The Petroleum Exploration of Ireland's Offshore Basins

Special Publication reviewing procedures

The Society makes every effort to ensure that the scientific and production quality of its books matches that of its journals. Since 1997, all book proposals have been refereed by specialist reviewers as well as by the Society's Books Editorial Committee. If the referees identify weaknesses in the proposal, these must be addressed before the proposal is accepted.

Once the book is accepted, the Society has a team of Book Editors (listed above) who ensure that the volume editors follow strict guidelines on refereeing and quality control. We insist that individual papers can only be accepted after satisfactory review by two independent referees. The questions on the review forms are similar to those for *Journal of the Geological Society.* The referees' forms and comments must be available to the Society's Book Editors on request.

Although many of the books result from meetings, the editors are expected to commission papers that were not presented at the meeting to ensure that the book provides a balanced coverage of the subject. Being accepted for presentation at the meeting does not guarantee inclusion in the book.

Geological Society Special Publications are included in the ISI Science Citation Index, but they do not have an impact factor, the latter being applicable only to journals.

More information about submitting a proposal and producing a Special Publication can be found on the Society's web site: www.geolsoc.org.uk.

It is recommended that reference to all or part of this book should be made in one of the following ways:

SHANNON, P. M., HAUGHTON, P. D. W. & CORCORAN, D. V. (eds) 2001 *The Petroleum Exploration of Ireland's Offshore Basins.* Geological Society, London, Special Publications, **188**.

SPENCER, A. M. & MACTIERNAN, B. 2001. Petroleum systems offshore western Ireland in an Atlantic margin context *In:* SHANNON, P. M., HAUGHTON, P. D. W. & CORCORAN, D. V. (eds) *The Petroleum Exploration of Ireland's Offshore Basins.* Geological Society, London, Special Publications, **188**, 9–29.

GEOLOGICAL SOCIETY SPECIAL PUBLICATION NO. 188

The Petroleum Exploration of Ireland's Offshore Basins

EDITED BY

P. M. SHANNON
University College Dublin, Ireland

P. D. W. HAUGHTON
University College Dublin, Ireland

and

D. V. CORCORAN
Statoil Exploration Ireland Ltd, Ireland

2001
Published by
The Geological Society
London

THE GEOLOGICAL SOCIETY

The Geological Society of London (GSL) was founded in 1807. It is the oldest national geological society in the world and the largest in Europe. It was incorporated under Royal Charter in 1825 and is Registered Charity 210161. The Society is the UK national learned and professional society for geology with a worldwide Fellowship (FGS) of 9000. The Society has the power to confer Chartered status on suitably qualified Fellows, and about 2000 of the Fellowship carry the title (CGeol). Chartered Geologists may also obtain the equivalent European title, European Geologist (EurGeol). One fifth of the Society's fellowship resides outside the UK. To find out more about the Society, log on to *www.geolsoc.org.uk*.

The Geological Society Publishing House (Bath, UK) produces the Society's international journals and books, and acts as European distributor for selected publications of the American Association of Petroleum Geologists (AAPG), the American Geological Institute (AGI), the Indonesian Petroleum Association (IPA), the Geological Society of America (GSA), the Society for Sedimentary Geology (SEPM) and the Geologists' Association (GA). Joint marketing agreements ensure that GSL Fellows may purchase these societies' publications at a discount. The Society's online bookshop (accessible from *www.geolsoc.org.uk*) offers secure book purchasing with your credit or debit card.

To find out about joining the Society and benefiting from substantial discounts on publications of GSL and other societies worldwide, consult *www.geolsoc.org.uk,* or contact the Fellowship Department at: The Geological Society, Burlington House, Piccadilly, London W1J 0BG: Tel. +44 (0)20 7434 9944; Fax +44 (0)20 7439 8975; Email: *enquiries@geolsoc.org.uk.* For information about the Society's meetings, consult *Events* on *www.geolsoc.org.uk.* To find out more about the Society's Corporate Affiliates Scheme, write to *enquiries@geolsoc.org.uk.*

Published by The Geological Society from:
The Geological Society Publishing House
Unit 7, Brassmill Enterprise Centre
Brassmill Lane
Bath BA1 3JN, UK
(*Orders:* Tel. +44 (0)1225 445046
Fax +44 (0)1225 442836)
Online bookshop: http://bookshop.geolsoc.org.uk

British Library Cataloguing in Publication Data
A catalogue record for this book is available from the British Library.

ISBN 1-86239-087-8

Typeset by Alden Bookset, Somerset, UK
Printed by The Alden Press, Oxford, UK.

Distributors

USA
AAPG Bookstore
PO Box 979
Tulsa
OK 74101-0979
USA
Orders: Tel. +1 918 584-2555
Fax +1 918 560-2652
E-mail *bookstore@aapg.org*

Australia
Australian Mineral Foundation Bookshop
63 Conyngham Street
Glenside
South Australia 5065
Australia
Orders: Tel. +61 88 379-0444
Fax +61 88 379-4634
E-mail *bookshop@amf.com.au*

India
Affiliated East-West Press PVT Ltd
G-1/16 Ansari Road, Daryaganj,
New Delhi 110 002
India
Orders: Tel. +91 11 327-9113
Fax +91 11 326-0538
E-mail *affiliat@nda.vsnl.net.in*

Japan
Kanda Book Trading Co.
Cityhouse Tama 204
Tsurumaki 1-3-10
Tama-shi
Tokyo 206-0034
Japan
Orders: Tel. +81 (0)423 57-7650
Fax +81 (0)423 57-7651

Contents

The petroleum exploration of Ireland's offshore basins: introduction

P. M. SHANNON[1], D. V. CORCORAN[2] & P. D. W. HAUGHTON[1]

[1]*Department of Geology, University College Dublin, Belfield, Dublin 4. Ireland (e-mail: p.shannon@ucd.ie)*

[2]*Statoil Exploration (Ireland) Ltd., Statoil House, 6 George's Dock, IFSC, Dublin 1. Ireland*

Ireland is virtually encircled by sedimentary basins (Fig. 1) that developed in response to a series of rift episodes interspersed with periods of thermal subsidence. A number of inversion episodes also played a role in the development of sediment source areas and in the structuring of the basins.

These basins can be categorized into two groups. The first comprises the basins of Northern Ireland, the Irish Sea and Celtic Sea areas, and the inboard basins (Slyne, Erris and Donegal basins) of the Atlantic margin. They generally have a NE–SW elongate morphology and typically lie within 100 km of the shore. Their sedimentary fill is predominantly of pre-Tertiary age and they have no major bathymetric expression. The second group, comprising the outboard basins of the Atlantic margin (Goban Spur, Porcupine, Rockall and Hatton basins), lies in deep water. These basins are characterized by having an extensive surface area, typically containing a predominantly Cretaceous and Tertiary succession and having an underfilled sedimentary character.

The Irish offshore basins have been the focus of intermittent phases of exploration since the first well was drilled in 1970. To date, a total of 136 wells has been drilled (Fig. 2), with 37 of these in the basins west of Ireland. The total cost of wells in the Irish offshore, in 2001 prices, is approximately IR £1500 million. A significant amount of 2D reflection seismic data has been acquired (Fig. 3), both as speculative and proprietary surveys. Two commercial gas fields (Kinsale Head and Ballycotton) are currently in production in the North Celtic Sea Basin but are nearing the end of their productive lives. The Corrib gas field in the Slyne Basin is currently undergoing the final stages of appraisal and field development will commence shortly. Some other gas and oil accumulations have been discovered but all of them appear to be relatively small and currently non-commercial to marginally commercial.

Reservoir and source rock horizons have been encountered in the various basins (Croker & Shannon 1995). Most of the drilling to date has concentrated on structural traps (e.g. inversion anticlines and tilted fault blocks) but recent exploration has begun to focus upon a variety of stratigraphic traps (Shannon & Naylor 1998).

The results of the exploration in the Irish basins have been generally disappointing. Several phases of exploration drilling in various basins have raised expectations only to see hopes dashed and exploration drilling dwindle for a time before the next phase of optimism and renewed exploration (Naylor 1996). These various phases have been influenced by a variety of factors, such as Ireland's exploration policy, the oil price, new exploration ideas and advances in drilling and production technology.

Exploration history

Five major periods of exploration have taken place during the past four decades. These blended into one another, reflecting major changes within the international oil industry.

The early era (pre-1973)

The first offshore well (48/25-1), targeted on a shallow Cretaceous prospect in the North Celtic Sea Basin, was spudded in 1970 and encountered gas shows. The following year a second well on this block discovered the Kinsale Head gas field (Colley *et al.* 1981; Naylor & Shannon 1982; Murray 1995; Taber *et al.* 1995). Drilling through the 1970s concentrated on shallow inversion structures. The geology of the region was very poorly known, due largely to the poor seismic data quality in the region, and the exploration results were generally disappointing.

From: SHANNON, P.M., HAUGHTON, P.D.W. & CORCORAN, D.V. (eds). 2001. *The Petroleum Exploration of Ireland's Offshore Basins*. Geological Society, London, Special Publications, **188**, 1–8. 0305-8719/01/$15.00

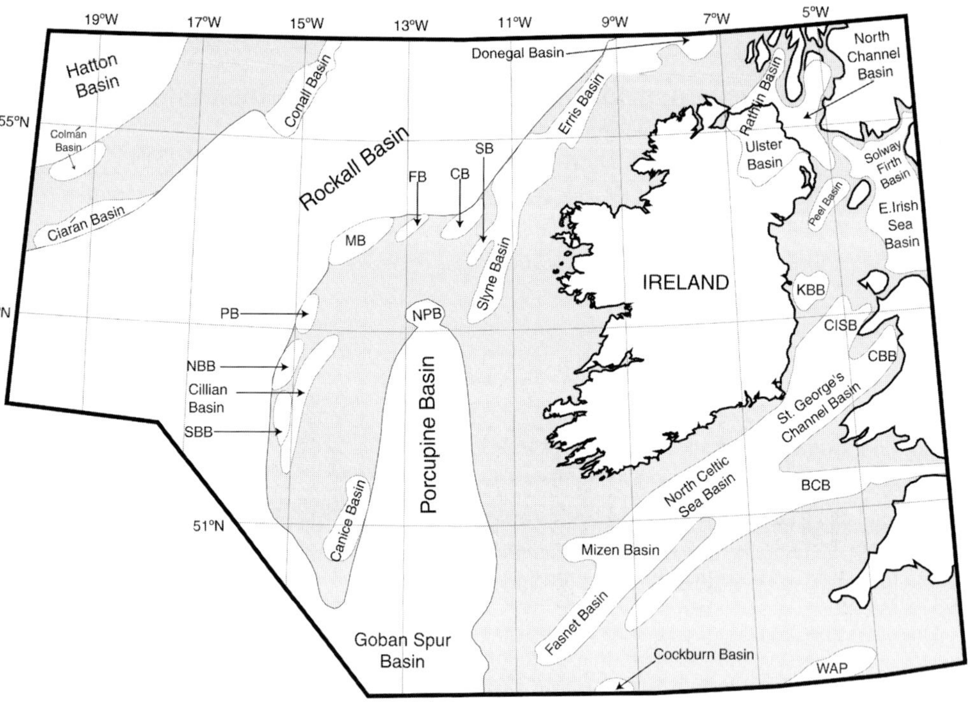

Fig. 1. Location map of sedimentary basins in the Irish offshore and adjacent areas. Abbreviations of basins are as follows: BCB = Bristol Channel Basin; CB = Colm Basin; CBB = Cardigan Bay Basin; CISB = Central Irish Sea Basin; FB = Fursa Basin; KBB = Kish Bank Basin; MB = Macdara Basin; NBB = North Bróna Basin, NPB = North Porcupine Basin; PB = Pádraig Basin; SBB = South Bróna Basin; SE = Slyne Embayment; WAP = Western Approaches Basin. Based largely on Naylor *et al.* (1999).

The era of oil shortages (1973–83)

The oil crisis of 1973 led to a large increase in oil prices and spurred non-OPEC countries into encouraging exploration for indigenous oil and gas. Major discoveries were made in the UK and Norwegian sectors of the North Sea. New Irish licensing terms, modelled on those applied to the Norwegian offshore, were issued in 1975 and reflected a belief that large fields awaited discovery in the Irish offshore. The terms incorporated the aspiration that the state should benefit by joint ownership of the fields (state participation) and also by the receipt of taxes and royalties. A select number of oil companies were granted acreage under the First Licensing Round in the Fastnet, Porcupine, Slyne, Erris, Donegal and Kish Bank basins and a phase of drilling commenced in 1976. The drilling levels reached a peak in 1978 when 15 wells were drilled (Fig. 2).

The oil crisis of 1979 served to renew exploration interest in the Irish offshore. Improved seismic acquisition techniques revealed deep, unexplored fault block structures at Jurassic level within the North Celtic Sea and Porcupine basins. The Second Licensing Round took place in 1981 and resulted in the entry of a large number of new companies to the Irish offshore. During the following years an 'open door' licensing policy was followed, whereby interested companies could either licence open acreage, or take a 'seismic option' for a fixed period on such acreage.

The era of economic recession (1983–93)

By *c.*1983 the effects of the oil price rises of the 1970s had taken a toll on the industrial countries and economic recession had begun to bite. The demand for oil and gas dropped and the oil industry began to turn away from the perceived high risk, unproven regions. While some discoveries were made (MacDonald *et al.* 1987; Shannon 1993*a*,*b*; Caston 1995), there was little evidence of large accumulations. In contrast to other countries the Irish fiscal terms looked

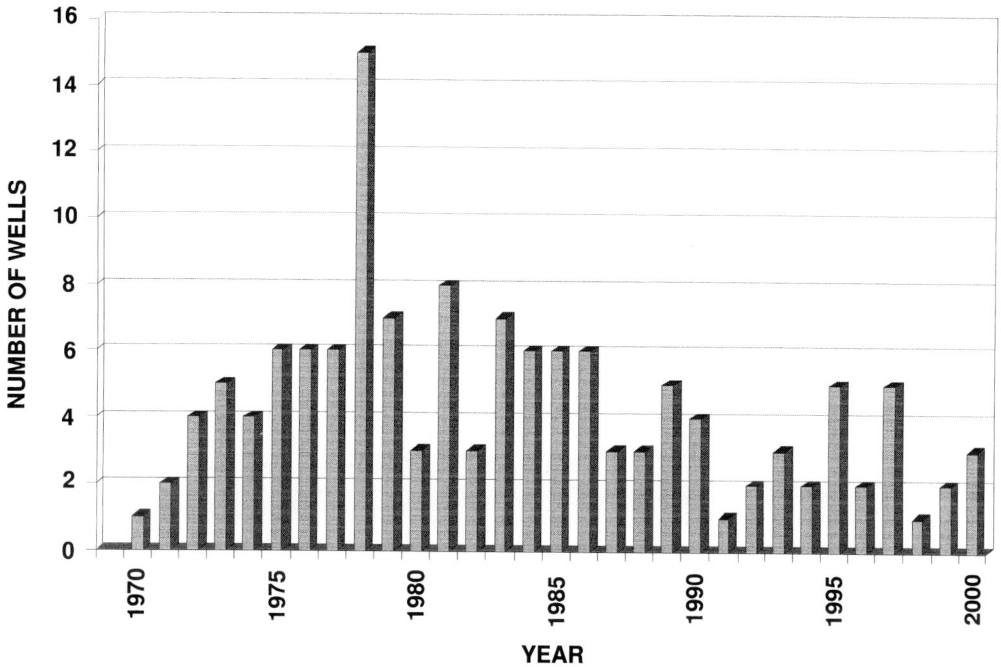

Fig. 2. Number of wells drilled per year, offshore Ireland.

unattractive, especially for marginal fields which were now regarded as the likely outcome in Irish acreage. The uptake of acreage in the Third Licensing Round of 1984 was low. Over the course of the next few years the Irish terms were modified several times in order to make them more attractive. Eventually the state participation and royalty components were abolished. The Licensing Terms of 1992 offered attractive tax incentives for exploration and production.

The era of Atlantic margin optimism (1993–9)

By the early 1990s exploration had slowed down in most of the Irish basins. The majority of the obvious structures in relatively shallow water had been drilled and the results had been generally disappointing. One exception was the discovery of the Ballycotton gas field, a small accumulation located close to the Kinsale Head gas field. Some exploration interest remained in the Kish Bank and Central Irish Sea basins. However, exploration in the UK and Norwegian parts of the Atlantic margin had provided encouraging results, and this region became an exploration 'hot spot'. New seismic data in the Slyne, Erris and Rockall basins added to the encouragement and indicated the presence of interesting structures and potential plays. The First Frontier Licensing Round in Irish waters in 1994 was very successful and brought the return of several companies who had previously left Irish waters, together with a number of new companies. The new frontier licensing terms afforded a relatively long lead time to companies before a decision was required to drill or drop acreage. The Second Frontier Licensing Round in 1995 offered blocks in the northern part of the Porcupine Basin and was again heavily subscribed. The Third Frontier Licensing Round of 1997 offered deep water blocks in the Rockall and Erris Basins and was also successful in terms of the acreage licensed. A novel aspect of this round was the instigation of the Petroleum Infrastructure Programme (PIP) whereby the licencees jointly contributed to funding research related to the Rockall region, largely in Irish academic and service institutions. Drilling during the period took place in the Kish Bank Basin and in the shallower waters of the Slyne Basin, with the latter drilling leading to the discovery of the Corrib gas field.

The present and the future (1999 onwards)

The Fourth Frontier Licensing Round, in the southern Porcupine Basin, took place in 1999 but only two exploration groups were awarded

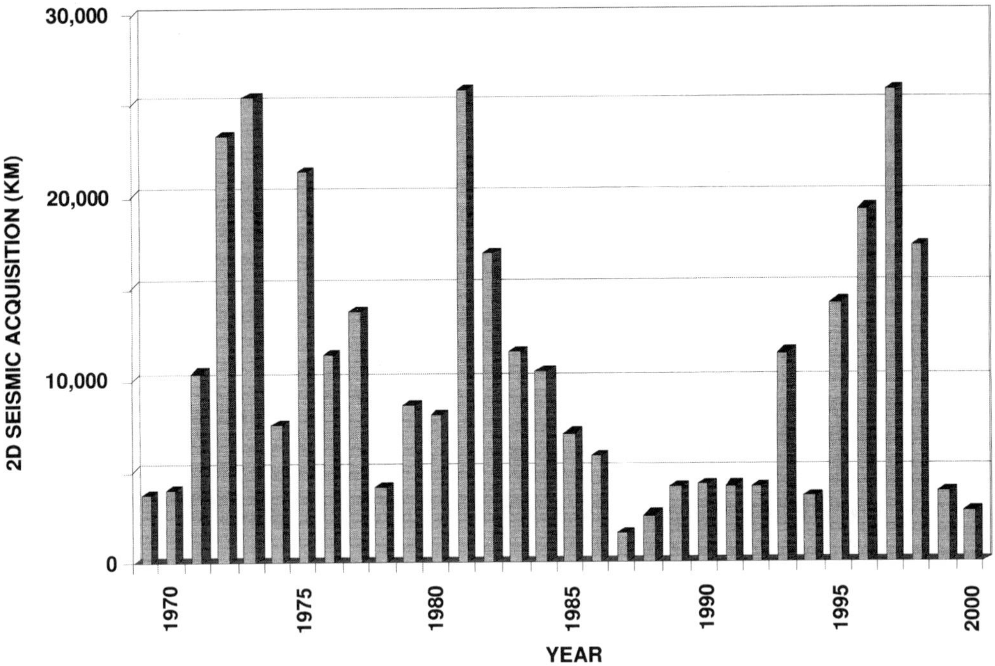

Fig. 3. Annual 2D seismic acquisition (km), offshore Ireland.

acreage. At the time of the announcement of the round the optimistic exploration air of the industry still prevailed. However, by the time the round closed for bidding, major changes in the nature and confidence of the industry had occurred. The results of appraisal drilling on the Connemara oil accumulation in the Porcupine Basin were disappointing. The oil price had dropped significantly. 'Merger mania' was in full swing, with a resultant re-focus on finding giant fields by the enlarged companies. There was an increasing interest in more attractive and prolific deep water margins and a number of companies began to feel an overexposure to the high risk, deep water Irish Atlantic margin.

The undoubted highlight of the last 5 years has been the discovery of the Corrib gas field in the Slyne Basin with the drilling of well 18/20-1 in 1996. Appraisal wells have been drilled and successfully tested. Well 18/20-2Z, drilled in 1998, tested gas from a Triassic reservoir at a stabilized flow rate of up to 64 MMSCFD. This is very timely, with the anticipated depletion of the Kinsale Head Field within the next few years. Nevertheless, efforts to prolong the life of the Kinsale Head gas field have continued in recent years, with incremental reserves now being produced through existing facilities. This additional production is the result of a successful re-evaluation of the SW Kinsale extension which culminated in the sub-sea completion and tieback of well 48/25-3 in 1999.

This development of Ireland's gas resources has occurred against the background of a buoyant indigenous energy market, fuelled by the burgeoning appetite of the 'Celtic Tiger' economy. Published forecasts predict that local demand for natural gas could triple by 2005. It is anticipated that further exploration for gas will be stimulated in this environment.

Presently there is little active exploration taking place in the Central Irish Sea and Kish Bank basins. Some exploration activity is anticipated in the south coast basins during the next couple of years with drilling expected in the North Celtic Sea and Fastnet basins. The main focus of exploration activity is currently in the basins west of Ireland – the Porcupine, Slyne, Erris and Rockall basins. Although the potential of the Irish frontier basins is still generally recognized by the industry, Ireland is at a disadvantage in having relatively unproven deep water plays. However, one or more commercial discoveries in the next couple of years could help revitalize exploration confidence.

Despite the relative lack of discoveries there are encouraging prospects, especially for the

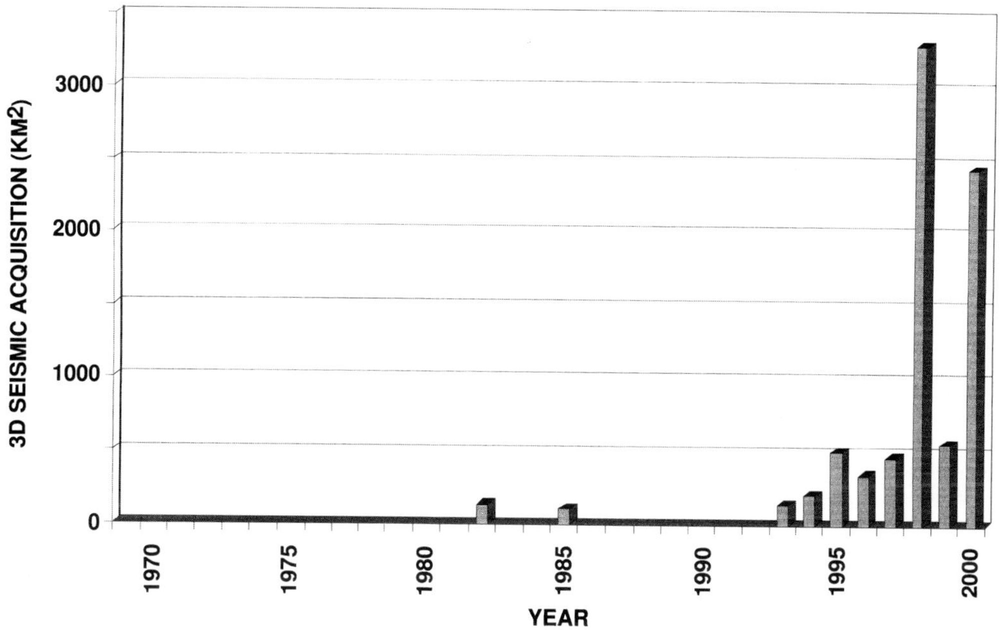

Fig. 4. Annual 3D seismic acquisition (km^2), offshore Ireland.

basins west of Ireland (Shannon *et al.* 1995; Spencer *et al.* 1999; Walsh *et al.* 1999). A number of gas and oil prospects remain in the Celtic Sea basins at various levels but any discoveries are likely to be small. Some remaining (mostly gas) prospects exist in the Central Irish Sea and the Kish Bank basins (Shannon & Naylor 1998). A number of rift episodes, important in the generation of oil- and gas-bearing structures, are recognized or suggested in all the basins west of Ireland and are comparable to those in other regions of the Atlantic margin. Jurassic source rocks are proven in some of the basins and can be speculated with moderate confidence in others. Structural and stratigraphic traps have been identified at various levels. However, there is very little control on the age or structure of the pre-Tertiary succession in the Rockall Basin, due to a combination of lack of wells and of the seismic masking effect of extensive shallow sills in the basin. In particular, the presence, extent and nature of Jurassic and older strata are uncertain. The unknown nature of source and reservoir rocks in the Rockall Basin, and the lack of information on likely reservoirs in the southern Porcupine Basin, represent major exploration risks (Shannon & Spencer 1999). The structural complexity of the Slyne and Erris basins, together with the problem of seismic data quality, represent the major challenges in this region.

The level of licensing in the Irish offshore remains encouragingly high at the time of writing (early 2001). A total of 104 blocks or part blocks (97 to the west, 7 in the basins to the south of Ireland) are held under 21 licences. Some 33 blocks/part blocks (14 to the south, 15 in the basins to the west) are held under 11 licensing options. A total of 5 blocks/part blocks in the North Celtic Sea Basin are held under petroleum leases. Fourteen blocks are held under licence options in basins to the south of Ireland, while 15 blocks are held under licence options in the west coast basins.

The recent exploration of Ireland's offshore basins has taken place in the context of rapid technological change. During the past decade, significant advances have occurred in the area of floating production systems, sub-sea completion technology and the acquisition, processing and visualization of 3D seismic data. Although the volume of acquisition is increasing, there has been a limited employment of 3D technology in exploration programmes in the Irish offshore basins (Fig. 4). For example, the present paltry coverage ($<2500\,km^2$) within the Porcupine Basin compares most unfavourably with the almost blanket 3D coverage ($>15000\,km^2$) of the Faroe–Shetland Basin.

This volume

The 27 papers presented in this thematic volume provide a significant amount of new information on the structural and stratigraphic evolution, thermal history, petroleum systems and reservoir geology of Ireland's offshore basins. The volume provides a companion to the 1995 Geological Society Special Publication (No. 93) which elucidated the petroleum geology of the Irish offshore. The papers in the present volume focus largely upon the petroleum exploration of the region and indicate exploration thinking and results. While the papers cover a broad spectrum of topics and areas, some common themes are identified. In addition, they highlight a number of issues that need to be addressed in order to reduce exploration risk in these basins.

1. All of the offshore Late Palaeozoic to Cenozoic basins have experienced a multiphase extensional and inversion history. Several rifting episodes are recognized or inferred in a number of these basins: Permo-Triassic, Middle to Late Jurassic and Early to middle Cretaceous. At least two periods of pervasive exhumation are interpreted, one during the Late Carboniferous to Late Permian and one during the Tertiary. Regional uplift and erosion is also interpreted for the Early to middle Cretaceous period. Seismic-stratigraphic analysis of the Paleogene, Neogene and Quaternary sediments of the Atlantic margin basins offers an eloquent testimony to the interlinkage between pulsed sedimentation and the uplift and denudation of inboard regions during the Cenozoic.
2. Petroleum systems are proven or possible in all of the basins. However, the impact of regional exhumation upon the pre-existing petroleum systems within Ireland's offshore basins remains to be fully evaluated. Initial studies from the Irish Sea basins indicate that this can have a radical effect on the source rocks, reservoirs, seal integrity, trap morphology and hydrocarbon displacement patterns within the basins.
3. Geochemical studies of oils and shales suggest that the Lower and Middle Jurassic shales have considerable potential as effective oil-prone source rocks, where preserved in these offshore basins. Oil shows have been typed to Lower Jurassic shales in the Slyne, North Celtic Sea and St George's Channel basins. Biomarker evidence suggests that Middle Jurassic sources have contributed to the hydrocarbon budget in the Porcupine Basin, in addition to the hydrocarbon charge provided by the Upper Jurassic Kimmeridge Clay Formation equivalents. However, our understanding of the Jurassic-sourced petroleum systems requires better constraints on the distribution of individual source rock units within each of the basins. Constraining the timing of maturation and the hydrocarbon drainage pattern from these source units also remains a significant challenge. The source rock potential of the Atlantic margin outboard basins is largely unknown and continues to be a significant exploration risk factor in these basins.
4. The thermal evolution of these sedimentary basins is poorly constrained. Evidence from vitrinite reflectance (VR), apatite fission track analysis (AFTA) and fluid inclusion data suggests that advective heat transfer by hydrodynamic systems occurred during the development of these basins, and may be responsible for both long and short duration heating events recorded by these palaeotemperature indicators. Interestingly, VR data do not record a regionally elevated basal heat flux coincident with Paleogene igneous activity in Atlantic margin basins.
5. The presence, quality and maturation history of Carboniferous source rocks is an issue of concern with respect to the Irish Sea basins. Speculative palaeogeographical reconstructions suggest that the Namurian Holywell Shale, the prolific oil and gas source in the East Irish Sea Basin, may be present in the Kish Bank Basin but was never deposited in the Central Irish Sea area.
6. While there has been a significant amount of speculation on the likely age, facies and hydrocarbon habitat of the pre-Neogene succession in the Rockall Basin, there is very little direct geological information on these aspects. Likewise, the nature of the succession in the southern Porcupine Basin remains conjectural until wells are drilled.
7. Some of the Irish Atlantic margin basins are characterized by extensive, near-surface Tertiary lava flows, igneous dykes and sills. Signal penetration beneath this near-surface high velocity layer is a particular problem due to a variable morphology, a heterogeneous velocity structure and the absorption of P- and

S-wave energy. The application of 3D seismic technology, combined with low-frequency acquisition, judicious multiple attenuation and velocity picking offers a way forward to improved sub-basalt imaging in this environment.

8. The Neogene and Recent successions in the Atlantic margin basins contain evidence for major changes in oceanic current circulation patterns, sediment transport, slope development and mass failure features. Exciting evidence for the growth of carbonate mound clusters, and their possible linkage with gas escape features, is coming to light. These features are likely to present opportunities for fruitful study into aspects of climate change in late Neogene to Recent times, demonstrating that there are significant scientific research offshoots of the exploration effort in Ireland's offshore basins.

These themes reflect some of the questions being addressed by explorationists and others working in Ireland's offshore basins. This present volume is not intended to present the definitive statement on the petroleum geology of these basins, but rather to present a snapshot of current understanding which will hopefully provide a stimulus for new ideas and a template for further exploration. The quest to understand the Irish offshore basins and their hydrocarbon habitat continues.

This volume arose from the proceedings of a two-day conference held in Dublin in April 1999. The conference was organized by the Department of the Marine and Natural Resources, the Institute of Petroleum and the Irish Offshore Operators' Association. We would like to take this opportunity to thank the members of the PEIOB conference organizing committee – Peter Croker, Pat Shannon and Geirr Haarr, whose collective energies ensured that the conference was a successful and enjoyable event. In producing this volume we are indebted to a large number of individuals and organizations: the authors who gave freely of their time and expertise to produce these papers, and the companies, universities and institutions who permitted publication. Finally, we wish to thank all the referees who thoroughly reviewed the initial manuscripts and greatly contributed to the quality of the final papers. They are: Nigel Ainsworth, Philip Allen, Morten Sparre Andersen, Ray Archer, Ken Baxter, Chris Bean, Doug Boyd, Richard Bray, Andrew Brock, Glen Cayley, Tim Chapman, Geoff Clayton, Dermot Corcoran, John Conroy, Peter Croker, Bryan Cronin, Alex Densmore, Ian Duncan, Robin Dyer, Richard England, Martin Feely, Delwyn Geraghty, Ken Glennie, Paul Green, Paul Griffiths, Adrian Hartley, Stuart Haszeldine, Geir Ultveit Haugen, Peter Haughton, Steve Hay, Amy Heath, Ken Higgs, Deepak Inamdar, Sarah Johnston, Gareth Jones, Steve Jones, Jan Sverre Laberg, Xiang-Yang Li, Brian MacTiernan, Angela McDonnell, Neil Meadows, John Moore, Noel Murphy, Dave Naylor, Phil Newman, Brian O'Reilly, John O'Sullivan, Adrian Phillips, Daniel Praeg, Peter Readman, Jonathan Redfern, Dave Roberts, Adrian Robinson, Pat Shannon, Iain Sinclair, George Sevastopulo, Steve Smith, Mike Stephenson, Dave Tappin, Michael Tate, Alastair Thomson, Vikram Unnithan, Anne Walsh, Tjeerd van Weering, Andy Wheeler. Thanks are also due to Peter Croker and Michael Hanrahan of the Petroleum Affairs Division, Department of the Marine and Natural Resources, for providing the data for Figures 2–4.

References

CASTON, V.N.D. 1995. The Helvick oil accumulation, Block 49/9, North Celtic Sea Basin. *In*: CROKER, P.F. & SHANNON, P.M. (eds) *The Petroleum Geology of Ireland's Offshore Basins*. Geological Society, London, Special Publications, **93**, 209–225.

COLLEY, M.G., MCWILLIAMS, A.S.F. & MYERS, R.C. 1981. Geology of the Kinsale Head gas field, Celtic Sea, Ireland. *In*: ILLING, L.V. & HOBSON, G.D. (eds) *Petroleum Geology of the Continental Shelf of North-West Europe*. Heyden, London, 504–510.

CROKER, P.F. & SHANNON, P.M. 1995. The petroleum geology of Ireland's offshore basins: introduction. *In*: CROKER, P.F. & SHANNON, P.M. (eds) *The Petroleum Geology of Ireland's Offshore Basins*. Geological Society, London, Special Publications, **93**, 1–8.

MACDONALD, H., ALLAN, P.M. & LOVELL, J.P.B. 1987. Geology of oil accumulation in Block 26/28, Porcupine Basin, offshore Ireland. *In*: BROOKS, J. & GLENNIE, K.W. (eds) *Petroleum Geology of North West Europe*. Graham & Trotman, London, 643–651.

MURRAY, M.V. 1995. Development of small gas fields in the Kinsale Head area. *In*: CROKER, P.F. & SHANNON, P.M. (eds) *The Petroleum Geology of Ireland's Offshore Basins*. Geological Society, London, Special Publications, **93**, 259–260.

NAYLOR, D. 1996. History of oil and gas exploration in Ireland. *In*: GLENNIE, K. & HURST, A. (eds) *AD 1995: NW Europe's Hydrocarbon Industry*. The Geological Society, London, 43–52.

NAYLOR, D., SHANNON, P.M. 1982. *The Geology of Offshore Ireland and West Britain*. Graham & Trotman Ltd, London.

NAYLOR, D., SHANNON, P., MURPHY, N. 1999. *Irish Rockall region – a standard structural nomenclature system*. Petroleum Affairs Division, Dublin, Special Publication, **1/99**.

SHANNON, P.M. 1993*a*. Submarine Fan Types in the Porcupine Basin, Ireland. *In*: SPENCER, A.M. (ed.) *Generation, Accumulation and Production of Europe's hydrocarbons. III*. Special Publication of the European Association of Petroleum

Geoscientists, No. 3, Springer-Verlag, Berlin, 111–120.

SHANNON, P.M. 1993*b*. Oil and gas in Ireland – exploration, production and research. *First Break*, **11**, 429–433.

SHANNON, P.M. & NAYLOR, D. 1998. An assessment of Irish offshore basins and petroleum plays. *Journal of Petroleum Geology*, **21**, 125–152.

SHANNON, P.M. & SPENCER, A.M. 1999. Atlantic margin: offshore Norway to offshore Ireland. Introduction and review. *In*: FLEET, A.J & BOLDY, S.A.R. (eds) *Petroleum Geology of Northwest Europe: Proceedings of the 5th Conference*. Geological Society, London, 229–230.

SHANNON, P.M., JACOB, A.W.B., MAKRIS, J., O'REILLY, B., HAUSER, F. & VOGT, U. 1995. Basin development and petroleum prospectivity of the Rockall and Hatton region. *In*: CROKER, P.F. & SHANNON, P.M. (eds) *The Petroleum Geology of Ireland's Offshore Basins*. Geological Society, London, Special Publications, **93**, 435–457.

SPENCER, A.M., BIRKELAND, Ø., KNAG, G.Ø. & FREDSTED, R. 1999. Petroleum systems of the Atlantic margin of northwest Europe. *In*: FLEET, A.J. & BOLDY, S.A.R. (eds) *Petroleum Geology of Northwest Europe: Proceedings of the 5th Conference*. Geological Society, London, 231–246.

TABER, D.R., VICKERS, M.K. & WINN, R.D. Jr. 1995. The definition of the Albian 'A' Sand reservoir fairway and aspects of associated gas accumulations in the North Celtic Sea Basin. *In*: CROKER, P.F. & SHANNON, P.M. (eds) *The Petroleum Geology of Ireland's Offshore Basins*. Geological Society, London, Special Publications, **93**, 227–244.

WALSH, A., KNAG, G., MORRIS, M., QUINQUIS, H., TRICKER, P., BIRD, C. & BOWER, S. 1999. Petroleum geology of the Irish Rockall Trough – a frontier challenge. *In*: FLEET, A.J. & BOLDY, S.A.R. (eds) *Petroleum Geology of Northwest Europe: Proceedings of the 5th Conference*. Geological Society, London, 433–444.

Petroleum systems offshore western Ireland in an Atlantic margin context

A. M. SPENCER[1] & B. MacTIERNAN[2]

[1]Statoil, 4035 Stavanger, Norway (e-mail: ams@statoil.com)

[2]Statoil Exploration (Ireland) Ltd, 6 George's Dock, IFSC, Dublin 2, Ireland

Abstract: The Rockall, Slyne, Erris and Porcupine basins on the Atlantic margin off Ireland belong to a family of geologically similar basins stretching from offshore mid-Norway to offshore Newfoundland. Jurassic sequences act as reservoir and source rocks in many of the basins. Cretaceous extensional faulting was widespread and major subsidence affected several basins. Cretaceous submarine fan and shallow marine sandstones and Paleocene submarine fan sandstones often provide reservoir targets. Cretaceous and Paleocene to Eocene volcanic rocks are widespread and Eocene to Recent net subsidence has resulted in water depths which generally exceed 200 m and reach over 2000 m in the south. Passive uplift affected the land areas to the east in Neogene times.

Proven Jurassic-sourced petroleum systems occur in six basins from the Halten Terrace to the Jeanne d'Arc Basin, including the northern Porcupine Basin and the Slyne–Erris basins. In the latter the Jurassic petroleum system has been destroyed by uplift, but a Carboniferous petroleum system has proved successful there for gas. West and northwest of these proven basins, on the 'outboard' side of the Atlantic margin, are large frontier areas. Recent gas discoveries in the deep Vøring Basin prove the existence of petroleum systems there but the source is not known. On both margins of the Rockall Basin and in the southern Porcupine Basin petroleum systems may exist but are not yet proven.

In this article we review the petroleum geology of the Irish Atlantic basins (Rockall, Slyne, Erris and Porcupine basins) in the context of the wider family of basins along the Atlantic margin, analysing them from a petroleum system point of view. 'A petroleum system is ... a natural system that encompasses a pod of active source rock and all the related oil and gas and which includes all the geological elements and processes that are essential if a hydrocarbon accumulation is to exist' (Magoon & Dow 1994, p.10). The present article builds largely upon an earlier review of the petroleum systems of the Atlantic margin basins (Spencer *et al.* 1999). This article does not describe the Celtic Sea basins, which do not belong to the Atlantic margin family, nor the little known Hatton Basin in the far west.

The part of the Atlantic margin of NW Europe described here stretches 2500 km from the Porcupine Basin to the Lofoten Islands (Figs 1, 2). On the landward side of the margin are basins in which exploration started in the 1970s – the Porcupine Basin, the Slyne and Erris basins, the Faroe–Shetland Basin, the northern North Sea and the Halten Terrace. To the northwest of these are four large frontier areas – the Rockall Basin, the Faroes Shelf and the Møre and Vøring basins – and in most of these areas petroleum exploration is just starting, following licensing rounds in 1996 and 2000 in Norway, 1997 in the UK, 2000 in the Faroes and 1997 and 1998 in Ireland. This is, therefore, an exciting time in the early exploration history of these frontier areas, especially in view of the large petroleum finds in the Faroe–Shetland Basin in 1992 (Foinaven oil field), in the Vøring Basin in 1997 (the Nyk High and Ormen Lange gas finds) and in 1998 in the Slyne Basin (Corrib gas field).

Five areas on the landward side of the margin contain proven petroleum systems: the northern Porcupine Basin, the Slyne and Erris basins, the Faroe–Shetland Basin, the northern North Sea and the Halten Terrace. The hydrocarbons in most of these basins are in Jurassic sandstone reservoirs in late Jurassic fault traps. A direct analogue is the Jeanne d'Arc Basin on the Newfoundland Shelf, where Jurassic-sourced hydrocarbons are reservoired in Upper Jurassic and Lower Cretaceous sandstones in traps affected by early Cretaceous faulting.

On the 'outboard' side of the European Atlantic margin, the frontier areas have many features in common. Some have Cretaceous strata which are many kilometres thick: the

From: Shannon, P.M., Haughton, P.D.W. & Corcoran, D.V. (eds) 2001. *The Petroleum Exploration of Ireland's Offshore Basins*. Geological Society, London, Special Publications, **188**, 9–29.

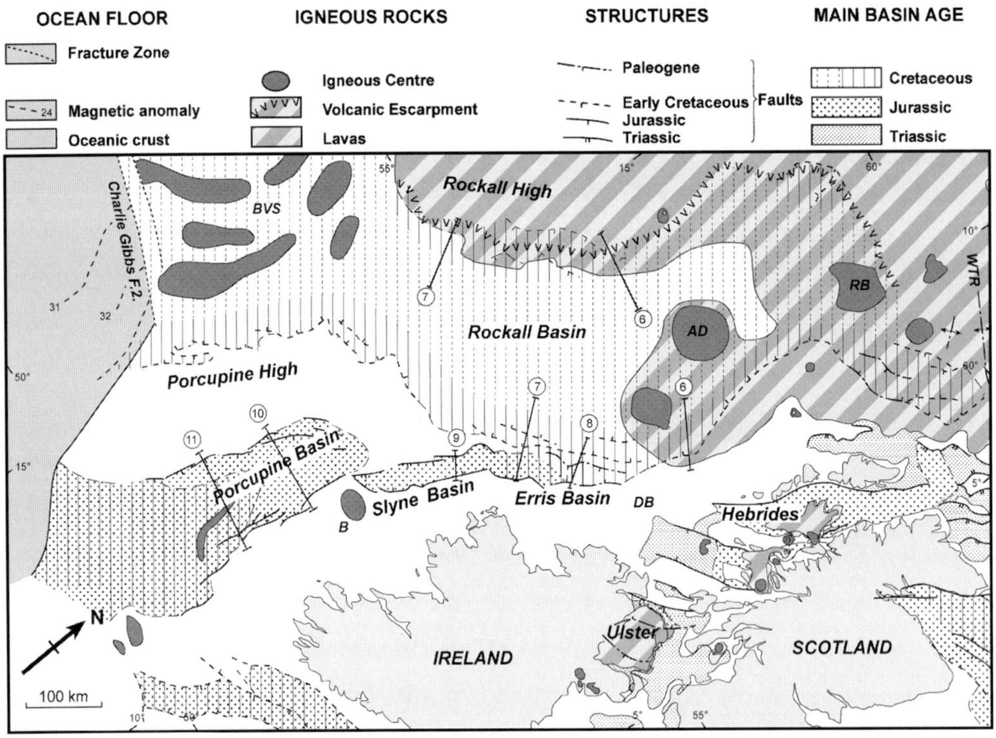

Fig. 1. Tectonic map showing the main basins of the Atlantic margin: Ireland to Scotland. AD, Anton Dohrn Seamount; B, Brendon Igneous Centre; DB, Donegal Basin; RB, Rosemary Bank; BVS, Barra Volcanic Ridge system; WTR, Wyville–Thomson Ridge.

southern Porcupine, Møre and Vøring basins. Tertiary strata are thick in some basins (eastern Faroes Shelf) but thin (starved) in others (Rockall). Cretaceous extensional faulting created fault traps in many of the basins and mid-Tertiary compressional domes provide important traps in some. The least known aspect of the frontier basins is the presence and maturity history of any source rocks. Any Jurassic source rocks present may have reached maturity in some of the basins in Cretaceous times, before deposition of Paleogene reservoirs. Cretaceous source rocks, if present, will have matured later. The distribution of Cretaceous and Paleogene reservoirs is little known. In the north, for example, Cretaceous turbidite reservoirs may have been derived from the northwest (Greenland). Along the flanks of the basins submarine fan systems can be mapped at Paleogene levels but, in many places, these are now in zones of monoclinal basinward dip and will therefore depend on stratigraphic trapping for prospectivity. For the frontier areas the most important question is whether a petroleum system is present and, if so, over what area.

Geological development

The period of the geological development of the Atlantic margin region which is of most relevance to the petroleum geology is that from Triassic times onwards. This article, therefore, does not review older events and structures. Triassic grabens, filled with thick continental sequences, occur irregularly on the present platform areas of the margin from western Ireland to mid-Norway (Figs 1, 2). Jurassic marine and deltaic rocks with broadly similar stratigraphic columns, often 1–2 km thick, occur in the Porcupine, Slyne, Erris, Hebrides, North Sea and Halten Terrace basins. Late Jurassic rifting affected most of these basins and created their fault-block traps. It was accompanied by deposition of the most important source rock interval, Volgian to Ryazanian marine shales, which are strongly transgressive and stretch throughout and beyond the rift basins. In early Cretaceous time, throughgoing extensional faulting, from the Rockall Basin to Lofoten, first created the NE-trend characteristic of the Atlantic margin (Doré *et al.* 1997; Roberts *et al.*

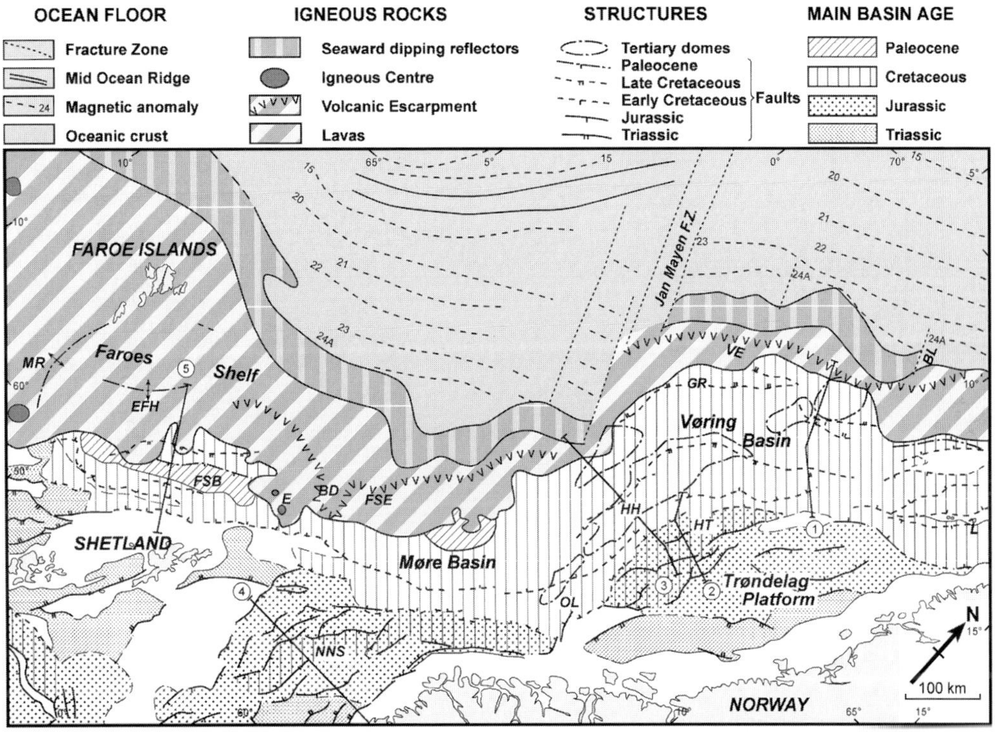

Fig. 2. Tectonic map showing the main basins of the Atlantic margin: Faroe–Shetland to Norway. BD, Brendans Dome; BL, Bivrost Lineament; E, Erlend Centres; EFH, East Faroe High; FSB, Faroe Shetland Basin; FSE, Faroe Shetland Escarpment; GR, Gjallar Ridge; HH, Helland Hansen Arch; HT, Halten Terrace; L, Lofoten Islands; MR, Munkagrunnar Ridge; NNS, northern North Sea; OL, Ormen Lange Dome; VE, Vøring Escarpment.

1999). This younger rift trend cuts across the older basins, truncating the N–S late Jurassic fault system in the northern North Sea, for example. Late Cretaceous and minor Paleocene extensional faulting occurs further northwest in the Vøring and Faroe–Shetland basins (Fig. 2), closer to the line of lithospheric break-up between Europe and Greenland at which, in Eocene times, the North Atlantic Ocean was initiated. Two later tectonic episodes are important: compressional doming in mid Tertiary times (Doré & Lundin 1996) and 'passive' Neogene uplift of the Norwegian (Doré *et al.* 1999) and British and Irish landmasses. During Eocene to Recent times, the Atlantic margin basins have undergone net subsidence so that they are now largely in water depths greater than 200 m and which reach over 2000 m in the Rockall and southern Porcupine basins.

The Atlantic margin basins are bordered to the northwest by a zone of subaerial Paleocene–Eocene volcanic lavas which are over 5 km thick in the Faroe Islands. This volcanic pile may conceal eroded but prospective Paleocene and Mesozoic strata. Other major volcanic provinces, of probably early Cretaceous age, are known to occur in the southern Porcupine and southern Rockall basins. On first reflection, the abundance of these igneous rocks would seem to be both detrimental to the presence of petroleum and a hindrance to finding sub-basalt petroleum accumulations. Further analysis, however, shows that the wealth of knowledge of the igneous rocks, studied for over 150 years in western Scotland, can aid interpretation (e.g. P.H. Naylor *et al.* 1999).

Atlantic margin basins of Norway, UK and the Faroes

Vøring Basin

In the north, the Vøring Basin contains three broad synclines separated by narrower faulted highs (Blystad *et al.* 1995; Fig. 3). In the south a central anticline is flanked by two synclines (Fig. 4). The most important feature of the basin is the thickness of the Cretaceous section, which

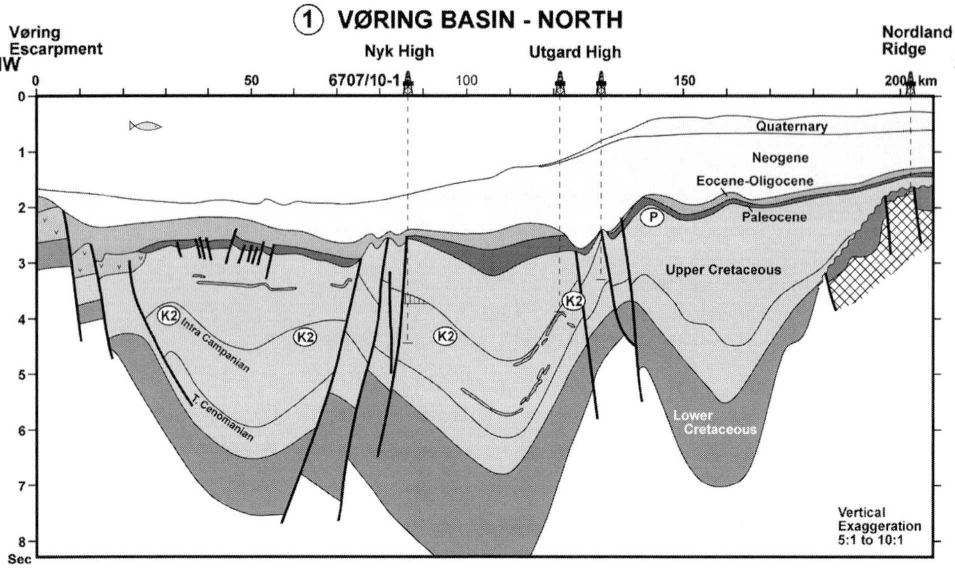

Fig. 3. Geological profile showing the plays of the Vøring Basin – north. Profile based on Blystad *et al* (1995, profile D). See Figure 2 for location. For Figures 3–12 and 14 the hydrocarbon plays are identified by letter: E, Eocene; P, Paleocene; K2, Upper Cretaceous; K1, Lower Cretaceous; J3, Upper Jurassic; J1–2, Lower–Middle Jurassic; T, Triassic; D, Devonian.

reaches 6 s TWT (*c*. 10 km). Seismic imaging of the deepest Cretaceous is poor and so interpretations of the sub-Cretaceous section and of early Cretaceous events are tentative. We follow Lundin & Doré (1997) in inferring that the Vøring Basin was first created as a result of early Cretaceous, throughgoing, rifting and that the second main rifting phase occurred in latest Cretaceous to Paleocene times, particularly in the northwest of the basin (Hjelstuen *et al.* 1999).

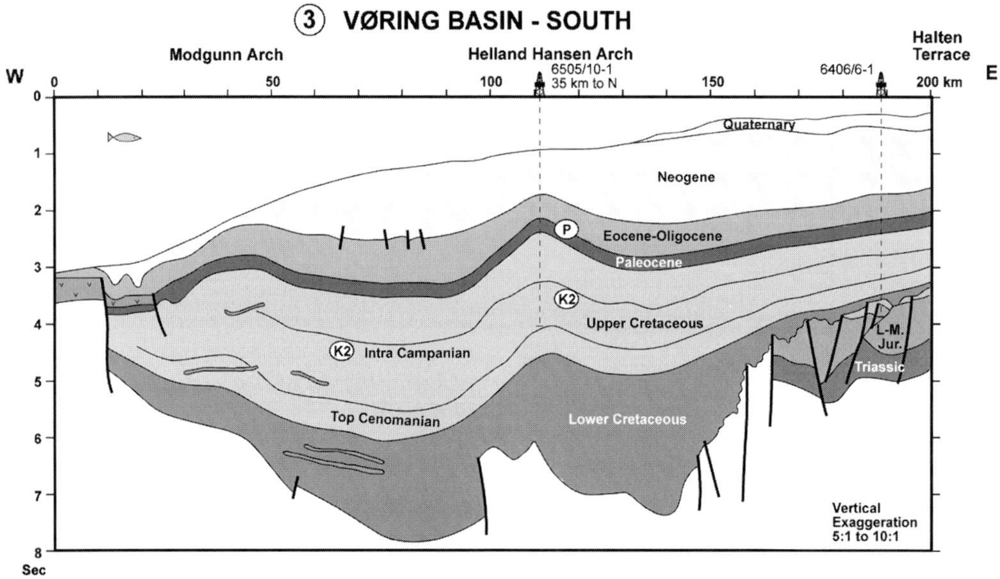

Fig. 4. Geological profile showing the plays of the Vøring Basin – south. Profile based on Blystad *et al.* (1995, profile J). See Figure 2 for location and Figure 3 for abbreviations.

The third main tectonic episode was the formation of compressional anticlines and domes in the Cenozoic, particularly in Miocene time, related to sinistral wrenching due to plate reorganizations in the Norwegian Sea (Doré & Lundin 1996).

Two main plays are envisaged – at Upper Cretaceous and Paleocene levels (Brekke *et al.* 1999) – and major gas finds have been made in both. In the Nyk High discovery to the north (Fig. 3), gas is contained in an Upper Cretaceous reservoir trapped in a latest Cretaceous–Paleocene age fault block (Kittilsen *et al.* 1999). Further south, drilling for Upper Cretaceous sandstones (Sanchez-Ferrer *et al.* 1999) has revealed the sandstones to be almost absent there. Provenance studies have suggested that the Cretaceous sandstones were not just derived from Scandinavia, but came also from source areas in Greenland (Morton & Grant 1998). The second find, the Ormen Lange discovery, occurs to the south on the margins of the Vøring and Møre basins in Paleocene sandstones in a Tertiary compressional anticline.

The source rocks for these hydrocarbons is problematical. Upper Jurassic source rocks, if present, had been buried so deeply in most areas that by late Cretaceous times they would have been over-mature. Shallower, 'frontier' source rocks are unproven but possible: at Lower Cretaceous, Cenomanian–Turonian or Paleocene levels (Doré *et al.* 1997). Sourcing hydrocarbons in the Vøring Basin must thus rely on re-migration from older, hypothetical hydrocarbon pools, themselves fed from Jurassic (or Lower Cretaceous) sources, or on the presence of speculative, shallower, source rocks.

Halten Terrace

This area is defined as the easternmost part of the Vøring Basin. It has a block-faulted structure at Jurassic level as a result of late Jurassic rifting (Fig. 5). Faulting continued into the early Cretaceous. The Cretaceous period was mainly characterized by the passive subsidence of the eastern Vøring Basin, so that the Cretaceous section steadily thickens westwards across the Halten Terrace. Minor inversion movements occurred in mid-Tertiary times. A Neogene tectonic episode affected the whole Mid-Norway shelf, with passive uplift and erosion onshore and along the coast and accelerated subsidence further west. The proven hydrocarbon system here is related mostly to the Upper Jurassic oil-prone source rock, but a gas-prone coal sequence also occurs in the Lower Jurassic. The rank of these sources increases from immature in the east to over-mature in the west (Koch & Heum 1995). Most discoveries occur in the pre-rift play in Lower–Middle Jurassic reservoirs. Some wells have drilled syn-rift Upper Jurassic targets with one success (Draugen) and a few wells have been aimed at post-rift Cretaceous structural and stratigraphic traps with limited success.

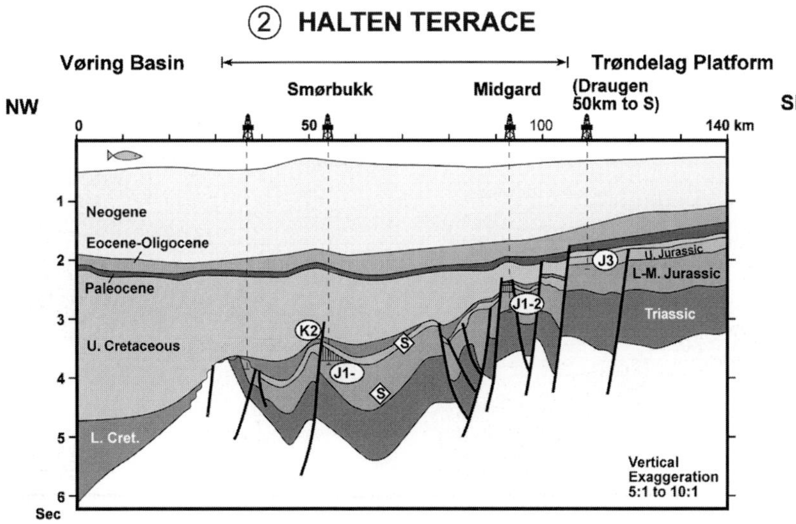

Fig. 5. Geological profile showing the proven plays of the Halten Terrace. Profile based on Blystad *et al.* (1995, profile F). See Figure 2 for location and Figure 3 for abbreviations.

Møre Basin

The Møre Basin comprises two regions. First, on the southeast side of the basin, a NE-trending fault-block province appears to cut across the north-trending structures of the northern North Sea (Fig. 2). The NE-trending faulting appears to have started in Volgian times but major offsets in the base Cretaceous horizon confirm that faulting continued into early Cretaceous times. The geometry of these fault blocks has been interpreted in terms of giant footwall collapse into the deeply subsiding Møre Basin (Graue 1992, fig. 6; Jongepier *et al.* 1996, fig. 9). The second, main, part of the basin has a much gentler structure and a huge thickness of Cretaceous section, like the Vøring Basin. Faulting related to the latest Cretaceous to Paleocene rifting episode of the Vøring Basin is not observed in the Møre Basin. Mid-Cenozoic compression resulted in a few large domes and anticlines (e.g. Ormen Lange). The Møre marginal high to the west of the Faroe–Shetland Escarpment was probably uplifted and eroded in latest Cretaceous to Paleocene times before being covered by late Paleocene–Eocene subaerial basalts.

All of the 17 exploration wells drilled are in the southeastern fault-block province in the UK and Norwegian sectors. This exploration has focused on pre-rift Jurassic targets in the steeply tilted fault blocks (Graue 1992; Jongepier *et al.* 1996). A few wells have been drilled to explore for onlapping Lower Cretaceous submarine fan sandstones. Upper Cretaceous and Paleocene submarine fan sandstones are almost unexplored and westerly-derived Paleocene sands may occur (Doré *et al.* 1997, fig. 7). Over the main, deep, part of the basin, the sourcing of hydrocarbons will again be problematical. One potential hazard for drilling and development is a recent feature on the sea bed known as the Storegga Slide; it is one of the largest submarine slides known in the world (Bugge *et al.* 1988).

Northern North Sea

This area is adjacent to, but not one of, the Atlantic margin basins. It is included here because its extremely well known petroleum system can form a reference with which to compare the Atlantic margin basins. Late Jurassic rifting created fault-block traps that are now buried beneath a sealing Cretaceous and Tertiary post-rift cover (Fig. 6). The principal source rocks are 'Kimmeridgian' shales of late Jurassic to earliest Cretaceous age, which achieved maturity during continuous later subsidence and burial. Oil generation began over wide areas in Eocene time and gas generation began in the areas of deepest subsidence during the Neogene. The pre-rift play (Triassic and Lower–Middle Jurassic reservoirs) is the most important and comprises 75% of the discoveries. The syn-rift play is defined as including Upper Jurassic and Lower Cretaceous reservoirs. The post-rift play is a minor play in the northern part of the northern North Sea.

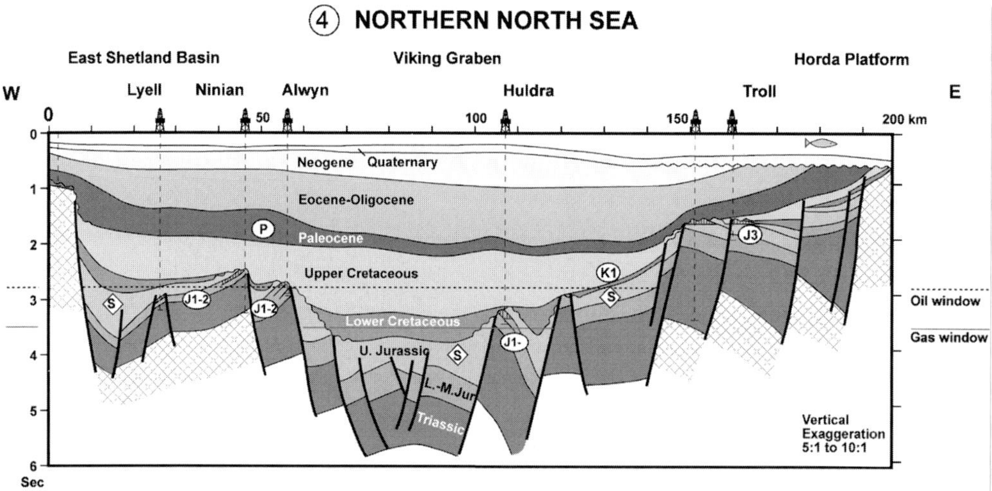

Fig. 6. Geological profile showing the proven plays of the northern North Sea. Profile based on Nopec NNST-84-6 seismic line. See Figure 2 for location and Figure 3 for abbreviations.

Faroe–Shetland Basin

The Faroe–Shetland Basin (FSB) is the continuation of the fault-block province of the Møre Basin (Fig. 2) and, on its eastern flank, comprises a series of NE- or north-trending faulted ridges (Fig. 7). Only wells on the southeastern ridges have reached Jurassic strata, which rarely exceed 100 m in thickness, often comprise Upper Jurassic shales and commonly rest unconformably on Triassic or basement rocks (Stoker *et al.* 1993, fig. 38). To the NW, Cretaceous strata reach several kilometres in thickness and are succeeded by Paleocene sediments up to 3 km thick. These thick successions are the results of two major rifting phases (Duindam, 1987; Earle *et al.* 1989): early Cretaceous, especially in the southeast, and Campanian to Paleocene. Inversion took place in Paleogene and Miocene times (Earle *et al.* 1989; Ebdon *et al.* 1995). Paleocene–Eocene volcanism produced the lava pile of the Faroes Shelf to the west, a major sill complex intruding Upper Cretaceous shales (Stoker *et al.* 1993) and the Erlend (Gatliff *et al.* 1984) and other intrusive complexes.

The FSB has been intensively explored in the UK sector. Drilling began in 1972 and was followed in 1977 by the discovery of the giant Clair oil field, reservoired in Devonian sandstones and Precambrian basement (Ridd 1981). The next large oil discovery was in Paleocene reservoirs at Foinaven in 1992 (Cooper *et al.* 1999; Lamers & Carmichael 1999). Numerous exploration plays have been proven: footwall traps with old reservoirs on the eroded ridges (Clair); Triassic and Jurassic sandstones (Herries *et al.* 1999); Lower Cretaceous sandstones (Goodchild *et al.* 1999) and Paleocene stratigraphic traps (Foinaven, Schiehallion etc).

The latter discoveries have led to a surge in understanding of the Paleogene sequences which built out from the eroding Shetland Platform (Mitchell *et al.* 1993; Ebdon *et al.* 1995). The deep marine basin formed in late Cretaceous times began to accumulate submarine fans in early Paleocene times and was progressively infilled, so that by the early Eocene (at the time of the main volcanism), deltaic conditions were achieved. The petroleum generation and migration history appears to have been complex. Bailey *et al.* (1987) inferred that the Kimmeridgian to Ryazanian shales could have sourced major petroleum accumulations. Recent studies have confirmed this, emphasizing that this sourcing system appears very oil prone and that Middle Jurassic lacustrine source rocks are also involved (Scotchman *et al.* 1998). As a result of the major Cretaceous and Paleocene burial, Jurassic sources will have become mature early. Also, the oils commonly contain both biodegraded and 'fresher' components, suggesting a complex charging history, sometimes involving re-migration from Mesozoic to Paleogene traps (Holmes *et al.* 1999; Iliffe *et al.* 1999).

Faroes Shelf

The Faroes Shelf is almost completely covered by Paleocene–Eocene basalts. It is limited northwards by a zone of seaward-dipping

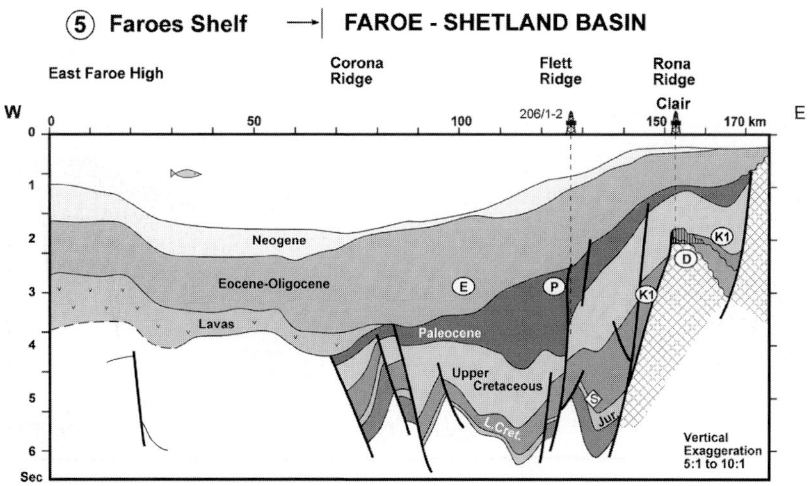

Fig. 7. Geological profile showing the proven plays of the Faroe–Shetland Basin. See Figure 2 for location and Figure 3 for abbreviations.

reflectors and abuts the Rockall Basin to the SW (Figs 1, 2). The Paleocene–Eocene basalt series is at least 5.5 km thick on the Faroe Islands (3.5 km exposed and >2 km drilled; Waagstein 1988) and may total up to 7 km. The lavas covering most of the shelf were erupted subaerially and volcanic escarpments mark the contemporary shorelines in the east (Smythe *et al.* 1983) and close to the margin of Rockall Basin in the south (Boldreel & Andersen 1994). The lavas are affected by anticlines with varied trends (Wyville–Thomson and Munkagrunnar ridges and East Faroe High) resulting from compressional phases in Eocene to Miocene times (Boldreel & Andersen 1993).

The structure and stratigraphy of the sub-basalt interval is enigmatic but is likely to include a thick sedimentary interval below the shelf in the east (Richardson *et al.* 1999). Paleocene and Mesozoic sedimentary rocks may be present extending from the FSB and could contain petroleum plays. The late Cretaceous to Paleocene sedimentary section below the basalts in the Kangerlussuaq area of East Greenland may provide an analogue (Larsen *et al.* 1999). Note that traces of hydrocarbons have been recorded in the basalts from the Faroe Islands (Laier *et al.* 1997), in the form of waxy coatings in vesicles and as tiny amounts of gas in the Lopra 1 well.

UK Rockall Basin

The northeastern part of the Rockall Basin lies in UK-designated waters and it is this area that is described here. The basin is bordered to the east by the Hebrides Platform which contains deep Triassic fault basins and, in the Skye region, a Jurassic sequence (Hettangian to Callovian) which is 800 m thick (Fyfe *et al.* 1993; Morton 1993). On the east flank of the basin (Fig. 8) Musgrove & Mitchener (1996) interpreted early Cretaceous rifting based on well 132/15-1. In the northeast a suite of fault blocks is clearly seen on seismic data (e.g. Tate *et al.* 1999) and is interpreted here as Cretaceous and Paleocene in age. Throughout most of the basin, however, the Mesozoic section is obscured by extensive Paleocene–Eocene lavas (Wood & Hall, 1987; Wood *et al.*, 1988) and by intrusive seamounts and major igneous centres (e.g. Evans *et al.* 1989; Abraham & Ritchie 1991). Two of these centres, Anton Dohrn and Rosemary Bank, have been dated as Maastrichtian or older (Jones *et al.* 1974; Morton *et al.* 1995).

Only five exploration wells have been drilled: three just outside the NE flank of the basin (1988–91); one in 1992 (132/15-1) just inside the SE flank (Fig. 8; Musgrove & Mitchener 1996); and a deep water stratigraphic test in 1980 (163/6-1A), which found 1200 m of Tertiary strata above 1045 m of Paleocene basalt and dacite (Morton *et al.* 1988). Seismic interpretations suggest that Jurassic strata may be present at least in the eastern part of the basin, providing possible sources for hydrocarbons. Lower Cretaceous sandstones may act as reservoirs in the fault traps and Paleocene and Eocene basin floor fans may also be present (Waddams & Cordingley 1999).

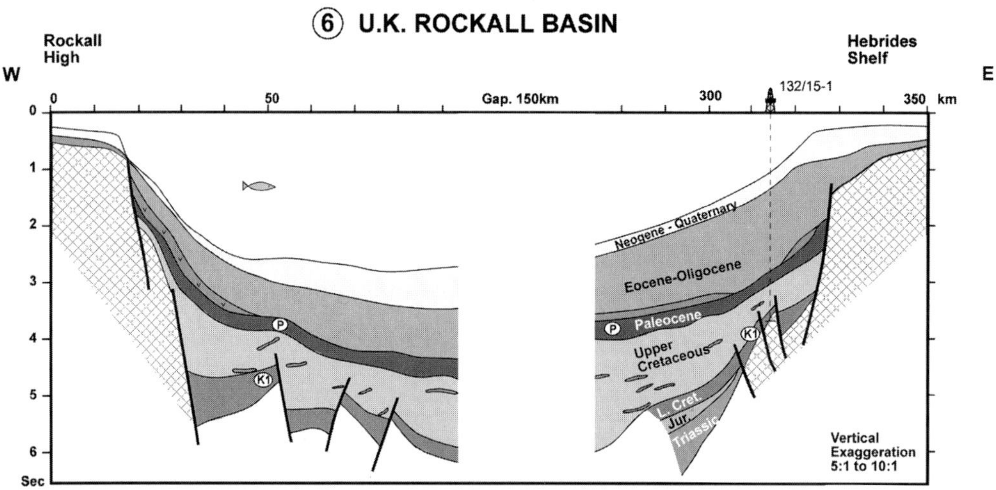

Fig. 8. Geological profile showing the possible plays of the south of the UK Rockall Basin. Profile based on seismic lines M89-WB-2 (west) and BP132-91-287 (east). See Figure 1 for location and Figure 3 for abbreviations.

Irish Atlantic basins

Irish Rockall Basin

The major part of the Rockall Basin lies in Irish-designated waters. Here the basin is bounded to the east by faulted margins against the Erris Ridge, the Slyne and Erris basins and Porcupine Bank. A faulted margin against the Rockall High forms the western limit of the basin and the Charlie–Gibbs Fracture Zone delimits the southern extent. The southernmost part of the basin is occupied by a major Cretaceous volcanic province (Scrutton & Bentley 1988). Three main structural trends appear to have influenced the evolution of the basin: NE–SW, N–S and NW–SE (Corfield *et al.* 1999). The first two are the most important in terms of Mesozoic basin evolution and define the trends of the basin-bounding faults. The NW–SE lineaments are interpreted as transfer zones similar to those seen in the Faroe–Shetland and Vøring regions. A structural nomenclature for the Rockall Basin has recently been proposed by D. Naylor *et al.* (1999).

No exploration wells have been drilled in the Irish Rockall Basin. A single exploration well is located in the UK sector just to the north (132/15-1) and proved early Cretaceous strata resting on crystalline basement (Musgrove & Mitchener 1996). In 1997, eleven licenses, each consisting of three or more blocks, were awarded in the Irish Rockall Round, mostly in the NE and SE but with one on the west flank of the basin.

Crustal structure and evolution. Interpretations of the crustal structure of the Rockall Basin have long been uncertain. Roberts (1975) and Scrutton (1986) made the first interpretations that the thin crust below the basin was oceanic and of late Jurassic to early Cretaceous age or of Cretaceous age, respectively. Later seismic refraction and reflection studies revealed a highly stretched continental crust (Roberts *et al.* 1988; O'Reilly *et al.* 1995). In detail, the structure below the basin comprises a three-part sedimentary sequence up to 5 km thick, overlying a crustal layer that is only 5–6 km thick (O'Reilly *et al.* 1995). The three-part sedimentary succession has been interpreted as comprising a deep section (?Palaeozoic to Jurassic rocks) rifted in Cretaceous times, overlain by post-rift Cretaceous to Paleocene and Eocene to Recent layers (Shannon *et al.* 1995, 1999). The highly attenuated crust (5–6 km thick compared with *c.* 30 km below Ireland) implies a stretching factor of as much as 6, suggesting that multiple rifting events may have affected the basin.

Petroleum geology. Half-grabens and tilted fault blocks are present in parts of the basin and are most clearly seen on its margins (Fig. 9).

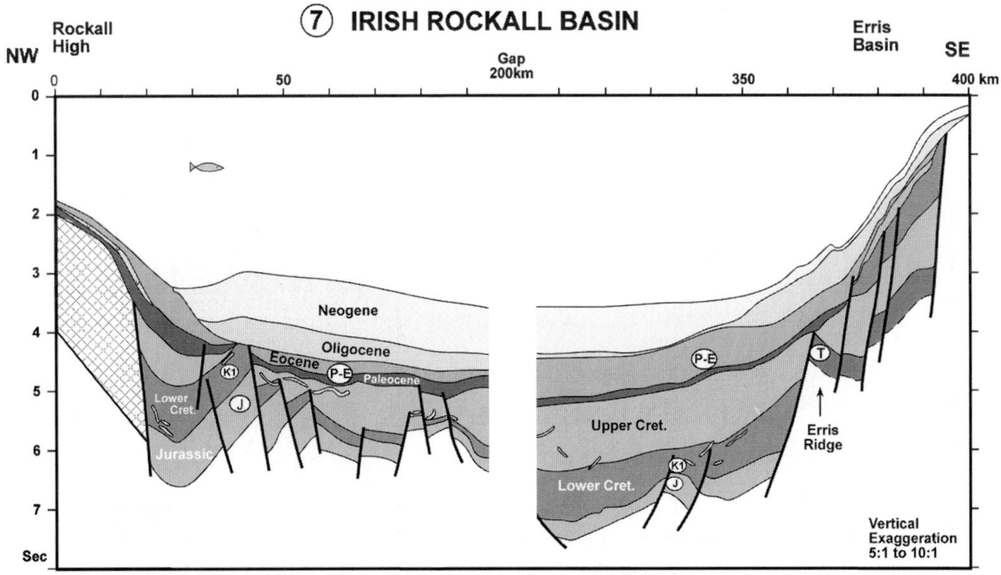

Fig. 9. Geological profile showing the possible plays of the Irish Rockall Basin. Profile based on seismic lines GSR96-0116-2116 (west) and WRM96-103 (east). See Figure 1 for location and Figure 3 for abbreviations.

Pre-Cretaceous strata are considered to be similar to those found in the Slyne, Erris, Hebrides and Faroe–Shetland basins. Upper Jurassic rocks will be the most significant source rocks and Jurassic and possibly older rocks may be important reservoirs (Walsh *et al.* 1999). The fault blocks seen on both margins of the basin are interpreted to be of early Cretaceous age. Cretaceous and Cenozoic strata thin rapidly away from the margins, reflecting increased thermal subsidence exceeding sedimentation. Both Lower Cretaceous and Paleocene to Lower Eocene intervals may contain submarine fan sandstone reservoirs. Assuming a petroleum system is present in the Rockall Basin, the main exploration challenges are to understand these plays, define the traps and predict the hydrocarbon phase in the prospects.

Slyne and Erris basins

The Slyne Basin is a 35 km wide eroded graben system which trends NNE for 200 km across the Palaeozoic and Precambrian basement rocks of the shelf off northwest Ireland (Fig. 1). The graben is deepest in the south (Fig. 10). To the north it joins with the 150 km long Erris Basin (Figs 9, 10). To the NE of the Erris Basin lies the Donegal Basin (Dobson & Whittington 1992) which, though largely filled with Palaeozoic strata (e.g. Carboniferous section in well 13/3-1), could contain Triassic or Cretaceous strata to the west or SW where the boundary with the Erris Basin is ill-defined. The Erris Basin is separated from the NE-trending margin fault of the Rockall Basin by the narrow 'Erris Ridge' (Cunningham & Shannon 1997). The existence of these basins was first revealed by early 'sparker' reflection surveys (Bailey *et al.* 1977). Six petroleum exploration wells were drilled in the period 1978–99, resulting in one gas discovery (Corrib).

Structural evolution. The Slyne and Erris basins preserve a thick Jurassic section (over 2200 m in well 27/13-1; Scotchman & Thomas 1995), overlying Triassic and Zechstein strata which rest unconformably on Carboniferous rocks. In the north, in the Erris Basin, the unconformably overlying Cretaceous strata reach over 1 km in thickness (Chapman *et al.* 1999). A Triassic rift episode has been inferred. Major fault activity is suggested by the huge thickness of the Jurassic section, estimated at up to 2.5 km, which has been dated as Aalenian to Bathonian (Dancer *et al.* 1999). Late Jurassic strata have not been encountered, but in early Cretaceous time, major footwall uplift on the NE-trending Rockall Basin margin fault led to erosion of the Erris Basin in the 'Erris Ridge' zone (Chapman *et al.* 1999, fig. 7c). The area of the Slyne and Erris basins has been affected by regional uplifts in mid-Cretaceous and Oligocene/Miocene times which together amount to 1–2 km (Scotchman & Thomas 1995).

Petroleum geology. The Jurassic lithostratigraphy is similar to that of Skye, so the same formation names may be applied (Trueblood 1992). Source rocks and maturity have been carefully investigated in well 27/13-1 (Scotchman & Thomas 1995). Two oil-prone Liassic source rocks are present in the well (the Toarcian

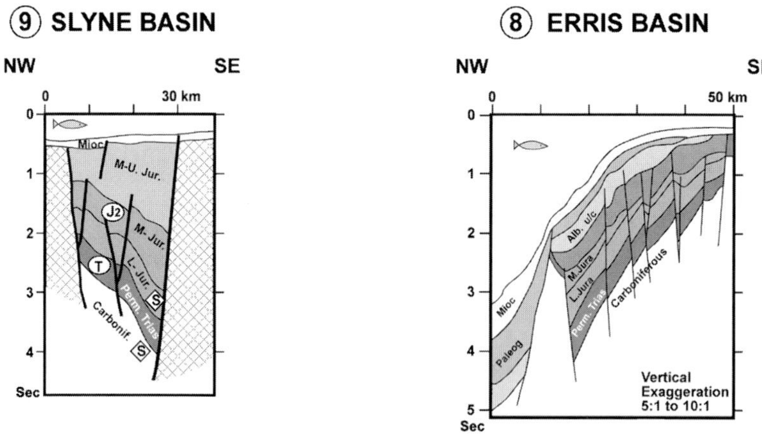

Fig. 10. Geological profiles showing the plays of the Slyne and Erris basins. See Figure 1 for location and Figure 3 for abbreviations.

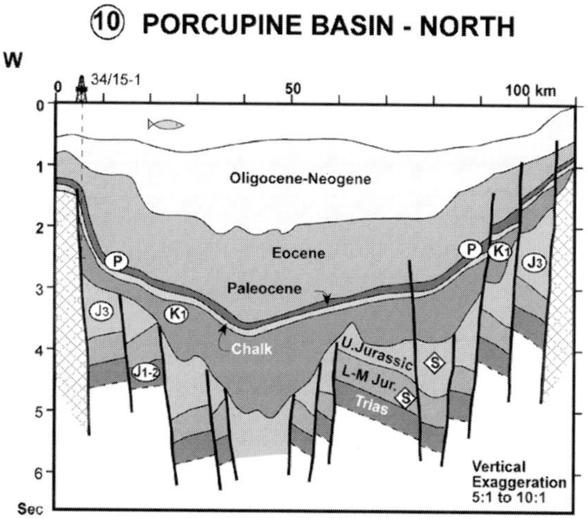

Fig. 11. Geological profile showing the proven plays of the Porcupine Basin – north. See Figure 1 for location and Figure 3 for abbreviations.

'Portree Shale' and the Sinemurian–Hettangian 'Pabba Shale'). They appear to be the source for the oil shows in the Middle Jurassic sandstones. Vitrinite studies and basin modelling suggest that the top of the palaeo-oil window occurs today at 2.6 km, corresponding to 3.6 to 4.2 km at maximum burial. Hydrocarbon generation occurred in late Jurassic and again in late Cretaceous to early Tertiary times. The oil shows in the well are biodegraded and water-washed, probably due to breaching of the structure during the later uplifts. These results of Scotchman & Thomas (1995) indicate that a Jurassic (Liassic) petroleum system is present in the Slyne Basin. However, it is not known whether any petroleum accumulations belonging to this petroleum system have survived the Cretaceous and Tertiary uplift phases.

Deeper, gas-prone source rocks have been penetrated in the 19/5-1 well in thick Namurian–Westphalian claystones (Murphy & Croker 1992). The Corrib gas field in Triassic sandstones demonstrates a second working petroleum system, probably sourced from such Carboniferous strata.

Porcupine Basin

The Porcupine Basin is bordered on the east, north and west by platform areas composed of Palaeozoic and Precambrian rocks. To the south it widens and its sedimentary fill thickens, with Cenozoic and Cretaceous strata reaching 4 km and 6 km in thickness, respectively (Tate 1993). The first well was drilled in 1977 and oil was first tested in 1978. A total of 26 exploration wells have been drilled, mostly in the northern third of the basin, of which four wells have recorded flows of hydrocarbons (see Croker & Shannon 1987; Naylor 1996).

Structural evolution. The wells prove a Middle to Upper Jurassic section that rests unconformably on thick Permo-Carboniferous strata (Tate & Dobson 1989). Triassic and Liassic strata are almost always absent except in the extreme north. Middle Jurassic sediments are commonly non-marine. Upper Jurassic transgressive sediments, with rapidly varying facies, accumulated during the east–west rifting that created the symmetrical, block-faulted structure clearly seen in the north (Fig. 11). Greater extension is inferred in the south (Tate *et al.* 1993). Cretaceous and Tertiary sediments passively filled the rift trough, with shallow marine facies belts being replaced by deep marine sediments southwards. A median ridge in the south (Fig. 12) was inferred to be of volcanic origin and of early Cretaceous age (Tate & Dobson 1988); a recent interpretation, however, proposes a non-volcanic origin as a peridotite ridge (Pennell *et al.* 1999). Other, scattered, volcanic rocks include: sills intruded at Lower Cretaceous levels in some wells; Danian lavas in wells 35/2-1 and 26/29-1; and the Brendan Igneous Centre (presumed to be late Cretaceous to Paleocene) located NE of the basin (see Naylor 1998).

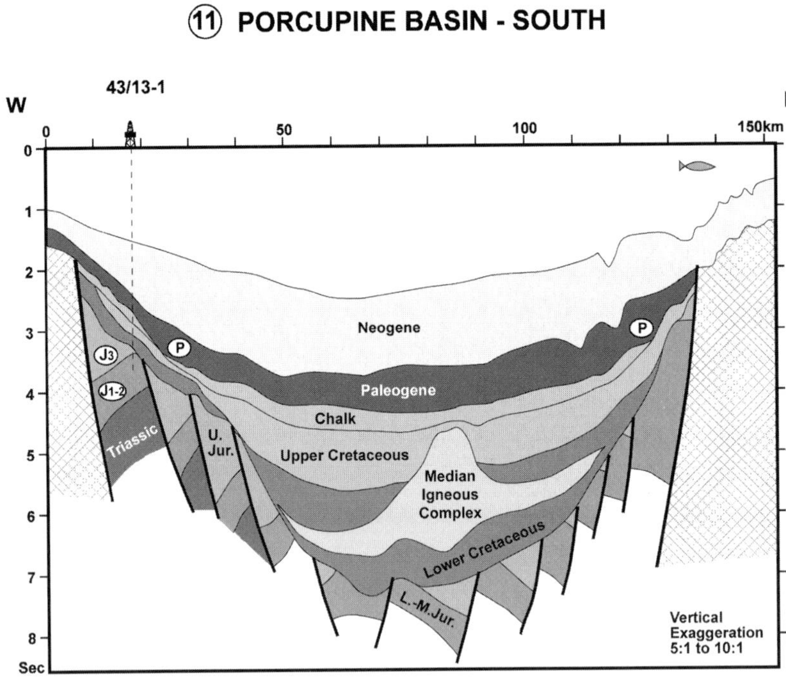

Fig. 12. Geological profile showing the plays of the Porcupine Basin – south. See Figure 1 for location and Figure 3 for abbreviations.

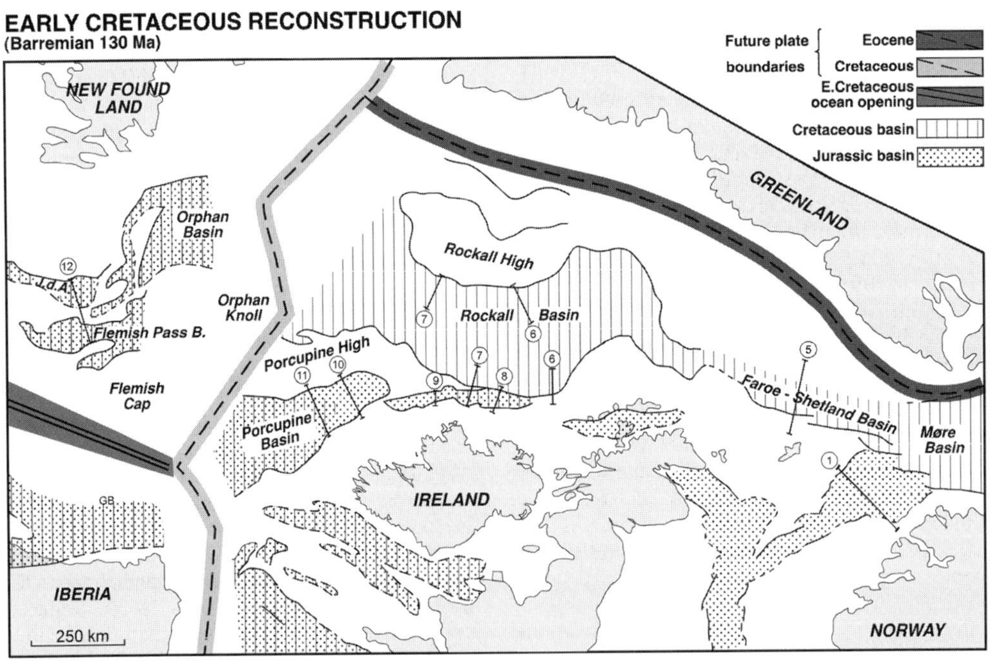

Fig. 13. Early Cretaceous continental reconstruction (Barremian, 130 Ma) of the Newfoundland to Ireland area showing the Atlantic margin basins.

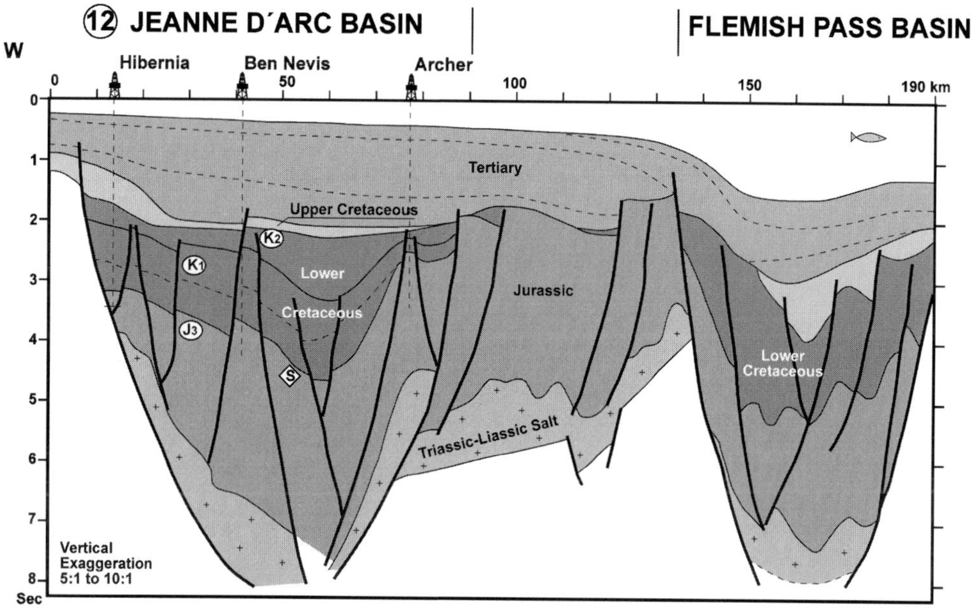

Fig. 14. Geological profile showing the plays of the Jeanne d'Arc Basin. Profile based on Enachescu (1987). See Figure 13 for location and Figure 3 for abbreviations.

Petroleum geology. Most of the exploration drilling has been directed at the Jurassic tilted fault blocks in the north. The Connemara accumulation has such a trap and contains oil in thin Upper Jurassic sandstones which are laterally impersistent (MacDonald *et al.* 1987). That oil was inferred to have been sourced from late Jurassic mudstones down-dip, with the oil-generation threshold lying today at approximately 2.2 km. The Upper Jurassic, especially the Kimmeridgian, was originally regarded as the single most important source rock interval in the basin (Croker & Shannon 1987). Butterworth *et al.* (1999) suggested, however, that the Porcupine oils have been derived from a mix of Middle Jurassic lacustrine source and an atypical marine Upper Jurassic source type. Lower Cretaceous deltaic to deep marine sandstones provide one post-rift play and oil shows have been found in Barremian sandstones in two wells (Moore & Shannon 1995). Paleocene to Eocene submarine fans interpreted on seismic lines (Shannon 1992) represent a second, largely untested, play type with possible stratigraphic traps.

Newfoundland Atlantic basins

The continental margin NE of Newfoundland contains several sedimentary basins of which the Jeanne d'Arc Basin has seen the most petroleum exploration and discovery. Prior to Atlantic Ocean opening, this area lay adjacent to the southern Porcupine and Rockall basins (Fig. 13). Rifting began between Africa and North America in the Middle Jurassic, separating Iberia and Newfoundland in early Cretaceous time and propagating north to separate Newfoundland and the Irish region in mid- to late Cretaceous times (Johnston *et al.* 2001).

The northeast Newfoundland continental margin has therefore undergone a complex Mesozoic extensional history, involving late Triassic to early Jurassic, late Jurassic to early Cretaceous and Aptian–Albian phases (Enachescu 1987; Tankard *et al.* 1989). Salt was deposited widely in late Triassic to early Jurassic times and has undergone halokinetic movements, further complicating the structural development (Fig. 14). The main source rock is a restricted marine Kimmeridgian shale. Multiple reservoirs are present – fluvial sandstones in the Upper Jurassic, deltaic sandstones at the base of the Cretaceous and shallow marine sandstones in the Barremian to Aptian (DeSilva 1999). Some of the hydrocarbon finds are also segmented structurally, with over 30 fault blocks identified in one reservoir in the giant Hibernia Field (Sinclair *et al.* 1999).

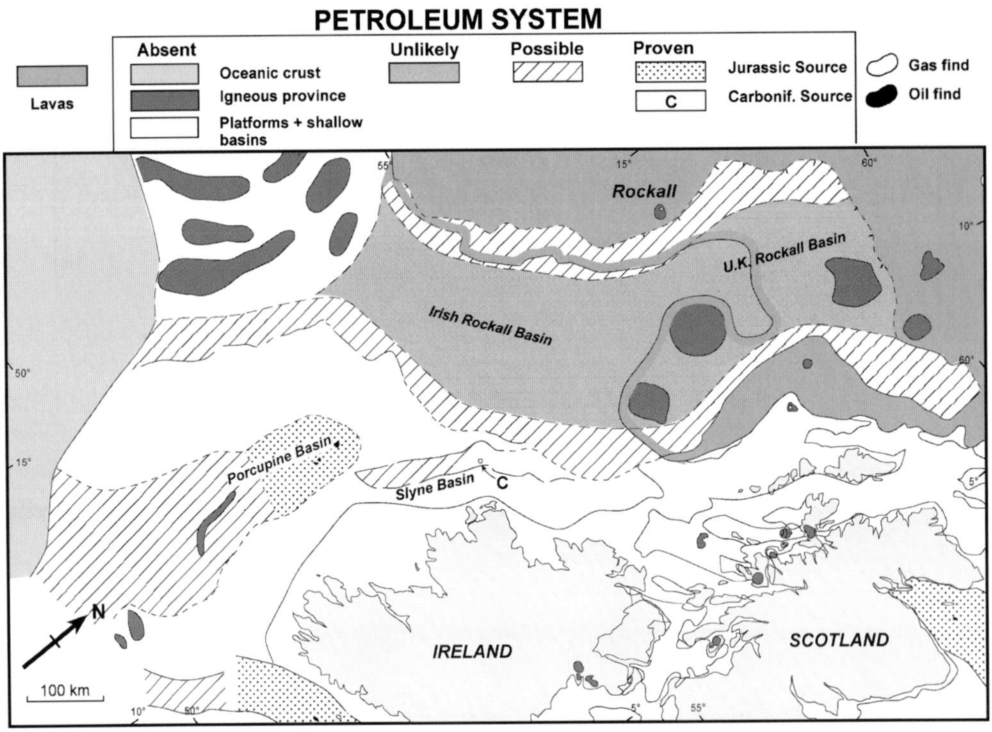

Fig. 15. Map of petroleum systems of the Atlantic margin: Ireland to Scotland.

Petroleum systems

The basins of the Atlantic margin reviewed here can be grouped according to whether a petroleum system is proven, possible, unlikely or absent (Figs 15, 16). Areas where the presence of a petroleum system has been proven include the established petroleum provinces of the Halten Terrace, the northern North Sea, the northern Porcupine Basin and the Jeanne d'Arc Basin. In these areas petroleum systems involve generation from known Jurassic source rocks. In the first three areas the maturity history is largely understood and the hydrocarbon finds result from direct, one-step, migration from mature kitchen areas to the traps.

In the Faroe–Shetland Basin there has been much drilling and the 13 discoveries there confirm the existence of a petroleum system which is believed to be fed from Jurassic source rocks. Large Cretaceous and Paleogene subsidence led to early oil generation with the result that finds in Paleogene reservoirs may have been fed by re-migration (two-step) from deeper hydrocarbon pools.

In the Slyne Basin, a Jurassic petroleum system is possible but may have been largely destroyed as a result of Cenozoic uplift and erosion. The Corrib gas field in the Slyne Basin belongs to a quite separate petroleum system sourced almost certainly from Carboniferous coals. A great Carboniferous petroleum system extends from the Ukraine and Poland to the southern North Sea and the Irish Sea. The Corrib source beds may, however, represent a continuation of the Scottish Midland Valley setting, rather than the 'foreland' setting of the southern North Sea.

On the 'outboard' side of the Atlantic margin are the mostly frontier status areas (the Vøring and Møre basins, the Faroes Shelf and the Rockall and southern Porcupine basins). The Nyk High and Ormen Lange discoveries prove the existence of petroleum systems in the Vøring Basin, but the source rocks and migration routes that fed them are unknown. If sourced from Jurassic source rocks, re-migration would be needed. Although the extent of the petroleum systems away from the two discovery wells is unknown, the wide extent of the petroleum systems to the east and south and the (destroyed) petroleum system onshore in East Greenland (Price & Witham 1997) give encouragement that they will prove to be widespread within the

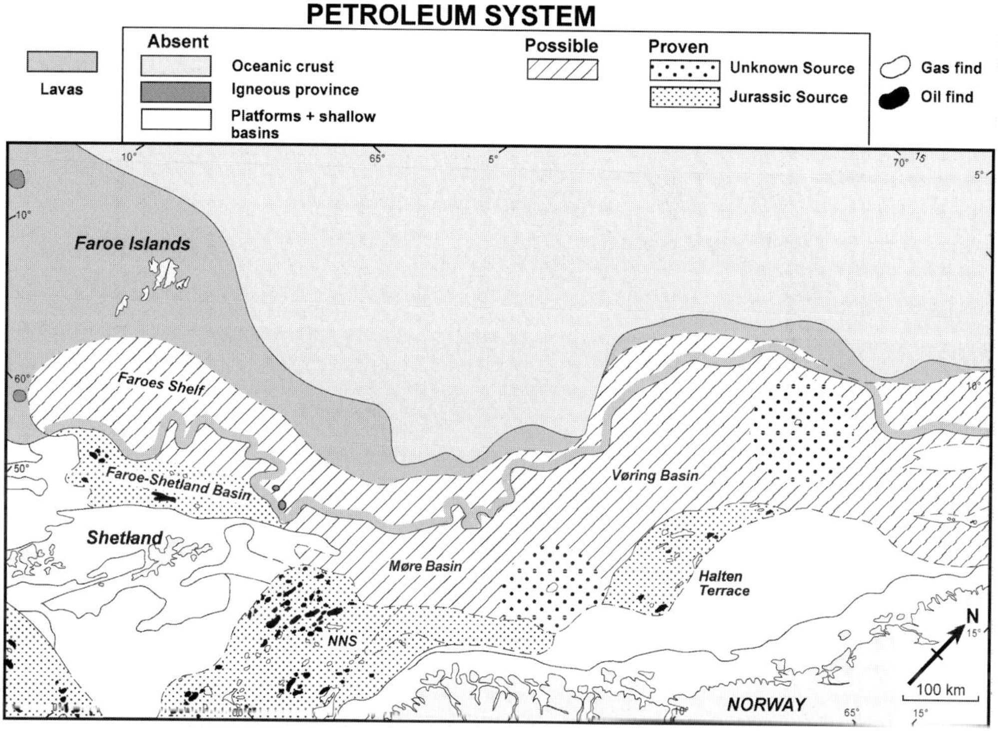

Fig. 16. Map of petroleum systems of the Atlantic margin: Faroe–Shetland to Norway.

Vøring and Møre basins. On the Faroes Shelf the presence of a petroleum system is unknown, apart from the hydrocarbon traces in the basalts described by Laier *et al.* (1997). Future exploration will reveal whether the proven petroleum system of the FSB extends westwards beneath the Faroes Shelf.

In the Rockall Basin there has been almost no drilling in deep water but the inferred geology is favourable to the existence of a petroleum system along the flanks of the basin. The central zone of the basin has probably undergone so much crustal stretching that deep source potential will have been destroyed and so the existence of a petroleum system is judged unlikely there. In the southern Porcupine Basin, the existence of a petroleum system is unproven but possible. Proven Jurassic petroleum systems exist in the north of the basin and in the Jeanne d'Arc Basin which, prior to Atlantic opening, lay along trend to the south.

Exploration potential of the Irish Atlantic margin basins

The Rockall, Slyne, Erris and Porcupine basins and the related basins along the Atlantic margin form a family of basins that have many similar features in their geological development and petroleum systems. The basins can also be compared in their exploration status. Table 1 gives statistics on the prospective areas, exploration wells and discoveries in the basins and of the presence of petroleum systems and exploration plays. For the frontier basins, Table 1 lists the elements of the petroleum geology that are unknown. Also shown are the major technical challenges facing future exploration in all of the basins. Can the knowledge and lessons from the other basins assist in assessing the exploration potential of the Irish basins?

The eastern flank of the Rockall Basin (Figs 8, 9) is a continuation of a major tectonic trend along the Atlantic margin: the southeastern flanks of the Faroe–Shetland (Fig. 7) and Møre basins and the eastern flank of the Vøring Basin against the Halten Terrace area (Figs 3–5). Faulted ridges and terraces along this trend have yielded discoveries in Mesozoic and Devonian reservoirs, all supplied from Jurassic sources. Along the same trend, the Paleocene play has proved successful in submarine fan sandstones in the Faroe–Shetland Basin and in the southern Vøring Basin, with the discoveries reservoired

Table 1. *Exploration status of the Atlantic margin basins*

Basin	Area ($\times10^3$ km^2)	Exploration wells	Finds	Petroleum system	Plays	Unknowns	Challenges
North Vøring	50	4	1	Proven	P, K2	Source and supply of hydrocarbons	Reservoir distribution Phase prediction
South Vøring	25	4	1	Proven	P, K2	Source and supply of hydrocarbons	Reservoir distribution Phase prediction
Halten Terrace	15	115	22	Jurassic, proven	K2, J3, J1–2		K2 reservoir distribution High pressure/high pressure prediction
Møre	40	17	2	SE – Jurassic, proven; Main basin – possible	P, K2, K1, J1–2	Existence of plays: source, reservoirs	Phase prediction: remigration, trap definition
Northern North Sea	40	600	110	Jurassic, proven	P, K1, J3, J1–2, T		Definition of subtle traps
Faroe–Shetland	20	150	13	Jurassic, proven	E, P, K1, J, D	Hydrocarbon supply history	Reservoir distribution Trap definition
Faroes Shelf	15	0	0	Possible	P, K1–J	Subbasalt stratigraphy Existence of plays	Seismic imaging below basalts
UK Rockall	W: 10 E: 13	W=0 E=6	0	Possible	E,P, K1, J	Existence of plays: source, reservoirs	Seismic imaging below basalts Phase prediction
Irish Rockall	W: 15	0	0	Possible	P–E, K1, J, T	Existence of plays: source, reservoirs	Trap definition Phase prediction
Slyne & Erris	Erris: 7 Slyne: 5	6	1	Jurassic possible; ?Carboniferous proven	J2, T		Effect of uplift on trapping Extents of plays
Porcupine Basin – North	15	26	3	Jurassic, proven	P–E, K1,J3, J1–2		Reservoir distribution
Porcupine Basin – South	27	1	0	Possible	P, J3, J1–2	Existence of plays and reservoirs	Reservoir distribution
Jeanne d'Arc	15	80	16	Proven	K2, K1, J3		Fault segmentation of traps Multiple reservoirs

Key for plays: E, Eocene; P, Paleocene; K2, Upper Cretaceous; K1, Lower Cretaceous; J3, Upper Jurassic; J1–2, Lower–Middle Jurassic; J, Jurassic; T, Triassic; D, Devonian.

in, respectively, stratigraphic–structural traps on a regional monocline and a major inversion anticline. Both Mesozoic fault blocks and Paleogene monoclinal stratigraphic traps are possible on the east flank of the Rockall Basin but prospectivity will depend on the presence of source and reservoir rocks there.

The western flank of the Rockall Basin (Fig. 9) is the least well known of the sub-basins. There are no clear analogues along-strike to the northeast, but the equivalent areas are covered by Paleocene–Eocene basalts northwards to the west flank of the Vøring Basin, where one well has been drilled on the northern Gjallar Ridge (1999, 6704/12-1).

In the narrow, grabenal Slyne and Erris basins (Fig. 10), the largely destroyed Jurassic petroleum system has no analogues along-strike to the NE. The destroyed Jurassic petroleum system in East Greenland (Price & Witham 1997) may be an analogue. The Carboniferous petroleum system of the Slyne–Erris area may prove prolific within that small basin system. In terms of analogues, however, this petroleum system may be unique in its combination of source setting, grabenal basin structure and Neogene uplift which is not repeated in the basins with proven Carboniferous petroleum systems to the east.

The northern Porcupine Basin has seen the most drilling of the Irish Atlantic basins, because of the numerous fault blocks and the relatively shallow water (Fig. 11). The results have been modest and future potential will depend upon the success of new plays. The southern Porcupine Basin is analogous, in its huge Cretaceous and Tertiary thicknesses (Fig. 12), to the Vøring and Møre basins (Figs 3, 4), suggesting that exploration focus should be at Tertiary or Cretaceous levels or on the basin flanks.

The main requirement for prospectivity in the Irish basins is the presence of viable source rocks. The prolific sources present in the Halten Terrace, northern North Sea and Faroe–Shetland basins have resulted in discoveries in those three basins totalling *c.* 50 billion BOE recoverable. The Irish Atlantic margin basins, which cover an even larger area, show many positive indicators of working hydrocarbon systems, giving hope that future exploration will yield commercial discoveries.

The authors wish to thank an anonymous referee and Peter Haughton for comments which have improved the article greatly. We thank Statoil for permission to publish.

References

ABRAHAM, D.A. & RITCHIE, J.D. 1991. The Darwin complex, a Tertiary igneous centre in the Northern Rockall Trough. *Scottish Journal of Geology*, **27**, 113–125.

BAILEY, R.J., JACKSON, P.D. & BENNELL, J.D. 1977. Marine geology of the Slyne Ridge. *Journal of the Geological Society, London*, **133**, 165–172.

BAILEY, N.J.L., WALKO, P. & SAUER, M.J. 1987. Geochemistry and source rock potential of the west of Shetlands. *In*: BROOKS, J. & GLENNIE, K.W. (eds) *Petroleum Geology of North West Europe*. Graham & Trotman, London, 711–721.

BLYSTAD, P., BREKKE, H., FÆRSETH, R.B., LARSEN, B.T., SKOGSEID, J. & TØRUDBAKKEN, B. 1995. Structural elements of the Norwegian continental shelf. Part II: the Norwegian Sea region. *Norwegian Petroleum Directorate Bulletin*, **8**.

BOLDREEL, L.O. & ANDERSEN, M.S. 1993. Late Paleocene to Miocene compression in the Faeroe–Rockall area. *In*: PARKER, J.R. (ed.) *Petroleum Geology of Northwest Europe: Proceedings of the 4th Conference*. Geological Society, London, 1025–1034.

BOLDREEL, L.O. & ANDERSEN, M.S. 1994. Tertiary development of the Faroe–Rockall Plateau based on reflection seismic data. *Geological Society, Denmark, Bulletin*, **41**, 162–180.

BREKKE, H., DAHLGREN, S., NYLAND, B. & MAGNUS, C. 1999. The prospectivity of the Vøring and Møre basins on the Norwegian Sea continental margin. *In*: FLEET, A.J. & BOLDY, S.A.R. (eds) *Petroleum Geology of NW Europe: Proceedings of the 5th Conference*. Geological Society, London, 261–274.

BUGGE, T., BELDERSON, R.H. & KENYON, N.H. 1988. The Storegga Slide. *Philosophical Transactions of the Royal Society, London*, **A325**, 357–388.

BUTTERWORTH, P., HOLBA, A., HERTIG, S., HUGHES, W. & ATKINSON, C. 1999. Jurassic non-marine source rocks and oils of the Porcupine Basin and other North Atlantic margin basins. *In*: FLEET, A.J. & BOLDY, S.A.R. (eds) *Petroleum Geology of NW Europe: Proceedings of the 5th Conference*. Geological Society, London, 471–486.

CHAPMAN, T.J., BROKS, T.M., CORCORAN, D.V., DUNCAN, L.A. & DANCER, P.N. 1999. The structural evolution of the Erris Trough, offshore northwest Ireland, and implications for hydrocarbon generation. *In*: FLEET, A.J. & BOLDY, S.A.R. (eds) *Petroleum Geology of NW Europe: Proceedings of the 5th Conference*. Geological Society, London, 455–469.

COOPER, M.M., EVANS, A.C., LYNCH, D.J., NEVILLE, G. & NEWLEY, T. 1999. The Foinaven Field: managing reservoir development uncertainty prior to start-up. *In*: FLEET, A.J. & BOLDY, S.A.R. (eds) *Petroleum Geology of NW Europe: Proceedings of the 5th Conference*. Geological Society, London, 675–682.

CORFIELD, S., MURPHY, N. & PARKER, S. 1999. The structural and stratigraphic framework of the Irish Rockall Trough. *In*: FLEET, A.J. & BOLDY, S.A.R.

(eds) *Petroleum Geology of NW Europe: Proceedings of the 5th Conference*. Geological Society, London, 407–420.

Croker, P.F. & Shannon, P.M. 1987. The evolution and hydrocarbon prospectivity of the Porcupine Basin, offshore Ireland. *In*: Brooks, J. & Glennie, K.W. (eds) *Petroleum Geology of North West Europe*. Graham & Trotman, London, 633–642.

Cunningham, G.A. & Shannon, P.M. 1997. The Erris Ridge: a major geological feature in the NW Irish offshore basins. *Journal of the Geological Society, London*, **154**, 503–508.

Dancer, P.N., Algar, S.T. & Wilson, I.R. 1999. Structural evolution of the Slyne Trough. *In*: Fleet, A.J. & Boldy, S.A.R. (eds) *Petroleum Geology of NW Europe: Proceedings of the 5th Conference*. Geological Society, London, 445–453.

DeSilva, N.R. 1999. Sedimentary basins and petroleum systems offshore Newfoundland and Labrador. *In*: Fleet, A.J. & Boldy, S.A.R. (eds) *Petroleum Geology of NW Europe: Proceedings of the 5th Conference*. Geological Society, London, 501–515.

Dobson, M.R. & Whittington, R.J. 1992. Aspects of the geology of the Malin Sea area. *In*: Parnell, J. (ed.) *Basins on the Atlantic Seaboard: petroleum geology, sedimentology and basin evolution*. Geological Society, London, Special Publications, **62**, 291–311.

Doré, A.G. & Lundin, E.R. 1996. Cenozoic compressional structures on the NE Atlantic margin: nature, origin and potential significance for hydrocarbon exploration. *Petroleum Geoscience*, **2**, 299–311.

Doré, A.G., Lundin, E.R., Birkeland, Ø., Eliassen, P.E. & Jensen, L.N. 1997. The NE Atlantic margin: implications of late Mesozoic and Cenozoic events for hydrocarbon prospectivity. *Petroleum Geoscience*, **3**, 117–131.

Doré, A.G., Lundin, E.R., Jensen, L.N., Birkeland, Ø., Eliassen, P.E. & Fichler, C. 1999. Principal tectonic events in the evolution of the NW European Atlantic margin. *In*: Fleet, A.J. & Boldy, S.A.R. (eds) *Petroleum Geology of NW Europe: Proceedings of the 5th Conference*. Geological Society, London, 41–61.

Duindam, P. 1987. Structural evolution of the West Shetland continental margin. *In*: Brooks, J. & Glennie, K.W. (eds) *Petroleum Geology of North West Europe*. Graham & Trotman, London, 765–773.

Earle, M.M., Jankowski, E.J. & Vann, I.R. 1989. Structural and stratigraphic evolution of the Faroe–Shetland Channel and northern Rockall Trough. *In*: Tankard, A.J.. & Balkwill, H.R. (eds) *Extensional tectonics and Stratigraphy of the North Atlantic Margins*. American Association of Petroleum Geologists Memoir, **46**, 461–469.

Ebdon, C.C., Granger, P.J., Johnson, H.D. & Evans, A.M. 1995. Early Tertiary evolution and sequence stratigraphy of the Faroe–Shetland Basin: implications for hydrocarbon prospectivity. *In*: Scrutton, R.A., Stoker, M.S., Shimmield, G.B. & Tudhope, A.W. (eds) *The Tectonics, Sedimentation and Palaeoceanography of the North Atlantic Region*. Geological Society, London, Special Publications, **90**, 51–69.

Enachescu, M.E. 1987. Tectonic and structural framework of the northeast Newfoundland continental margin. *In*: Beaumont, C. & Tankard, A.J., *Sedimentary basins and basin- forming mechanisms*. Canadian Society of Petroleum Geologists, Memoir, **12**, 117–146.

Evans, D., Abraham, D.A. & Hitchen, K. 1989. The Geikie igneous centre, west of Lewis: its structure and influence on Tertiary geology. *Scottish Journal of Geology*, **25**, 339–352.

Fyfe, J.A., Long, D. & Evans, D. 1993. *United Kingdom Offshore Regional Report: The geology of the Malin–Hebrides area.* British Geological Survey, HMSO, London.

Gatliff, R.W., Hitchen, K., Ritchie, J.D. & Smythe, D.K. 1984. Internal structure of the Erlend Tertiary volcanic complex, north of Shetland, revealed by seismic reflection. *Journal of the Geological Society, London*, **141**, 555–562.

Goodchild, M.W., Henry, K.L., Hinkley, R.J. & Imbus, S.W. 1999. The Victory gas field, west of Shetland. *In*: Fleet, A.J. & Boldy, S.A.R. (eds) *Petroleum Geology of NW Europe: Proceedings of the 5th Conference*. Geological Society, London, 713–724.

Graue, K. 1992. Extensional tectonics in the northernmost North Sea: Rifting, uplift, erosion and footwall collapse in Late Jurassic to Early Cretaceous times. *In*: Spencer, A.M. (ed.) *Generation, Accumulation and Production of Europe's Hydrocarbons II*. EAPG Special Publication, No. 2, Springer-Verlag, Berlin, 23–34.

Herries, R., Poddubiuk, R. & Wilcockson, P. 1999. Solan, Strathmore and the back basin play, west of Shetland. *In*: Fleet, A.J. & Boldy, S.A.R. (eds) *Petroleum Geology of NW Europe: Proceedings of the 5th Conference*. Geological Society, London, 693–712.

Hjelstuen, B.O., Eldholm, O. & Skogseid, J. 1999. Cenozoic evolution of the northern Vøring margin. *Bulletin of the Geological Society of America*, **111**, 1792–1807.

Holmes, A.J., Griffith, C.E. & Scotchman, I.C. 1999. The Jurassic petroleum system of the west of Britain Atlantic margin – an integration of tectonics, geochemistry and basin modelling. *In*: Fleet, A.J. & Boldy, S.A.R. (eds) *Petroleum Geology of NW Europe: Proceedings of the 5th Conference*. Geological Society, London, 1351–1365.

Iliffe, J.E., Robertson, A.G., Ward, G.H.F., Wynn, C., Pead, S.D.M. & Cameron, N. 1999. The importance of fluid pressures and migration to the hydrocarbon prospectivity of the Faroe–Shetland White Zone. *In*: Fleet, A.J. & Boldy, S.A.R. (eds) *Petroleum Geology of NW Europe: Proceedings of the 5th Conference*. Geological Society, London, 601–611.

Johnston, S., Doré, A.G. & Spencer, A.M. 2001. The Mesozoic evolution of the Southern North Atlantic region and its relationship to basin development in the southern Porcupine Basin, offshore Ireland. *In*: SHANNON, P.M., HAUGHTON, P.D.W. & CORCORAN, D.V. (eds) *The Petroleum Exploration of Ireland's Offshore Basins*. Geological Society, London, Special Publications, **188**, 237–263.

JONES, E.J.W., RAMSAY, A.T.S., PRESTON, N.J. & SMITH, A.C.S. 1974. A Cretaceous guyot in the Rockall Trough. *Nature*, **251**, 129–131.

JONGEPIER, K., RUI, J.C. & GRUE, K. 1996. Triassic to early Cretaceous stratigraphic and structural development of the northeastern Møre Basin margin, off Mid-Norway. *Norsk Geologisk Tidsskrift*, **76**, 199–214.

KITTILSEN, J.E., OLSEN, R.R., MARTEN, R.F., HANSEN, E.K. & HOLLINGSWORTH, R.R. 1999. The first deepwater well in Norway and its implications for the Upper Cretaceous play, Vøring Basin. *In*: FLEET, A.J. & BOLDY, S.A.R. (eds) *Petroleum Geology of NW Europe: Proceedings of the 5th Conference*. Geological Society, London, 275–280.

KOCH, J-O. & HEUM, O.R. 1995. Exploration trends of the Halten Terrace. *In*: HANSLIEN, S. (ed.) *Petroleum exploration and exploitation in Norway*. NPF Special Publication 4, Elsevier, Amsterdam, 235–251.

LAIER, T., NYTOFT, H.P., JØRGENSEN, O. & ISAKSEN, G.H. 1997. Hydrocarbon traces in the Tertiary basalts of the Faroe Islands. *Marine and Petroleum Geology*, **14**, 257–266.

LAMERS, E. & CARMICHAEL, S.M.M. 1999. The Paleocene deep water sandstone hydrocarbon play West of Shetland. *In*: FLEET, A.J. & BOLDY, S.A.R. (eds) *Petroleum Geology of NW Europe: Proceedings of the 5th Conference*. Geological Society, London, 645–659.

LARSEN, M., HAMBERG, L., OLAUSSEN, S., PREUSS, T. & STEMMERIK, L. 1999. Sandstone wedges of the Cretaceous–Lower Tertiary Kangerlussuaq Basin, East Greeland – outcrop analogues to the offshore North Atlantic. *In*: FLEET, A.J. & BOLDY, S.A.R. (eds) *Petroleum Geology of NW Europe: Proceedings of the 5th Conference*. Geological Society, London, 337–348.

LUNDIN, E.R. & DORÉ, A.G. 1997. A tectonic model for the Norwegian passive margin with implications for the NE Atlantic: early Cretaceous to break-up. *Journal of the Geological Society, London*, **154**, 545–550.

MACDONALD, H., ALLAN, P.M. & LOVELL, J.P.B. 1987. Geology of oil accumulation in Block 26/28, Porcupine Basin, offshore Ireland. *In*: BROOKS, J. & GLENNIE, K.W. (eds) *Petroleum Geology of North West Europe*. Graham & Trotman, London, 643–651.

MAGOON, L.B. & DOW, W.G. 1994. The petroleum system. *In*: MAGOON, L.B. & DOW, W.G. (eds) *The petroleum system – from source to trap*. American Association of Petroleum Geologists Memoir, **60**, 3–24.

MITCHELL, S.M., BEAMISH, G.W.J., WOOD, M.V., MALACEK, S.J., ARMENTROUT, J.A., DAMUTH, J.E. & OLSON, H.C. 1993. Paleogene sequence stratigraphic framework of the Faeroe Basin. *In*: PARKER, J.R. (ed.) *Petroleum Geology of Northwest Europe: Proceedings of the 4th Conference*. Geological Society, London, 1011–1023.

MOORE, J.G. & SHANNON, P.M. 1995. The Cretaceous succession in the Porcupine Basin, offshore Ireland: facies distribution and hydrocarbon potential. *In*: CROKER, P.F. & SHANNON, P.M. (eds) *The Petroleum Geology of Ireland's Offshore Basins*. Geological Society, London, Special Publications, **93**, 345–370.

MORTON, A.C. & GRANT, S. 1998. Cretaceous depositional systems in the Norwegian Sea: heavy mineral constraints. *AAPG Bulletin*, **82**, 274–290.

MORTON, A.C., DIXON, J.E., FITTON, J.G., MACINTYRE, R.M., SMYTHE, D.K. & TAYLOR, P.N. 1988. Early Tertiary volcanic rocks in well 163/6-1A, Rockall Trough. *In*: MORTON, A.C. & PARSON, L.M. (eds) *Early Tertiary volcanism and the opening of the NE Atlantic*. Geological Society, London, Special Publications **39**, 293–308.

MORTON, A.C., HITCHEN, K., RITCHIE, J.D., HINE, N.M., WHITEHOUSE, M. & CARTER, S.G. 1995. Late Cretaceous basalts from Rosemary Bank, northern Rockall Trough. *Journal of the Geological Society, London*, **152**, 947–952.

MORTON, N. 1993. Potential reservoir and source rocks in relation to Upper Triassic to Middle Jurassic sequence stratigraphy, Atlantic margin basins of the British Isles. *In*: PARKER, J.R. (ed.) *Petroleum Geology of Northwest Europe: Proceedings of the 4th Conference*. Geological Society, London, 285–297.

MURPHY, N.J. & CROKER, P.F. 1992. Many play concepts seen over wide area in Erris, Slyne troughs off Ireland. *Oil and Gas Journal*, **90**, 92–97.

MUSGROVE, F.W. & MITCHENER, B. 1996. Analysis of the pre-Tertiary rifting history of the Rockall Trough. *Petroleum Geoscience*, **2**, 353–360.

NAYLOR, D. 1996. History of oil and gas exploration in Ireland. *In*: GLENNIE, K. & HURST, A. (eds) *AD 1995: NW Europe's hydrocarbon industry*. Geological Society, London, 43–52.

NAYLOR, D. 1998. *Irish shorelines through geological time*. Occasional papers in Irish Science and Technology. Royal Dublin Society, No. **17**.

NAYLOR, D., SHANNON, P. & MURPHY, N. 1999. *Irish Rockall Basin region – a standard structural nomenclature system*. Petroleum Affairs Division, Dublin. Special Publication **1/99**.

NAYLOR, P.H., BELL, B.R., JOLLEY, D.W., DURNALL, P. & FREDSTED, R. 1999. Palaeogene magmatism in the Faeroe–Shetland Basin: influences on uplift history and sedimentation. *In*: FLEET, A.J. & BOLDY, S.A.R. (eds) *Petroleum Geology of NW Europe: Proceedings of the 5th Conference*. Geological Society, London, 545–558.

O'REILLY, B.M., HAUSER, F., JACOB, A.W.B., SHANNON, P.M., MAKRIS, J. & VOGT, U. 1995. The Erris and eastern Rockall Troughs: structural and sedimentological development. *In*: CROKER, P.F. & SHANNON, P.M. (eds) *The Petroleum Geology of*

Ireland's Offshore Basins. Geological Society, London, Special Publications, **93**, 413–421.

PENNELL, J., RESTON, T. & GUSSINYE, M.P. 1999. From rift basin to rifted margin – the effect of increasing extension: a case study from the Porcupine Basin (Paper T51 I01). *Supplement to Eos, Transactions AGU*, **80** (46), F1057.

PRICE, S.P. & WITHAM, A.G. 1997. Exhumed hydrocarbon traps in East Greenland: analogs for the Lower–Middle Jurassic play of NW Europe. *AAPG Bulletin*, **81**, 196–221.

RICHARDSON, K.R., WHITE, R.S., ENGLAND, R.W. & FRUEHN, J. 1999. Crustal structure east of the Faroe Islands: mapping sub-basalt sediments using wide-angle seismic data. *Petroleum Geoscience*, **5**, 161–172.

RIDD, M.F. 1981. Petroleum geology west of the Shetlands. *In*: ILLING, L.V. & HOBSON, G.D. (eds) *Petroleum Geology of the continental Shelf of North-west Europe*. Heyden, London, 414–425.

ROBERTS, D.G. 1975. Marine geology of the Rockall Plateau and Trough. *Philosophical Transactions of the Royal Society of London*, **A278**, 447–509.

ROBERTS, D.G., GINZBERG, A., NUNN, K. & MCQUILLIN, R. 1988. The structure of the Rockall Trough from seismic refraction and wide-angle reflection measurements. *Nature*, **332**, 632–635.

ROBERTS, D.G., THOMPSON, M., MITCHENER, B., HOSSACK, J., CARMICHAEL, S. & BJØRNSETH, H-M. 1999. Paleozoic to Tertiary rift and basin dynamics: mid-Norway to the Bay of Biscay – a new context for hydrocarbon prospectivity in the deep water frontier. *In*: FLEET, A.J. & BOLDY, S.A.R. (eds) *Petroleum Geology of NW Europe: Proceedings of the 5th Conference*. Geological Society, London, 7–40.

SANCHEZ-FERRER, F., JAMES, S.D., LAK, B. & EVANS, A.M. 1999. Techniques used in the exploration for turbidite reservoirs in a frontier setting – Helland Hansen licence, Vøring Basin, offshore mid Norway. *In*: FLEET, A.J. & BOLDY, S.A.R. (eds) *Petroleum Geology of NW Europe: Proceedings of the 5th Conference*. Geological Society, London, 281–292.

SCOTCHMAN, I.C. & THOMAS, J.R.W. 1995. Maturity and hydrocarbon generation in the Slyne Trough, northwest Ireland. *In*: CROKER, P.F. & SHANNON, P.M. (eds) *The Petroleum Geology of Ireland's Offshore Basins*. Geological Society, London, Special Publications, **93**, 385–411.

SCOTCHMAN, I.C., GRIFFITH, C.E., HOLMES, A.J. & JONES, D.M. 1998. The Jurassic petroleum system north and west of Britain: a geochemical oil-source correlation study. *Organic Geochemistry*, **29**, 671–700.

SCRUTTON, R.A. 1986. The geology, crustal structure and evolution of Rockall Trough and the Faroe–Shetland channel. *Proceedings of the Royal Society, Edinburgh*, **88B**, 7–26.

SCRUTTON, R.A. & BENTLEY, P.A.D. 1988. Major Cretaceous volcanic province in southern Rockall Trough. *Earth and Planetary Science Letters*, **91**, 198–204.

SHANNON, P.M. 1992. Early Tertiary submarine fan deposits in the Porcupine Basin, offshore Ireland. *In*: PARNELL, A.J. (ed.) *Basins on the Atlantic Seaboard: petroleum geology, sedimentology and basin evolution*. Geological Society, London, Special Publications, **62**, 351–373.

SHANNON, P.M., JACOB, A.W.B., MAKRIS, J., O'REILLY, B., HAUSER, F. & VOGT, U. 1995. Basin development and petroleum prospectivity of the Rockall and Hatton region. *In*: CROKER, P.F. & SHANNON, P.M. (eds) *The Petroleum Geology of Ireland's Offshore Basins*. Geological Society, London, Special Publications, **93**, 435–457.

SHANNON, P.M., JACOB, A.W.B., O'REILLY, B.M., HAUSER, F., READMAN, P.W. & MAKRIS, J. 1999. Structural setting, geological development and basin modelling in the Rockall Trough. *In*: FLEET, A.J. & BOLDY, S.A.R. (eds) *Petroleum Geology of NW Europe: Proceedings of the 5th Conference*. Geological Society, London, 421–431.

SINCLAIR, I.K., EVANS, J.E., ALBRECHTSONS, E.A. & SYDORA, L.J. 1999. The Hibernia oilfield – effects of episodic tectonism on structural character and reservoir compartmentalization. *In*: FLEET, A.J. & BOLDY, S.A.R. (eds) *Petroleum Geology of NW Europe: Proceedings of the 5th Conference*. Geological Society, London, 517–528.

SMYTHE, D.K., CHALMERS, J.A., SKUCE, A.G., DOBINSON, A. & MOULD, A.S. 1983. Early opening history of the North Atlantic – structure and origin of the Faeroe – Shetland escarpment. *Geophysical Journal of the Royal Astronomical Society*, **72**, 373–398.

SPENCER, A.M., BIRKELAND, Ø., KNAG, G.Ø. & FREDSTED, R. 1999. Petroleum systems of the Atlantic margin of NW Europe. *In*: FLEET, A.J. & BOLDY, S.A.R. (eds) *Petroleum Geology of NW Europe: Proceedings of the 5th Conference*. Geological Society, London, 231–246.

STOKER, M.S., HITCHEN, K. & GRAHAM, C.C. 1993. *The geology of the Hebrides and West Shetland shelves, and adjacent deep-water areas*. British Geological Survey, HMSO, London.

TANKARD, A.J., WELSINK, H.J. & JENKINS, W.A.M. 1989. Structural styles and stratigraphy of the Jeanne d'Arc Basin, Grand Banks of Newfoundland. *In*: TANKARD, A.J. & BALKWILL, H.R. (eds) *Extensional tectonics and stratigraphy of the North Atlantic margins*. American Association of Petroleum Geologists Memoir, **46**, 265–282.

TATE, M.P. 1993. Structural framework and tectono-stratigraphic evolution of the Porcupine Seabight Basin, offshore western Ireland. *Marine and Petroleum Geology*, **10**, 95–123.

TATE, M.P. & DOBSON, M.R. 1988. Syn- and post- rift igneous activity in the Porcupine Seabight Basin and adjacent continental margin, west of Ireland. *In*: MORTON, A.C. & PARSON, L.M. (eds) *Early Tertiary volcanism and the opening of the NE Atlantic*. Geological Society, London, Special Publications, **39**, 309–334.

TATE, M.P. & DOBSON, M.R. 1989. Pre-Mesozoic geology of the western and north-western Irish continental shelf. *Journal of the Geological Society of London*, **146**, 229–240.

TATE, M.P., WHITE, N. & CONROY, J-J. 1993. Lithospheric extension and magmatism in the Porcupine Basin west of Ireland. *Journal of Geophysical Research*, **98**, 13905–13923.

TATE, M.P., DODD, C.D. & GRANT, N.T. 1999. The Northeast Rockall Basin and its significance in the evolution of the Rockall–Faeroes/East Greenland rift system. *In*: FLEET, A.J. & BOLDY, S.A.R. (eds) *Petroleum Geology of NW Europe: Proceedings of the 5th Conference*. Geological Society London, 391–406.

TRUEBLOOD, S. 1992. Petroleum geology of the Slyne Trough and adjacent basins. *In*: PARNELL, J. (ed.) *Basins on the Atlantic Seaboard: petroleum geology, sedimentology and basin evolution*. Geological Society, London, Special Publications, **62**, 315–326.

WAAGSTEIN, R. 1988. Structure, composition and age of the Faroe basalt plateau. *In*: MORTON, A.C. & PARSON, L.M. (eds) *Early Tertiary volcanism and the opening of the NE Atlantic*. Geological Society, London, Special Publications, **39**, 225–238.

WADDAMS, P. & CORDINGLEY, T. 1999. Regional geology and exploration potential of the NE Rockall Basin. *In*: FLEET, A.J. & BOLDY, S.A.R. (eds) *Petroleum Geology of NW Europe: Proceedings of the 5th Conference*. Geological Society, London, 379–390.

WALSH, A., KNAG, G., MORRIS, M., QUINQUIS, H., TRICKER, P., BIRD, C. & BOWER, S. 1999. Petroleum geology of the Irish Rockall Trough – a frontier challenge. *In*: FLEET, A.J. & BOLDY, S.A.R. (eds) *Petroleum Geology of NW Europe: Proceedings of the 5th Conference*. Geological Society, London, 433–444.

WOOD, M.V., HALL, J. & VAN HOORN, B. 1987. Post-Mesozoic differential subsidence in the north-east Rockall Trough related to volcanicity and sedimentation. *In*: BROOKS, J. & GLENNIE, K.W. (eds) *Petroleum Geology of North West Europe*. Graham & Trotman, London, 677–687.

WOOD, M.V., HALL, J. & DOODY, J.J. 1988. Distribution of early Tertiary lavas in the NE Rockall Trough. *In*: MORTON, A.C. & PARSON, L.M. (eds) *Early Tertiary volcanism and the opening of the NE Atlantic*, Geological Society, London, Special Publication, **39**, 283–292.

Petroleum geochemistry of the Lower and Middle Jurassic in Atlantic margin basins of Ireland and the UK

IAIN C. SCOTCHMAN

Statoil (UK) Ltd., Statoil House, 11a, Regent Street, London, SW1Y 4ST, UK (e-mail: iain.scotchman@statoil.com)

Abstract: Potential hydrocarbon source rocks of Lower and Middle Jurassic age have been reported from outcrop, shallow boreholes and exploration wells in Atlantic margin basins of the UK (Hebrides, West of Shetlands and flanking the NE Rockall Trough) and, recently, in the continuation of this trend offshore Ireland (Slyne, Erris and Porcupine basins). Previously these organic-rich mudrocks were considered to be of little economic importance, due largely to their perceived limited areal distribution and low maturity. However, recent geochemical studies of oils and shales from exploration drilling of these basins shows the Lower and Middle Jurassic to have considerable potential as effective hydrocarbon source rocks, supplanting the Late Jurassic–Early Cretaceous Kimmeridge Clay Formation equivalents as the only viable oil source rock in the region.

Flanking the Atlantic margin in the Irish and UK sectors, rich oil source potential occurs in two transgressive mudrock cycles of Lower Jurassic age. These are the Sinemurian–Pliensbachian interval and the overlying Toarcian section, present in basins such as the Solan, Minch, Hebrides, Slyne, North Celtic Sea, St George's Channel and Central English Channel. The Middle Jurassic source rocks have a more limited areal distribution and occur in the Faroe–Shetland, Solan, West Lewis, West Flannan, Hebrides, Slyne and North Porcupine basins with oil source potential in regressive marginal marine to lacustrine facies mudrocks.

Geochemical studies were undertaken on mudrocks from the Lower and Middle Jurassic sections in Atlantic margin basins (outcrop, shallow borehole core and exploration well cores and cuttings samples) and on oils from drill stem test and shows (core and cuttings extracts). Detailed analyses using GC, GC–MS and carbon isotopes allowed both characterization of the source rocks and oil-to-source correlation. Biomarker and carbon isotope studies of oils from the Faroe–Shetland Basin (Foinaven and Schiehallion fields), the Porcupine Basin (Connemara accumulation), the Wessex Basin (Wytch Farm and Kimmeridge oil fields) and wells in the Slyne Basin show strong correlations to the various source rock developments in the Lower and Middle Jurassic. The mixed biodegraded Foinaven and Schiehallion oils have a major waxy component and correlate with lacustrine Middle Jurassic source rocks in the Solan and West Lewis/West Flannan basins. Middle Jurassic sourcing of the Connemara oils is also suggested, while oils in the Slyne Basin appear to have been largely sourced by the Lower Jurassic Pabba Shale Formation. Oils in the Wessex Basin (Wytch Farm and Kimmeridge) appear to have been sourced by Hettangian–Sinemurian mudrocks and those in the North Celtic Sea Basin by Toarcian source rocks.

The results from this study, in combination with previously published data, show that rich, effective oil-prone source rocks occur in both the Lower and Middle Jurassic of the Atlantic margin basins offshore Ireland and the UK. These source rocks can be correlated with indigenous oils, indicating the existence of a previously under-evaluated petroleum system.

Geochemical oil-to-source correlation studies of oils from the limited number of discoveries and shows in the Atlantic margin basins flanking western Ireland and the UK (Fig. 1) indicate a source in Jurassic rocks other than the ubiquitous Kimmeridge Clay Formation and its lateral equivalents. This paper documents the occurrences of Lower and Middle Jurassic source rocks and demonstrates their correlation to oils in these basins, indicating them to be effective source rocks.

Jurassic petroleum systems

Three Jurassic petroleum systems can be recognized in the Atlantic margin basins north-

From: SHANNON, P.M., HAUGHTON, P.D.W. & CORCORAN, D.V. (eds) 2001. *The Petroleum Exploration of Ireland's Offshore Basins*. Geological Society, London, Special Publications, **188**, 31–60.
0305-8719/01/$15.00

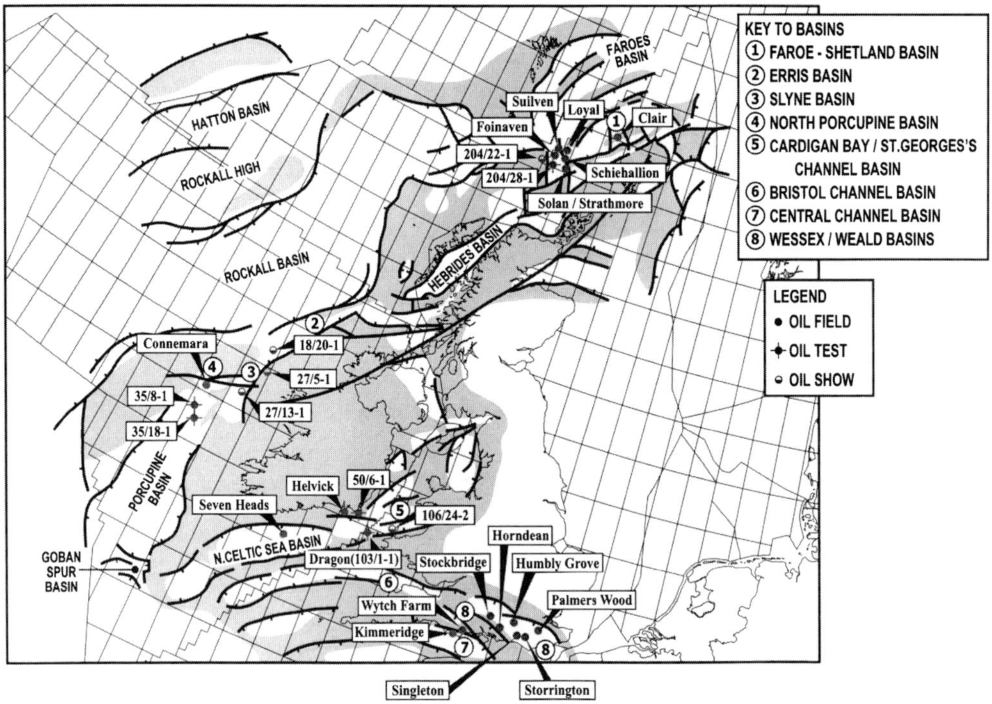

Fig. 1. Location map of the Atlantic margin basins offshore Ireland and the UK, showing significant oil fields, hydrocarbon discoveries and wells with oil shows.

west of Ireland and Britain (Fig. 2), encompassing Lower Jurassic, Middle Jurassic and Upper Jurassic source rocks (Scotchman & Doré 1995). The most prolific of these systems over NW Europe is that of the Upper Jurassic (Holmes *et al.* 1999; Spencer *et al.* 1999). Sourced by the Kimmeridge Clay Formation and its equivalents (Doré *et al.* 1985), the system appears ubiquitous to the Atlantic margin Jurassic basin system (Scotchman & Doré 1995), from the Jeanne d'Arc Basin in the southwest (von der Dick *et al.* 1989), through the Porcupine Basin (MacDonald *et al.* 1987), the West of Shetlands (Bailey *et al.* 1987; Scotchman *et al.* 1998; Jowitt *et al.* 1999), the northern North Sea (Barnard & Cooper 1981; Pegrum & Spencer 1990), to Haltenbanken (Koch & Heum 1995) and the Barents Sea in the north, extending round to the West Siberian Basin in Russia (Kontorovich *et al.* 1997).

A less extensively developed Lower Jurassic petroleum system is recognized in the Slyne Basin (Scotchman & Thomas 1995), in the Celtic Sea/Cardigan Bay/Central Channel/Wessex/Weald basin systems (Colter & Harvard 1981; Butler & Pullan 1990; Caston 1995) and in the Paris Basin (Espitalié *et al.*, 1987) and southern North Sea/West Netherlands basins (Bodenhausen & Ott 1981), with oils sourced by marine shales of Hettangian–Sinemurian and Pleinsbachian–Toarcian ages (Fleet *et al.* 1987; Cornford 1998).

Local developments of marine and non-marine, often lacustrine, mudrocks provide the source for the Middle Jurassic petroleum system, which appears to have a very restricted regional distribution. Middle Jurassic source rocks and generated oils have been recognized in the Porcupine and Jeanne d'Arc basins (Cornford 1998; Butterworth *et al.* 1999), while source rocks occur in basins in the northeast Rockall Basin/West of Shetlands areas (Hitchen & Stoker 1993; Scotchman *et al.* 1998; Holmes *et al.* 1999; Lamers & Carmichael 1999) and in the Hebrides Basin (Vincent & Tyson 1999).

This paper will concentrate on the development and distribution of the Lower and Middle Jurassic petroleum systems. Geochemical analyses of source rock and oil samples were made by the Newcastle Research Group in Fossil Fuels and Environmental Geochemistry at Newcastle University and other contractors using methods as described in Scotchman *et al.* (1998). Analytical data appear in tables 1 to 5 of Scotchman *et al.* (1998) and tables 3 to 8 of

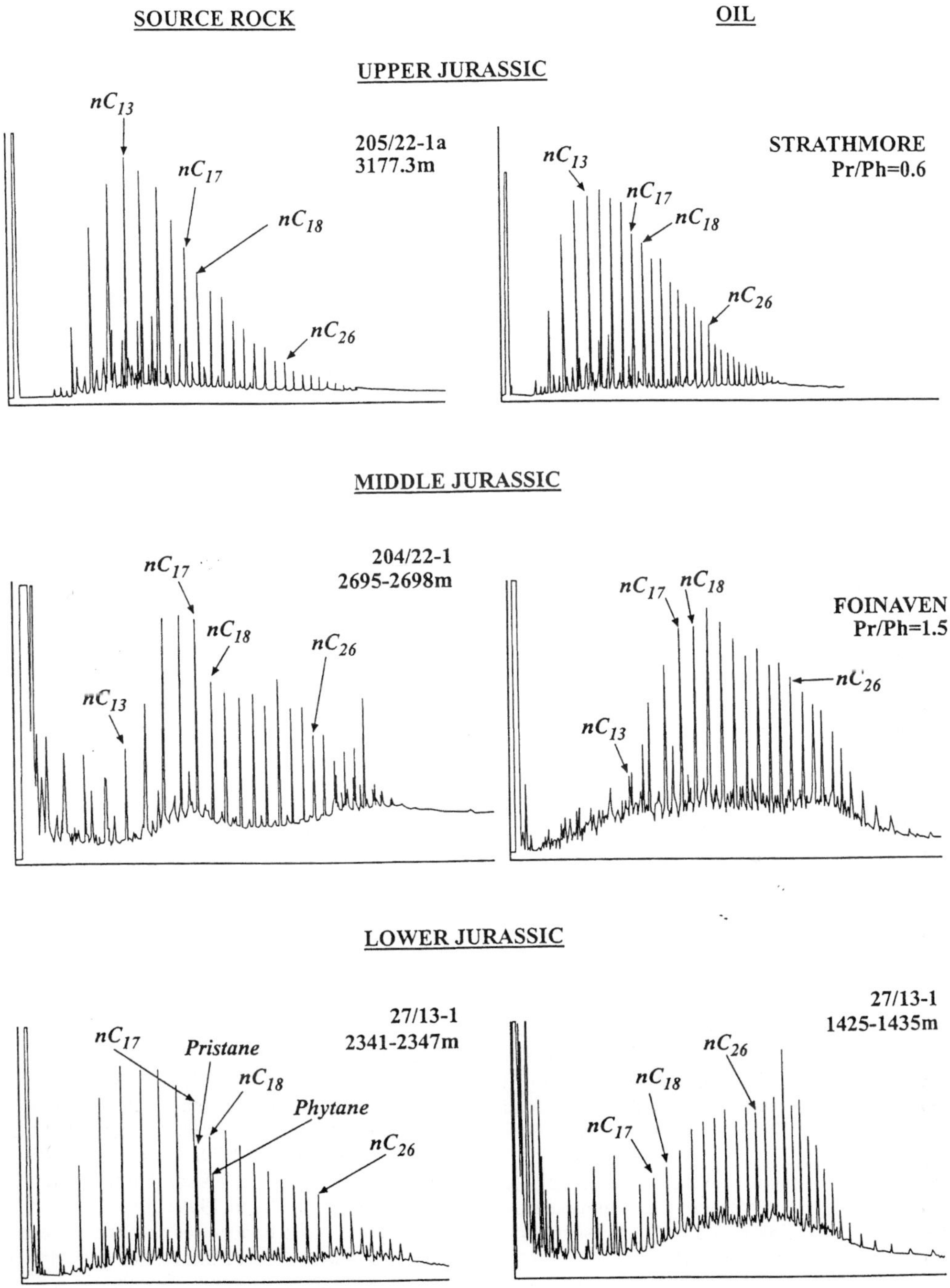

Fig. 2. Key characteristics of typical source rocks and oils from the Upper, Middle and Lower Jurassic petroleum systems.

Table 1. *Source rock samples*

Location/Depth	Sample type	Formation	TOC (wt%)	S_1 (kg t^{-1})	S_2 (kg t^{-1})	T_{max} (°C)	HI	VR (%Ro)
Kimmeridge-5								
1660.5 m	Core	Blue Lias	1.5	0.24	0.62	443	40	0.29
1661.0 m	Core	Blue Lias	1.2	0.27	0.61	440	50	0.29
1661.6 m	Core	Blue Lias	1.1	0.19	0.45	nm	40	0.29
1662.2 m	Core	Blue Lias	1.3	0.33	1.01	504	80	0.28
1662.6 m	Core	Blue Lias	1.2	0.36	1.39	452	110	0.27
1663.4 m	Core	Blue Lias	1.2	0.33	0.78	442	60	0.59
Charmouth								
DC/88-1	O/C	Black Ven Marl	2.3	0.05	2.95	426	130	0.61
DC/88-12	O/C	Black Ven Marl	5.9	0.43	36.28	419	620	0.51
DC/88-17	O/C	SWB	6.5	0.27	33.18	414	510	0.54
DC/88-18	O/C	SWB	7.2	0.46	38.61	417	530	0.57
DC/88-21	O/C	SWB	4.6	0.16	18.65	416	410	0.47
DC/88-22	O/C	SWB	5.6	0.25	26.69	413	480	0.59
Brora								
BC-1	O/C	Brora Coal Fm.	65.0	1.29	63.73	421	90	0.43
BRA-1	O/C	Brora Coal Fm.	21.2	0.29	26.92	429	120	nm
BRA-2	O/C	Brora Coal Fm.	18.2	0.36	24.06	420	130	nm
BAF-1	O/C	BAF	1.9	0.02	5.37	414	290	0.48
BAF-2	O/C	BAF	15.7	0.55	22.59	439	140	0.49
BAF-3	O/C	BAF	45.4	1.59	213.99	434	470	0.36
Hallaig, Raasay								
SK/95-1	O/C	Pabba Shale Fm	1.0	0.11	0.21	nm	21	0.71
SK/95-2	O/C	Cullaidh Shale	3.1	0.25	2.6	nm	84	0.57
SK/95-3	O/C	Dun Caan Shale	4.7	0.67	12.41	nm	264	0.40
Elgol, Skye								
SK/95-5	O/C	Lealt Shale Fm	2.2	0.07	2.95	nm	134	2.58
Upper Glen-1								
2950′	SWC	Cullaidh Shale	3.98	0.48	14.75	441	371	0.79
2990′	SWC	Cullaidh Shale	2.92	0.15	9.47	438	324	nm
4330′	SWC	Portree Shale	6.79	1.30	25.57	429	377	0.93
4450′	SWC	Portree Shale	6.88	1.55	25.47	429	370	nm
4620′	SWC	Portree Shale	2.89	0.52	10.64	432	368	nm
5330′	SWC	Pabba Shale Fm	0.89	nm	nm	nm	nm	1.2
5520′	SWC	Pabba Shale Fm	1.11	0.09	1.04	439	94	nm
5700′	SWC	Pabba Shale Fm	1.19	0.21	0.32	470	27	nm
L134/5-1								
2060–2080′	Ctgs	Pabba Shale Fm	3.1	0.53	10.03	432	328	nm
2140–2160′	Ctgs	Pabba Shale Fm	2.0	0.88	10.43	423	532	nm
4500–4520′	Ctgs	Broadford Beds	0.8	0.38	0.36	424	46	nm
4750–4800′	Ctgs	Broadford Beds	1.3	0.59	1.85	434	142	nm
'Slyne Trough'								
2049.0 m	SWC	Garantiana Clay	2.63	0.33	6.22	436	237	nm
2098.0 m	SWC	Garantiana Clay	2.20	0.29	7.14	436	325	0.42
2634.0 m	SWC	Dun Caan Shale	1.88	0.50	6.87	438	365	0.48
2656.0 m	SWC	Portree Shale	3.94	0.34	20.63	437	524	nm
2838.0 m	SWC	Portree Shale	7.02	0.70	33.07	436	471	0.48
2867.5 m	SWC	Portree Shale	7.42	0.67	36.24	441	488	nm
3194.0 m	SWC	Pabba Shale Fm	3.33	0.91	4.65	447	140	0.74
3207.0 m	SWC	Pabba Shale Fm	3.58	0.51	4.27	447	119	0.73

Data for West of Shetlands source rocks appear in Scotchman *et al.* (1998, tables 1 and 2) and for Slyne Trough well 27/13-1 in Scotchman & Thomas (1995, tables 3 and 4).

O/C, outcrop; nm, not measured/measurable; Ctgs, cuttings; SWB, Shales With Beef; SWC, sidewall core; BAF, Brora Argillaceous Formation

Table 2. *Oil samples*

Location/ DST/depth	Formation	Sample	API gravity (°)
Kimmeridge-1			
1791.0′	Cornbrash	DST	46
Wytch Farm			
3038–3225′	Bridport Sst.	DST	38
Bran Point			
DC/88-24A	Bencliff Grit	Seep	—
DC/88-24B	Bencliff Grit	Seep	—
Mupe Bay			
DC/88-25	Wealden	Seep	—
DC/88-26	Wealden	Seep	—
Schiehallion			
Oil	Palaeocene	DST	27
'Slyne Trough'			
683 m	M. Jurassic	SWC Extract	—
1967.5 m	M. Jurassic	MDT	—
Connemara			
26/28-1			
2339.5–2328.0 m	M. Jurassic	DST-1	31.8
2245–2268 m	M. Jurassic	DST-2	34.8
1960–1970 m	M. Jurassic	DST-3	31.8
2247 m	M. Jurassic	RFT 1	31.1
26/28-2			
2134.0–2154.1 m	M. Jurassic	DST-2	33.4
2089.0–2122.1 m	M. Jurassic	DST-3	36.6
26/28-A1			
2223.5 m	M. Jurassic	MDT	30.8
2236.3 m	M. Jurassic	MDT	33.5
26/28-A1z			
2154.4 m	M. Jurassic	MDT	35.1
26/28-A2			
1959.8 m	U. Jurassic	MDT	31.5

Data for West of Shetlands oils appear in Scotchman *et al.* (1998, table 3) and for Slyne Basin well 27/13-1 in Scotchman & Thomas (1995, table 1).

Scotchman & Thomas (1995), with additional data reported here in Tables 1–4.

Hydrocarbon occurrences west and south of Ireland and Britain

Hydrocarbon discoveries west and south of Ireland and Britain comprise large, heavy oil fields and small gas accumulations in the UK West of Shetlands area; oil and gas discoveries in the Slyne and Porcupine basins, offshore western Ireland; and oil, heavy oil, condensate and gas in the Celtic Sea/English Channel basin systems south of Ireland and the UK.

The main oil discoveries in the West of Shetlands are located in the Faroe–Shetland Basin, comprising the Clair, Foinaven and Schiehallion/Loyal oil fields and the Suilven oil discovery (Fig. 1), and the West Shetland Basin, comprising the Solan/Strathmore and 204/28-1 oil discoveries. The Devonian–Carboniferous reservoired Clair Field located on the Rona Ridge (Coney *et al.* 1993) is the largest accumulation on the UK continental shelf with 3–5 billion barrels in-place (Haszeldine *et al.* 1987) but only 200–300 million barrels of oil

Table 3. *Source rock extract and biomarker data*

Sample	PR/PH	nC_{17}/PR	nC_{18}/PH	CPI	Ts/Tm	C_{31} 22S/(22R+22S) hopane ratio	% C_{27} Steranes	% C_{28} Steranes	% C_{29} Steranes	C_{29} 20S/(20S+20R) sterane ratio	$\delta^{13}C$ sat.HC	$\delta^{13}C$ aro. HC	$\delta^{13}C$ kerogen
Kimmeridge-5													
1660.5 m	1.76	4.92	7.29	1.06	nm	nm	nm	nm	nm	nm	nm	nm	nm
1661.0 m	1.93	5.34	9.18	1.07	nm	nm	nm	nm	nm	nm	nm	nm	nm
1661.1 m	1.83	6.27	9.54	1.10	nm	nm	nm	nm	nm	nm	nm	nm	nm
1662.2 m	1.74	6.06	9.32	1.06	nm	nm	nm	nm	nm	nm	nm	nm	nm
1662.6 m	1.83	5.94	9.28	1.07	nm	nm	nm	nm	nm	nm	nm	nm	nm
1663.4 m	1.60	3.50	5.15	1.08	nm	nm	nm	nm	nm	nm	nm	nm	nm
Charmouth													
DC/88-1	0.89	1.91	0.89	2.29	nm	nm	nm	nm	nm	nm	nm	nm	nm
DC/88-12	0.38	1.78	0.28	2.89	nm	nm	nm	nm	nm	nm	nm	nm	nm
DC/88-17	0.45	1.09	0.29	2.93	nm	nm	nm	nm	nm	nm	nm	nm	nm
DC/88-18	0.35	1.06	0.16	2.92	nm	nm	nm	nm	nm	nm	nm	nm	nm
DC/88-21	0.38	0.90	0.16	2.63	nm	nm	nm	nm	nm	nm	nm	nm	nm
DC/88-22	0.34	0.95	0.15	2.64	nm	nm	nm	nm	nm	nm	nm	nm	nm
Brora													
BC-1	nm	nm	nm	nm	nm	nm	nm	nm	nm	nm	−26.0	−25.0	nm
BRA-1	2.44	0.57	1.21	1.30	nm	nm	nm	nm	nm	nm	nm	nm	nm
BRA-2	1.82	1.11	1.55	1.96	nm	nm	nm	nm	nm	nm	nm	nm	nm
BAF-1	2.70	1.52	2.80	2.62	nm	nm	nm	nm	nm	nm	nm	nm	−26.5
BAF-2	1.32	0.38	0.32	1.09	nm	nm	nm	nm	nm	nm	nm	nm	−23.6
BAF-3	nm	nm	nm	nm	nm	nm	nm	nm	nm	nm	−29.5	−26.8	nm
L134/5-1													
2060–2080′	nm	nm	nm	nm	0.28	0.42	26.1	30.9	43.0	0.30	−29.31	−28.77	nm
2140–2160′	nm	nm	nm	nm	0.21	0.47	21.9	31.8	46.2	0.16	−29.50	−28.86	nm
4500–4520′	nm	nm	nm	nm	0.24	0.55	29.0	31.9	39.1	0.37	−28.75	−28.01	nm
4750–4800′	nm	nm	nm	nm	0.45	0.57	31.0	27.1	42.0	0.39	−29.00	−28.37	nm
'Slyne Basin'													
2634.0 m	nm	nm	nm	nm	1.00	0.62	39.2	24.7	35.4	0.43	−27.70	−28.42	nm
2838.0 m	nm	nm	nm	nm	1.08	0.62	37.2	23.9	39.0	0.46	−26.90	−29.08	nm
3207.0 m	nm	nm	nm	nm	0.54	0.61	40.6	24.9	34.4	0.44	−28.78	nm	nm

Data for West of Shetlands source rocks appear in Scotchman *et al.* (1998, table 3) and for Slyne Trough well 27/13-1 in Scotchman & Thomas (1995, table 7).
nm, not measured; PR, pristane; PH, phytane; CPI, Carbon Preference Index; Ts/Tm, 17α(H)-22,29,30-trisnorhopane/18α(H)-22,29,30-trisnorneohopane ratio; sat. HC, saturated hydrocarbon fraction, per mil.; aro. HC, aromatic hydrocarbon fraction, per mil.

Table 4. *Oil fraction biomarker data*

Location/ DST/depth	PR/PH	nC_{17}/PR	nC_{18}/PH	CPI	Ts/Tm	C_{31} 22S/ (22R+22S) hopane ratio	% C_{27} steranes	% C_{28} steranes	% C_{29} steranes	C_{29} 20S/ (20S+20R) sterane ratio	$\delta^{13}C$ sat.HC	$\delta^{13}C$ aro. HC
Kimmeridge-1												
1791.0′	1.53	2.17	2.75	1.04	1.5	0.58	38.2	23.6	38.2	0.42	−30.0	nm
Wytch Farm												
3038–3225′	1.62	2.05	2.77	1.04	0.89	0.57	39.1	26.1	34.8	0.45	−28.8	nm
Bran Point												
DC/88-24A	nm	nm	nm	nm	1.78	0.57	37.4	24.3	38.3	0.56	−29.4	nm
DC/88-24B	nm	nm	nm	nm	2.13	0.56	38.2	23.6	38.2	0.50	−29.3	nm
Mupe Bay												
DC/88-25	nm	nm	nm	nm	1.6	0.55	35.5	24.8	39.7	0.74	−29.4	nm
DC/88-26	nm	nm	nm	nm	1.14	0.58	35.6	25.4	39.0	0.45	−29.4	nm
Schiehallion												
Oil	1.52	0.70	0.50	1.93	1.14	0.73	33.1	29.3	37.6	0.48	nm	nm
'Slyne Trough'												
683 m	nm	nm	nm	nm	0.52	0.61	33.8	26.2	40.1	0.49	−29.09	−28.06
1967.5 m	nm	nm	nm	nm	0.59	0.59	37.8	25.6	36.6	0.44	−29.04	−28.33
Connemara												
26/28-1												
2339.5–2328.0 m	1.82	1.75	2.70	1.24	0.90	0.60	42	26	32	0.39	−27.91	−26.22
2245–2268 m	1.41	1.96	2.50	1.12	0.98	0.59	41	27	32	0.36	−30.30	−26.57
1960–1970 m	1.34	1.67	2.00	1.16	1.41	0.59	40	27	33	0.31	−30.71	−27.83
2247 m	1.51	1.89	2.44	1.19	0.94	0.60	42	26	32	0.37	−29.31	−28.02
26/28-2												
2134.0–2154.1 m	1.41	2.17	2.70	1.12	0.99	0.59	41	27	32	0.38	−30.55	−27.70
2089.0–2122.1 m	1.53	2.63	3.70	1.10	1.02	0.60	42	26	32	0.38	−30.72	−26.95
26/28-A1												
2223.5 m	2.93	1.54	3.13	2.28	0.77	0.62	43	27	29	0.45	−26.15	−26.59
2236.3 m	1.45	2.00	2.44	1.14	0.96	0.59	41	26	32	0.37	−28.62	−27.01
26/28-A1z												
2154.4 m	1.49	1.92	2.56	1.1	0.96	0.57	41	26	32	0.36	−30.18	−27.4
26/28-A2												
1959.8 m	1.26	1.67	1.67	1.2	1.49	0.60	40	27	34	0.29	−29.88	−27.81

Data for West of Shetlands oils appear in Scotchman *et al.* (1998, table 5) and for Slyne Trough well 27/13-1 in Scotchman & Thomas (1995, table 7).
nm, not measured; PR, pristane; PH, phytane; CPI, Carbon Preference Index; Ts/Tm, 17α(H)-22,29,30-trisnorhopane/18α(H)-22,29,30-trisnorneohopane ratio; sat. HC, saturated hydrocarbon fraction, per mil.; aro. HC, aromatic hydrocarbon fraction, per mil.

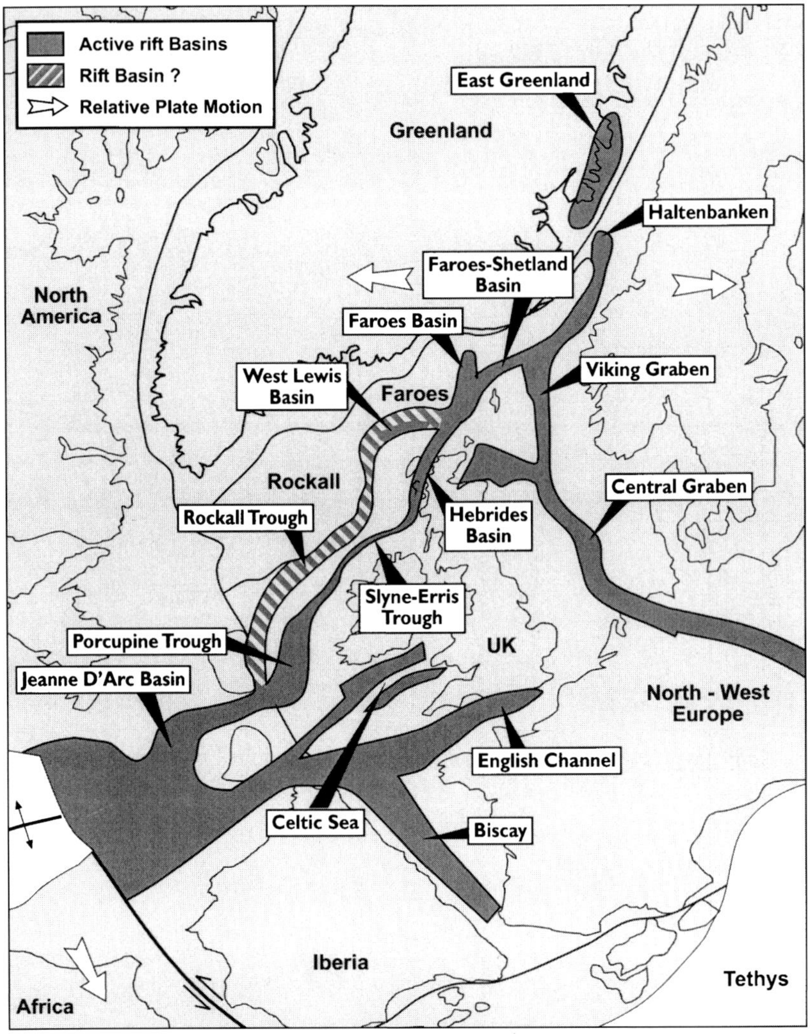

Fig. 3. Early Jurassic plate reconstruction illustrating the rift basin system initiated in the Permo-Triassic which controlled deposition of Jurassic, particularly Liassic, source rocks.

(MMBO) recoverable reserves (Potter 1998). To the southwest, the Foinaven, Alligin, Schiehallion and Loyal fields, with combined reserves in excess of 625 MMBO (Cooper *et al.* 1999; Leach *et al.* 1999), are located over the Westray Ridge in Paleocene deep-water sandstone reservoirs. A further Paleocene oil and gas discovery, Suilven, with reserves of about 100 MMBO (Beckman 1998), is located on the northern flank of the Westray Ridge. In the West Shetland Basin the Upper Jurassic 204/28-1 discovery on the North Rona High on the southwest flank of the basin contains in excess of 100 million barrels of very heavy oil (Scout data), while the Triassic–Upper Jurassic-reservoired Strathmore/Solan fields (Herries *et al.* 1999) have combined reserves of about 60 MMBO. All of these oils are heavy, with API gravity ranging from 9° (204/28-1) to 27° (Schiehallion), biodegraded to variable degrees, acidic (Clair) or waxy (Foinaven) and are undersaturated with low gas-to-oil ratios (GOR) of 85 to 320 SCF/BBL. Small gas caps overlie the oils of the Clair and Foinaven fields.

To the west of Ireland, oil has been discovered in Jurassic reservoirs in the Connemara accumulation in the northern Porcupine Basin, with about 195 million barrels of light, waxy oil in-place (MacDonald *et al.* 1987) and reserves of 40–80 MMBO. Oil shows also occur in the

Upper Jurassic and Lower Cretaceous rocks of the Porcupine Basin, with wells 35/8-1 and 35/8-2 testing 730 and 925 BOPD respectively (Shannon & Naylor 1998). Biodegraded oil shows also occur in the Slyne Basin, in the Middle Jurassic of wells 27/13-1 (Scotchman & Thomas, 1995), 27/5-1 and 18/20-1.

In the Celtic Sea, Cardigan Bay, Central Channel, Wessex and Weald basins stretching from southern Ireland across southern England, oil and gas discoveries occur in reservoirs of Triassic to Lower Cretaceous age. The largest discovery is the Wytch Farm oil field in the Wessex Basin with reserves of 428 million

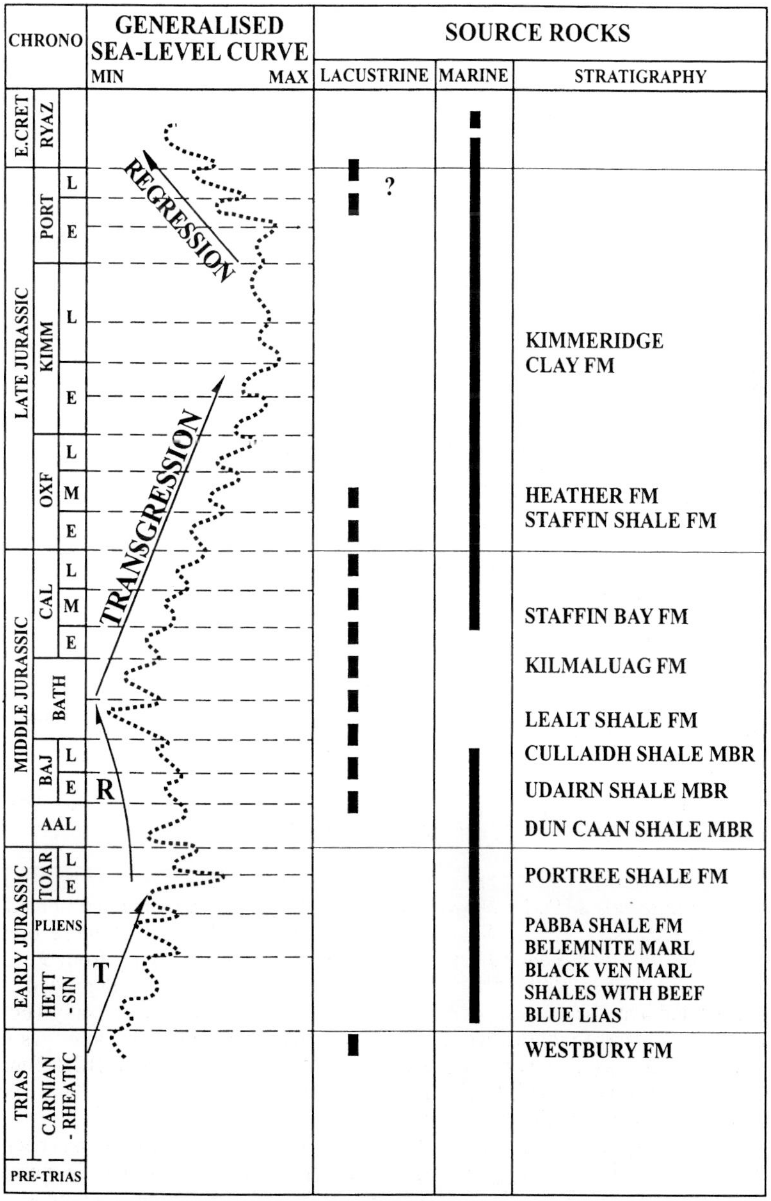

Fig. 4. Stratigraphic development of marine and lacustrine source rocks in the Jurassic of northwest of Ireland and Britain, in relation to transgressive and regressive cycles. The development of lacustrine source rocks during the Middle Jurassic, related to regressive and early transgressive phases, is illustrated.

barrels of light 38–42° API gravity oil in reservoirs of Lower Triassic (Sherwood Sandstone Group), Lower Jurassic (Bridport Sandstone Formation) and Middle Jurassic Frome Clay Member (Colter & Harvard 1981; Underhill & Stoneley 1998). In the North Celtic Sea Basin, the Helvick accumulation of 2–5 MMBO 44° API gravity oil is reservoired in Middle–Upper Jurassic sandstones (Caston 1995). Condensate of 42° API occurs in similar-aged reservoirs of the Dragon discovery (well 103/1-1) in the St George's Channel Basin. Medium API gravity oils (35–42° API) occur in small fields in the Wessex and Weald basins, reservoired in Middle Jurassic Great Oolite Group limestones such as Humbly Grove, Stockbridge, Horndean, Storrington fields and in younger sandstones such as Palmers Wood Field (Butler & Pullan 1990). Reserves are less than 10 MMBO. The 3.2 MMBO Kimmeridge oil field with 46° API gravity oil is in Middle Jurassic fractured limestones of the Cornbrash (Evans *et al.* 1998).

Lower Jurassic petroleum system

The Jurassic began with the progressive development of fully marine conditions across the Atlantic margin rift basins (Fig. 3) as a transgression flooded the late Triassic continental area (Ziegler 1990). The timing of this transgression is variable, ranging from the late Triassic in southern England (Cope 1995), where it deposited the Rhaetic section including the Westbury Formation source rock (MacQuaker *et al.* 1986), to the Hettangian in the Slyne Basin, west of Ireland and in the Hebrides Basin (Morton 1992, 1993; Morton & Hudson 1995). The early Jurassic section generally comprises a transgressive – regressive cyclic depositional sequence of shales, sandstones and limestones, reflecting extensional tectonic and eustatic influences (Cope 1995). The main developments of source rocks are in the Hettangian/early Sinemurian, late Sinemurian/Pliensbachian and the Toarcian (Fig. 4) and were primarily associated with transgressive cycles, with deposition occurring under relatively deep-water conditions.

Hettangian–early Sinemurian source rocks and oils

These source rocks are primarily developed in the basins south and west of Ireland and Britain: in the North and South Celtic Sea, Cardigan Bay, Central Channel, Wessex and Weald basins.

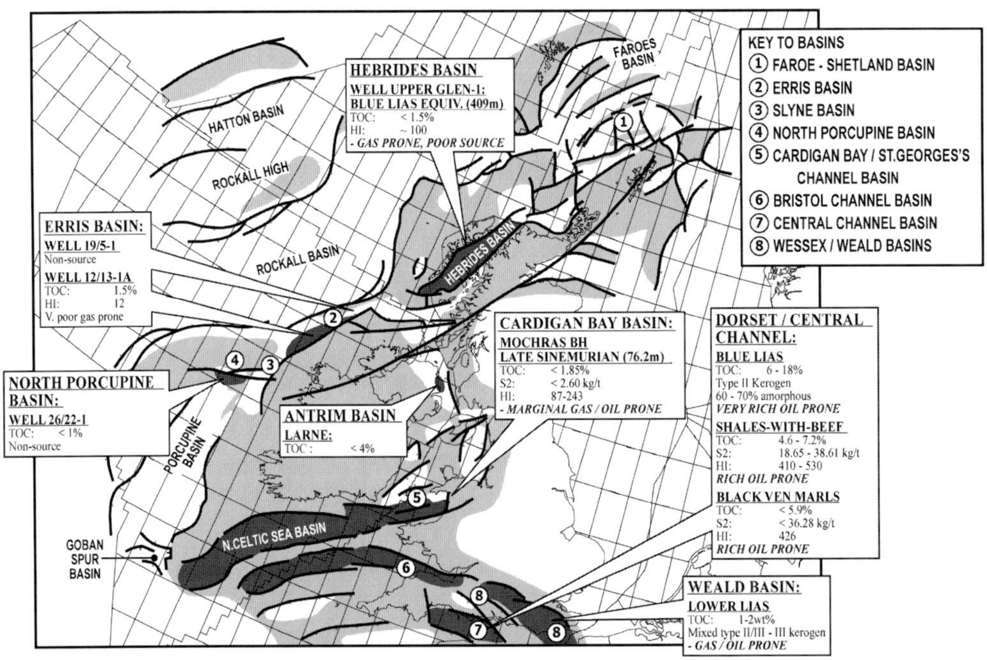

Fig. 5. The distribution of Hettangian–early Sinemurian source rocks in the Atlantic margin basins of Ireland and the UK.

However, they also occur on the northwest margins in the northern Porcupine Basin, Erris Basin and the Antrim and Hebrides basins (Fig. 5).

The most well-known source rock development of this age is the Blue Lias of the Wessex Basin (Selley & Stoneley 1987; Fleet *et al.* 1987), which comprises a 26 m thick succession of alternating organic-rich mudstones and limestones, the latter largely of early diagenetic origin, which extends up into the early Sinemurian (House 1993; Hesselbo & Jenkyns 1995). Deposition was generally under anoxic conditions with high surface water productivity. Overlying the Blue Lias are the early Sinemurian Shales with Beef and the lower part of the Black Ven Marls, 25 m and 28 m thick, respectively. These comprise dark laminated shales and marls with occasional bands of concretions and tabular limestones, the Shales with Beef containing abundant diagenetic fibrous calcites.

The Hettangian–early Sinemurian mudrocks are very rich, oil-prone source rocks. Total organic carbon (TOC) ranges between less than 1 wt% for the limestone bands to 18 wt% in the richest shales (Cornford 1998). TOC values up to 7.2 wt% occur in organic-rich 'paper shales' within the overlying Shales with Beef and lower Black Ven Marls (Table 1), with average values of 5–6 wt%. Pyrolysis source potential (S2) is similarly variable (see Fleet *et al.* 1987, fig. 3), with values up to 38.6 kg t^{-1}. Kerogens largely comprise marine amorphous Type II material (60–70%) with varying amounts of terrestrially derived material, the proportion of the latter increasing towards the margins of the basin, where mixed Type II/III and terrestrial Type III kerogens predominate.

Gas chromatograph (GC) and gas chromatograph–mass spectrometer (GC–MS) data (Fig. 6) show the source rock extracts to have a waxy character with a slight odd carbon preference index (CPI) and nC_{30}+ 'hump' in the *n*-alkane distribution. This is particularly apparent in the oil-window maturity Blue Lias Formation in well Kimmeridge-5.

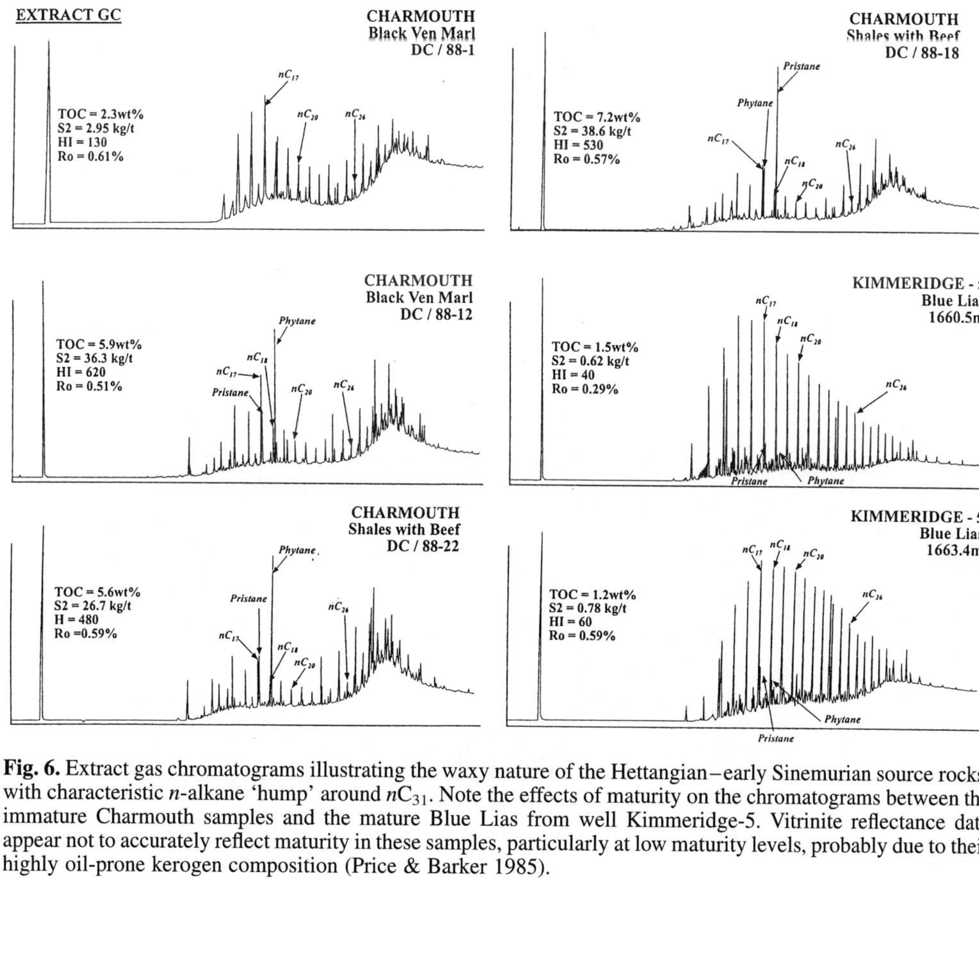

Fig. 6. Extract gas chromatograms illustrating the waxy nature of the Hettangian–early Sinemurian source rocks, with characteristic *n*-alkane 'hump' around nC_{31}. Note the effects of maturity on the chromatograms between the immature Charmouth samples and the mature Blue Lias from well Kimmeridge-5. Vitrinite reflectance data appear not to accurately reflect maturity in these samples, particularly at low maturity levels, probably due to their highly oil-prone kerogen composition (Price & Barker 1985).

Oils from from the Wytch Farm and Kimmeridge oil fields and bitumen extracts from the biodegraded Dorset coast oil seeps at Bran Point and Mupe Bay (Stoneley & Selley 1986; Cornford *et al.* 1988; Miles *et al.* 1993, 1994; Kinghorn *et al.* 1994; Wimbledon *et al.* 1996; Parfitt & Farrimond 1998) all show similar geochemical characteristics to these source rocks (Fig. 7). In particular, the sterane distribution is distinctive with the nC_{27} sterane content being relatively greater than the nC_{29} steranes, but significantly greater than the nC_{28} steranes (Fig. 8), while the waxy character, 'hump' at nC_{30}, CPI, pristane/phytane (PR/PH) ratio and saturate fraction carbon isotope values all further indicate sourcing by these mudrocks. This unit has also been demonstrated to be the source for oils in the Weald Basin (Ebukanson & Kinghorn 1986; Burwood *et al.* 1991), including the Humbly Grove Field (Hancock & Mithen 1987).

In the Weald Basin, more marginal facies kerogens (mixed Type II/III and terrestrial Type III) predominate, with generally lower TOCs of 1–2 wt% (Ebukanson & Kinghorn 1985) and mixed gas–oil source potential. Similar organic facies occur in mudrocks of this age to the southwest in the North and South Celtic Sea basins, with largely gas source potential.

In the basins north and west of Ireland and Britain, only poor, largely gas-prone source rocks are developed in the Hettangian to early Sinemurian. In the northern Porcupine and Erris basins (wells 26/22-1 and 19/5-1 and 12/13-1a respectively), poor gas-prone, terrestrial kerogens occur in shales with TOCs less than 1 wt%. Along-trend in the Hebrides Basin, the Blue Lias Formation equivalent is up to 409 m thick in well Upper Glen-1, but is generally non- to poor source rock with TOCs less than 1.5 wt% and hydrogen indices (HI) less than 100, indicating only gas potential (Butterworth *et al.* 1999). However, to the south in the Antrim Basin at Larne, Hettangian to early Sinemurian-age black shales in outcrop are over 100 m thick and have TOCs up to 4 wt% (Parnell *et al.*1992).

Late Sinemurian–Pliensbachian source rocks and oils

Late Sinemurian to Pliensbachian source rock deposition appears to have been very widespread. Deposition of organic-rich mudrocks occurred in the Central Channel and Wessex basins (House 1993; Hesselbo & Jenkyns 1995), throughout the Celtic Sea basins (Fleet *et al.* 1987), to the Slyne

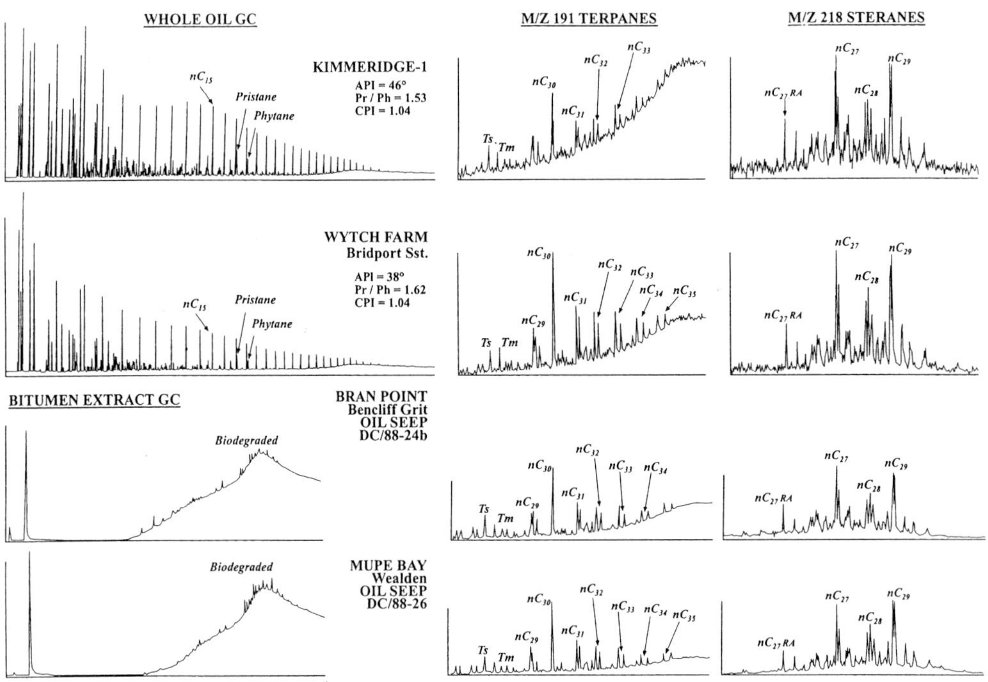

Fig. 7. Geochemical data from Central English Channel Basin oil fields and seeps, showing good correlation both between the oils and seeps, notwithstanding the severe biodegradation of the latter, and with the Hettangian–Sinemurian source rocks.

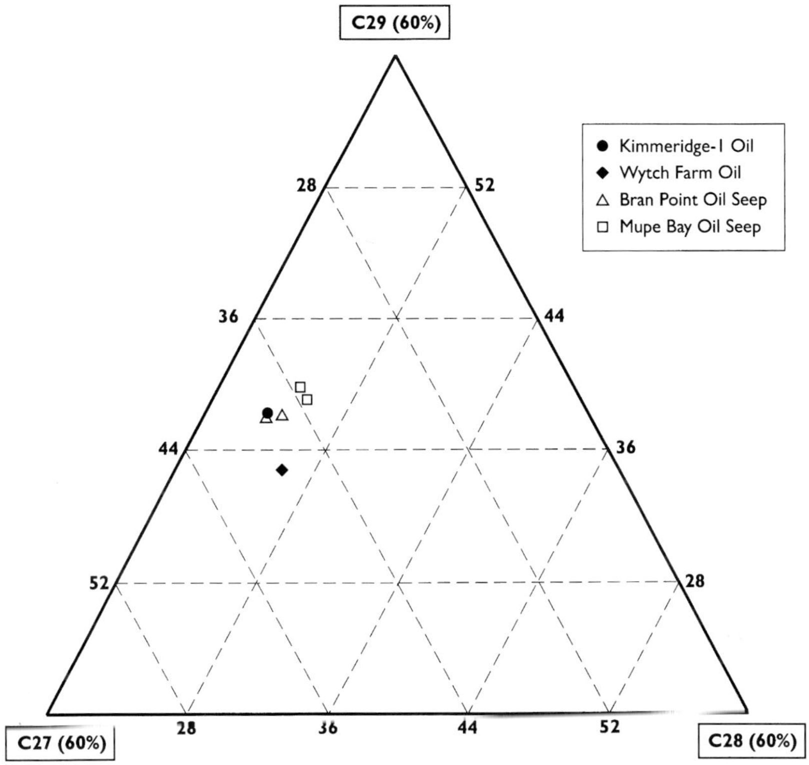

Fig. 8. Sterane composition of oils and seeps from the Central English Channel Basin, showing a close correlation suggesting a similar source rock. The Kimmeridge-1 oil and the Bran Point oil seeps appear to have a very similar composition.

Basin (Scotchman & Thomas 1995; Dancer *et al.* 1999) and the Hebrides Basin (Morton & Hudson 1995), extending northeastwards to the West of Shetlands Solan Basin and the northern North Sea North Viking Graben and East Shetland Basin (Cornford 1998; Fig. 9).

In Dorset, this source unit is represented by the upper part of the Black Ven Marls (13.7 m) and the 23 m thick Belemnite Marls with similar TOC values and oil-prone kerogens to the units below. In the Celtic Sea basins, offshore southern Ireland, this unit has poorer quality, mixed source potential for oil and gas, with predominantly Type II/III kerogens (Cornford 1998). In well 49/9-1, TOC is reported as 1.52–2.11 wt%, the kerogen comprising 10–15% Type II sapropel and 65–75% terrestrially derived vitrinite, suggesting a nearshore depositional environment (Caston 1995). To the southwest in the Goban Spur Basin, the source quality is even poorer with some 400 m of section in well 62/7-1 having average TOC of 1.5 wt%, comprising mainly inertinitic and woody gas-prone kerogen (Cook 1987). Northeastwards into the Cardigan Bay Basin, the mixed oil–gas-prone nature of the kerogens continues, with thin, richer oil-prone units within a generally gas-prone shale sequence, as illustrated by well 107/21-1 and the Mochras Borehole (Barr *et al.* 1981).

In the Slyne Basin, the late Sinemurian–Pliensbachian is represented by the Pabba Shale Formation equivalent (Dancer *et al.* 1999) which is up to 253 m thick. Some 127 m of organic-rich shales occur in well 27/13-1, with TOC up to 6.5 wt% (average 3.7 wt%) and maximum S2 of 24.5 kg t^{-1} (mean 11.1 kg t^{-1}) which, with a HI of 205 to 377 indicates a rich, oil–gas-prone source rock (Scotchman & Thomas 1995). The kerogen comprises up to 90% amorphous organic matter (AOM) which is finely disseminated and partly degraded, giving the unit considerable gas potential. Along-strike in the Hebrides Basin, the age-equivalent Pabba Shale Formation occurs both in wells and in outcrop (Morton & Hudson 1995). The 375 m thick unit in well Upper Glen-1

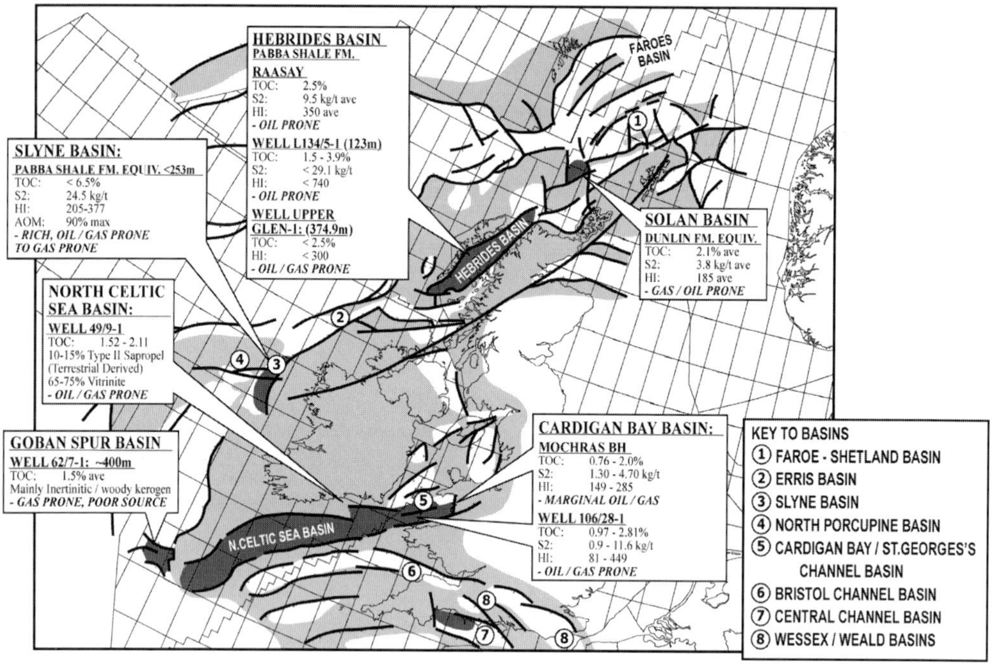

Fig. 9. The distribution of late Sinemurian–Pliensbachian source rocks in the Atlantic margin basins of Ireland and the UK.

is oil–gas prone with TOC less than 2.5 wt% and HI less than 300 (Butterworth *et al.* 1999), while in well L134/5-1 the unit is 123 m thick with oil-prone kerogens with TOCs of 1.5 to 3.9 wt%, S2 up to 29.1 kg t^{-1} and HI less than 740. Shales with TOC averaging 2.5%, S2 9.5 kg t^{-1} and HI 350 occur in outcrop from Raasay, indicating a similar oil-prone source rock. Further along the basin trend in the Solan Basin, the Sinemurian–Pliensbachian-aged Dunlin Group equivalent has gas–oil potential in well 202/3a-3, with average TOC of 2.1 wt%, S2 of 3.8 kg t^{-1} and HI of 185.

Geochemically, the late Sinemurian–Pliensbachian source rock interval is characterized by a waxy *n*-alkane signature with a slight odd over even CPI greater than *n*C23 (Fig. 10). Isotopically, the aromatic fraction $\delta^{13}C$ values are relatively heavier than those of the Portree Shale Formation (Fig.11), indicating a more terrestrial and less marine derivation. The kerogens are generally mixed marine Type II sapropels and terrestrially derived Type III material, with mixed oil and gas potential. The general dispersed, pyritic and degraded nature of the amorphous organic matter (AOM) which, in the Slyne Basin, comprises 90% of the total, indicates deposition under very poorly oxygenated but not anoxic bottom waters.

Oil shows from wells in the Slyne Basin show sourcing from both the Portree Shale Formation equivalent in well 27/13-1 (Scotchman & Thomas 1995) and the Pabba Shale Formation equivalent in other wells (Figs 11 to 13). Oils from the Pabba Shale Formation have a higher 18α(H)-22,29,30-trisnorneohopane/ 17α(H)-22,29,30-trisnorhopane (Ts/Tm) ratio, a more prominent C_{30} pentacyclic terpane peak, a higher C_{25}/C_{26} tricyclic terpanes ratio and heavier aromatic fraction $\delta^{13}C$ values than those from the Portree Shale Formation, while the Portree Shale Formation sourced-oils have a higher C_{27-28} vs. C_{29} sterane content (Fig. 13).

Toarcian source rocks and oils

The youngest of the early Jurassic anoxic events occurred in the Toarcian and resulted in the widespread deposition of organic-rich mudrocks across the Atlantic margin basins, eastern England, the southern North Sea, west Netherlands and the Paris Basin (Fleet *et al.* 1987; Cornford 1998). In the Atlantic margin basins,

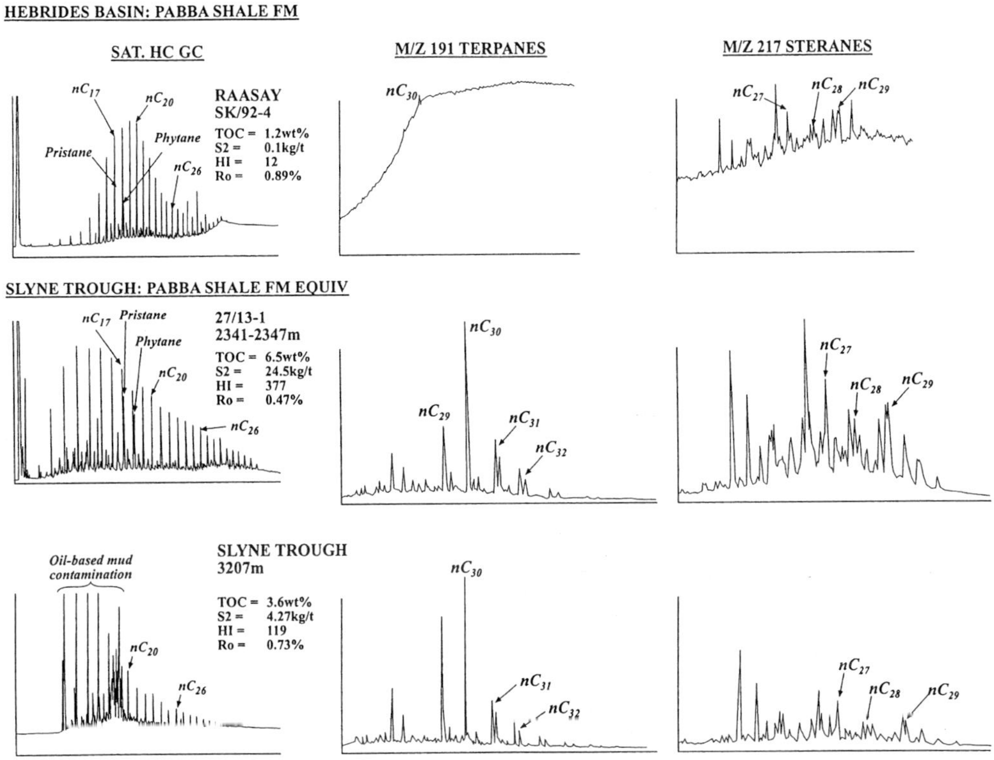

Fig. 10. Pliensbachian Pabba Shale Formation source rocks from the Hebrides Basin and its equivalents from the Slyne Trough. The saturated hydrocarbon fraction GCs show the general waxy nature of the hydrocarbons, with the dominant nC_{30} terpane peak and the relatively low nC_{28} sterane content.

source rock facies are known from the Celtic Sea, Slyne Basin and Hebrides basins and were probably deposited throughout the early Mesozoic rift-basin system (Fig. 14). In the Hebrides Basin the unit comprises the Portree Shale Formation which occurs at outcrop on the eastern Isle of Skye and on Raasay and in several wells from Skye and the Sea of the Hebrides. Geochemical analyses indicate a rich, oil-prone source rock (Scotchman *et al.* 1998, table 2) which reaches a maximum thickness of 61.6 m in well Upper Glen-1. Average data are: TOC 4.6 wt% (range up to 5.5 wt%); S2 13.8 kg t^{-1} and HI 429 (range up to 550). Kerogen data indicate over 80% marine AOM with prasinophyte algae and microhystrydid acritarchs, with the palynomorph assemblage dominated by small sphaeromorph pollen. Along the basin strike to the southwest, Slyne Basin well 27/13-1 contains some 84 m of Portree Shale Formation equivalent with richer oil source characteristics (Scotchman & Thomas 1995). In the well, TOC reaches a maximum of 7.5 wt%, averaging 4.7 wt% while the maximum recorded S2 is 29.1 kg t^{-1} (average 24.8 kg t^{-1}) and HI ranges from 330 to 540. The kerogen is a rich oil-prone Type II–II/III mix, comprising 90% AOM with acritarchs, dinocysts and prasinophyte algae plus degraded waxy humic or cuticular material.

In the basins south and west of Ireland, the Toarcian source rocks become generally less rich and oil prone and more variable in facies, with mixed gas–oil potential and oil potential. The poorest source rocks are reported from the Goban Spur Basin where some 600 m of Toarcian section in well 62/7-1 contains gas–oil-prone shales with average TOC of 1 wt%, the marine kerogens being diluted by a large woody, terrestrial input (Cook 1987). Similar source potential occurs on the flanks of the North Celtic Sea Basin in well 48/19-1, where the unit has TOCs of 1.5–2.3 wt%, S2 of 0.5–5.1 kg t^{-1} and HI of 35–222 (Murphy *et al.* 1995), indicating poor to good gas to oil potential. Source richness and oil proneness increase into the depocentres of the Celtic Sea basins (Murphy *et al.* 1995) due largely to reduced dilution by terrestrial, more gas-prone kerogens and increased productivity

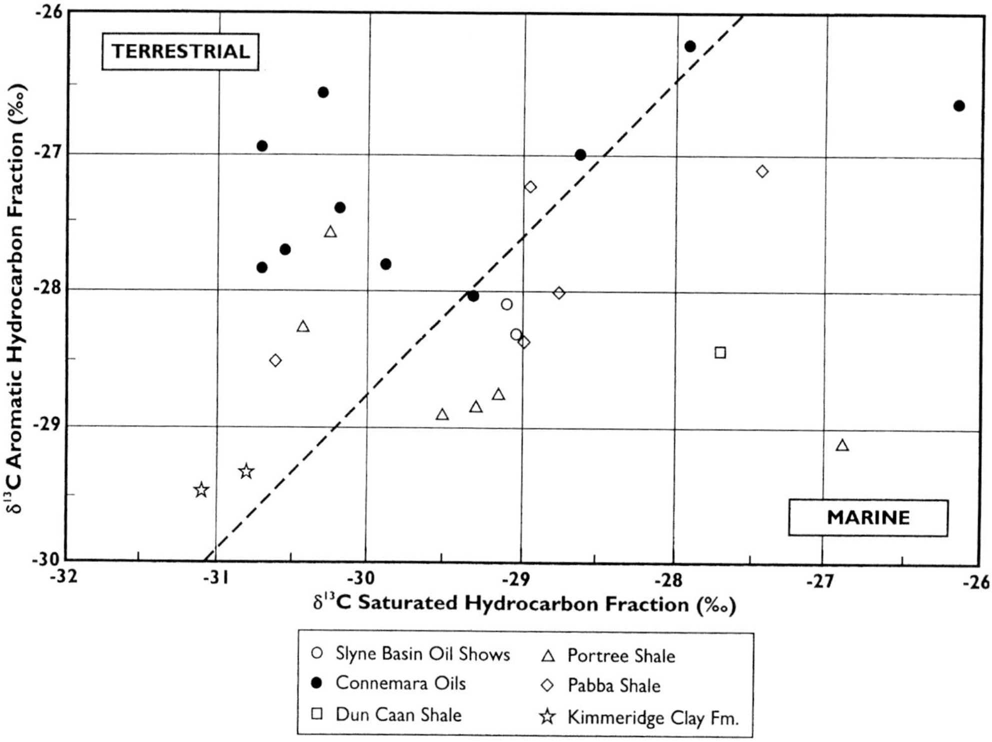

Fig. 11. Cross-plot of aromatic and saturate hydrocarbon fraction $\delta^{13}C$ data, with marine source rocks and oils plotting to the lower right of the line (Sofer 1984) and terrestrial/non-marine source rocks and oils to the upper left. In particular, the Connemara Field oils appear to be largely non-marine in origin.

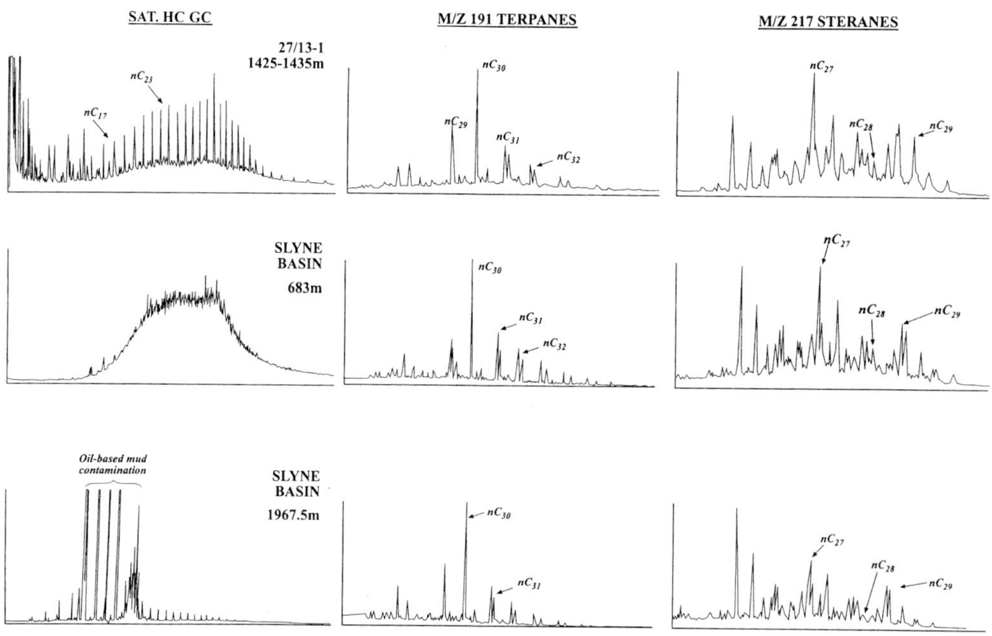

Fig. 12. Oil shows from the Slyne Trough which, although either biodegraded or contaminated by oil-based drilling mud, show that the biomarkers are unaffected and indicate sourcing from the Portree Shale Formation equivalent in the case of 27/13-1 and the Pabba Shale equivalent in the other Slyne Trough samples.

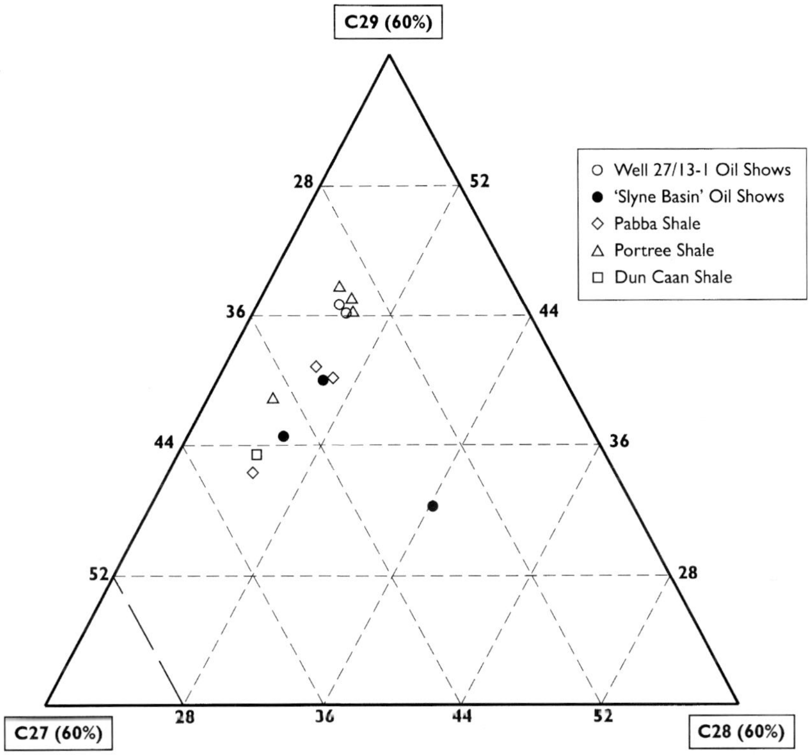

Fig. 13. Sterane composition of Lower–Middle Jurassic source rocks and oil shows from the Slyne Trough. The 27/13-1 oil shows appear to have been sourced by the Portree Shale Formation equivalent while those from the 'Slyne Trough' wells correlate closer with the Pabba Shale Formation equivalent.

and preservation of marine oil-prone kerogens. Murphy *et al.* (1995) report wells from more basinal locations with TOCs upto 3.0 wt%, S2 up to 13.9 kg t^{-1} and maximum HI of 525, indicating rich, oil-prone source potential. These authors also note that the oil potential is limited in the early Toarcian to the northeastern part of the North Celtic Sea Basin, while it increases southwestwards in the middle and late Toarcian. Oils in the Helvick accumulation are correlated with Liassic (Toarcian) source rocks, due to their waxy nature and similar biomarker and carbon isotope composition (see Caston 1995, Fig. 10). No source rocks of this age occur in the Wessex or Weald basins of southern England.

The Toarcian source rocks predominantly contain marine Type II kerogens diluted towards the basin flanks by land-derived Type II sapropel and varying proportions of Type III herbaceous material. In the Slyne and Hebrides basins, the oil-prone kerogens contain predominantly algal marine AOM, generally greater than 80–90%. These have a waxy nC_{25}+ alkane signature, with a slight odd over even CPI predominance greater than nC_{23}. The sterane distribution is characterized by a relatively low nC_{28} and high nC_{29} contents (Fig. 15), as is also seen in the oil shows from Slyne Trough well 27/13-1 (Fig. 13).

Middle Jurassic petroleum system

Middle Jurassic source rocks occur under very different conditions to those of the Lower Jurassic, with deposition in a regressive setting during the Aalenian–Bathonian and they have a restricted areal distribution. Lacustrine or marginal marine source rocks typify this interval (Fig. 4). The succeeding Callovian–Oxfordian interval marks the return to transgressive, marine conditions, the timing of this non-marine to marine transition varying considerably along the Atlantic margin basins and occurring latest in the northern Porcupine Basin. These source rocks have very distinctive geochemical characteristics, as discussed below, having generated oils which share these properties. The oils occur in the Foinaven and Schiehallion fields West of Shetlands (Scotchman *et al.* 1998; Lamers &

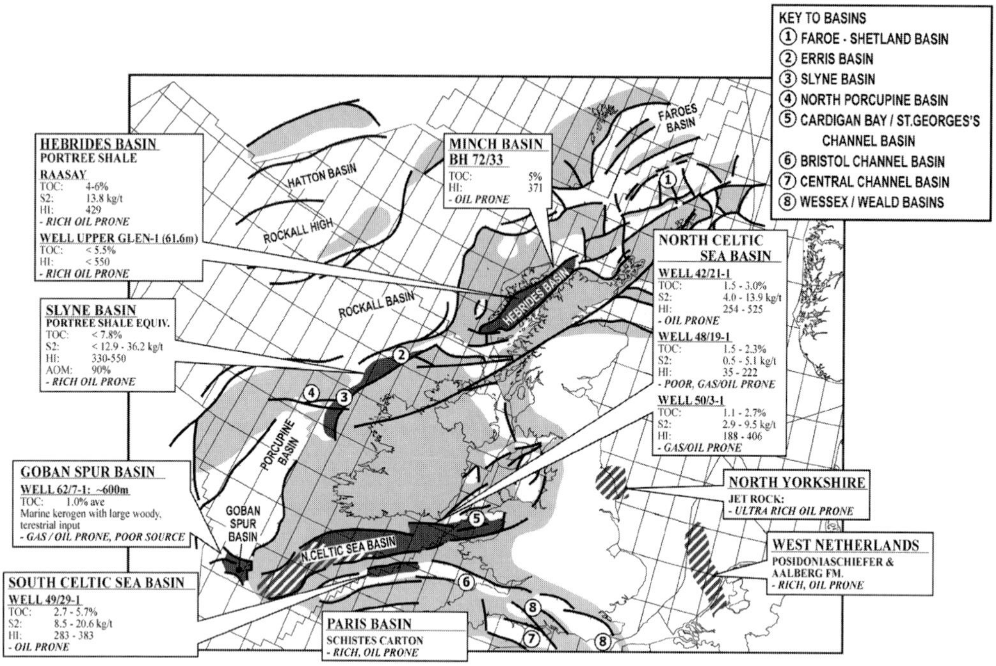

Fig. 14. The distribution of Toarcian source rocks in the Atlantic margin basins of Ireland and the UK.

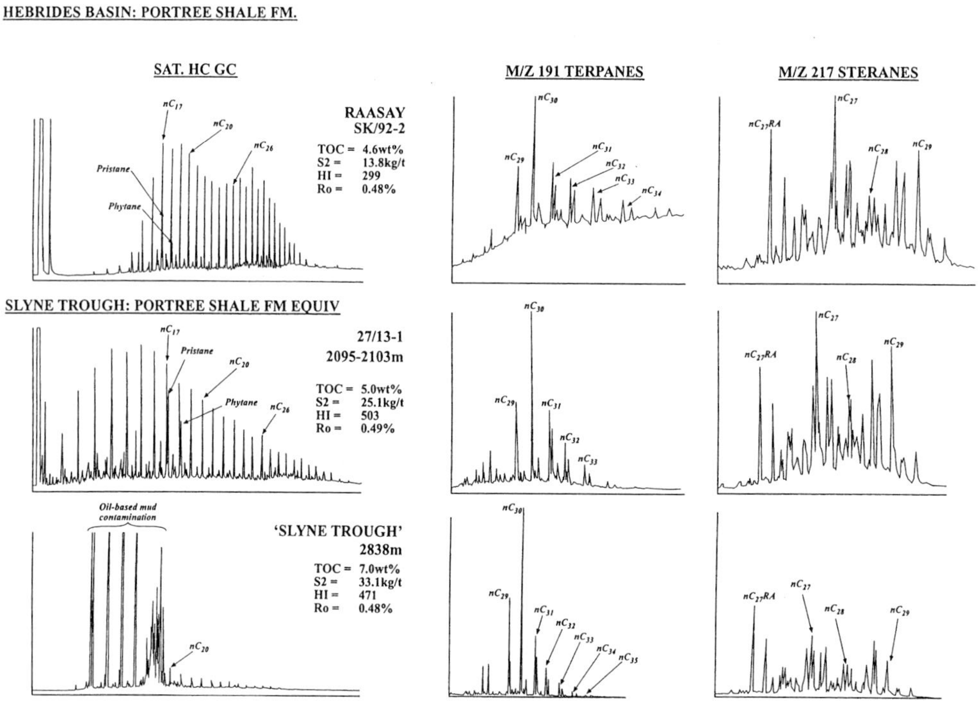

Fig. 15. Geochemistry of Toarcian Portree Shale Formation source rocks from the Hebrides Basin and its equivalents from the Slyne Trough. The saturated hydrocarbon fraction GCs show the general waxy nature of the hydrocarbons with 'hump' centred on nC_{29}, with the dominant nC_{30} terpane peak and the relatively low nC_{28} sterane content.

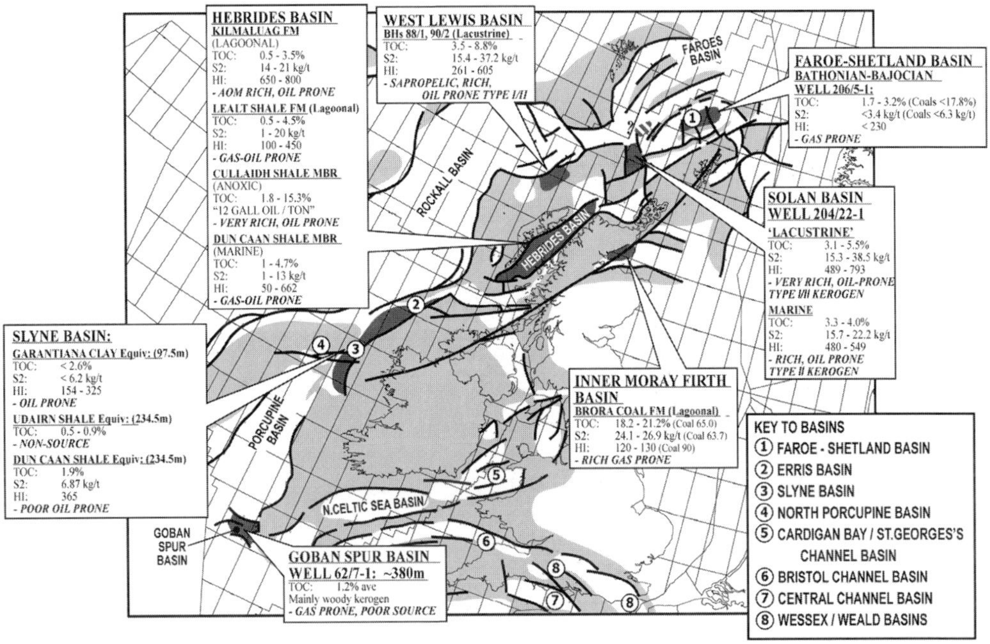

Fig. 16. The distribution of Aalenian–Bathonian source rocks in the Atlantic margin basins of Ireland and the UK.

Carmichael 1999) and in the Connemara, accumulation in the Porcupine Basin (Cornford 1998; Butterworth *et al.* 1999). These oils show clear geochemical evidence of mixed sourcing from both marine Upper Jurassic and Middle Jurassic non-marine, lacustrine rocks.

Aalenian–Bathonian source rocks and oils

The generally organic-rich Bathonian shallow marginal marine to lacustrine mudrocks of the Great Estuarine Group of eastern and central England are typical of this interval, with thin, very rich oil shales in the Atlantic margin area (Fig. 16). In the Hebrides Basin, the sequence is represented by a cyclic series of shales and sandstones deposited under varying marine to lagoonal settings (Morton & Hudson 1995 and references therein). The shallow-marine Aalenian to upper Bajocian Bearreraig Sandstone Formation contains the Aalenian Dun Caan Shale Member (12 m thick) at the base (Vincent & Tyson 1999), with the Aalenian–Bajocian Udairn Shale Member (72 m thick) above the Ollach Sandstone Member. At the top, marine dark shales of the Garantiana Clay Member (upper Bajocian, 2 m thick) pass up into the brackish water–lagoonal Cullaidh Shale Formation at the base of the Great Estuarine Group. The Great Estuarine Group comprises three upward-regressive cycles with the brackish–lagoonal Cullaidh Formation shales, the lagoonal Lealt Formation and the marine–brackish–lagoonal sediments of the Duntulm Formation respectively at the base of each coarsening-up cycle. Of these units, the Bathonian Cullaidh Shale Formation of the Hebrides Basin, deposited under anoxic conditions and outcropping on Raasay and the Isle of Skye, is the most widespread and comprises a thin oil-prone unit 3–6 m thick with TOC up to 6.3 wt% (Table 1). TOCs of up to 15.3 wt% and reported pyrolysis yields of 12 gallons ton^{-1} and 74 kg t^{-1} (Bjorøy *et al.* 1988; Thrasher 1992) clearly show the very rich, oil-prone nature of the unit which is often masked by the effects of local high maturity on account of proximity to Tertiary igneous intrusions (Thrasher 1992). The brackish–lagoonal shales in the overlying cycles from northern Skye are gas–oil prone in the case of the Lealt Shale Formation (44 m thick) and rich oil prone with a high AOM content in the Kilmaluag Formation (25 m thick) with Type II and Type I/II kerogens respectively (Morton & Hudson 1995; Vincent & Tyson 1999).

A further section of potential source rock occurs in the older Aalenian marine Dun Caan Shale Member in Raasay, where it is about 27 m thick (Morton & Hudson 1995). Generally poorer oil–gas source potential is indicated by the S2 of $1–13\,kg\,t^{-1}$ and HI of 50–662 (Table 1): TOC ranges from 1.0 to 4.7 wt%, Thrasher (1992) quoting an average of 2 wt%. Marine Garantiana Clay Formation shales in Hebrides Basin Borehole 88/6 are poor, gas-prone source rocks with maximum TOC of 1.63 wt%. S2 of $11.70\,kg\,t^{-1}$ and HI of 81. The humic and inertinitic kerogens present are predominantly terrestrial in origin.

Middle Jurassic source rocks also occur in the basins on the southeastern flank of the Rockall Trough in the West Lewis and West Flannan basins, where Bathonian-aged Great Estuarine Group shales with TOCs of up to 8.8 wt% occur

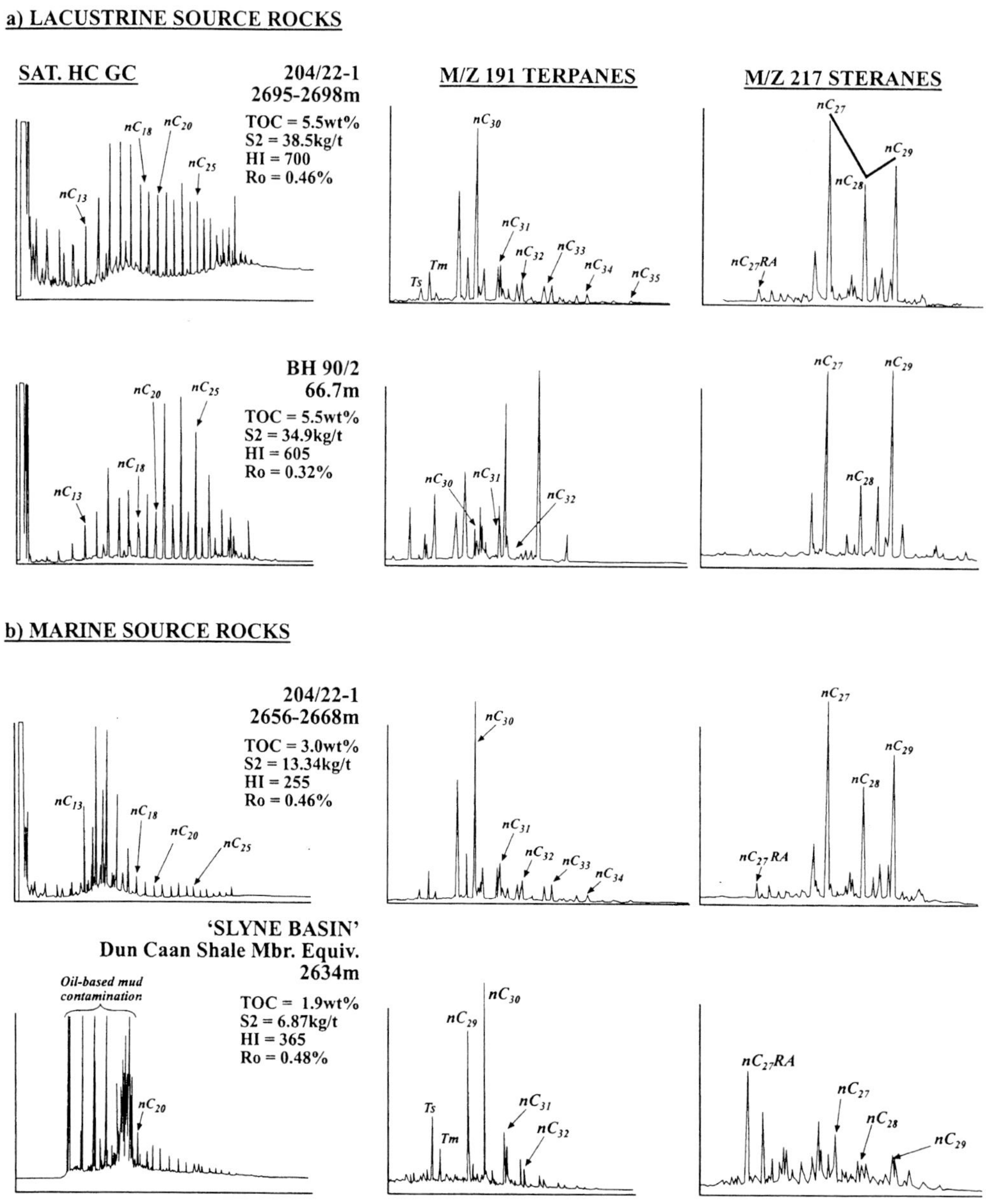

Fig. 17. Geochemistry of Aalenian–Bathonian age lacustrine and marine facies source rocks, showing the very waxy nature of the former samples with *n*-alkane distributions with 'humps' developed at nC_{17} and nC_{30}. Again, the nC_{28} sterane content is relatively low.

in the British Geological Survey's 88/1 and 90/2 cored boreholes (Hitchen & Stoker 1993). Excellent oil source potential in these lacustrine to marginal marine rocks is indicated by S2 yields of 15 to 37 kg t^{-1} and HI of up to 605, containing hydrocarbons that are dominated by either C_{21} or C_{23} *n*-alkanes, with a very high odd carbon number predominance above *n*-C_{20}, and have a PR/PH ratio <1, suggesting these very immature rocks were deposited under anoxic conditions. Sterane distributions from these samples are dominated by 14(α)H,17(α)H isomers and show distinct patterns (Figs 17 and 18) with a relatively lower C_{28} sterane content than the C_{29} steranes which suggests a terrestrial source (Huang & Meinschein 1979). Kinetic data (Holmes *et al.* 1999) indicate these samples to contain Type I and Type I/II kerogens.

Further occurrences of Middle Jurassic shales are present along the strike of the Mesozoic basin system to the northeast in the West of Shetlands area as thick pods in what, on the current limited database, appear to be separate sub-basins. One such pod of Middle Jurassic occurs in the Solan Basin, on the flank of the North Rona High where 200 m of late Bajocian to late Bathonian shales penetrated by well 204/22-1 have oil source potential. Here the kerogens are of variable composition, with a 13 m thick basal marine section of low source potential overlain by lacustrine mudstones which contain rich, oil-prone source rocks. These are richest at the base, with TOCs of 10–12 wt% up to 20 wt%, becoming poorer (TOC 2–3 wt%) and less oil prone up-hole as the facies becomes increasingly marine (Scotchman *et al.* 1998, fig. 7). The lacustrine facies samples contain waxy hydrocarbons with an *n*-alkane distribution up to *n*-C_{34} (Fig. 17). Similar-aged lacustrine source rock developments are inferred to the north in the Quadrant 204 area from the geochemical characteristics of reservoired oils as discussed below.

The West of Shetland oils from the Foinaven and Schiehallion oil fields (Fig. 19) have a high wax content, as shown by the predominance of nC_{20}+ molecular weight components, with an odd/even *n*-alkane preference (high CPI) and a

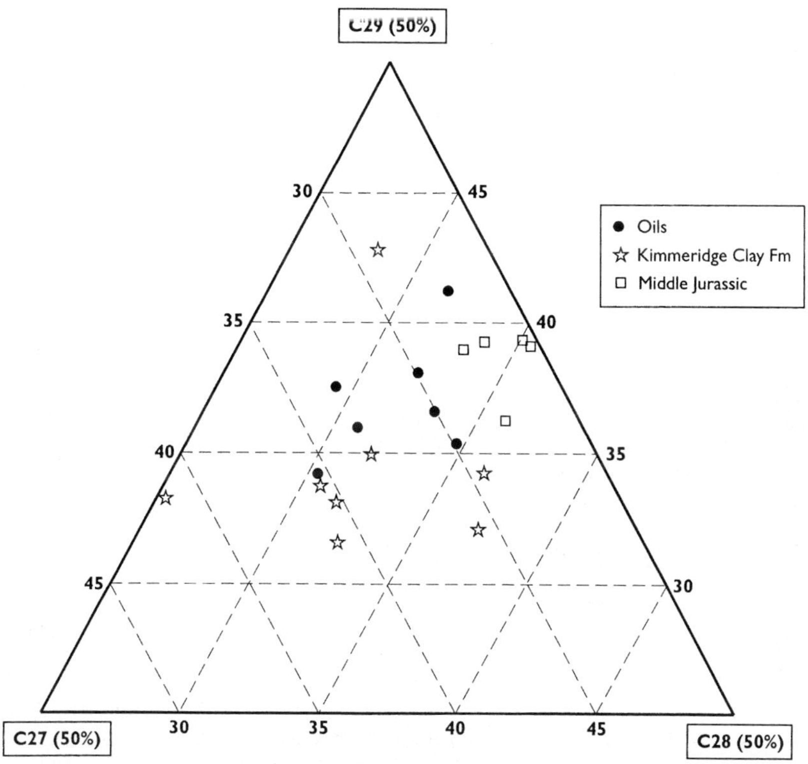

Fig. 18. Sterane composition of oils and Middle–Upper Jurassic source rocks of the West of Shetlands, showing the variable degree of co-sourcing of the oils by both the Middle Jurassic and the Kimmeridge Clay Formation equivalent source rocks.

low C_{28} sterane content (Figs 18 and 19; Scotchman *et al.* 1998, fig. 20). These features indicate sourcing from the lacustrine facies Middle Jurassic rocks as seen in well 204/22-1, while the presence of 17α(H),18α(H),21β(H) 28,30-bisnorhopane indicates an Upper Jurassic Kimmeridge Clay Formation sourced component (Fig. 20; Scotchman *et al.* 1998). Traces of gammacerane are also present in these oils, probably from the Middle Jurassic component, suggesting a highly saline lacustrine depositional environment (Peters & Moldowan 1993).

Well 206/5-1, located on the northwestern flank of the Faroe–Shetland Basin, contains an anomalously thick Middle to Upper Jurassic section in what is interpreted to be a large pod of sediment slumped off the Rona Ridge (Haszeldine *et al.* 1987). Some 348 m of Bajocian to Bathonian-aged marine shales with minor sandstones have TOCs ranging from 1.7 to 3.2 wt%; S2 yield of up to 3.4 kg t^{-1} and HI of less than 230, indicating rich, gas-prone source potential. Thin coals in the section have TOC and S2 up to 17.3 wt% and 6.3 kg t^{-1} respectively.

In the Irish offshore area, thick developments of equivalent-age source rocks occur in the Slyne Basin, although hydrocarbon potential appears poorer than in the Hebrides and Rockall flank basins (Fig. 16). The richest, most oil-prone rocks occur in the 97.5 m thick Garantiana Clay Member equivalent, with TOC and S2 up to 2.6 wt% and 6.2 kg t^{-1} respectively with a HI of 154–325, indicating mixed Type III–II/III kerogen with good potential for oil and gas generation. These kerogens are of mixed marine and terrestrial origin and were deposited under anoxic conditions. The thick Udairn Shale Member equivalent is non-source, while the 235 m thick basal Dun Caan Shale Member equivalent (Fig. 17) is a poor (average TOC of 1.9 wt%) but oil-prone (S2 of 6.87 kg t^{-1} and HI of 365) source with Type II/III kerogen of a mixed terrestrial/marine origin from an anoxic environment.

Southwest of Ireland, in the Goban Spur Basin, some 380 m of Middle Jurassic shales in well 62/7-1 are a gas-prone, poor source, containing mainly woody kerogens with an average TOC of 1.2 wt% (Cook 1987).

Callovian–Oxfordian source rocks and oils

Callovian to Oxfordian marine source rocks occur in and on the flanks of the Faroe–Shetland Basin and in the Hebrides Basin. These mudrocks are equivalent to the Heather Formation of the northern North Sea, which demonstrates a significant local and regional variation in both thickness and in the development of organic facies due to syn-rift deposition (Johnson *et al.* 1993). In the Heather Formation, the most transgressive, oil-prone shale facies appears to be restricted to the rift basins, such as the North Viking Graben (Gormly *et al.* 1994), whereas on the basin flanks the unit is largely a

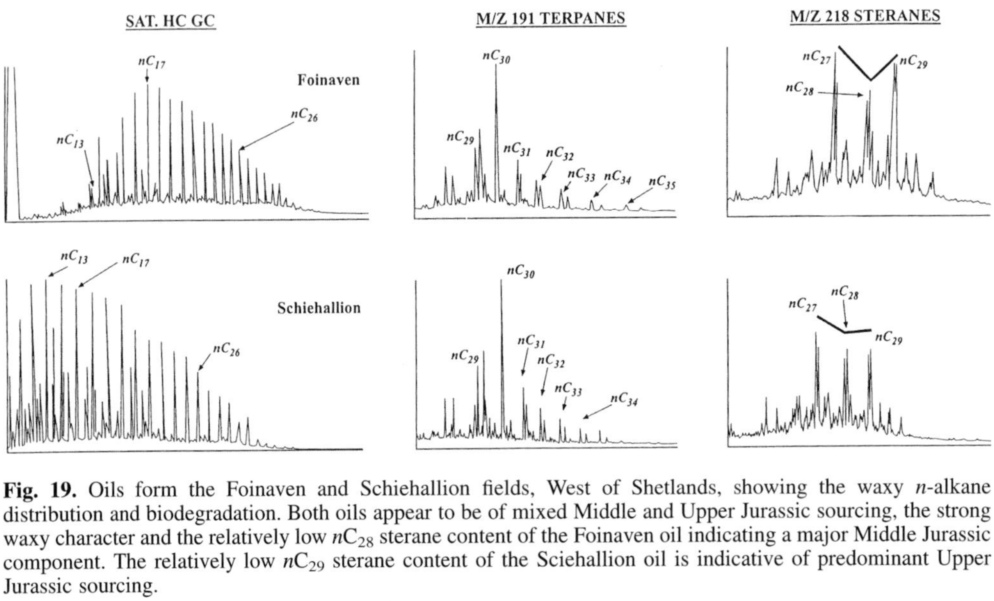

Fig. 19. Oils form the Foinaven and Schiehallion fields, West of Shetlands, showing the waxy *n*-alkane distribution and biodegradation. Both oils appear to be of mixed Middle and Upper Jurassic sourcing, the strong waxy character and the relatively low nC_{28} sterane content of the Foinaven oil indicating a major Middle Jurassic component. The relatively low nC_{29} sterane content of the Sciehallion oil is indicative of predominant Upper Jurassic sourcing.

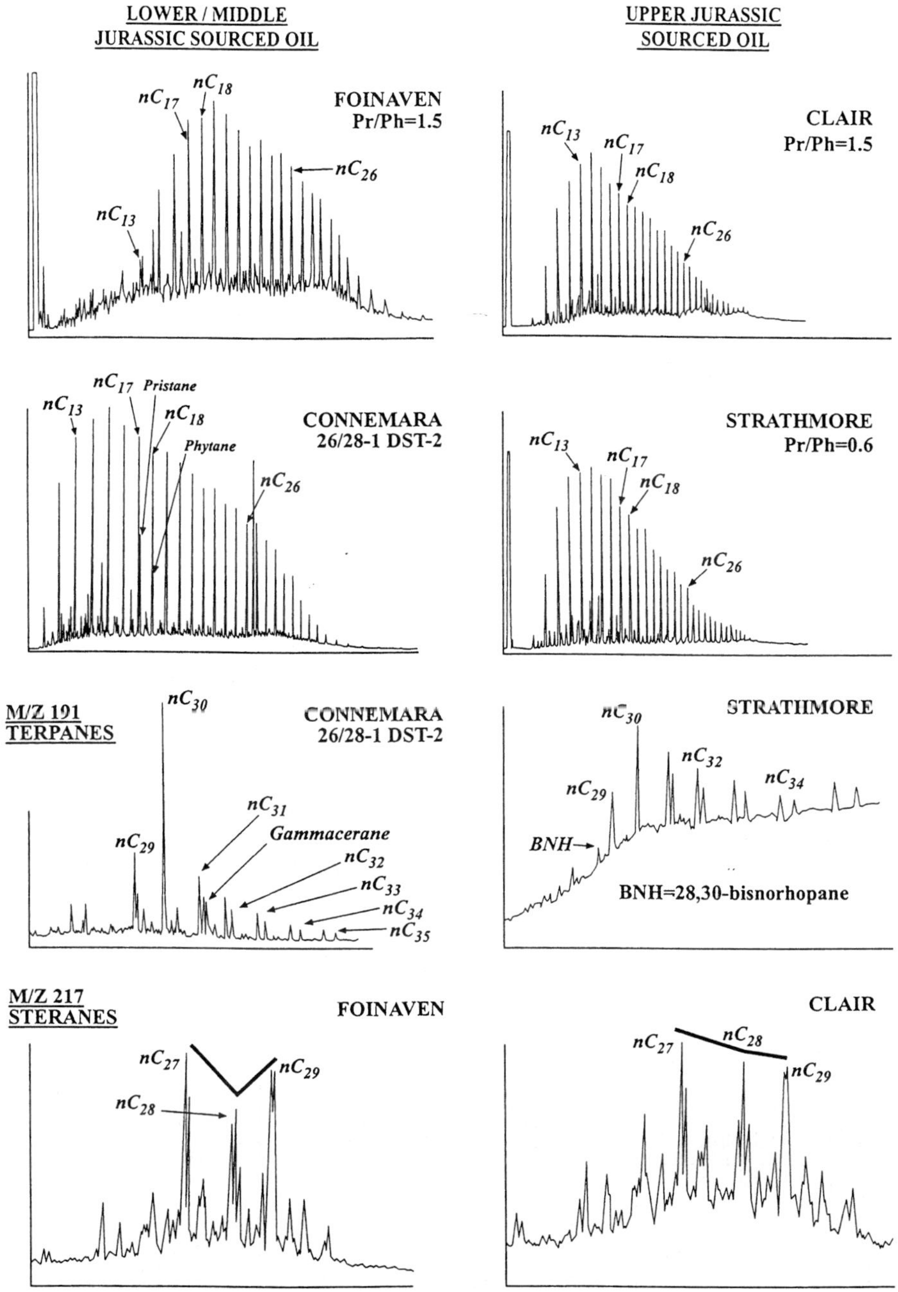

Fig. 20. Geochemical comparison of Middle and Upper Jurassic-sourced oils, showing the waxy nature of the former and low wax content of the latter. Gammacerane may occur in the lacustrine-sourced Middle Jurassic oils while 28,30-bisnorhopane occurs only in Upper Jurassic oils. The sterane distributions are also distinctive, with the former having a relatively low nC_{28} sterane content compared to the latter.

gas-prone source rock (Barnard & Cooper 1981). Ultra-rich, oil-prone marine source rocks are also developed in localized, probably very anoxic basin settings, such as the Brora Argillaceous Formation of the Inner Moray Firth (Andrews *et al.* 1990; Cornford 1998).

In the Atlantic margin basins (Fig. 21) the Heather Formation equivalent occurs in the

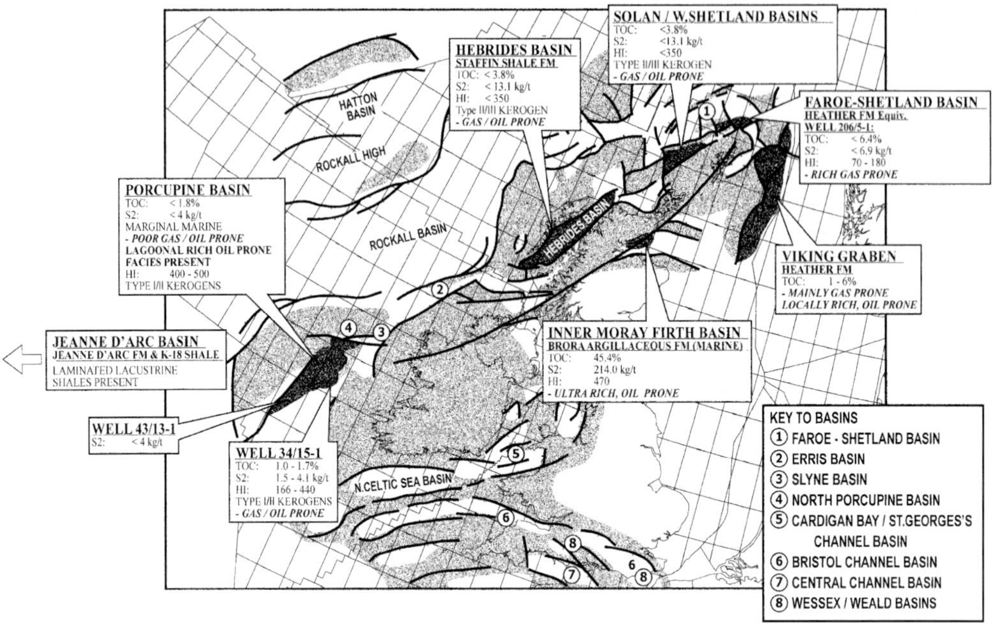

Fig. 21. The distribution of Callovian–Oxfordian source rocks in the Atlantic margin basins of Ireland and the UK.

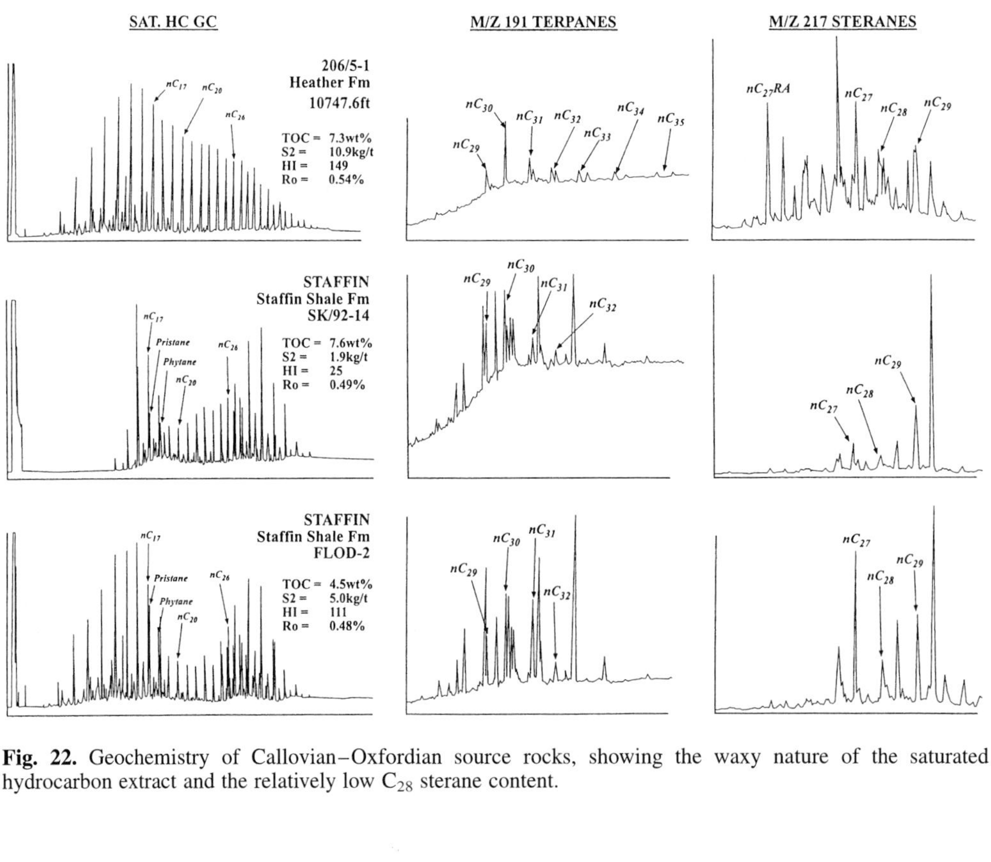

Fig. 22. Geochemistry of Callovian–Oxfordian source rocks, showing the waxy nature of the saturated hydrocarbon extract and the relatively low C_{28} sterane content.

Faroe–Shetland and Solan basins. The formation occurs at the top of the Middle–Upper Jurassic sections in wells such as 204/22-1 and 205/6-1, or as thin units, often overlying the basal late Jurassic unconformity, as in well 205/26a-3. In the Solan Basin wells the unit contains Type II/III oil–gas-prone kerogens with TOCs up to 3.8 wt% and S2 up to 13.1 kg t^{-1}. However, in the Faroe–Shetland Basin, these source rocks are much richer and oil–gas-prone in well 206/5-1, with TOCs up to 7.3 wt% and a typical waxy character (Fig. 22).

In the Hebrides area, organic-rich shales of the Callovian–Oxfordian Staffin Bay and Staffin Shale formations outcrop in northeast Skye with TOC and S2 less than 3.8 wt% and 13.1 kg t^{-1} respectively. The saturated hydrocarbon fraction has a waxy *n*-alkane composition and relatively low C_{28} sterane content (Fig. 22). The hydrogen index of less than 350 indicates Type II/III kerogens of mixed oil–gas potential.

Further to the southwest, along-strike of the Mesozoic basin system, equivalent-aged shales in the northern Porcupine Basin are of marginal marine facies, deposited on the flank of the Upper Jurassic basin and represent only a poor gas–oil source rock. TOCs are correspondingly poor, averaging less than 1.8 wt% with S2 less than 4 kg t^{-1}. However, lagoonal facies units present within the basin have HIs of 400–500 and contain Type I/Type II kerogens in wells 34/15-1 and 35/6-1 (Butterworth *et al.* 1999), where they form rich, waxy, oil-prone source rocks. Laminated lacustrine shales are also reported in the similar-aged Jeanne d'Arc Shale and K-18 Shale of the formerly contiguous Jeanne d'Arc Basin (Fig. 21).

Oils from the Connemara accumulation in the northern Porcupine Basin (Figs 23 and 24) show mixed marine and lacustrine Middle Jurassic sourcing, with a significant gammacerane content but no 28,30-bisnorhopane. The oils are waxy with a high CPI and relatively heavy aromatic hydrocarbon fraction $\delta^{13}C$ values, indicating a strong non-marine component (Fig. 11). The oils have a distinctive C_{27} sterane content (Fig. 24), with low C_{28} steranes and slightly higher C_{29} steranes. A highly saline, lacustrine source of Middle Jurassic age is suggested for these oils by Butterworth *et al.*

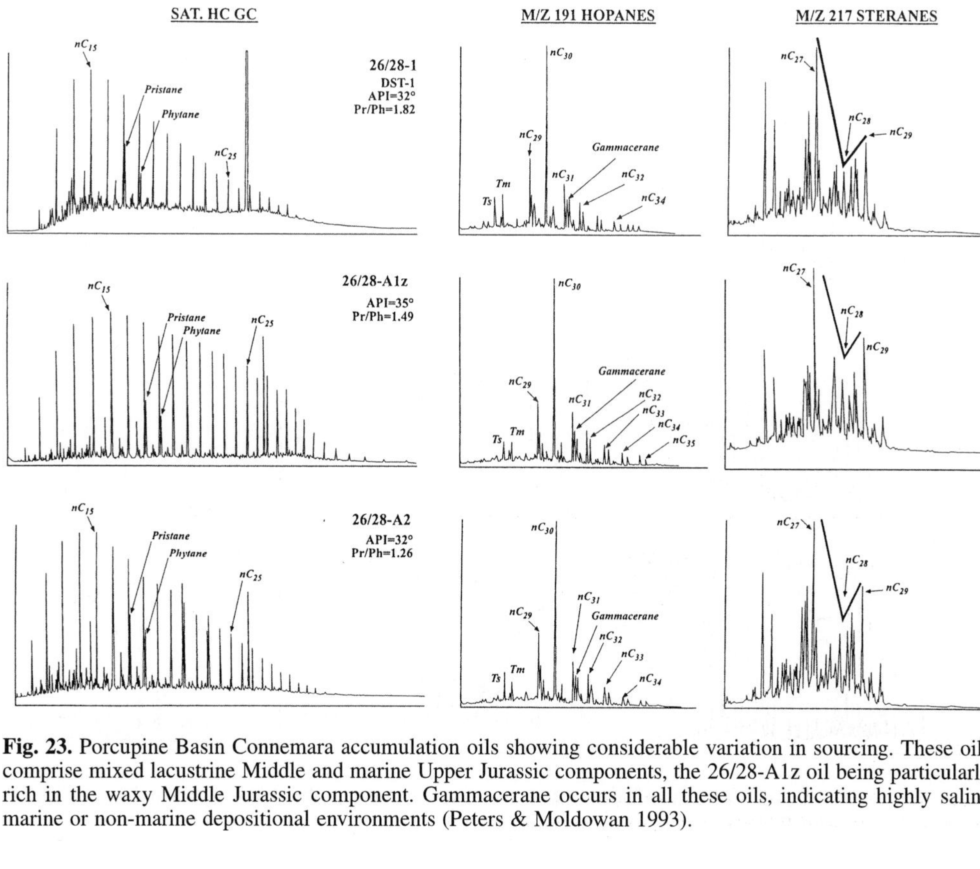

Fig. 23. Porcupine Basin Connemara accumulation oils showing considerable variation in sourcing. These oils comprise mixed lacustrine Middle and marine Upper Jurassic components, the 26/28-A1z oil being particularly rich in the waxy Middle Jurassic component. Gammacerane occurs in all these oils, indicating highly saline marine or non-marine depositional environments (Peters & Moldowan 1993).

(1999), with a variable marine Upper Jurassic sourced component. The sterane data (Fig. 24), however, show affinity to the marine Middle Jurassic Dun Caan Shale equivalent from the Slyne Basin, while the saturate and aromatic hydrocarbon isotope data (Fig. 11) additionally show the marine Lower Jurassic Pabba and Portree Shale formations as potential sources for the marine component. No data are available for the Upper Jurassic of the Porcupine Basin. Co-sourcing from both non-marine lacustrine and marine rocks can therefore be concluded, with a Middle Jurassic lacustrine component mixed with marine-sourced oil from Lower or Middle Jurassic shales or, on regional grounds, the Upper Jurassic. Further south in the Porcupine Basin, the Callovian–Oxfordian is developed in a more open marine depositional facies with consequent poorer, more gas-prone source potential, as seen in well 43/13-1, where the kerogens are mixed Type II/III with S2 below $4\,kg\,t^{-1}$.

In summary, the lacustrine and lagoonal facies Middle Jurassic source rocks are very oil prone and contain predominantly Type I/Type II kerogens, with a strong algal AOM content (80–90%) and PR/PH <1. Hydrocarbons from these rocks are waxy with high $nC_{20}+$ content with odd/even predominance and, characteristically, a low C_{28} and high C_{27} sterane content. In the Connemara accumulation the oils have a relatively heavy aromatic hydrocarbon fraction $\delta^{13}C$ values and light saturate hydrocarbon fraction values, indicating a terrestrial source (Sofer 1984). Marine source rocks of this age have a lower AOM and higher phytoclast content, with variable input of marine and terrestrial kerogens, resulting in variable oil–gas potential

Summary

Data presented in this study indicate Lower and Middle Jurassic-sourced oils to be volumetrically important in the Atlantic margin basins north and west of Ireland and the UK and that these oils are geochemically distinct from the regionally extensive Upper Jurassic Kimmeridge Clay

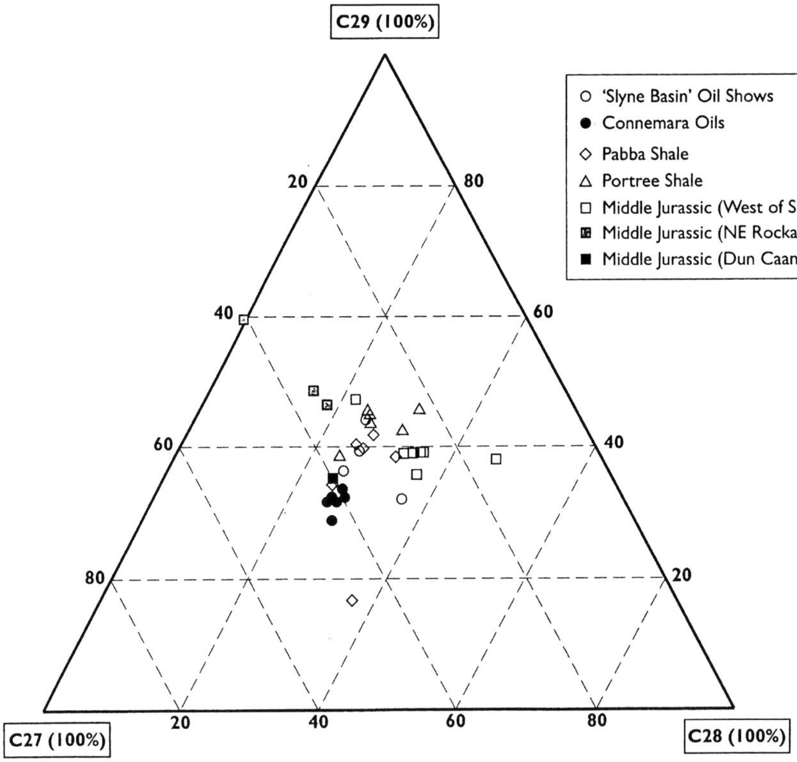

Fig. 24. Sterane composition of Lower–Middle Jurassic source rocks and oils from the Slyne Trough and Connemara Field, indicating a Middle Jurassic Dun Caan Shale equivalent source for the Connemara oils and a Lower Jurassic source (Portree and Pabba Shales) for those from the Slyne Trough.

Formation-sourced oils (Fig. 20). In particular, Lower–Middle Jurassic oils are waxier than their Upper Jurassic counterparts with an odd/even carbon preference in the $nC_{20}+$ alkanes (Table 5), due to their non-marine or lacustrine kerogen sourcing. C_{28} sterane contents in the Lower–Middle Jurassic oils are generally low and C_{29} steranes high, with the reverse in the Upper Jurassic oils. Gammacerane, indicative of highly saline, non-marine conditions, may be present in the former and 28,30-bisnorhopane in the latter. Finally, the Middle Jurassic sourced oils generally have relatively heavy aromatic hydrocarbon fraction $\delta^{13}C$ values and light saturate hydrocarbon fraction values (Fig. 11) while the oils from Upper Jurassic sources generally have the opposite, with heavy saturate and light aromatic hydrocarbon fraction values. Non-marine and marine sourcing respectively is indicated by these isotopic ratios (Sofer 1984).

Conclusions

A number of conclusions can be drawn from this study.

(1) Lower and Middle Jurassic mudrocks are effective source rocks in Atlantic margin basins north and west of Ireland and the UK, where they have sourced significant volumes of oil.

(2) Upper Jurassic Kimmeridge Clay Formation source rocks are volumetrically important only in the West of Shetlands where they are the major oil source. Elsewhere along the margin the Upper Jurassic equivalents comprise relatively lean mudrocks with only poor source potential, although they have good source potential in parts of the Porcupine Basin (Scotchman & Doré 1995).

(3) Marine Lower Jurassic oil-prone source rocks appear widely distributed through the Atlantic margin basins and have generated considerable amounts of oil in the Central Channel, North Celtic Sea and the Slyne basins.

(4) Marginal marine to lacustrine Middle Jurassic source rocks have a limited distribution in several basins (northern Porcupine, West Flannan, West Lewis, Hebrides, Solan and Judd) where they are extremely rich and oil prone. Oils generated by these source rocks, although waxy, are volumetrically important in the West of Shetlands and in the Porcupine Basin.

(5) Many oils are of mixed origin, with co-sourcing from lacustrine/non-marine Middle Jurassic and marine Upper Jurassic source rocks in the Porcupine Basin and West of Shetlands. A marine component from the Lower Jurassic cannot be ruled out in the former case.

(6) The wax content, $\delta^{13}C$ values of the saturate and aromatic hydrocarbon fractions, the C_{27}:C_{28}:C_{29} sterane ratio and prescence/absence of biomarkers such as gammacerane and 28,30-bisnorhopane in oils can be used to determine sourcing from either non-marine Middle Jurassic or marine Lower and Upper Jurassic rocks.

Table 5. *Comparison between Lower–Middle Jurassic- and Upper Jurassic-sourced oils*

Lower–Middle Jurassic-sourced oils	Upper Jurassic-sourced oils
Waxy	Low wax content
Predominant $nC_{20}+$ alkane distribution	nC_{15}–nC_{18} alkane peak distribution
$nC_{20}+$ odd/even carbon preference (Terrestrial or non-marine source)	Smooth $nC_{18}+$ alkane envelope (Marine, algal source)
$nC_{30}+$ USM ('hump')	Not present
28,30-bisnorhopane absent	28,30-bisnorhopane usually present
Gammacerane may be present (hypersaline environment)	Gammacerane absent (marine environment)
Low C_{28}, high C_{29} steranes	High C_{28}, low C_{29} steranes
Heavy aromatic and light saturate hydrocarbon fraction $\delta^{13}C$	Light aromatic and heavy saturate hydrocarbon fraction $\delta^{13}C$

Statoil (UK), Statoil Ireland Exploration and their licence co-venturers are thanked for permission to publish data used in this paper. John Kipps and Roy Rees-Williams draughted the figures.

References

ANDREWS, I.J., LONG, D., RICHARDS, P.C., THOMSON, A.R., BROWN, S., CHESHER, J.A. & MCCORMAC, A. 1990. *United Kingdom offshore regional report: the geology of the Moray Firth*. HMSO for British Geological Survey, London.

BAILEY, N.J.L., WALKO, P. & SAUER, M.J. 1987. Geochemistry and source rock potential of the west of Shetlands. *In*: BROOKS, J. & GLENNIE, K.W. (eds) *Petroleum Geology of North West Europe*. Graham and Trotman Ltd., London, 711–721.

BARNARD, P.C. & COOPER, B.S. 1981. Oils and source rocks of the North Sea area. *In*: ILLING, L.V. & HOBSON, G.D. (eds) *Petroleum Geology of the Continental Shelf of North-west Europe*. Heyden, London, 169–175.

BARR, K.W., COLTER, V.S. & YOUNG, R. 1981. The geology of the Cardigan Bay – St George's Channel Basin. *In*: ILLING, L.V. & HOBSON, G.D. (eds) *Petroleum Geology of the Continental Shelf of North-west Europe*. Heyden, London, 432–443.

BECKMAN, J. 1998. West of Shetland limps into action. *Offshore*, **57** (1), 14.

BJORØY, M., HALL, P.B., LOBERG, R., MCDERMOTT, J.A. & MILLS, N. 1987. Hydrocarbons from non-marine source rocks. *In*: MATTAVELLI, L. & NOVELLI, L. (eds) *Advances in Organic Geochemistry 1987, Part 1*. Pergamon, Oxford, 221–224.

BODENHAUSEN, J.W.A. & OTT, W.F. 1981. Habitat of the Rijswijk oil province, onshore The Netherlands. *In*: ILLING, L.V. & HOBSON, G.D. (eds) *Petroleum Geology of the Continental Shelf of North-west Europe*. Heyden, London, 301–309.

BURWOOD, R., STAFFURTH, J., DE WALQUE, L. & DE WITTE, S.M. 1991. Petroleum geochemistry of the Wessex–Weald basin of Southern England: a problem in source–oil correlation. *In*: MANNING, D. (ed.) *Organic Geochemistry, Advances and Applications in Energy and Natural Environment*. Ext. Abstracts, 15th Meeting EAOG, Manchester University Press, Manchester, 22–27.

BUTLER, M. & PULLAN, C.P. 1990. Tertiary structures and hydrocarbon entrapment in the Weald basin of southern England. *In*: HARDMAN, R.F.P. & BROOKS, J. (eds) *Tectonic Events Responsible for Britain's Oil and Gas Reserves*. Geological Society, London, Special Publications, **55**, 371–391.

BUTTERWORTH, P., HOLBA, A., HERTIG, S., HUGHES, W. & ATKINSON, C. 1999. Jurassic non-marine source rocks and oils of the Porcupine Basin and other North Atlantic margin basins. *In*: FLEET, A.J. & BOLDY, S.A.R. (eds) *Petroleum Geology of NW Europe: Proceedings of the 5th Conference*. Geological Society, London, 471–486.

CASTON, V.N.D. 1995. The Helvick oil accumulation, Block 49/9, North Celtic Sea Basin. *In*: CROKER, P.F. & SHANNON, P.M. (eds) *The Petroleum Geology of Ireland's Offshore Basins*. Geological Society, London, Special Publications, **93**, 209–225.

COLTER, V.S. & HARVARD, D.J. 1981. The Wytch Farm oil field, Dorset. *In*: ILLING, L.V. & HOBSON, G.D. (eds) *Petroleum Geology of the Continental Shelf of North-west Europe*. Heyden, London, 494–503.

CONEY, D., FYFE, T.B., RETAIL, P. & SMITH, P.J. 1993. Clair Appraisal: the benefits of a co-operative approach. *In*: PARKER, J.R. (ed.) *Petroleum Geology of Northwest Europe: Proceedings of the 4th Conference*. Geological Society, London, 1409–1420.

COOK, D.R. 1987. The Goban Spur – exploration in a deep water frontier basin. *In*: BROOKS, J. & GLENNIE, K.W. (eds) *Petroleum Geology of North West Europe*. Graham and Trotman Ltd., London, 623–632.

COOPER, M.M., EVANS, A.C., LYNCH, D.J., NEVILLE, G. & NEWLEY, T. 1999. The Foinaven Field: Managing reservoir development uncertainty prior to start-up. *In*: FLEET, A.J. & BOLDY, S.A.R. (eds) *Petroleum Geology of NW Europe: Proceedings of the 5th Conference*. Geological Society, London, 675–682.

COPE, J.C.W. 1995. Introduction to the British Jurassic. *In*: TAYLOR, P.D. (ed.) *Field Geology of the British Jurassic*. Geological Society, London, 1–7.

CORNFORD, C. 1998. Source rocks and hydrocarbons of the North Sea. *In*: GLENNIE, K.W. (ed.) *Petroleum Geology of the North Sea: Basic Concepts and Recent Advances*. 4th edn, Blackwell, London, 376–462.

CORNFORD, C., CHRISTIE, O., ENDRESEN, U., JENSEN, P. & MYHR, M.-B. 1988. Source rock and seep oil maturity in Dorset, southern England. *Organic Geochemistry*, **13**, 399–409.

DANCER, P.N., ALGAR, S.T. & WILSON, I.R. 1999. Structural evolution of the Slyne Trough. *In*: FLEET, A.J. & BOLDY, S.A.R. (eds) *Petroleum Geology of NW Europe: Proceedings of the 5th Conference*. Geological Society, London, 445–453.

DORÉ, A.G., VOLLSET, J. & HAMAR, G.P. 1985. Correlation of the offshore sequences referred to the Kimmeridge Clay Formation: relevance to the Norwegian sector. *In*: THOMAS, B.M., DORÉ, A.G., EGGEN, S.S., HOME, P.C. & MAGNE LARSEN, R. (eds) *Petroleum Geochemistry and Exploration of the Norwegian Shelf*. NPS, Graham and Trotman, London, 27–37.

EBUKANSON, E.J. & KINGHORN, R.R.F. 1985. Kerogen facies in the major Jurassic mudrock formations of southern England and the implications on the depositional environments of their precursors. *Journal of Petroleum Geology*, **8**, 435–462.

EBUKANSON, E.J. & KINGHORN, R.R.F. 1986. Oil and gas accumulations and their possible source rocks in southern England. *Journal of Petroleum Geology*, **9**, 413–428.

ESPITALIÉ, J., MARQUIS, F. & SAGE, L. 1987. Organic geochemistry of the Paris Basin. *In*: BROOKS, J. & GLENNIE, K.W. (eds) *Petroleum Geology of North

West Europe. Graham and Trotman Ltd, London, 71–86.

EVANS, J., JENKINS, D. & GLUYAS, J.G. 1998. The Kimmeridge Oil Field: an enigma demystified. *In*: UNDERHILL, J.R. (ed.) *Development, Evolution and Petroleum Geology of the Wessex Basin*. Geological Society, London, Special Publications, **133**, 407–414.

FLEET, A.J., CLAYTON, C.J., JENKYNS, H.C. & PARKINSON, D.N. 1987. Liassic source rock deposition in western Europe. *In*: BROOKS, J. & GLENNIE, K.W. (eds) *Petroleum Geology of North West Europe*. Graham and Trotman Ltd, London, 59–70.

GORMLY, J.R., BUCK, S.P. & CHUNG, M. 1994. Oil–source rock correlation in the North Viking Graben. *In*: TELNÆS, N., VAN GRAAS, G. & ØYGARD, K. (eds) *Advances in Organic Geochemistry 1993: Proceedings of the 16th International Meeting on Organic Geochemistry, Stavanger, Norway, 20–24 September 1993. Organic Geochemistry*, **22**, 403–413.

HANCOCK, F.R.P. & MITHEN, D.P. 1987. The geology of the Humbly Grove Oilfield. *In*: BROOKS, J. & GLENNIE, K.W. (eds) *Petroleum Geology of North West Europe*. Graham and Trotman Ltd, London, 161–170.

HASZELDINE, R.S., RITCHIE, J.D. & HITCHEN, K. 1987. Seismic and well evidence for the early development of the Faeroe–Shetland Basin. *Scottish Journal of Geology*, **23**, 283–300.

HERRIES, R., PODDUBIUK, R. & WILCOCKSON, P. 1999. Solan, Strathmore and the back basin play, West of Shetland. *In*: FLEET, A.J. & BOLDY, S.A.R. (eds) *Petroleum Geology of NW Europe: Proceedings of the 5th Conference*. Geological Society, London, 693–712.

HESSELBO, S.P. & JENKYNS, H.C. 1995. A comparison of the Hettangian to Bajocian successions of Dorset and Yorkshire. *In*: TAYLOR, P.D. (ed.) *Field Geology of the British Jurassic*. Geological Society, London, 105–150.

HITCHEN, K. & STOKER, M.S. 1993. Mesozoic rocks from the Hebrides shelf and the implications for hydrocarbon prospectivity in the northern Rockall Trough. *Marine and Petroleum Geology*, **10**, 246–254.

HOLMES, A.J., GRIFFITH, C.E. & SCOTCHMAN, I.C. 1999. The Jurassic petroleum system of the west of Britain Atlantic margin – an integration of tectonics, geochemistry and basin modelling. *In*: FLEET, A.J. & BOLDY, S.A.R. (eds) *Petroleum Geology of NW Europe: Proceedings of the 5th Conference*. Geological Society, London, 1351–1365.

HOUSE, M.R. 1993. *Geology of the Dorset Coast*. Field Guide, 2nd edn, Geologists' Association, London, **22**.

HUANG, W.-Y. & MEINSCHEIN, W.G. 1979. Sterols as ecological indicators. *Geochimica et Cosmochimica Acta*, **43**, 739–745.

JOHNSON, H., RICHARDS, P.C., LONG, D. & GRAHAM, C.C. 1993. *United Kingdom offshore regional report: the geology of the northern North Sea*. HMSO for British Geological Survey, London.

JOWITT, R., HINDLE, A., JONES, D. & ROSE, P. 1999. Petroleum systems analysis of the Palaeocene play in the West of Shetlands area. *In*: FLEET, A.J. & BOLDY, S.A.R. (eds) *Petroleum Geology of NW Europe: Proceedings of the 5th Conference*. Geological Society, London, 1367–1381.

KINGHORN, R.R.F., SELLEY, R.C. & STONELEY, R. 1994. The Mupe Bay oil seep demythologized? *Marine and Petroleum Geology*, **11**, 124.

KOCH, J.-O. & HEUM, O.R. 1995. Exploration trends of the Halten Terrace. *In*: HANSLIEN, S. (ed.) *Petroleum exploration and exploitation in Norway*. NPF Special Publication, Elsevier, Amsterdam vol **4**, 235–251.

KONTOROVICH, A.E., MOSKVIN, V.I., BOSTRIKOV, O.I., DANILOVA, V.P., FOMIN, A.N., FOMICHEV, A.S., KOSTYREVA, E.A. & MELENEVSKY, V.N. 1997. Main oil source formations of the West Siberian Basin. *Petroleum Geoscience*, **3**, 343–358.

LAMERS, E. & CARMICHAEL, S.M.M. 1999. The Palaeocene deep water sandstone play West of Shetland. *In*: FLEET, A.J. & BOLDY, S.A.R. (eds) *Petroleum Geology of NW Europe: Proceedings of the 5th Conference*. Geological Society, London, 645–659.

LEACH, H.M., HERBERT, N., LOS, A. & SMITH, R.L. 1999. The Schiehallion development. *In*: FLEET, A.J. & BOLDY, S.A.R. (eds) *Petroleum Geology of NW Europe: Proceedings of the 5th Conference*. Geological Society, London, 683–692.

MACDONALD, H., ALLAN, P.M. & LOVELL, J.P.B. 1987. Geology of oil accumulation in Block 26/28, Porcupine Basin, offshore Ireland. *In*: BROOKS, J. & GLENNIE, K.W. (eds) *Petroleum Geology of North West Europe*. Graham and Trotman Ltd, London, 643–651.

MACQUAKER, J.H.S., FARRIMOND, P. & BRASSEL, S.C. 1986. Biological markers in the Rhaetian black shales of South West Britain. *In*: LEYTHAEUSER, D. & RULLKOTTER, J. (eds) *Advances in Organic Geochemistry*. Pergamon, Oxford, 93–100.

MILES, J.A., DOWNES, C.J. & COOK, S.E. 1993. The fossil oil seep in Mupe Bay. Dorset: a myth investigated. *Marine and Petroleum Geology*, **10**, 58–70.

MILES, J.A., DOWNES, C.J. & COOK, S.E. 1994. Reply to 'The Mupe Bay oil seep demythologized?' by Kinghorn, R.R.F., Selley, R.C. & Stoneley, R. *Marine and Petroleum Geology*, **11**, 125–126.

MORTON, N. 1992. Late Triassic to Middle Jurassic stratigraphy, palaeogeography and tectonics west of the British Isles. *In*: PARNELL, J. (ed.) *Basins on the Atlantic Seaboard: Petroleum Geology. Sedimentology and Basin Evolution*. Geological Society, London, Special Publications, **62**, 53–68.

MORTON, N. 1993. Potential reservoir and source rocks in relation to Upper Triassic to Middle Jurassic sequence stratigraphy, Atlantic margin basins of the British Isles. *In*: PARKER, J.R. (ed.) *Petroleum Geology of Northwest Europe: Proceedings of the 4th Conference*. Geological Society, London, 285–297.

MORTON, N. & HUDSON, J.D. 1995. Field Guide to the Jurassic of the Isles of Raasay and Skye, Inner Hebrides, NW Scotland. *In*: TAYLOR, P.D. (ed.) *Field Geology of the British Jurassic*. Geological Society, London, 209–280.

MURPHY, N.J., SAUER, M.J. & ARMSTRONG, J.P. 1995. Toarcian source rock potential in the North Celtic Sea Basin, offshore Ireland. *In*: CROKER, P.F. & SHANNON, P.M. (eds) *The Petroleum Geology of Ireland's Offshore Basins*. Geological Society, London, Special Publications, **93**, 193–207.

PARFITT, M.A. & FARRIMOND, P. 1998. The Mupe Bay oil seep: a detailed organic geochemical study of a controversial outcrop. *In*: UNDERHILL, J.R. (ed.) *Development, Evolution and Petroleum Geology of the Wessex Basin*. Geological Society, London, Special Publications, **133**, 387–397.

PARNELL, J., MONSON, B. & BUCKMAN, J. 1992. Excursion Guide: Basins and petroleum geology in the north of Ireland. *In*: PARNELL, J. (ed.) *Basins on the Atlantic Seaboard: Petroleum Geology, Sedimentology and Basin Evolution*. Geological Society, London, Special Publications, **62**, 449–464.

PEGRUM, R.M. & SPENCER, A.M. 1990. Hydrocarbon plays in the North Sea. *In*: BROOKS, J. (ed.) *Classic Petroleum Provinces*. Geological Society, London, Special Publications, **50**, 441–470.

PETERS, K.E. & MOLDOWAN, J.M. 1993. *The Biomarker Guide*. Prentice Hall, New Jersey.

POTTER, N. 1998. 3D, horizontal drilling changing Clair development economics. *Offshore*, **57**, 110–114.

PRICE, L.C. & BARKER, C.E. 1985. Suppression of vitrinite reflectance in amorphous rich kerogen – a major unrecognised problem. *Journal of Petroleum Geology*, **8**, 59–84.

SCOTCHMAN, I.C. & DORÉ, A.G. 1995. A Regional Appraisal of Source Rocks North and West of Britain (Abstract). *AAPG Bulletin*, **79**, 1247.

SCOTCHMAN, I.C. & THOMAS, J.R.W. 1995. Maturity and Hydrocarbon Generation in the Slyne Trough, Northwest Ireland. *In*: CROKER, P.F. & SHANNON, P.M. (eds) *The Petroleum Geology of Ireland's Offshore Basins*. Geological Society, London, Special Publications, **93**, 385–411.

SCOTCHMAN, I.C., GRIFFITH, C.E., HOLMES, A.J. & JONES, D.M. 1998. The Jurassic petroleum system north and west of Britain: a geochemical oil-source correlation study. *In*: HORSFIELD, B., RADKE, M., SCHAEFER, R.G. & WILKES, H. (eds) *Advances in Organic Geochemistry 1997, Organic Geochemistry*, **29**, 671–700.

SELLEY, R.C. & STONELEY, R. 1987. Petroleum habitat in south Dorset. *In*: BROOKS, J. & GLENNIE, K.W. (eds) *Petroleum Geology of North West Europe*. Graham and Trotman Ltd, London, 139–148.

SHANNON, P.M. & NAYLOR, D. 1998. An assessment of Irish offshore basins and petroleum plays. *Journal of Petroleum Geology*, **21**, 125–152.

SOFER, Z. 1984. Stable carbon isotope comparisons of crude oils: application to source depositional environment and petroleum alteration. *AAPG Bulletin*, **68**, 31–49.

SPENCER, A.M., BIRKELUND, Ø., KNAG, G.Ø. & FREDSTED, R. 1999. Petroleum systems of the Atlantic margin of northwest Europe. *In*: FLEET, A.J. & BOLDY, S.A.R. (eds) *Petroleum Geology of NW Europe: Proceedings of the 5th Conference*. Geological Society, London, 231–246.

STONELEY, R. & SELLEY, R.C. 1986. *A Field Guide to the Petroleum Geology of the Wessex Basin*. Imperial College, London.

THRASHER, J. 1992. Thermal effect of the Tertiary Cuillins Intrusive Complex in the Jurassic of the Hebrides: an organic geochemical study. *In*: PARNELL, J. (ed.) *Basins on the Atlantic Seaboard: Petroleum Geology, Sedimentology and Basin Evolution*. Geological Society, London, Special Publications, **62**, 35–49.

UNDERHILL, J.R. & STONELEY, R. 1998. Introduction to the development, evolution and petroleum geology of the Wessex Basin. *In*: UNDERHILL, J.R. (ed.) *Development, Evolution and Petroleum Geology of the Wessex Basin*. Geological Society, London, Special Publications, **133**, 1–18.

VINCENT, A.J. & TYSON, R.V. 1999. Organic facies of the Middle Jurassic of the Inner Hebrides, Scotland. *Petroleum Geoscience*, **5**, 83–92.

VON DER DICK, H., MELOCHE, J.D., DWYER, J. & GUNTER, P. 1989. Source rock geochemistry and hydrocarbon generation in the Jeanne d'Arc Basin, Grand Banks, offshore eastern Canada. *Journal of Petroleum Geology*, **12**, 51–68.

WIMBLEDON, W.A., ALLEN, P. & FLEET, A.J. 1996. Penecontemporaneous oil-seep in the Wealden (early Cretaceous) at Mupe Bay, Dorset, U.K. *Sedimentary Geology*, **102**, 213–220.

ZIEGLER, P.A. 1990. *Geological Atlas of Western and Central Europe*. 2nd edn, Shell Internationale Petroleum Maatschappij BV, The Hague.

Interpretation of vitrinite reflectance profiles in sedimentary basins, onshore and offshore Ireland

D. V. CORCORAN & G. CLAYTON

Department of Geology, Trinity College Dublin, Dublin 2, Ireland

(e-mail: Dermot.Corcoran@statoil.com)

Abstract: Vitrinite reflectance (VR) data (R_m%) have been compiled from 77 Irish offshore wells and 17 onshore boreholes. This database has facilitated the analysis of vitrinite reflectance v. depth relationships by both basin and stratigraphic interval. In general, VR gradients from the Carboniferous sections are defined by less scattered trends than those from Mesozoic and Cenozoic sections, reflecting the less complex vitrinite populations within Carboniferous coals and shales. A composite approach (display of profiles from a number of wells together) to the interpretation of vitrinite reflectance profiles has been utilized to characterize the thermal history and the prevalent heat transfer mechanisms within the various basins. Calculated peak palaeotemperatures from the wells are used to compute palaeogeothermal gradients and to estimate the magnitude of net exhumation at selected locations. Average palaeogeothermal gradients in the onshore Carboniferous basins range from less than 3 °C km^{-1} at well IIP-2 in the Clare Basin to 119 °C km^{-1} at well N998 in the Navan area of the Dublin Basin. Lateral variations in palaeogeothermal gradients recorded in the Carboniferous sections are consistent with a gravity-driven hydrothermal system discharging heated fluids, along fault systems, in a foreland platform area. In general, palaeogeothermal gradients are substantially higher in the Carboniferous sections (mean 60 °C km^{-1}) than in the Mesozoic or Cenozoic sections (mean 32 °C km^{-1}). Maturation levels in many of the Carboniferous sections are considered to be the consequence of burial, elevated heat flows and a regional advective system during late Carboniferous to early Permian times rather than Mesozoic or Cenozoic processes. Empirically derived methods of calculating peak palaeotemperature from VR are compared with kinetic models and, although differing in detail, within the resolution of this dataset are shown to produce similar trends.

There is a considerable body of evidence to suggest that the extensional evolution of Ireland's Late Palaeozoic to Cenozoic sedimentary basins has been punctuated by a multiphase inversion history. Regional stratigraphic evidence, combined with VR and apatite fission-track data, suggests at least two periods of pervasive exhumation occurred; one during Late Carboniferous–Late Permian time and another during Tertiary time. Both of these phases are characterized by a component of compressional inversion and the widespread occurrence of extrusive and intrusive igneous rocks. However, the contrasting thermal signature of these regional uplift events suggests that both the basin setting and the mechanism of regional exhumation exerted a fundamental control on processes that determined heat flow distribution within a basin. In terms of the hydrocarbon exploration of Ireland's sedimentary basins the model presented here has important implications for the timing of maturation of Carboniferous source rocks in these basins. Where Carboniferous source rocks are present, they will make a significant contribution to the hydrocarbon budget only in those basins that experienced relatively low heat flow during Late Carboniferous–Early Permian time and where sufficient Mesozoic burial has occurred to subsequently expose the kerogen to higher temperatures. This observation is consistent with the presence of gas accumulations, which are postulated to have been derived from Carboniferous source rocks, in both the Slyne Basin and the Northwest Carboniferous Basin.

Vitrinite reflectance (VR) is one method of evaluating the thermal alteration of sedimentary rocks (Héroux *et al.* 1979). It has been successfully demonstrated to be a reliable indicator of organic maturation in sedimentary successions and is widely used in the oil industry to define potential areas of oil and gas generation in a basin (Peters & Cassa 1994). In addition, lateral and vertical trends in vitrinite reflectance result from a range of basin-scale geological processes and can offer important constraints with respect to thermal

From: SHANNON, P.M., HAUGHTON, P.D.W. & CORCORAN, D.V. (eds). 2001. *The Petroleum Exploration of Ireland's Offshore Basins*. Geological Society, London, Special Publications, **188**, 61–90. 0305-8719/01/$15.00

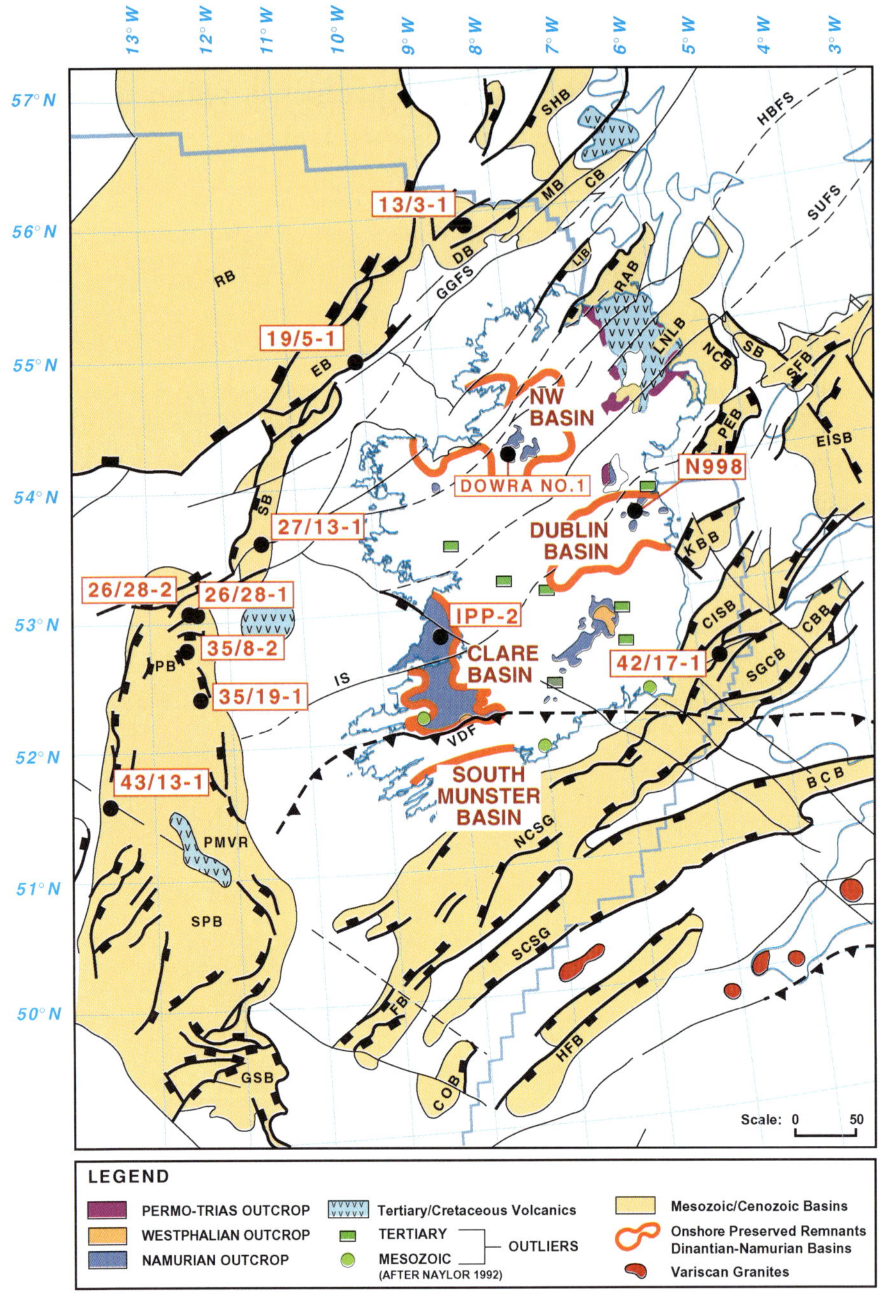

Fig. 1. Location map showing post-Dinantian outcrop map of Ireland, location of onshore and offshore basins, some major tectonic lineaments and exploration wells referred to in text. BCB, Bristol Channel Basin; CB, Colonsay Basin; CBB, Cardigan Bay Basin; CISB, Central Irish Sea Basin; COB, Cockburn Basin; DB, Donegal Basin; EISB, East Irish Sea Basin; EB, Erris Basin; FB, Fastnet Basin; GGFS, Great Glen Fault System; GSB, Goban Spur Basin; HBFS, Highland Boundary Fault System; HFB, Haig Fras Basin; IS, Iapetus Suture; KBB, Kish Bank Basin; LIB, Lough Indaal Basin; LNLB, Lough Neagh–Larne Basin; MB, Malin Basin; NCB, North Channel Basin; NCSG, North Celtic Sea Graben; PEB, Peel Basin; PB, Porcupine Basin; PMVR, Porcupine Median Volcanic Ridge; RAB, Rathlin Basin; RB, Rockall Basin; SB, Stranraer Basin; SHB, Sea of Hebrides Basin; SCSG, South Celtic Sea Graben; SFB, Solway Firth Basin; SGCB, St George's Channel Basin; SPB, South Porcupine Basin; SB, Slyne Basin; SUFS, Southern Uplands Fault System; VDF, Variscan Deformation Front.

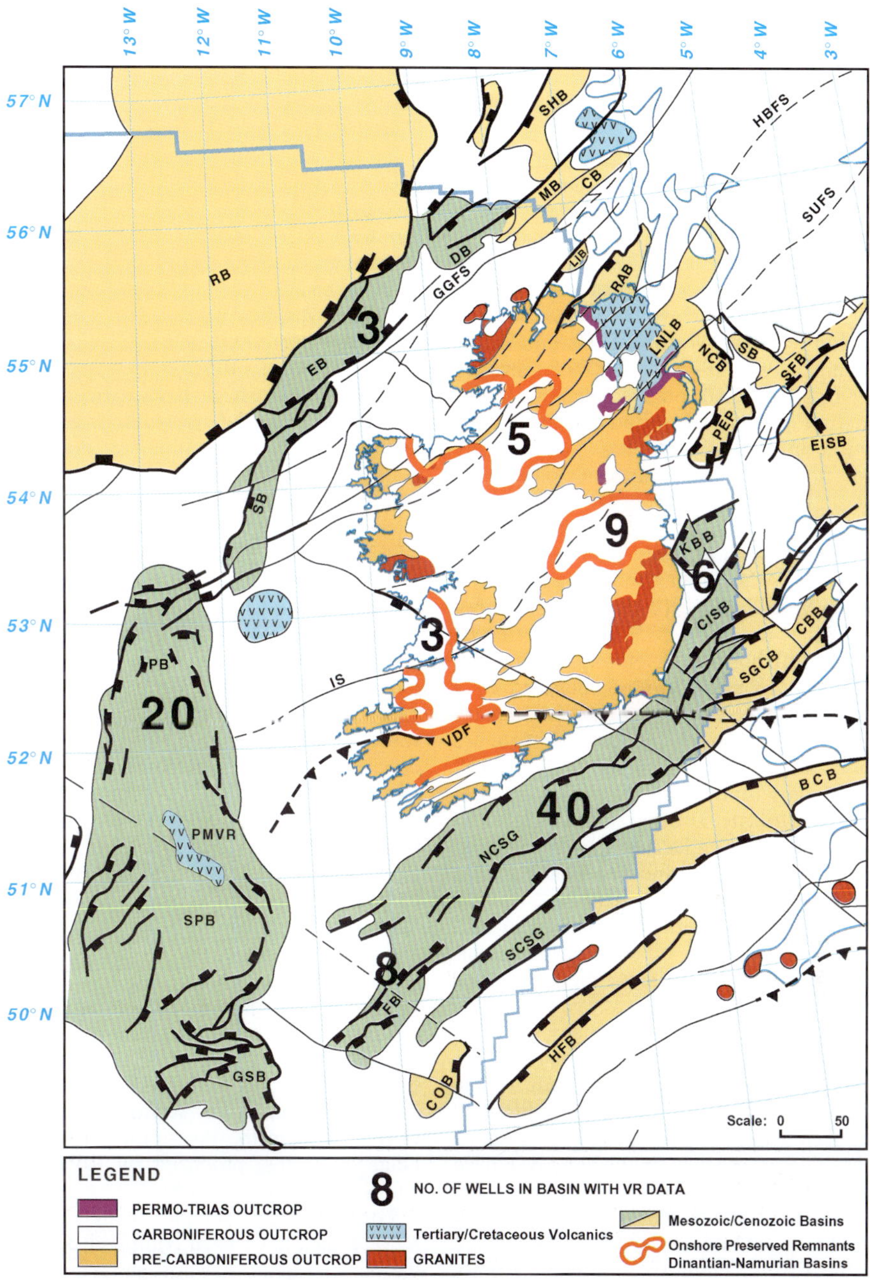

Fig. 2. Simplified geological map of Ireland showing distribution of wells, incorporated in the VR database, by basin. Tectonic elements and basins labelled as in Figure 1.

evolution and exhumation history. VR data are of particular importance in exhumed basin settings where a quantitative assessment of the peak palaeotemperatures achieved by potential source rocks, before uplift, is a critical factor in determining the hydrocarbon prospectivity. Furthermore, by establishing a peak palaeotemperature profile at a given well location, a palaeogeothermal gradient (at peak palaeotemperature exposure) can be determined. This may provide an insight into the mechanism of heating and cooling of the sedimentary rocks at that location (Bray *et al.* 1992; Duddy *et al.* 1994). In some cases, stratigraphic relationships can be used to constrain the timing of peak palaeotemperature exposure (Evans & Clayton 1998; Corcoran & Clayton 1999).

Post-Palaeozoic rocks are rarely preserved onshore in Ireland, except in the northeastern part of the island (Naylor 1992). The remaining onshore sedimentary succession is dominated by thick sequences of Carboniferous rocks including the preserved remnants of Dinantian–Namurian basins, such as the Dublin, Clare, South Munster and Northwest Carboniferous basins (Philcox *et al.* 1992; Croker 1995; Strogen *et al.* 1996; Fig. 1), and some isolated outliers of Westphalian rocks. Limited hydrocarbon exploration in these basins has yielded only minor gas flows from Dinantian sandstones in the Northwest Basin. Carboniferous sedimentary successions have also been drilled in a number of offshore exploration wells in the Irish Sea, Celtic Sea and Atlantic margin basins (Robeson *et al.* 1988; Maddox *et al.* 1995; Newman 1999). However, the morphology of Carboniferous basins in the offshore area is poorly understood.

In contrast, Ireland is almost encircled by a series of Mesozoic to Cenozoic sedimentary basins (Fig. 2). Although the general structure and stratigraphy of these offshore basins is well understood, only limited exploration success has been achieved to date. Drilling of more than 130 exploration wells has yielded only two producing gas fields in the North Celtic Sea Basin, with a third commercial gas accumulation currently under appraisal in the Slyne Basin. In addition, a number of undeveloped oil and gas accumulations have been discovered in the Celtic Sea and Porcupine basins.

One observation common to many of these basins is that they appear to have experienced a multiphase exhumation history. In spite of the well-documented impact that the uplift and removal of overburden has on petroleum systems within basins (Nyland *et al.* 1992; Doré & Jensen 1996), few published studies have been dedicated to this aspect of the evolution of these Irish offshore basins. Murdoch *et al.* (1995) were among the first to employ a multidisciplinary approach to assess the magnitude and timing of Tertiary uplift events and their impact on the maturity of the Jurassic source rocks in the North Celtic Sea Basin. Scotchman & Thomas (1995) identified a number of consequences, for the Jurassic petroleum system in the Slyne Basin, of 1–2 km of late Mesozoic to early Cenozoic exhumation in this area. Duncan *et al.* (1998) addressed the impact of a multiphase inversion history on the Carboniferous to Triassic petroleum system in the Central Irish Sea area and highlighted evidence for an Early Cretaceous exhumation event in this area, in addition to describing uplift phases at the end of the Carboniferous period and during Tertiary time.

The central theme of this paper is to present an overview of a compiled VR database for these offshore and onshore basins and to identify the implications of these data with regard to the thermal evolution and hydrocarbon prospectivity of these basins. It is suggested that a composite approach (display of profiles from a number of wells together) to the analysis and interpretation of the VR data offers some new insights into the thermal history of Irish sedimentary basins, which need to be reconciled with regional tectonic and basin evolution models.

Vitrinite reflectance database

VR data (mean random reflectance; $R_m\%$) have been compiled from 77 Irish offshore wells and 17 onshore boreholes (Fig. 2). This database has facilitated the analysis of VR v. depth behaviour by basin and by stratigraphic interval. The VR data have been compiled from a number of sources. For the onshore basins this paper draws upon the extensive VR datasets generated over the past decade by a number of research students at Trinity College Dublin: Baily (1992), Fitzgerald (1994), Goodhue (1996) and associated publications including those by Goodhue & Clayton (1999) and Clayton & Baily (2000). For the offshore basins, available organic geochemical and maturation reports for 77 wells were reviewed. These reports have been generated by various oil companies and contractors based on samples recovered from exploration wells drilled in these basins before 1996. In addition, this study incorporates the large VR dataset, pertaining to Carboniferous sections encountered in wells drilled in the Porcupine, Slyne, Erris and Donegal basins, produced by Robeson (1987) and Robeson *et al.* (1988).

For each well, sample depth and measured $R_m\%$ values have been compiled, together with the stratigraphic age of the samples where available. Sample type and number of vitrinite particles have also been recorded to assist with quality control of the database. All depths have been converted to metres true vertical depth below sea bed (TVD Sub SB) or true vertical depth below ground level (TVD Sub GL) in the case of the onshore boreholes. In addition, peak palaeotemperature data have been computed from the measured $R_m\%$ for each sample, by utilizing the three different empirically based

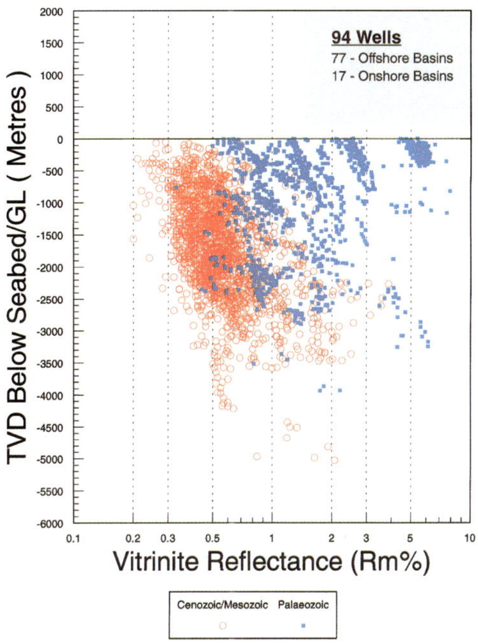

Fig. 3. Semi-log plot of $R_m\%$ (mean random vitrinite reflectance) v. depth for all 94 wells in the database, with assigned stratigraphic ages of the sample data. Noteworthy features are the wide scatter in the dataset (values range from 0.2 to 7.56 $R_m\%$) and a weak correlation of $R_m\%$ with depth, which suggests substantial differences in the thermal and structural evolution of the Irish sedimentary basins (for comparison see Fig. 7). Two generic VR families can be identified: VR recorded in Mesozoic–Cenozoic sections and VR recorded in Palaeozoic (mainly Carboniferous) sections. For a given present-day burial depth, organic maturity levels of the Palaeozoic sections are generally higher than for Mesozoic–Cenozoic sections.

schemes of Barker & Pawliewicz (1986), Barker (1988) and Barker & Goldstein (1990).

VR v. depth relationships are commonly presented in the literature as either linear or semi-log plots. Although linear VR v. depth profiles were favoured by Stach *et al.* (1982), Suggate (1998) and others, the 'semi-log' approach of Dow & O'Connor (1982) has been adopted here because of the large range of $R_m\%$ values (0.2–7.56 $R_m\%$) contained within this VR database (Fig. 3). Initial inspection of these data suggests a weak correlation of $R_m\%$ with depth. However, by integrating the stratigraphic ages of the sample data some generic subsets of the database can be identified and characterized.

The stratigraphic information permits the differentiation of at least two VR populations in the data; VR recorded in Mesozoic–Cenozoic sections v. VR recorded in Palaeozoic (mainly Carboniferous) sections (Fig. 3). As might be expected, for a given present-day burial depth, organic maturity levels of the Palaeozoic sections are generally higher than for the Mesozoic–Cenozoic sections. This is consistent with the observation that many of the Palaeozoic $R_m\%$ values pertain to the exhumed remnants of the onshore Carboniferous basins. In addition, these Carboniferous sections are characterized by more discrete, better-defined, VR profiles relative to Mesozoic–Cenozoic sections. This observation reflects the less complex vitrinite populations within the vitrinite-rich, paralic, sequences of coals and shales of Carboniferous age (Durand *et al.* 1986). In contrast, many of the Mesozoic–Cenozoic VR profiles exhibit a wide scatter. Some of this scatter may be due to the presence of reworked vitrinite populations within the Mesozoic sedimentary column caused by the denudation of a Westphalian–Stephanian blanket that once covered the Irish landmass. A further contribution to this scatter may be provided by the presence of vitrinite-like 'bituminite' populations, within Mesozoic–Cenozoic marine sediments. These 'bituminite' populations have been identified as a major problem with respect to the determination of organic maturation trends within the Liassic sediments of the Paris Basin (Alpern & Cheymol 1978).

The Carboniferous subset of the data gives an indication of the range of individual maturity gradients within Carboniferous sections (Fig. 4). At present, the offshore sections are buried below a Mesozoic–Cenozoic cover of variable thickness. Maturity levels for this offshore subset of the data range widely from 0.44 $R_m\%$ to 6.1 $R_m\%$. Discounting profile deviations associated with igneous intrusions, average maturity gradients, inferred from Robeson *et al.* (1988), range from 0.19 $R_m\%$ km^{-1} for Erris Basin well 19/5-1 to 1.93 $R_m\%$ km^{-1} for wells in the north of the Porcupine Basin. In general, maturity gradients for these offshore sections are high, with the exception of well 19/5-1, which manifests a maturity gradient more typical of Mesozoic sections. This anomalous gradient has important implications for the timing of Carboniferous source rock maturation in the Erris Basin. It may represent overprinting by Mesozoic and Cenozoic burial following relatively low heat flow in this area during Late Carboniferous–Early Permian time.

In contrast, the Mesozoic–Cenozoic subset of the VR database (Fig. 5) is characterized by lower levels of maturity (0.2–3.9 $R_m\%$) and dominantly low-magnitude, poorly defined

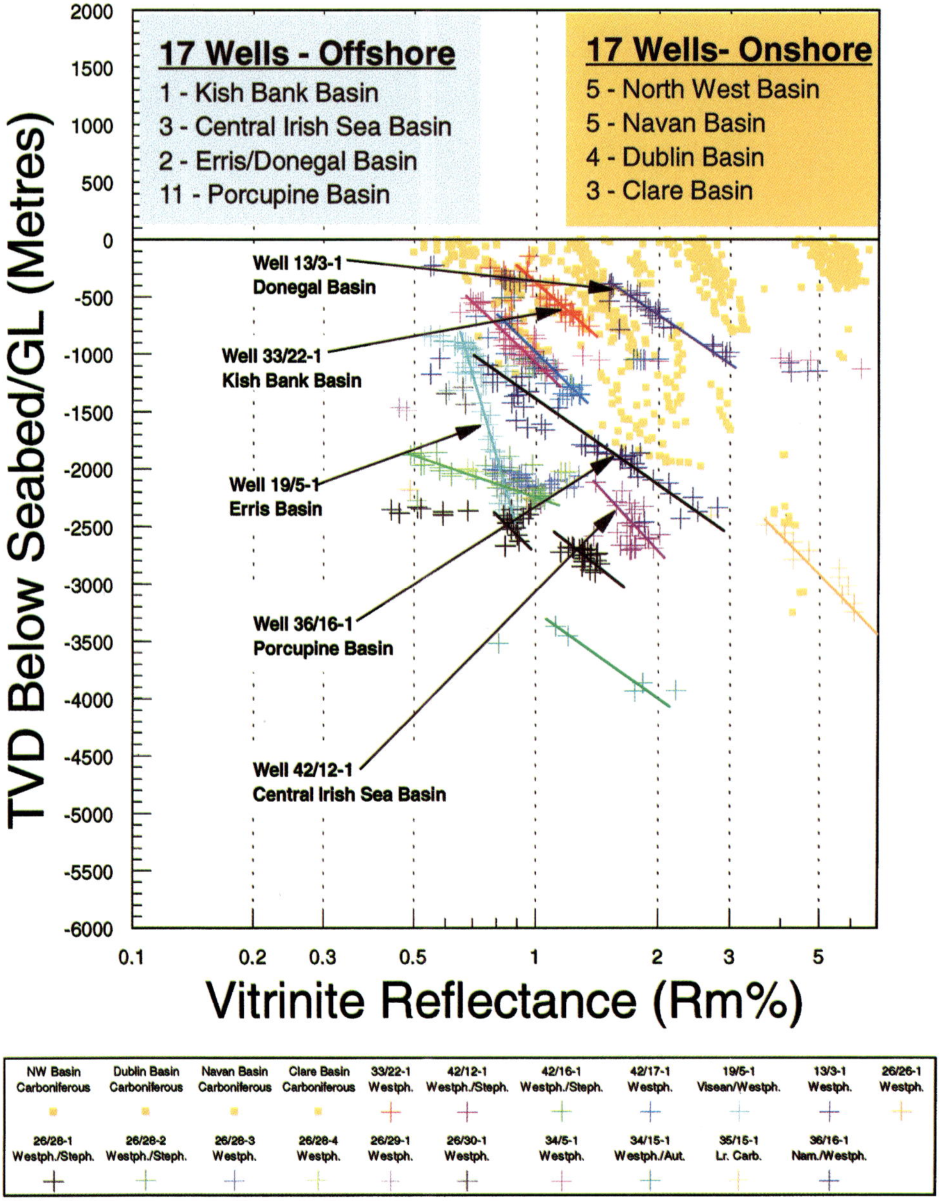

Fig. 4. R_m% v. depth for all Carboniferous sections encountered in wells offshore and onshore. Data from onshore wells are plotted in background as yellow squares. Data from offshore wells from the Irish Sea and the Atlantic margin basins are plotted as colour crosses with best-fit maturity gradient lines displayed. In general, maturity gradients for these offshore sections are high (range 0.19–1.93 R_m% km^{-1}), with the exception of well 19/5-1, which manifests a maturity gradient more typical of Mesozoic sections.

maturity gradients (0.13–0.78 R_m% km^{-1}). By identifying the stratigraphic age and basin setting of the VR samples, this subset of the data can be analysed in terms of the well-documented late Mesozoic to Cenozoic exhumation history of NW Europe (Roberts 1989; Brodie & White 1995; Menpes & Hillis 1995). Although the VR data manifest considerable scatter, it is possible to differentiate between those basins that are characterized by continual subsidence, such as the Porcupine Basin, and those basins that have experienced exhumation during Cenozoic time,

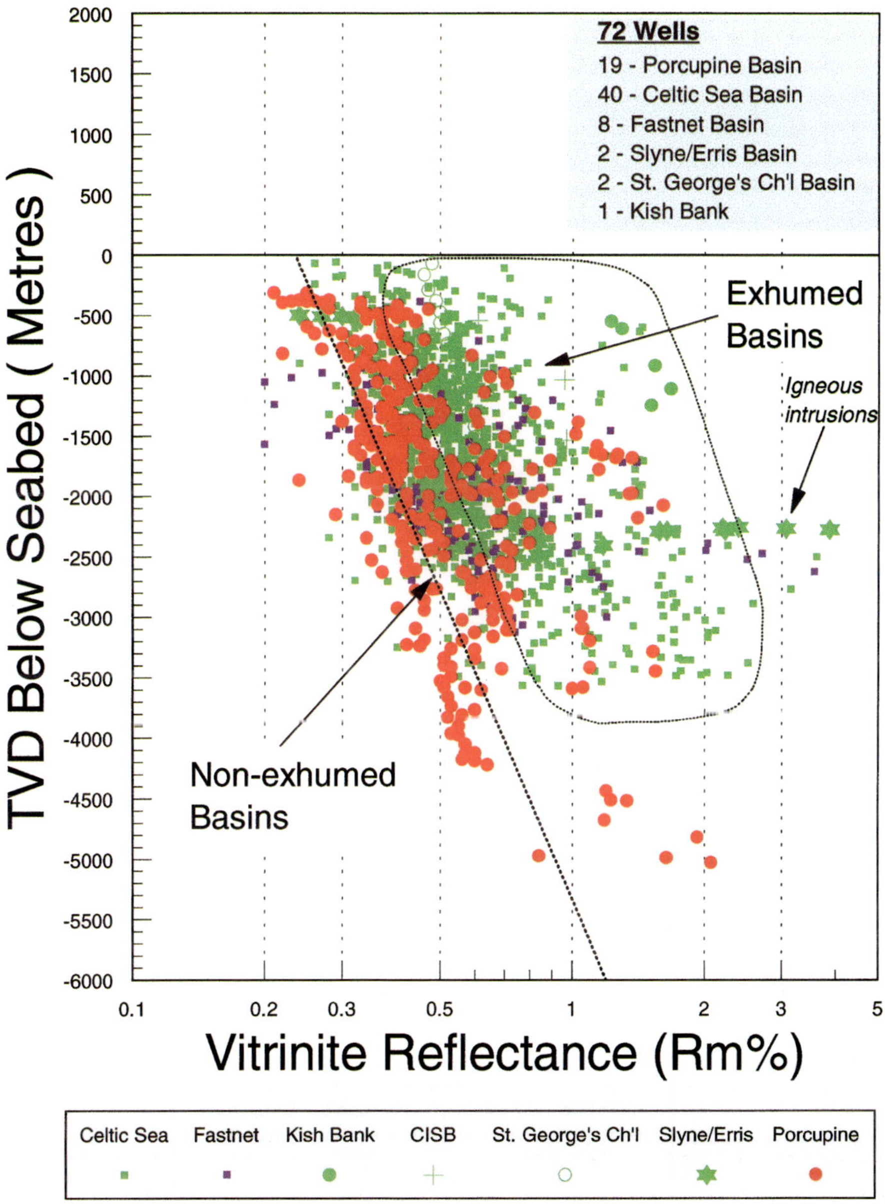

Fig. 5. $R_m\%$ v. depth for all Mesozoic and Cenozoic sections encountered in offshore wells. Although data are noisy, because of the presence of igneous intrusions, reworked vitrinite populations and inherent scatter, they do differentiate between those basins characterized by continual subsidence (e.g. Porcupine Basin) and those basins that have experienced exhumation (e.g. Celtic Sea, St George's Channel, Slyne and Erris basins).

e.g. Slyne, St George's Channel and Celtic Sea basins (Fig. 5). Confirmation that a major phase of exhumation is of post-Cretaceous age is provided by the $R_m\%$ data that have been recorded in Cretaceous rocks in some of these offshore basins (Fig. 6). Displacement of the Cretaceous vitrinite maturity levels, away from the 'normal' depth trend towards higher maturity levels, suggests they have experienced post-Cretaceous exhumation. A familiar pattern is observed with higher $R_m\%$ values present at equivalent present-day depth in basins, such as the Celtic Sea Basin, that have undergone exhumation during Cenozoic time (Fig. 6).

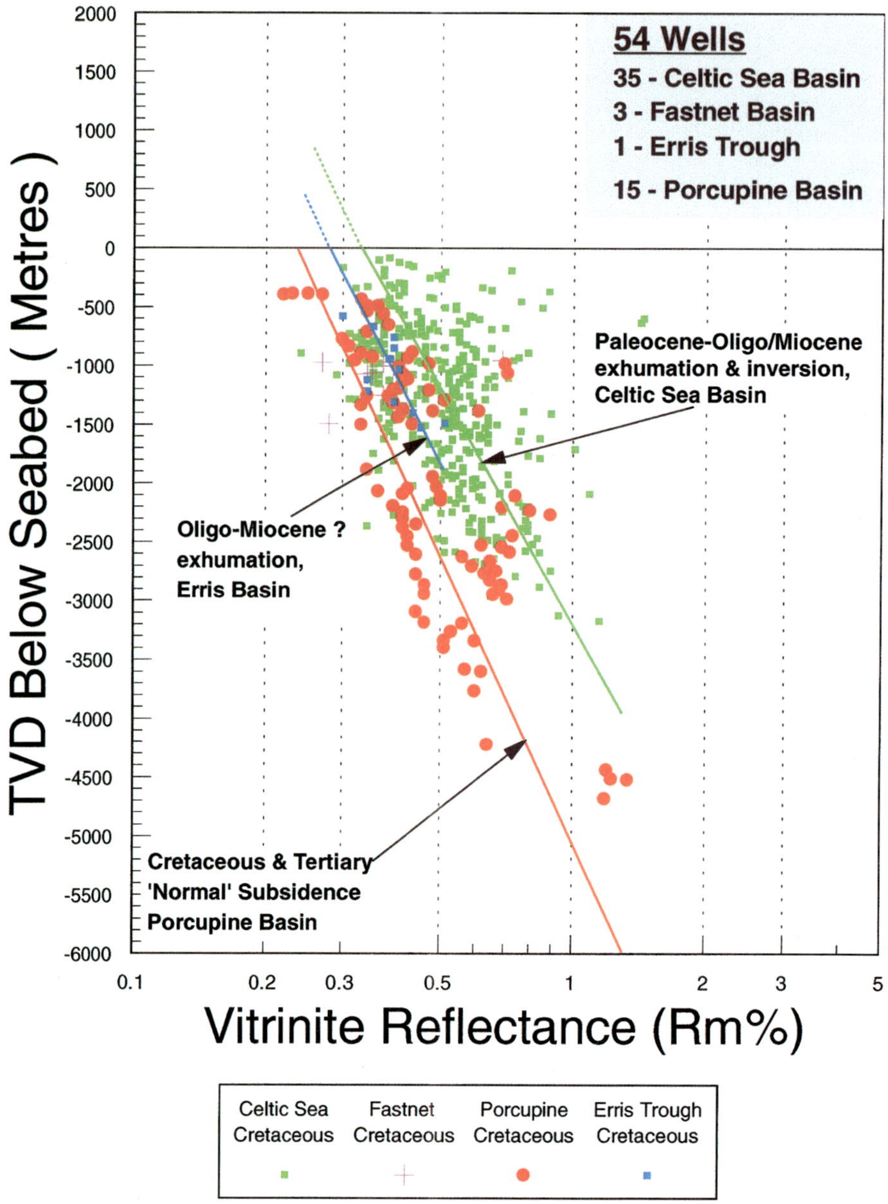

Fig. 6. R_m% v. depth for all Cretaceous sections encountered in offshore wells (subset of data plotted in Fig. 5). Data are scattered but the displacement of the Cretaceous R_m% levels in some basins, such as the Celtic Sea Basin, away from the 'normal' subsidence trend towards higher maturity levels, suggests that the timing of basin exhumation is post-Cretaceous.

Pitfalls and sources of uncertainty with VR data

Some of the scatter in the VR data may be related to the inherent difficulties in the comparison of VR determinations made by different workers in various laboratories. Logistical factors that affect the quality of VR data include the problem of caving (cuttings samples) in well bores, consistent identification of vitrinite, differentiation of primary vitrinite and insufficient grains to make a reliable determination of the reflectance of the sample. Dembicki (1984) highlighted the inconsistencies in the analytical results reported for common samples.

Geological conditions during burial and maturation can influence the reflectance level attained by the vitrinite. Compaction state (porosity), pore pressure and hydrodynamic flow through the sediments containing the vitrinite can alter the thermal conductivity and permeability of the sediments and thus affect the dominant heat transfer mechanism (Law *et al.* 1989; Deming 1994a; McTavish 1998). In addition, variations in the chemical composition of vitrinite may lead to invalid comparison of VR gradients. Early workers suggested that vitrinite reflectance is a maturity measurement independent of kerogen composition and organic facies (Stach *et al.* 1982). However, there is a gathering consensus that variation in the initial hydrogen content of the vitrinite macerals is significant, with hydrogen-poor, oxygen-rich vitrinite maturing at an enhanced rate compared with hydrogen-enriched vitrinite (Price & Barker 1985; Hao & Chen 1992; Goodarzi *et al.* 1994). Suppressed and retarded vitrinite reflectance (deviations of VR profile towards lower values of R_m%) can result from the presence of hydrogen-rich vitrinites and overpressuring, respectively, in a sedimentary column. Nevertheless, in spite of these pitfalls, the generally consistent picture observed in these data suggests that the underlying thermal maturity 'signal' rises above the ambient 'noise' of the dataset, at least when the data are considered at a regional scale.

Translation of VR to palaeotemperature

The thermal alteration of vitrinite proceeds via a series of irreversible chemical reactions. Consequently, VR is one of a number of organic thermal maturation indicators that provide a means of ascertaining the maximum temperature exposure of sedimentary rocks (Hood *et al.* 1975; Héroux *et al.* 1979). Nevertheless, conversion of VR values to absolute palaeotemperatures is still problematic (Price 1983). Much debate has centred on the relative influence of time and temperature on the increase of vitrinite reflectance. However, both recent kinetic models (Burnham & Sweeney 1989; Sweeney & Burnham 1990) and field studies in areas where the duration of heating is well constrained (Barker 1988, 1991, 1993), indicate that VR is most sensitive to maximum temperature, under conditions of uniform heating rates. Barker (1991, 1993) demonstrated that in geologically young systems VR adjusts to changes in temperature over a geologically brief time period (10^3-10^5 years). However, Suggate (1982) suggested that 10^6-10^7 years were required to allow stabilization of vitrinite reflectance at peak temperature. Morrow & Issler (1993) compared VR values, computed via a number of common forward modelling techniques, for known temperature histories under constant heating rates. They concluded that, within the range of organic maturity asssociated with oil and gas generation, the EASY%Ro algorithm of Sweeney & Burnham (1990) is the most appropriate method for the prediction of VR. However, they suggested that the algorithm slightly overestimates VR in strata of low to medium maturity (<0.9 R_m%).

A number of published empirically based and kinetic schemes are available to translate VR data into peak palaeotemperature, including those by Barker & Pawliewicz (1986), Barker (1988), Burnham & Sweeney (1989), Barker & Goldstein (1990) and Sweeney & Burnham (1990). The current study utilized the well-delineated global dataset (from 10 non-inverted basins) of Rowley & White (1998), to derive some model data points allowing a comparison of the different VR to palaeotemperature translation schemes (Fig. 7). Assumption of a best-fit line, representing on average how VR behaves with depth, in a basin characterized by continual subsidence, permits the examination of the peak palaeotemperature v. depth relationship derived from each translation scheme (Fig. 8).

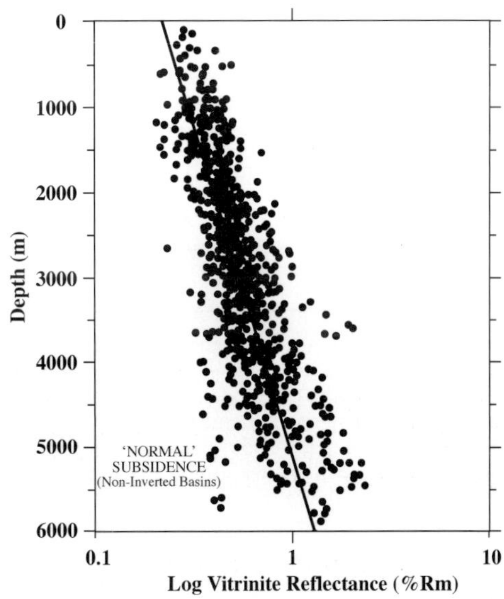

Fig. 7. Semi-log plot of R_m% v. depth for a global database of 10 non-inverted basins (adapted from Rowley & White 1998). Best-fit line is used to generate model data points to compare a number of VR to peak palaeotemperature translation schemes (see Fig. 8).

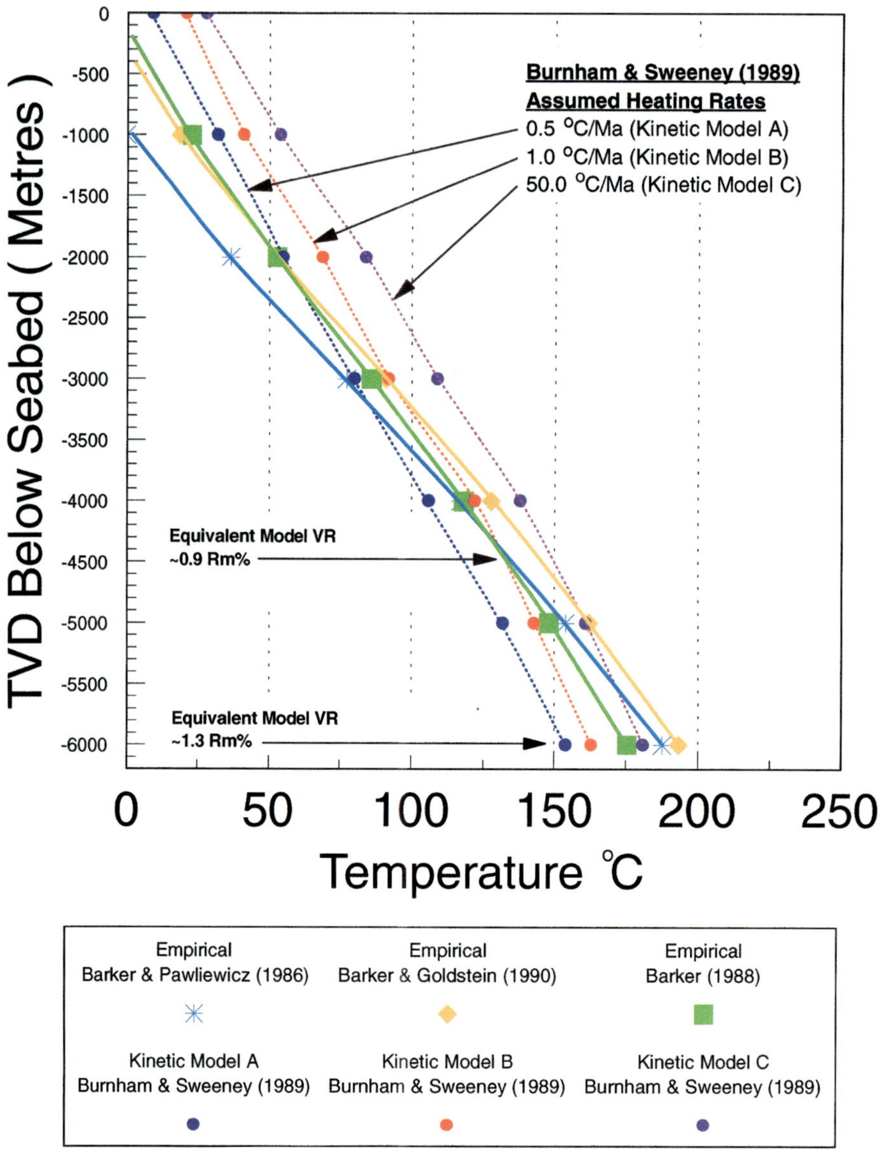

Fig. 8. Comparison of some empirical v. kinetic VR to palaeotemperature translation schemes. Plot of computed palaeotemperature v. depth for some model data points derived from Fig. 7. For example, a depth of *c.* 4500 m below sea bed equates to a model VR data point of 0.9 R_m%. Palaeotemperatures and palaeogeothermal gradients derived via the empirical scheme of Barker (1988) most closely approximate the kinetic scheme of Burnham & Sweeney (1989). However, at VR values of <0.9 R_m% the Barker (1988) scheme predicts lower palaeotemperatures than the Burnham & Sweeney (1989) kinetic model. At VR values between 0.9 and 1.3 R_m%, higher palaeotemperatures are predicted via the Barker (1988) scheme (*c.* 13 °C higher at 6000 m below sea bed, assuming a heating rate of 1 °C Ma^{-1}).

For example, at VR values <0.9 R_m% the kinetic model (assumed heating rate 1 °C Ma^{-1}) of Burnham & Sweeney (1989) will yield higher palaeotemperatures but lower palaeogeothermal gradients than the Barker (1988) empirical scheme. Between VR values of 0.9 and 1.3 R_m% the Burnham & Sweeney model will result in lower palaeotemperatures and lower palaeogeothermal gradients than the Barker (1988) scheme (Fig. 8). In general, over the

Table 1. *Well 35/15-1, Porcupine Basin: sample depth, type, age, mean random vitrinite reflectance (R_m%) and number of vitrinite reflectance determinations (n) for each sample*

KB DEPTH *Midpoint (FT)*	TVD Sub SB *Midpoint (M)*	SAMPLE *Type*	*Age*	Determ. *(n)*	VR *(Rm%)*	Bark. & Paw. (1986) *Peak Temp. °C*	Bark. & Gold. (1990) *Peak Temp. °C*	Barker (1988) *Peak Temp. °C*	Burnham & Sweeney (1989) *Peak Temp. °C (assumed heating rate 1°C/Ma)*
5525	**−1346.54**	**Ctgs.**	**Tert.**	**21**	**0.37**	**26**	**44**	**45**	**60**
5775	**−1422.73**	**Ctgs.**	**Tert.**	**20**	**0.39**	**33**	**50**	**50**	**65**
6075	**−1514.17**	**Ctgs.**	**Tert.**	**20**	**0.41**	**40**	**56**	**55**	**70**
6275	**−1575.13**	**Ctgs.**	**Tert.**	**22**	**0.46**	**54**	**70**	**67**	**79**
6525	**−1651.32**	**Ctgs.**	**Tert.**	**14**	**0.48**	**60**	**75**	**72**	**82**
6525	−1651.32	Ctgs.	Tert.	3	0.63	95	107	100	105
6525	−1651.32	Ctgs.	Tert.	5	1.54	209	213	193	174
7025	**−1803.72**	**Ctgs.**	**Tert.**	**20**	**0.48**	**60**	**75**	**72**	**82**
7025	−1803.72	Ctgs.	Tert.	2	0.92	143	152	139	137
7495	**−1946.96**	**Ctgs.**	**E. Cret.**	**11**	**0.48**	**60**	**75**	**72**	**82**
7495	−1946.96	Ctgs.	E. Cret.	3	0.61	90	103	97	101
7775	**−2032.30**	**Ctgs.**	**E. Cret.**	**12**	**0.49**	**62**	**77**	**74**	**84**
7775	−2032.30	Ctgs.	E. Cret.	2	1.01	155	163	149	142
8025	**−2108.50**	**Ctgs.**	**E. Cret.**	**4**	**0.5**	**65**	**80**	**76**	**87**
8025	−2108.50	Ctgs.	E. Cret.	3	0.96	149	157	144	141
8275	−2184.70	Ctgs.	E. Carb.	12	0.49	62	77	74	84
8275	−2184.70	Ctgs.	E. Carb.	5	0.99	153	161	147	143
8525	−2260.89	Ctgs.	E. Carb.	13	0.95	147	156	143	140
8525	**−2260.89**	**Ctgs.**	**E. Carb.**	**14**	**2.5**	**271**	**271**	**243**	**208**
8775	**−2337.09**	**Ctgs.**	**E. Carb.**	**5**	**2.8**	**286**	**285**	**255**	**217**
9025	**−2413.29**	**Ctgs.**	**E. Carb.**	**4**	**2.7**	**281**	**280**	**251**	**214**
9275	**−2489.48**	**Ctgs.**	**E. Carb.**	**20**	**3.7**	**322**	**318**	**284**	**253**
9525	**−2565.68**	**Ctgs.**	**E. Carb.**	**11**	**4.1**	**335**	**330**	**295**	**267**
9775	**−2641.87**	**Ctgs.**	**E. Carb.**	**14**	**4.3**	**341**	**336**	**300**	**276**
9775	−2641.87	Ctgs.	E. Carb.	2	5.7	377	369	329	
10025	**−2718.07**	**Ctgs.**	**E. Carb.**	**7**	**4.7**	**352**	**346**	**309**	**300**
10025	−2718.07	Ctgs.	E. Carb.	9	6.1	386	377	336	
10280	**−2795.79**	**Ctgs.**	**E. Carb.**	**15**	**4.3**	**341**	**336**	**300**	**276**
10280	−2795.79	Ctgs.	E. Carb.	4	6	384	375	334	
10525	**−2870.46**	**Ctgs.**	**E. Carb.**	**20**	**5.6**	**375**	**367**	**327**	
10775	**−2946.66**	**Ctgs.**	**E. Carb.**	**22**	**5.6**	**375**	**367**	**327**	
11030	**−3024.38**	**Ctgs.**	**E. Carb.**	**23**	**5.9**	**381**	**373**	**333**	
11280	**−3100.58**	**Ctgs.**	**E. Carb.**	**20**	**5.7**	**377**	**369**	**329**	
11530	**−3176.77**	**Ctgs.**	**E. Carb.**	**18**	**6.1**	**386**	**377**	**336**	
11775	**−3251.44**	**Ctgs.**	**E. Carb.**	**18**	**6.1**	**386**	**377**	**336**	

A comparison is presented of VR to peak palaeotemperature translation via the empirically based translation schemes of Barker & Pawliewicz (1986), Barker & Goldstein (1990) and Barker (1988) with the kinetic translation scheme of Burnham & Sweeney (1989), simplified by Sweeney & Burnham (1990) and presented by Suzuki *et al.* (1993) for a heating rate of 1 °C Ma^{-1}. It should be noted that the kinetic model of Burnham & Sweeney (1989) is not calibrated above VR values >4.7 R_m% and consequently, kinetically derived peak palaeotemperatures are not available for some of the deeper samples in well 35/15-1. A graphical comparison of the Barker (1988) v. Burnham & Sweeney (1989) translation schemes is presented in Fig. 9.

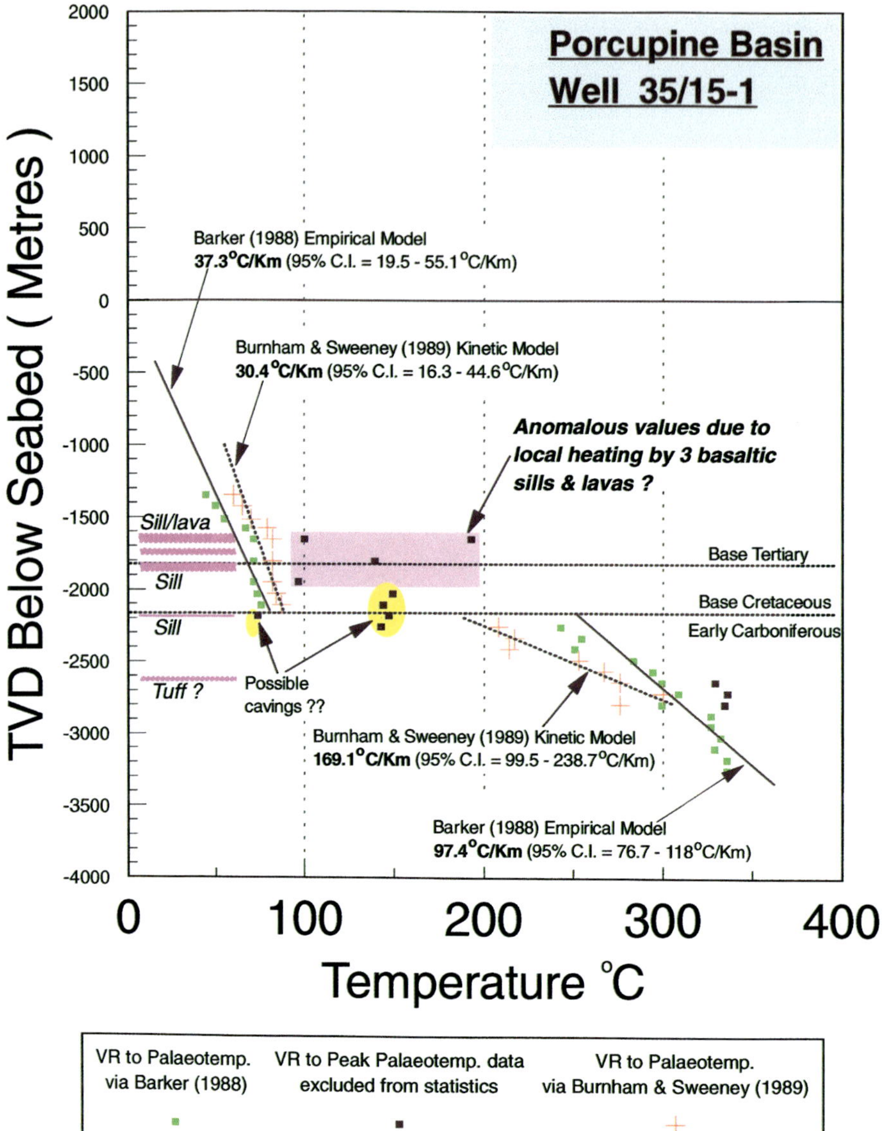

Fig. 9. VR to palaeotemperature translation schemes and impact on estimated palaeogeothermal gradients: a comparison of the Barker (1988) empirical scheme v. the kinetic model of Burnham & Sweeney (1989) for some real data points. Plot of computed palaeotemperature v. depth for sample data from well 35/15-1, Porcupine Basin. Only data points in bold in Table 1 were incorporated in the statistical analysis and used to estimate the average palaeogeothermal gradients for the Mesozoic–Cenozoic and Palaeozoic segments. The filtered data points (■) are interpreted as anomalous owing to the presence of igneous intrusions or extrusions and/or caved material. A comparison of the estimated palaeogeothermal gradients for the Mesozoic–Cenozoic section indicates that the Barker (1988) scheme yields a *higher* palaeogeothermal gradient (37.3 °C km^{-1}) than the Burnham & Sweeney model (30.4 °C km^{-1}). The reverse is true for the Palaeozoic section, with the Barker (1988) scheme yielding a palaeogeothermal gradient of 97.4 °C km^{-1}, which is significantly *lower* than the palaeogeothermal gradient (169.1 °C km^{-1}) estimated via the Burnham & Sweeney model. It should be noted that the kinetic model of Burnham & Sweeney (1989) is not calibrated above VR values $>4.7\ R_m\%$ and, consequently, kinetically derived peak palaeotemperatures are not available for some of the deeper samples in well 35/15-1.

Table 2 *Computed best-fit palaeogeothermal gradients and summary statistics for Carboniferous sections encountered in 28 wells*

Well	Stratigraphic Interval	Avg. Palaeogeothermal Gradient °C/km	95% Confidence Interval °C/km lower	upper	R^2	n	Depth Range km
BHB 1	Dinantian	43.3	24.8	61.7	0.59	19	0.378
Drumkeeran 1	Dinantian	49.9	45.5	54.3	0.952	29	2.399
Kilcoo Cross	Dinantian	45.2	39.8	50.5	0.914	30	1.344
Macnean 2	Dinantian	48.9	42.4	55.4	0.884	33	1.155
Slisgarrow 1	Dinantian	34.7	30.2	39.2	0.89	33	1.92
1327/1	Dinantian	30.2	25.7	34.7	0.835	39	0.791
1328/19	Dinantian	39.8	32.9	46.6	0.766	44	0.804
1450/1	Dinantian	29.9	21.6	38.2	0.679	28	0.761
1453/5	Dinantian-Namurian	41.2	33.3	49.1	0.79	32	0.396
Athboy	Dinantian	50.7	41.5	59.9	0.71	52	1.78
N852	Dinantian	81.6	59.2	103.8	0.802	17	0.433
N996	Dinantian	66.4	25.5	107.4	0.464	16	0.62
N998	Dinantian	119.4	86.4	152.5	0.698	26	0.4
N1001	Dinantian	100.7	67.5	133.9	0.721	18	0.605
Doonbeg (Upp. section)	Namurian	11.9	−5.7	29.5	0.23	10	1.036
Doonbeg (Lr. section)	Dinantian	6.4	0.31	12.5	0.68	6	1.375
Lissycassy	Namurian	46.7	38.2	55.2	0.669	62	0.35
IPP 2	Namurian	3.3	−14.3	21	0.003	52	0.373
13/3-1	Westphalian B–C	105.2	89.7	120.7	0.865	32	0.637
19/5-1	Dinantian—Westphalian A–B	20.9	17.3	24.6	0.677	64	1.588
26/28-1 (Upp. section)	Westphalian D—Stephanian	78.2	−12.2	168.6	0.147	21	0.34
26/28-1 (Lr. section)	Westphalian B–C	51.4	13.7	89.1	0.3	21	0.22
26/28-2	Westphalian C—Stephanian	81.4	18.4	144.4	0.188	32	0.39
34/5-1	Westphalian B–D	92.5	57.2	127.8	0.366	50	0.655
34/15-1	Westphalian—Stephanian	105.2	46.7	163.6	0.916	5	0.564
35/15-1	Dinantian	97.4	76.7	118	0.875	13	0.914
36/16-1	Namurian-Westphalian D	104.5	94.2	114.9	0.91	43	0.917
33/22-1	Westphalian B–D	80.1	60.8	99.5	0.745	27	0.613
42/12-1	Westphalian D—Stephanian	33.2	7.6	58.9	0.195	31	0.5
42/17-1	Westphalian C–D	65.3	53.5	77.2	0.826	29	0.884

The SPSS statistical analysis package was used to determine a best-fit, least-squares, linear regression line for each section with depth as the independent variable and temperature as the dependent variable. R^2 (the success of the best-fit line in predicting temperature from depth), 95% Confidence Interval (95% probability that actual average palaeogeothermal gradient falls within this interval), n (number of palaeotemperature points used to define the gradient) and the depth range for these points in kilometres were also recorded for each well.

narrow maturity range examined, these model curves indicate that the use of empirically based translation schemes will result in the derivation of higher magnitude palaeogeothermal gradients than might be predicted by kinetic translation schemes. In addition, these model curves indicate that palaeotemperatures derived via the Barker (1988) translation scheme most closely approximate to the palaeotemperature output from the kinetic schemes of Burnham & Sweeney (1989). These results are consistent with the conclusions reached by Johnston *et al.* (1993) from their analysis of an extensive VR dataset from Alaska.

The impact on estimated palaeotemperatures and palaeogeothermal gradients of the Barker (1988) and Burnham & Sweeney (1989) schemes has also been examined over a broader maturity range (Table 1; Fig. 9). VR data from well 35/15-1 in the Porcupine Basin have been used to compare these two palaeotemperature translation schemes. This well encountered Tertiary and Cretaceous sediments resting unconformably upon rocks of Early Carboniferous age. Recorded VR values range from 0.37 R_m% in the Tertiary section to 6.1 R_m% in the pre-Mesozoic section (Table 1). Although the interpretation of the palaeotemperature profile for this well is complicated by the local presence of igneous intrusions and/or caved material, a significant break in the palaeotemperature profile and change in the palaeogeothermal gradient is observed at the Base Cretaceous unconformity (Fig. 9). The stratigraphic evidence from this well indicates that peak palaeotemperature exposure of the early Carboniferous section must have been achieved before the deposition of the Early Cretaceous (Barremian) section, which directly overlies the Base Cretaceous unconformity.

A comparison of the estimated palaeogeothermal gradients for the Mesozoic–Cenozoic section indicates that the Barker (1988) scheme yields a higher palaeogeothermal gradient (37.3 °C km^{-1}, with 95% confidence interval (CI) 19.5–55.1 °C km^{-1}) than the Burnham & Sweeney model (30.4 °C km^{-1}, with 95% CI 16.3–44.6 °C km^{-1}). The reverse is true for the Palaeozoic section, with the Barker (1988) scheme yielding a palaeogeothermal gradient of 97.4 °C km^{-1} (95% CI 76.7–118 °C km^{-1}), which is significantly lower than the palaeogeothermal gradient (169.1 °C km^{-1}, with 95% CI 99.5–238.7 °C km^{-1}) estimated via the Burnham & Sweeney model. It should be noted that the kinetic model of Burnham & Sweeney (1989) is not calibrated above VR values >4.7 R_m% and, consequently, kinetically derived peak palaeotemperatures are not available for some of the deeper samples in the well (Table 1). It is concluded that both the kinetic and empirical VR to palaeotemperature translation schemes indicate a significantly higher palaeogeothermal gradient in the Palaeozoic section of well 35/15-1, relative to that of the Mesozoic–Cenozoic section.

Consequently, in the context of the VR database examined here, a considerable difference in palaeogeothermal gradients between Mesozoic–Cenozoic and Palaeozoic sections is evident, regardless of which of the two translation schemes is used. Both kinetic and empirical models rely upon some underlying assumptions. The advantage of the more theoretically rigorous, multiple-reaction, Arrhenius-based simulation methods of Burnham & Sweeney (1989) is somewhat undermined in this case by a limited knowledge of heating rates in Irish sedimentary basins throughout the *c.* 400 Ma history represented by the samples in this VR database and by the absence of kinetic model calibration at VR values >4.7 R_m%. For this reason the empirically based equation of Barker (1988) has been used to convert VR into peak palaeotemperature in this study.

Peak palaeotemperatures and palaeogeothermal gradients

The graphical techniques of Bray *et al.* (1992) have been adopted to interpret palaeotemperature profiles derived from the VR database. A statistical analysis package (SPSS) has been utilized to derive the best-fit average palaeogeothermal gradient for each well with at least five VR data points spread over a minimum vertical interval of 200 m. Both R^2 (the success of the best-fit line in predicting temperature from depth) and 95% confidence intervals (95% probability that actual average palaeogeothermal gradient falls within this interval) were determined for each best-fit line. An overview of these best-fit average gradients and summary statistics for Carboniferous sections is presented in Table 2. The best-fit (least-squares regression) line for each well was computed with depth as the independent variable, and temperature as the dependent variable. Some filtering of the data was necessary in a number of wells to remove data points associated with igneous intrusions and outlying data points. Because of the observed offsets in palaeotemperature profiles at the base of the Mesozoic section, 'segmented piecewise' regression was performed for those wells where both Mesozoic–Cenozoic and Carboniferous sections were present.

Using the palaeotemperature conversion, subsets of the database can be characterized by the magnitude of peak palaeotemperatures and/or palaeogeothermal gradients. For example, examination of the subset of the database that pertains to onshore Carboniferous sections reveals that, by plotting peak palaeotemperature v. depth for these sections, an estimate of the magnitude of the palaeogeothermal gradient at peak temperature exposure for these well locations can be obtained (Fig. 10). A broad range in average palaeogeothermal gradients is observed, ranging from less than the 3 °C km^{-1} gradient at well IPP-2 in the Clare Basin to 119 °C km^{-1} at well N998 in the Navan area of the Dublin Basin. Low to negative palaeogeothermal gradients are identified in some wells in the Clare Basin, suggesting advective heat transfer consistent with a hydrothermal circulation system in a foreland basin setting.

Carboniferous sections encountered in the offshore area manifest a similar pattern of dominantly high palaeogeothermal gradients, with gradients reaching 105 °C km^{-1} observed in well 13/3-1 in the Donegal Basin. A notable exception to these high trends is the low palaeogeothermal gradient (*c.* 21 °C km^{-1}) recorded in the Viséan–Westphalian section

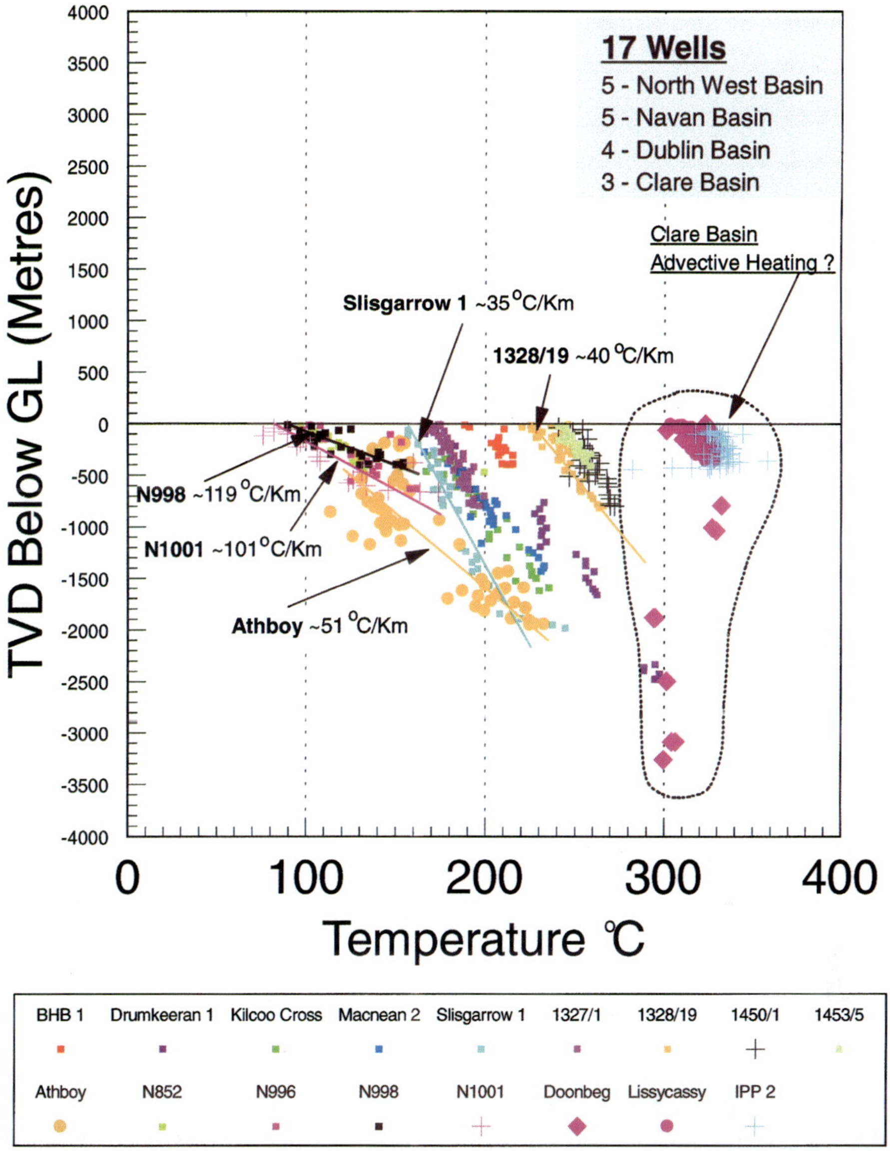

Fig. 10. Peak palaeotemperature v. depth for Carboniferous sections encountered in onshore boreholes. $R_m\%$ has been translated to maximum palaeotemperature via the Barker (1988) empirical scheme. Average palaeogeothermal gradients range from less than *c.* 3 °C km^{-1} at well IPP-2 in the Clare Basin to *c.* 119 °C km^{-1} at well N998 in the Navan area of the Dublin Basin.

encountered in well 19/5-1 in the Erris Basin (Fig. 11).

The variation in palaeogeothermal gradients recorded in Mesozoic–Cenozoic sections is explored using palaeotemperature profiles for three deep wells (35/8-2, 35/19-1, 43/13-1) in the Porcupine Basin (Fig. 12). In excess of 4 km of Mesozoic–Cenozoic overburden was encountered at each well location, with total depth (TD) reached in rocks of Late Jurassic age. No major discontinuity was observed in the palaeotemperature profiles at the Base Cretaceous unconformity. The average present-day geothermal gradient in these three wells is *c.* 34 °C km^{-1}, based upon eight Horner-corrected bottom hole temperatures (BHTs) and an assumed average sea-bed temperature of 5 °C. A number of interpretations of the VR data are possible but

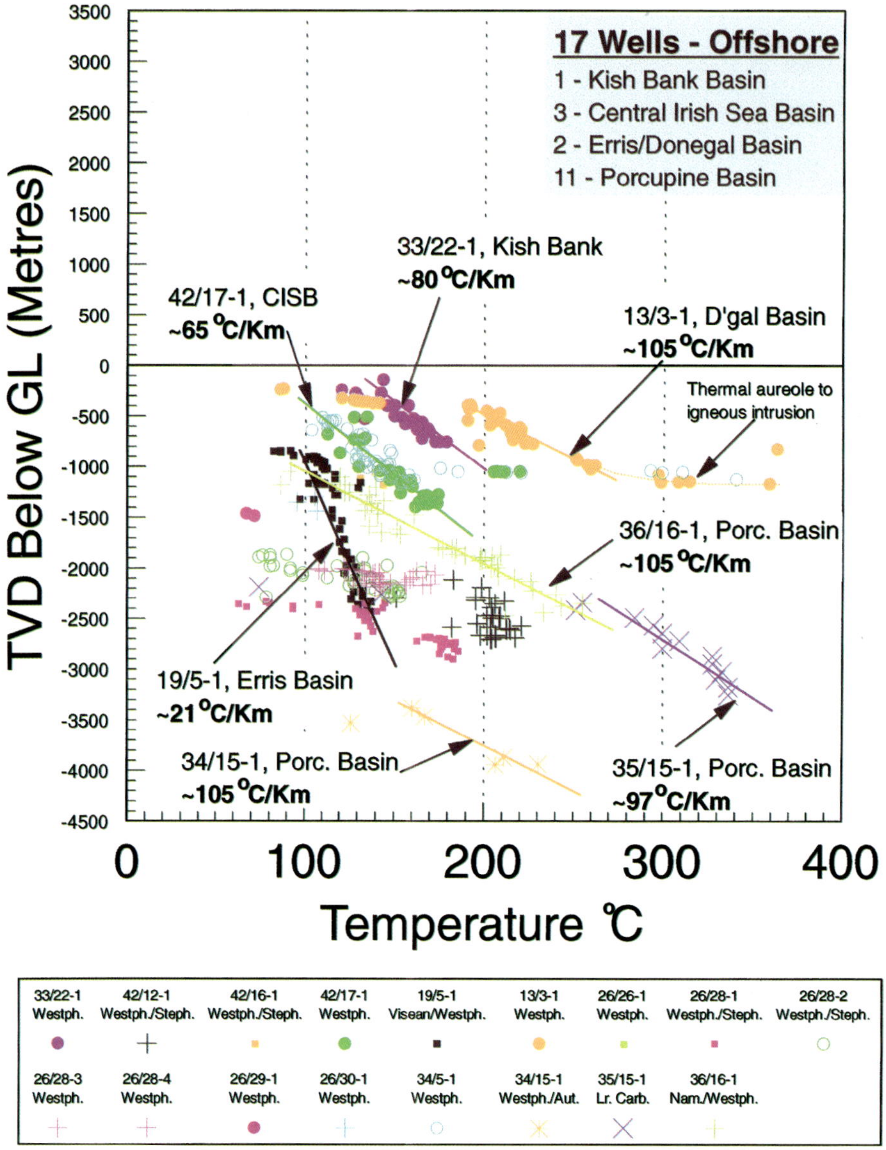

Fig. 11. Peak palaeotemperature v. depth for Carboniferous sections encountered in offshore wells. $R_m\%$ has been translated to maximum palaeotemperature via the Barker (1988) empirical scheme. Discounting profile deviations caused by igneous intrusions, the average palaeogeothermal gradients range from anomalously low (*c.* 21 °C km^{-1}) at well 19/5-1 in the Erris Basin to *c.* 105 °C km^{-1} at well 13/3-1 in the Donegal Basin and wells 34/15-1 and 36/16-1 in the Porcupine Basin.

in all cases the evidence suggests that the rocks encountered in these wells are at their maximum temperature exposure at the present day. Consequently, the VR-derived temperature profiles for these wells represent the present-day temperature profiles and geothermal gradients for these locations. The best-fit average geothermal gradient, derived from utilizing all the VR data in well 35/19-1, is *c.*35 °C km^{-1} (95% CI 29–42 °C km^{-1}). An alternative interpretation suggests that the high palaeotemperatures recorded (three sidewall cores and three cuttings samples) towards the base of well 35/19-1 may be the result of anomalous heat flow associated with a high thermal conductivity salt diapir intruded into the

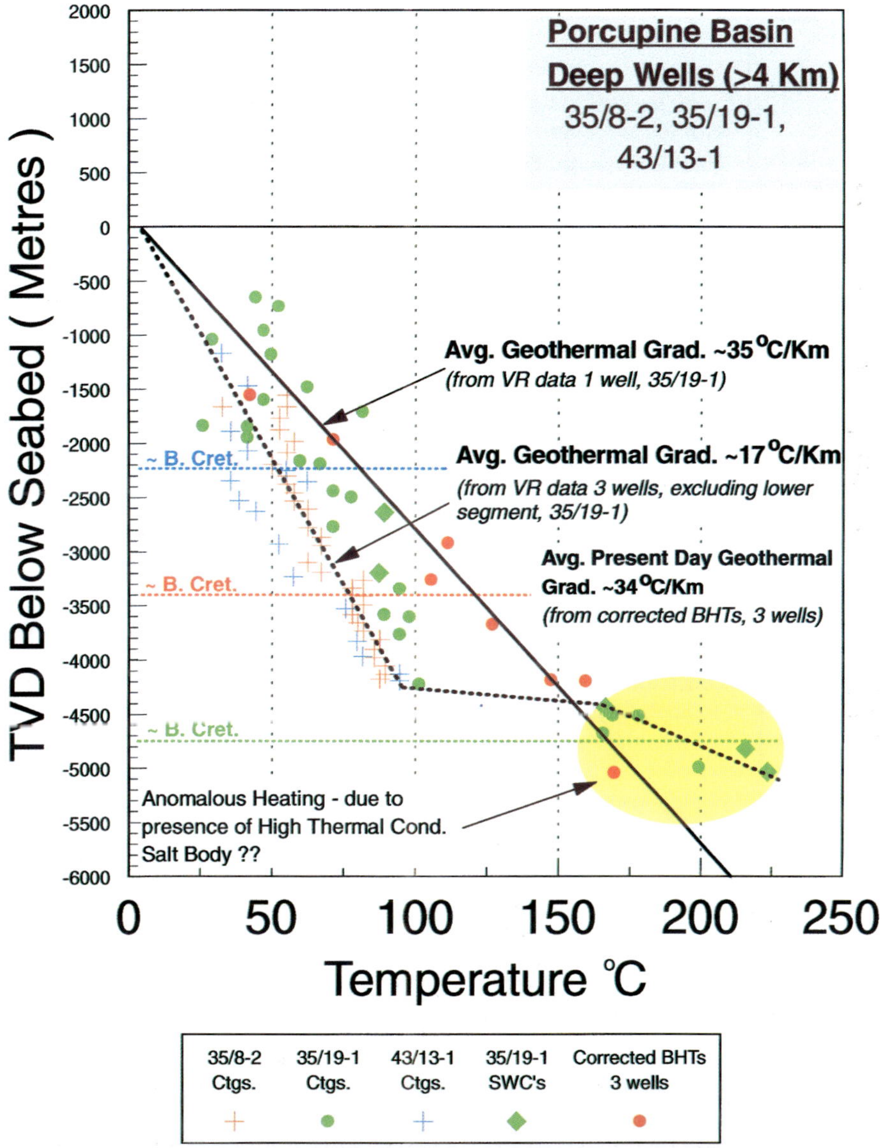

Fig. 12. Palaeogeothermal gradients recorded in Mesozoic–Cenozoic sections in three wells from the Porcupine Basin with Mesozoic–Cenozoic sections in excess of 4 km. $R_m\%$ has been translated to maximum palaeotemperature via the Barker (1988) empirical scheme. The average present-day geothermal gradient in these three wells is *c.* 34 °C km^{-1}, based upon eight Horner-corrected BHTs and an assumed average sea-bed temperature of 5 °C. A number of interpretations of the VR data are possible but in all cases the evidence suggests that the sediments encountered in these wells are at their maximum temperature exposure present day. The best-fit average, present-day, geothermal gradient, derived from utilizing all the VR data in well 35/19-1, is *c.* 35 °C km^{-1} (95% CI 29–42 °C km^{-1}).

Late Jurassic sediments in this well. In this scenario, although the shallower data manifest considerable scatter and are dominated by cuttings samples (only two sidewall cores), an average geothermal gradient of *c.*17 °C km^{-1} (95% CI 13–24 °C km^{-1}) is estimated for the Cretaceous–Tertiary section in these three wells (excluding the lower segment in 35/19-1). This interpretation would conflict with the corrected BHT estimates for these wells.

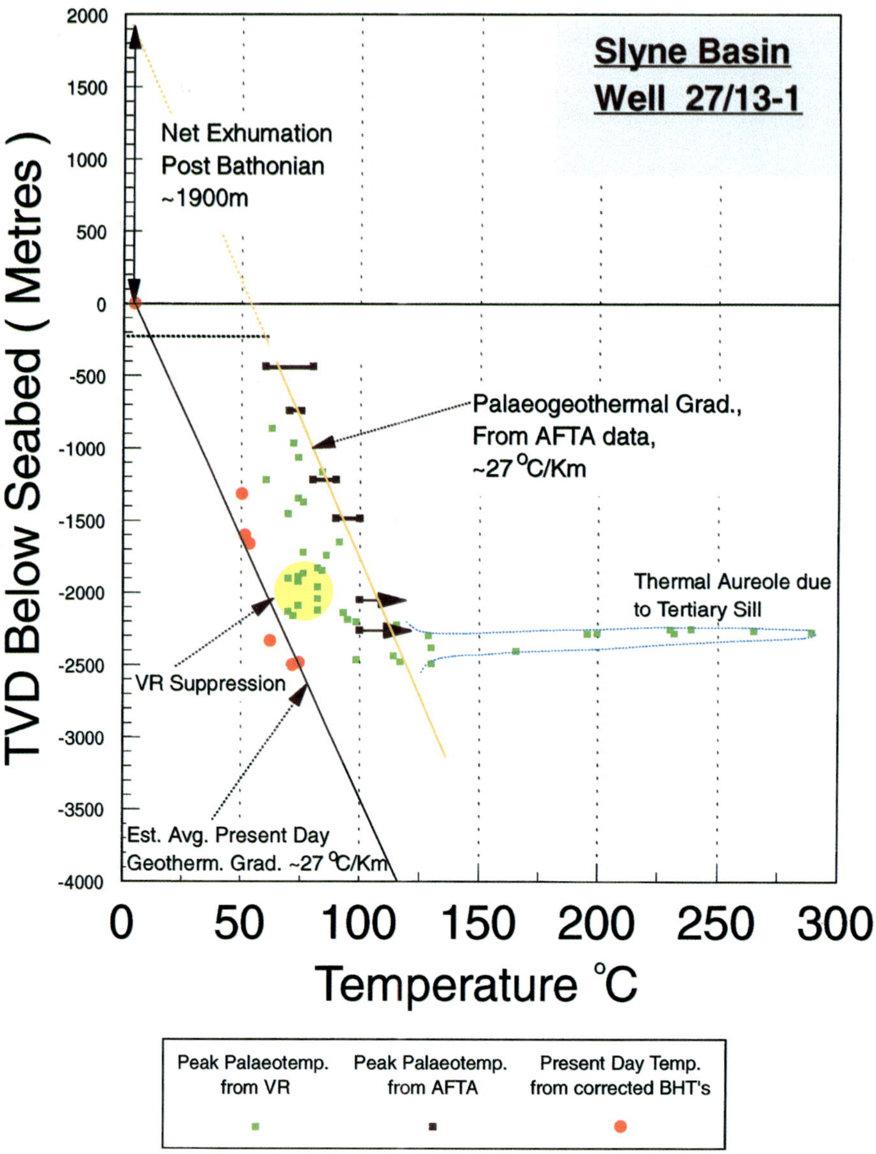

Fig. 13. Peak palaeotemperatures, from VR and AFTA data, v. TVD below sea bed for well 27/13-1 in the Slyne Basin. R_m% has been translated to maximum palaeotemperature via the Barker (1988) empirical scheme. The palaeogeothermal gradient (*c.* 27 °C km^{-1}; 95% CI 17–37 °C km^{-1}) derived from filtered VR data is consistent with the gradient-derived from AFTA data. Black filled arrows indicate minimum estimates of peak palaeotemperature from AFTA data as fission tracks are totally annealed above *c.* 110 °C. AFTA data indicate cooling to present temperatures via *c.* 1900 m of exhumation during late Mesozoic to early Cenozoic time. VR and AFTA data have been adapted from Scotchman & Thomas (1995).

Further north, in the Slyne Basin, published apatite fission-track analysis and VR data indicate an average palaeogeothermal gradient of *c.*27.5 °C km^{-1} in a Hettangian–Bathonian section in well 27/13-1 (Scotchman & Thomas 1995). Again, the palaeotemperature profile is noisy because of VR suppression, caused by oil saturation in the kerogen-rich Liassic intervals, and elevated VR measurements as a result of the presence of a thermal aureole associated with an igneous sill (Fig. 13). However, the VR profile is consistent with independent estimates of peak

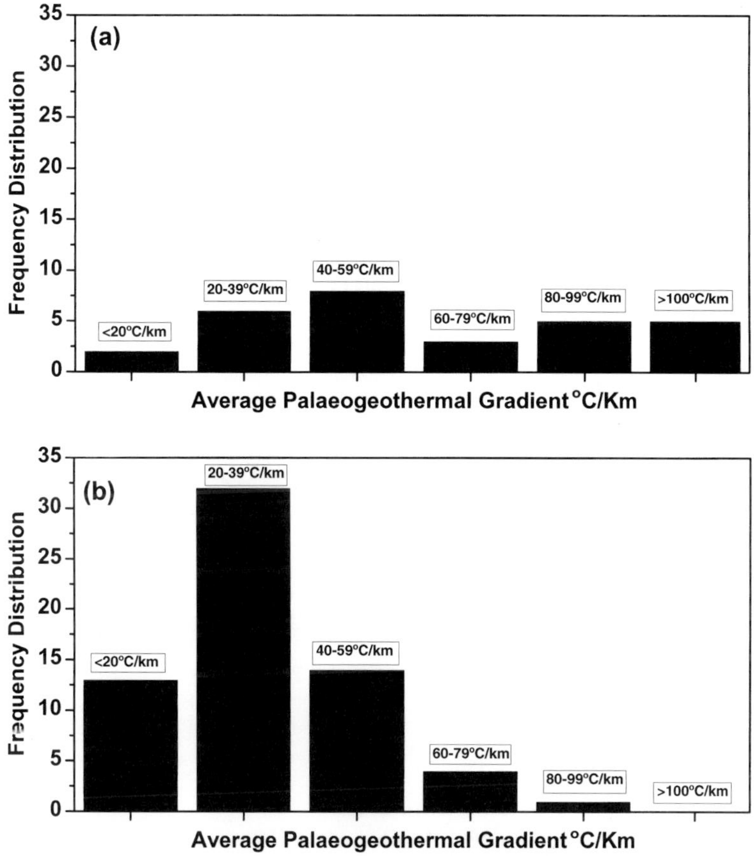

Fig. 14. Frequency distributions comparing (**a**) the average palaeogeothermal gradients for Carboniferous sections (28 wells) v. (**b**) the average palaeogeothermal gradients for Mesozoic–Cenozoic sections (64 wells) derived from the VR database. In general, palaeogeothermal gradients are substantially higher in the Carboniferous sections (mean 60 °C km^{-1}) than in the Mesozoic–Cenozoic sections (mean 32 °C km^{-1}).

palaeotemperature derived from apatite fission-track data and confirms an average palaeogeothermal gradient of *c.* 27 °C km^{-1} (95% CI 17–37 °C km^{-1}), at peak palaeotemperature exposure of this section, similar to the present-day average geothermal gradient (*c.* 27 °C km^{-1}) derived from Horner-corrected BHT data (Fig. 13). In addition, the apatite fission-track data indicate cooling to present temperatures via *c.* 1900 m of exhumation during late Mesozoic to early Cenozoic time.

A general conclusion is that palaeogeothermal gradients are substantially higher in the Carboniferous sections (estimated mean 60 °C km^{-1}) than in the Mesozoic–Cenozoic sections (estimated mean 32 °C km^{-1}). Figure 14 illustrates a frequency distribution comparison of average palaeogeothermal gradients for Carboniferous sections (28 wells) v. Cenozoic–Mesozoic sections (64 wells) derived from the VR database. Furthermore, there is stratigraphic evidence in the database suggesting that many of the high palaeogeothermal gradients observed in Carboniferous sections were established before the deposition of Mesozoic sediments. VR- and apatite fission-track-derived palaeotemperature data for two wells in the northern part of the Porcupine Basin offer support for this hypothesis (Fig. 15).

Wells 26/28-1 and 26/28-2 have been drilled within 2 km of each other on the same Mesozoic tilted fault-block structure (MacDonald *et al.* 1987). Both wells encountered a similar stratigraphy, with major unconformities at the base of the Cretaceous and the base of the Mesozoic sequences, where rocks of Mid-Jurassic age rest unconformably upon Westphalian–Stephanian sediments. In each well, a significant break in the palaeotemperature profile is observed at the Base

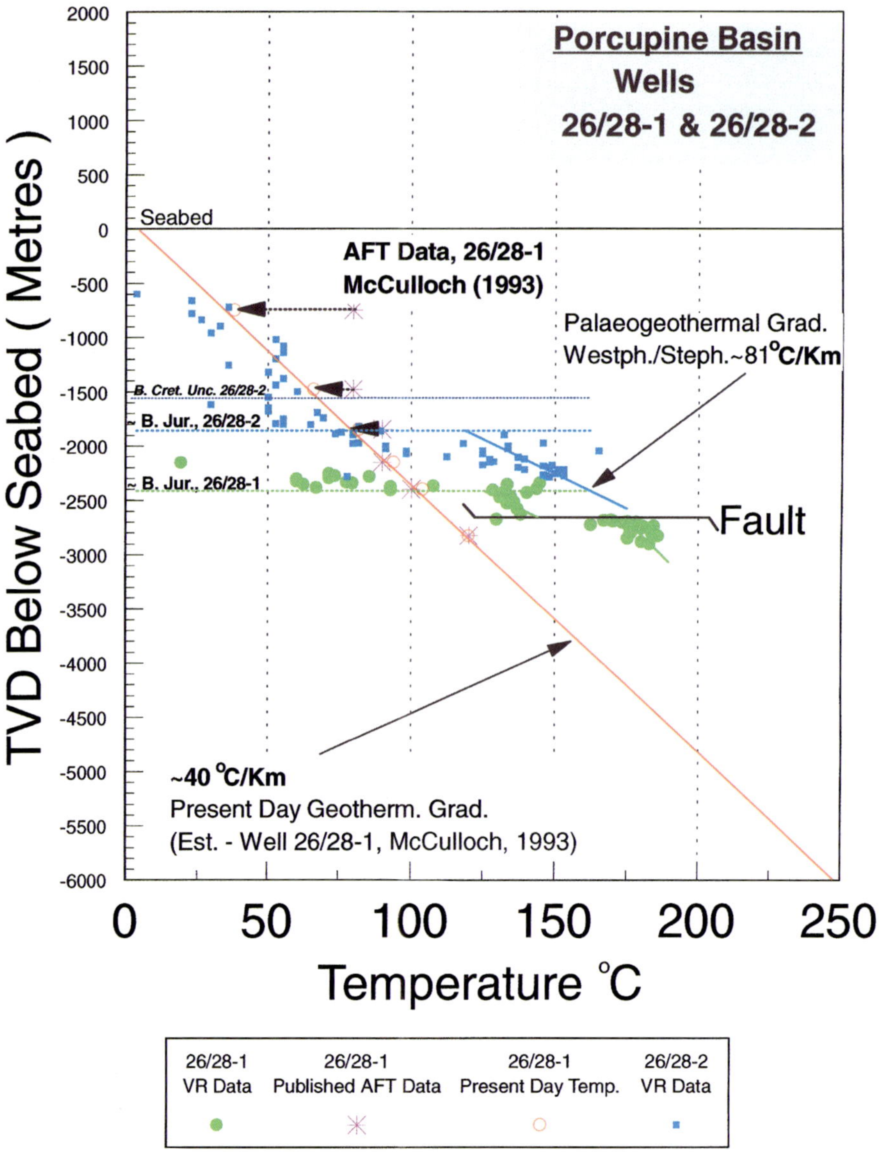

Fig. 15. Evidence for elevated heat flows during Late Carboniferous–Early Permian time. Peak palaeotemperature, from VR and apatite fission-track data, v. TVD below sea bed for wells 26/28-1 and 26/28-2 in the northern part of the Porcupine Basin. R_m% has been translated to maximum palaeotemperature via the Barker (1988) empirical scheme. Fission-track and VR data confirm that the Mesozoic–Cenozoic sections are at peak temperature exposure present day, whereas the Carboniferous section has experienced significantly higher temperatures at some time before deposition of the Mid-Jurassic sediments.

Mesozoic, but not at the Base Cretaceous unconformity (Fig. 15). Independent assessment of downhole ostracod colour changes recorded in well 26/28-1 is consistent with this profile (Ainsworth *et al.* 1990). Computed palaeogeothermal gradients for Carboniferous sections are in the range of 51–81 °C km^{-1} for both wells; significantly higher than the present-day geothermal gradient (*c.*40 °C km^{-1}) estimated by McCulloch (1993). The computed average geothermal gradient for the Mesozoic–Cenozoic section in well 26/28-2 is *c.* 35 °C km^{-1} (95% CI 25–46 °C km^{-1}). The derived palaeotemperatures for this Mesozoic–Cenozoic section plot close to the present-day geothermal gradient, suggesting that this section is experiencing maximum temperature exposure at the present day. This interpretation is consistent with the results of fission-track analysis of six samples from well 26/28-1 published by McCulloch (1993). The upper two apatite samples have fission-track ages significantly older than stratigraphic age and show no evidence of post-depositional annealing at temperatures >80 °C. In contrast, the bottom four samples, including one from the Westphalian–Stephanian section, are dominated by annealing at present-day temperatures >90 °C.

These observations suggest that the Mesozoic–Cenozoic sections in this area are at peak temperature at the present day, whereas the Carboniferous section has experienced significantly higher temperatures in the past. Stratigraphic relationships indicate that this higher temperature exposure must pre-date Mid-Jurassic deposition. A major discontinuity in the palaeotemperature profile between the Mesozoic and Palaeozoic sections in the Central Irish Sea–St George's Channel basins has previously been reported by Corcoran & Clayton (1999). This evidence supports the hypothesis that maturation levels in many Carboniferous sections are a consequence of burial and elevated heat flows during Late Carboniferous–Early Permian time, rather than a consequence of Mesozoic–Cenozoic processes. Green *et al.* (2001, fig. 7) have presented apatite fission-track evidence from Carboniferous and Lower Palaeozoic outcrop samples recovered onshore Ireland and north Wales. These samples clearly indicate cooling from a major palaeothermal episode that occurred between 300 and 250 Ma. However, Green *et al.* (2001) have offered an alternative hypothesis with respect to the Carboniferous sections in the offshore wells in the Central Irish Sea Basin, which they suggest reached their maximum palaeotemperatures during Early Cretaceous time.

Implications for thermal and tectonic evolution

There is a considerable body of evidence that suggests that the extensional evolution of Ireland's Late Palaeozoic to Cenozoic sedimentary basins has been punctuated by a multiphase exhumation history. Stratigraphic evidence, combined with VR and apatite fission-track data, suggests at least two periods of regional exhumation have occurred; during Late Carboniferous–Late Permian time and again during Tertiary time. In addition, published evidence from apatite fission-track work suggests that a major episode of exhumation occurred during Early Cretaceous time, at least in the Central Irish Sea Basin (Duncan *et al.* 1998; Green *et al.* 2001). Both the Late Carboniferous–Late Permian and Tertiary phases of exhumation are characterized by a component of compressional inversion and the widespread occurrence of extrusive and intrusive igneous rocks. However, the contrasting thermal signatures of these regional uplift events suggests that basin setting and the mechanism of regional exhumation both exert a fundamental control on processes that determine heat flow distribution within a basin.

Although the driving force for Tertiary exhumation is a source of some debate, it is now generally accepted that many sedimentary basins in NW Europe have experienced exhumation during Palaeogene to Neogene times. Evidence cited for Tertiary exhumation of offshore Ireland includes the complete or partial absence from some basins of the post-rift sediments predicted by the McKenzie lithospheric stretching model (Brodie & White 1995), the stratigraphic restoration of Cretaceous Chalk and Tertiary isopachs (Murdoch *et al.* 1995) and the observation of late Cretaceous anticlinal flexures on seismic data (Roberts 1989). In addition, the analysis of compaction trends derived from sonic velocities, palaeotemperature profiles derived from fission-track and VR data and the analysis of normalized drilling rates offers further evidence for exhumation events during the Tertiary period (Hillis 1995).

Direct evidence for Tertiary exhumation is also observed onshore. In Northern Ireland, the erosion and karstification of Santonian–Maastrichtian chalks, before extrusion of voluminous Paleocene basalt flows, is consistent with a phase of regional uplift at this time. Furthermore, Evans & Clayton (1998) interpreted from VR data 1–1.5 km of post-Campanian exhumation for the Ballydeenlea Chalk Breccia outlier in County Kerry.

Although less well documented, a similar geological rationale can be applied to support the hypothesis of pervasive Late Carboniferous–Late Permian exhumation of onshore and offshore Ireland. Corcoran & Clayton (1999) summarized the local evidence for igneous activity during the Stephanian–Early Permian time. These observations are consistent with the extensive calc-alkaline to subalkaline volcanism of the Permo-Carboniferous Pangaean province, reviewed by Doblas *et al.* (1998) and interpreted as evidence of a mantle plume impinging upon the base of the lithosphere at this time.

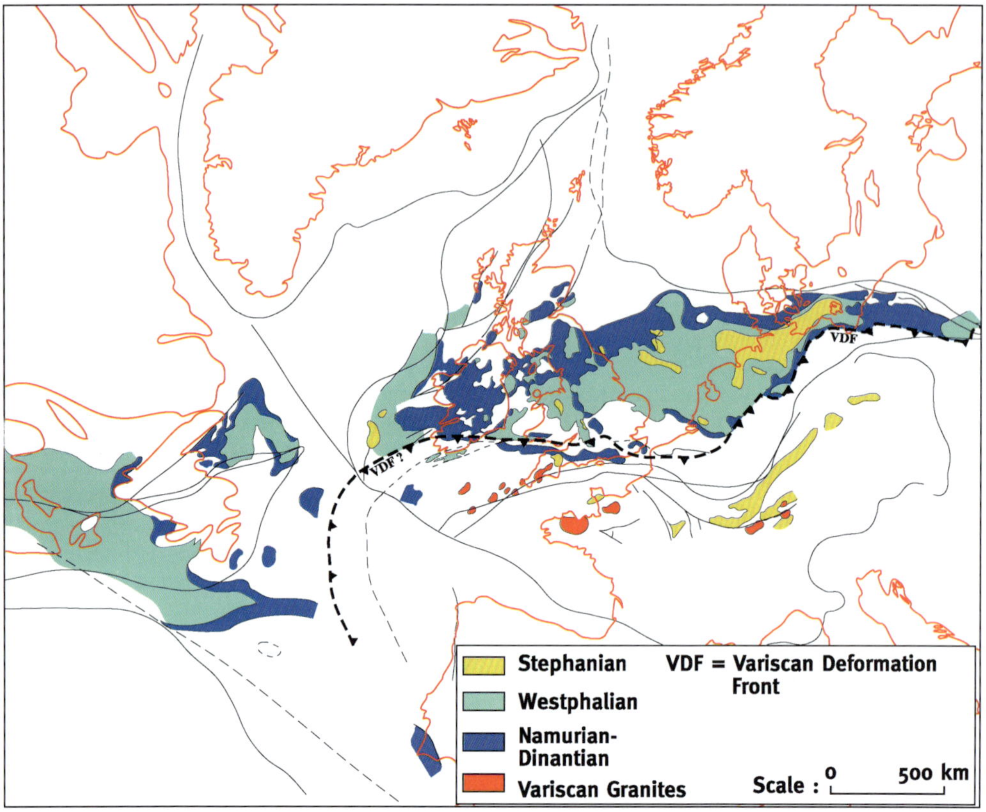

Fig. 16. Pre-Permian subcrop map, distal zone of Variscan foreland basins on a pre-Atlantic plate reconstruction (adapted from Maynard *et al.* 1997). The Late Palaeozoic stratigraphic record preserved in basins along-strike suggests that a Westphalian–Stephanian–Autunian cover existed across Ireland before denudation at the Saalian Unconformity.

With the exception of one isolated outlier of Westphalian D sedimentary rocks in Wexford (Clayton *et al.* 1986), the youngest preserved Carboniferous succession onshore Ireland is of Westphalian A age. However, stratigraphic evidence for missing Late Carboniferous–Early Permian cover can be deduced from the examination of a pre-Permian subcrop map constructed on a pre-Atlantic plate reconstruction (Fig. 16). A review of the stratigraphic record in those basins along-strike, within a similar distal setting of the Variscan foreland, suggests that Late Westphalian–Stephanian–Autunian sedimentary cover could have been deposited over Ireland and subsequently removed to produce the Saalian unconformity. In NE Germany, Glennie (1997) and others have described >600 m of Stephanian sediments locally overlain by >2 km of Autunian basalts. To the west of Ireland, *c*.450 m of Stephano-Autunian sediments are preserved in two wells on the western flank of the Porcupine Basin (Robeson *et al.* 1988). Further along-strike in the Maritime Province of Canada, Bell & Howie (1990) interpreted well and seismic data that suggested that up to 5 km of Late Westphalian–Early Permian sediments are preserved in the eastern part of the Magdalen Basin.

The primary mode of heat transport in the crust is conduction, with basal heat flow and thermal conductivity being the critical determinants of temperature distribution in the sedimentary pile (Deming 1994a). In addition, it is generally acknowledged that fluid flow is capable of transporting significant quantities of heat within the permeable strata of sedimentary basins, which can result in the distortion of conductive heat distribution (Jessop & Majorowicz 1994). The major physical and thermal driving forces for fluid flow in a sedimentary basin are topographic gradients, sediment compaction, diagenesis and buoyancy

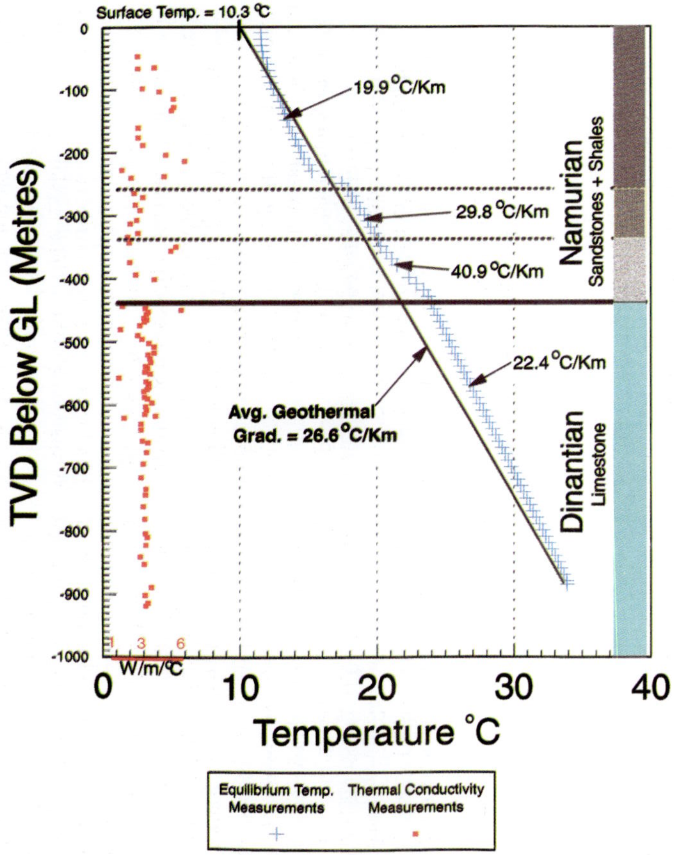

Fig. 17. Equilibrium temperature and thermal conductivity measurements from well IPP-2, Clare Basin. Fourier's Law ($Qz = K(dT/dZ)$) states that under conditions of a constant heat flow (Qz) the thermal conductivity (K) and the temperature gradient (dT/dZ) are inversely related. Observed geothermal gradient suggests that conductive heat flow is the dominant heat transfer mechanism in the Clare Basin, present day. Data from Brock & Barton (1984).

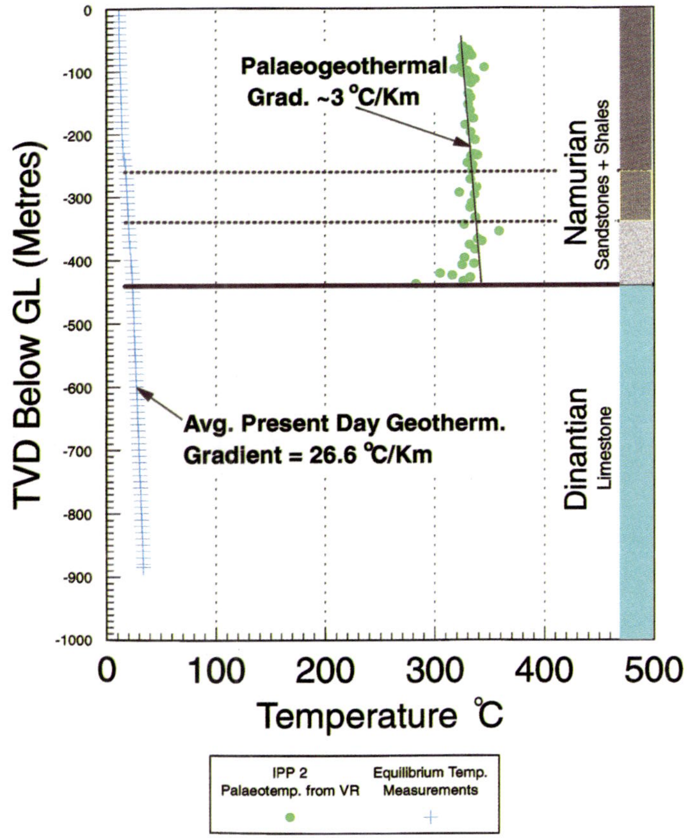

Fig. 18. Present-day geothermal gradient compared with palaeogeothermal gradient derived from VR data, well IPP-2 in the Clare Basin. R_m% has been translated to maximum palaeotemperature via the Barker (1988) empirical scheme. The very low to negative palaeogeothermal gradient suggests that advective heating was the dominant heat transfer mechanism at this location, at the time of peak palaeotemperatures.

forces (Deming 1994b). The nature and distribution of these forces changes throughout the evolution of a basin and can alter the dominant mode of heat transfer through time (Hitchon 1984; Majorowicz *et al.* 1985). For example, the large number of equilibrium temperature and thermal conductivity measurements for well IPP-2 in the Clare Basin, reported by Brock & Barton (1984), demonstrate that conductive heat transfer is the dominant control on the present-day geothermal gradient (average *c.*26 °C km^{-1}) in this basin (Fig. 17). However, examination of the peak palaeotemperatures derived from the VR data for this borehole indicates the presence of a very low palaeogeothermal gradient (average *c.* 3 °C km^{-1}, with 95% CI −14.3–21 °C km^{-1}) at the peak palaeotemperature exposure of the Namurian sediments in this borehole (Fig. 18). Goodhue & Clayton (1999) suggested that the reversal of the palaeogeothermal gradient towards the base of the Namurian section indicates a complex thermal regime that involves advective heating and fluid circulation through fractures in the Clare Shale.

Contrasting thermal signatures are associated with the Late Carboniferous–Late Permian and Early Tertiary palaeo-thermal events and the concomitant or subsequent phases of uplift and erosion. Tertiary exhumation is characterized by moderately elevated palaeotemperatures and variable, but generally low, palaeogeothermal gradients, although McCulloch (1993) invoked a post-Paleocene geothermal gradient in excess of 50 °C km^{-1} to account for palaeotemperatures derived from apatite fission-track data in well 26/26-1, in the northern Porcupine Basin. Green *et al.* (1999) have documented a very low palaeogeothermal gradient (little change in peak palaeotemperatures derived from apatite fission-track data over a thickness of 3 km of Permo-Triassic sediments and lavas) in the Larne No. 2 borehole. This is in spite of the close proximity to the Tertiary Volcanic Province and the consequent presence of numerous Tertiary intrusions in the penetrated section. They interpreted this gradient as the consequence of hydrothermal fluid circulation within the porous and permeable Permo-Triassic sandstones below the Tertiary lava pile.

The Late Carboniferous–Early Permian thermal event is characterized by high palaeotemperatures and in general very high palaeogeothermal gradients (estimated mean gradient *c.*60 °C km^{-1}). These palaeogeothermal gradients are consistent with the elevated maturity gradients found in other Variscan foreland settings (Robert 1989). Estimated palaeogeothermal gradients in excess of 80 °C km^{-1} for late Palaeozoic time have been reported from northern Switzerland (Schegg & Leu 1998). These values are consistent with high heat flows induced by extensional collapse, post-kinematic plutonism and hydrothermal activity during the final phase of the Central European Variscan orogeny. Intrusion of the Cornubian batholith and the Haig Fras granitic plutons to the south of Ireland (Gardiner & Sheridan 1981) established a mountain belt, which acted as a recharge area for Carboniferous and Devonian aquifers and provided the hydraulic head to drive fluids northwards through the foreland basin. Vertical discharge of these heated fluids, along major fault systems in a foreland platform setting, offers a mechanism to account for two observations with respect to VR trends in Carboniferous rocks of the Irish sedimentary basins (Fig. 19). First, the maturation levels (R_m% levels) in Carboniferous rocks manifest a general decrease in a northern direction indicating a fall in Variscan palaeotemperatures to the north. Although some of the observed R_m% trend may be due to different erosion levels, as the preserved Palaeozoic stratigraphy, of onshore Ireland, generally youngs to the north, the evidence is consistent with a model of progressive heat dissipation from gravity-driven fluids discharging towards the north. Second, the rapid lateral variation observed in Carboniferous palaeogeothermal gradients suggests that active fault systems and hydraulically induced fracture zones acted as vertical conduits for fluid flow, resulting in locally complex, vertical heat distribution patterns. For example, low palaeogeothermal gradients (<3 °C km^{-1}) are locally observed in an area of high palaeotemperatures (>300 °C) in the Clare Basin, whereas low to high palaeogeothermal gradients (30–119 °C km^{-1}) are observed, at relatively lower palaeotemperatures (75–280 °C), in close proximity in the Navan area of the Dublin Basin (Figs 10 and 19). Furthermore, it has been suggested that lateral fluid flow in this foreland basin setting offers a mechanism for the emplacement of large hydrothermal sulphide ore bodies along fault systems active during Variscan deformation (Johnston 1999).

Implications for hydrocarbon generation and hydrocarbon prospectivity

Proven Jurassic-sourced petroleum systems are present in four of Ireland's offshore basins: the Slyne, Porcupine, North Celtic Sea and the St George's Channel basins (Caston 1995; Taber *et al.* 1995; Spencer & MacTiernan 2001). The recent discovery of gas in Triassic sandstones in the Slyne Basin indicates a second working petroleum system in this basin, probably sourced from Dinantian–Westphalian shales and coals (Spencer & MacTiernan 2001). This petroleum system is possibly present, although not yet proved, in those offshore Permo-Triassic basins around Ireland that are stratigraphically analogous to the prolific East Irish Sea hydrocarbon province. Published drilling results indicate that the timing of maturation of Carboniferous source rocks is one of the major exploration risk factors in these basins, as discussed by Duncan *et al.* (1998) and Corcoran & Clayton (1999).

Considerable lateral variation is observed in the organic maturity levels and magnitude of palaeogeothermal gradients recorded in Carboniferous sections (Fig. 19). In the onshore area, Palaeozoic sediments at outcrop in the south and west of the island are dominantly post-mature (R_m% >3.0) for gas generation present day (Clayton *et al.* 1989). In contrast, maturation levels at outcrop in the Northwest Carboniferous Basin are in the dry gas generation window (R_m% 1.2–3.0) and Carboniferous outcrops in Northern Ireland are immature to early mature for gas generation present day. The relatively low palaeogeothermal gradients observed in wells in the onshore Northwest Basin, combined with the low gradient recorded in well 19/5-1 on the margin of the Erris Basin, suggest that this area experienced relatively low heat flows during Late Carboniferous–Early Permian time. The well-constrained, linear palaeotemperature–depth profiles for these wells are consistent with a conductive heat transfer mechanism with little evidence for modification by advective heating (Duddy *et al.* 1994).

Exhumation of the Slyne, Erris and Northwest basins during Late Carboniferous–Early Permian time may have temporarily arrested the maturation of the Carboniferous source rocks. However, subsequent reburial during Mesozoic time (and exposure to temperatures that exceed those reached in the earlier episode of burial) would have produced renewed maturation of these source rocks and provided a gas charge for the available Mesozoic hydrocarbon traps, before exhumation during Late Mesozoic–Early Cenozoic time. These observations are consistent with the presence of the Corrib gas accumulation in the Slyne Basin and the presence of gas in Dinantian sandstones in the Dowra No. 1 well in the Northwest Basin.

The above observations suggest that, where Carboniferous source rocks are present, they will make a significant contribution to the hydrocarbon budget only in those basins that have experienced relatively low heat flow during Late Carboniferous–Early Permian time, and where sufficient Mesozoic burial has occurred to subsequently expose the kerogen to higher palaeotemperatures. There is little evidence that elevated heat flows, at the basin scale, resulted from the emplacement of sills and dykes into Atlantic margin basins, during Early Tertiary time. England (1992) has suggested that heat derived from rising Paleocene magma would have little effect upon the thermal maturation of sediments in these basins. Experience along the UK Atlantic margin indicates that source rock maturation and hydrocarbon generation histories are rarely dominated by elevated Early Tertiary heat flow (Green *et al.* 1999).

Conclusions

VR data (R_m%) have been compiled from 77 Irish offshore wells and 17 onshore boreholes. This database has facilitated the analysis of VR v. depth behaviour by basin and by stratigraphic interval. A composite approach to the display and interpretation of VR and peak palaeotemperature profiles has been utilized to present an overview of this database and to gain insights into the thermal history of some of the Irish sedimentary basins. The following are the principal conclusions of this study:

(1) VR profiles manifest considerable scatter. In general, VR profiles from Carboniferous sections are better delineated than those from Mesozoic and Cenozoic sections, reflecting the relatively less complex vitrinite populations within Carboniferous coals and shales.

(2) For non-exhumed basins, kinetic VR to palaeotemperature translation models suggest lower palaeogeothermal gradients than might be predicted from empirically based VR to palaeotemperature translation schemes.

(3) Considerable inter- and intra-basin variation in thermal maturity and palaeogeothermal gradient patterns is observed. In general, palaeogeothermal gradients are substantially higher in Carboniferous sections than in the Mesozoic–Cenozoic sections and provide evidence of elevated heat flows during Late Carboniferous to Early Permian time.

(4) There is no evidence, from VR data, of elevated heat flows during Tertiary time in Irish Atlantic margin basins.

(5) A model of northwards-focused hydrodynamic flow of heated fluids, with vertical discharge along major fault systems in a foreland platform setting, offers a mechanism

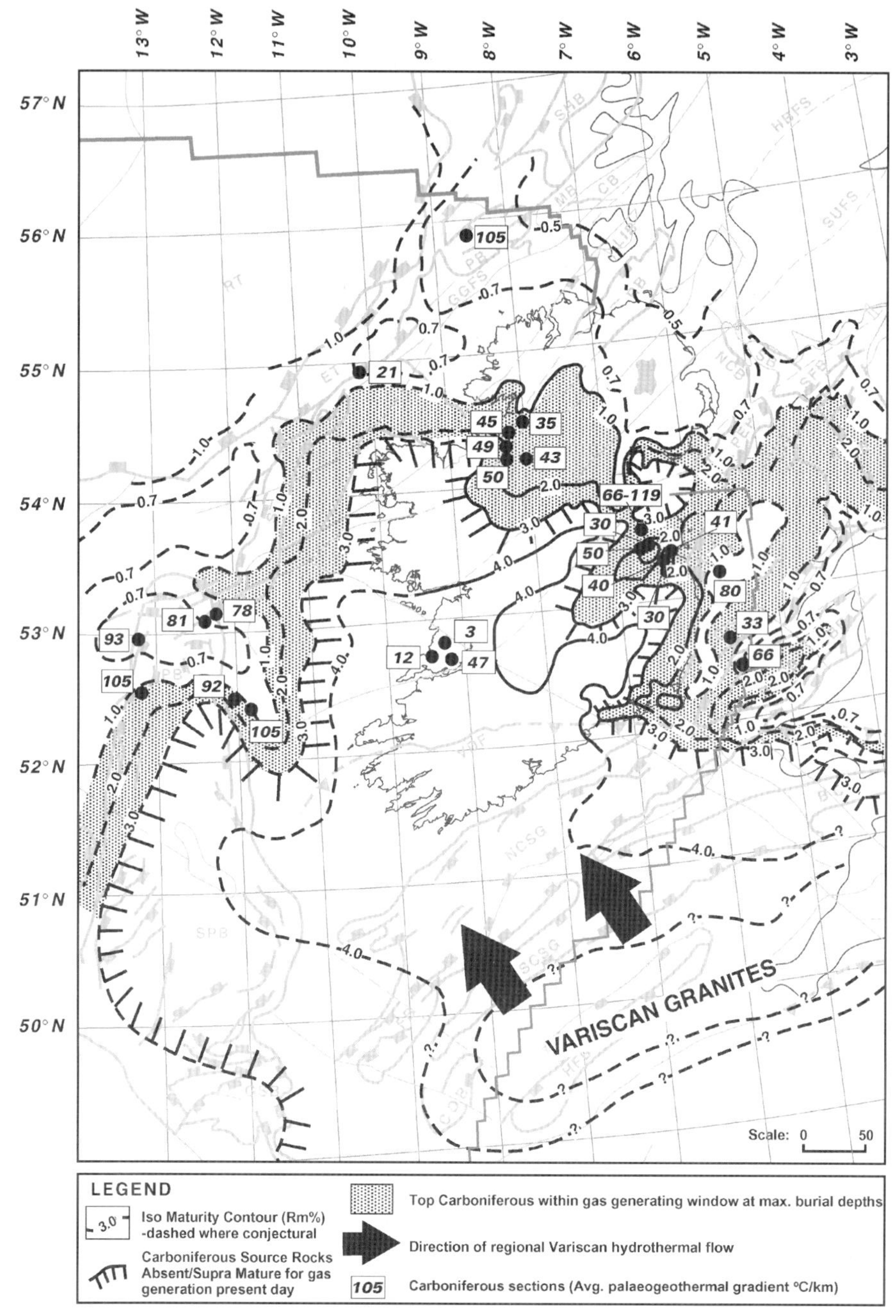

Fig. 19. Peak maturity levels of Carboniferous rocks: generalized $R_m\%$ contour map for Carboniferous and older sediments that subcrop the Saalian Unconformity surface. $R_m\%$ levels in Carboniferous rocks manifest a general decrease towards the north indicating a fall in Variscan palaeotemperatures in this direction. Rapid lateral variations in palaeogeothermal gradients are consistent with a gravity-driven hydrothermal system discharging heated fluids, along fault systems and fracture zones, in a foreland platform area. With respect to organic maturity levels and timing, the southern area is dominated by early maturity and high Variscan heat flows as a result of intrusion of Variscan granites and a regional advective system. The northern area is dominated by later maturity because of relatively lower Variscan heat flows at the distal end of the regional advective system and is overprinted by Mesozoic burial. Modified after Clayton *et al.* (1989), Maddox *et al.* (1995), Newman (1999) and Middleton *et al.* (2001).

to account for two observations with respect to VR trends in Carboniferous rocks of the Irish sedimentary basins. First, Variscan palaeotemperatures fall up-stratigraphy and to the north. Second, rapid lateral variation in Carboniferous palaeogeothermal gradients suggests that active fault systems and hydraulically induced fracture zones acted as vertical conduits for fluid flow, resulting in locally complex, vertical heat distribution patterns during Late Carboniferous to Early Permian time.

(6) Many of the onshore and offshore basins have experienced a multiphase exhumation history. The VR database indicates that extensive regional uplift and erosion occurred across Ireland during Late Carboniferous to Late Permian times. In addition, Paleocene to Oligo-Miocene exhumation is pervasive in the offshore area, with the exception of the Porcupine Basin.

(7) Where Carboniferous source rocks are present, they will make a significant contribution to the hydrocarbon budget only in those basins where sufficient Mesozoic burial has occurred to expose the kerogen to higher palaeotemperatures than might have been achieved prior to the Late Carboniferous–Early Permian uplift.

The authors would like to thank the Petroleum Affairs Division, Department of the Marine and Natural Resources, for permission to publish this paper. In particular, the authors express their sincere thanks to K. Robinson, P. Croker and N. Murphy for access to organic geochemical and maturation well records. This paper is based upon many reports and studies undertaken by past operators (Amoco, BP, British Gas, Britoil, Conoco, Elf, Enterprise, Esso, Cities Services, Fina, Gulf, Marathon, Shell and Total), their contractors and consultants. The authors would also like to thank P. Green and P. Haughton for their constructive reviews of the original manuscript. Finally, the authors would like to thank Shell UK Exploration and Production for their generous sponsorship of the colour printing costs.

References

AINSWORTH, N.R., BURNETT, R.D. & KONTROVITZ, M. 1990. Ostracod colour change by thermal alteration, offshore Ireland and Western UK. *Marine and Petroleum Geology*, **7**, 288–297.

ALPERN, B. & CHEYMOL, D. 1978. Reflectance et fluorescence des organoclastes du Toarcian du Bassin de Paris en fonction de la profondeur et de la temperature. *Revue de l'Institut Français du Pétrole*, **33**, 515–535.

BAILY, H. 1992. *Organic petrology of Dinantian and Namurian rocks of the Northwest Carboniferous Basin, onshore Ireland*. PhD thesis, University of Dublin.

BARKER, C.E. 1988. Geothermics of petroleum systems: implications of the stabilisation of kerogen thermal maturation after a geologically brief heating duration at peak temperature. *In*: MAGOON, L.B. (ed.) *Petroleum Systems of the United States*. US Geological Survey Bulletin, **1870**, 26–29.

BARKER, C.E. 1991. Implications to organic maturation studies of evidence for a geologically rapid increase and stabilisation of vitrinite reflectance at peak temperature: Cerro Prieto geothermal system, Mexico. *AAPG Bulletin*, **75**, 1852–1863.

BARKER, C.E. 1993. Implications to organic maturation studies of evidence for a geologically rapid increase and stabilisation of vitrinite reflectance at peak temperature: Cerro Prieto geothermal system, Mexico: Reply. *AAPG Bulletin*, **77**, 673–678.

BARKER, C.E. & GOLDSTEIN, R.H. 1990. Fluid-inclusion technique for determining maximum temperature in calcite and its comparison to vitrinite reflectance. *Geology*, **18**, 1003–1006.

BARKER, C.E. & PAWLIEWICZ, M.J. 1986. The correlation of vitrinite reflectance with maximum heating in humic organic matter. *In*: BUNTEBARTH, G. & STEGENA, L. (eds) *Palaeogeothermics*. Springer, New York, 79–93.

BELL, J.S. & HOWIE, R.D. 1990. Palaeozoic geology. *In*: KEEN, M.J. & WILLIAMS, G.L. (eds) *Geology of the Continental Margin of Eastern Canada*. Geological Survey of Canada, Geology of Canada, **2**, 141–165.

BRAY, R.J., GREEN, P.F. & DUDDY, I.R. 1992. Thermal history reconstruction using apatite fission track analysis and vitrinite reflectance: a case study from the UK East Midlands and the Southern North Sea. *In*: HARDMAN, R.P.F. (ed.) *Exploration Britain: Geological Insights for the Next Decade*. Geological Society, London, Special Publications, **67**, 3–25.

BROCK, A. & BARTON, K.J. 1984. *Equilibrium temperature and heat flow measurements in Ireland*. Commission of European Communities, Brussels, **EUR 9517**.

BRODIE, J. & WHITE, N. 1995. The link between sedimentary basin inversion and igneous underplating. *In*: BUCHANAN, J.G. & BUCHANAN, P.G. (eds) *Basin Inversion*. Geological Society, London, Special Publications, **88**, 21–38.

BURNHAM, A.K. & SWEENEY, J.J. 1989. A chemical kinetic model of vitrinite reflectance maturation. *Geochimica et Cosmochimica Acta*, **53**, 2649–2657.

CASTON, V.N.D. 1995. The Helvick oil accumulation, Block 49/9, North Celtic Sea Basin. *In*: CROKER, P.F. & SHANNON, P.M. (eds) *The Petroleum Geology of Ireland's Offshore Basins*. Geological Society, London, Special Publications, **93**, 209–225.

CLAYTON, G. & BAILY, H. 2000. Maturation levels in the onshore and offshore pre-Westphalian rocks in Ireland. *In*: WHITICAR, M. (ed.) The Search for Deep Gas. Proceedings of the First International

Deep Gas Workshop, Hannover, 1990. *Geologisches Jahrbuch*, **D107**, 25–41.

CLAYTON, G., HAUGHEY, N., SEVASTOPULO, G.D., BURNETT, R. 1989. *Thermal Maturation Levels in Devonian and Carboniferous Rocks of Ireland.* Geological Survey of Ireland, Dublin.

CLAYTON, G., SEVASTOPULO, G.D. & SLEEMAN, A.G. 1986. Carboniferous (Dinantian and Silesian) and Permo-Triassic rocks in south County Wexford, Ireland. *Geological Journal*, **21**, 355–374.

CORCORAN, D. & CLAYTON, G. 1999. Interpretation of vitrinite reflectance profiles in the Central Irish Sea area: implications for the timing of organic maturation. *Journal of Petroleum Geology*, **22**, 261–286.

CROKER, P.F. 1995. The Clare Basin: a geological and geophysical outline. *In*: CROKER, P.F. & SHANNON, P.M. (eds) *The Petroleum Geology of Ireland's Offshore Basins*. Geological Society, London, Special Publications, **93**, 327–339.

DEMBICKI, H. Jr 1984. An interlaboratory comparison of source rock data. *Geochimica et Cosmochimica Acta*, **48**, 2641–2649.

DEMING, D. 1994*a*. Overburden rock, temperature and heatflow. *In*: MAGOON, L.B. & DOW, W.G. (eds) *The Petroleum System—From Source to Trap*. American Association of Petroleum Geologists, Memoirs, **60**, 165–186.

DEMING, D. 1994*b*. Fluid flow and heat transport in the upper continental crust. *In*: PARNELL, J. (ed.) *Geofluids: Origin, Migration and Evolution of Fluids in Sedimentary Basins*. Geological Society, London, Special Publications, **78**, 27–42.

DOBLAS, M., OYARZUN, R., LOPEZ-RUIZ, J. & 6 OTHERS, 1998. Permo-Carboniferous volcanism in Europe and northwest Africa: a superplume exhaust valve in the centre of Pangaea? *Journal of African Earth Science*, **26**, 89–99.

DORÉ, A.G. & JENSEN, L.N. 1996. The impact of late Cenozoic uplift and erosion on hydrocarbon exploration: offshore Norway and some other uplifted basins. *Global and Planetary Change*, **12**, 415–436.

DOW, W.G. & O'CONNOR, D.I. 1982. Kerogen maturity and type by reflected light microscopy applied to petroleum exploration. *In*: STAPLIN, F., DOW, W.G., MILNER, C.W.D., O'CONNOR, D.I., VAN GIJZEL, P., WELTE, P.H. & YUKLER, M.A. (eds) *How to Assess Maturation and Palaeotemperatures*. Society of Economic Paleontologists and Mineralogists, Short Course, **7**, 133–157.

DUDDY, I.R., GREEN, P.F., BRAY, R.J. & HEGARTY, K.A. 1994. Recognition of the thermal effects of fluid flow in sedimentary basins. *In*: PARNELL, J. (ed.) *Geofluids: Origin, Migration and Evolution of Fluids in Sedimentary Basins*. Geological Society, London, Special Publications, **78**, 325–345.

DUNCAN, W.I., GREEN, P.F. & DUDDY, I.R. 1998. Source rock burial history and seal effectiveness: key facets to understanding hydrocarbon exploration in the East and Central Irish Sea Basins. *AAPG Bulletin*, **82**, 1401–1415.

DURAND, B., ALPERN, B., PITTION, J.L. & PRADIER, B. 1986. Reflectance of vitrinite as a control of thermal history of sediments. *In*: BURRUS, J. (ed.) *Thermal Modelling in Sedimentary Basins*. Technip, Paris, 441–474.

ENGLAND, R.W. 1992. The role of Palaeocene magmatism in the tectonic evolution of the Sea of Hebrides Basin: implications for basin evolution on the NW Seaboard. *In*: PARNELL, J. (ed.) *Basins on the Atlantic Seaboard: Petroleum Geology, Sedimentology and Basin Evolution*. Geological Society, London, Special Publications, **62**, 163–174.

EVANS, A. & CLAYTON, G. 1998. The geological history of the Ballydeenlea Chalk Breccia, County Kerry, Ireland. *Marine and Petroleum Geology*, **15**, 299–307.

FITZGERALD, L. G. 1994. *Thermal maturation history of the Lower Carboniferous sequence in the vicinity of the Navan Zn/Pb orebody and the Athboy area, Co. Meath, Ireland.* PhD thesis, University of Dublin.

FITZGERALD, E., FEELY, M., JOHNSTON, J.D., CLAYTON, G., FITZGERALD, L.J. & SEVASTOPULO, G.D. 1994. The Variscan thermal history of west Clare, Ireland. *Geological Magazine*, **131**, 545–558.

GARDINER, P.R.R. & SHERIDAN, D.J.R. 1981. Tectonic framework of the Celtic Sea and adjacent areas with special reference to the location of the Variscan Front. *Journal of Structural Geology*, **3**, 317–331.

GLENNIE, K.W. 1997. Recent advances in understanding the southern North Sea Basin: a summary. *In*: ZIEGLER, K., TURNER, P. & DEINES, S.R. (eds) *Petroleum Geology of the Southern North Sea: Future Potential*. Geological Society, London, Special Publications, **123**, 17–29.

GOODARZI, F., SNOWDON, L., GENTZIS, T. & PEARSON, D. 1994. Petrological and chemical characteristics of liptinite-rich coals from Alberta, Canada. *Marine and Petroleum Geology*, **11**, 307–319.

GOODHUE, R. 1996. *A palynofacies, geochemical and maturation investigation of the Namurian rocks of County Clare*. PhD thesis, University of Dublin.

GOODHUE, R. & CLAYTON, G. 1999. Organic maturation levels, thermal history and hydrocarbon source rock potential of the Namurian rocks of the Clare Basin, Ireland. *Marine and Petroleum Geology*, **16**, 667–675.

GREEN, P.F., DUDDY, I.R., BRAY, R.J., DUNCAN, W.I. & CORCORAN, D.V. 2001. The influence of thermal history hydrocarbon prospectivity in the Central Irish Sea Basin. *In*: SHANNON, P.M., HAUGHTON, P.D.W. & CORCORAN, D.V. (eds) *The Petroleum Exploration of Ireland's Offshore Basins*. Geological Society, London, Special Publications, **188**, 171–188.

GREEN, P.F., DUDDY, I.R., HEGARTY, K.A. & BRAY, R.J. 1999. Early Tertiary heat flow along the UK Atlantic margin and adjacent areas. *In*: FLEET, A.J. & BOLDY, S.A.R. (eds) *Petroleum Geology of Northwest Europe: Proceedings of the 5th Conference*. Geological Society, London, 349–357.

HAO, Fang & CHEN, Jianyu 1992. The cause and mechanism of vitrinite reflectance anomalies. *Journal of Petroleum Geology*, **15**, 419–434.

HÉROUX, Y.A., CHAGNON, A. & BERTRAND, R. 1979. Compilation and correlation of major thermal maturation indicators. *AAPG Bulletin*, **63**, 2128–2144.

HILLIS, R.R. 1995. Regional Tertiary exhumation in and around the United Kingdom. *In*: BUCHANAN, J.G. & BUCHANAN, P.G. (eds) *Basin Inversion*. Geological Society, London, Special Publications, **88**, 167–190.

HITCHON, B. 1984. Geothermal gradients, hydrodynamics and hydrocarbon occurrences, Alberta, Canada. *AAPG Bulletin*, **68**, 713–743.

HOOD, A., GUTJAHR, C.C.M. & HEACOCK, R.L. 1975. Organic metamorphism and the generation of petroleum. *AAPG Bulletin*, **59**, 986–996.

JESSOP, A.M. & MAJOROWICZ, J.A. 1994. Fluid flow and heat transfer in sedimentary basins. *In*: PARNELL, J. (ed.) *Geofluids: Origin, Migration and Evolution of Fluids in Sedimentary Basins*. Geological Society, London, Special Publications, **78**, 43–54.

JOHNSTON, J.D. 1999. Regional fluid flow and the genesis of Irish Carboniferous base metal deposits. *Mineralium Deposita*, **34**, 571–598.

JOHNSTON, M.J., HOWELL, D.G. & BIRD, K.J. 1993. Thermal maturity patterns in Alaska: implications for tectonic evolution and hydrocarbon potential. *AAPG Bulletin*, **77**, 1874–1903.

LAW, B.E., NUCCIO, V.F. & BARKER, C.E. 1989. Kinky vitrinite reflectance well profiles: evidence of palaeopore pressure in low-permeability, gas-bearing sequences in Rocky Mountain foreland basins. *AAPG Bulletin*, **73**, 999–1010.

MACDONALD, H., ALLAN, P.M. & LOVELL, J.P.B. 1987. Geology of oil accumulation in Block 26/28, Porcupine Basin, offshore Ireland. *In*: BROOKS, J. & GLENNIE, K.W. (eds) *Petroleum Geology of North West Europe*. Graham & Trotman, London, 643–651.

MADDOX, S.J., BLOW, R. & HARDMAN, M. 1995. Hydrocarbon prospectivity of the Central Irish Sea Basin with reference to Block 42/12, offshore Ireland. *In*: CROKER, P.F. & SHANNON, P.M. (eds) *The Petroleum Geology of Ireland's Offshore Basins*. Geological Society, London, Special Publications, **93**, 59–77.

MAJOROWICZ, J.A., RAHMAN, M., JONES, F.W. & MCMILLEN, N.J. 1985. The palaeogeothermal and present thermal regimes of the Alberta Basin and their significance for petroleum occurrences. *Bulletin of Canadian Petroleum Geology*, **33**, 12–21.

MAYNARD, J.R., HOFMANN, W., DUNAY, R.E., BENTHAM, P.N., DEAN, K.P. & WATSON, I. 1997. The Carboniferous of western Europe: the development of a petroleum system. *Petroleum Geoscience*, **3**, 97–115.

MCCULLOCH, A.A. 1993. Apatite fission track results from Ireland and the Porcupine basin and their significance for the evolution of the North Atlantic. *Marine and Petroleum Geology*, **10**, 572–590.

MCTAVISH, R.A. 1998. The role of overpressure in the retardation of organic matter maturation. *Journal of Petroleum Geology*, **21**, 153–185.

MENPES, R.J. & HILLIS, R.R. 1995. Quantification of Tertiary exhumation from sonic velocity data, Celtic Sea/South-Western Approaches. *In*: BUCHANAN, J.G. & BUCHANAN, P.G. (eds) *Basin Inversion*. Geological Society, London, Special Publications, **88**, 191–207.

MIDDLETON, D.W.J., PARNELL, J., GREEN, P.F., XU, G. & MCSHERRY, M. 2001. Hot fluid flow events in Atlantic margin basins: an example from the Rathlin Basin. *In*: SHANNON, P.M., HAUGHTON, P.D.W. & CORCORAN, D.V. (eds) *The Petroleum Exploration of Ireland's Offshore Basins*. Geological Society, London, Special Publications, **188**, 91–106.

MORROW, D.W. & ISSLER, D.R. 1993. Calculation of vitrinite reflectance from thermal histories: a comparison of some methods. *AAPG Bulletin*, **77**, 610–624.

MURDOCH, L.M., MUSGROVE, F.W. & PERRY, J.S. 1995. Tertiary uplift and inversion history in the North Celtic Sea Basin and its influence on source rock maturity. *In*: CROKER, P.F. & SHANNON, P.M. (eds) *The Petroleum Geology of Ireland's Offshore Basins*. Geological Society, London, Special Publications, **93**, 297–319.

NAYLOR, D. 1992. The post-Variscan history of Ireland. *In*: PARNELL, J. (ed.) *Basins on the Atlantic Seaboard*. Geological Society, London, Special Publications, **62**, 255–275.

NEWMAN, P.J. 1999. The geology and hydrocarbon potential of the Peel and Solway Basins, East Irish Sea. *Journal of Petroleum Geology*, **22**, 305–324.

NYLAND, B., JENSEN, L.N., SKAGEN, J., SKARPNES, O. & VORREN, T. 1992. Tertiary uplift and erosion in the Barents Sea: magnitude, timing and consequences. *In*: LARSEN, R.M., BREKKE, H., LARSEN, B.T. & TALLERAAS, E. (eds) *Structural and Tectonic Modelling and its Application to Petroleum Geology*. Norwegian Petroleum Society, Special Publication, **1**, 153–162.

PETERS, K.E. & CASSA, M.R. 1994. Applied source rock geochemistry. *In*: MAGOON, L.B. & DOW, W.G. (eds) *The Petroleum System—from Source to Trap*. American Association of Petroleum Geologists, Memoirs, **60**, 93–120.

PHILCOX, M.E., BAILY, H., CLAYTON, G. & SEVASTOPULO, G.D. 1992. Evolution of the Carboniferous Lough Allen Basin, Northwest Ireland. *In*: PARNELL, J. (ed.) *Basins of the Atlantic Seaboard. Petroleum Geology, Sedimentology and Basin Evolution*. Geological Society, London, Special Publications, **62**, 203–215.

PRICE, L.C. 1983. Geologic time as a parameter in organic metamorphism and vitrinite reflectance as an absolute palaeogeothermometer. *Journal of Petroleum Geology*, **6**, 5–38.

PRICE, L.C. & BARKER, C.E. 1985. Suppression of vitrinite reflectance in amorphous rich kerogen—a major unrecognized problem. *Journal of Petroleum Geology*, **8**, 59–84.

ROBERT, P. 1989. The thermal setting of Carboniferous basins in relation to the Variscan Orogeny in Central and Western Europe. *International Journal of Coal Geology*, **13**, 171–206.

ROBERTS, D.G. 1989. Basin inversion in and around the British Isles. *In*: COOPER, M.A. & WILLIAMS, G.D. (eds) *Inversion Tectonics*. Geological Society, London, Special Publications, **44**, 131–150.

ROBESON, D. 1987. *Palynostratigraphy and thermal maturity of Carboniferous strata Offshore Western Ireland*. PhD thesis, University of Dublin.

ROBESON, D., BURNETT, R.D. & CLAYTON, G. 1988. The Upper Palaeozoic geology of the Porcupine, Erris and Donegal Basins, Offshore Ireland. *Irish Journal of Earth Sciences*, **9**, 153–175.

ROWLEY, E.J. & WHITE, N.J. 1998. Inverse modelling of extension and denudation in the East Irish Sea and surrounding areas. *Earth and Planetary Science Letters*, **161**, 57–71.

SCHEGG, R. & LEU, W. 1998. Analysis of erosion events and palaeogeothermal gradients in the North Alpine Foreland Basin of Switzerland. *In*: DÜPPENBECKER, S.J. & ILIFFE, J.E. (eds) *Basin Modelling: Practice and Progress*. Geological Society, London, Special Publications, **141**, 137–155.

SCOTCHMAN, I.C. & THOMAS, J.R.W. 1995. Maturity and hydrocarbon generation in the Slyne Trough, northwest Ireland. *In*: CROKER, P.F. & SHANNON, P.M. (eds) *The Petroleum Geology of Ireland's Offshore Basins*. Geological Society, London, Special Publications, **93**, 385–411.

SPENCER, A.M. & MACTIERNAN, B. 2001. Petroleum systems offshore western Ireland in an Atlantic margin context. *In*: SHANNON, P.M., HAUGHTON, P.D.W. & CORCORAN, D.V. (eds) *The Petroleum Exploration of Ireland's Offshore Basins*. Geological Society, London, Special Publications, **188**, 9–29.

STACH, E., MACHOWSKY, M.T., TEICHMÜLLER, M., TAYLOR, G.H., CHANDRA, D., TEICHMÜLLER, R. 1982. *Stach's Textbook of Coal Petrology*. Borntraeger, Stuttgart.

STROGEN, P., SOMERVILLE, I.D., PICKARD, N.A.H., JONES, G.L.L. & FLEMING, M. 1996. Controls on ramp, platform and basinal sedimentation in the Dinantian of the Dublin Basin and Shannon Trough, Ireland. *In*: STROGEN, P., SOMERVILLE, I.D. & JONES, G.L.L. (eds) *Recent Advances in Lower Carboniferous Geology*. Geological Society, London, Special Publications, **107**, 263–279.

SUGGATE, R.P. 1982. Low rank sequences and scales of organic metamorphism. *Journal of Petroleum Geology*, **4**, 377–392.

SUGGATE, R.P. 1998. Relations between depth of burial, vitrinite reflectance and geothermal gradient. *Journal of Petroleum Geology*, **21**, 5–32.

SUZUKI, N., MATSUBAYASHI, H. & WAPLES, D.W. 1993. A simpler kinetic model of vitrinite reflectance. *AAPG Bulletin*, **77**, 1502–1508.

SWEENEY, J.J. & BURNHAM, A.K. 1990. Evaluation of a simple model of vitrinite reflectance based on chemical kinetics. *AAPG Bulletin*, **74**, 1559–1570.

TABER, D.R., VICKERS, M.K. & WINN, R.D. JR 1995. The definition of the Albian 'A' Sand reservoir fairway and aspects of associated gas accumulations in the North Celtic Sea Basin. *In*: CROKER, P.F. & SHANNON, P.M. (eds) *The Petroleum Geology of Ireland's Offshore Basins*. Geological Society, London, Special Publications, **93**, 227–244.

Hot fluid flow events in Atlantic margin basins: an example from the Rathlin Basin

DAVID W. J. MIDDLETON[1], JOHN PARNELL[1], PAUL F. GREEN[2], GUOJIAN XU[3] & MARIE McSHERRY[4]

[1]*Department of Geology and Petroleum Geology, University of Aberdeen, Aberdeen AB24 3UE, UK (e-mail: d.middleton@abdn.ac.uk)*

[2]*Geotrack International Pty Ltd, 37 Melville Road, Brunswick West, Vic. 3055, Australia*

[3]*Geology Department, University of Papua New Guinea, Box 414, University P.O. NCD, Papua New Guinea*

[4]*Mobil North Sea Limited, Grampian House, Union Row, Aberdeen AB10 1SA, UK*

Abstract: An understanding of the thermal and tectonic evolution of sedimentary basins is essential to the effective modelling of source rock maturation and hydrocarbon charge and entrapment histories of potential hydrocarbon systems. A growing body of data suggests that a number of basins on the Atlantic margin to the west of Britain and Ireland have suffered short-lived episodes of migration of anomalously hot fluids through reservoir intervals. These events leave higher temperature signatures in affected basins than predicted from burial under conditions of vertical conductive heat transfer, and should be considered during hydrocarbon appraisal of a prospective basin. The Rathlin Basin displays a thermal history influenced by one or more such hot fluid flow events, with fluid palaeotemperatures in excess of 170 °C recorded in the Permo-Triassic and Carboniferous section, and is typical of other Atlantic margin basins affected in this way.

The NE Atlantic margin has become one of Europe's most active regions for hydrocarbon exploration in the last decade. Both proven and prospective plays occur in offshore basins stretching from Portugal to northern Norway. Hydrocarbon accumulations have been discovered in reservoirs ranging in age from Devonian (e.g. Clair Field) to Tertiary time (e.g. Foinaven and Schiehallion Fields), and in a range of structural and stratigraphic traps. In many of these basins the hydrocarbon expulsion and migration history is complex, with oil and gas accumulations often having a multi-stage charge history (Doré *et al.* 1997).

Understanding the thermal and tectonic evolution of these Atlantic margin basins is a significant challenge, particularly in the light of continuing research, which suggests that many of these basins have had a complex palaeothermal history (Green *et al.* 1999; Parnell *et al.* 1999). Temperature profiles as defined by thermal indicators often fluctuate markedly through the geological section, suggesting input of heat along certain horizons (Duddy *et al.* 1994). This phenomenon may be attributed to the effect of advecting hot fluids, with the fluctuating temperature profiles representing the flow of heat into reservoir intervals and along fracture pathways, thus indicating a dependence on the porosity and permeability characteristics of the host rock.

This work presents data from the Rathlin Basin (Fig. 1) as an example of an Atlantic margin basin affected by one or more hot fluid flow events. The probable temperature, source, timing and duration of the hot fluids are discussed, as quantified by the application of fluid inclusion studies in combination with apatite fission-track analysis and vitrinite reflectance.

Geological setting of the Rathlin Basin

The Rathlin Basin (Fig. 1) is one of a system of elongate NE–SW-trending basins along the NE Atlantic margin, which extends from offshore Iberia to the Arctic (Naylor & Shannon 1982). Many of these basins began to form during Late Palaeozoic time, accommodation space being provided by the extensional reactivation of Caledonian thrusts and strike-slip faults that had formed during the closure of the Iapetus

From: SHANNON, P.M., HAUGHTON, P.D.W. & CORCORAN, D.V. (eds) 2001. *The Petroleum Exploration of Ireland's Offshore Basins*. Geological Society, London, Special Publications, **188**, 91–105.
0305-8719/01/$15.00

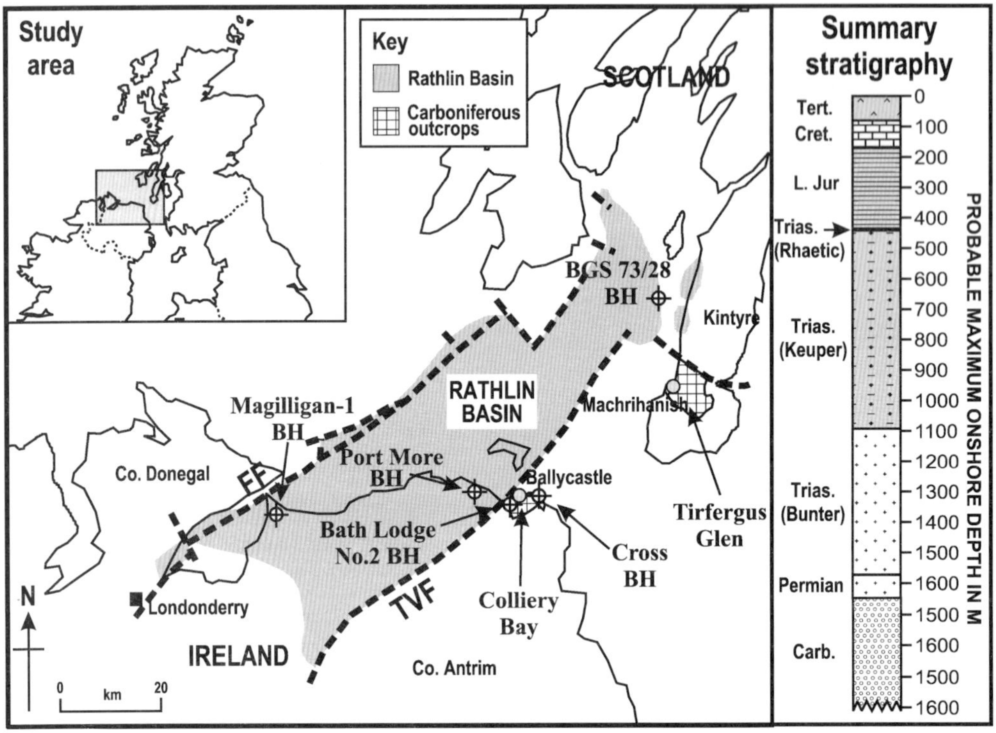

Fig. 1. Location of the Rathlin Basin based on the Permo-Triassic and Carboniferous outcrop and subcrop. Additional Carboniferous outcrops at Ballycastle and Machrihanish also shown (modified from Anderson *et al.* 1995). Summary stratigraphic column is based on a composite of Port More and Magilligan-1 borehole data, and represents probable maximum onshore thicknesses. TVF, Tow Valley Fault; FF, Foyle Fault.

Ocean (Anderson *et al.* 1995). Many of these Atlantic margin basins subsequently experienced repeated episodes of subsidence and basin inversion. This tectonism is manifested in the presence of multiple unconformities in the Late Palaeozoic to Cenozoic stratigraphy of these basins.

The preserved fill of the Rathlin Basin is dominated by Permo-Triassic continental deposits resting upon mixed marine and non-marine deposits of Carboniferous age. Interpretation of local seismic and gravity data suggests that the Permo-Triassic and Carboniferous sections reach thicknesses of about 2 km and 1 km, respectively (Evans *et al.* 1980). Both sections are dominated by sandstones, with subordinate mudrocks. Several coal seams of Dinantian and Namurian age occur in the Carboniferous section, and were formerly exploited in the Ballycastle and Machrihanish coalfields (McCallien & Anderson 1930; Wilson & Robbie 1966). A Westphalian section is present in Kintyre but missing in Co. Antrim, reflecting a major unconformity between Permo-Triassic and Carboniferous sequences. Outcrop evidence indicates that both the Palaeozoic and Mesozoic rocks are cross-cut by Paleocene dykes (Wilson & Robbie 1966).

A generalized stratigraphic evolution and burial history for the Rathlin Basin has been summarized by Parnell (1992). This interpretation is based upon a combination of stratigraphic data from the Carboniferous succession in the Ballycastle Coalfield, south of the Tow Valley Fault (Wilson & Robbie 1966), and the Mesozoic–Cenozoic succession north of the Tow Valley Fault as exemplified in the Port More borehole (Wilson & Manning 1978). The Carboniferous facies outcropping in the Ballycastle Coalfield suggest continuation at depth north of the Tow Valley Fault, such that the combined stratigraphy is a reasonable representation of the preserved sequences in the Rathlin Basin (Fig. 1). On the west side of the basin, the Magilligan-1 borehole encountered Permo-Triassic rocks resting upon Namurian sediments. However, regional palaeogeographical reconstructions indicate that sedimentation in the area continued into Westphalian time (Eagar 1974). This suggests that substantial Variscan uplift and denudation occurred in the Rathlin Basin.

Sandstone samples were collected from boreholes and exposures in the Carboniferous and Permo-Triassic sections (Fig. 1). The sandstone samples come from coal exploration boreholes at Bath Lodge and Cross, geothermal boreholes at Magilligan and Port More, and a shallow offshore borehole drilled by the British Geological Survey (BGS) west of Kintyre (BGS 73/28). Additional samples were collected from outcrop at Colliery Bay near Ballycastle and Tirfergus Glen, Kintyre. The samples were generally quartzose sandstones, with carbonate or anhydrite cements in the Permo-Triassic sandstones. The sandstone from the Cross borehole exhibits patchy alteration marked by iron oxide staining.

Methods

Fluid inclusion study

Fluid inclusions have a great diversity of applications to exploration geology, and are invaluable in the reconstruction of fluid flow histories in reservoir rocks (De Vivo & Frezzotti 1994; Goldstein & Reynolds 1994). Their value lies in the information that they can yield about temperature, salinity, pressure and composition of fluids that have migrated through basins in the geological past. Fluid inclusions represent encapsulated vacuoles of fluid trapped within diagenetic mineral cement phases and vein and fracture fills. It is assumed that they have remained as a closed, static system since their formation, which means they provide micron-scale samples of reservoir fluids at certain points through geological time.

The first stage in a fluid inclusion study is to identify inclusion populations based on occurrence, host minerals, homogenization temperature and final ice melting temperature of inclusions within the suite. A distinction can be made in fluid inclusion populations between primary and secondary occurrences. Primary inclusions are those trapped at grain surfaces, at grain–overgrowth boundaries, and at growth zone boundaries within overgrowths. Such inclusions became trapped at the time of cement precipitation and represent samples of the fluid that produced the host cement phase in the reservoir. Secondary inclusions form in annealed microfractures across grains, often cross-cutting overgrowth zones that themselves bear primary inclusions. These micron-width cracks form as a result of event-driven fluid migration and are subsequently annealed by dissolution–reprecipitation of the host phase. During this process, fluid inclusions are trapped in the annealing microfracture and appear as trails across grains, representing samples of reservoir fluids close to the time of fluid migration.

Wafers for fluid inclusion analysis were produced from sandstone samples from the outcrops and boreholes noted above (Fig. 1). The samples were examined using a Linkam THM600 heating–freezing stage attached to a Nikon Optiphot2-POL microscope. Ultraviolet fluorescence microscopy was carried out using a Nikon Eclipse E600 instrument with a Y-FL Epi-fluorescence attachment and a UV-2A filter.

Thermal history reconstruction using AFTA and VR

Thermal history reconstruction (THR) is a technique that relies on the application of apatite fission-track analysis (AFTA) and vitrinite reflectance (VR) in the identification of major episodes of heating and cooling in the evolution of a sedimentary basin. As such, it is a very powerful tool in basin analysis, and helps to constrain the timing of heating and cooling events and allows quantification of palaeotemperatures. The application of THR in hydrocarbon prospectivity appraisal provides controls on source rock maturation and timing of likely charge from source kitchens in the sedimentary succession. The techniques involved in THR have been described by Bray *et al.* (1992), Duddy *et al.* (1994) and Green *et al.* (1995).

AFTA is a technique that relies on the analysis of radiation damage trails (fission tracks) in detrital apatite grains (Green *et al.* 1989, 2001). Each track is created by spontaneous fission of an atom of uranium, with new tracks constantly being created throughout geological time. These tracks form at an initial length with little variability, and immediately begin to anneal at a rate dependent on the ambient temperature. If the temperature falls, any individual track is frozen at the shortened length obtained at maximum palaeotemperature. As new tracks are continuously being formed as a result of continuing fission from uranium, a sample that cools after having reached a high temperature in the past will have a population of both short and long tracks. Long tracks are those that formed after cooling, and shortened tracks represent those that formed before the onset of cooling, and will have a length proportional to the maximum palaeotemperature achieved. The ratio of short to long tracks then gives some information about the timing of the cooling event in the evolution of the basin. At a temperature of *c.* 110–120 °C, all tracks are completely annealed to zero length, and any tracks in the apatite grain must have formed after cooling below this critical temperature. In a case such as this, the number of tracks records the time of cooling of the interval to below 110–120 °C (Green *et al.* 1989). The chlorine content of the apatite exerts a control on the rate of annealing, so that if track distributions and lengths are analysed as a function of chlorine content, the absolute timing of palaeothermal events can be determined.

Analysis of a number of samples over a depth range yields information about the palaeotemperature profile with depth. Heating as a result of continuous burial, in conditions that give rise to a constant geothermal gradient, should produce a linear profile, the gradient of which reflects the palaeogeothermal gradient at maximum burial in the basin history. Conversely, a fluctuating palaeotemperature profile indicates localized heating by igneous intrusions (Summer & Verosub 1989) and/or input of anomalously hot fluids along certain horizons (Zaigos & Blackwell 1986; Duddy *et al.* 1994).

Vitrinite reflectance is a technique for determining the thermal maturity of rocks in a sedimentary basin. Vitrinite and other macerals become more reflective to light in a kinetically dependent manner (i.e. dependent on time as well as temperature) so that prolonged heating of vitrinite in its host geological interval increases the vitrinite reflectance (Burnham & Sweeney 1989). The reflectance is quantifiable as a parameter $\%R_o$, which can be modelled using the Easy$\%R_o$ algorithm (Sweeney & Burnham 1990). This algorithm takes into account the kinetic nature of thermal maturation of vitrinite by using an Arrhenius first-order parallel-reaction approach with a distribution of activation energies. The Easy$\%R_o$ program thus enables us to reconstruct maximum palaeotemperatures with respect to a vast range of heating rates, from simulation of a slowly subsiding basin (1 °C per 10 Ma) to laboratory experiments lasting a few days (1 °C per week).

Vitrinite reflectance measurements have been made in Carboniferous rocks at Ballycastle (Clayton *et al.* 1989), Colliery Bay, Tirfergus Glen and at Bath Lodge No. 2 and Magilligan-1 boreholes (authors' unpublished data). These measurements were made using samples in close proximity to the sandstone samples used for fluid inclusion studies. The samples from the Permo-Triassic red beds did not yield any vitrinite.

Table 1. *Summary of data obtained from fluid inclusions in the Rathlin Basin*

Sample	Depth (m)	Stratigraphic age	Occurrence	Host minerals	Nature	Range T_h (°C)	*n*	Mean T_h (°C)	Population type
Port More BH	1191	Permo-Triassic	P	Quartz	Aqueous	146.8–171.2	12	164	1
Port More BH	1398	Permo-Triassic	P	Quartz/Anhydrite	Aqueous	145.2–182.5	15	149(Anh) and 168 (Qtz)	3
Port More BH	1456	Permo-Triassic	P and S	Quartz	Aqueous	132.7–162.9	20	149	1
BGS 73/28 BH	29.27	Permo-Triassic	S	Quartz	Aqueous	157.8–184.4	20	173	1
Magilligan-1 BH	1312	Carboniferous	P and S	Quartz	Aqueous	152.6–181.6	20	169	1
Bath Lodge No. 2 BH	57	Carboniferous	P and S	Quartz	Aqueous	149.2–176.2	20	160	1
Cross BH	426	Carboniferous	P and S	Quartz	Aqueous	120.2–185.3	20	177 (early) and 133 (late)	2
Colliery Bay	Outcrop	Carboniferous	S	Quartz	Aqueous	145.4–169.1	6	154	1
Tirfergus	Outcrop	Carboniferous	S	Quartz	Aqueous	124.1–162.8	20	144	1

Occurrence data 'P' and 'S' refer to the presence of primary and secondary inclusions respectively. T_h, homogenization temperature (°C). Description of Population Types 1–3 is given in text. Cross borehole late T_h is associated with secondary alteration.

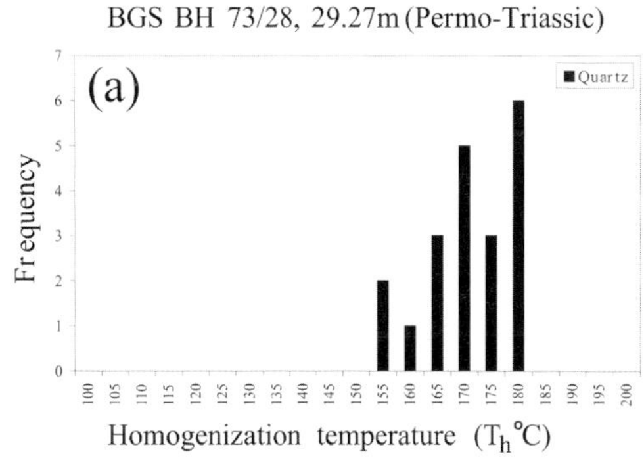

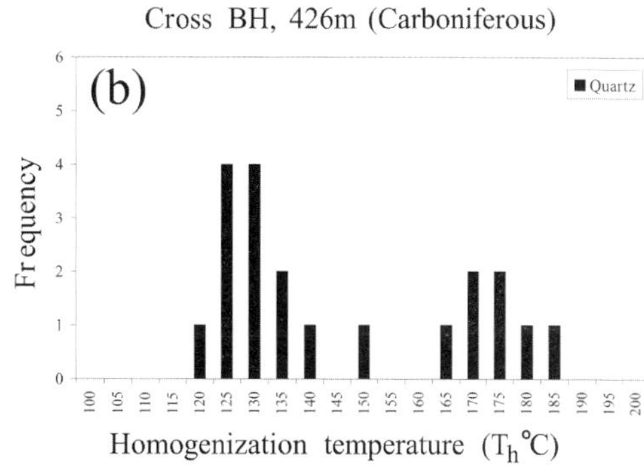

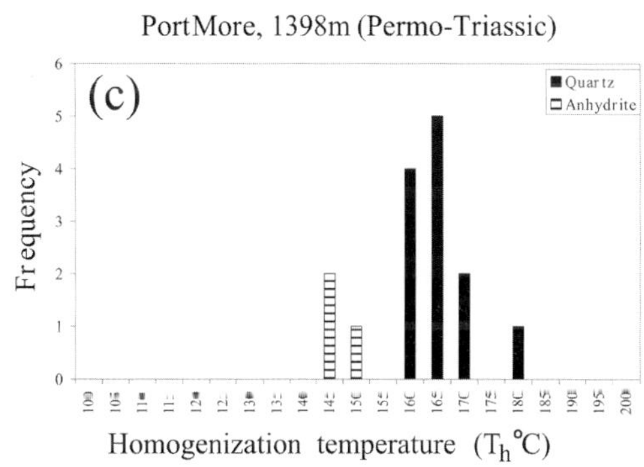

Fig. 2. Histograms of homogenization temperature (T_h) distribution in fluid inclusion population types, for selected samples, from the Rathlin Basin. (**a**) Type 1, unimodal T_h distribution, single host phase. (**b**) Type 2, bimodal T_h distribution, single host phase. (**c**) Type 3, bimodal T_h distribution, dual host phases (quartz and anhydrite). (See text for description of population types.)

Results

Fluid inclusions

Samples recovered from five boreholes and two outcrop localities were examined for fluid inclusions. Homogenization temperatures (T_h °C), occurrence types (primary or secondary) and host minerals were recorded for each sample and a summary of these data is presented in Table 1.

Three classes of inclusion populations were identified in the Rathlin Basin samples. Each sample has been assigned to Population Type 1, 2 or 3 based on the definition of these population types in the Rathlin Basin, as described below.

Population Type 1. Quartz-hosted population displaying primary and secondary two-phase aqueous inclusions, with unimodal distribution of T_h (Fig. 2a). Vapour/liquid ratios of 0.05–0.25 are recorded. Fluid inclusions are rounded to subrounded and of high sphericity.

Population Type 2. Quartz-hosted population displaying primary and secondary two-phase aqueous inclusions, with bimodal distribution of T_h (Fig. 2b). The high-T_h modal group occurs as primary and secondary inclusions, with vapour/liquid ratios of 0.05–0.25. The low-T_h modal group occurs as secondary inclusions, with vapour/liquid ratios of 0.1–0.2. Fluid inclusions are rounded to subrounded and of high sphericity.

Population Type 3. Quartz- and anhydrite-hosted population displaying primary two-phase aqueous inclusions, with bimodal distribution of T_h correlating with host mineral type (Fig. 2c). Vapour/liquid ratios of 0.05–0.2 occur in the quartz phase, whereas ratios of 0.2–0.5 occur in the anhydrite phase. Fluid inclusions are rounded to subrounded and of high sphericity.

Vitrinite reflectance

Vitrinite reflectance data for samples from the Rathlin Basin are recorded in Table 2. Mean random vitrinite reflectance (R_m%) varies from 0.52 to 0.92 in the study area.

Using the modelling program Easy%R_o (Sweeney & Burnham 1990), a temperature indicative of a given VR value can be estimated with respect to the assumed heating rate in the Rathlin Basin during subsidence. This requires effective modelling of the basin subsidence

Table 2. *Summary of palaeotemperatures derived from fluid inclusion (T_h) and vitrinite reflectance data ($T(R_m)$)*

Sample	Depth (m)	Stratigraphic age	Mean T_h (°C)	R_m%	n	$T(R_m)$ (°C)
Magilligan-1 BH	1312	Carboniferous	169	0.92	66	148
Bath Lodge No. 2 BH	57	Carboniferous	160	0.55	34	102
Colliery Bay	Outcrop	Carboniferous	154	0.61	100	108
Tirfergus	Outcrop	Carboniferous	144	0.52	74	96

history reconstructed from local borehole and outcrop evidence (Fig. 3) and consideration of regional stratigraphy (see section on geological setting, above).

Reconstruction of the burial history was used to assess the rate of heating of vitrinite-bearing rocks in the Rathlin Basin. Episodes of significant basin subsidence are interpreted in Carboniferous, Permo-Triassic and Tertiary time. Initial subsidence during Carboniferous time created the accommodation space for an infill of mainly marine and non-marine sandstones, claystones and coals. Subsidence continued into Permo-Triassic time and the basin was predominantly infilled with continental red bed clastic deposits. The subsidence history of the Jurassic–Cretaceous periods is somewhat more speculative. However, regional outcrop evidence indicates that marine conditions were established in the basin during Late Triassic time and continued into Early Jurassic time with the deposition of a marine mudstone sequence. Relatively minor subsidence and sedimentation is inferred to have occurred during Mid- to Late Jurassic time, with all of these sediments removed by erosion during Early Cretaceous time (Fig. 3). Subsidence of the area resumed during Cenomanian time with the recurrence of marine sedimentation, which culminated in the widespread deposition of limestone and chalk sequences during mid-Santonian to Maastrichtian time (Wilson & Manning 1978). Basin development in the Early Tertiary period was, however, less quiescent, and was characterized by the extensive extrusion of basaltic lavas and the emplacement of hypabyssal intrusions in parts of the Rathlin Basin. Magmatism was widespread along the NE Atlantic margin at this time and has been linked to rifting associated with the opening of the NE Atlantic in late Paleocene time (Stoker *et al.* 1993).

The modelled heating rates for the Rathlin Basin through geological time are displayed in Fig. 4. Increased heating rates are interpreted to be the result of rifting during Permo-Triassic time and rapid magma emplacement during the Early Tertiary period. However, Green *et al.* (1999) argued against a significantly elevated geothermal gradient for the adjacent Hebridean basins during Early Tertiary time, and THR indicates that the prevailing geothermal gradients in the Rathlin Basin during the Tertiary period were close to the present-day values of 30 °C km^{-1} and 37 °C km^{-1} in the Magilligan-1 and Port More boreholes, respectively (Green 1996).

This quantification of heating rate allows modelling of palaeotemperature from R_m% measurements using the Easy%R_o program. The peak palaeotemperatures indicated by the VR data, under the assumed heating rates, are shown in Table 2. It is evident that palaeotemperatures derived from fluid inclusion data are consistently higher than those recorded by VR data by up to 58 °C.

Apatite fission-track analysis

AFTA data from the Rathlin Basin (Table 3) yield information on regional cooling episodes, and help to constrain the thermal history of the basin when interpreted in conjunction with VR data. VR modelling from the Magilligan-1 borehole (Fig. 5) yields palaeotemperatures of 128 °C at 1189 m and 143 °C at 1344 m, which are consistent with AFTA analysis of core samples, which show cooling from >110 °C at some time before 250 Ma. AFTA also indicates cooling from 100 to 110 °C at some time between 230 and 135 Ma. Similarly, AFTA data from the Port More borehole show cooling from >105 °C at some time between 200 and 100 Ma. Two episodes of cooling are identified in the BGS 73/28 borehole, the first from a maximum temperature of 75–90 °C beginning some time between 180 and 35 Ma, and subsequently from a peak temperature of 30–70 °C occurring some time between 45 and 0 Ma.

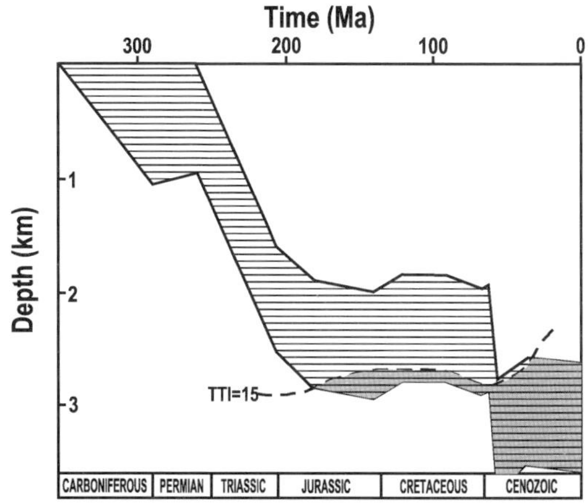

Fig. 3. Total subsidence plot for the vitrinite-bearing Carboniferous interval (ruled) in the Rathlin Basin, assuming high rates of subsidence in Mesozoic (Triassic sedimentation) and early Tertiary time (extrusive vulcanism). TTI (time–temperature index) of 15 marks onset of oil generation according to Lopatin-type models (after Parnell 1992). (See text for discussion on reconstruction of burial history.)

Discussion

Fluid inclusion populations

Three distinct fluid inclusion population types are described in this study (Fig. 2) and are here interpreted to record distinct fluid flow histories. Population Type 1 (Fig. 2a) is the most prevalent in the study area, and probably represents a single hot aqueous fluid pulse through the interval (see below). Thermal waters would have been trapped as high-T_h aqueous fluid inclusions during rapid cement precipitation and microfracture annealing, as silica dropped out of solution when mineral solubility decreased with falling temperature. This would have occurred when hot, silica-saturated water came into contact with relatively cold country rock.

Population Type 2 (Fig. 2b), recorded at 426 m in the Cross borehole, consists of two aqueous inclusion suites displaying different mean T_h values. This indicates at least two aqueous fluid migration events, possibly from the same source, but at distinctly different temperatures. This is interpreted as representing basinal fluids being heated at depth and being vented upwards through the basin periodically. The bimodal distribution of fluid palaeotemperatures could indicate a dissipation of energy from the heat source over time. This is further supported by paragenetic studies of the sample, as the cooler fluids (mean T_h 133 °C) are recorded in fluid inclusions that are associated with patchy alteration as a result of iron oxide staining. This alteration, which is manifest as a reddening at outcrop, is documented throughout the Carboniferous sequences in Northern Ireland (Wilson & Robbie 1966; Wang 1992). The alteration is attributed to the liberation of iron during dissolution of a ferroan dolomite cement combined with the incursion of oxidizing meteoric fluids (Wang 1992). This is significant in identifying the relative ages of the two fluid inclusion suites, and also imposes a timing constraint on fluid migration, as discussed below.

Population Type 3 (Fig. 2c), recorded at 1398 m in the Port More borehole, consists of aqueous inclusions, with distinct T_h characteristics, hosted in two cement phases. This indicates that at least two aqueous fluid pulses occurred with different fluid chemistries (i.e. saturated with respect to different mineral phases) and temperatures, possibly derived from different sources at depth.

Application of VR in combination with fluid inclusion studies

Palaeotemperatures in the Rathlin Basin indicated by fluid inclusion studies are consistently higher than those derived from VR (Table 2). This difference is significant, and is up to 58 °C in the recorded samples. The differences in fluid inclusion palaeotemperatures (T_h) and palaeotemperatures derived from VR ($T(R_m)$) are minimum values, as $T(R_m)$ is modelled at its highest possible value, and no pressure corrections were applied to the T_h data (which may add 20–30 °C to the recorded values). Figure 6 shows

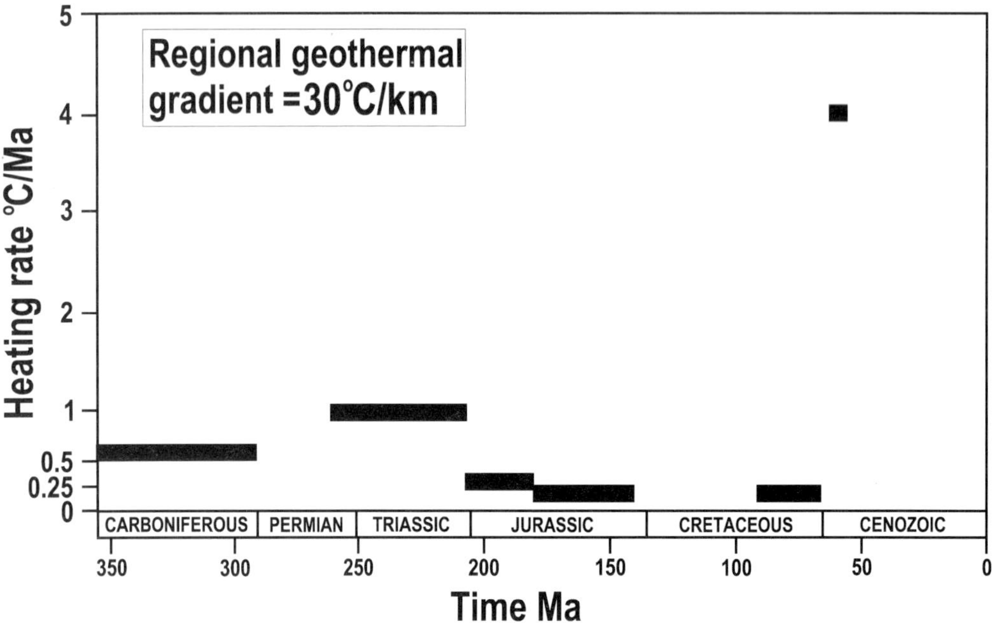

Fig. 4. Modelled heating rates throughout the subsidence history of the Rathlin Basin. A geothermal gradient of 30 °C km^{-1} has been assumed throughout the burial history, based on THR data derived from the Port More and Magilligan-1 boreholes (Green 1996).

the spatial distribution of sample points in the Rathlin Basin, highlighting the differences in $T(R_m)$ and T_h.

This phenomenon is attributed to the fact that the increasing reflectance of vitrinite on heating is dependent on kinetically driven chemical reactions, i.e. for R_m to increase, heating must continue for a prolonged period of time (Burnham & Sweeney 1989). In theory, a short duration heating event may have little or no effect on the pre-existing reflectance. Conversely, fluid inclusions represent a geologically instantaneous trapping of fluids passing through the section, and are not dependent on the duration of the fluid flow event (Goldstein & Reynolds 1994).

The presence of high fluid inclusion homogenization temperatures in the Rathlin Basin is evidence for one or more phases of hot fluid flow in the Rathlin Basin, which must have transferred an amount of heat to the sedimentary rocks, including the vitrinite-bearing Carboniferous sandstones. However, the fact that the temperatures recorded by VR are significantly lower than those recorded by fluid inclusions indicates that heating as a result of hydrothermal fluid migration must have been on a geologically brief time scale.

Table 3. *Summary of apatite fission-track data derived from the Port More and Magilligan-1 boreholes*

Sample source	Depth (m)	Stratigraphic age	Present temperature (°C)	Mean track length (μm)	Apatite fission-track age (Ma)
Magilligan-1	319	Permo-Triassic	20	12.956±0.19	297.76±23.0
Magilligan-1	601	Permo-Triassic	28	12.526±0.25	238.46±18.4
Magilligan-1	1250	Carboniferous	47	10.256±0.62	201.96±19.7
Port More	711	Permo-Triassic	41	11.146±0.79	143.76±62.4
Port More	1097	Permo-Triassic	57	11.956±0.19	193.96±11.7
Port More	1271	Permo-Triassic	65	11.716±0.24	124.06±16.7
Port More	1791	Permo-Triassic	87	11.066±0.48	101.66±17.7

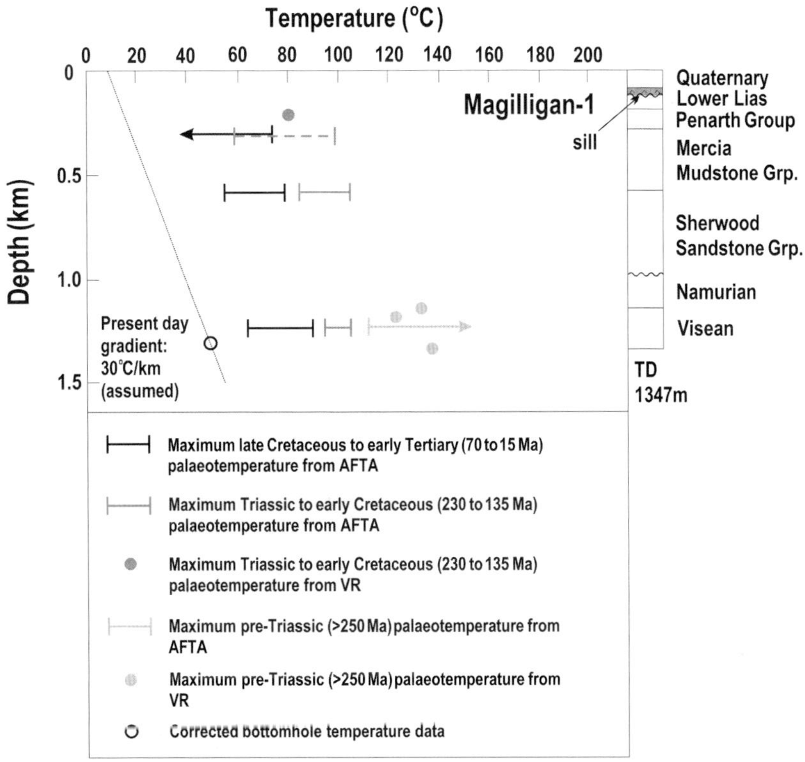

Fig. 5. Palaeotemperatures for Magilligan-1 borehole, derived from combined AFTA and VR data. This analysis indicates that Carboniferous sediments achieved peak palaeotemperature exposure before Triassic time (>250 Ma). The most recent episode of cooling occurred in late Cretaceous to Tertiary time, between 70 and 15 Ma.

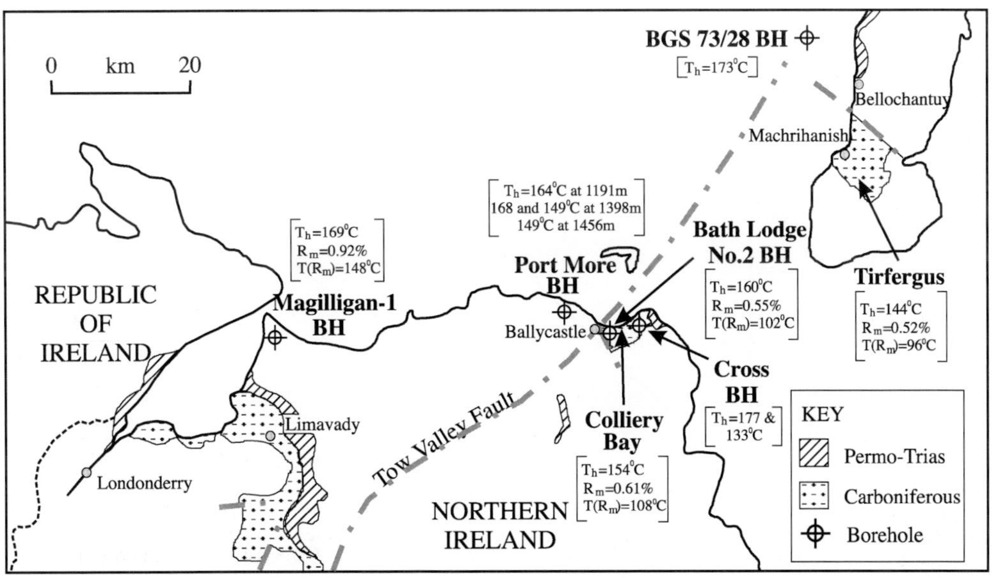

Fig. 6. Spatial distribution of sample points in the Rathlin Basin highlighting the differences in palaeotemperatures derived from fluid inclusions (T_h) and modelled from vitrinite reflectance data ($T(R_m)$).

Figure 7 is a crossplot of values of T_h against $T(R_m)$ in the study area. Where thermal maturity derived from both T_h and R_m reflects the maximum temperature to which the rock has been exposed, all points should lie on or close to the line $m = 1$. However, the effect of short-lived pulses of hot fluids is to create data points that lie above this line. The implication of this work is that an influence from brief hot fluid flow events may be suspected in any dataset that has numerous data points above the line $m = 1$.

Duration of fluid-driven heating

The duration of hot fluid flow events can be constrained by modelling how VR increases when high ambient temperatures are applied over various time scales. Figure 8 shows how reflectance increases from an ambient level of $0.5\%R_{max}$ for heating to temperatures of 200 and 250 °C during the burial history recorded from a confidential well on the Atlantic margin (authors' unpublished data). The model accounts for the kinetic nature of vitrinite maturation by using the Arrhenius-based kinetic algorithm of Sweeney & Burnham (1990). On the basis of this modelling, it can be seen that heating for up to $10^0–10^2$ a duration has a negligible effect on R_{max}. A similar effect is observed in modelled AFTA data, which is also kinetically dependent (Duddy *et al.* 1994.) The increase in VR over longer time periods suggests that it may be possible to use fluid inclusion data in combination with VR data to assess the longevity of a given fluid-driven heating event.

The apparent brevity of hydrothermal fluid flow in the Rathlin Basin is supported by other studies. For example, Barker (1991) has demonstrated that VR responds to fluid-driven heating over brief time scales in open, fluid-rich, hydrothermal systems. This supports the hypothesis that for a large discrepancy to exist between T_h and $T(R_m)$ (Fig. 7), fluid-driven heating must have been on a time scale of <1 ka.

One of the implications of a short duration heating event, via the emplacement of hot fluids, is that fluid migration must have been very rapid. This, in turn, suggests that fluid flow was probably accommodated by fracture pathways rather than via intergranular porosity. Fluid migration via the latter medium takes longer and allows rapid dissipation of heat into the host rock. Evidence for fracture-controlled fluid movement is displayed in the Colliery Bay sample, wherein all the T_h data were obtained from fluid inclusions in annealed microfractures close to a granulation seam (Fig. 9a). The granulation seam has acted as a fluid pathway, allowing the relatively rapid transportation of fluids along the plane of the seam. These hot

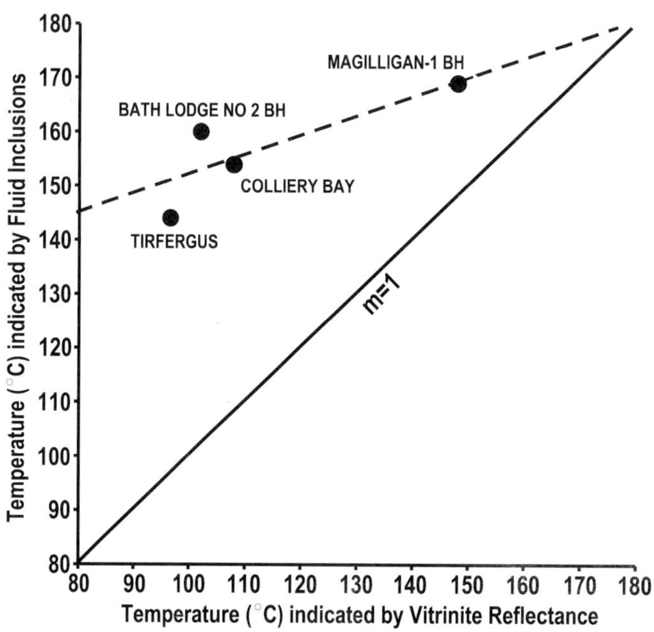

Fig. 7. Crossplot to show the discrepancy between values of T_h and $T(R_m)$ derived from the same sample. The palaeotemperatures derived from T_h data are consistently higher than those indicated by the measured R_m values.

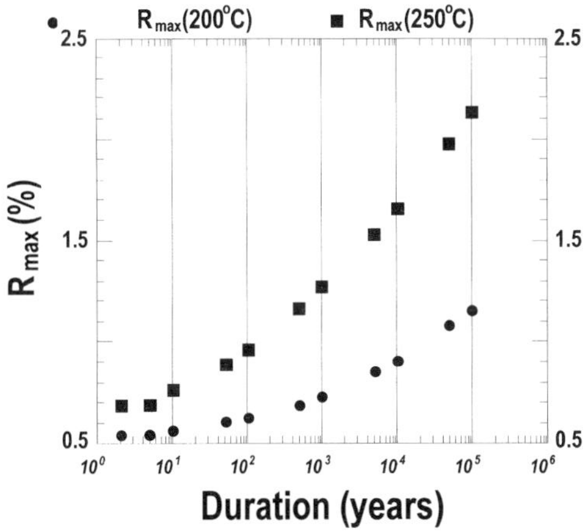

Fig. 8. Modelling of reflectance increase from an ambient value of 0.5% R_{max} with heating to temperatures of 200 and 250 °C for a given (but unstated) burial history. The diagram is constructed using the Arrhenius-based kinetic parallel-reaction algorithm of Sweeney & Burnham (1990).

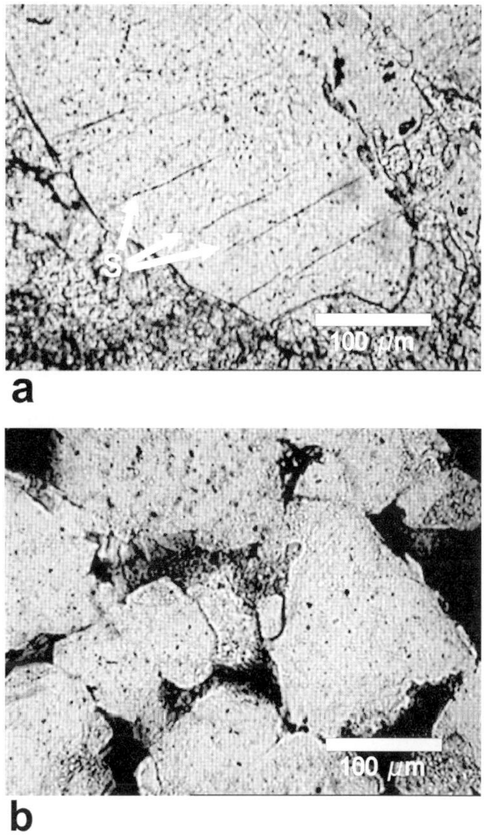

Fig. 9. Photomicrographs to demonstrate the role of granulation seams as fracture pathways for fluid flow in the Carboniferous sediments at Colliery Bay: (**a**) in proximity to a granulation seam, showing secondary fluid inclusions in annealed microfractures (S); (**b**) at distance from the granulation seam, showing a quartz-dominated matrix with good intergranular porosity and a scarcity of fluid inclusions.

fluids have become encapsulated as secondary monophase and two-phase fluid inclusions in parallel trails through numerous quartz grains within 10 mm of the fracture (Fig. 9a), but are absent in the porous sandstone at a distance from the granulation seam (Fig. 9b). This is evidence that hot advecting fluids exploited fracture pathways in preference to intergranular pathways in the Rathlin Basin.

Potential heat sources

Conduction is the primary mode of heat transport in the crust (Deming 1994). Rocks in sedimentary basins are exposed to rising temperature through subsidence and burial. Formation temperatures are also raised when heat flux is elevated as a result of crustal extension and attenuation or igneous activity. However, conduction of basal heat flow is inadequate to explain the T_h values of >150 °C observed in the Carboniferous and Permo-Triassic section of the Rathlin Basin. This is because the maximum burial depth of the Carboniferous units was probably close to 3.5 km (Fig. 3), which, under conditions of a constant geothermal gradient through time of *c*. 30 °C km^{-1} (Green 1996), would yield a maximum palaeotemperature for the Carboniferous section of 110 °C owing solely to conductive heat transfer in the sedimentary pile.

Another obvious source of heat is that derived from igneous intrusions. However, there is little evidence, geophysical or otherwise, for the presence of a regional (batholithic) intrusion underlying the Rathlin Basin, although hypabyssal intrusions are present. Rocks within the thermal aureoles of such dykes and sills tend to reflect thermal effects in standard maturation parameters (e.g. VR, spore colour index, conodont alteration index, etc.). However, the thermal aureoles of such intrusions tend to be very small, usually no more than twice the width of the intrusion itself (Briggs 1935; Thrasher 1992). Furthermore, many of the high-temperature T_h data obtained from the Rathlin Basin occur at a distance from any known intrusions.

The Rathlin Basin is presented here as an example of an Atlantic margin basin affected by hot fluids. Advective heating events are recognized in other Atlantic margin basins such as the Rockall and West of Shetland basins (authors' unpublished data). This indicates that the heating source may be regional, even extrabasinal, or that the heat is generated by some common mechanism in these Atlantic margin basins. One possibility is that Tertiary plume-related activity, associated with a major hotspot of 2000 km width under East Greenland (Ebdon *et al.* 1995), may have had a major effect on thermal and hydrodynamic regimes along the NE Atlantic margin. Such a regional scale heat source could explain the large spatial distribution of basins affected by hot fluids, but does not account for the possibility that advective heating events may have occurred in the pre-Tertiary period, as discussed below.

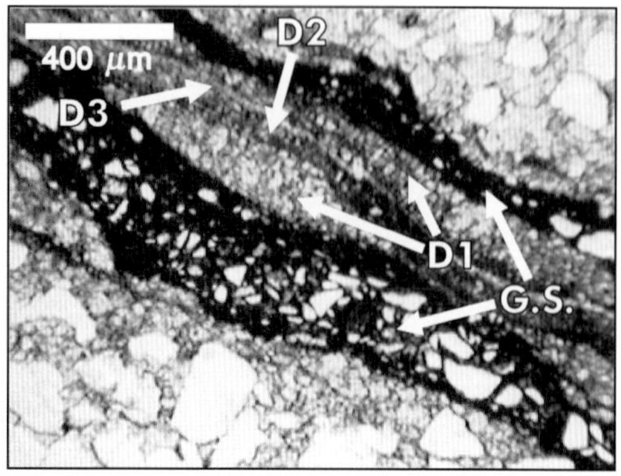

Fig. 10. Photomicrograph of a Carboniferous sample from Colliery Bay, showing successive generations of fracture fill, indicative of multi-phase fluid flow. Successive phases of fluid flow are recorded as sequential dolomite vein fills with distinct textural and compositional differences (D1–D3, D1 being the earliest fill). All the high-T_h fluid inclusion data obtained from Colliery Bay occurs in close proximity to these fractures, without actually forming part of the fracture fill. G.S., granulation seam boundary.

Timing of hot fluids

There are several plausible time periods when anomalously hot fluids may have been generated on the Atlantic margin. Very high palaeogeothermal gradients obtained from VR profiles in Carboniferous rocks of onshore and offshore Ireland, which show no expression in Jurassic reflectance profiles, indicate a pre-Jurassic heating episode (Fitzgerald *et al.* 1994; Corcoran & Clayton 1999). Late Carboniferous–early Permian igneous activity is a likely origin for this heating. It should be noted, however, that there is no evidence from combined AFTA and VR data for elevated geothermal gradients during Carboniferous time in the Rathlin Basin (Green 1996). Crustal extension during Triassic time, which resulted in widespread basin growth, may also have resulted in hot fluid pulses being generated at that time.

In the West of Shetlands region high (>150 °C) fluid inclusion temperatures, in rocks as young as Paleocene, suggest advective heating possibly related to Paleocene–Eocene igneous activity (Parnell *et al.* 1999).

The measurement of high fluid inclusion palaeotemperatures in the Permo-Triassic sandstones of the Port More and BGS 73/28 boreholes precludes these T_h data from reflecting a late Carboniferous–early Permian event as inferred for many Carboniferous sections in Ireland (Corcoran & Clayton 1999). The thermal maturity of the Carboniferous sequence in Co. Antrim is much lower than elsewhere in Ireland (Clayton *et al.* 1989), so it has clearly experienced a different thermal history. Late Carboniferous–early Permian igneous activity was important in the region, as evidenced by lavas of this age in the Larne No. 2 borehole (Penn *et al.* 1983), the Sound of Islay (Upton *et al.* 1986) and Arran (Leitch 1942). However, this activity did not result in elevated reflectance values in the Carboniferous sequence of Co. Antrim. Distance from the Variscan Orogen may have been a more important factor in controlling the maturation levels of the Carboniferous units in this area (Corcoran & Clayton 2001).

All the fluid inclusions in the Colliery Bay sample occur in annealed microfractures closely related to granulation seams, which have been attributed a late Carboniferous age of formation (Evans *et al.* 1998). However, these structures appear to have had a multi-stage history of reactivation (authors' unpublished data), having acted as conduits for multiple phases of fluid flow (Fig. 10). As a result, these granulation seams cannot be used to constrain the timing of inclusion entrapment.

The presence of two suites of inclusions in the Cross borehole sample (Population Type 2, Fig. 2b), suggests that at least two fluid flow events have occurred. Paragenetic studies indicate that entrapment of the high-T_h inclusion suite must have occurred during a relatively early fluid migration phase, whereas the low-T_h inclusions suite represents a later fluid event associated with iron oxide alteration. The timing of this alteration in the Carboniferous sandstones is difficult to constrain. Major alteration has been associated with deep oxidative weathering of the Carboniferous sandstones during Permian time (Wang 1992). However, further alteration of these sandstones probably also occurred during Mesozoic and Tertiary uplift. The presence of high-temperature inclusions in Permo-Triassic sandstones suggests that entrapment of this inclusion suite is unlikely to have occurred before iron oxide alteration during Permian time. This suggests that there may be a record of two hot fluid events in the sample. The earlier high-T_h modal population (mean T_h 177 °C) in this Carboniferous sample from the Cross borehole is probably the correlative of the high-T_h (mean T_h 161 °C) quartz-hosted inclusions in the other Permo-Triassic and Carboniferous samples.

Modern analogues

There are numerous modern analogues indicating that these hot fluid pulses have a relatively short lifespan. For example, present-day hydrothermal systems are documented around the East Pacific Rise, where hot fluids are venting at the sea bed (Lowell *et al.* 1995). The lifespan of this hydrothermal system is estimated to be *c.* 40 ka (Converse *et al.* 1984). In a continental setting, the Grant Canyon and Bacon Flat oil fields in Nevada, USA, are part of an active, relatively young (0.6–2.5 Ma) hydrothermal system (Hulen *et al.* 1994). In this case, the hot fluids are in part responsible for the hydrocarbon prospectivity of the system, as an upwelling hot-water plume has promoted oil migration, and has heated the reservoir rocks to temperatures usually encountered some 3 km deeper.

There is, however, an important distinction to be made here regarding hydrocarbon generation. Evidence from VR suggests that hot fluid flow events in the Rathlin Basin have occured on time scales of thousands to tens of years, considerably less than the time scales shown by these modern systems. Although fluid-driven heating may stimulate hydrocarbon generation from source rocks at shallow crustal levels if the heating is long lived, short-lived hydrothermal events are unlikely to influence the maturation of organic matter. This is because hydrocarbon generation is also a kinetic process, and is more appropriately modelled by changes in VR than by palaeotemperatures recorded from fluid inclusions.

Conclusions

The Rathlin Basin is an example of an Atlantic margin basin that has had a palaeothermal history affected by one or more phases of hot fluid flow through parts of the sedimentary succession. The hot fluids are recorded by fluid inclusion minimum trapping temperatures (T_h) of up to 177 °C in Carboniferous and Permo-Triassic sandstones. However, vitrinite reflectance data indicates temperatures much lower than those recorded from fluid inclusions. This discrepancy arises from the fact that vitrinite maturation is a kinetic process, requiring prolonged heating, whereas fluid inclusions record geologically instantaneous entrapment of advecting fluids.

The influence of advective heat transfer may be suspected in any basin that displays the following features:

(1) palaeotemperatures that are much higher than predicted from the burial history under conditions of vertical conductive heat transfer;

(2) palaeotemperature profiles that fluctuate markedly, suggesting fluid-driven heat transfer along certain horizons;

(3) discrepancies between palaeotemperatures derived from fluid inclusion studies and kinetically dependent thermal maturity indicators such as VR and AFTA.

The rapid emplacement of hydrothermal fluids indicates that fluid flow was mainly accommodated by fracture-controlled fluid migration, which has the potential to progress more quickly than intergranular fluid flow. Timing of hydrothermal fluid flow is difficult to constrain because of the multi-stage flow history of the fracture pathways, although potential periods of hot fluid flow in the Rathlin Basin include Late Carboniferous–Early Permian, Triassic and Paleocene–Eocene times. The source of the heat appears to be regional, as a result of the large spatial distribution of Atlantic margin basins affected by hot fluids, and may have been related to mantle plume activity during the Tertiary period.

Hot fluid flow events have important implications for hydrocarbon prospectivity in a basin. Short-lived hydrothermal fluid flow may not be detected by kinetically dependent thermal maturity parameters such as VR and AFTA because of the brevity of these events, leading to inaccurate reconstruction of palaeothermal histories of affected basins. However, hot fluid flow events, as identified by fluid inclusion studies, may not significantly affect hydrocarbon generation where the hydrothermal activity is short lived. By combining VR and AFTA with fluid inclusion studies, the timing, source, duration and palaeotemperatures of short-lived hot fluid flow events can be constrained.

We would like to thank D. Corcoran, M. Feely and P. Allen for their constructive reviews of an earlier draft of this manuscript. This work was undertaken as part of a PhD research studentship sponsored by the European Social Fund.

References

ANDERSON, T.B., PARNELL, J. & RUFFELL, A.H. 1995. Influence of basement on the geometry of Permo-Triassic Basins in the northwest British Isles. *In*: BOLDY, S.A.R. (ed.) *Permian and Triassic Rifting in northwest Europe*. Geological Society, London, Special Publications, **91**, 103–122.

BARKER, C.E. 1991. Implications to organic maturation studies of evidence for a geologically rapid increase and stabilisation of vitrinite reflectance at peak temperature: Cerro Prieto Geothermal System, Mexico. *AAPG Bulletin*, **75**, 1852–1863.

BRAY, R.J., GREEN, P.F. & DUDDY, I.R. 1992. Thermal history reconstruction using apatite fission track analysis and vitrinite reflectance: a case study from the UK East Midlands and the Southern North Sea. *In*: HARDMAN, R.F.P. (ed.) *Exploration Britain: Geological Insights for the Next Decade*. Geological Society, London, Special Publications, **67**, 3–25.

BRIGGS, H. 1935. Alteration of coal-seams in the vicinity of igneous intrusions, and associated problems. *Transactions of the Institution of Mining Engineers*, **89**, 187–219.

BURNHAM, A.K. & SWEENEY, J.J. 1989. A chemical kinetic model of vitrinite reflectance maturation. *Geochimica et Cosmochimica Acta*, **53**, 2649–2657.

CLAYTON, G., HAUGHEY, N., SEVASTOPULO, G.D., BURNETT, R. 1989. *Thermal Maturation Levels in the Devonian and Carboniferous Rocks of Ireland*. Geological Survey of Ireland, Dublin.

CONVERSE, D.R., HOLLAND, H.D. & EDMOND, J.M. 1984. Flow rates in the axial hot springs of the East Pacific Rise (21°N): implications for the heat budget and the formation of massive sulphide deposits. *Earth and Planetary Science Letters*, **69**, 159–175.

CORCORAN, D.V. & CLAYTON, G. 1999. Interpretation of vitrinite reflectance profiles in the Central Irish Sea area: implications for the timing of organic maturation. *Journal of Petroleum Geology*, **22**, 261–286.

CORCORAN, D.V. & CLAYTON, G. 2001. Interpretation of vitrinite reflectance profiles in sedimentary basins onshore and offshore Ireland. *In*: SHANNON, P.M., HAUGHTON, P.D.W. & CORCORAN, D.V. (eds) *The Petroleum Exploration of Ireland's Offshore Basins*. Geological Society, London, Special Publications, **188**, 61–90.

DEMING, D. 1994. Overburden rock, temperature and heat flow. *In*: MAGOON, L.B. & DOW, W.G. (eds) *The Petroleum System—from Source to Trap*. American Association of Petroleum Geologists, Memoirs, **60**, 165–185.

DE VIVO, B. & FREZZOTTI, M.F. 1994. *Fluid Inclusions in Minerals: Methods and Applications*. Short Course of the Working Group (IMA) 'Inclusions in minerals', Virginia Tech, Blacksburg, VA.

DORÉ, A.G., LUNDIN, E.R., BIRKELAND, Ø., ELIASSEN, P.E. & JENSEN, L.N. 1997. The NE Atlantic Margin: implications of late Mesozoic and Cenozoic events for hydrocarbon prospectivity. *Petroleum Geoscience*, **3**, 117–131.

DUDDY, I.R., GREEN, P.F., BRAY, R.J. & HEGARTY, K.A. 1994. Recognition of the thermal effects of fluid flow in sedimentary basins. *In*: PARNELL, J. (ed.) *Geofluids: Origin, Migration and Evolution of Fluids in Sedimentary Basins*. Geological Society, London, Special Publications, **78**, 325–345.

EAGAR, R.M.C. 1974. Neuere Arbeiten uber das Westfal in Irland. *Zentralblatt für Geologie und Palaontologie*, **1**, 291–308.

EBDON, C.C., GRANGER, P.J., JOHNSON, H.D. & EVANS, A.M. 1995. Early Tertiary evolution and sequence stratigraphy of the Faeroe–Shetland Basin: implications for hydrocarbon prospectivity. *In*: SCRUTTON, R.A., STOKER, M.S., SHIMMIELD, G.B. & TUDHOPE, A.W. (eds) *The Tectonics, Sedimentation and Palaeoceanography of the North Atlantic Region*. Geological Society, London, Special Publications, **90**, 51–69.

EVANS, R., HENDRY, J.P., PARNELL, J. & KALIN, R.M. 1998. Origin and significance of fracture-related dolomite in porous sandstones: an example from the Carboniferous of County Antrim, Northern Ireland. *In*: MORAD, S. (ed.) *Carbonate Cementation in Sandstones*. International Association of Sedimentologists, Special Publication, **26**, 409–435.

EVANS, D., KENOLTY, N., DOBSON, M.R. & WHITTINGTON, R.J. 1980. *The Geology of the Malin Sea*. Report of the Institute of Geological Sciences, **79/15**.

FITZGERALD, E., FEELY, M., JOHNSTON, J.D., CLAYTON, G., FITZGERALD, L.J. & SEVASTOPULO, G.D. 1994. The Variscan thermal history of West Clare, Ireland. *Geological Magazine*, **131**, 545–558.

GOLDSTEIN, R.H., REYNOLDS, T.J. 1994. *Systematics of Fluid Inclusions in Diagenetic Minerals*. Society of Economic Paleontologists and Mineralogists, Short Course, **31**.

GREEN, P.F. 1996. *West of Scotland: Thermal and Tectonic Development and Hydrocarbon Generation History Assessed using AFTA and VR*. Non-proprietary report, Geotrack Report **523**, 285–316.

GREEN, P.F., DUDDY, I.R. & BRAY, R.J. 1995. Applications of thermal history reconstruction in inverted basins. *In*: BUCHANAN, J.G. & BUCHANAN, P.G. (eds) *Basin Inversion*. Geological Society, London, Special Publications, **88**, 149–165.

GREEN, P.F., DUDDY, I.R., BRAY, R.J., DUNCAN, W.I. & CORCORAN, D.V. 2001. The influence of thermal history on hydrocarbon prospectivity in the

Central Irish Sea Basin. *In*: SHANNON, P.M., HAUGHTON, P.D.W. & CORCORAN, D.V. (eds) *The Petroleum Exploration of Ireland's Offshore Basins*. Geological Society, London, Special Publications, **188**, 171–188.

GREEN, P.F., DUDDY, I.R., GLEADOW, A.J.W. & LOVERING, J.F. 1989. Apatite fission track analysis as a palaeotemperature indicator for hydrocarbon exploration. *In*: NAESER, N.D. & MCCOLLOH, T. (eds) *Thermal History of Sedimentary Basins – Methods and Case Histories*. Springer, Berlin, 181–195.

GREEN, P.F., DUDDY, I.R., HEGARTY, K.A. & BRAY, R.J. 1999. Early Tertiary heat flow along the UK Atlantic continental margin and adjacent areas. *In*: FLEET, A.J. & BOLDY, S.A.R. (eds) *Petroleum Geology of Northwest Europe: Proceedings of the 5th Conference*. Geological Society, London, 349–357.

HULEN, J.B., GOFF, F., ROSS, J.R., BORTZ, L.C. & BERESKIN, S.R. 1994. Geology and geothermal origin of Grant Canyon and Bacon Flat Oil Fields, Railroad Valley, Nevada. *AAPG Bulletin*, **78**, 596–623.

LEITCH, D. 1942. The Upper Carboniferous rocks of Arran. *Transactions of the Geological Society of Glasgow*, **20**, 141–154.

LOWELL, R.P., RONA, P.A. & VON HERZEN, R.P. 1995. Seafloor hydrothermal systems. *Journal of Geophysical Research*, **100** (B1), 327–352.

MCCALLIEN, W.J. & ANDERSON, R.B. 1930. The Carboniferous sediments of Kintyre. *Transactions of the Royal Society of Edinburgh*, **56**, 599–619.

NAYLOR, D., SHANNON, P.M. 1982. *The Geology of Offshore Ireland and West Britain*. Graham & Trotman, London.

PARNELL, J. 1992. Burial histories and hydrocarbon source rocks on the North West Seaboard. *In*: PARNELL, J. (ed.) *Basins on the Atlantic Seaboard: Petroleum Geology, Sedimentology and Basin Evolution*. Geological Society, London, Special Publications, **62**, 3–16.

PARNELL, J., CAREY, P.F., GREEN, P. & DUNCAN, W. 1999. Hydrocarbon migration history, West of Shetland: integrated fluid inclusion and fission track studies. *In*: FLEET, A.J. & BOLDY, S.A.R. (eds) *Petroleum Geology of Northwest Europe: Proceedings of the 5th Conference*. Geological Society, London, 613–625.

PENN, I.E., HOLLIDAY, D.W., KIRBY, G.A. & 5 OTHERS, 1983. The Larne No. 2 Borehole: discovery of a new Permian volcanic centre. *Scottish Journal of Geology*, **19**, 333–346.

STOKER, M.S., HITCHEN, K. & GRAHAM, C.C. 1993. *The Geology of the Hebrides and West Shetland Shelves, and Adjacent Deep-Water Areas*. British Geological Survey, UK Offshore Regional Report. HMSO, London.

SUMMER, N.S. & VEROSUB, K.L. 1989. A low temperature hydrothermal maturation mechanism for sedimentary basins associated with volcanic rocks. *In*: PRICE, P.A. (ed.) *Origin and Evolution of Sedimentary Basins and their Economic Potential*. Geophysical Monographs, American Geophysical Union, **48**, 129–136.

SWEENEY, J.J. & BURNHAM, A.K. 1990. Evaluation of a simple model of vitrinite reflectance based on chemical kinetics. *AAPG Bulletin*, **74**, 1559–1570.

THRASHER, J. 1992. Thermal effect of the Tertiary Cuillins Intrusive Complex in the Jurassic of the Hebrides: an organic geochemical study. *In*: PARNELL, J. (ed.) *Basins on the Atlantic Seaboard: Petroleum Geology, Sedimentology and Basin Evolution*. Geological Society, London, Special Publications, **62**, 35–49.

UPTON, G.J., FITTON, J.G. & MACINTYRE, R.M. 1986. The Glas Eilean lavas: evidence of a Lower Permian volcano tectonic basin between Islay and Jura, Inner Hebrides. *Transactions of the Royal Society of Edinburgh: Earth Sciences*, **77**, 289–293.

WANG, W.H. 1992. Origin of reddening and secondary porosity in Carboniferous sandstones, Northern Ireland. *In*: PARNELL, J. (ed.) *Basins on the Atlantic Seaboard: Petroleum Geology, Sedimentology and Basin Evolution*. Geological Society, London, Special Publications, **62**, 243–254.

WILSON, H.E. & MANNING, P.I. 1978. *Geology of the Causeway Coast. Memoir of the Geological Survey of Northern Ireland, Sheet 7*. HMSO, Belfast.

WILSON, H.E., ROBBIE, J.A. 1996. *Geology of the Country around Ballycastle. Memoir of the Geological Survey of Northern Ireland, Sheet 8*. HMSO, Belfast.

ZAIGOS, J.P. & BLACKWELL, D.D. 1986. A model for the transient temperature effect of horizontal fluid flow in geothermal systems. *Journal of Volcanology and Geothermal Research*, **27**, 371–397.

Hydrocarbon prospectivity in the Irish Sea area: insights from recent exploration of the Central Irish Sea, Peel and Solway basins

JONATHAN FLOODPAGE[1], PHIL NEWMAN & JASON WHITE

TotalFinaElf Exploration UK PLC, Crawpeel Rd, Altens Industrial Estate, Aberdeen AB12 3FG, UK

[1]*Present address: TotalFinaElf S.A., Tour Coupole, Place de la Coupole, La Defense 6, 92078 Paris, France*

(e-mail: jonathan.floodpage@totalfinaelf.com)

Abstract: Compared with the prolific success of the Triassic play in the East Irish Sea Basin (EISB) the lack of hydrocarbon discovery in neighbouring Permo-Triassic basins of the Irish Sea has been an enigma. However, recent exploration of the Peel, Solway and Central Irish Sea basins has provided new insights into the geology of these basins and the controls upon hydrocarbon prospectivity in the Irish Sea area. Regional seismic interpretation suggests that 12 of the 15 exploration wells drilled in the basins adjacent to the EISB tested valid structural closures at top Triassic reservoir level. Re-evaluation of the Irish Sea petroleum system reveals that, although effective reservoirs occur in the Lower–Middle Triassic Ormskirk Sandstone Formation, and evaporites in the Middle–Upper Triassic Mercia Mudstone Group provide a regional top seal, the major factor controlling hydrocarbon prospectivity is the limited presence of effective source rocks in the underlying Carboniferous section. A further control upon prospectivity is the timing of hydrocarbon migration, from those areas where Carboniferous source rocks were deposited and preserved. The Namurian basinal marine oil- and gas-prone shales, which form the principal source of hydrocarbons for the Triassic play in the EISB, are restricted to an east–west fairway extending from the EISB into the Kish Bank Basin. Rocks of this age are absent from the Peel and Solway basins as a result of Variscan uplift and erosion. However, palaeogeographical reconstructions based on well and outcrop data suggest that, even if preserved, the depositional environment was not conducive to the formation of marine oil- and gas-prone source rocks. Well and seismic data suggest that rocks of Namurian age were not deposited in the Central Irish Sea area, which remained high during much of Dinantian and Namurian time. Potential source rock development in the Central Irish Sea area is therefore limited to the Westphalian section, which is organically lean and dominated by inertinitic kerogens. Potential hydrocarbon traps in the Central Irish Sea, Peel and Solway basins formed largely as a result of Early Cretaceous tectonism and were subsequently modified by fault reactivation during Tertiary uplift phases. Trap formation appears to post-date the most likely timing of hydrocarbon charge, which this study suggests would have occurred in Late Triassic and Jurassic time. Hydrocarbon entrapment in the EISB may have been favoured by limited Early Cretaceous uplift, coupled with renewed hydrocarbon generation and re-migration during Early Tertiary time. It is concluded that the remaining prospectivity of the Triassic play in the Irish Sea area is likely to be restricted to the proven play fairway within the EISB.

Five Permo-Triassic basins are preserved in the Irish Sea area that is bounded to the north by the Longford Down and Southern Uplands Caledonide massifs and to the south by the Mid Irish Sea High (Fig. 1). The largest of these Permo-Triassic basins is the East Irish Sea Basin (EISB), which lies to the south and east, respectively, of the smaller Solway and Peel basins. The southern part of the study area comprises the Kish Bank Basin and the Central Irish Sea Basin (CISB) (Fig. 1). These basins are the preserved remnants of a previously more extensive cover of Permo-Triassic and younger rocks, which are today separated by intra-basinal highs, resulting from post-Triassic phases of uplift and erosion.

The EISB is an established hydrocarbon province with reserves of 7.6 TCF (trillion cubic feet) of gas and 200 MMBBL (million barrels) of oil proven in 12 fields (Quirk *et al.* 1999). However, exploration success has so far been elusive in the Solway, Peel, Kish Bank and Central Irish Sea basins. The main exploration

From: SHANNON, P.M., HAUGHTON, P.D.W. & CORCORAN, D.V. (eds) 2001. *The Petroleum Exploration of Ireland's Offshore Basins*. Geological Society, London, Special Publications, **188**, 107–134.

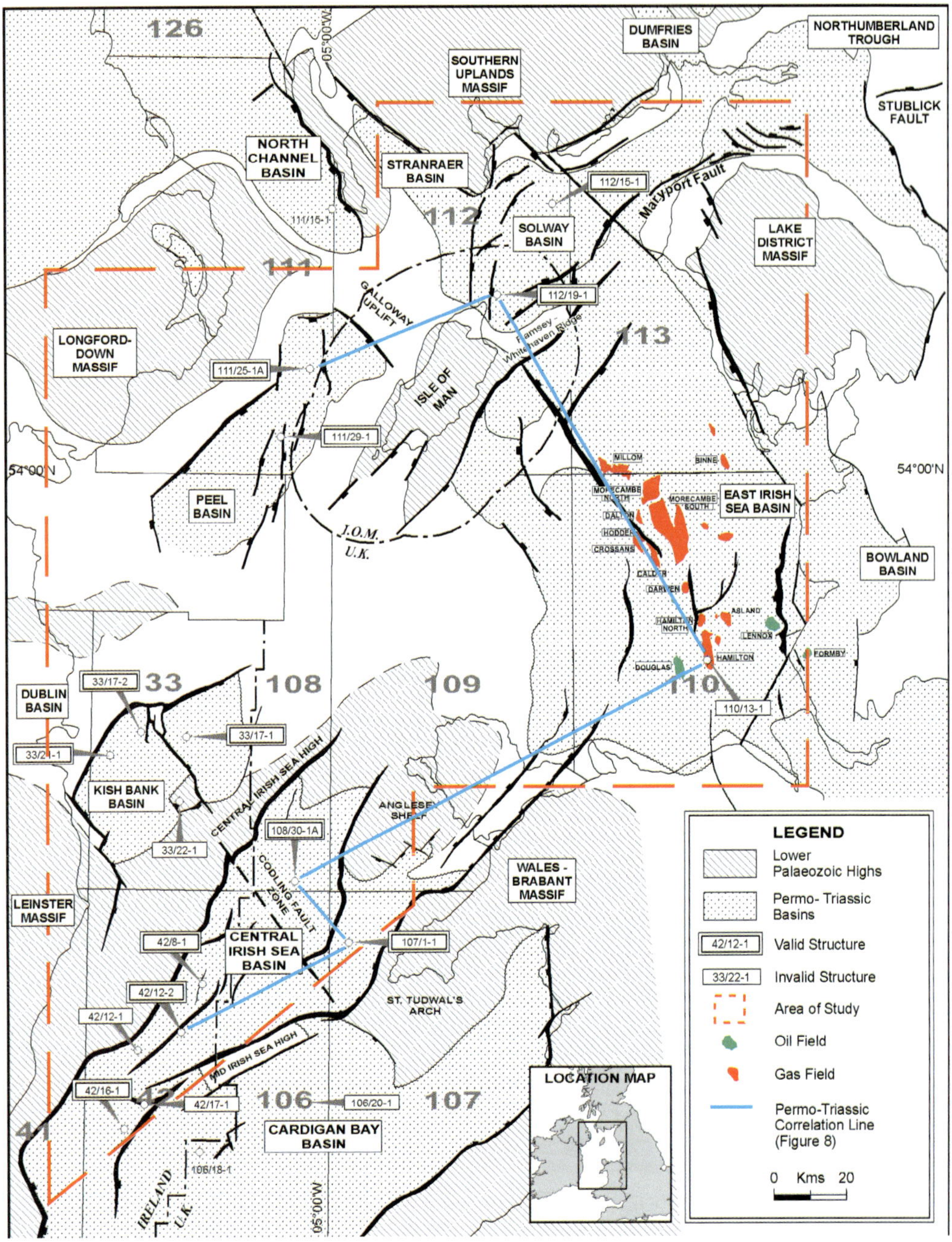

Fig. 1. Tectonic elements map of the Irish Sea area with key exploration wells and study area highlighted.

play in the EISB consists of a Lower Triassic sandstone reservoir (Sherwood Sandstone Group) in structural traps, sealed vertically and laterally by Upper Triassic evaporites and shales (Mercia Mudstone Group) and charged from source rocks in the underlying Carboniferous units (Hardman *et al.* 1993). Thick marine shales of Pendleian to Yeadonian age, known from outcrop in North Wales and the Bowland Basin and from wells in the southern part of the EISB, have good to rich oil- and gas-prone source rock potential and have been geochemically matched to the oils recovered from the Lennox and Douglas fields in the south of the basin (Armstrong *et al.* 1997).

The discovery of oil in the Douglas and Lennox accumulations in the early 1990s (Haig *et al.* 1997; Yaliz 1997) provided a fresh impetus for exploration in the previously unexplored Solway and Peel basins and the poorly explored basins of the Central Irish Sea area. Following the UKCS 14th and 16th Rounds of Licensing in 1993 and 1995 and the First Isle of Man Licensing Round in 1995, new licences were awarded in the CISB, Peel Basin and Solway Basin. Elf Exploration UK plc and partners were particularly active in exploring these basins and drilled four exploration wells between 1994 and 1998. Two of these wells are located in the Peel Basin (UK111/25-1A and UK111/29-1), one in the Solway Basin (IOM112/19-1) and one in the CISB (UK108/30-1A) (Fig. 1).

The aim of this paper is to integrate the new geological data available from these exploration wells with seismic interpretations based on 2D and 3D data in a re-evaluation of the hydrocarbon potential of the Irish Sea area. Comparisons are drawn with the EISB and a post-mortem of drilling results in the Peel, Solway and Central Irish Sea basins is discussed. Systematic mapping at top Sherwood Sandstone level suggests that 12 of the 15 exploration wells drilled, in the basins peripheral to the EISB (within the study area defined in Fig. 1), were located on valid structural closures. Consequently, the main reason for failure appears to be related to lack of hydrocarbon charge, which could be explained either by the absence of source rocks or the unfavourable timing of source rock maturation and hydrocarbon migration, relative to trap formation. These possibilities are explored in some detail in this paper.

Seismic and well database

This study had access to most of the wells drilled in the EISB together with all of the exploration wells drilled, pre-1998, in basins outlying the EISB. These data have been integrated with published information and the results of extensive outcrop sampling around the Solway Firth, Isle of Man, eastern Ireland and western Wales (Geochem 1992). The seismic dataset, covering the CISB, Solway and Peel basins, comprises regional speculative 2D surveys acquired by Western Geophysical, Nopec, Jebco and Geoteam, and block-specific 2D and 3D surveys, provided courtesy of Elf Exploration UK, BHP and BG E&P. In total, *c.* 1000 km^2 of 3D and 7000 km of 2D data have been integrated in this study.

Hydrocarbon occurrence

All the wells drilled within the study area, in basins adjacent to the EISB, have been abandoned as dry holes (Fig. 1). Minor gas shows were encountered in the basal sandstones of the Mercia Mudstone Group, in the Sherwood Sandstone and in the Carboniferous sequence of CISB well IR42/16-1, but at ditch gas concentrations of <0.9% C1 (methane). Examination of thin sections from a core cut at the top of the Sherwood Sandstone Group, in the Solway Basin well IOM112/19-1, revealed traces of bitumen, which were volumetrically insignificant (Newman 1999). The most significant hydrocarbon shows occur in the Kish Bank Basin, where weak oil shows were recorded in the Sherwood Sandstone of well IR33/17-1 and in the Westphalian sequence of IR33/22-1. In the latter well the oil shows were associated with ditch gas readings of up to 3.4% C1. In addition, evidence for gas seepage has been reported from surface geochemical surveys (Croker & Power 1996), and Seepfinder surveys (Dunford & Dancer 2001), in the Kish Bank Basin area.

Tectono-stratigraphic setting

During Early Palaeozoic time much of the Irish Sea formed part of the Iapetus Ocean. The Longford Down and Southern Uplands massifs lay on the northern margin of this ocean (Laurentian plate), whereas the Lake District and Wales lay on the southern margin (East Avalonian plate) (Soper *et al.* 1987; Jackson *et al.* 1995). Closure of the Iapetus Ocean occurred during the Caledonian Orogeny, with the formation of NE–SW-trending thrust faults and NW–SE transfer faults. These two fault trends form the main structural framework of the area and have been repeatedly reactivated during subsequent basin evolution.

Three main post-Early Palaeozoic megasequences are identified in the Irish Sea area, based on well and seismic data (Fig. 2). The first of these is the Carboniferous Megasequence, which may locally contain important source rock intervals. The second is the Permo-Triassic Megasequence, which contains the main reservoir and seal intervals for the Triassic play. The third, the post-Triassic Megasequence, exerts important controls on source rock maturation and the timing of hydrocarbon migration from the Carboniferous source rocks.

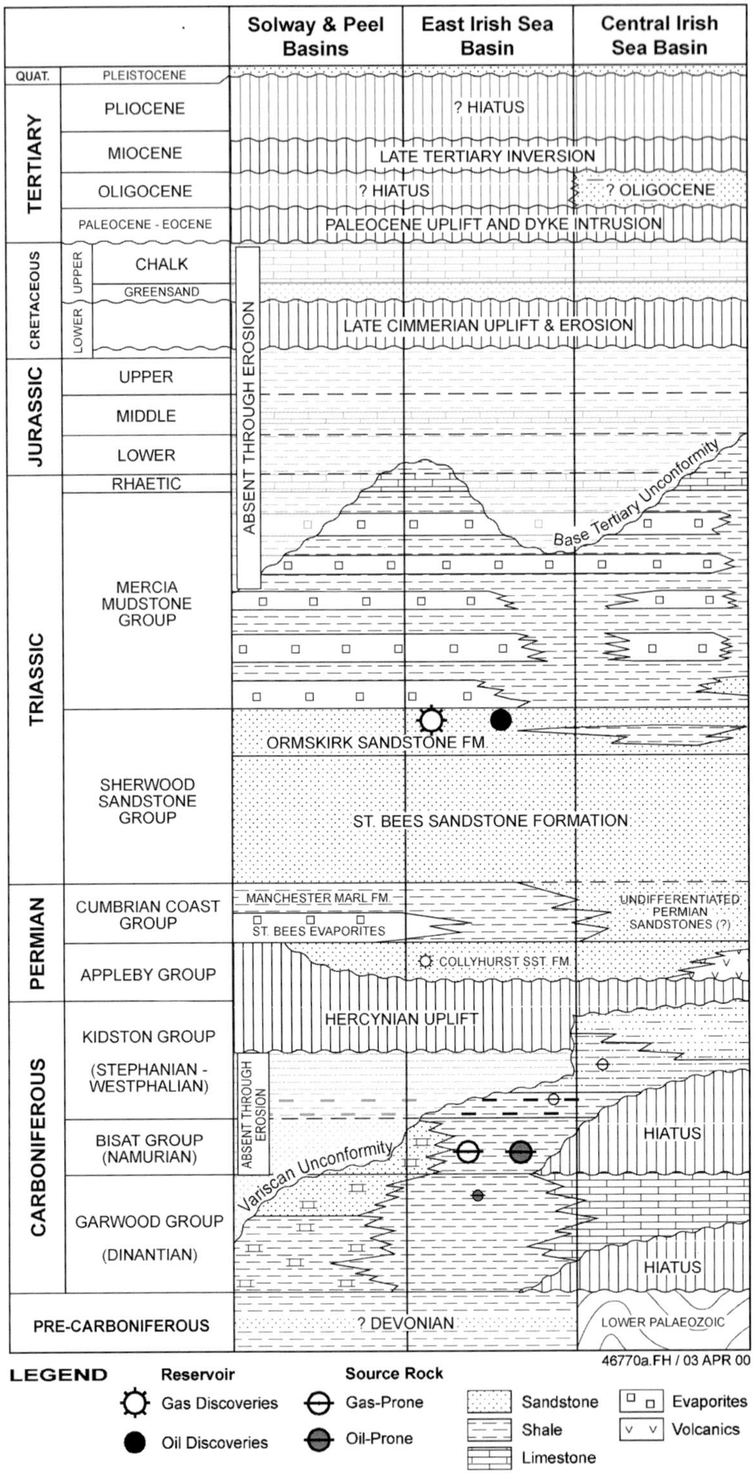

Fig. 2. Comparative tectono-stratigraphy of the Irish Sea basins with the major hydrocarbon source rock and reservoir intervals highlighted.

Carboniferous Megasequence

In the Irish Sea area, Early Carboniferous time marked a change to extensional tectonics in response to north–south extension (Fraser *et al.* 1990). In the north, extension across the NE–SW-trending faults generated a basin system extending SW from the Northumberland Trough through the Solway and Peel basins (Fig. 1) (Newman 1999). These basins are separated from the EISB to the south by the Ramsey–Whitehaven Ridge and Lake District Massif. During Early Carboniferous time, the EISB formed an offshore extension of the Bowland Basin (Fraser *et al.* 1990) and may have continued further west to connect with the Dublin Basin (Fig. 1). This basinal system appears to have experienced rapid subsidence during Carboniferous time with >3000 m of strata deposited. The northern margin of the EISB is fault bounded, whereas the southern margin is formed by a northerly-dipping ramp that constituted the northern limit of the Wales–Brabant Massif. This massif forms a regional high of Lower Palaeozoic rocks, which extends in an easterly direction from the Leinster Massif, through the Central Irish Sea area, into Wales and southern England. This regional high influenced deposition in the Irish Sea area during much of the Carboniferous period.

Well and outcrop data indicate that Early Carboniferous sedimentation in the Solway Basin and Northumberland Trough comprised cyclical deltaic to shallow-water sandstones, shales, limestones and coals (Fig. 3a and b) (Newman 1999). In contrast, limited well data from the EISB suggest that this basin was starved of clastic deposits and that deposition was dominated by marine shales and calci-turbidites (Fig. 3) (Jackson *et al.* 1995). This suggests that the Ramsey–Whitehaven Ridge and Lake District Massif formed a barrier to the influx of coarse clastic deposits into the EISB at this time. Further south, seismic and well data suggest that Early Carboniferous sediments may be thin to locally absent in the Central Irish Sea area (Fig. 4).

Palaeo-facies reconstructions (Fraser *et al.* 1990; Newman 1999) suggest that by early Namurian times major delta systems prograding from the NE had encroached upon the Northumberland Trough and Solway Basin, whereas deep-water sedimentation persisted over the southern part of the EISB (Fig. 3b). In contrast to the Solway Basin and Northumberland Trough, the prevailing environment in the southern part of the EISB was conducive to the deposition of marine shales with good to rich source rock potential (Armstrong *et al.* 1997). Seismic and well data suggest that Namurian sediments are absent through non-deposition over much of the CISB. Westphalian strata lie unconformably upon Precambrian rocks at outcrop in Anglesey (Jackson *et al.* 1995). Neither of the two Central Irish Sea area wells (IR33/22-1 and IR42/17-1) that penetrated the base of the Westphalian sequence encountered Namurian rocks, although Maddox *et al.* (1995) have suggested that Namurian sediments could be present in the undrilled, basinal, part of the CISB.

Regional seismic data indicate that the Kish Bank Basin is underlain by a thick package of reflectors (*c.* 1300 ms two-way travel time (TWT)), which onlap the Lower Palaeozoic Central Irish Sea High (Fig. 5a). The lower part of this package is likely to be of Dinantian–Namurian age and probably represents a synrift succession similar to that beneath the EISB (Fig. 3). There is no evidence for the development of similar seismic facies of equivalent age beneath the CISB (Fig. 5a and b). In contrast, a thin package of reflectors, interpreted to be of Dinantian age, is directly overlain by a characteristically transparent seismic facies, which correlates with the Westphalian D sediments encountered in IR42/17-1 (Izatt *et al.* 2001).

Structural restoration, to pre-Variscan uplift, indicates that during Late Carboniferous time much of the underlying topography was buried and that the Irish Sea was covered by a succession of sandstones, shales and coals of Westphalian to Stephanian age, which gradually transgressed the CISB from the north (Fig. 4). Seismic data, tied to wells IR42/17-1 and IR42/16-1, indicate that at least 1000 m of Westphalian to Stephanian rocks were deposited in this area (Figs 5 and 6).

At the end of Carboniferous time, north–south compression associated with the Variscan Orogeny resulted in basin inversion and erosion (Fraser *et al.* 1990; Newman 1999). This inversion and erosion appears to have been most intense along the Northumberland Trough and the Peel and Solway basins, where denudation of the entire Upper Carboniferous section has occurred (Fig. 6). Seismic and well data indicate complex patterns of inversion and erosion in the EISB, with variable degrees of erosion of the Carboniferous basin fill. Seismic data from the Peel and Solway basins and EISB indicate a strong angular unconformity between the Carboniferous and overlying Permo-Triassic rocks (Fig. 7). In contrast, over the CISB, there is no marked angular unconformity observed

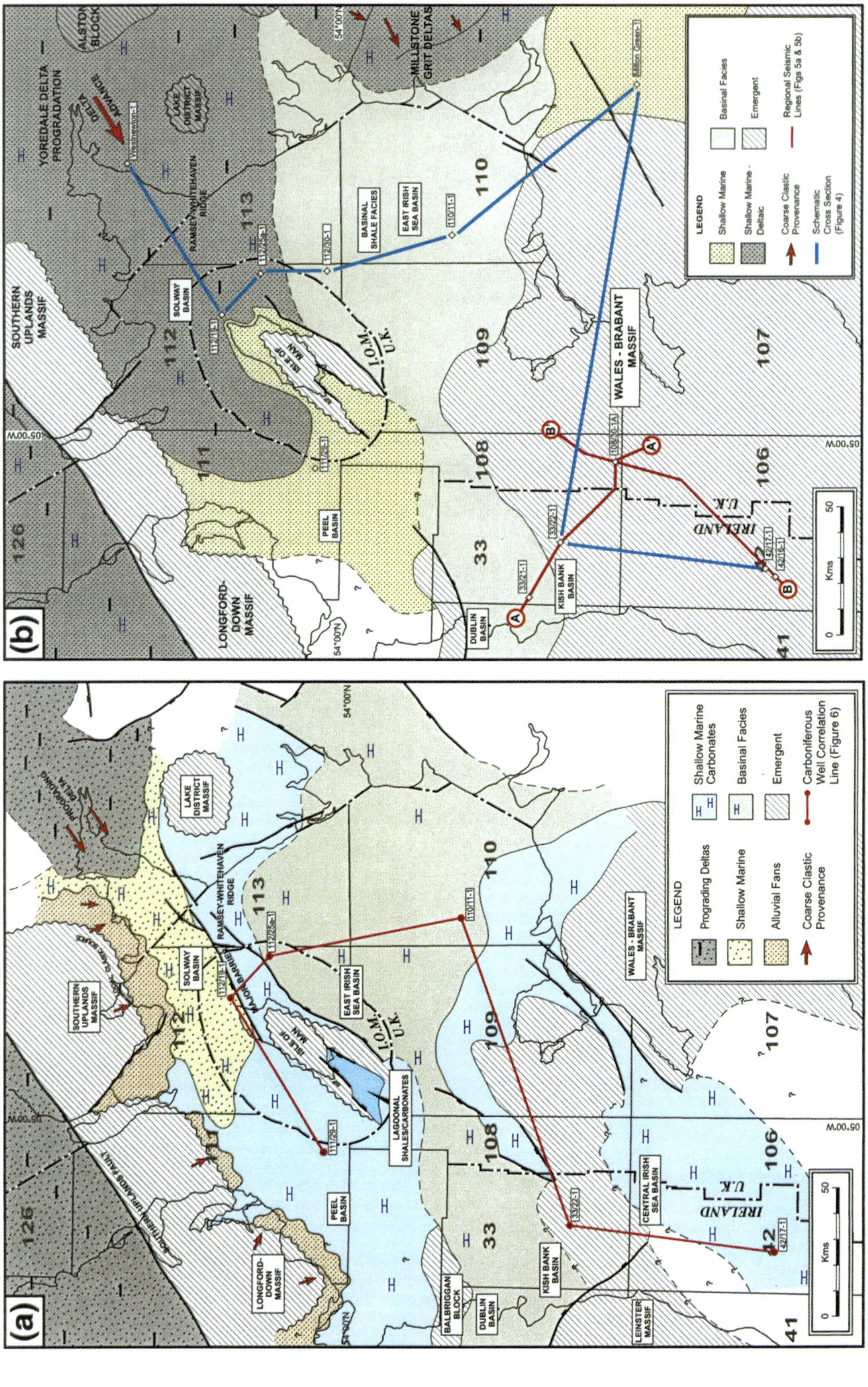
(a)
(b)
LEGEND
Prograding Deltas
Shallow Marine
Alluvial Fans
Coarse Clastic Provenance
Shallow Marine Carbonates
Basinal Facies
Emergent
Carboniferous Well Correlation Line (Figure 6)
Shallow Marine - Deltaic
Schematic Cross Section (Figure 4)
Regional Seismic Lines (Figs 5a & 5b)
SOUTHERN UPLANDS MASSIF
LAKE DISTRICT MASSIF
SOLWAY BASIN
RAMSEY-WHITEHAVEN RIDGE
EAST IRISH SEA BASIN
LAGOONAL SHALES/CARBONATES
PEEL BASIN
ISLE OF MAN
LONGFORD-DOWN MASSIF
KISH BANK BASIN
CENTRAL IRISH SEA BASIN
DUBLIN BASIN
BALBRIGGAN BLOCK
LEINSTER MASSIF
WALES - BRABANT MASSIF
PROGRADING DELTA
SOUTHERN UPLANDS FAULT
YOREDALE DELTA PROGRADATION
DELTA ADVANCE
ALSTON BLOCK
MILLSTONE GRIT DELTAS
BASINAL SHALE FACIES
I.O.M. U.K.
IRELAND U.K.
54°00'N
05°00'W
Kms 0 50
126 112 113 111 110 109 108 107 106 33 42 41
112/19-1
112/25a-1
112/30-1
110/11-1
111/29-1
33/21-1
33/22-1
42/16-1
42/17-1
Westnewton-1
Milton Green-1

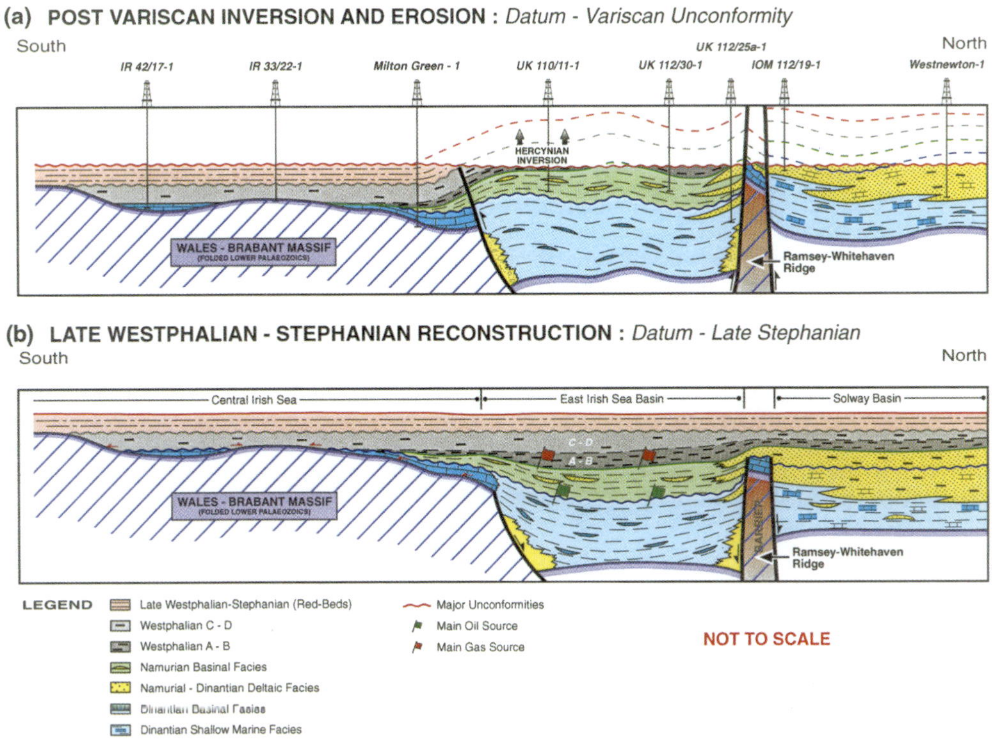

Fig. 4. (**a**) Schematic north–south stratigraphic section through the Irish Sea area illustrating structural control of regional Carboniferous facies trends. Datum is Variscan unconformity. (**b**) Schematic north–south stratigraphic section through the Irish Sea area restored to pre-Variscan uplift and erosion (shortening is not restored). Datum is Late Stephanian time. Pre-Variscan deformation configuration indicates that the Westphalian C–D and Stephanian section onlaps the Wales–Brabant Massif and Dinantian carbonate platform to the south.

between the Carboniferous and Permo-Triassic successions, suggesting limited Variscan uplift and erosion in this area (Fig. 5). As has been demonstrated in the English Midlands (Fraser *et al.* 1990), latest Carboniferous inversion activity was localized in areas of Dinantian synsedimentary growth faulting (such as the EISB, Peel and Solway basins), whereas adjacent Dinantian platform areas (such as the CISB) were relatively undisturbed.

Permo-Triassic Megasequence

Evidence, particularly in the vicinity of the Isle of Man, suggests that Early Permian rifting may have occurred in response to east–west extension (Quirk & Kimbell 1997). This extension, which appears to have reactivated the pre-existing NW–SE-trending faults, created a series of half-grabens infilled by thick coarse clastic deposits, e.g. North Channel, Stranraer and Dumfries basins (Fig. 1). In contrast, the Late Permian to Early Triassic sediments appear to have been deposited in response to regional subsidence (Quirk & Kimbell 1997; Maingarm *et al.* 1999; Newman 1999). This interpretation is supported by well correlation work and seismic interpretation, which indicates relatively uniform stratigraphic thicknesses, with facies maintained over long distances, in the Irish Sea area (Figs 7 and 8). This broad regional subsidence facilitated the northward progradation of a major river system during Early Triassic time (Jackson *et al.* 1995; Jackson *et al.* 1997) and the deposition of

Fig. 3. Palaeogeography of the Irish Sea area during (**a**) Dinantian, (**b**) Namurian time. These reconstructions suggest that organic-rich marine shales, of Namurian age, were restricted to an east–west basin extending from the Dublin Basin through the Kish Bank Basin into the EISB.

the Sherwood Sandstone Group. These sediments were in turn overlain by evaporites and shales of the Mercia Mudstone Group.

Deformational styles observed in the Triassic sediments are strongly controlled by the marked lithological contrasts between the Mercia Mudstone Group and the Sherwood Sandstone Group. In general, the relatively incompetent lithologies of the Mercia Mudstone Group show complex, ductile deformation with faults detaching within the Triassic evaporites and the development of low-amplitude halokinetic structures. In contrast, the competent lithologies of the Sherwood Sandstone Group manifest brittle behaviour with a mixture of basement-involved and basement-detached deformation styles observed. In the EISB, Peel and Solway basins seismic data indicate that these normal faults, which displace the top Sherwood Sandstone Group reflector, often sole out within shales and evaporites of Late Permian age (Fig. 7). However, in the CISB, where the Late Permian evaporites are probably absent, many of these faults have propagated into the underlying Carboniferous section (Fig. 5).

Post-Triassic Megasequence

The tectono-stratigraphic history of this megasequence is difficult to reconstruct because of the limited preservation of post-Triassic rocks caused by multiple phases of Mesozoic and Cenozoic uplift and erosion. However, the occurrence of isolated outliers of Lower Jurassic rocks in a number of basins, e.g. Peel, Solway, EISB, Cheshire and Kish Bank basins (Cope 1997; Warrington 1997; Newman 1999), suggests that widespread deposition of Early Jurassic sediments occurred in the Irish Sea area. Middle to Upper Jurassic strata have not been documented in the Irish Sea area, although rocks of this age are known to the south in the St George's Channel Basin and to the north in the Hebridean Basin. This evidence, when combined with the absence of strong faunal provinciality at this time, suggests that a marine connection may have existed through the Irish Sea during Mid- to Late Jurassic time (Cope 1997).

The Permo-Triassic succession of the Irish Sea is disrupted by numerous normal faults, which record at least one phase of extension during Mesozoic time. Regional geological constraints (Quirk & Kimbell 1997; Newman 1999) suggest that significant east–west extension probably occurred during Mid- to Late Jurassic time (Izatt *et al.* 2001). Reactivation of the underlying Caledonian basement faults influenced the development of these basins and the intervening basin highs, e.g. Mid Irish Sea High, Central Irish Sea High, Ramsey–Whitehaven Ridge and the Galloway Uplift (Fig. 1). Seismic mapping at top Sherwood Sandstone level indicates that at least three major fault trends were developed, throughout the Irish Sea area, post-Early Triassic: NE–SW, NW–SE and north–south. The first two trends are inherited from the underlying Caledonian framework, and oblique slip on these faults as a result of east–west extension is thought to have produced the north–south-trending faults (Maingarm *et al.* 1999; Newman 1999; Izatt *et al.* 2001).

There is no evidence for the preservation of Early Cretaceous strata in the Irish Sea area and palaeogeographical reconstructions indicate that deposition of marine sediments is unlikely to have occurred in this area, at this time (Cope 1997). The emergence of the Irish Sea area during Early Cretaceous time may have been the result of a well-documented global sea-level fall (Haq *et al.* 1987) but may also have been partly caused by basin-scale uplift.

Regional stratigraphic evidence from Northern Ireland suggests that marine sedimentation recommenced in Cenomanian time (Cope 1997), followed by the deposition of 200–300 m of chalk across the area during Late Cretaceous time. Apatite fission-track analysis (AFTA) data suggest that two phases of Tertiary uplift occurred; a rapid phase of cooling during Paleocene time (*c.* 60 Ma) and a second, slower phase of cooling during Miocene time (*c.* 20 Ma) (Newman 1999). The first phase of Tertiary ('Laramide') uplift was probably caused by epeirogenic uplift associated with the development of the Icelandic plume (Brodie & White 1994) and the emplacement of the Tertiary

Fig. 5. (**a**) NW–SE regional seismic line through the Kish Bank Basin and northern Central Irish Sea Basin illustrating seismic tie to wells IR33/21-1, IR33/22-1 and UK108/30-1A. Thick package of reflectors, onlapping the Central Irish Sea High to the south, suggests that a Dinantian–Namurian synrift succession is present in the Kish Bank Basin (seismic line courtesy of Western Geophysical, Nopec and Jebco). (**b**) SW–NE regional seismic line along the axis of the Central Irish Sea Basin illustrating seismic tie to wells IR42/16-1, IR42/17-1 and UK108/30-1A. Dinantian section is directly overlain by a transparent seismic facies, which correlates with the coal-poor Westphalian D succession encountered in well IR42/17-1 (seismic line courtesy of Western Geophysical and Jebco).

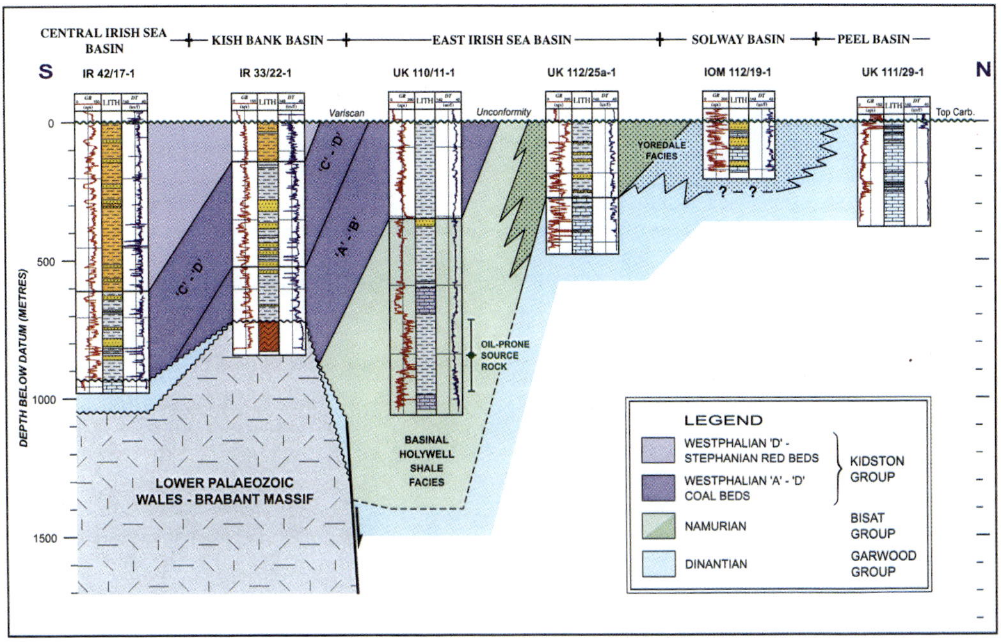

Fig. 6. Regional stratigraphic well log correlation for the Carboniferous sequences of the Irish Sea area illustrating: Namurian facies variations, with basinal Holywell Shale facies restricted to the East Irish Sea Basin; onlap of the Wales–Brabant Massif to the south by the Westphalian A–B section; increasing magnitude of Variscan uplift towards the north resulting in complete removal of the Namurian and Westphalian sections from the Solway and Peel basins. Datum is Variscan unconformity. GR, gamma-ray log in API units); DT, sonic log (in μs per foot).

igneous province of NW Britain. The second phase, during Miocene time, may be related to Alpine compressional movements and associated uplift (Fig. 2).

Hydrocarbon prospectivity evaluation

Reservoir potential

The Ormskirk Sandstone Formation, which forms the principal reservoir rock for the Triassic play, lies at the top of the Sherwood Sandstone Group, immediately below the regional seal provided by the Mercia Mudstone Group. Regional well log and seismic correlation indicates that the Ormskirk Sandstone Formation was deposited throughout the Irish Sea area and is only missing by erosion from intra-basinal highs (Fig. 8). Reservoir quality of the Ormskirk Sandstone Formation is generally good and constitutes a low exploration risk over much of the Irish Sea study area. In Peel Basin wells UK111/25-1A and UK111/29-1 and Solway Basin well IOM112/19-1 this formation has mean porosity values ranging from 15 to 17% with net to gross values ranging from 65 to 95% (Newman 1999), equivalent to the better reservoir qualities reported from the EISB (Haig *et al.* 1997; Yaliz 1997).

The exception to this general rule is found in the northern part of the CISB, where well data suggest a change of facies and a deterioration in reservoir quality (Figs 9 and 10). In well UK108/30-1A, the Ormskirk Sandstone Formation was found to comprise 351 m of argillaceous sandstones interbedded with mudstone, siltstone and anhydrite with an overall net to gross value of just 7%. Net sandstone porosity in this well averages just 13%, reflecting high clay content and the effects of anhydrite cementation (Fig. 9). Highly argillaceous sandstones were also encountered in the Ormskirk Sandstone Formation in well UK107/1-1. The deterioration in reservoir quality in this area appears to reflect the development of a playa lake influenced environment of deposition, in the NE of the CISB, during early Triassic time (Fig. 10). Reservoir quality improves as the formation thins towards the south of the basin, with fluvial and aeolian sandstones predominating in wells IR42/12-1 and IR42/12-2 (Maddox *et al.* 1995). Within the uppermost Ormskirk Sandstone Formation of the EISB, playa-influenced deposition is reported to have occurred only in the more actively subsiding parts of that basin (Jackson *et al.* 1995; Thompson & Meadows 1997). An analogous facies relationship is observed in the Wessex-Channel Basin of southern England, where the development of mudstone and anhydrite-rich sandstones of the Sherwood Sandstone Group coincides with an isopach thick in the region of the Seaborough-1 well (Butler 1998).

Seal potential

Evaporites and shales of the Mercia Mudstone Group were deposited over the entire Irish Sea area during Mid- to Late Triassic time. These lithologies act as the regional seal to the Ormskirk Sandstone Formation, although the presence of halite beds is thought to be critical to seal efficiency because of the multiple periods of fault reactivation during Early Cretaceous time (Late Cimmerian phase) and the Tertiary period (Mikkelsen & Floodpage 1997). With the exception of well UK111/29-1, in the western part of the Peel Basin, all available wells in the study area penetrated at least 20 m of halite beds in the Mercia Mudstone Group, suggesting that seal is a relatively low risk factor for exploration, throughout much of the Irish Sea area. However, the seal effectiveness of the Mercia Mudstone Group is locally compromised around the truncated periphery of the Permo-Triassic basins, and over intra-basinal highs, where Mesozoic and Cenozoic uplift has removed the halite-bearing units. With respect to the EISB, Seedhouse & Racey (1997) have suggested that there is an increased risk of top seal failure, where the Ormskirk Sandstone Formation is shallower than 600 m sub-sea.

In CISB wells IR42/12-1, IR42/12-2 and IR42/16-1 a reduction in the thickness of the intra-Mercia halite beds, coupled with the presence of porous sandstone units 100–200 m above the base of the Mercia Mudstone Group, may compromise cross-fault seal potential. These wells are interpreted to lie close to the margins of the Late Triassic evaporitic basin, where the influence of coarse clastic deposition was greater. Consequently, the sealing capacity of the Mercia Mudstone Group also deteriorates towards the south of the CISB (Maddox *et al.* 1995).

In spite of the severe uplift experienced by the Irish Sea area, the risk of seal failure is relatively low within the northern CISB, EISB and Solway basins. Seal failure is considered to be a significant exploration risk parameter in the Peel Basin and the southern CISB.

Source rock development

The principal source rocks for hydrocarbons in the EISB are marine Namurian shales of Pendleian to Yeadonian age (Armstrong *et al.* 1997). Palaeogeographical reconstructions suggest that basinal marine shales of this age were restricted to an east–west basin extending from the Dublin Basin through the Kish Bank Basin into the EISB (Fig. 3b). To the south of this axis, potential Namurian source rocks are likely to be absent, through non-deposition, over much of the CISB. In contrast, to the north, in the Peel and Solway basins, shallow marine and deltaic conditions had already been established in Dinantian time, suggesting that the subsequent environment in Namurian time was not favourable for the deposition of oil-prone source rocks (Newman 1999). Even if source rocks were deposited in this area, severe Variscan uplift and erosion has resulted in the removal of virtually all of the Namurian strata from the Peel and Solway basins.

Geochemical analysis of the Dinantian platform carbonates encountered in Peel Basin wells UK111/25-1A and UK111/29-1 reveal very low levels of total organic carbon (TOC), indicating minimal source rock potential (Table 1). Geochemical data from exploration well IOM112/19-1 and from outcrop show that the Dinantian sediments deposited in the Solway Basin have poor gas-prone source potential, primarily related to the presence of minor coal interbeds. Oil-prone shales of Dinantian age,

Fig. 7. (**a**) Regional NW–SE seismic line across the Solway Basin showing the synclinal form of Triassic structure and detached fault blocks of the Sherwood Sandstone Group. Thick preserved Upper Carboniferous section on the margin of the Lake District Massif and the strong angular unconformity between the Carboniferous and overlying Permo-Triassic rocks in the centre of the basin should be noted. (**b**) Regional NE–SW seismic line along the Solway Basin reveals thinning of the Sherwood Sandstone Group to SW and development of "collapse trenches" on the flanks of the basin. A detached deformation style predominates in the basin centre with faults in the Sherwood Sandstone Group apparently detaching on a Late Permian evaporite layer. (**c**) Regional NW–SE seismic line across the Peel Basin indicating half-graben form of the basin and Tertiary igneous activity associated with the northwestern faulted margin (adapted from Newman 1999; all three seismic lines are courtesy of Jebco Seismic UK Ltd.).

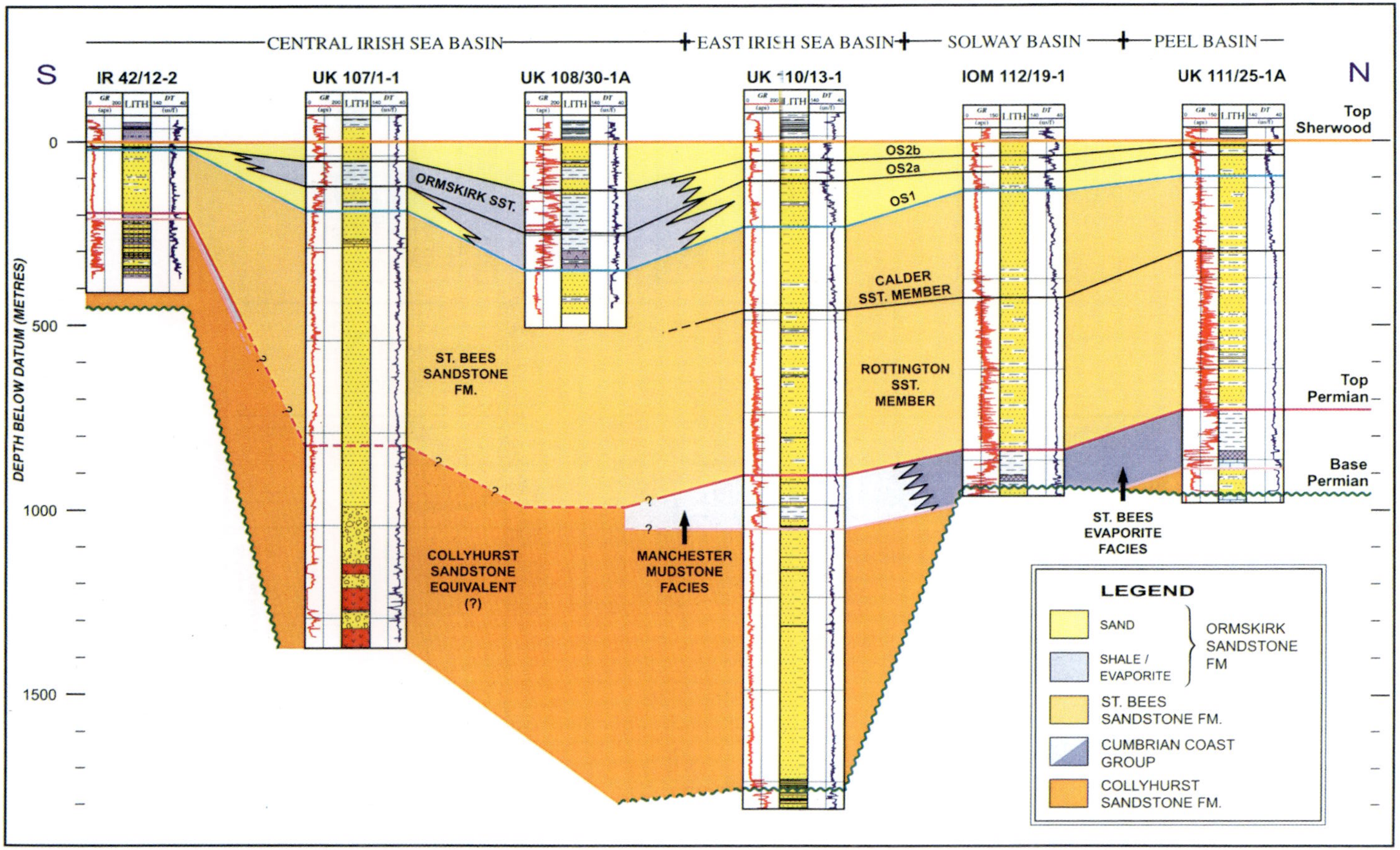

Fig. 8. Regional stratigraphic well log correlation for the Permian and Lower Triassic sequences of the Irish Sea area. Datum is top Sherwood Sandstone Group. Intra-Triassic events can be correlated across the Irish Sea area, including the tripartite subdivision of the Ormskirk Sandstone Formation (OS1, OS2a, OS2b), suggesting that the Upper Permian to Lower Triassic sediments were deposited in response to regional subsidence. In the northern part of the CISB a deterioration in the reservoir quality of the Ormskirk Sandstone Formation is observed as a result of the development of a playa lake facies in the area of well UK108/30-1A.

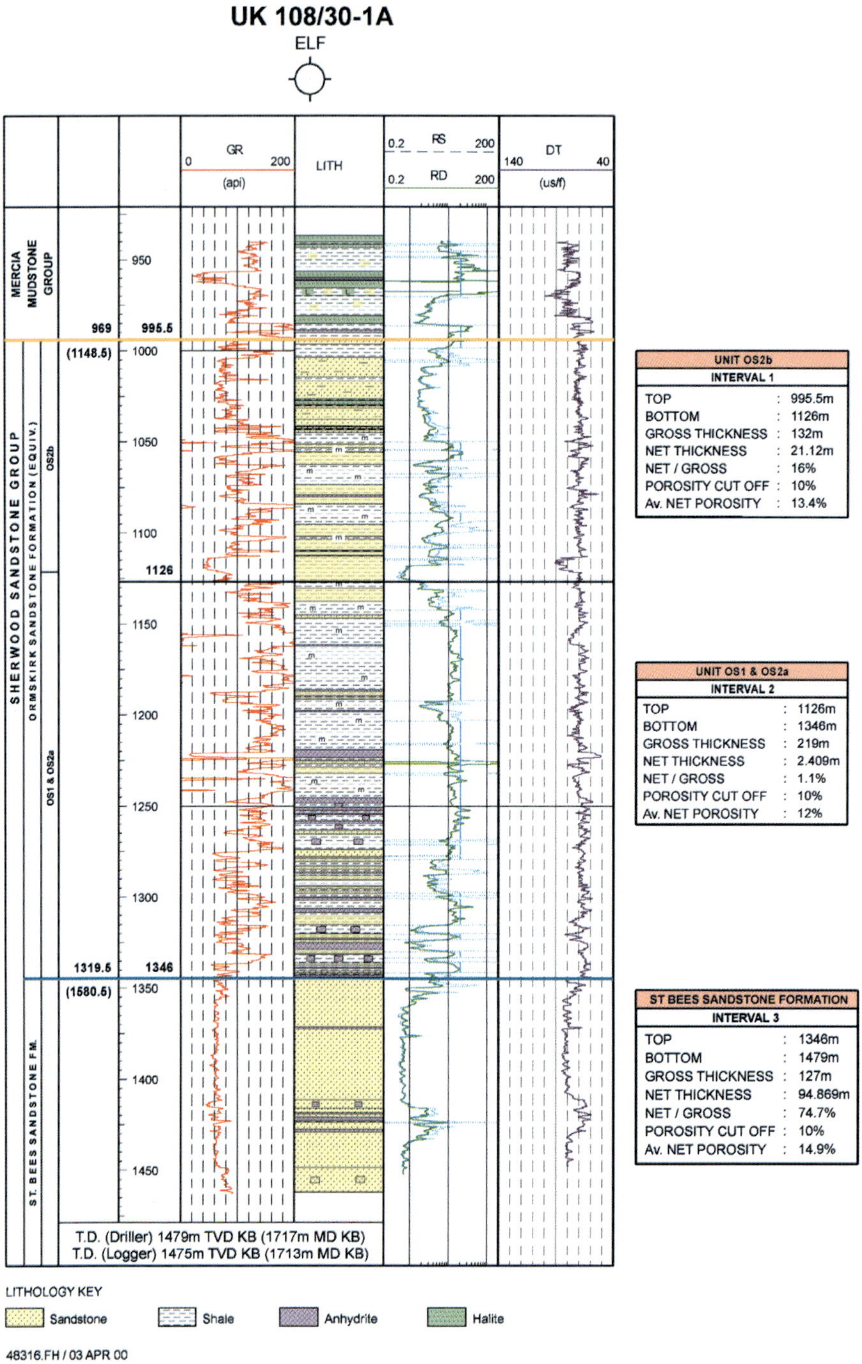

Fig. 9. Summary log of the Sherwood Sandstone section encountered in well UK108/30-1A. Petrophysical parameters for three reservoir intervals are summarized.

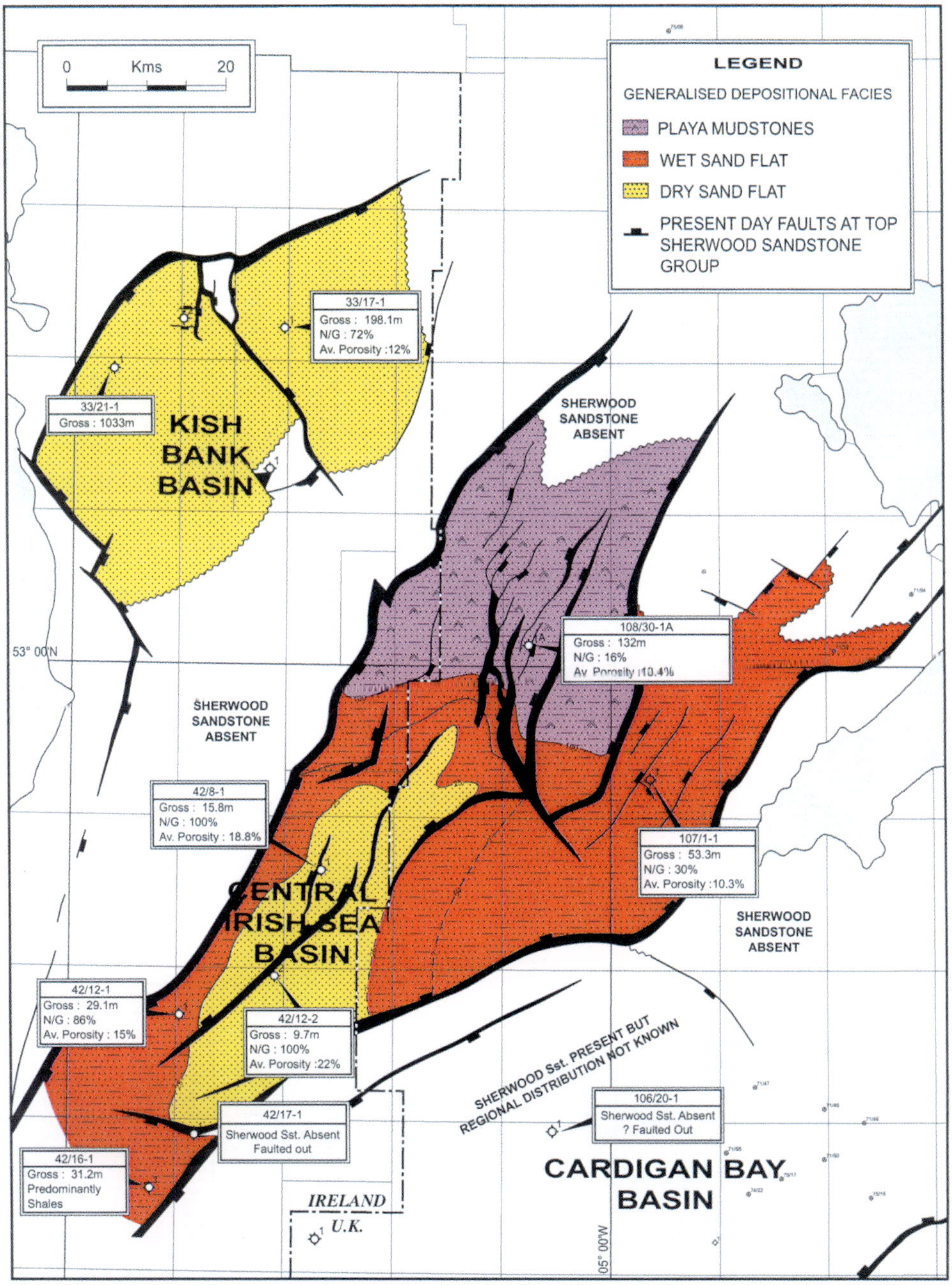

Fig. 10. Generalized depositional facies map, for the uppermost unit (OS2b) of the Ormskirk Sandstone Formation, CIS area. Well data indicate that net to gross value and average porosity of this unit decrease towards the NE of the CISB.

found on the southern coast of the Isle of Man, are thought to be lagoonal deposits and are of limited lateral extent (Newman 1999). Proprietary kerogen facies mapping of the Namurian sequence, based on available well data and outcrop analyses, indicates that in the vicinity of the Solway Basin the kerogen type is generally inertinitic, becoming more vitrinitic and sapropelic southwards towards the North Wales coast.

Table 1 *Summary of source rock potential of the Carboniferous sections encountered in the Irish Sea area*

Basin/sample location (data source)	Potential source horizon	TOC (av.) (wt %)	Pyrolysis S2 (av.) (mg g^{-1} rock)	Cumulative coal seam thickness (m)	Source quality/type
EISB/N Wales Coast (Armstrong ***et al.*** 1997)	Namurian Holywell Shales	3–5	7–13	n/a	Very good, oil-prone (Type II)
Solway/IOM112/19-1 (TFEEUK, proprietary)	Dinantian clastic deposits	<1	0.3	n/a	Very poor, gas-prone (Type IV)
Peel/UK111/25-1A (TFEEUK, proprietary)	Dinantian carbonates	0.04	n/a	n/a	None
Kish Bank/IR33/22-1 (Geochem 1992)	Westphalian B–D	1–2	1–2	11	Poor gas source with richer coaly horizons (Type III–IV)
Central Irish Sea IR42/12-1 (Robertson Research International Limited 1986)	Westphalian C–D	1–2	<2	<2	Poor gas source (Type IV)
IR42/17-1 (Geochem 1992)	Westphalian C–D	1–2	1–2	2.7	Poor gas source (Type IV)

Cumulative coal seam thickness is calculated for Westphalian section only.

The only potential source rocks in the Central Irish Sea area are coals developed in the Westphalian strata, which were encountered in Kish Bank Basin well IR33/22-1 (Westphalian B–D) and CISB wells IR42/12-1 and IR42/17-1 (Westphalian C–D). Previous workers have referred to the 'good quality' source rock potential of the Westphalian in these wells (Maddox *et al.* 1995). However, re-examination of the geochemical data from the above wells suggests that the 'encouraging' TOCs of 3–74% and extractable hydrocarbons (S2s) of 1–46 mg g^{-1} rock, reported by Maddox *et al.* (1995) from IR42/17-1, probably relate to individual coal fragments. Bulk cuttings samples from this well (Geochem 1992) show a maximum present-day TOC of 2.82% and S2 of 2.03 mg g^{-1} rock. The main lithology is shale, which is organically lean and dominated by inertinitic kerogens (Corcoran & Clayton 1999). Consequently, the source rock potential of the Westphalian C–D section, in this well, is limited to a few thin coal seams.

The thickness and organic content of the coal seams encountered in CISB wells is lower than in Kish Bank Basin well IR33/22-1 (Table 1). Onlap of the Wales–Brabant Massif during Late Carboniferous time has resulted in the absence of the organically richer Westphalian A–B interval from the Central Irish Sea wells (Figs 4 and 6). Furthermore, the Late Carboniferous environment of deposition became progressively more subaerial, and therefore coal poor, towards the south. Similar trends have been reported in the English Midlands, where the Coal Measures pass southwards into alluvial barren red bed facies (Besly 1988; Jackson *et al.* 1995).

It is possible that well IR42/17-1 is not representative of the CISB as a whole because of its position on the Mid Irish Sea High (Maddox *et al.* 1995). Seismic line B–B′ (Fig. 5b) does show some thickening of the Westphalian and Dinantian strata across the pair of faults to the NE of well IR42/17-1. However, the generally opaque seismic facies in the Westphalian section corresponds to the coal-poor seismic character documented at IR42/17-1. Also, the underlying bright seismic reflectors closely resemble those calibrated as Dinantian strata in IR42/17-1 and there is no suggestion of a major synsedimentary wedge of Early Carboniferous age, as observed to the NE in the Kish Bank Basin (Fig. 5a) and in the EISB. It is concluded that the presence of a viable source rock is the major exploration risk factor in the Peel, Solway and Central Irish Sea basins.

Source rock maturation and migration

Several workers have discussed the timing of tectonic uplift and hydrocarbon migration in the Irish Sea area (Hardman *et al.* 1993; Maddox *et al.* 1995; Green *et al.* 1997; Haig *et al.* 1997; Yaliz 1997; Duncan *et al.* 1998; Corcoran & Clayton 1999; Newman 1999; Green *et al.* 2001). Previous work by Hardman *et al.* (1993) concluded that the main phase of uplift in the EISB occurred during the Early Tertiary period and that an Early Cretaceous ('Late Cimmerian') phase of uplift was less significant. These workers suggested that hydrocarbon generation and migration, from Carboniferous source rocks, may have started during Jurassic time, with a final phase of expulsion occurring during Early Tertiary time, as a result of elevated heat flows induced by igneous activity. Maddox *et al.* (1995) reached similar conclusions for the CISB.

Recently published AFTA and vitrinite reflectance (VR) data from the EISB area (Green *et al.* 1997) have confirmed the likelihood that maximum palaeo-temperatures and peak maturity were reached in the EISB during Early Tertiary time, before regional Early Tertiary uplift (Fig. 11). However, Green *et al.* (1997) suggested that peak maturity may have been achieved before Tertiary time, in areas to the west and north of the EISB, during either Early Cretaceous or Late Carboniferous time (Fig. 11). Duncan *et al.* (1998) concluded, from the interpretation of the AFTA data for CISB well IR42/12-1, that maximum palaeo-temperature and peak maturity had occurred, in that area, during Early Cretaceous time (120–100 Ma), before Late Cimmerian uplift.

New data presented here (Figs 11 and 12) indicate that maximum palaeo-temperatures were also experienced by Permo-Triassic sediments, in CISB well UK107/1-1 (Fig. 12) and Peel Basin well UK111/29-1, during Early Cretaceous time. This evidence suggests that peak maturity would have been reached earlier in the CISB and Peel Basin than in the EISB. AFTA data from Kish Bank Basin wells (Wilson, pers. comm.) and proprietary AFTA data from CISB well UK108/30-1A and Solway Basin well IOM112/19-1 suggest that maximum palaeo-temperatures were reached, at these locations, during the Tertiary period. However, in all cases, the apatite grains have been completely annealed by the Tertiary event, thereby obscuring evidence of the earlier thermal history.

Corcoran & Clayton (1999) have recently presented interesting arguments suggesting that the palaeo-temperatures experienced by

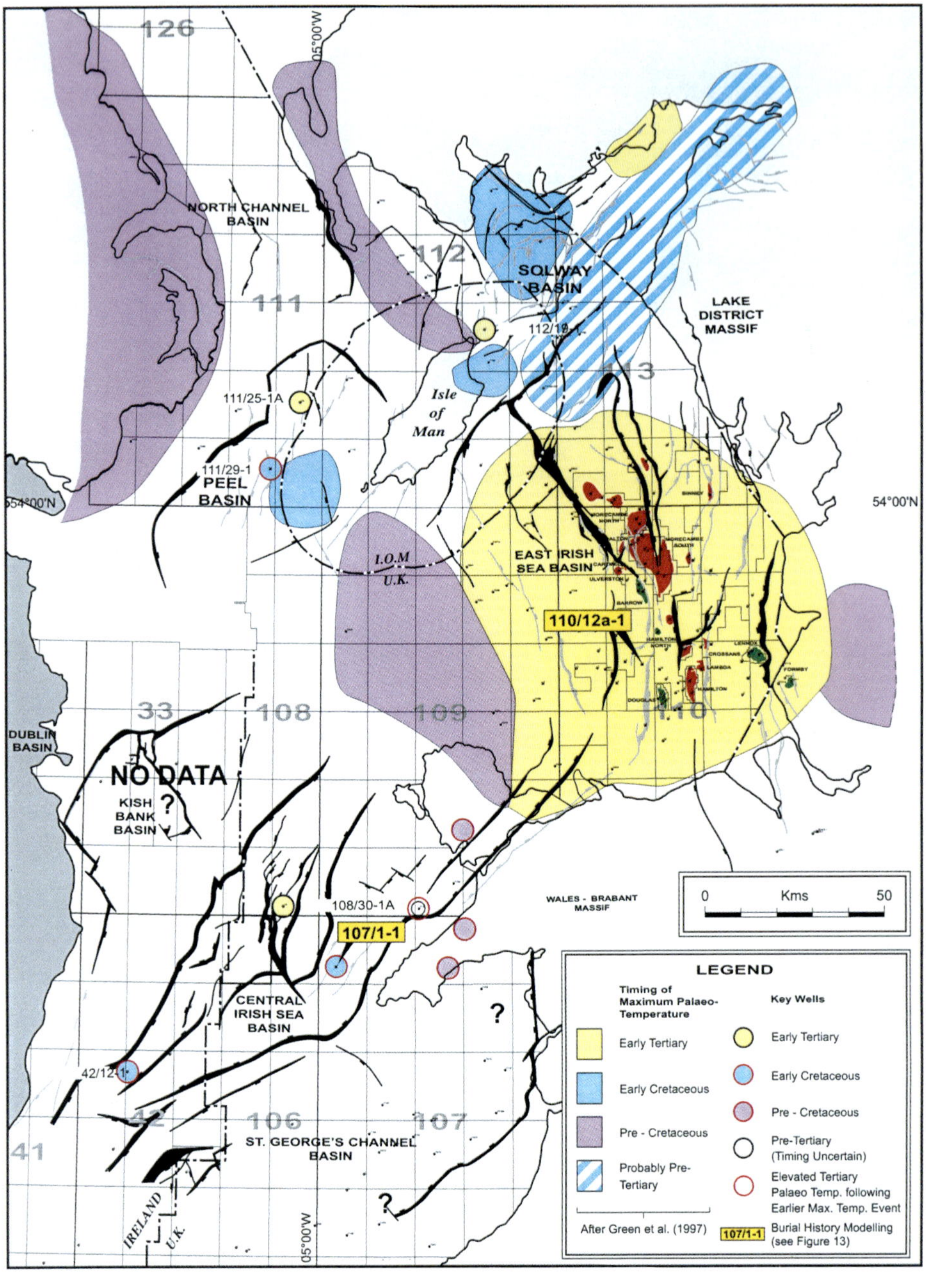

Fig. 11. Timing of cooling from maximum palaeo-temperatures deduced from AFTA. Maximum palaeo-temperatures were reached in the EISB during Early Tertiary time but to the west and north of the EISB peak palaeo-temperatures may have been achieved during Early Cretaceous or Late Carboniferous time. (Adapted from Green *et al.* 1997.)

Carboniferous sediments were reached before Permian uplift in the CISB and adjacent areas, based on the apparently high geothermal gradients indicated by VR data from the Carboniferous sections. Whereas some areas of the Irish Sea almost certainly reached peak maturity in Carboniferous times (Fig. 11), it is difficult to reconcile the conclusions of those

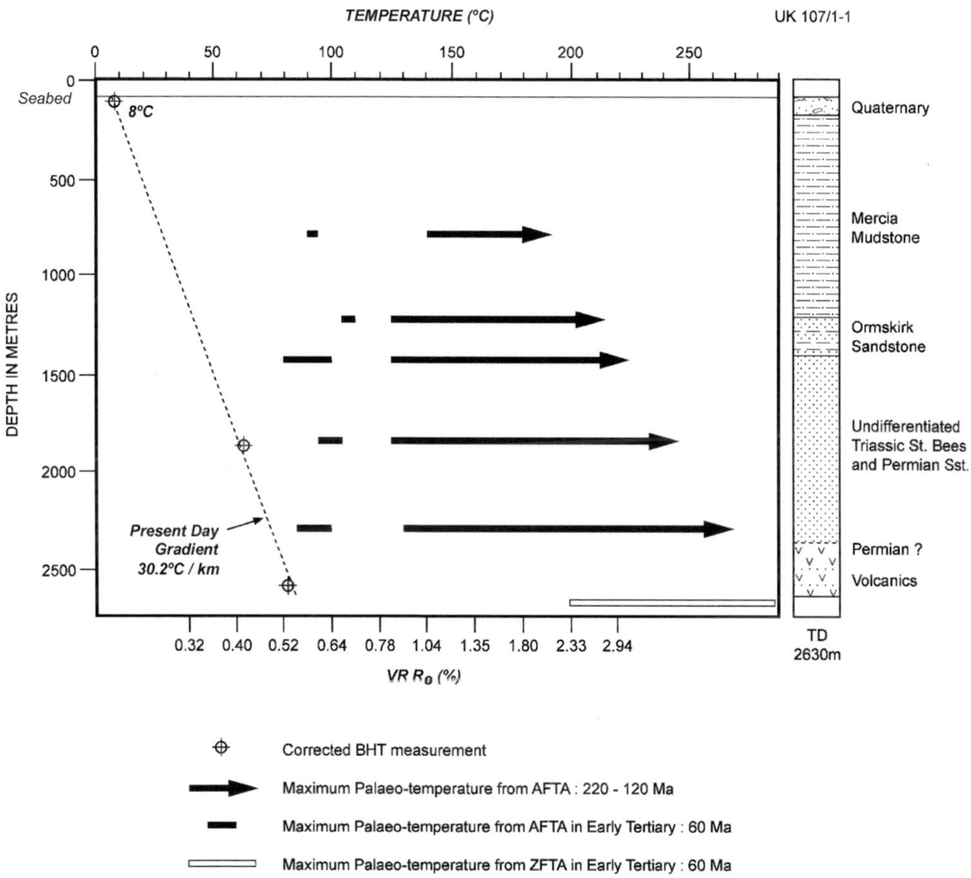

Fig. 12. Thermal history calibration, well UK107/1-1. Maximum palaeo-temperatures and timing interpreted from AFTA and ZFTA. A present-day geothermal gradient of 30.2 °C km^{-1} has been calculated using Horner-corrected BHTs. It should be noted that conversion of AFTA temperature estimates to VR R_o equivalent is based on the equation of Burnham & Sweeney (1989), assuming a heating duration of 10 Ma.

workers with the AFTA data from the Permo-Triassic rocks or with the preservation of Stephanian sediments and the lack of angular unconformity evident on seismic data in the Central Irish Sea area (Fig. 5).

Burial history modelling. To evaluate the timing of source rock maturation and hydrocarbon migration, burial history modelling (using the IFP-Beicip Genex software) was performed for a number of wells. Burial history modelling of Solway Basin well IOM112/19-1 was previously reviewed by Newman (1999). Data presented here from CISB well UK107/1-1 have been compared with a published model from EISB well UK110/12a-1 (Hardman *et al.* 1993) (Fig. 13).

(1) Well UK107/1-1 reached a total depth (TD) of 2630 m in volcanic rocks of probable Permian age (Figs 12 and 13a). These volcanic rocks were overlain by a succession of Permo-Triassic sandstones including the St Bees and Ormskirk Sandstone formations. A truncated Mercia Mudstone Group is unconformably overlain by Quaternary sediments. The post-Triassic stratigraphic and denudation history at this location has been estimated from published regional trends (Murdoch *et al.* 1995; Cope 1997).

Five apatite fission-track analyses and one zircon fission-track analysis (ZFTA) were made on samples from well UK107/1-1 (Fig. 12). The fission tracks from the apatite grains proved to be completely annealed, indicating palaeo-temperatures in excess of 110 °C. However, the fission

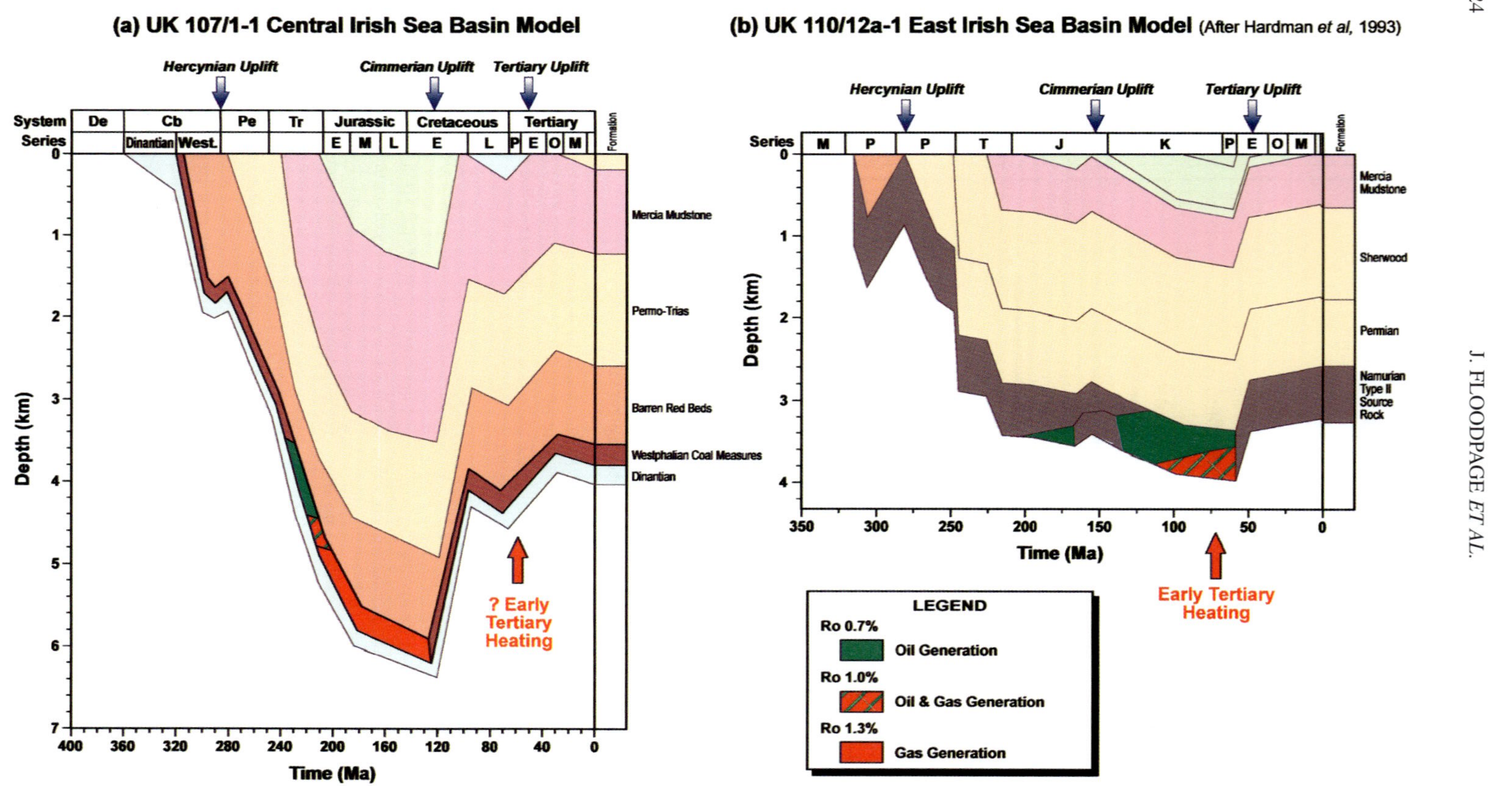

Fig. 13. Comparison of burial history model for (**a**) Central Irish Sea well UK107/1-1 with published model for (**b**) EISB well 110/12a-1 (after Hardman *et al.* 1993). The thicknesses of Dinantian–Westphalian sediments have been estimated for both locations from seismic interpretation and isopach mapping. These models suggest that the onset of hydrocarbon generation is earlier in the CISB (Mid-Triassic v. Early Jurassic time) and that limited Early Cretaceous uplift occurred in the EISB relative to the large magnitude (2000 m) of Early Cretaceous uplift modelled for the CISB.

tracks from these samples show clear evidence of cooling between 220 and 120 Ma, spanning Early Jurassic to Early Cretaceous time (Fig. 12). Traces of a later cooling episode (from temperatures of 90–110 °C), during Early Tertiary time (60 Ma), are also recorded in these samples. Partial annealment of the fission tracks in zircons, obtained from the deepest sample, indicates palaeo-temperatures in excess of 200 °C between 62 and 53 Ma, and could relate to heating caused by proximity to igneous bodies in the well.

Complete annealment of the apatite fission tracks in well UK107/1-1 prevents estimation of the geothermal gradient at the time of maximum palaeo-temperature. However, combined AFTA and VR data from the less deeply buried well IR42/12-1 (Duncan *et al.* 1998), defined a palaeo-geothermal gradient of 37.1 °C km^{-1} during Early Cretaceous time. The relatively low magnitude of the Early Tertiary palaeo-geothermal gradients, defined by AFTA data in both wells, strongly suggests that advective heat transfer has occurred in the CISB, during Early Tertiary time. The vertical migration of heated fluids, associated with the emplacement of sills and dykes, has been suggested as a mechanism for heat transfer in the EISB, during Early Tertiary time (Hardman *et al.* 1993; Green *et al.* 1997).

The burial history model for well UK107/1-1 indicates that Carboniferous source rocks reached the oil window during Mid-Triassic time and the gas window from Early Jurassic time (Fig. 13a). The AFTA data are consistent with up to 2000 m of uplift during Early Cretaceous time, which would have resulted in the 'switching off' of hydrocarbon generation at that time. Reburial during Late Cretaceous and Early Tertiary time was probably insufficient to permit resumption of hydrocarbon generation, even allowing for the effects of an Early Tertiary heating event. Early Tertiary AFTA palaeo-temperatures for the deepest sample (80–100 °C at 2271 m) (Fig. 12a) are only slightly higher than at present day (*c.* 76 °C), suggesting that limited (*c.* 700 m) post-Early Tertiary uplift has occurred at this location.

(2) Well UK110/12a-1 is located on the SW flank of the EISB. The interpreted burial history at this location is significantly different from the model employed at well UK107/1-1 in the CISB. Reduced Late Triassic (Mercia Mudstone Group) and Jurassic burial has been modelled at well UK110/12a-1 and the timing of Cimmerian uplift is inferred to be of Late Jurassic rather than Early Cretaceous age (Hardman *et al.* 1993). The onset of hydrocarbon generation occurred during Early Jurassic time, somewhat later than modelled for UK107/1-1, reflecting the shallower depth of burial of the SW flank of the EISB. However, the critical difference from the CISB model is that the AFTA data from UK110/12a-1 are consistent with only a limited Cimmerian uplift (Hardman *et al.* 1993). Consequently, although hydrocarbon generation in the EISB was interrupted by Cimmerian uplift, the generation of oil, and subsequently gas, resumed under the influence of Cretaceous reburial, possibly augmented by an Early Tertiary heating event.

It is concluded that the thermal history of the Central Irish Sea, Solway and Peel basins differs significantly from that of the EISB. The Irish Sea area was probably subjected to a widespread igneous heating event, during Early Tertiary time, which resulted in the pervasive circulation of hot fluids. Late-stage condensate and gas generation in the EISB may have been associated with this heating event but is primarily the result of maximum burial being achieved, by the Carboniferous source rocks, during Late Cretaceous to Early Tertiary time (Hardman *et al.* 1993; Yaliz 1997). The absence of this critical late generation phase in the CISB, Peel and Solway basins probably resulted from the Early Cretaceous (Late Cimmerian) uplift of these basins. The magnitude of this uplift, and the subsequent limited reburial during Cretaceous time, prevented further maturation of any Carboniferous source rocks in these basins. It is likely that the Kish Bank Basin has experienced a similar evolution to that of the CISB. However, our understanding of the pre-Cenozoic thermal history of this basin is limited by the paucity of vitrinite in the preserved Mesozoic and Cenozoic sections and the presence of fully annealed apatite fission tracks caused by Tertiary heating.

Timing of structure formation

Seismic interpretation and structural mapping have been used to develop an understanding of the timing of structure formation in the CISB, Peel and Solway basins.

Central Irish Sea Basin. Basin-wide mapping at top Sherwood Sandstone Group level (Fig. 14) indicates that the CISB is divided into separate western and eastern fault-bounded synclines by a central axial high trend (areas shallower than 1500 m), which continues along the length of the basin. All the major structural culminations are located along this axial trend, including the closures tested by wells IR42/8-1, IR42/12-2, UK107/1-1 and UK108/30-1A.

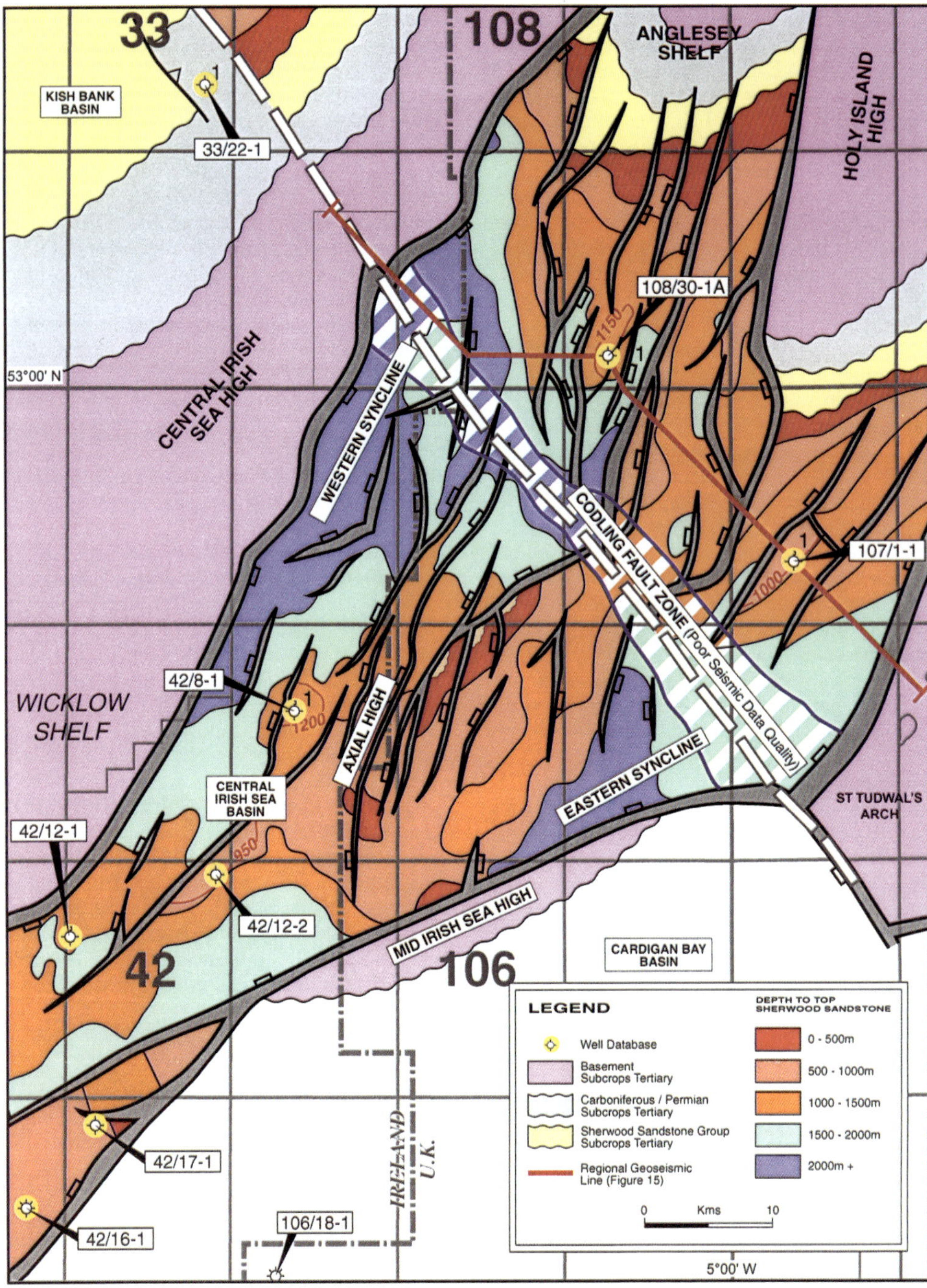

Fig. 14. Simplified depth structure map at top Sherwood Sandstone Group for the CISB. This map illustrates the major structural units in the CISB, which is dominated by a NE–SW-trending axial high (areas shallower than 1500 m), which extends along the length of the basin. All the major structural culminations are located along this axial trend including the valid structural closures tested by wells IR42/8-1, IR42/12-2, UK107/1-1 and UK 108/30-1A.

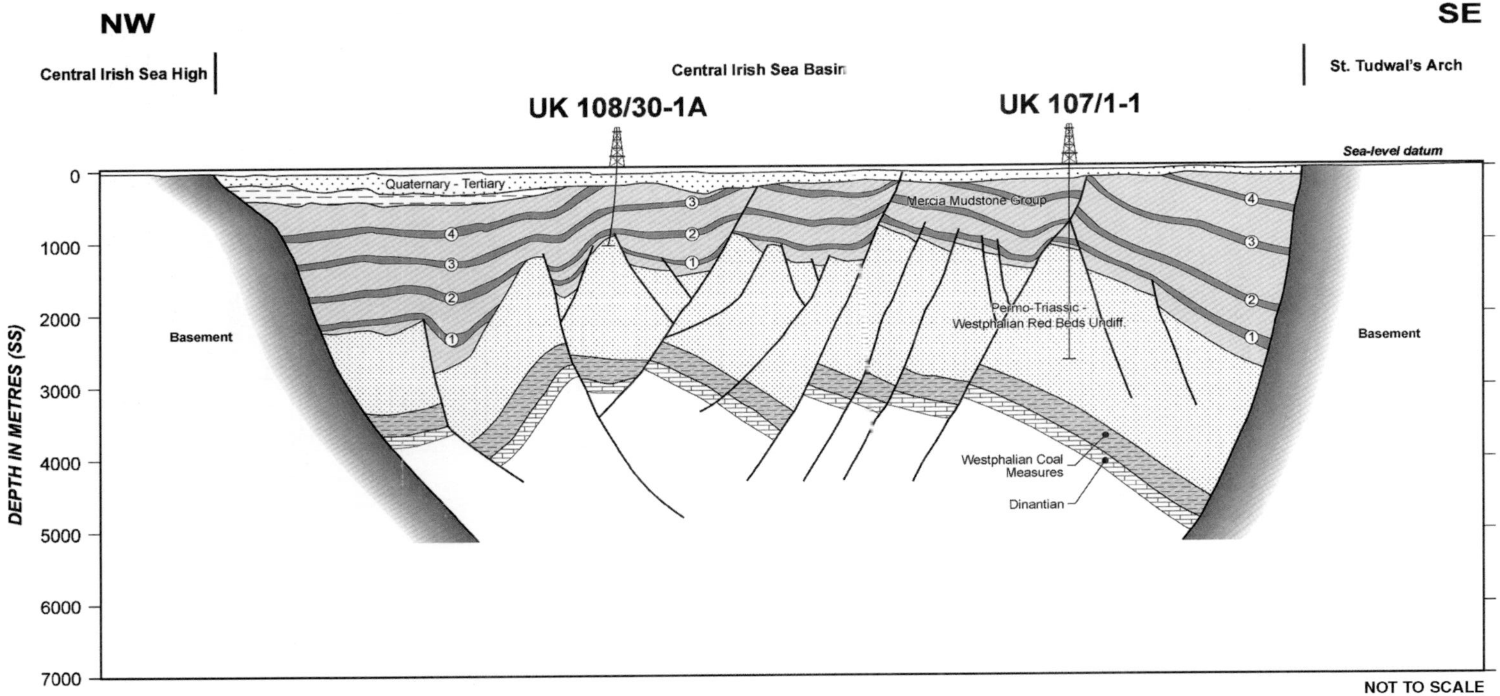

Fig. 15. NW–SE geoseismic line extending across the northern part of the CISB (see Fig. 14 for location). This line illustrates the structural morphology of the Central Irish Sea Basin, which consists of opposing fault-bounded synclines separated by a central axial high from which the younger halite beds (3 and 4) of the Mercia Mudstone Group have been removed by erosion.

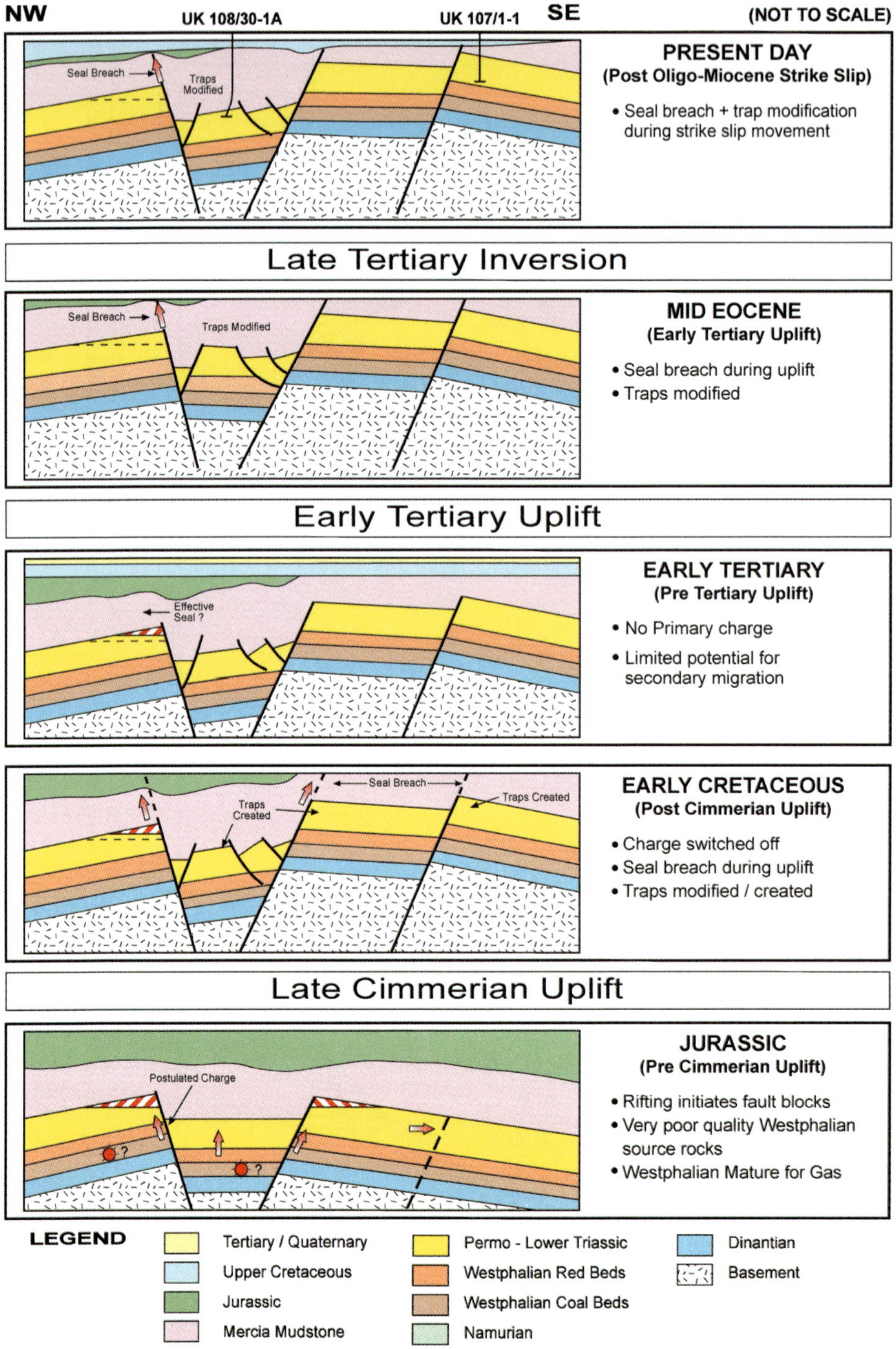

Fig. 16. Structural evolution model for the northern CISB area (based upon geoseismic line presented in Fig. 15). The section illustrates the effect of Cretaceous and Tertiary uplift on basin structure and hydrocarbon traps, with uplift resulting in the likely breaching of early-formed (Jurassic) traps.

A geoseismic section (Fig. 15) clearly shows truncation of younger intra-Mercia Mudstone Group reflectors towards the crest of the axial high, indicating that the formation of this anticlinal flexure post-dates the Triassic age of these sediments. Locally, within the CISB, the entire Mercia Mudstone Group has been removed by erosion and Tertiary sediments (?Oligo-Miocene) rest directly upon the Sherwood Sandstone Group (Fig. 14). However, the anticlinal morphology of the CISB axial high suggests an origin related to basin-scale exhumation, rather than localized footwall uplift caused by rotation of individual fault blocks.

Unfortunately, the seismic data do not constrain the timing of uplift and erosion more precisely than to post-Triassic and pre-?Oligocene age. Regional stratigraphic evidence indicates that the main phases of regional uplift were during Early Cretaceous (Late Cimmerian) and Early Tertiary time. Evidence from AFTA data suggests that the Early Cretaceous phase of uplift was the more severe of the two in the CISB. Figure 16 is a schematic 2D reconstruction of the CISB (along the geoseismic line presented in Fig. 15) illustrating the possible effect of uplift on basin structure and trap integrity. This interpretation suggests that the CISB axial high formed during Early Cretaceous time. Furthermore, most of the relief, on the mapped fault-block closures, was formed during Early Cretaceous and Tertiary uplift phases and therefore post-dates hydrocarbon charge in the CISB. It is possible that tilted fault-block closures were formed during Late Jurassic extension, before uplift, and that these traps could have been charged by gas sourced from the poor-quality Westphalian C–D Coal Measures. However, severe Late Cimmerian uplift would have greatly modified these early traps and would probably have resulted in fault seal breach, causing loss of gas to the surface.

One possible explanation for the Late Cimmerian development of the axial high is that it formed in response to extensional faulting at a time of regional Early Cretaceous exhumation (Hawkes *et al.* 1998). In this scenario, maximum erosion occurred in the footwall blocks located along the axial high, with limited erosion occurring in the adjacent fault-bounded synclinal areas. Subsequent, Early Tertiary compression may have folded the axial high, causing some additional inversion and erosion, as suggested by Maingarm *et al.* (1999).

Solway and Peel basins. The Peel and Solway basins, as defined by mapping of the top Sherwood Sandstone Group, are the isolated remnants of a previously more extensive cover of Permo-Triassic rocks (Fig. 17). The Solway Basin is a NE–SW-trending syncline at Permo-Triassic level, which was formed by the uplift of the surrounding structural highs (Fig. 7a and b). Extensional faulting around the perimeter of the basin, probably resulting from uplift of the margins, combined with dip towards the basin centre, created the structural traps tested by wells UK112/15-1 and IOM112/19-1. Burial history modelling (Newman 1999), together with evidence from uraninite inclusions found in bitumens in Carboniferous outcrops along the northern margin of the Solway Basin (Parnell 1995, 1997), suggests that limited migration of hydrocarbons may have taken place during Jurassic time. AFTA data from the Isle of Man and northern coast of the Solway Basin (Green *et al.* 1997) indicate probable uplift during Early Cretaceous time. However, Quirk & Kimbell (1997) proposed Late Jurassic faulting to explain the major fault displacements observed to the east of the Isle of Man. Potential hydrocarbon traps were therefore probably initiated at this time, but were further modified during Early Cretaceous and Tertiary time. These data suggest that, even if a significant source rock had been deposited and preserved in the Solway Basin, trap formation is likely to have post-dated hydrocarbon migration.

In contrast, the Peel Basin manifests an elongate NE–SW-trending half-graben morphology (Fig. 17). Proprietary AFTA data from well UK111/29-1 suggest that the main phase of uplift probably occurred during Early Cretaceous time and resulted in the tilting of the basin towards the NW as observed in the seismic data (Fig. 7c). All the evidence suggests that, even if a significant source rock had been deposited and preserved in the Peel Basin, trap formation is likely to have post-dated hydrocarbon migration.

Discussion. As a result of the absence of a preserved Cretaceous sequence in the Irish Sea area, local evidence for an Early Cretaceous uplift event relies primarily on indirect inferences from AFTA data. However, there is widespread regional evidence for a major Late Cimmerian unconformity of Early Cretaceous age, in the Wessex-Channel and Celtic Sea basins (McMahon & Turner 1998), and along the Atlantic margin (Roberts *et al.* 1999). Hawkes *et al.* (1998) have related widespread Late Cimmerian uplift to the initiation of rifting and subsequent development of a 'break-up unconformity' in the Rockall Trough and Bay of

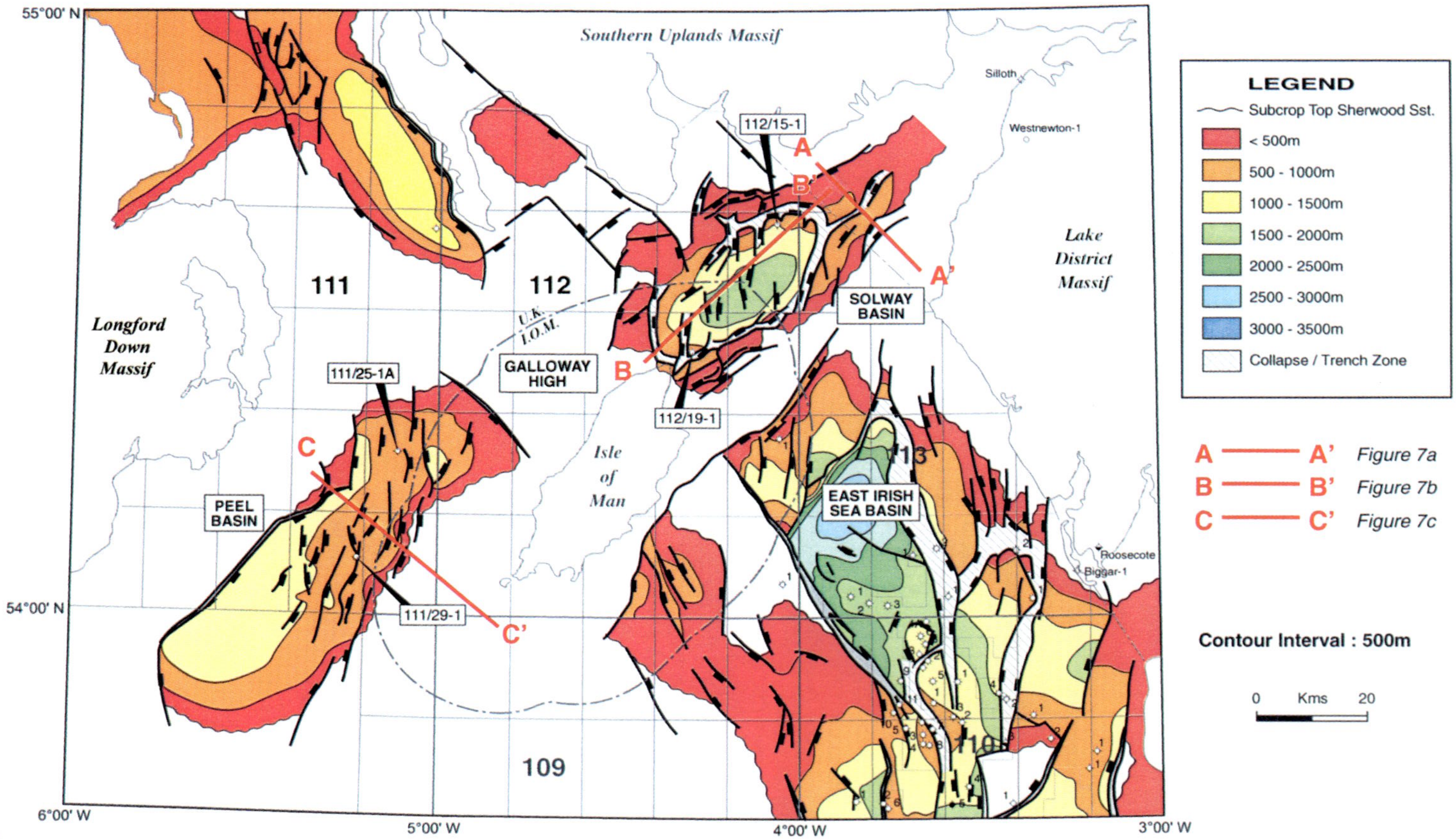

Fig. 17. Regional depth structure map at top Sherwood Sandstone Group for the Peel, Solway and adjacent basins. The synclinal form of the Solway Basin contrasts with the half-graben morphology developed in the Peel Basin. These basins are effectively the preserved remnants of a previously more extensive cover of Permo-Triassic rocks.

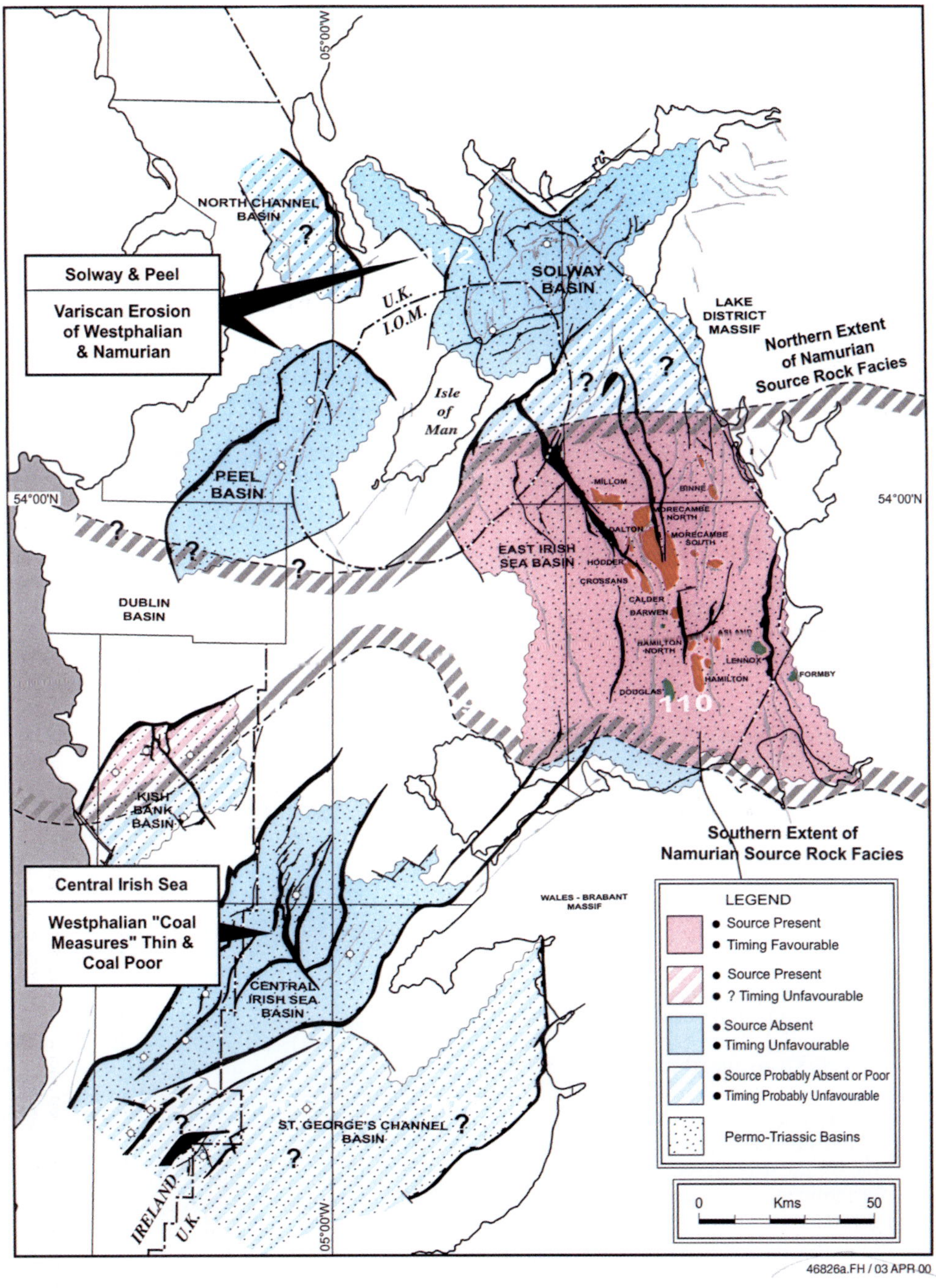

Fig. 18. Generalized hydrocarbon charge risk assessment map for the Triassic play of the Irish Sea area. This mapping indicates that source presence and quality plus the timing of hydrocarbon charge are significant risk factors in all Irish Sea basins with the exception of the EISB.

Biscay, and have suggested that a massive thermal pulse, during Early Cretaceous time, might have uplifted both basinal and marginal areas at this time.

Conclusions

In spite of the prolific success of the Triassic hydrocarbon play in the EISB, exploration elsewhere in the Irish Sea area has so far proved to be unsuccessful. Post-well evaluation suggests that 12 of the 15 wells drilled in basins adjacent to the EISB tested valid present-day structural closures at top Triassic reservoir level.

A regional assessment of the remaining hydrocarbon potential in the study area indicates that both reservoir presence (Lower Triassic Ormskirk Sandstone Formation) and seal presence (Mercia Mudstone Group) are low risk exploration parameters in these basins. However, reservoir effectiveness is significantly reduced in the northern part of the CISB, where a facies change has resulted in the deposition of argillaceous and anhydritic sandstones at the top of the Ormskirk Sandstone Formation. The sealing capacity of the Mercia Mudstone Group is locally reduced in the western part of the Peel Basin and the southern part of the CISB, in response to an increased coarse clastic influx and corresponding reduction in evaporite content, possibly reflecting a more proximal depositional environment in these areas during Late Triassic time.

This study suggests that the most significant factor controlling Triassic prospectivity in the Irish Sea area is the presence of an effective source rock in the Carboniferous sequences (Fig. 18). Palaeogeographical reconstructions indicate that the oil-prone Namurian marine shales are probably restricted to a east–west fairway, extending from the Bowland Basin through the EISB to the Dublin Basin. To the north of this axis, well and outcrop data reveal that shallow marine and deltaic conditions prevailed during Namurian time and were not favourable for the deposition of oil-prone source rocks. Furthermore, Variscan uplift and erosion was so severe in the Solway and Peel basins that rocks of this age have been removed by erosion. Geochemical analysis of the preserved Dinantian sediments in these basins indicates very limited source rock potential. Well and seismic data, together with regional onshore trends, suggest that the Namurian source rocks are absent over the CISB through non-deposition. The Westphalian sequence of the CISB is generally a shale-prone succession, which is organically lean and dominated by inertinitic kerogen. Gas-prone source potential is restricted to interbedded coal seams, but these are not well developed in the CIS area.

The relative timing of hydrocarbon generation and trap formation is also of critical importance to the hydrocarbon prospectivity of the Irish Sea area (Fig. 18). Recently acquired AFTA data indicate that the Peel, CIS and, probably, the Solway basins ceased generating hydrocarbons during Early Cretaceous time. Subsequent Late Cretaceous and Early Tertiary reburial was insufficient to achieve the temperatures necessary for the resumption of hydrocarbon generation, despite the effects of Tertiary heating. Seismic interpretation and structural analysis, complemented by AFT data, suggest that the main present-day structural closures mapped in the Peel, Solway and CIS basins were probably generated during successive Early Cretaceous and Tertiary uplift phases, thereby post-dating any hydrocarbon generation. The failure of three apparently valid structures in the Kish Bank Basin indicates that similar exploration risks may pertain to this basin. The success of the Triassic play in the EISB can therefore be partly attributed to limited Cimmerian uplift, which permitted (1) the preservation of earlier Jurassic structures and (2) the resumption of hydrocarbon generation during the Late Cretaceous and Early Tertiary period. The EISB is the only basin, in the Irish Sea area, where a functioning petroleum system has been proven. This suggests that the remaining prospectivity of the Irish Sea area is likely to be concentrated within the EISB.

The authors would like to thank TotalFinaElf Exploration UK plc and partners Enterprise Oil plc and Amerada Hess for permission to publish this paper. We are grateful to TGS Nopec, Western Geophysical and Jebco for allowing reproduction of speculative seismic lines, to Geochem for access to proprietary geochemical data, and to BHP for permission to publish AFT data from well UK107/1-1. We would also like to acknowledge the contribution of other Elf geoscientists to the work presented in this paper, particularly J. Mather, G. Desrousseau and L. Sides. Finally, we would like to acknowledge the constructive reviews of J. Redfern, G. Cayley and D. Corcoran.

References

ARMSTRONG, J.P., SMITH, J., D'ELIA, V.A.A. & TRUEBLOOD, S.P. 1997. The occurrence and correlation of oils and Namurian source rocks in the Liverpool Bay–North Wales area. *In*: MEADOWS, N.S., TRUEBLOOD, S.P., HARDMAN, M. & COWAN, G. (eds) *Petroleum Geology of the Irish Sea and Adjacent Areas.* Geological Society, London, Special Publications, **124**, 195–211.

BESLY, B.M. 1988. Palaeogeographic implications of late Westphalian to early Permian red beds, Central England. *In*: BESLY, B.M. & KELLING, G. (eds) *Sedimentation in a Synorogenic Basin Complex: the Carboniferous of North West Europe*. Blackie, Glasgow, 200–221.

BRODIE, J. & WHITE, N. 1994. Sedimentary basin inversion caused by igneous underplating: Northwest European continental shelf. *Geology*, **22**, 147–150.

BURNHAM, A.K. & SWEENEY, J.J. 1989. A chemical kinetic model of vitrinite reflectance maturation. *Geochimica et Cosmochimica Acta*, **53**, 2649–2657.

BUTLER, M. 1998. The geological history of the southern Wessex Basin—a review of new information from oil exploration. *In*: UNDERHILL, J.R. (ed.) *Development, Evolution and Petroleum Geology of the Wessex Basin*. Geological Society, London, Special Publications, **133**, 67–86.

COPE, J.C.W. 1997. The Mesozoic and Tertiary history of the Irish Sea. *In*: MEADOWS, N.S., TRUEBLOOD, S.P., HARDMAN, M. & COWAN, G. (eds) *Petroleum Geology of the Irish Sea and Adjacent Areas*. Geological Society, London, Special Publications, **124**, 47–59.

CORCORAN, D. & CLAYTON, G. 1999. Interpretation of vitrinite reflectance profiles in the Central Irish Sea Area: implications for the timing of organic maturation. *Journal of Petroleum Geology*, **22**, 261–286.

CROKER, P.F. & POWER, K.T. 1996. Revising the solid geology of the western Irish Sea—evidence from shallow gas, seismic, bathymetry, gravity and high resolution aeromagnetic data. *Abstracts, Annual Irish Geological Research Meeting, University College, Dublin, 23–25 February 1996*.

DUNCAN, W.I., GREEN, P.F. & DUDDY, I.R. 1998. Source rock burial history and seal effectiveness: key facets to understanding hydrocarbon exploration potential in the East and Central Irish Sea basins. *AAPG Bulletin*, **82**, 1401–1415.

DUNFORD, G.M. & DANCER, P.N. 2001. Hydrocarbon potential of the Kish Bank Basin: integration within a regional model for the Greater Irish Sea Basin. *In*: SHANNON, P.M., HAUGHTON, P.D.W. & CORCORAN, D.V. (eds) *The Petroleum Exploration of Ireland's Offshore Basins*. Geological Society, London, Special Publications, **188**, 135–154.

FRASER, A.J., NASH, D.F., STEELE, R.P. & EBDON, C.C. 1990. A regional assessment of the intra-Carboniferous play of northern England. *In*: BROOKS, J. (ed.) *Classic Petroleum Provinces*. Geological Society, London, Special Publications, **50**, 417–440.

GEOCHEM, 1992. *The Hydrocarbon Prospectivity of the Caernarvon and Cardigan Bay Basin Areas*. Proprietary report.

GREEN, P.F., DUDDY, I.R. & BRAY, R.J. 1997. Variation in thermal history styles around the Irish Sea and adjacent areas: implications for hydrocarbon occurrence and tectonic evolution. *In*: MEADOWS, N.S., TRUEBLOOD, S.P., HARDMAN, M. & COWAN, G. (eds) *Petroleum Geology of the Irish Sea and Adjacent Areas*. Geological Society, London, Special Publications, **124**, 73–93.

GREEN, P.F., DUDDY, I.R., BRAY, R.J., DUNCAN, W.I. & CORCORAN, D.V. 2001. The influence of thermal history on hydrocarbon prospectivity in the Central Irish Sea Basin. *In*: SHANNON, P.M., HAUGHTON, P.D.W. & CORCORAN, D.V. (eds) *The Petroleum Exploration of Ireland's Offshore Basins*. Geological Society, London, Special Publications, **188**, 171–188.

HAIG, D.B., PICKERING, S.C. & PROBERT, R. 1997. The Lennox oil and gas field. *In*: MEADOWS, N.S., TRUEBLOOD, S.P., HARDMAN, M. & COWAN, G. (ed.) *Petroleum Geology of the Irish Sea and Adjacent Areas*. Geological Society, London, Special Publications, **124**, 417–436.

HAQ, B.U., HARDENBOL, J. & VAIL, P.R. 1987. Chronology of fluctuating sea-levels since the Triassic. *Science*, **235**, 1156–1167.

HARDMAN, M., BUCHANAN, J., HERRINGTON, P. & CARR, A. 1993. Geochemical modelling of the East Irish Sea Basin: its influence on predicting hydrocarbon type and quality. *In*: PARKER, J.R. (ed.) *Petroleum Geology of Northwest Europe: Proceedings of the 4th Conference*. Geological Society, London, 809–821.

HAWKES, P.W., FRASER, A.J. & EINCHCOMB, C.C.G. 1998. The tectono stratigraphic development and tectonic history of the Weald and Wessex basins, southern England, UK. *In*: UNDERHILL, J.R. (ed.) *Development, Evolution and Petroleum Geology of the Wessex Basin*. Geological Society, London, Special Publications, **133**, 39–65.

IZATT, C., MAINGARM, S. & RACEY, A. 2001. Fault distribution and timing in the Central Irish Sea Basin. *In*: SHANNON, P.M., HAUGHTON, P.D.W. & CORCORAN, D.V. (eds) *The Petroleum Exploration of Ireland's Offshore Basins*. Geological Society, London, Special Publications, **188**, 155–169.

JACKSON, D.I., JACKSON, A.A., EVANS, D., WINGFIELD, R.T.R., BARNES, R.P. & ARTHUR, M.J. 1995. *United Kingdom Offshore Regional Report: the Geology of the Irish Sea*. HMSO, London.

JACKSON, D.I., JOHNSON, H. & SMITH, N.J.P. 1997. Stratigraphical relationships and a revised lithostratigraphical nomenclature for the Carboniferous, Permian and Triassic rocks of the offshore East Irish Sea Basin. *In*: MEADOWS, N.S., TRUEBLOOD, S.P., HARDMAN, M. & COWAN, G. (eds) *Petroleum Geology of the Irish Sea and Adjacent Areas*. Geological Society, London, Special Publications, **124**, 11–32.

MADDOX, S.J., BLOW, R. & HARDMAN, M. 1995. Hydrocarbon prospectivity of the Central Irish Sea Basin with reference to Block 42/12, offshore Ireland. *In*: CROKER, P.F. & SHANNON, P.M. (eds) *The Petroleum Geology of Ireland's Offshore Basins*. Geological Society, London, Special Publications, **93**, 59–77.

MAINGARM, S., IZATT, C., WHITTINGTON, R.J. & FITCHES, W.R. 1999. Tectonic evolution of the Southern–Central Irish Sea Basin. *Journal of Petroleum Geology*, **22**, 287–304.

McMAHON, N.A. & TURNER, J. 1998. The documentation of a latest Jurassic–earliest Cretaceous uplift throughout southern England and adjacent offshore areas. *In*: UNDERHILL, J.R. (ed.) *Development, Evolution and Petroleum Geology of the Wessex Basin*. Geological Society, London, Special Publications, **133**, 215–240.

MIKKELSEN, P.W. & FLOODPAGE, J.B. 1997. The hydrocarbon potential of the Cheshire Basin. *In*: MEADOWS, N.S., TRUEBLOOD, S.P., HARDMAN, M. & COWAN, G. (eds) *Petroleum Geology of the Irish Sea and Adjacent Areas*. Geological Society, London, Special Publications, **124**, 161–183.

MURDOCH, L.M., MUSGROVE, F.W. & PERRY, J.S. 1995. Tertiary uplift and inversion history in the North Celtic Sea Basin and its influence on source rock maturity. *In*: CROKER, P.F. & SHANNON, P.M. (eds) *The Petroleum Geology of Ireland's Offshore Basins*. Geological Society, London, Special Publications, **93**, 297–319.

NEWMAN, P.J. 1999. The geology and hydrocarbon potential of the Peel and Solway basins, East Irish Sea. *Journal of Petroleum Geology*, **22**, 305–324.

PARNELL, J. 1995. Hydrocarbon migration in the Solway Basin. *Geological Journal*, **30**, 25–38.

PARNELL, J. 1997. Fluid migration history in the north Irish Sea–North Channel region. *In*: MEADOWS, N.S., TRUEBLOOD, S.P., HARDMAN, M. & COWAN, G. (eds) *Petroleum Geology of the Irish Sea and Adjacent Areas*. Geological Society, London, Special Publications, **124**, 213–228.

QUIRK, D.G. & KIMBELL, G.S. 1997. Structural evolution of the Isle of Man and central part of the Irish Sea. *In*: MEADOWS, N.S., TRUEBLOOD, S.P., HARDMAN, M. & COWAN, G. (eds) *Petroleum Geology of the Irish Sea and Adjacent Areas*. Geological Society, London, Special Publications, **124**, 135–159.

QUIRK, D.G., ROY, S., KNOTT, I., REDFERN, J. & HILL, L. 1999. Petroleum geology and future hydrocarbon potential of the Irish Sea. *Journal of Petroleum Geology*, **22**, 243–260.

ROBERTS, D.G., THOMPSON, M., MICHENER, B., HOSSACK, J., CARMICHAEL, S. & BJØRNSETH, H.-M. 1999. Palaeozoic to Tertiary rift and basin dynamics: mid-Norway to the Bay of Biscay—a new context for hydrocarbon prospectivity in the deep water frontier. *In*: FLEET, A.J. & BOLDY, S.A.R. (eds) *Petroleum Geology of Northwest Europe: Proceedings of the 5th Conference*. Geological Society, London, 7–40.

ROBERTSON RESEARCH INTERNATIONAL LIMITED 1986. *The petroleum evaluation of the interval 4500′–9265′ of the 42/12-1 well drilled offshore Republic of Ireland*. Unpublished report. Robertson Research International, Llandudno.

SEEDHOUSE, J.K. & RACEY, A. 1997. Sealing capacity of the Mercia Mudstone Group in the East Irish Sea Basin: implications for petroleum exploration. *Journal of Petroleum Geology*, **20**, 261–286.

SOPER, N.J., WEBB, B.C. & WOODCOCK, N.H. 1987. Late Caledonian (Acadian) transpression in north-west England: timing, geometry and geotectonic significance. *Proceedings of the Yorkshire Geological. Society*, **46**, 175–192.

THOMPSON, J. & MEADOWS, N.S. 1997. Clastic sabkhas and diachroneity at the top of the Sherwood Sandstone Group: East Irish Sea Basin. *In*: MEADOWS, N.S., TRUEBLOOD, S.P., HARDMAN, M. & COWAN, G. (eds) *Petroleum Geology of the Irish Sea and Adjacent Areas*. Geological Society, London, Special Publications, **124**, 237–251.

WARRINGTON, G. 1997. The Penarth Group–Lias Group succession (Late Triassic–Early Jurassic) in the East Irish Sea Basin and neighbouring areas: a stratigraphical review. *In*: MEADOWS, N.S., TRUEBLOOD, S.P., HARDMAN, M. & COWAN, G. (eds) *Petroleum Geology of the Irish Sea and Adjacent Areas*. Geological Society, London, Special Publications, **124**, 33–46.

YALIZ, A.M. 1997. The Douglas Oil Field. *In*: MEADOWS, N.S., TRUEBLOOD, S.P., HARDMAN, M. & COWAN, G. (eds) *Petroleum Geology of the Irish Sea and Adjacent Areas*. Geological Society, London, Special Publications, **124**, 399–416.

Hydrocarbon potential of the Kish Bank Basin: integration within a regional model for the Greater Irish Sea Basin

G. M. DUNFORD[1], P. N. DANCER[2] & K. D. LONG[3]

[1]*Dunford Exploration Limited, Durham Cottage, Hewelsfield Common GL15 6US, UK (e-mail: G.Dunford@g-dunford.demon.co.uk)*

[2]*Enterprise Energy Ireland, Embassy House, Herbert Park Lane, Ballsbridge, Dublin 4, Ireland*

[3]*Enterprise Energy Norge Ltd, Løkkeveien 103, PO Box 399, N-4002 Stavanger, Norway*

Abstract: The Kish Bank Basin lies in the western Irish Sea *c.* 20 km east of Dublin. It is one of a number of remnants of a larger Permo-Triassic basin system that may have extended across the whole of the Irish Sea. It has a geological history similar to that of the East Irish Sea Basin, initially developing by the reactivation of Caledonian faults that controlled subsequent deposition during Dinantian and Namurian time, with Westphalian deposition in a sag-basin that overstepped the adjacent basement highs. Variscan dextral transpression resulted in the formation of the Codling and Bray faults, and Permian to Jurassic extension formed a set of north–south-trending faults. Liassic outliers are preserved in the hanging walls of the basin margin faults. Early Cretaceous uplift was followed by chalk deposition. Tertiary movements reactivated older faults, isolating the Kish Bank Basin, and producing 9 km of dextral strike-slip along the Codling Fault Zone. The main reservoir in the hydrocarbon play is provided by the Sherwood Sandstone Group, as successfully exploited in the East Irish Sea. Three wells have been drilled to test this reservoir. These encountered high-quality Sherwood Sandstone reservoirs beneath the good potential seal of the Mercia Mudstone Group (which included thick halites). Source rock potential is from either the Westphalian Coal Measures, as penetrated in well 33/22-1, or from inferred Dinantian to Namurian basinal shales. There is good evidence of an active source system, with oil shows in wells 33/17-1 and 33/22-1, data from geochemical analysis of sea-bed cores, a 'Seepfinder' survey, sea-bed mounds and seismic evidence of shallow gas. The main risks of the play are the migration pathway and the timing of trap formation with respect to migration. Migration favours the eastern side of the basin, and many of the tilted fault blocks that formed during Permian to Jurassic time have been modified by Early Cretaceous inversion and by Tertiary strike-slip compression. All of the structures that have been drilled to date have been either formed or modified after the time of peak hydrocarbon generation and migration.

The Kish Bank Basin is in the Irish Sea, east of Dublin, in water depths of 3–100 m. It is an erosional remnant of an inferred larger linked Permo-Triassic basin system that covered the northern Irish Sea, North Channel and western England (Fig. 1). This basin system, hereafter called the Greater Irish Sea Basin, probably originally extended further north and west, as shown by onshore successions near Kingscourt (Visscher 1971), and in Antrim (Manning & Wilson 1975).

Triassic facies, despite great thickness variations, were similar across this basinal system, allowing correlation of the Triassic sequences in the Kish Bank and East Irish Sea basins. Enterprise's regional evaluation suggests that the depositional and tectonic history of the Kish Bank Basin was linked to that of the East Irish Sea Basin, and hence the Kish Bank Basin has a similar potential for both oil and gas. The main play is the Early Triassic Sherwood Sandstone Group reservoir sealed by the Mercia Mudstone Group and sourced with oil and/or gas from the upper Dinantian–Namurian deposits, or with gas from the Westphalian Coal Measures. Secondary reservoirs are the Permian Collyhurst Sandstone and Carboniferous sandstones.

The Triassic play (Fig. 2) was tested by well 33/21-1, drilled by Shell in 1979 on an inversion anticline in the hanging wall of the Dalkey Fault (Figs 3 and 4), and by well 33/17-1, drilled by Charterhouse in 1986, at the north end of a horst to the east of the Codling Fault (Fig. 3). Oil shows were recorded in well 33/17-1 and in Carboniferous sandstones in well 33/22-1, drilled on the high to the south (Fig. 2).

From: SHANNON, P.M., HAUGHTON, P.D.W. & CORCORAN, D.V. (eds) 2001. *The Petroleum Exploration of Ireland's Offshore Basins*. Geological Society, London, Special Publications, **188**, 135–154.
0305-8719/01/$15.00

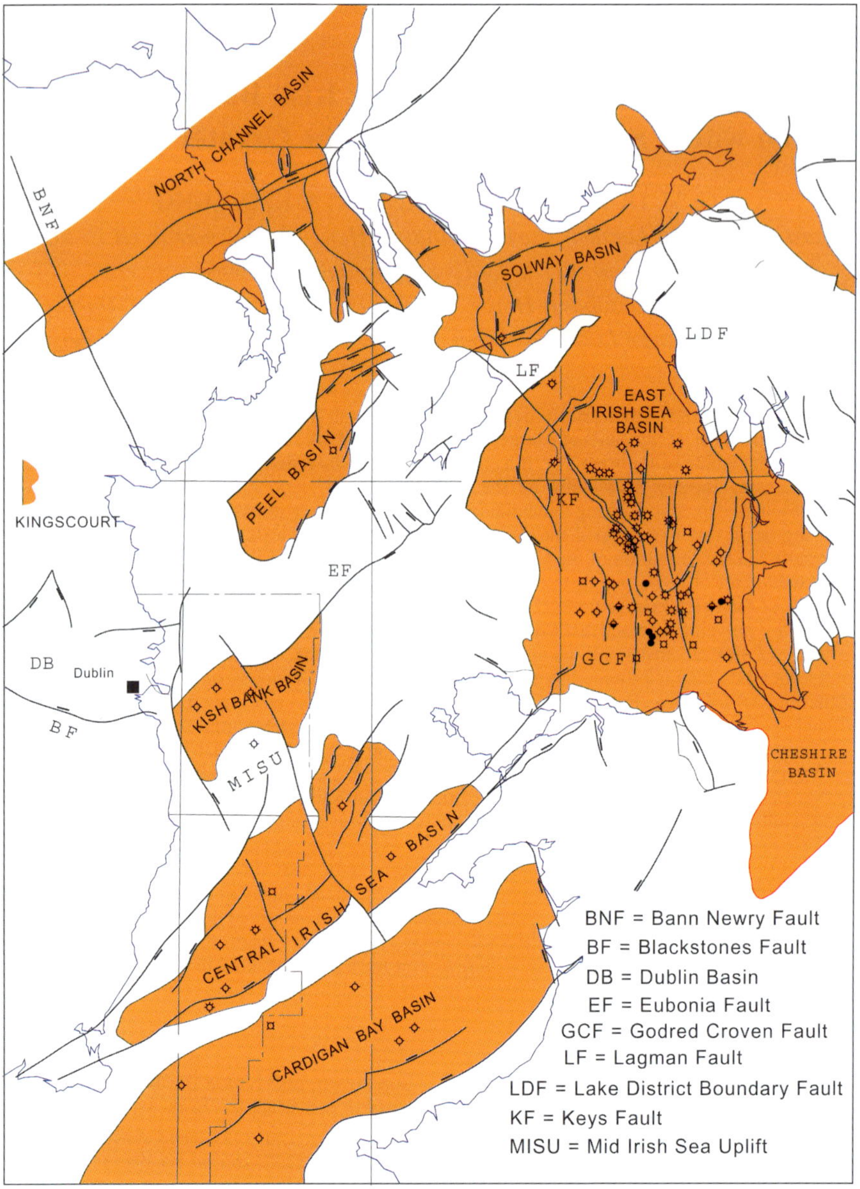

Fig. 1. Location of the Kish Bank Basin and the other remnants of the Greater Irish Sea Basin. Orange denotes areas where Permo-Triassic rocks are preserved. Major faults are shown, including those bounding the Kish Bank Basin to its west that are on trend with the Eubonia and Lagman faults (EF and LF).

Enterprise Oil re-evaluated the basin using speculative and proprietary seismic data shot in 1992–1996 and reprocessed older data. This led to the drilling of well 33/17-2A by Enterprise in 1997. It tested a combined anticline and tilted fault-block closure, to the west of the Codling Fault (Fig. 4).

Structure of the Kish Bank Basin

The Kish Bank Basin is a NW-dipping half-graben (Fig. 2) divided in two by the NNW-trending Codling Fault Zone (Bott & Young 1971; Dobson & Whittington 1979; Jenner 1981; Naylor *et al.* 1993). New seismic interpretation

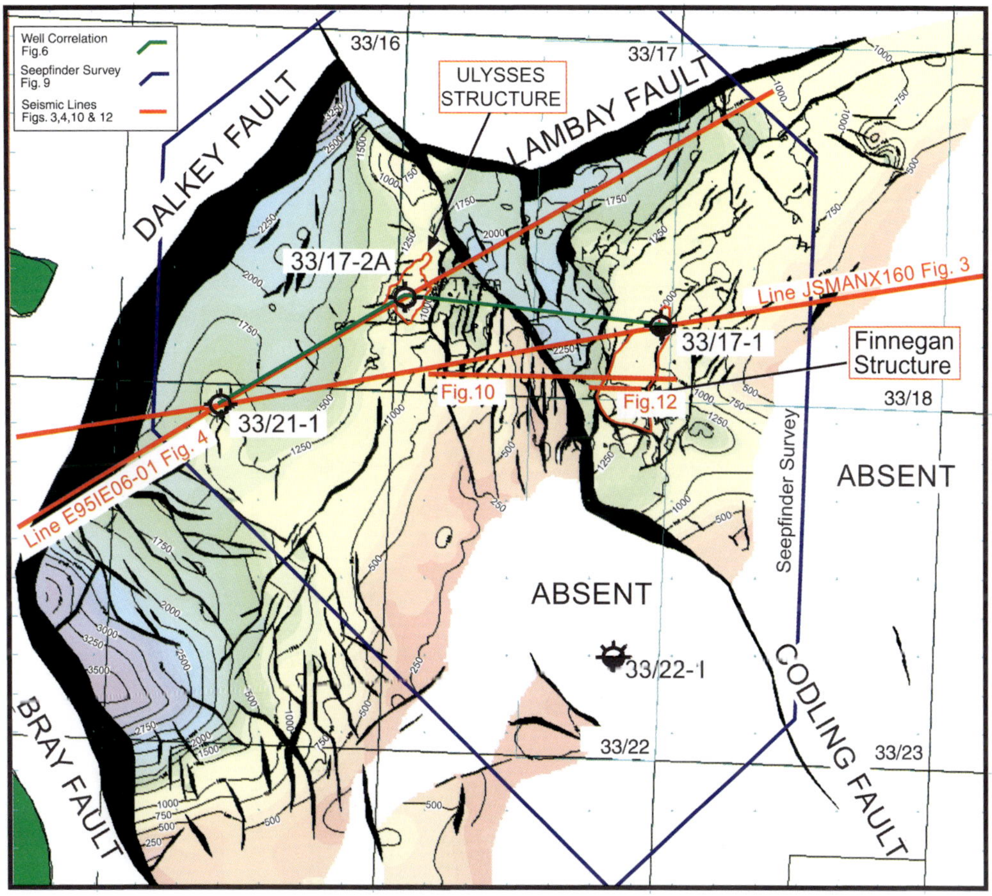

Fig. 2. Top Ormskirk Sandstone Formation depth structure map. Deep blue denotes depths >3000 m. The basin is a NW-dipping half-graben divided by the Codling Fault. The Bray Fault in the SW is parallel to the Codling Fault. The Dalkey and Lambay faults have been offset from each other by 9 km. The location of the wells, seismic lines, structures, Seepfinder survey and correlation diagram are shown.

suggests 9 km of post-Triassic (probably Early Tertiary) dextral strike-slip movement on the Codling Fault Zone. The basin is bounded to the NW by the NE–SW-trending Dalkey and Lambay faults and to the SW by the NNW-trending Bray Fault. In the deepest, southwestern part of the basin, the Top Sherwood event is at 3000 m subsea (Fig. 2). Enterprise's gravity modelling and seismic interpretation (Figs 3 and 4) suggests that the top of the Dinantian carbonates (or Top Basement) event is at >6000 m subsea in the hanging wall of the Dalkey Fault.

To the SE, the limit of the basin is formed by Triassic, Permian and Carboniferous subcrop beneath the Base Tertiary unconformity on the NW-dipping flank of the Mid-Irish Sea Uplift. This Caledonian (NE–SW)-trending, NW-dipping fault block separates the Kish Bank Basin from the Central Irish Sea Basin to the south. The SW part of the basin has numerous minor faults that approximately parallel, and are antithetic to, the Bray Fault. The central part of the basin is relatively unfaulted, except for rare faults antithetic to the Dalkey Fault.

The eastern part of the basin has closely spaced, generally north–south-trending faults either side of the Codling Fault. Tilted fault-block closures are formed by these faults, two of which have been partially tested by wells 33/17-2A and 33/17-1. NW–SE faults, such as the southern bounding fault of the Finnegan structure (Fig. 2), are subparallel to the Codling Fault and probably had a similar history, including Tertiary strike slip movement.

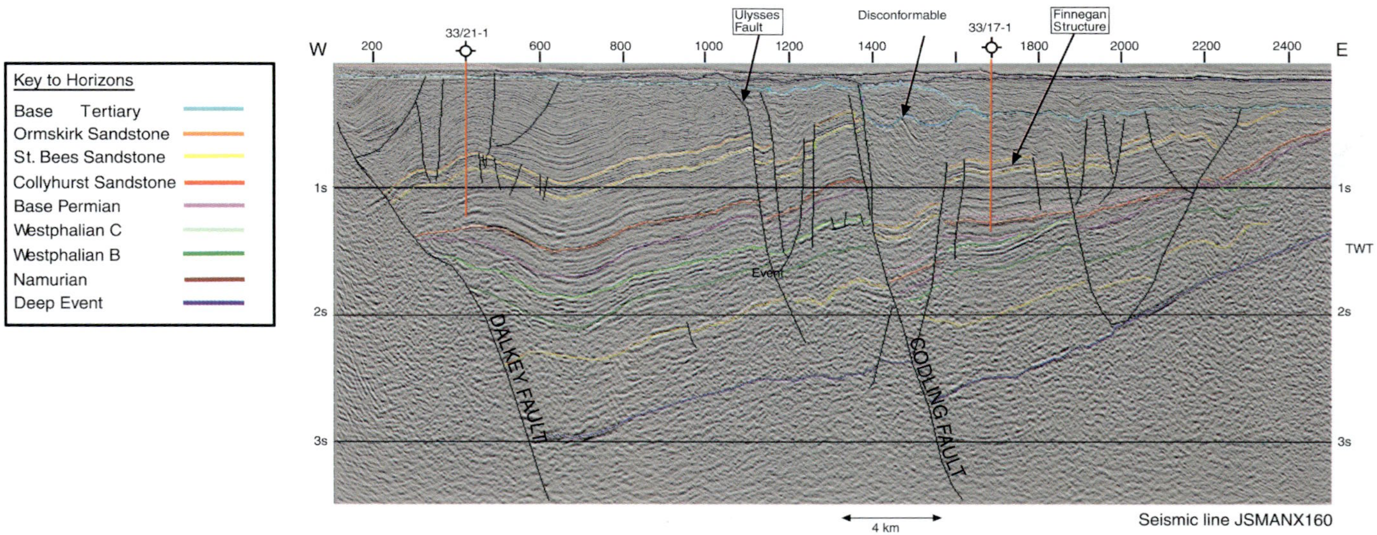

Fig. 3. Seismic line JSMANX160 (location shown in Fig. 2). The line shows the structure and stratigraphy and the wells 33/21-1 and 33/17-1. The disconformable event in the Codling Fault Zone may be caused by salt movement in the MMG.

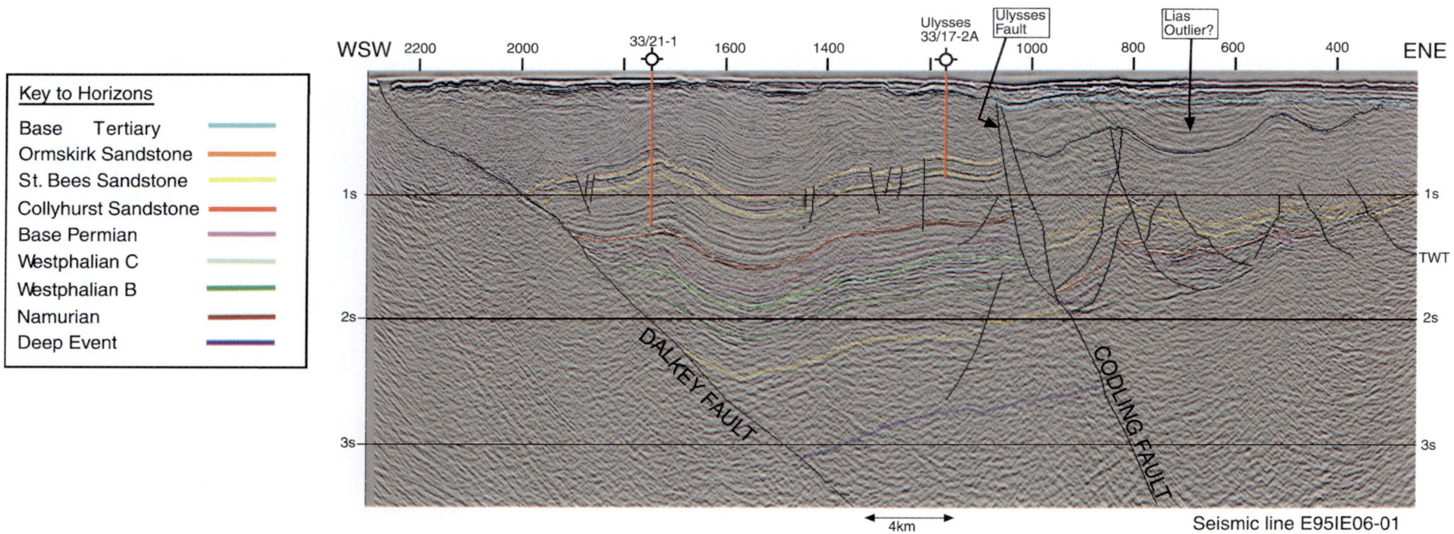

Fig. 4. Seismic line E95IE06-01 (location shown in Fig. 2). The line shows the structures tested by wells 33/21-1 and the Ulysses structure tested by well 33/17-2A. A small untested independent closure is present on the footwall of the Ulysses Fault. A possible Liassic outlier is present beneath the Tertiary sequence in the ENE. This is in the hanging wall of the Lambay Fault, which is subparallel to the line.

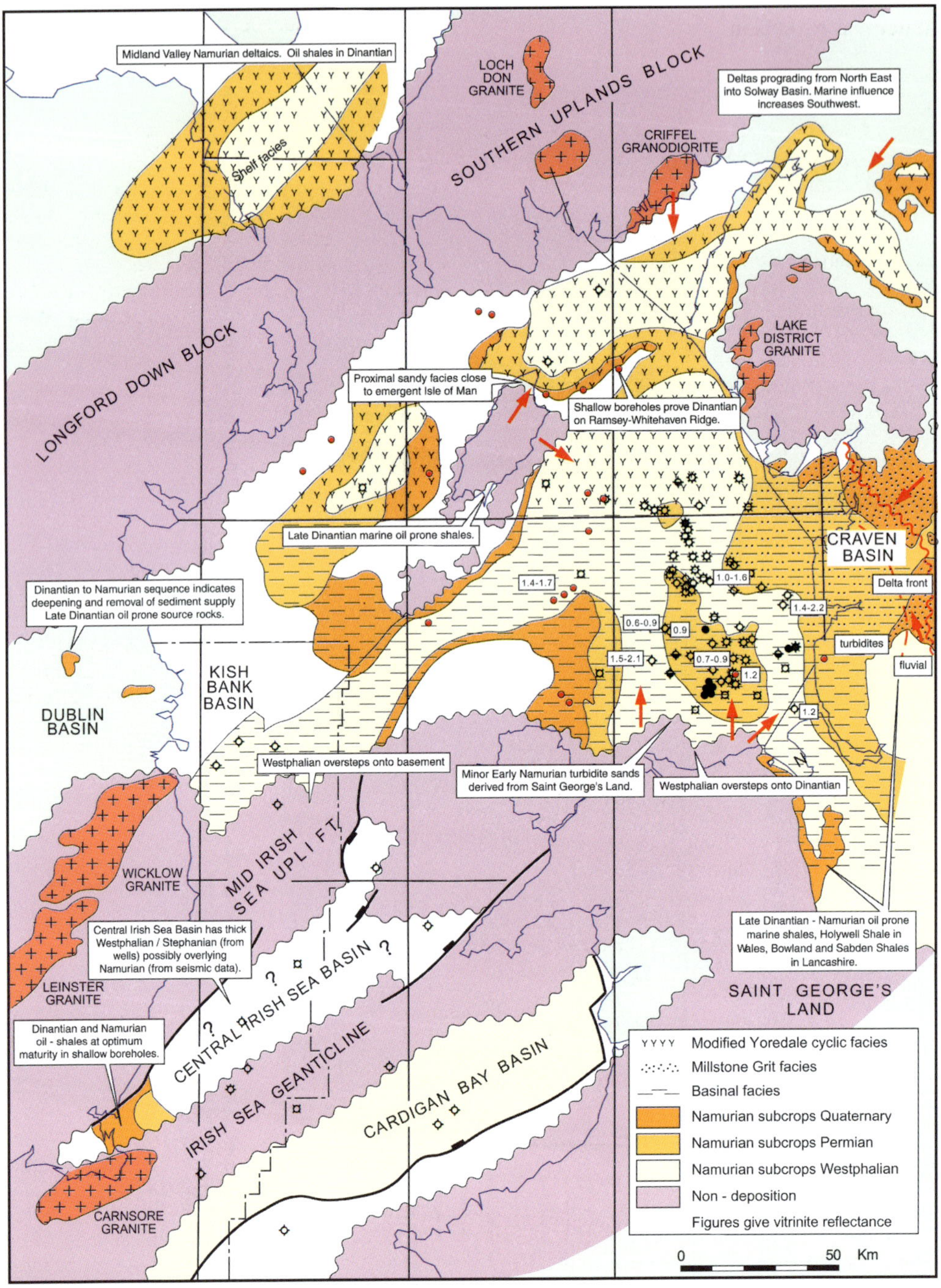

Fig. 5. Namurian outcrop, subcrop and inferred palaeogeography for the Irish Sea and adjacent areas. Coloured areas show preservation of Namurian rocks. The Kish Bank Basin is expected to have basinal shales, as it lies between the Dublin and Craven basins in the same Carboniferous basin as the southern, oil-prone part of the East Irish Sea Basin.

Structural evolution

Caledonian

The Caledonian structural framework of Ireland, Scotland and the Irish Sea was produced by the NW–SE-directed closure of the Iapetus Ocean during Silurian–Devonian time. This framework forms the major structural control on the orientation of post-Caledonian faulting and basin formation in the Kish Bank area. NE–SW-trending Carboniferous basins such as the Northumberland, Solway and Peel basins (Chadwick & Holliday 1991; Chadwick *et al.* 1995; Corfield *et al.* 1996) developed along the line of the NW-dipping Iapetus Suture (England & Soper 1997), across the Solway Firth and the Irish Sea. The Kish Bank Basin is separated from the Peel Basin by a NE–SW ridge extending SW from the Isle of Man (Fig. 1).

Caledonian sinistral transpression characterized this part of the Irish Sea (Quirk & Kimbell 1997) and continued into Early Devonian time. Mountain building and major erosion provided considerable amounts of sediment, which was deposited into subsequently formed Devonian and Carboniferous basins.

The NE–SW Dalkey and Lambay faults (Fig. 2) were probably initiated during Caledonian compression and later reactivated as normal faults. These faults are on trend with the Eubonia and Lagman faults to the NE (which separate the Isle of Man and Ramsey–Whitehaven Ridge from the East Irish Sea Basin) and the Blackstones Fault south of the Dublin Basin (Fig. 1). The NNW–SSE fault trend, represented by the Bray and Codling faults in the Kish Bank Basin, is common in the East Irish Sea Basin (e.g. Keys Fault and Lake District Boundary Fault). These faults may have been active as strike-slip faults during Caledonian movements, although they may have formed as early as Charnian times (Jackson & Mulholland 1993).

Carboniferous

Post-tectonic relaxation, possibly related to subduction south of the British Isles (Leeder 1988), began in Mid-Devonian time, and continued into Dinantian–Namurian times. Throughout the Greater Irish Sea area (shown by Fig. 5), Dinantian–Namurian deposition was partially controlled by reactivation of the Caledonian structural framework (and partly by new faults, as Dinantian extension was oblique to the Caledonian trend). The major faults ceased to be active during Late Namurian time and the Westphalian sequence was deposited in a sag basin.

Variscan

The NNW–SSE-trending Codling, Bray and Keys faults may have been reactivated as strike-slip faults during the Variscan orogeny. In the Kish Bank Basin, new seismic interpretation suggests up to 3 km of Hercynian dextral strike slip along the Codling Fault (as there is a 12 km offset of the Westphalian–Stephanian depocentres compared with the 9 km of post-Triassic dextral offset).

In the adjacent Dublin Basin, inversion produced NE–SW-trending thrust anticlines and reverse movement on Early Carboniferous normal faults. Inversion (and later periods of uplift) has removed the Upper Carboniferous post-rift section and much of the Lower Carboniferous synrift section from the Dublin Basin (Nolan 1989). The bounding highs of the Kish Bank Basin were also probably rejuvenated by reverse movements on their bounding faults. In the Kish Bank Basin, inversion and erosion was less severe and the Westphalian to Stephanian post-rift section has been preserved over most of the basin.

Permian–Triassic

By analogy to the East Irish Sea Basin (Jackson & Mulholland 1993), several periods of extension with intervening periods of thermal sag probably occurred during Permian to Jurassic time. The Kish Bank Basin and East Irish Sea Basin formed part of the Clyde Belt trending NNW–SSE through the Irish Sea, Cheshire Basin, Worcester Graben and Wessex Basin, linking the Boreal and Tethyan Oceans (Ziegler 1988; Jackson & Mulholland 1993).

Extension directions varied either side of east–west (Jackson & Mulholland 1993; England & Soper 1997). Extension used previously existing faults and new north–south faults almost perpendicular to the likely Permian extension direction. The preserved Permo-Triassic section in the Kish Bank Basin provides no strong evidence for the Permian age of these faults. However, the authors' seismic interpretation suggests that the dominant north–south faults in the southern part of the East Irish Sea Basin (e.g. the Godred Croven Fault) were active during Permian (Collyhurst Sandstone) deposition. Some north–south faults may be Tertiary faults formed as a result of dextral strike slip along the Codling and Bray faults.

Extension would have also exploited Caledonian and Variscan faults. Caledonian trending faults, such as the Lambay and Dalkey faults, may have had a sinistral strike-slip component as well as a normal component. The NW-trending Codling and Bray faults were probably active normal faults during Permian–Jurassic time. As with the similarly oriented Keys Fault, in the East Irish Sea Basin, the WSW–ENE extension direction is almost perpendicular to their trend. The Codling Fault has a similar hade (50°) to the Keys Fault. These similarities argue against a purely Tertiary strike-slip fault with associated north–south splays. This has important implications for hydrocarbon prospectivity, because if the Codling Fault system had formed during Tertiary time, then all structures formed by these faults would carry a high risk to the timing of trap formation with respect to migration timing (see discussion below).

Lack of well control in the hanging wall of the Codling Fault precludes using Triassic facies variations as evidence for the timing of movement. The lack of appreciable thickening of Permian and Sherwood Sandstone sections across the Codling Fault (once post-Triassic strike-slip has been restored) suggests that movement was probably minor. Seismic data in the hanging wall of the Codling Fault are too poor to prove normal movement during Mercia Mudstone Group deposition.

Jurassic

The Lower Jurassic sequence preserved in the hanging walls of the bounding faults was probably deposited in a sag basin that extended beyond these present-day bounding faults. Liassic units thin towards the Dalkey, Lambay and Bray faults, suggesting that they were not basin-bounding faults during Liassic deposition (Figs 3 and 4). This contrasts with the interpretation of Cope (1997). By analogy with the East Irish Sea Basin (where much of the fault movement is of post-Liassic age, according to Jackson & Mulholland 1993), extensional faulting probably recommenced during Mid- and Late Jurassic times (Cope 1997) as a result of rapid opening of the central Atlantic Ocean.

The lack of post-Liassic section makes difficult any assessment of the effects of Mid–Late Jurassic rifting. Post-Liassic history has been deduced from apatite fission-track analysis (AFTA), vitrinite reflectance (VR) data, shale velocities and analogies with other basins. Several periods of uplift and erosion may have occurred from Mid-Jurassic times onwards. In the centre of the Cardigan Bay Basin (e.g. well 106/24-1) there was no major hiatus throughout Bajocian–Purbeckian times (Barr *et al.* 1981). A similar history is postulated for the Kish Bank Basin.

There is evidence for late Jurassic rifting on the Keys Fault near the Isle of Man (Quirk & Kimbell 1997) and elsewhere in the East Irish Sea Basin (Jackson & Mulholland 1993). In the Kish Bank Basin, NE–SW late Jurassic rifting probably produced extensional (down to the east) normal movement on the Bray and Codling faults. The Dalkey and Lambay faults lie subparallel to the extension direction and may have experienced sinistral transtension with little extension.

Cretaceous

Major uplift at the end of Jurassic or beginning of Cretaceous time is suggested by AFTA data from other Irish Sea basins (Hardman *et al.* 1993) and pre-Greensand compression in Northern Ireland (Wilson 1972). The amount of Early Cretaceous erosion in the Kish Bank Basin is uncertain, as the thermal effects of this period have been masked by the effects of the Early Tertiary uplift. The amount of Early Cretaceous uplift and erosion used in burial history modelling has been conservatively estimated as *c.* 500 m (from the authors' seismic interpretation of the NE Celtic Sea, the nearest basin with a preserved Cretaceous section).

Tertiary

AFTA data from all three wells (unpublished work for Enterprise Oil) suggest that Tertiary uplift (commencing 60 Ma) and denudation was greater than in earlier periods. Shale velocity analysis (Ware, pers. comm.) suggests that the amount of erosion during Tertiary time was greater in the SW than in the NE, with 1500 m removed from the 33/21-1 inversion structure, 875 m from the 33/17-2A location, and 350 m from the 33/17-1 area. About 960 m has been removed from the Mid Irish Sea Uplift at the location of well 33/22-1. These values are 500–1000 m less than those derived from AFTA data. These data are compatible with a model of hydrothermal heating of the Sherwood Sandstone Group during a Paleocene heat pulse coeval with dyke emplacement in Scotland, Northern Ireland, Anglesey, the East Irish Sea Basin (Arter & Fagin 1993) and along the Codling Fault Zone in the Central Irish Sea Basin and Kish Bank Basin (magnetic data also suggest a dyke at the south end of the Finnegan structure, Fig. 2). Early Tertiary heating and uplift was probably caused by the inception of the Iceland mantle plume. Ridge push, associated with sea-floor spreading in the North Atlantic, may have also contributed to compression and uplift throughout Tertiary time.

The Tertiary stress regime in the Kish Bank Basin was NNW-directed compression with WSW–ENE extension. The compression was oblique to the NNW-trending Codling Fault, which thus became active as a dextral strike-slip fault. There was 6 km of post-Triassic movement along the fault itself, but 9 km spread over a shear zone of 10 km width on either side (see Fig. 2 for the displacement of structural features).

The strike-slip movements were probably mainly of Eocene–Oligocene age, by analogy with similar NW–SE strike-slip faults elsewhere in the UK such as the Bann–Newry Fault Zone in Northern Ireland (Jenner 1981) and the Sticklepath Fault in Devon (Dearman 1963; Jenner 1981). Some older north–south faults were probably reactivated during Early Tertiary strike slip, and new north–south splays may have formed either side of the Codling Fault.

Uplift probably continued intermittently throughout mid–late Tertiary time owing to changes in sea-floor spreading geometry in the North Atlantic (Boldreel & Andersen 1993) and Alpine movements to the SE (Ziegler *et al.* 1995). Most faults were probably reactivated throughout Tertiary time. Drainage patterns in eastern Ireland (Cope 1997) suggest Recent activity on north–south faults and there has been Recent seismic activity on the Codling Fault and other faults in the Irish Sea (Cope 1997).

The current remnant basin outlines are controlled by the reactivation of the basin-bounding faults and erosion of their footwalls. The westerly tilt of the basin was caused by Tertiary uplift of the Rhinns of Galloway–Isle of Man–Anglesey Axis (Jackson & Mulholland 1993), which now separates the Kish Bank Basin from the East Irish Sea Basin.

Stratigraphy

Aspects of the stratigraphy have been previously discussed by Jenner (1981) and Naylor *et al.* (1993). The recent well 33/17-2A (Fig. 6)

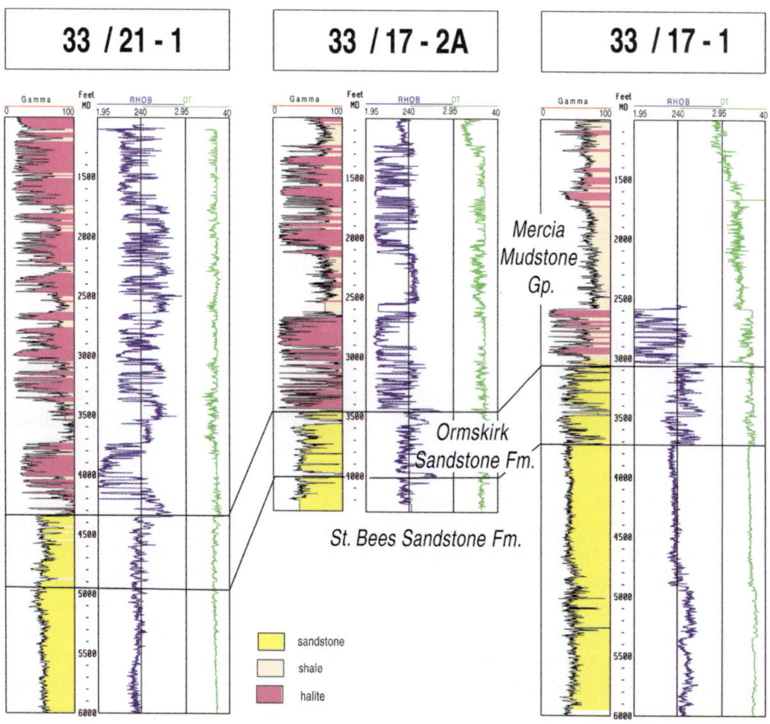

Fig. 6. Log correlation of the Mercia Mudstone and Ormskirk Sandstone sections from the three wells. The Mercia Mudstone has about 30% halite in wells west of the Codling Fault and 16% in well 33/17-1. Shale content in the Ormskirk Sandstone increases from west to east. RHOB, bulk density (g cm^{-3}); DT, sonic interval transit time (μs per foot).

reached TD (total depth) at the top of the St. Bees Sandstone and thus provides new information on the Mercia Mudstone Group and Ormskirk Sandstone Formation in this area. This information has been used, as much as possible, to try and date movement of the Codling Fault, as this has such important timing implications. The thickness of Jurassic–Tertiary units that have been eroded must also be estimated to attempt any burial history reconstruction.

Dinantian–Namurian

A new seismic interpretation (Figs 3 and 4) suggests that there is up to 2000 m of Dinantian–Namurian section in the Kish Bank Basin, a value consistent with the results of gravity modelling. The Dalkey and Lambay faults probably controlled deposition, as the interpreted Namurian section thickens into their hanging walls. The Codling Fault shows no thickening and thus does not appear to have been active. The succession, which thins southeastward, onto the NW flank of the Mid Irish Sea Uplift, corresponds to a zone of low seismic amplitudes (Figs 3 and 4) similar to that in the southern part of the East Irish Sea Basin. It has not, however, been drilled and thus the facies must be inferred from areas outside the basin.

Well control from the East Irish Sea Basin and outcrop data from NW England, NE Wales and eastern Ireland, all suggest the presence of a continuous Dublin–Craven Basin (Fig. 5) in which basinal shales were deposited during Late Dinantian to Early Namurian times (Fraser & Gawthorpe 1990; Lawrence *et al.* 1997). The Dinantian–Early Namurian synrift infill in the Dublin Basin is similar to that in the Bowland Trough, Widmerpool Gulf and North Staffordshire basins (Gawthorpe *et al.* 1989). Deposition was dominated by basinal mudstones with little coarse-grained clastic sediment input and significant thickening towards the basin-bounding faults (Corfield *et al.* 1996). A similar depositional pattern is anticipated in the Kish Bank Basin (Fig. 5) with deposition of thick mudstones similar to the organic-rich Bowland Shale and Edale Shales. This contrasts with the Peel and Solway basins (Newman 1999), where the Upper Dinantian sequence was deposited in a shallow-water 'Yoredale' facies. Data from the Solway and Northumberland basins also suggest Yoredale facies deposition in Namurian time but the succession is only locally preserved.

In the Central Irish Sea Basin, south of the Mid-Irish Sea Uplift, the Namurian section has not been penetrated by exploration wells, but Namurian and Upper Dinantian oil-prone source rocks were found in boreholes in County Wexford (SE Ireland) at the SW end of the basin (Clayton *et al.* 1986). The absence of Namurian section in well 42/17-1 (Maingarm *et al.* 1999) on the flank of the Caledonian Lleyn Ridge, may not be representative of the deeper parts of the basin.

Westphalian–Stephanian

Seismic interpretation of the Westphalian–Stephanian thickness (up to 1500 m in the basin depocentre) suggests that the major faults in the Kish Bank Basin were inactive during the later part of the Carboniferous period, and the Westphalian section was deposited in a sag-basin. On the southern margin of the Kish Bank Basin, the Westphalian section oversteps the Namurian–Dinantian units, on the NW flank of the Mid Irish Sea Uplift, to rest unconformably on Cambrian basement at the location of well 33/22-1 (Fig. 2). This well encountered a 740 m Westphalian B–D to Stephanian succession.

Permian

Well 33/17-1 (Fig. 3) partially penetrated the Early Permian Collyhurst Sandstone Formation. The sandstones were fine to coarse grained with rounded to well-rounded grains and good visible porosity (Naylor *et al.* 1993). The Collyhurst Sandstone in the Irish Sea Basin is largely aeolian (Jenner 1981) and a similar depositional environment is likely in the Kish Bank Basin. The overlying Late Permian Manchester Marl Formation (nomenclature of Naylor *et al.* 1993) is 97 m thick and consists of red–brown calcareous mudstones. A similar sequence is present in the Kingscourt oulier (Visscher 1971; Jenner 1981). The presence of thin sandstones and siltstones and the thinness of the unit degrade its potential as a seal for the Collyhurst Sandstone reservoir.

Early Triassic

Sherwood Sandstone Group facies are similar across the whole Greater Irish Sea Basin (Fig. 7) because deposition kept pace with faulting and the whole area had a similar climatic and subsidence history. Lithostratigraphic correlations can thus be made between the Kish Bank Basin and the other sub-basins of the Irish Sea. As in the East Irish Sea Basin, the Sherwood Sandstone Group can be divided into a lower St. Bees Sandstone Formation and an overlying Ormskirk Formation using differences in log

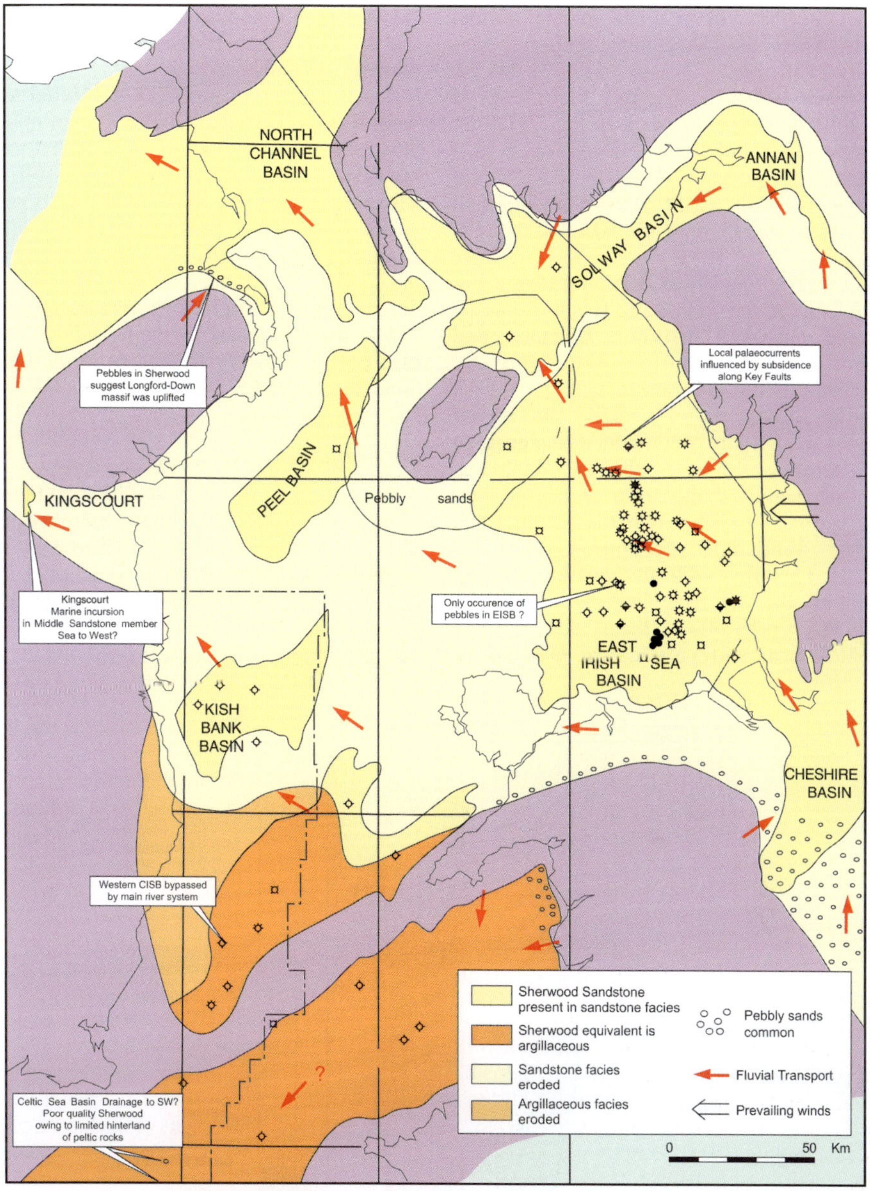

Fig. 7. Sherwood Sandstone palaeogeography and distribution map covering the Irish Sea and adjacent areas. The inferred original extent of the Ormskirk Sandstone fluvial depositional system is shown and (in bolder colours) the areas of present-day preservation. The present-day basins are erosional remnants of a Greater Irish Sea linked basin system.

(Fig. 6) and seismic (Figs 3 and 4) character reflecting the overall more uniform lithology of the St. Bees Sandstone.

The St. Bees Sandstone was fully penetrated by well 33/17-1, almost fully by well 33/21-1, and just reached by well 33/17-2A. There is little difference in facies between the St. Bees Sandstone in these wells. In the Greater Irish Sea Basin, pebbles and conglomerates are found in the St. Bees Sandstone and the laterally equivalent Chester Pebble Beds in proximal areas, close to massifs, such as in the Cheshire Basin and in isolated wells in the southern East Irish Sea Basin (e.g. released well 110/7-2,

Fig. 7). The lack of pebbles in the almost complete Sherwood Sandstone section penetrated by well 33/21-1, only 4 km from the Dalkey Fault, suggests that this fault, even if active, was unlikely to have been the bounding fault of the Greater Irish Sea Basin. This is also suggested by the presence of 400 m of Sherwood Sandstone (probably all St. Bees equivalent) in the Kingscourt outlier, onshore, 80 km to the NW (Fig. 7).

The Ormskirk Sandstone is similar (Fig. 6) in all three wells in the Kish Bank Basin. It can be subdivided into the Thurstaston, Delamere and Frodsham members (from base to top, Fig. 6), on the basis of log correlation with East Irish Sea wells. It consists of aeolian and fluvial sandstones with more interbedded shales than the Ormskirk Sandstone of the East Irish Sea Basin. Log analysis suggests a higher percentage of aeolian sandstones in the 171 m thick Ormskirk Sandstone Formation of well 33/17-2A. This fits with its position on the footwall of the (active) Codling Fault (by analogy with the East Irish Sea Basin). Sedimentological and petrological analysis of sidewall cores showed strong bimodal sorting and a lack of fine-grained components, typical of aeolian sandstones.

The shales were probably deposited as overbank fines in interchannel areas or in ephemeral playa lakes. The axes of Ormskirk Sandstone braided stream systems in the East Irish Sea Basin tend to follow the hanging walls of major faults (Meadows & Beach 1993). Channelling of a major stream system along the hanging wall of the Dalkey Fault may explain the SW to NE decrease in net to gross ratio from 80% (in the 265 m section found in well 33/21-1) to 75%, then 66% (log analysis, Fig. 6) in the Ormskirk Sandstone in the Kish Bank Basin.

Late Triassic

The wells west of the Codling Fault (Figs 4 and 6) have about 30% halite in the Mercia Mudstone Group, compared with only 16% in well 33/17-1 in the east. There is no seismic evidence for halokinesis affecting these ratios in the areas around the wells but seismic interpretation suggests that a disconformable event in the

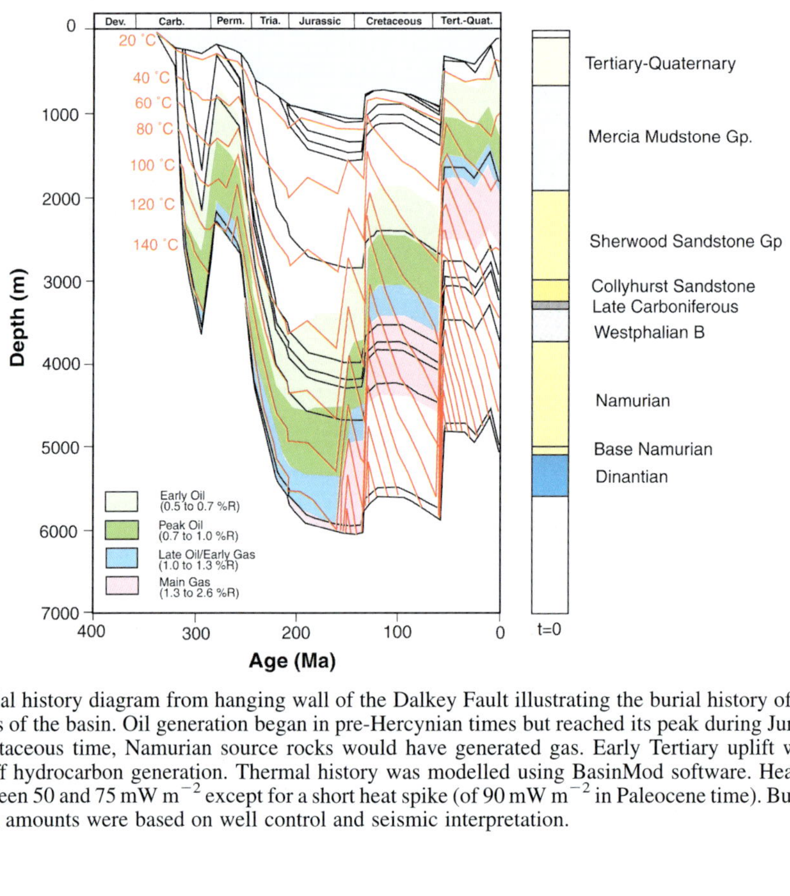

Fig. 8. Burial history diagram from hanging wall of the Dalkey Fault illustrating the burial history of one of the deeper parts of the basin. Oil generation began in pre-Hercynian times but reached its peak during Jurassic time. During Cretaceous time, Namurian source rocks would have generated gas. Early Tertiary uplift would have switched off hydrocarbon generation. Thermal history was modelled using BasinMod software. Heat flow was varied between 50 and 75 mW m^{-2} except for a short heat spike (of 90 mW m^{-2} in Paleocene time). Burial history and erosion amounts were based on well control and seismic interpretation.

hanging wall of the Codling Fault (Fig. 3) may be a surface caused by salt withdrawal during Tertiary strike-slip movements. The southwesterly increase in the proportion of halite indicates that during Late Triassic time, the deepest part of the basin was in the west. This suggests that the present basin-bounding faults were active during Late Triassic times, even if they did not form the edge of the depositional basin. Palynological dating supports this interpretation, as the untruncated Anisian and Ladinian sequences thickened westwards.

Correlation of halites between the three wells is difficult, and it is even more difficult to correlate them with the type section in the East Irish Sea Basin. This suggests that several depocentres existed within the Greater Irish Sea linked basin complex during Late Triassic time. There is no well control in the hanging wall of the Codling Fault to ascertain if it influenced Mercia Mudstone Group facies. Sonic velocities in the Mercia Mudstone Group show a NE–SW increase across the three wells. Uplift and denudation values derived from these velocities do not fall on a linear trend across the basin (after correcting for strike slip) and suggest 300–400 m of normal throw on the Codling Fault at the time of maximum burial (latest Jurassic or latest Cretaceous time; see Fig. 8).

Jurassic

Liassic outliers, in the hanging walls of the bounding faults, have been interpreted from seismic data and confirmed by Liassic-aged dredge samples. The Liassic sequence conformably overlies the Triassic units. There are no known pre-Tertiary sediments younger than Liassic age preserved in the Kish Bank Basin (although a Mid-Jurassic section was suggested by Broughan *et al.* 1989).

Cretaceous

The North Celtic Sea Basin contains thick Wealden sediments (released wells, e.g. 50/11-1), generally derived from the NE, i.e. the Irish Sea area. It is thus likely that the Kish Bank Basin was eroded during Early Cretaceous time. If any Lower Cretaceous section accumulated, it was probably eroded before Greensand deposition, which in Northern Ireland began in Cenomanian time (Cope 1997). Widespread deposition of thin glauconitic sandstones was followed by a Chalk succession over large areas of NW Europe. In Antrim, up to 150 m of White Limestone and 14 m of Hibernian Greensand are preserved beneath Tertiary lavas (Wilson 1972). A similar section may have been deposited over the Kish Bank Basin before its removal by Tertiary erosion.

Tertiary–Recent

Eocene–Oligocene pull-apart basins developed along the Bann–Newry Fault zone (Jenner 1981), Sticklepath Fault (Jenner 1981) and Codling Fault Zone in the Central Irish Sea Basin (authors' seismic interpretation). If similar basins developed along the Codling Fault in the Kish Bank Basin, they were probably uplifted and eroded later in the Tertiary period, as there is no Tertiary isopach thickness in the fault zone. The Tertiary sequence thickens eastwards along a hinge line only approximately following the Codling Fault.

Hydrocarbon system

Reservoir

The Ormskirk Sandstone Formation has good reservoir parameters in all three wells. Net-to-gross ratios are high (66–80%) with no evidence of a change to a facies dominated by playa lake and interchannel shales as in Central Irish Sea Basin well 108/30-1 (Floodpage *et al.* 1999). As in the East Irish Sea Basin, porosity is influenced by depositional facies (Meadows & Beach 1993). The porosity of the Ormskirk Sandstone Formation is best (averaging 17.7%) in well 33/17-2A, probably because of more aeolian sandstones in the 171 m interval drilled. Results from a repeat formation test (RFT) indicate good permeability. Well 33/21-1 (Fig. 6) has a 265 m Ormskirk Sandstone Formation interval with a higher net-to-gross ratio (80%). The average porosity is lower (12%) possibly because of its greater maximum depth of burial (*c.* 2800 m compared with 1900 m for well 33/17-2A) but this may also be due to a different proportion of facies types. Coarsening-upwards sequences at the top of this interval (see Fig. 6) may suggest a proximal fluvial setting for the top part adjacent to the Dalkey Fault, although Naylor *et al.* (1993) suggested mixed fluvial and aeolian sandstones for the whole Ormskirk Sandstone Formation in this well.

The 215 m thick Ormskirk Sandstone Formation in well 33/17-1 is also of poorer quality than in well 33/17-2A, with an average (log) porosity of 14% and a net to gross value of 66%. The better porosity in this well compared with that in well 33/21-1 is probably because of the lack of burial in well 33/17-1 (maximum burial depth of 900 m was similar to the present burial depth, on the basis of sonic velocities). Permeabilities for these porosities should be good (>250mD) given data from elswhere in the East Irish Sea. The Ormskirk Sandstone Formation is thus of good reservoir quality for gas, and reasonable quality for oil, throughout the Kish Bank Basin.

The St. Bees Sandstone is not sealed from the Ormskirk Sandstone Formation and thus can be a reservoir only in structures with relief greater than the thickness of the Ormskirk Sandstone Formation (average 217 m). The top part (Calder Sandstone of Jackson & Mulholland 1993) is of good reservoir quality. The lower half (Silicified Zone or St. Bees Sandstone of Barnes *et al.* 1994), as in the East Irish Sea Basin (Colter & Ebbern 1978), is tightly cemented by quartz overgrowths (Naylor *et al.* 1993) and would be a poor reservoir.

Seal

The thickness and halite content (33% in well 33/21-1, 30% in well 33/17-2A, and 16% in well 33/17-1) of the Mercia Mudstone Group compare favourably with those of the East Irish Sea Basin and thus seal is not considered a significant risk. In well 33/17-1, which contained the least halite, this was concentrated towards the base of the section (Fig. 6) and thus the Mercia Mudstone Group is likely to be a good seal even here.

Source rocks

Upper Dinantian to Lower Namurian organic-rich basinal shales, probably deposited throughout the Dublin–Craven Basin (including the Kish Bank Basin; see Fig. 5), are the source of oil in the East Irish Sea Basin (e.g. Douglas and Lennox oil fields; see Armstrong *et al.* 1997; Haig *et al.* 1997; Yaliz 1997) and Lancashire (e.g. Formby oil field; see Armstrong *et al.* 1997) and have good oil-source potential in the Dublin Basin. In the East Irish Sea, they contain small but significant amounts of oil-prone sapropel and have total organic carbon (TOC) values (Hardman *et al.* 1993) of 0.2–8.3% (averaging 1.93%). Namurian and Dinantian strata have not

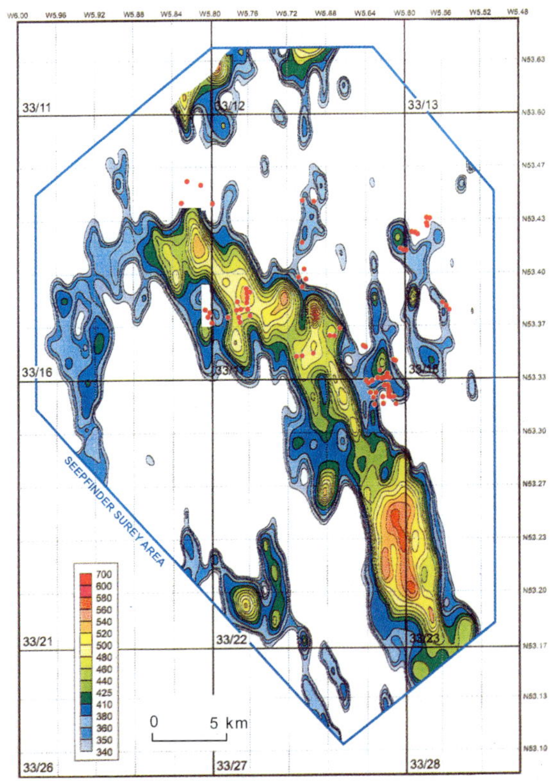

Fig. 9. Seepfinder anomalies and geochemical sample sites. The strongest anomalies indicate leakage of liquid hydrocarbons along the Codling Fault Zone. Weaker anomalies are present along the Dalkey and Lambay faults. A strong anomaly is present on the Mid Irish Sea Uplift, where the source rock may subcrop.

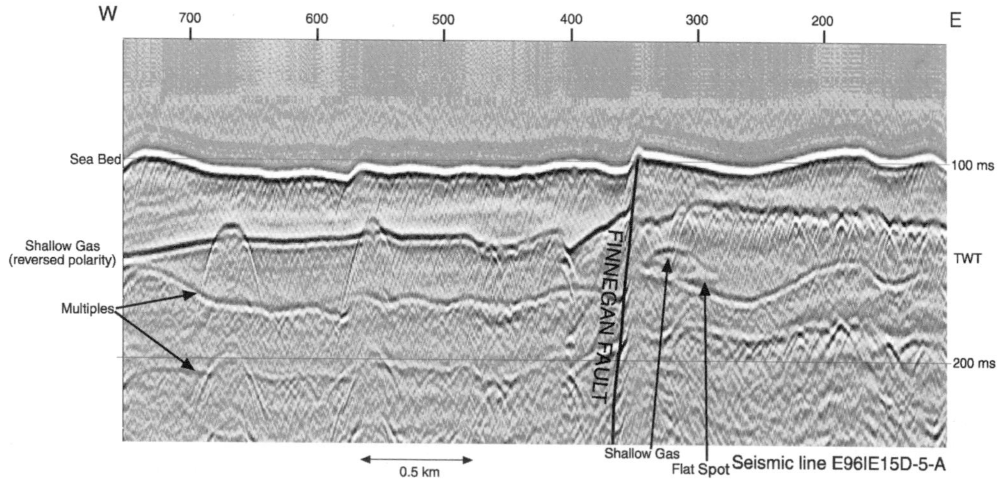

Fig. 10. Site survey line from Finnegan structure showing seismic evidence of shallow gas.

been proved in the Kish Bank Basin, but gravity modelling and seismic interpretation (Figs 3 and 4) suggest the presence of up to 2000 m of this section.

Westphalian coals are poorly developed in well 33/22-1, which was drilled outside the basin on the high to the SE, with only 11 m of coal in the 329 m Westphalian C section (Jenner 1981). The kerogen composition of this coal-poor section indicated poor to moderate potential for gas generation, which may not be representative of the basin as a whole. In the East Irish Sea Basin (Hardman *et al.* 1993), the Coal Measures have high TOC values (3.8–77.3%) but most of the kerogen is inertinite. Although the section is rather poor in well 33/22-1, seismic data suggest that the Coal Measures thicken northwestwards into the basin, where they may be up to 1500 m thick. The Westphalian C section, which has the most coal, has a distinctive seismic character that can be traced over the whole Kish Bank Basin (contrasting with the conclusion drawn by Naylor *et al.* 1993).

Dead oil shows were recorded in the top 5 m of the Sherwood Sandstone in well 33/17-1, and oil staining and fluorescence in the Westphalian section of well 33/22-1. Geochemical analysis of sea-bed cores gave evidence of a hydrocarbon seep from the Ulysses structure (Figs 2 and 4) with wet gases and $C_{12}+$ extracts. Although well 33/17-2A (Figs 2 and 4) proved part of the structure to be dry, the fault-block part (with a seismic flat spot) remains untested.

A 'Seepfinder' survey (Fig. 9) showed strong evidence of seepage along the Codling Fault Zone, weaker anomalies along the Dalkey and Lambay faults, and a strong anomaly on the Mid Irish Sea Uplift, SW of well 33/22-1, where the source rock may subcrop. There are many near-surface gas anomalies shown by high-resolution (site-survey) seismic data over the Finnegan structure (Fig. 10), south of well 33/17-1. These include a strong reverse polarity seismic event, gas chimneys, flat spots and phase reversals. Sea-bed mounds may also be evidence of hydrocarbon seepage (Croker 1995). Many are sited over faults and may represent bioherms living off leaking hydrocarbons. Sediment plumes over some of these mounds also suggest gas leakage.

Timing of hydrocarbon generation

The Kish Bank Basin had a similar structural history to the southern East Irish Sea Basin, in terms of the timing and relative magnitudes of burial and uplift events (see Hardman *et al.* 1993). Most of the oil-prone southern part of the East Irish Sea Basin has, however, had less pre-Variscan and Mesozoic burial than the Kish Bank Basin. This has led to a more favourable, later maturity with oil generation continuing after the Early Cretaceous inversion (Yaliz 1997).

Oil generation began (Fig. 8) in Late Carboniferous times over most of the Kish Bank Basin (before Variscan uplift) and recommenced during Late Triassic (during deposition of the seal) and Jurassic time. Oil potential would have been largely exhausted before Early Cretaceous uplift. Only gas would have been generated during subsequent burial during Late Cretaceous time and possibly during the

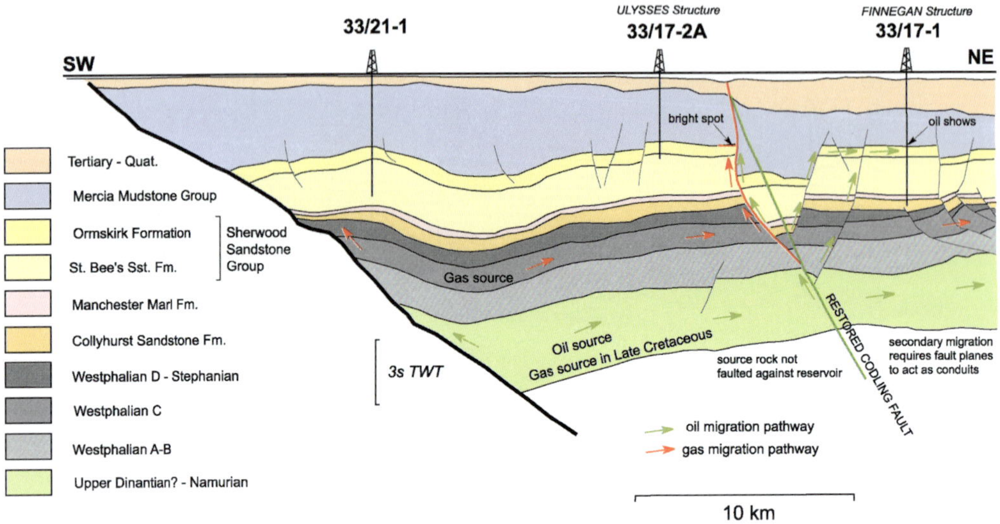

Fig. 11. Pre strike-slip cross-section showing migration routes. Reconstructed cross-section after removing 9 km dextral offset. Migration routes favour the eastern part of the basin. The Ulysses and Codling faults would have facilitated secondary migration into the Sherwood Sandstone Group.

Paleocene heating event. Hydrocarbon generation would have been largely turned off by uplift and cooling during the Tertiary period. Structures formed during, or significantly modified by, Tertiary uplift and strike-slip faulting thus have a high timing risk, as they rely on re-migration of previously trapped hydrocarbons.

Migration

Hydrocarbon migration probably occurred before Early Tertiary dextral strike slip along the Codling Fault Zone. Primary migration of hydrocarbons generated in the deeply buried source rocks in the west, in the hanging walls of the major faults, would have been to the SE within the Carboniferous section. This is towards the location of well 33/22-1, which had shows, and the subcrop of the Carboniferous units, where there is a Seepfinder anomaly (Fig. 9).

In the East Irish Sea Basin, secondary migration into and through the Permo-Triassic sandstone section is often facilitated by cross-fault juxtaposition of Carboniferous source rocks with Permo-Trias reservoirs. This situation pertains to the Douglas oil field, a hanging-wall structure sourced from the footwall of its controlling fault (Yaliz 1997). This situation exists in places along the Codling Fault, but it has been produced only by (post-migration) dextral strike slip, rather than by (pre-migration) normal faulting. Triassic prospects thus rely on vertical secondary migration into the Permo-Triassic sandstones, facilitated by faults. The first faults encountered by hydrocarbons on their primary migration route from the Dalkey hanging-wall kitchen are the Ulysses Fault (Figs 4 and 11) and the Codling Fault. Migration would then follow the Top Ormskirk interface, generally updip to the SE, towards the structure tested by well 33/22-1.

The untested fault-block part of the Ulysses structure (Fig. 4) could have been sourced via the Ulysses Fault. The tested anticlinal part of the closure relies on fill-and-spill from the fault block. Structures east of the Codling Fault could have been sourced via the Codling Fault, or via the numerous north–south faults forming these features.

There is a very limited drainage area where Carboniferous dips are up towards the bounding faults. Thus only small volumes of hydrocarbons could have moved up the Dalkey Fault into the Sherwood Sandstone and thence towards the structure tested by well 33/21-1, even if it existed at the right time. Primary and secondary migration paths generally favour prospects in the eastern part of the basin.

Post-migration structural alteration

Most of the closed tilted fault blocks (including that drilled by well 33/17-2A), are situated in a zone of 10 km width either side of the Codling

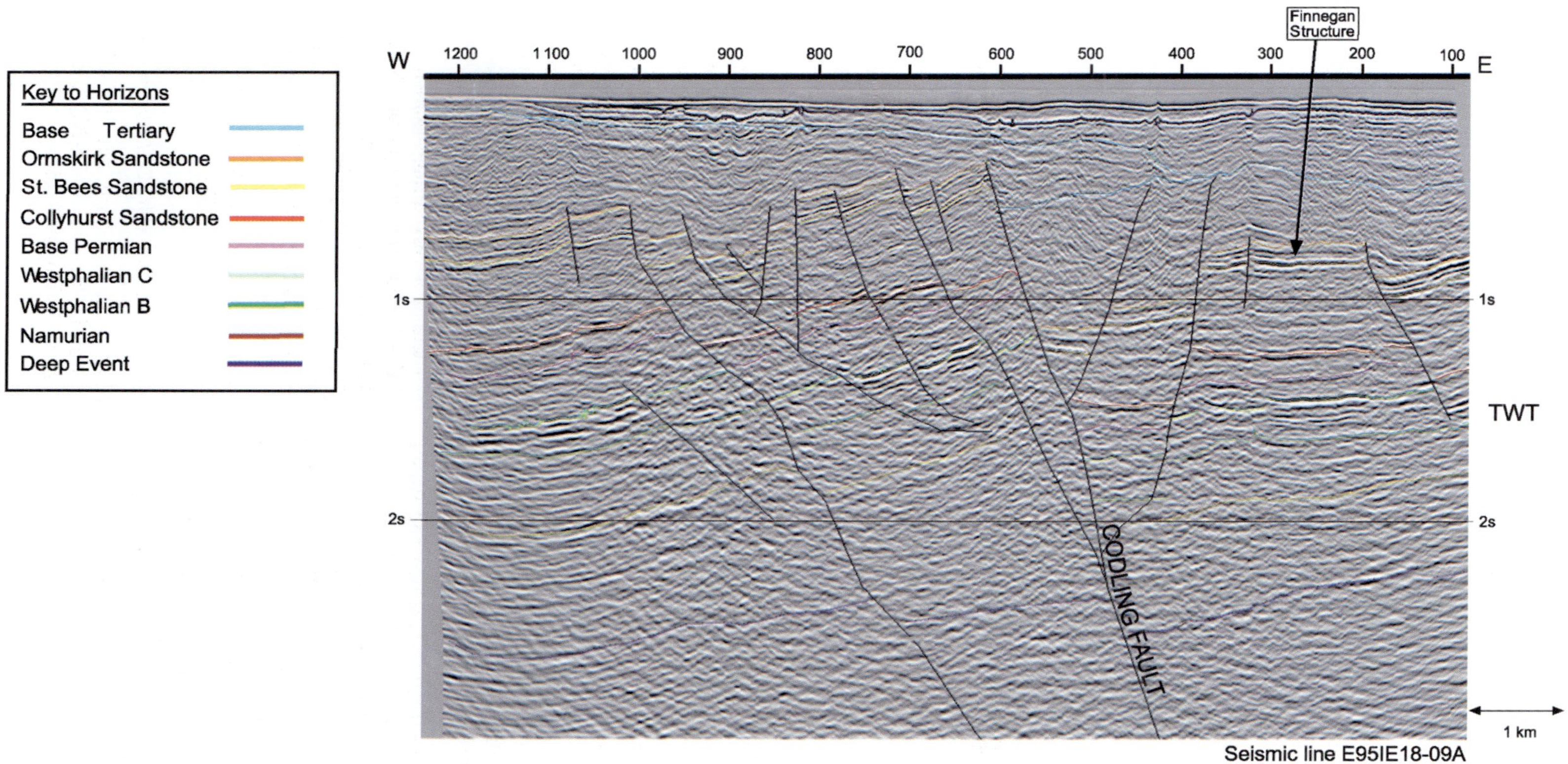

Fig. 12. Seismic line E95IE18-09A, showing Finnegan structure and Permian and Carboniferous plays. Permian and Upper Carboniferous reservoirs are faulted against the Mercia Mudstone across the Codling Fault. This juxtaposition was effected by Tertiary strike-slip movement.

Fault, and are formed by north–south faults. This faulted zone is divided, and dextrally offset by the Codling Fault. The north–south faults were, by analogy with the East Irish Sea Basin, probably initiated in Permian time and were active as normal faults during Triassic and Jurassic time.

Tilted fault-block closures formed during the Permian–Jurassic extensional phase have then been modified during Early Cretaceous inversion, and during the Tertiary period, by compression and strike-slip. The Codling Fault was mainly transtensional with evidence of dextral offset up to 4 km away from the main fault. These movements may have breached some previously existing closures and tilted others (causing re-migration of previously trapped hydrocarbons). Some closures (such as the 33/17-2A structure) may have been partly formed by Tertiary movements and thus post-date hydrocarbon migration. The graben forming the southern closure of the Ulysses structure (Fig. 2) and the Ulysses Fault appear to have been dextrally offset 1 km in a SSE direction, parallel to the Codling Fault. The strike closure of the anticlinal part of the structure may thus have formed after the main period of hydrocarbon migration. The 33/21-1 anticline was probably formed by compression during Tertiary time (although an Early Cretaceous age is also possible) rather than extensional roll-over, as Liassic to Permian units generally (with the exception of the Ormskirk Sandstone) thin towards the Dalkey Fault. The north–south (Finnegan) horst tested by well 33/17-1 probably formed during a Permian–Jurassic extensional phase. The well was dry even though it is just within closure. Alteration of the structure during Tertiary strike-slip faulting could have caused re-migration of previously trapped hydrocarbons to a new culmination at the southern end of the horst. This could explain the dead oil shows in the water-wet Sherwood Sandstone in well 33/17-1.

Collyhurst Sandstone play

The Manchester Marl is a potential top-seal to the Collyhurst Sandstone, although in well 33/17-1 it was sandy. Cross-fault seal is also a risk for most Collyhurst fault blocks, as any fault with a throw >100 m may juxtapose the Collyhurst Sandstone with the St. Bees Sandstone Formation. The Collyhurst Sandstone is, however, prospective in roll-over anticlines only cut by smaller faults. The Collyhurst Sandstone could also be prospective if juxtaposed with the Mercia Mudstone Group across a large fault (Fig. 12). This juxtaposition has, however, usually been effected by strike-slip movement of the Codling Fault (compare Figs 4 and 11). This has adverse timing implications.

Carboniferous play

The Carboniferous section in well 33/22-1 included porous sandstones that could be prospective in fault-block and roll-over anticlinal structures if there was an effective cross-fault seal and an intra-Carboniferous top-seal. Well 33/22-1 probably failed owing to the lack of an intra-Westphalian seal, which allowed escape of hydrocarbons to the Base-Tertiary surface, and severe biodegradation. There may be no effective seal below the Manchester Marl Formation. In the southern North Sea, fields are generally defined (and closed) by the base of the Permian Silverpit Formation (Bailey *et al.* 1993).

Conclusions

The Kish Bank Basin has proven Triassic reservoir and seal facies and there is good evidence for a hydrocarbon source. Migration and the timing of migration of the hydrocarbons are the most significant risks. Migration pathways favour the eastern part of the basin, which was updip from the source kitchen during the time of potential migration. Timing is a major risk, as most fault-block structures of economic size are within the Codling Fault dextral shear zone, and thus have been partly formed, or altered, after hydrocarbon migration. Deeper secondary plays, such as the Permian and Carboniferous plays, present higher risks because of cross-fault and vertical seal problems.

The authors are grateful to Enterprise Oil and Centrica UK for permission to publish this paper and to include data from the recently drilled well 33/17-2A. The authors would also like to thank I. Wilson for his contribution to the paper, and H. O'Reilly and S. Coffey for drafting the figures.

References

ARMSTRONG, J.P., SMITH, J., D'ELIA, V.A.A. & TRUEBLOOD, S.P. 1997. The occurrence and correlation of oils and Namurian source rocks in the Liverpool Bay–North Wales area. *In*: MEADOWS, N.S., TRUEBLOOD, S.P., HARDMAN, M. & COWAN, G. (eds) *Petroleum Geology of the Irish Sea and Adjacent Areas*. Geological Society, London, Special Publications, **124**, 195–211.

ARTER, G. & FAGIN, S.W. 1993. The Fleetwood Dyke and the Tynwald Fault zone, Block 113/27, East Irish Sea Basin. *In*: PARKER, J.R. (ed.) *Petroleum*

Geology of Northwest Europe: Proceedings of the 4th Conference. Geological Society, London, 835–843.

BAILEY, J.B., ARBIN, P., DAFFINOTI, O., GIBSON, P. & RITCHIE, J.S. 1993. Permo-Carboniferous plays of the Silver Pit Basin. *In*: PARKER, J.R. (ed.) *Petroleum Geology of Northwest Europe: Proceedings of the 4th Conference*. Geological Society, London, 707–715.

BARNES, R.P., AMBROSE, K., HOLLIDAY, D.W. & JONES, N.S. 1994. Lithostratigraphical subdivision of the Triassic Sherwood Sandstone Group in west Cumbria. *Proceedings of the Yorkshire Geological Society*, **50**, 51–60.

BARR, K.W., COLTER, V.S. & YOUNG, R. 1981. The geology of the Cardigan Bay–St. George's Channel Basin. *In*: ILLING, L.V. & HOBSON, G.D. (eds) *Petroleum Geology of the Continental Shelf of North-west Europe*. Heyden, London, 432–443.

BOLDREEL, L.O. & ANDERSEN, M.S. 1993. Late Paleocene to Miocene compression in the Faeroe Rockall area. *In*: PARKER, J.R. (ed.) *Petroleum Geology of Northwest Europe: Proceedings of the 4th Conference*. Geological Society, London, 1025–1034.

BOTT, M.H.P. & YOUNG, D.G.G. 1971. Gravity measurements in the north Irish Sea. *Quarterly Journal of the Geological Society of London*, **126**, 413–434.

BROUGHAN, F.M., NAYLOR, D. & ANSTEY, N.A. 1989. Jurassic rocks in the Kish Bank Basin. *Irish Journal of Earth Sciences*, **10**, 99–106.

CHADWICK, R.A. & HOLLIDAY, D.W. 1991. Deep crustal structure and Carboniferous basin development within the Iapetus suture zone, northern England. *Journal of the Geological Society, London*, **148**, 41–53.

CHADWICK, R.A., HOLLIDAY, D.W., HOLLOWAY, S., HULBERT, A.G. 1995. *The Northumberland–Solway Basin and Adjacent Areas. Subsurface Memoir.* British Geological Survey, Keyworth.

CLAYTON, G., SEVASTOPULO, G.D. & SLEEMAN, A.G. 1986. Carboniferous (Dinantian and Silesian) and Permo-Triassic rocks in south County Wexford, Ireland. *Geological Journal*, **21**, 355–374.

COLTER, V.S. & EBBERN, J. 1978. The petrography and reservoir properties of some Triassic sandstones of the Northern Irish Sea Basin. *Journal of the Geological Society, London*, **135**, 57–62.

COPE, J.C.W. 1997. The Mesozoic and Tertiary history of the Irish Sea. *In*: MEADOWS, N.S., TRUEBLOOD, S.P., HARDMAN, M. & COWAN, G. (eds) *Petroleum Geology of the Irish Sea and Adjacent Areas*. Geological Society, London, Special Publications, **124**, 47–59.

CORFIELD, S.M., GAWTHORPE, D.L., GAGE, M., FRASER, A.J. & BESLY, B.M. 1996. Inversion tectonics of the Variscan foreland of the British Isles. *Journal of the Geological Society, London*, **153**, 17–32.

CROKER, P.F. 1995. Shallow gas accumulation and migration in the western Irish Sea. *In*: CROKER, P.F. & SHANNON, P.M. (eds) *The Petroleum Geology of Ireland's Offshore Basins*. Geological Society, London, Special Publications, **93**, 41–58.

DEARMAN, W.R. 1963. Wrench-faulting in Cornwall and South Devon. *Proceedings of the Geologists' Association*, **74**, 265–287.

DOBSON, M.R. & WHITTINGTON, R.J. 1979. The Geology of the Kish Bank Basin. *Journal of the Geological Society London*, **1363**, 243–249.

ENGLAND, R.W. & SOPER, N.J. 1997. Lower crustal structure of the East Irish Sea from deep seismic reflection data. *In*: MEADOWS, N.S., TRUEBLOOD, S.P., HARDMAN, M. & COWAN, G. (eds) *Petroleum Geology of the Irish Sea and Adjacent Areas*. Geological Society, London, Special Publications, **124**, 61–72.

FLOODPAGE, J., WHITE, J. & NEWMAN, P. 1999. The hydrocarbon prospectivity of the Irish Sea: insights from recent exploration of the Central Irish Sea, Peel and Solway Basins. *In*: CROKER, P.F. & O'LOUGHLIN, O. (eds) *The Petroleum Exploration of Ireland's Offshore Basins, Extended Abstracts*. Petroleum Affairs Division, Department of the Marine and Natural Resources, Dublin, 28–31.

FRASER, A.J. & GAWTHORPE, R.L. 1990. Tectono-stratigraphic development and hydrocarbon habitat of the Carboniferous in northern England. *In*: HARDMAN, R.F.P. & BROOKS, J. (eds) *Tectonic Events Responsible for Britain's Oil and Gas Reserves*. Geological Society, London, Special Publications, **55**, 87–105.

GAWTHORPE, R.L., GUTTERIDGE, P. & LEEDER, M.R. 1989. Late Devonian and Dinantian basin evolution in northern England and North Wales. *In*: ARTHURTON, R.S., GUTTERIDGE, P. & NOLAN, S.C. (eds) *The Role of Tectonics in Devonian and Carboniferous Sedimentation in the British Isles*. Yorkshire Geological Society, Occasional Publications, **6**, 1–23.

HAIG, D.B., PICKERING, S.C. & PROBERT, R. 1997. The Lennox oil and gas field. *In*: MEADOWS, N.S., TRUEBLOOD, S.P., HARDMAN, M. & COWAN, G. (eds) *Petroleum Geology of the Irish Sea and Adjacent Areas*. Geological Society, London, Special Publications, **124**, 417–436.

HARDMAN, M., BUCHANAN, J., HERRINGTON, P. & CARR, A. 1993. Geochemical modelling of the East Irish Sea Basin: its influence on predicting hydrocarbon type and quality. *In*: PARKER, J.R. (ed.) *Petroleum Geology of Northwest Europe: Proceedings of the 4th Conference*. Geological Society, London, 809–821.

JACKSON, D.I. & MULHOLLAND, P. 1993. Tectonic and stratigraphic aspects of the East Irish Sea Basin and adjacent areas: contrasts in their post-Carboniferous structural styles. *In*: PARKER, J.R. (ed.) *Petroleum Geology of Northwest Europe: Proceedings of the 4th Conference*. Geological Society, London, 791–808.

JENNER, J.K. 1981. The structure and stratigraphy of the Kish Bank Basin. *In*: ILLING, L.V. & HOBSON, G.D. (eds) *Petroleum Geology of the Continental Shelf of North-west Europe*. Heyden, London, 426–431.

LAWRENCE, S.R., COSTER, P.W. & IRELAND, R.J. 1997. Structural development and petroleum potential of the northern flanks of the Bowland Basin (Carboniferous) North-West England. *In*: BROOKS, J. & GLENNIE, K.W. (eds) *Petroleum Geology of Northwest Europe*. Graham & Trotman, London, 225–233.

LEEDER, M.R. 1988. Recent developments in Carboniferous geology: a critical review with implications for the British Isles and NW Europe. *Proceedings of the Geologists' Association*, **99**, 73–100.

MAINGARM, S., IZATT, C., WHITTINGTON, R.J. & FITCHES, W.R. 1999. Tectonic evolution of the southern–central Irish Sea Basin. *Journal of Petroleum Geology*, **22**, 287–304.

MANNING, P.I. & WILSON, H.E. 1975. *Stratigraphy of the Larne Borehole, County Antrim*. Bulletin of the Geological Survey of Great Britain **50**.

MEADOWS, N.S. & BEACH, A. 1993. Controls on reservoir quality in the Triassic Sherwood Sandstone of the Irish Sea. *In*: PARKER, J.R. (ed.) *Petroleum Geology of Northwest Europe: Proceedings of the 4th Conference*. Geological Society, London, 823–833.

NAYLOR, D., HAUGHEY, N., CLAYTON, G. & GRAHAM, J.R. 1993. The Kish Bank Basin, offshore Ireland. *In*: PARKER, J.R. (ed.) *Petroleum Geology of Northwest Europe: Proceedings of the 4th Conference*. Geological Society, London, 845–855.

NEWMAN, P.J. 1999. The geology and hydrocarbon potential of the Peel and Solway Basins, East Irish Sea. *Journal of Petroleum Geology*, **22**, 305–324.

NOLAN, S.C. 1989. The style and timing of Dinantian syn-sedimentary tectonics in the eastern part of the Dublin Basin, Ireland. *In*: ARTHURTON, N.S., GUTTERIDGE, P. & NOLAN, S.C. (eds) *The Role of Tectonics in Devonian and Carboniferous Sedimentation in the British Isles*. Yorkshire Geological Society, Occasional Publications, **6**, 83–97.

QUIRK, D.G. & KIMBELL, G.S. 1997. Structural evolution of the Isle of Man and Central part of the Irish Sea. *In*: MEADOWS, N.S., TRUEBLOOD, S.P., HARDMAN, M. & COWAN, G. (eds) *Petroleum Geology of the Irish Sea and Adjacent Areas*. Geological Society, London, Special Publications, **124**, 135–159.

VISSCHER, H. 1971. *The Permian and Triassic of the Kingscourt Outlier, Ireland*. Geological Survey of Ireland, Special Paper No. 1.

WILSON, H.E. (ed.) *Regional Geology of Northern Ireland*. Ministry of Commerce, Geological Survey of Northern Ireland, Belfast.

YALIZ, A.M. 1997. The Douglas Oil Field. *In*: MEADOWS, N.S., TRUEBLOOD, S.P., HARDMAN, M. & COWAN, G. (eds) *Petroleum Geology of the Irish Sea and Adjacent Areas*. Geological Society, London, Special Publications, **124**, 399–416.

ZIEGLER, P.A. 1988. Post-Hercynian plate reorganisation in the Tethys and Arctic–North Atlantic domains. *In*: MANSPEIZER, W. (ed.) *Triassic–Jurassic Rifting, Continental Breakup and the Origin of the Atlantic Ocean and Passive Margins, Part B. Developments in Geotectonics 22, Parts A and B*. Elsevier, Amsterdam, 711–755.

ZIEGLER, P.A., CLOETINGH, S. & VAN WEES, J.D. 1995. Dynamics of intra-plate compressional deformation: the Alpine foreland and other examples. *Tectonophysics*, **252**, 7–59.

Fault distribution and timing in the Central Irish Sea Basin

CHRIS IZATT[1], SUVIMOL MAINGARM[2] & ANDREW RACEY[1]
[1]*BG International, 100 Thames Valley Park Drive, Reading RG6 1PT, UK*
(e-mail: chris.izatt@bg-group.com)
[2]*Institute of Geography and Earth Sciences, University of Wales, Aberystwyth SY23 3DB, UK*

Abstract: Well data analysis and the interpretation of 2D and 3D seismic reflection data provide valuable insights into the distribution and timing of fault activity within the Central Irish Sea Basin (CISB). Structural and stratigraphic relationships have been used to constrain the timing of fault movements and to interpret the mapped fault patterns in terms of the tectonic evolution of the area. Four main fault trends are identified at the Top Lower Triassic Sherwood Sandstone Group level: I, a NE–SW fault trend that parallels the basin-bounding faults and is believed to be of Mesozoic age; II, a pervasive system of north–south-trending faults that cross-cut the earlier NE–SW-trending faults, which manifests evidence of later Mesozoic extension followed by post-Oligocene transpressional fault reactivation; III, a NNE–SSW-trending, steeply dipping, fault set; IV, a WNW–ESE-trending conjugate extensional set that formed perpendicular to the NNE–SSW-trending transpressional faults during Late Tertiary dextral shearing. Early Tertiary axial centred basin inversion and regional exhumation have resulted in the elevation of the Sherwood Sandstone reservoir to shallow structural levels within the basin. Continued fault reactivation into Late Tertiary time has resulted in the compartmentalization of mapped structural closures and suggests that trap integrity is a major exploration risk factor in the CISB.

The Central Irish Sea Basin (CISB) is a NE–SW-trending graben located between the Kish Bank Basin to the NW and the St. George's Channel and Cardigan Bay basins to the SE (Fig. 1a). The CISB forms part of a linked system of Mesozoic basins between the UK and Ireland, which also includes the North Channel Basin (Shelton 1995), Solway Basin (Newman 1999), Peel Basin (Quirk *et al.* 1999), Kish Bank Basin (Naylor *et al.* 1993), the proven hydrocarbon province of the East Irish Sea Basin (Knipe *et al.* 1993), the St. George's Channel and Cardigan Bay basins (Tappin *et al.* 1994), and the North Celtic Sea Basin (Tucker & Arter 1987).

The CISB is *c.* 130 km long by 50 km wide and is bounded by the Mid-Irish Sea Uplift to the NW and by a southwestwards extension of the fault-bounded St. Tudwal's Arch to the SE (Fig. 1a and b). The NE part of the CISB is also known as the Caernarvon Bay Basin, and is separated from the rest of the CISB by a major NW–SE-trending lineament, the Codling Fault Zone (Fig. 1a). Carboniferous, Permo-Triassic, Tertiary and Quaternary sediments have been encountered by drilling in the basin (Fig. 2). In addition, interpretation of seismic data suggests that Jurassic and Cretaceous sediments could be locally preserved within the basin (Maddox *et al.* 1995). Two deep seismic reflection profiles, WINCH-4 and SWAT-2, acquired across the CISB and St. George's Channel and Cardigan Bay basins (BIRPS & ECORS 1986) indicate the graben geometry of the CISB with a preserved basin-fill consisting dominantly of Permo-Triassic sediments (Tappin *et al.* 1994).

The primary reservoir target for exploration wells drilled in the CISB has been the Lower Triassic Sherwood Sandstone Group, which comprises the Ormskirk Sandstone Formation and underlying St. Bees Sandstone Formation, sealed by halites and mudstones of the Upper Triassic Mercia Mudstone Group and sourced by possible Namurian or Westphalian claystones and coals (Fig. 2). Secondary targets include the Lower Permian Collyhurst Sandstone Formation and sands within the Westphalian section. All seven wells drilled in the basin have been plugged and abandoned as dry holes.

Results and a synthesis of these wells have been provided in a number of recent studies, which have discussed the hydrocarbon prospectivity of the CISB (Maddox *et al.* 1995;

From: SHANNON, P.M., HAUGHTON, P.D.W. & CORCORAN, D.V. (eds). 2001. *The Petroleum Exploration of Ireland's Offshore Basins*. Geological Society, London, Special Publications, **188**, 155–169. 0305-8719/01/$15.00

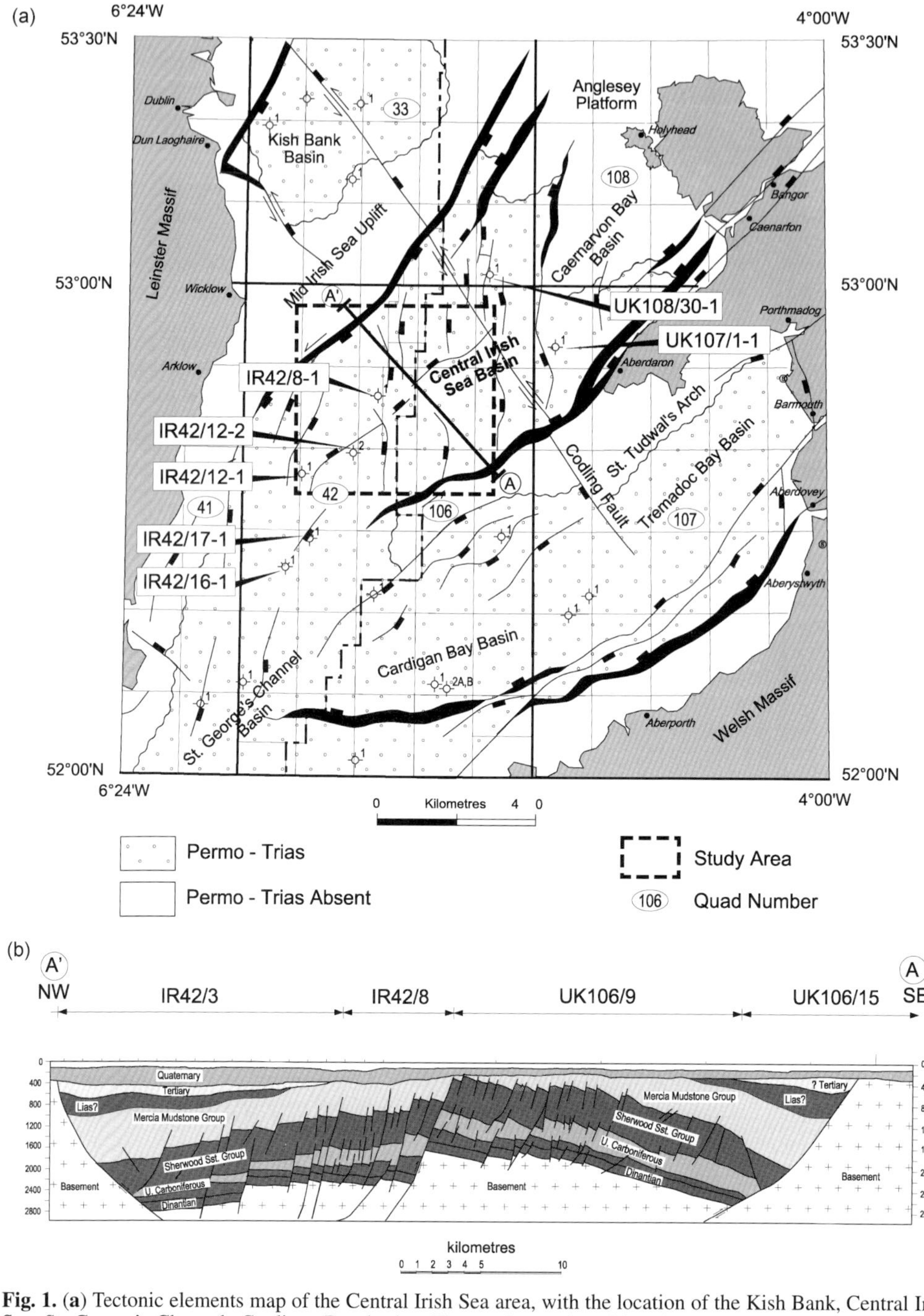

Fig. 1. (**a**) Tectonic elements map of the Central Irish Sea area, with the location of the Kish Bank, Central Irish Sea, St. George's Channel–Cardigan Bay basins. The north–south-striking faults within the Central Irish Sea Basin (CISB), and the NW–SE-trending Codling Fault Zone should also be noted. (**b**) Geoseismic section across the CISB showing graben-like geometry of the basin.

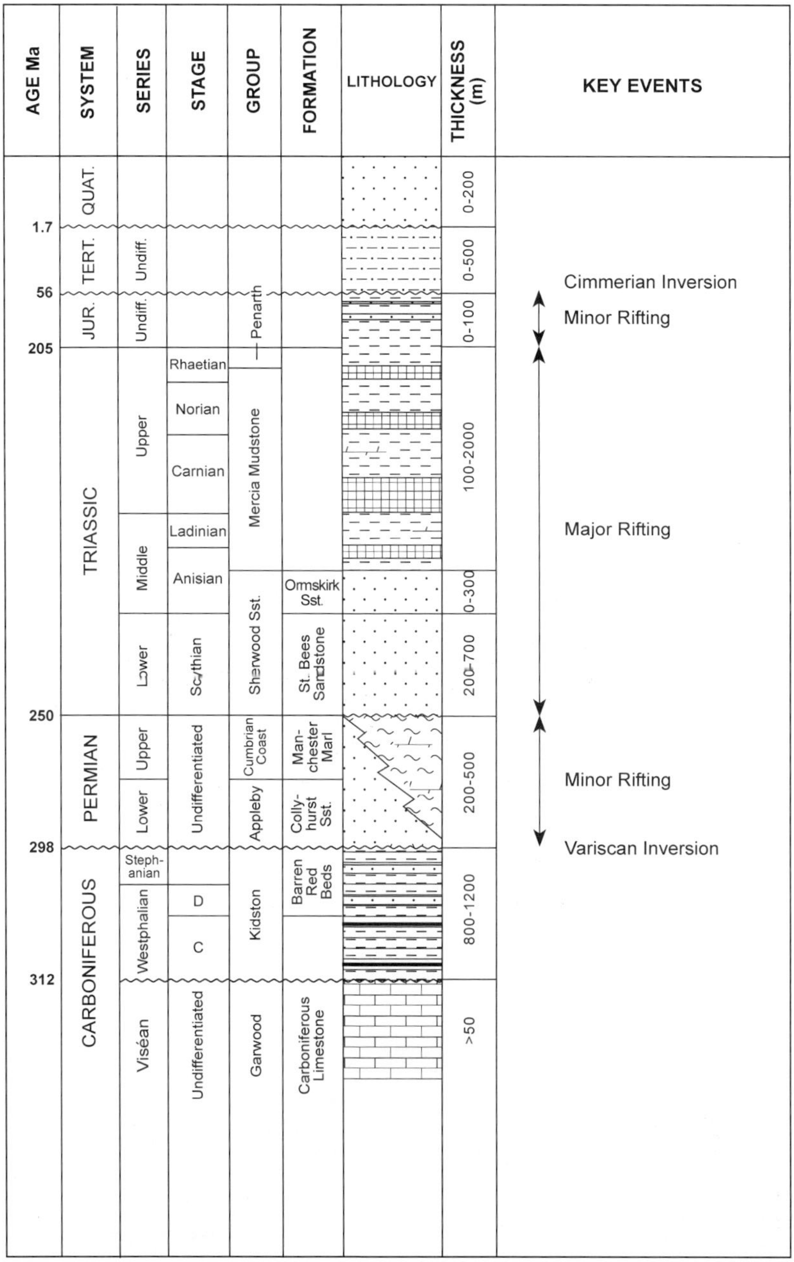

Fig. 2. Stratigraphic summary of the Central Irish Sea Basin. Major unconformities are recognized at the Base Westphalian, Base Permo-Triassic, Base Tertiary and Base Quaternary sequences.

Floodpage *et al.* 2001), the results of basin and thermal modelling (Duncan *et al.* 1998; Corcoran & Clayton 1999) and the structural evolution of the basin (Maingarm *et al.* 1999). The aim of this paper is to provide an insight into the distribution and timing of faulting in the CISB, as recorded by the preserved Palaeozoic to Cenozoic sediments, and to discuss the implications of this tectonic activity for the hydrocarbon prospectivity of the basin.

Well and seismic database

An extensive seismic and well database from the UK and Irish sectors of the CISB has been examined as part of a technical evaluation of UK exploration licence P939 (Fig. 3). The seven CISB wells, plus additional wells from the Kish Bank Basin (IR33/17-1, IR33/21-1 and IR33/22-1) and the St. George's Channel and Cardigan Bay basins (IR42/21-1, UK106/20-1, UK106/24-1, UK106/24-2, UK106/28-1, UK107/1-1 and UK107/16-1), have been used to establish a regional stratigraphic framework (Fig. 1a). The seismic interpretation was based on a regional grid of *c.* 1500 km of proprietary and speculative 2D seismic data in the southern part of the CISB plus 425 km^2 of proprietary 3D seismic data, acquired in 1996, over UK Blocks 106/4, 106/8 and 106/9 (Fig. 3). The 3D data quality is good to fair above 1.5 s two-way travel time (TWTT), except where data quality is reduced by the presence of sea-bed multiples. The primary Sherwood Sandstone Group reservoir target can be mapped with confidence around the basin, although mapping of deeper Permian and Carboniferous events is more problematic because of reduced data quality with increasing depth and limited well ties.

Tectonostratigraphic framework

The structural and stratigraphic evolution of the CISB has been documented in a number of publications (Maddox *et al.* 1995; Duncan *et al.* 1998; Corcoran & Clayton 1999; Maingarm *et al.* 1999; Floodpage *et al.* 2001). The oldest rocks penetrated in the basin to date are Dinantian

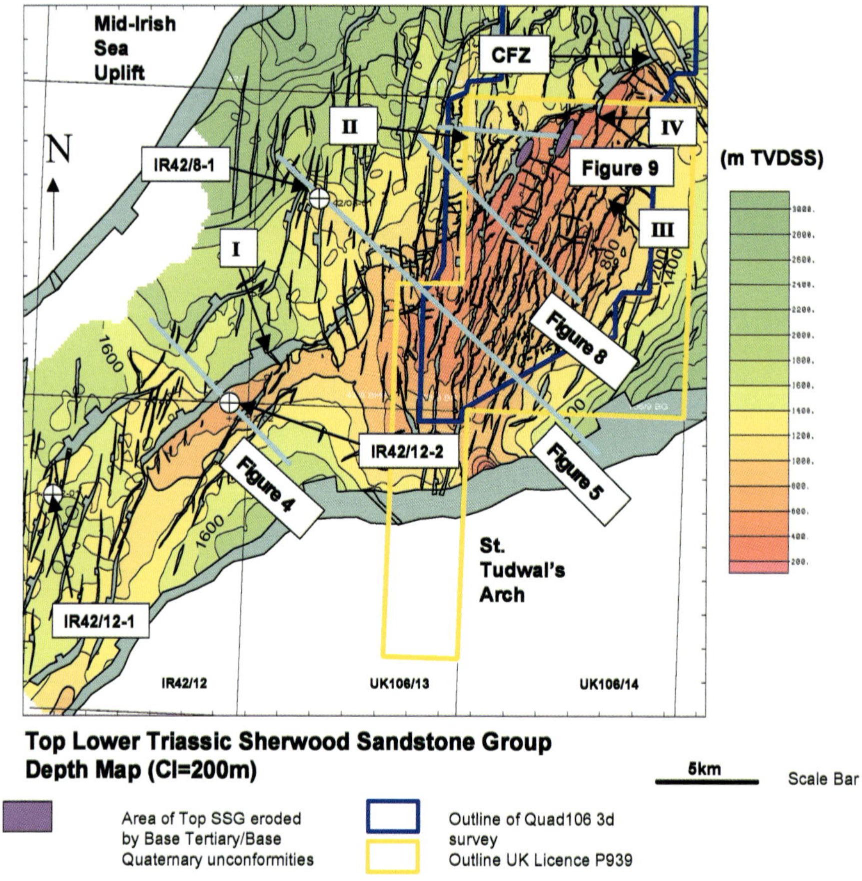

Fig. 3. Top Lower Triassic Sherwood Sandstone Group depth map of the southern part of the CISB. The presence of four main fault orientations should be noted: I, early NE–SE-trending faults; II, north–south-trending faults; III, NNE–SSW-trending faults; IV, WNW–ESE-trending faults; CFZ, Codling Fault Zone. Map also shows the location of the 3D seismic survey used in this study, and the location of Figs. 4, 5, 8 and 9.

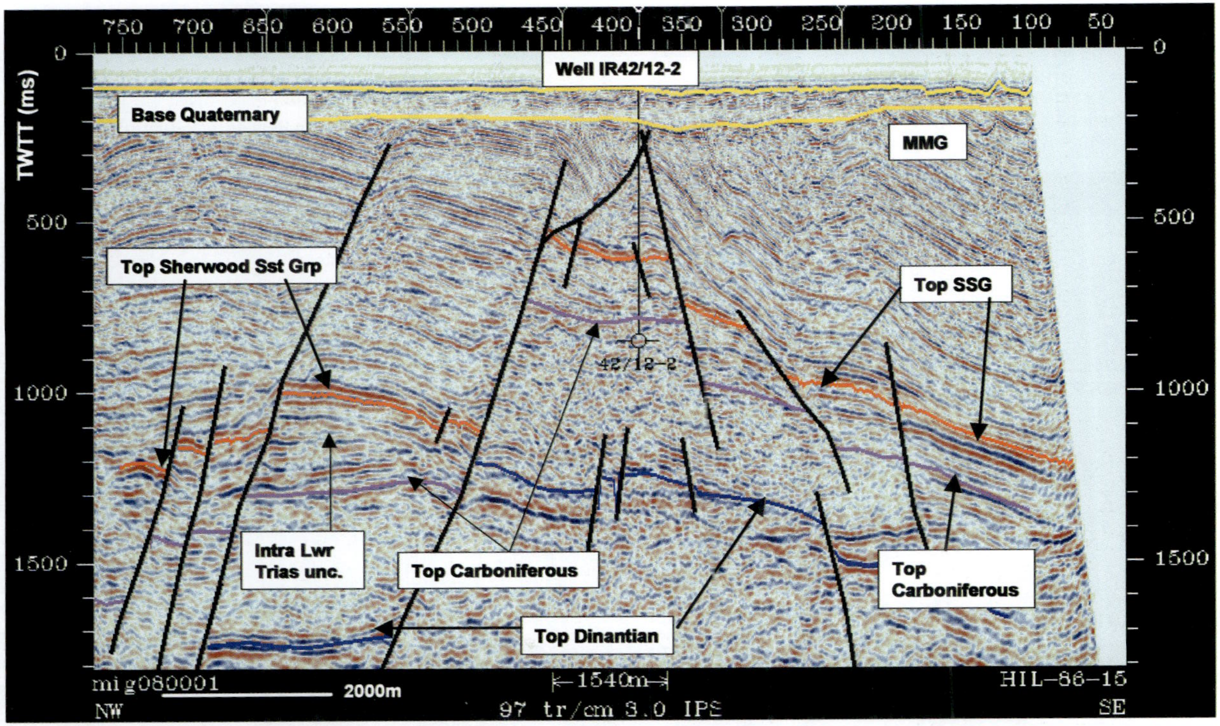

Fig. 4. Interpreted 2D seismic line HIL-86-15 through the fault terrace tested by well IR42/12-2 (see Fig. 3 for position of line). The relatively high-amplitude, continuous reflectors associated with the Top Dinantian and Top Sherwood Sandstone Group should be noted. The Westphalian C–D section is seismically transparent because of the low impedance contrasts between the sandstones and shales and the relative paucity of coal beds in this section. Interpreted horizon colour code: blue, Top Dinantian; purple, Top Carboniferous; red, Top Sherwood Sandstone Group (Top SSG); yellow ochre, Base Quaternary; bright yellow, sea bed.

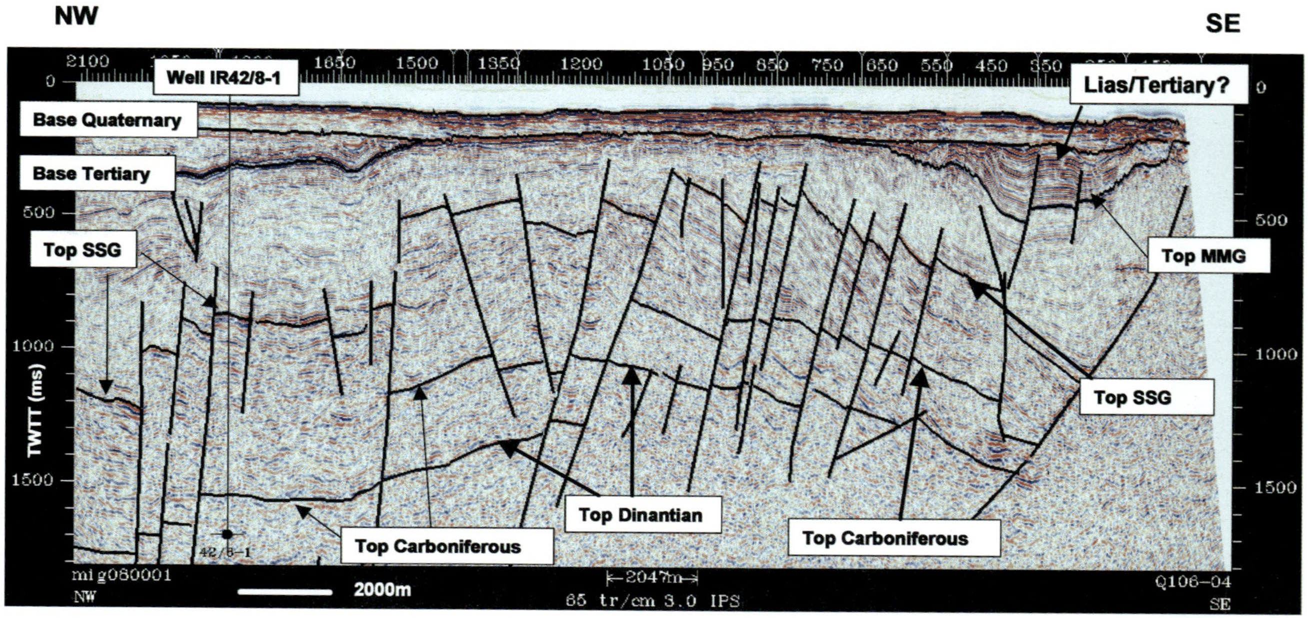

Fig. 5. Interpreted 2D seismic line Q106-04 through well IR42/8-1 (see Fig. 3 for position of line). The reflective package of Tertiary sediments preserved in the hanging wall of the basin-bounding fault, which is located to the NW of the end of this seismic line, should be noted. A highly reflective, but undated, package of reflectors is also observed in the hanging wall of the basin-bounding fault at the SW end of the line. Regional stratigraphic evidence suggests that this package is likely to be of Liassic or Tertiary age. Interpreted horizon colour code: blue, Top Dinantian; purple, Top Carboniferous; red, Top Sherwood Sandstone Group (Top SSG); beige, Base Tertiary; yellow ochre, Base Quaternary; bright yellow, sea bed.

marine carbonates, encountered in well IR42/17-1, which are unconformably overlain by interbedded coals, shales and sandstones of Westphalian C to Stephanian age (Maddox *et al.* 1995). No Namurian sediments have been drilled in the basin to date but the presence of a Namurian A–B section cropping out on the southern coast of the Isle of Anglesey (Racey, pers. obs.) suggests that Namurian deposition could have occurred, at least in the northeastern part of the CISB.

A thin, poorly dated, Permian section may be present in wells IR42/8-1 and IR42/12-2. However, in general, Upper Carboniferous sediments are unconformably overlain by interbedded sandstones, siltstones and claystones of the Lower Triassic Sherwood Sandstone Group (Fig. 2). Up to 2000 m of interbedded siltstones, claystones and halite beds of the Upper Triassic Mercia Mudstone Group overlie the Sherwood Sandstone Group. The post-Triassic stratigraphic record is severely truncated in the CISB. Jurassic and Cretaceous sediments have not been encountered by drilling to date, although these sediments may have been preserved in the hanging walls of basin-bounding faults. In addition, seismic and well evidence indicates that up to 500 m of Eocene to Oligocene interbedded sandstones and claystones are preserved. These Tertiary sediments are capped by a Quaternary cover, which generally comprises sands and gravels and is *c.* 200 m thick (Fig. 2).

A number of major unconformities are recognized in the CISB. The mapping of these erosive surfaces has been used to constrain the timing of fault activation and basin inversion in the CISB.

Seismic interpretation

A number of seismic reflectors can be mapped with confidence across the CISB including the Top Dinantian (Viséan), the Top Sherwood Sandstone Group and, locally, the Base Tertiary unconformity. The Top Dinantian event is a bright, high-amplitude, low-frequency reflector associated with a positive acoustic impedance contrast between the Westphalian clastic deposits and the underlying Dinantian carbonates (Fig. 4). Relative to the overlying Triassic section, the Westphalian section appears to be seismically transparent. In well IR42/17-1, the sonic log response of the Westphalian section confirms that there is a low acoustic impedance contrast between the interbedded sandstones and shales, which both manifest interval velocities of *c.* 4100–4900 m s^{-1}. The paucity of bright reflectors within the Westphalian section also suggests that there are fewer coals in the Westphalian C–D interval compared with the seismically more reflective Westphalian A–B coal measure sequences of the East Irish Sea and southern North Sea basins.

The Top Sherwood Sandstone Group event is seismically defined by two bright peaks and a trough (Fig. 4), associated with a negative acoustic impedance contrast between the siltstones, dolomites and halites of the Mercia Mudstone Group and the underlying Sherwood Sandstone Group. The Top Sherwood Sandstone Group event also marks the boundary between a relatively competent sequence of Permian to Lower Triassic strata dominated by sandstone, and the overlying incompetent fine-grained clastic deposits and halites of the Mercia Mudstone Group.

The Base Tertiary reflector, which is locally observed in the hanging wall of the northern boundary fault of the CISB, is a bright event that represents the positive acoustic impedance contrast between the low-velocity Tertiary section and the higher velocities of the underlying Upper Triassic Mercia Mudstone Group (Fig. 5). The Tertiary section, which has been penetrated by well IR42/8-1, is internally characterized by a number of bright events that represent the impedance contrast between sands and shales. A highly reflective package of reflectors is also identified in the hanging wall of the fault bounding the southern margin of the CISB. This reflective package cannot be correlated with any wells, although regional stratigraphic evidence suggests that this section is likely to be of Early Jurassic (Liassic) or Tertiary age.

A Top Sherwood Sandstone Group depth map (Fig. 3) illustrates the main faults that offset the Top Sherwood Sandstone Group horizon. In map view, a number of major tectonic elements can be recognized in the CISB. Well IR42/12-2 tested a structural closure on a central fault terrace that dominates the southern part of the basin (Figs 3 and 4). To the NE of this well location the Top Sherwood Sandstone Group becomes progressively shallower and rises to a series of culminations, in the north of UK Block 106/9, where it has been eroded by the Base Tertiary or Base Quaternary unconformity (Fig. 3). Within the centre of the basin, in Blocks IR42/8 and UK106/8 and 9, the Top Sherwood Sandstone Group is deformed by both NE–SW- and north–south-trending faults (Fig. 3). Seismic interpretation suggests that well IR42/8-1 tested a closure bounded by north–south-trending faults (Figs 3 and 5).

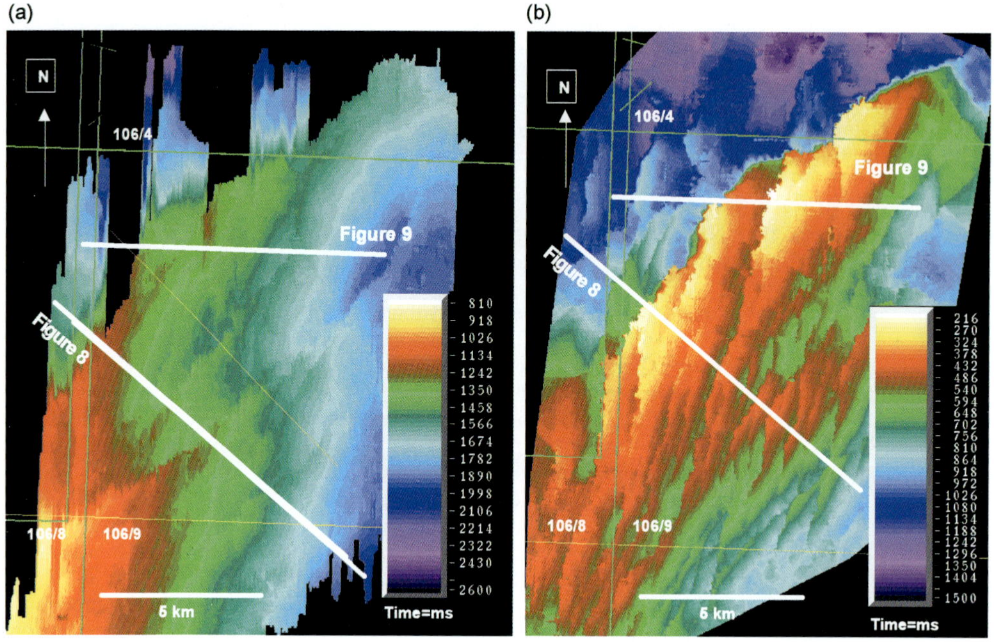

Fig. 6. Time structure maps showing the orientation of the main fault lineaments in UK Blocks 106/8 and 9: (**a**) Top Dinantian time map; (**b**) Top Lower Triassic Sherwood Sandstone Group time map. The higher density of faulting affecting the Top Sherwood Sandstone Group map, suggesting the presence of fault systems that do not involve the Lower Carboniferous section, should be noted. The area covered by these maps is essentially the outline of the 3D seismic survey shown in Fig. 3. The positions of Figs. 8 and 9 are shown.

Fault patterns

Fault patterns within the CISB have been evaluated previously using maps generated from 2D seismic reflection data (Maingarm *et al.* 1999). The fault trends described below have been identified following the interpretation of both 2D and 3D seismic data and the analysis of structure maps generated for the Top Dinantian and Top Lower Triassic Sherwood Sandstone Group horizons (Fig. 6a and b). In addition, fault trends are discussed with reference to an azimuth map of the Top Sherwood Sandstone Group, derived from the 3D seismic attribute mapping (Fig. 7). These fault trends are interpreted in terms of the regional tectonic evolution of the CISB.

Four main fault trends are identified at the Top Sherwood Sandstone Group level in the CISB (Figs 3 and 7), as follows.

NE–SW-trending faults (Late Triassic to Mid-Jurassic basin extension)

Fault trend I comprises a series of NE–SW-trending, basement involved, extensional faults, which includes the main basin-bounding faults and the fault bounding the central fault terrace tested by well IR42/12-2 (Figs 3, 4 and 8). The basin-bounding faults of the CISB are approximately parallel to the NE–SW-trending basin-bounding faults of the Kish Bank, St. George's Channel–Cardigan Bay and Celtic Sea basins (Fig. 1a). These faults have been interpreted as representing Mesozoic extensional reactivation of a pre-existing Caledonian structural fabric (Coward 1995; Musgrove *et al.* 1995).

The estimated cumulative extension across the CISB is *c.* 6000 m, consistent with a modelled β-factor of 1.12 for the basin (Coward & Ries, pers. comm.). This estimate is a minimum figure, as the Triassic and Jurassic footwall cut-offs are no longer preserved. The precise time of movement on these basin-bounding faults is unclear although the seismic interpretation indicates that significant extension has occurred post-Early Triassic time. Conglomerates and volcanic rocks interpreted as being of Permian age were encountered in the base of well UK107/7-1, indicating a possible earlier phase of extension at least within the Caernarvon Bay Basin. Chadwick & Evans (1995) have suggested the presence of an Early Triassic extensional phase in the Cheshire Basin and Worcester

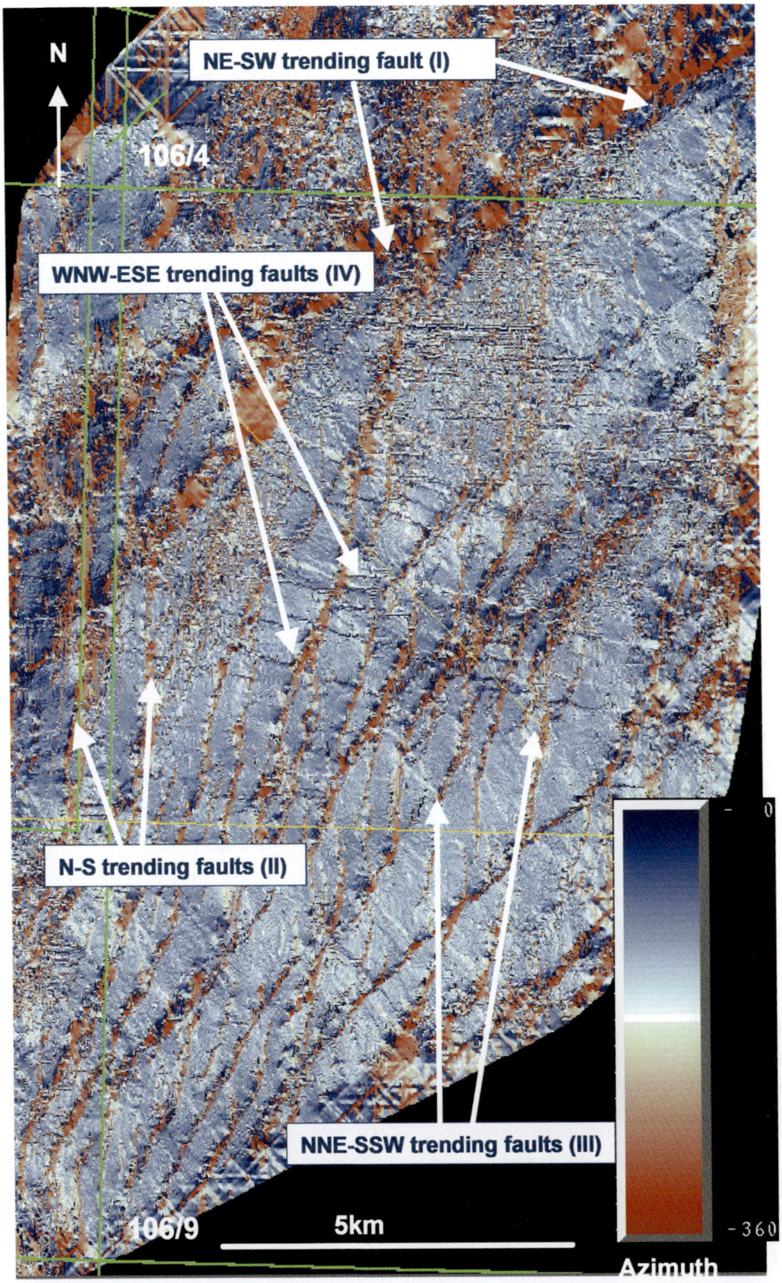

Fig. 7. Azimuth attribute from 3D seismic data in UK Block 106/9: Top Sherwood Sandstone Group azimuth map with illumination from the SE. Evidence for small-scale WNW–ESE-trending faults should be noted; these are interpreted to have formed as a conjugate extensional set perpendicular to the NNE–SSW-trending faults, under a Late Tertiary dextral shear couple.

Graben. Musgrove *et al.* (1995) offered evidence, from the EISB and the Celtic Sea Basin, in support of a regional phase of extension during Late Triassic time. The preservation of Middle–Upper Jurassic sequences to the south in the Celtic Sea and St. George's Channel and Cardigan Bay basins suggests that major extension in the CISB may have occurred at this time (Tappin *et al.* 1994).

North–south-trending faults (Mid- to Late Jurassic transtension)

Fault trend II consists of a series of faults striking north–south with fault throws dominantly to the west (Figs 3 and 8). These faults are basement involved and offset both Carboniferous and Permo-Triassic sequences, with fault displacements attenuating upwards into the overlying Upper Triassic Mercia Mudstone Group. These north–south-trending faults occur throughout the basin and form a critical component of a number of failed hydrocarbon traps in the CISB. Wells IR42/8-1, IR42/12-1 and possibly UK108/30-1 tested fault-dependent structural closures, at Top Sherwood Sandstone Group level, which were bounded by these north–south-trending faults.

The throw on these north–south-trending faults is *c.* 200–300 m, at the Top Sherwood Sandstone level, as compared with >1000 m on each of the NE–SW-trending basin-bounding faults. However, the density of north–south-trending faults is considerably higher than the density of distribution of the NE–SW-trending faults. The north–south-trending faults were initiated as extensional faults and cross-cut the earlier NE–SW-trending faults in the centre of the basin (Maingarm *et al.* 1999). This cross-cutting relationship gives rise to a series of sinistral offsets along the earlier NE–SW-striking extensional fault bounding the central fault terrace in Blocks IR42/8 and UK106/9 (Figs 3 and 7). These north–south-trending extensional or transtensional faults are consistent with a period of sinistral fault movement along the main NE–SW-trending basin-bounding faults. The age of this transtension is uncertain but appears to pre-date a possible period of Late Cretaceous–Early Tertiary uplift along the central axis of the CISB (Maingarm *et al.* 1999).

Extensional offsets across these north–south-trending faults, at the Top Sherwood Sandstone Group level, are clearly imaged on the 3D seismic dataset (Fig. 8). In addition, the seismic interpretation indicates that the Base Tertiary unconformity is deformed by a series of small north–south-trending folds. The axes of these folds are coincident with the extensional faults deforming the Top Sherwood Sandstone Group (Figs 8 and 9). This suggests that compressional reactivation has occurred along some of these faults during Tertiary time. The precise age of this folding and fault reactivation is unclear but it post-dates the deposition of Eocene sediments in the CISB. This folding also pre-dates a later regional inversion of the CISB along its NE–SW-trending basin axis. Minor compressional reactivation of these north–south-trending faults is consistent with simultaneous dextral shearing along the NE–SW-trending basin-bounding faults during Late Tertiary time.

NNE–SSW-trending faults (Late Tertiary steeply dipping faults)

Fault trend III consists of a relatively high-density set of NNE–SSW-trending transpressional faults, which appear to deform only the Permian to Lower Triassic section (Figs 8 and 9). This NNE–SSW fault trend is not prominent on the Top Dinantian time map (Fig. 6a) but is clearly identifiable on the Top Sherwood Sandstone Group time map and azimuth map (Figs 6b and 7). Fault spacing at the Top Sherwood Sandstone level is on average 500 m (Fig. 7). The majority of these fault planes are steeply dipping to vertical and have a maximum displacement of *c.* 200 m. Apparent displacement along these faults decreases with depth into a possible incompetent Permian section (Manchester Marl Formation equivalent) and decreases upwards into the incompetent mudstones and halites of the Mercia Mudstone Group.

The initiation of this set of steeply dipping faults may have been coincident with the compressional reactivation of the north–south-trending faults (fault trend II), which resulted from a phase of later Tertiary (post-Eocene) dextral shearing along the NE–SW-trending basin-bounding faults.

WNW–ESE-trending faults (Late Tertiary phase)

Fault trend IV comprises a relatively low-density population of WNW–ESE-trending extensional faults, which are observed, at Top Sherwood Sandstone Group level, on the 3D seismic data acquired over UK Blocks 106/8 and 9 (Figs 3 and 7). The offsets along these faults are relatively minor and at the limit of seismic resolution. These faults are interpreted as a conjugate set of extensional faults that possibly formed perpendicular to the NNE–

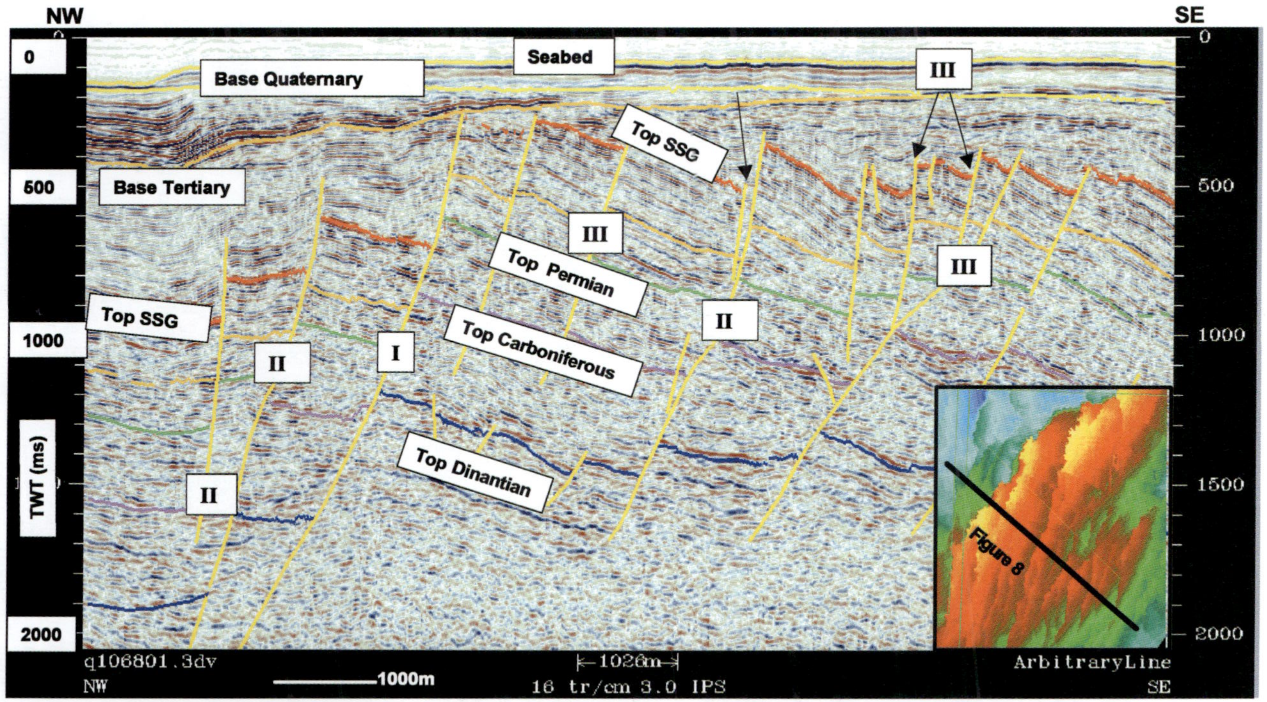

Fig. 8. Interpreted 3D seismic line across UK Block 106/9 with three of the main fault orientations labelled (see Figs. 3 and 6 for position of line); I, early NE–SE-trending faults; II, north–south-trending faults; III, NNE–SSW-trending faults. The small anticlinal flexures at Tertiary level should be noted, coincident with the north–south-trending faults, which indicates transpressional reactivation of this fault set in post-Eocene time. Interpreted horizon colour code: blue, Top Dinantian; purple, Top Carboniferous; green, Top Permian; brown, St. Bees Sandstone Formation; red, Top Sherwood Sandstone Group (Top SSG); beige, Base Tertiary; yellow ochre, Base Quaternary; bright yellow, sea bed.

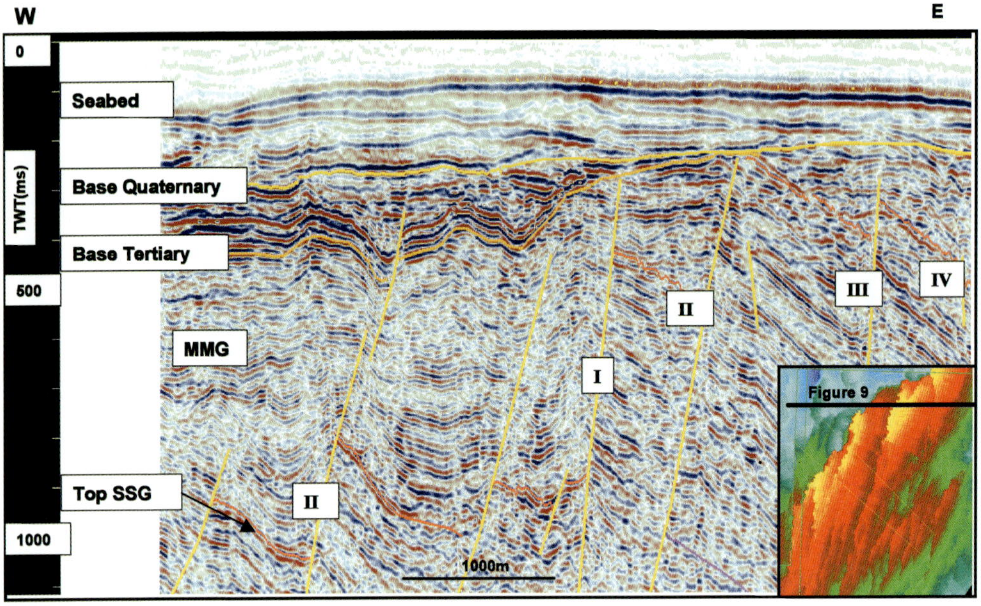

Fig. 9. Interpreted 3D seismic cross-line through the northern part of UK Block 106/9 with four main fault orientations labelled (see Figs. 3 and 6 for position of line). I, early NE–SE-trending faults; II, north–south-trending faults; III, NNE–SSW-trending faults; IV, WNW–ESE-trending faults. The evidence for transpressional reactivation of the north–south-trending faults, in post-Eocene time, should be noted. Interpreted horizon colour code: red, Top Sherwood Sandstone Group (Top SSG); beige, Base Tertiary; yellow ochre, Base Quaternary; bright yellow, sea bed.

SSW-trending transpressional faults, during the Late Tertiary dextral movements along NE–SW-trending basin-bounding faults.

In comparison with the NNE–SSW-trending faults the density of WNW–ESE-trending extensional faults is considerably reduced. Maximum displacement along these WNW–ESE-trending faults appears to be at the Top Sherwood Sandstone Group level, with displacement on these faults decreasing upwards into the more incompetent and ductile units of the Mercia Mudstone Group, and with depth into the structurally less competent Permian sequences. The high density of north–south- and NNE–SSW-trending faults, identified in the UK Block 106/9 area, is in part due to the higher resolution of the 3D seismic dataset. This densely faulted area is also located near a small bend in the trend of the southern bounding fault of the CISB basin (Figs 3 and 10), which may have acted as a buttress during the Tertiary dextral movement along the main basin-bounding fault.

NW–SE-trending fault: Codling Fault Zone

The Codling Fault Zone is a major NW–SE-trending lineament that cuts obliquely across the CISB in Blocks 106/4 and 106/9. This fault zone extends NW into the Kish Bank Basin, where it causes an apparent 4 km of dextral offset of the NE–SW-trending basin-bounding fault (Jenner 1981; Dunford & Dancer 2001). The Codling Fault Zone has a similar orientation to the Sticklepath–Lustleigh Fault Zone, described by Holloway & Chadwick (1986), along which there are preserved inliers of Eocene sediments within small pull-apart basins. Similar offsets are also located along the southern margins of the Cardigan Bay Basin (Turner 1997). In the NE part of UK Block 106/9 the Codling Fault Zone is marked by NW–SE-trending extensional faults that dip to the NE and form boundary faults to a transtensional basin containing a syntectonic basin fill of possible Oligocene sediment.

Discussion

Evidence of tectonic events affecting the CISB is recorded in fault patterns affecting the Top Dinantian and the Top Sherwood Sandstone Group. Analysis of the observed fault patterns suggests a complicated post-Triassic basin evolution (Fig. 10). An early series of NE–SW-trending faults involving basement and

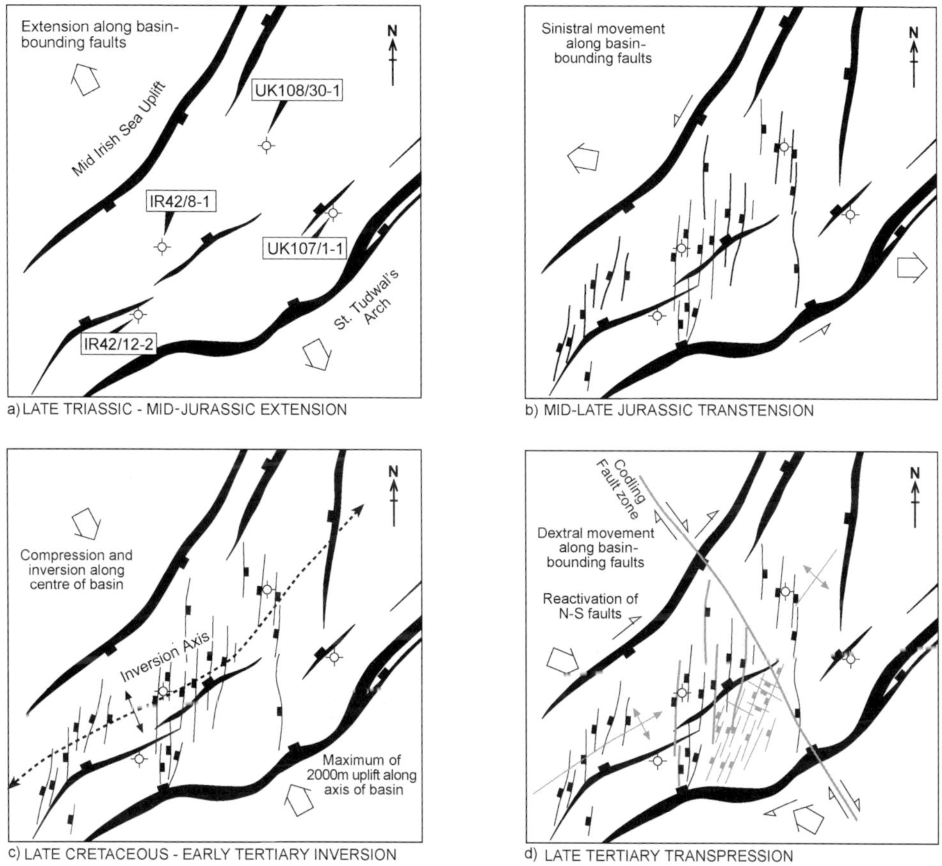

Fig. 10. Summary diagram showing fault development and change in stress field of the CISB through time. Some well locations are annotated for reference. (**a**) Late Triassic–Mid-Jurassic extension along NE–SW-striking faults (note the development of the IR42/12-2 fault terrace at this time). (**b**) Mid–Late Jurassic sinistral movement along the NE–SW-trending basin-bounding faults creating a series of north–south extensional faults that cross-cut the earlier extensional fabric. (**c**) Regional exhumation occurred during Late Cretaceous–Early Tertiary time followed by an Early Tertiary inversion along the NE–SW-trending central axis of the basin, which resulted in the development of a long-wavelength anticlinal flexure with maximum uplift of 2000 m along the axis of the CISB. (**d**) Dextral reactivation of the Codling Fault Zone during Eocene–Oligocene time resulted in the development of minor pull-apart basins along this fault zone. Late Tertiary dextral shearing, along the NE–SW-trending basin-bounding faults, caused partial reactivation of some of the earlier north–south-trending extensional faults, and the propagation of a high-density set of NNE–SSW-trending, steeply dipping faults.

Palaeozoic and Permo-Triassic sediments is interpreted as being of Late Triassic–Mid-Jurassic age (Maddox *et al.* 1995; Maingarm *et al.* 1999) and marks the initial basin extension (Fig. 10a). These early NE–SW-trending faults are subsequently cut by a later series of north–south-trending transtensional faults, which again involve both basement and Palaeozoic and Permo-Triassic cover. These north–south-trending faults could have propagated as a result of sinistral movements along the NE–SW-trending basin-bounding faults of the CISB (Fig. 10b). It is not possible to date the age of this faulting but it is believed to pre-date the Late Cretaceous–Early Tertiary basin inversion (Fig. 10c).

A Late Tertiary transpressional event, which locally reactivated some of the north–south-trending Mesozoic faults and resulted in localized folding of the Base Tertiary unconformity, is also recognized. This Late Tertiary transpressional event may be related to the propagation of a high density of NNE–SSW-trending transpressional faults deforming the competent Lower Triassic Sherwood Sandstone Group section (Fig. 10d). The inversion of the north–south-trending faults, together with the propagation of NNE–SSW-trending transpressional and WNW–ESE-trending extensional faults, is consistent with dextral fault displacements along the NE–SW-trending basin-bounding faults of the CISB, during Late Tertiary time.

An understanding of fault development in the CISB is important for evaluating the hydrocarbon prospectivity of the basin. All seven exploration wells drilled to date in the CISB have been dry holes with limited gas shows. Previous workers discussing the hydrocarbon prospectivity in the CISB (Maddox *et al.* 1995; Corcoran & Clayton 1999; Floodpage *et al.* 2001) have suggested that a primary reason for well failure has been either the lack of an effective source rock or the timing of source rock maturity. The present study indicates that trap integrity is another significant exploration risk factor in the CISB. Severely compartmentalized hydrocarbon traps, at the Top Sherwood Sandstone level, are a likely by-product of the complex fault patterns observed in the basin. In addition, continued reactivation, of selected fault trends into Late Tertiary time, increases the risk of tectonic breaching of hydrocarbon traps and fault seal failure.

Conclusions

The distribution and timing of fault activity in the CISB has been investigated, using the available well and seismic data, with the following conclusions:

(1) four main fault trends are identified at the Top Lower Triassic Sherwood Sandstone Group: trend I, NE–SW faults; trend II, north–south faults; trend III, NNE–SSW faults; trend IV, WNW–ESE faults. These fault patterns are interpreted as evidence for three main phases of fault activity within the CISB: Late Triassic–Mid-Jurassic basin extension, Late Jurassic transtension, and post-Eocene transpression.

(2) Axial centred basin inversion has occurred in the CISB, during Late Cretaceous to Early Tertiary time. This inversion pre-dates the post-Eocene transpressional reactivation of the north–south- and NNE–SSW-trending faults.

(3) Repeated uplift and erosion, combined with continued fault reactivation into Late Tertiary time, has resulted in the elevation of the Top Sherwood Sandstone Group to shallow levels with an increased risk of top seal failure.

The authors would like to thank BG International and their Licence P939 partners, Premier Oil plc and Talisman North Sea Ltd, for permission to publish this paper. Colleagues R. Blow, S. Maddox, P. Ellis and G. Oakes (BGI) are thanked for their contributions to this study. M. Coward and A. Ries are thanked for their comments. The cartography office at BG International is thanked for drafting the diagrams. We also thank W. Duncan, P. Newman and D. Corcoran for their reviews of an original draft of this paper.

References

BIRPS & ECORS 1986. Deep seismic reflection profiling between England, France and Ireland. *Journal of the Geological Society, London*, **143**, 45–52.

CHADWICK, R.A. & EVANS, D.J. 1995. The timing and direction of Permo-Triassic extension in southern Britain *In*: BOLDY, S.A.R. (ed.) *Permian and Triassic Rifting in Northwest Europe*. Geological Society, London, Special Publications, **91**, 161–192.

CORCORAN, D. & CLAYTON, G. 1999. Interpretation of vitrinite reflectance profiles in the Central Irish Sea area: implications for the timing of organic maturation. *Journal of Petroleum Geology*, **22**, 261–286.

COWARD, M.P. 1995. Structural and tectonic setting of the Permo-Triassic basins of Northwest Europe *In*: BOLDY, S.A.R. (ed.) *Permian and Triassic Rifting in Northwest Europe*. Geological Society, London, Special Publications, **91**, 7–39.

DUNFORD, G.M., DANCER, P.N. & LONG, K.D. 2001. Hydrocarbon potential of the Kish Bank Basin: integration within a regional model for the Greater Irish Sea Basin. *In*: SHANNON, P.M., HAUGHTON, P.D.W. & CORCORAN, D.V. (eds) *The Petroleum Exploration of Ireland's Offshore Basins*. Geological Society, London, Special Publications, **188**, 135–154.

DUNCAN, W.I., GREEN, P.F. & DUDDY, I.R. 1998. Source rock burial history and seal effectiveness: key facets to understanding hydrocarbon exploration potential in the East and Central Irish Sea Basins. *AAPG Bulletin*, **82**, 1401–1415.

FLOODPAGE, J., NEWMAN, P. & WHITE, J. 2001. Hydrocarbon prospectivity in the Irish Sea area: insights from recent exploration of the Central Irish Sea, Peel and Solway basins *In*: SHANNON, P.M., HAUGHTON, P.D.W. & CORCORAN, D.V. (eds) *The Petroleum Exploration of Ireland's Offshore Basins*. Geological Society, London, Special Publications, **188**, 107–134.

HOLLOWAY, S. & CHADWICK, R.A. 1986. The Sticklepath–Lustleigh fault zone: Tertiary sinistral reactivation of a Variscan dextral strike-slip fault. *Journal of the Geological Society, London*, **143**, 447–452.

JENNER, J.K. 1981. The structure and stratigraphy of the Kish Bank Basin *In*: ILLING, L.V. & HOBSON, G.D. (eds) *Petroleum Geology of the Continental Shelf of North-West Europe*. Heyden, London, 426–431.

KNIPE, R.J., COWAN, G. & BALENDRAN, V.S. 1993. The tectonic history of the East Irish Sea Basin with reference to the Morecambe Fields *In*: PARKER, J.R. (ed.) *Petroleum Geology of Northwest Europe:*

Proceedings of the 4th Conference. Geological Society, London, 857–866.

MADDOX, S.J., BLOW, R. & HARDMAN, M. 1995. Hydrocarbon prospectivity of the Central Irish Sea Basin with reference to Block 42/12, offshore Ireland *In*: CROKER, P.F. & SHANNON, P.M. (eds) *The Petroleum Geology of Ireland's Offshore Basins*. Geological Society, London, Special Publications, **93**, 59–77.

MAINGARM, S., IZATT, C., WHITTINGTON, R.J. & FITCHES, W.R. 1999. Tectonic evolution of the southern–Central Irish Sea Basin. *Journal of Petroleum Geology*, **22**, 287–304.

MUSGROVE, F.W., MURDOCH, L.M. & LENEHAN, T. 1995. The Variscan fold–thrust belt southeast of Ireland and its control on early Mesozoic extension and deposition: a method to predict the Sherwood Sandstone *In*: CROKER, P.F. & SHANNON, P.M. (eds) *The Petroleum Geology of Ireland's Offshore Basins*. Geological Society, London, Special Publications, **93**, 81–100.

NAYLOR, D., HAUGHEY, N., CLAYTON, G. & GRAHAM, J.R. 1993. The Kish Bank Basin, offshore Ireland *In*: PARKER, J.R. (ed.) *Petroleum Geology of Northwest Europe: Proceedings of the 4th Conference*. Geological Society, London, 845–855.

NEWMAN, P.J. 1999. The geology and hydrocarbon potential of the Peel and Solway Basins, East Irish Sea. *Journal of Petroleum Geology*, **22**, 305–324.

QUIRK, D.G., ROY, S., KNOTT, I., REDFERN, J. & HILL, L. 1999. Petroleum geology and future hydrocarbon potential of the Irish Sea. *Journal of Petroleum Geology*, **22**, 243–260.

SHELTON, R. 1995. Mesozoic basin evolution of the North Channel: preliminary results *In*: CROKER, P.F. & SHANNON, P.M. (eds) *The Petroleum Geology of Ireland's Offshore Basins*. Geological Society, London, Special Publications, **93**, 17–20.

TAPPIN, D. R., CHADWICK, R. A., JACKSON, A. A., Wingfield, R. T. R. & Smith, N. J. P. 1994. *The Geology of Cardigan Bay and the Bristol Channel*. British Geological Survey, United Kingdom Offshore Regional Report **8**.

TUCKER, R.M. & ARTER, G. 1987. The tectonic evolution of the North Celtic Sea and Cardigan Bay basins with special reference to tectonic inversion. *Tectonophysics*, **137**, 291–307.

TURNER, J.P. 1997. Strike-slip fault reactivation in the Cardigan Bay Basin. *Journal of the Geological Society, London*, **154**, 5–8.

The influence of thermal history on hydrocarbon prospectivity in the Central Irish Sea Basin

PAUL F. GREEN[1], IAN R. DUDDY[1], RICHARD J. BRAY[2], WILLIAM I. DUNCAN[3] & DERMOT V. CORCORAN[4]

[1]*Geotrack International, 37 Melville Road, West Brunswick, Vic. 3055, Australia (e-mail: mail@geotrack.com.au)*

[2]*Geotrack UK Office, 5 The Linen Yard, South Street, Crewkerne TA18 7HJ, UK*

[3]*Veba Oil and Gas UK Ltd, Bowater House, 114 Knightsbridge, London SW1X 7LD, UK*

[4]*Department of Geology, Trinity College, Dublin 2, Ireland*

Abstract: Thermal history reconstruction studies of four hydrocarbon exploration wells located in the Central Irish Sea Basin (CISB) reveal three major regional episodes of heating and cooling. Units throughout the pre-Quaternary section intersected in wells 42/12-1, 42/16-1 and 42/17-1 began to cool from their maximum post-depositional palaeotemperatures in Early Cretaceous time, between 120 and 115 Ma. Cooling from subsequent palaeotemperature peaks began in Late Cretaceous–Early Tertiary (70–55 Ma) and Late Tertiary (25–0 Ma) time. Results from well 42/21-1 are dominated by the two more recent episodes, and show no evidence of the Early Cretaceous episode. This is thought to reflect a different structural setting of this well, within a North Celtic Sea–Cardigan Bay trend. Palaeotemperature profiles suggest that heating in each episode was due largely to deeper burial, with subsequent cooling caused mainly by uplift and erosion. A maximum of *c.* 3 km of additional Late Triassic to Early Cretaceous section is required to explain the observed Early Cretaceous palaeotemperatures. Appropriate values for the Late Cretaceous–Early Tertiary and Late Tertiary episodes are *c.* 2 km and *c.* 1 km, respectively. All of these cooling episodes correlate closely with similar episodes recognized from previous studies in surrounding regions, from onshore Ireland, Scotland, South Wales and northern, eastern, central and SW England, and each appears to be of truly regional extent. Exploration risk in the CISB generation can be significantly reduced through recognition of the major palaeo-thermal episodes that have affected the region, and the variation in the magnitude of their effects across the region. The challenge for future exploration in the region is to identify regions where the main phase of hydrocarbon generation post-dated structuring.

Understanding the timing of hydrocarbon generation is a critical aspect of assessing regional hydrocarbon prospectivity. This is particularly important in sedimentary basins that have undergone a series of palaeo-thermal episodes, as a result of which a given source rock horizon may have reached maximum maturity at different times in various locations across the basin. The time at which a particular source rock cools from its maximum palaeotemperature effectively defines the time at which active hydrocarbon generation ceases. The relationship between this and the time at which various traps were formed can exert a critical control on hydrocarbon prospectivity, as only those traps formed before the main phase of generation will be available to be charged at that time. Accurate reconstruction of the thermal history of source rock sequences is therefore of major importance in reducing exploration risk in such regions. In addition, recognition of later tectonic episodes that might lead to breaching of seals, re-migration and loss of charge, is another important aspect of hydrocarbon prospectivity that can be investigated through thermal history studies.

In marked contrast to the East Irish Sea Basin, which contains significant hydrocarbon reserves (Colter 1997), the history of hydrocarbon exploration in the Central Irish Sea Basin (CISB) has been disappointing, despite many similarities in the geology of the two basins. Although a number of factors may be responsible for the differences in hydrocarbon prospectivity between these two provinces, here we focus on the role of thermal history of potential source rocks.

From: SHANNON, P.M., HAUGHTON, P.D.W. & CORCORAN, D.V. (eds) 2001. *The Petroleum Exploration of Ireland's Offshore Basins*. Geological Society, London, Special Publications, **188**, 171–188.
0305-8719/01/$15.00

In the Irish Sea region in general, source rocks are recognized within the Namurian section (Hardman *et al.* 1993; Armstrong *et al.* 1995, 1997), and Jurassic source rocks are also developed in surrounding regions (e.g. Scotchman & Thomas 1995). The area is characterized by a series of potential structuring episodes of latest Carboniferous, Mesozoic and Tertiary age (e.g. Stuart & Cowan 1991; Maddox *et al.* 1995). Previous thermal history studies from areas adjacent to the CISB, including the East Irish Sea Basin (Green *et al.* 1997) and onshore Ireland (Green *et al.* 2000), have revealed a series of palaeo-thermal episodes, in Late Carboniferous, Jurassic, Early Cretaceous, Early Tertiary and Late Tertiary times. The magnitude of peak palaeotemperatures during each episode varies significantly across these regions. The similarity in results from these regions, located to the east and west of the Central Irish Sea, suggests that the interplay between these various palaeo-thermal episodes and various structuring events is also likely to be crucial in understanding the history of hydrocarbon generation and accumulation in the CISB.

The Central Irish Sea Basin

The Central Irish Sea Basin (CISB) consists of a NE–SW-trending, Late Palaeozoic–Cenozoic, transtensional half-graben system that has experienced a multiphase inversion history. The basin is bounded to the north by the Mid Irish Sea Uplift and to the south it is separated from the St. George's Channel–Cardigan Bay Basin by the offshore extension of St. Tudwal's Arch (Fig. 1).

Five exploration wells have been drilled to date in the Irish sector of the CISB, all of which have been plugged and abandoned as dry holes. Potential hydrocarbon source rocks of Westphalian and Liassic (Early Jurassic) ages have been recognized in the Central Irish Sea area (Corcoran & Clayton 1999) although the efficacy of the Westphalian source rock system in the CISB has been questioned (Floodpage *et al.* 1999). The primary exploration target has been the Lower Triassic Sherwood Sandstone Group, sealed by evaporites and shales of the Upper Triassic Mercia Mudstone Group and sourced by Carboniferous shales and coals. Three of these wells appear to have tested valid hydrocarbon traps (Floodpage *et al.* 1999), and the absence of significant hydrocarbon accumulations in these wells strengthens the likelihood that the timing of hydrocarbon charge is a major exploration risk factor in the CISB.

However, maturation modelling of potential source rock horizons in the CISB is hampered by the severely truncated rock record. The challenge of thermal history reconstruction is to offer constraints to the thermal evolution of these potential source rock horizons in a multiphase inversion setting.

With this in mind, we report here results from four CISB wells (Fig. 1), as part of a continuing study designed to determine the thermal history of potential hydrocarbon source rocks across the region. Results from two of these wells have been published previously (Duncan *et al.* 1998), but some aspects of these results have been reassessed, and comparison with newer data provides tighter constraints on the interpretation of these older data than was previously possible.

Thermal history reconstruction using apatite fission-track analysis and vitrinite reflectance

Thermal history reconstruction (THR) is based on application of apatite fission-track analysis (AFTA®) and vitrinite reflectance (VR) data. Using THR, we can identify the timing of dominant episodes of heating and cooling that have affected a sedimentary section, quantify the palaeotemperatures through the section, and characterize mechanisms of heating and cooling (as described in detail by Bray *et al.* 1992; Duddy *et al.* 1994; Green *et al.* 1995).

AFTA is based on analysis of radiation damage trails (fission tracks) within the crystal lattice of detrital apatite grains, which are a common constituent of most sandstones and coarser sediments. The continuous production of new fission tracks through time, coupled to the reduction in track length as a function of temperature and time, provides the basis of the technique (Green *et al.* 1989a, 1989b). As temperature rises, track length is progressively reduced, as a result of partial repair of the radiation damage constituting the tracks ('partial annealing'). Once the temperature reaches some maximum value and begins to fall, track length is essentially frozen at the value reached at the thermal maximum. Temperature dominates over time in the kinetics of this process, such that a rise of 10 °C produces a similar change in length as an order of magnitude increase in time. Thus, most tracks are reduced to the same length regardless of when they formed. A sample that reached a maximum palaeotemperature of *c.* 90–100 °C at some time in the past, and then cooled and resided at lower temperatures until the present day, will therefore contain two populations of tracks: a shorter component, which represents tracks formed up to the time

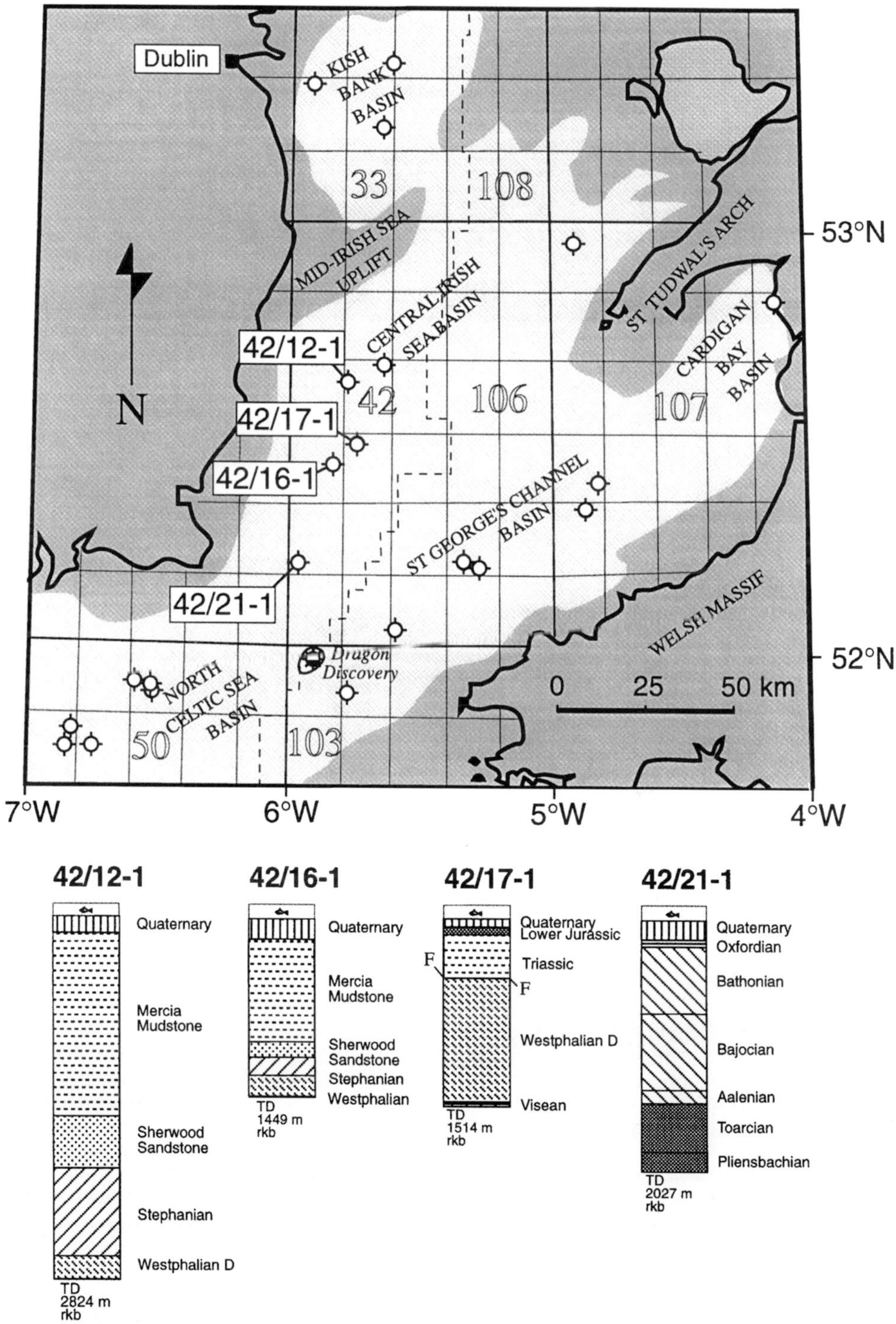

Fig. 1. Location map of the Central Irish Sea Basin and adjacent regions, showing locations of hydrocarbon exploration wells from which samples were analysed for this study. Stratigraphic columns are shown for each well, together with the total depth (TD) below rotary kelly bushing (RKB). The Triassic sequence is generally subdivided into Sherwood Sandstone and Mercia Mudstone groups.

at which cooling began, and a longer component representing tracks formed after the onset of cooling. The length of the shorter component is diagnostic of the maximum palaeotemperature reached before the onset of cooling, whereas that of the longer component is controlled by the history after the onset of cooling. The proportion of short to long tracks provides information on the time at which cooling began. For example, early cooling would result in few short tracks and mostly long tracks, whereas more recent cooling will produce more shorter lengths and fewer longer tracks. Thus, analysis of the distribution of track lengths provides estimates of both the time of cooling and the magnitude of the maximum palaeotemperature.

The number of tracks in a polished surface can also be used to measure a 'fission-track age'. In the absence of significant length reduction, this parameter would measure the time over which tracks have been retained. But because the probability of a track intersecting a surface depends on the track length, when length reduction is sufficiently severe the fission-track age is significantly reduced, and must be interpreted together with track length data in terms of thermal history rather than as an indicator of the timing of a discrete event.

In samples that reached sufficiently high values of maximum palaeotemperature, the track length is reduced to zero, because all of the radiation damage constituting the track is totally repaired ('total annealing'). Such samples begin to retain tracks only after cooling below this critical limit, which is typically *c.* 110–120 °C, depending on heating rate and apatite composition (chlorine content). In such cases AFTA provides only a minimum limit on the magnitude of the maximum palaeotemperature, but the fission-track age (combined with track length data, which record the post-cooling thermal history) provides key information on the time of cooling, and VR data from adjacent samples can still provide an estimate of the maximum palaeotemperature (as explained below).

Thermal history interpretation of AFTA and VR data is based on a detailed knowledge of the kinetic responses of both systems, which are well calibrated from studies in both geological and laboratory conditions. Thermal history information is extracted from the AFTA data by modelling measured AFTA parameters (fission-track age and track length distributions) through a variety of possible thermal history scenarios, varying the magnitude and timing of the maximum palaeotemperature so as to define the range of values of each parameter that give predictions consistent with the measured data within 95% confidence limits. The basics of this modelling procedure are well established for mono-compositional apatites (e.g. Green *et al.* 1989b), as a result of a series of laboratory experiments on Durango apatite (Green *et al.* 1986; Laslett *et al.* 1987; Duddy *et al.* 1988). However, the annealing kinetics of fission tracks in apatite are known to be affected by the chlorine content (Green *et al.* 1986), and in the studies described here, thermal history solutions have been extracted from the AFTA data using a 'multi-compositional' kinetic model that makes full quantitative allowance for the effect of Cl content on annealing rates of fission tracks in apatite (Green *et al.* 1996). This model is calibrated using a combination of laboratory and geological data from a variety of sedimentary basins around the world. Palaeotemperature estimates from AFTA are quoted as a range (corresponding to ±95% confidence limits) and have an absolute uncertainty of between ±5 and ±10 °C.

Observed VR values are converted to maximum palaeotemperatures using the kinetic model developed by Burnham & Sweeney (1989) and Sweeney & Burnham (1990). Information on the timing of these maximum palaeotemperatures is provided by the AFTA data. The VR-derived palaeotemperature estimates are shown as single values but probably have a precision of 5–10 °C. The kinetic response of VR as described by Burnham & Sweeney (1989) is very similar to the fission-track annealing kinetic model developed by Laslett *et al.* (1987) to describe the kinetics of fission-track annealing in Durango apatite. Total fission-track annealing in apatites with typical Cl content corresponds to a VR value of *c.* 0.7%, regardless of heating rate (Duddy *et al.* 1991, 1994).

Unlike VR data, which provide a value for only the maximum palaeotemperature, AFTA data also provide some control on the history after cooling from maximum palaeotemperatures through the lengths of tracks formed during this period, and it is often possible to resolve two discrete palaeo-thermal episodes from AFTA data in a single sample. This is most straightforward when an earlier event causes significant age and length reduction whereas a subsequent event produces only moderate length reduction. One additional episode during the cooling history is normally the limit of resolution from typical AFTA data. In rare instances, such as in one sample from well 42/16-1 described in this paper, information on three discrete episodes may be obtained from AFTA data in a single sample.

This sample was totally annealed in an early episode, cooled and was then reheated such that severe partial annealing was produced, cooled once more and then underwent moderate annealing relatively late in the history, such that the effects of all three events can be identified. Alternatively, integration of VR data with information from AFTA often allows three episodes to be resolved, by revealing the earliest episode, which may not be resolvable from AFTA alone.

Both AFTA and VR are dominated by the maximum palaeotemperature and preserve no information on the approach to the palaeo-thermal maximum. Therefore, in interpreting the data it is necessary to assume a value of heating rate, and the precise value of maximum palaeotemperature required to explain the data depends on the assumed value. Given the kinetics of the two processes, a change of an order of magnitude in the heating rate is equivalent to a change in the required palaeotemperature of *c.* 10 °C (Green *et al.* 1989b).

Wherever possible, AFTA data from each sample in this study have been interpreted in terms of two episodes of heating and cooling, using assumed heating and cooling rates of 1 °C Ma^{-1} and 10 °C Ma^{-1}, respectively, during each episode (with the maximum palaeotemperature reached during the earlier episode). Using statistical procedures, the timing of the onset of cooling and the peak palaeotemperatures during the two episodes are varied systematically, and by comparing predicted and measured parameters the range of conditions that are compatible with the data within 95% confidence limits can be defined.

We emphasize that the information derived from AFTA provides a direct estimate of the time at which the sample began to cool from its maximum post-depositional palaeotemperature (up to a maximum limit of *c.* 110 °C), which is generated from the AFTA data alone and is totally independent of any assumptions concerning the geological evolution of the region. This information is then used to infer the nature of the processes responsible for the observed palaeo-thermal effects, based on methods discussed in the next section.

Palaeotemperature profiles, palaeogeothermal gradients and removed section

Analysis of a series of samples using AFTA and VR over a range of depths reveals the variation of maximum palaeotemperature with depth, the 'palaeotemperature profile' characterizing each episode. From data that reveal multiple episodes of heating and cooling, separate palaeotemperature profiles can often be constructed for each episode, as in, for example, the discussion of Inner Moray Firth well 12/16-1 by Green *et al.* (1995).

The form of the palaeotemperature profile characterizing a particular palaeo-thermal episode provides key information on likely mechanisms of heating and cooling in that episode. Heating caused solely by deeper burial should produce a more or less linear palaeotemperature profile with a similar gradient to the present temperature profile. In contrast, heating caused primarily by increased basal heat flow (perhaps also with a minor component of deeper burial) should produce a more or less linear palaeo-temperature profile with a higher gradient than the present temperature profile.

Heating as a result of the passage of hot fluids can produce a variety of non-linear palaeo-temperature profiles, with different forms depending on the time scale of heating (see Ziagos & Blackwell 1986; Duddy *et al.* 1994). Heating effects caused by minor igneous intrusions can produce purely local anomalies, or may be more widespread if they cause circulation of heated fluids on a regional scale (e.g. Summer & Verosub 1989).

In sections where heating was due to deeper burial, either alone or possibly combined with elevated heat flow, fitting a line to the palaeotemperature profile provides an estimate of the palaeogeothermal gradient. Extrapolating this to an assumed palaeo-surface temperature then provides an estimate of the amount of section removed by erosion. This analysis depends critically on several assumptions, as discussed by Bray *et al.* (1992). Statistical techniques allow definition of the range of each parameter allowed by the palaeotemperature constraints within 95% confidence limits (Bray *et al.* 1992). Allowed values of palaeo-gradient and removed section are highly correlated, such that higher palaeogeothermal gradients require correspondingly lower values of removed section, and vice versa.

In summary, AFTA data are used to identify the timing of major cooling episodes, whereas AFTA and VR data provide estimates of the magnitude of maximum or peak palaeo-temperatures in each episode. From these results, palaeotemperature–depth profiles are constructed for each episode, and these provide insight into the mechanism of heating and cooling. If these profiles are linear, palaeo-geothermal gradients can be determined, and

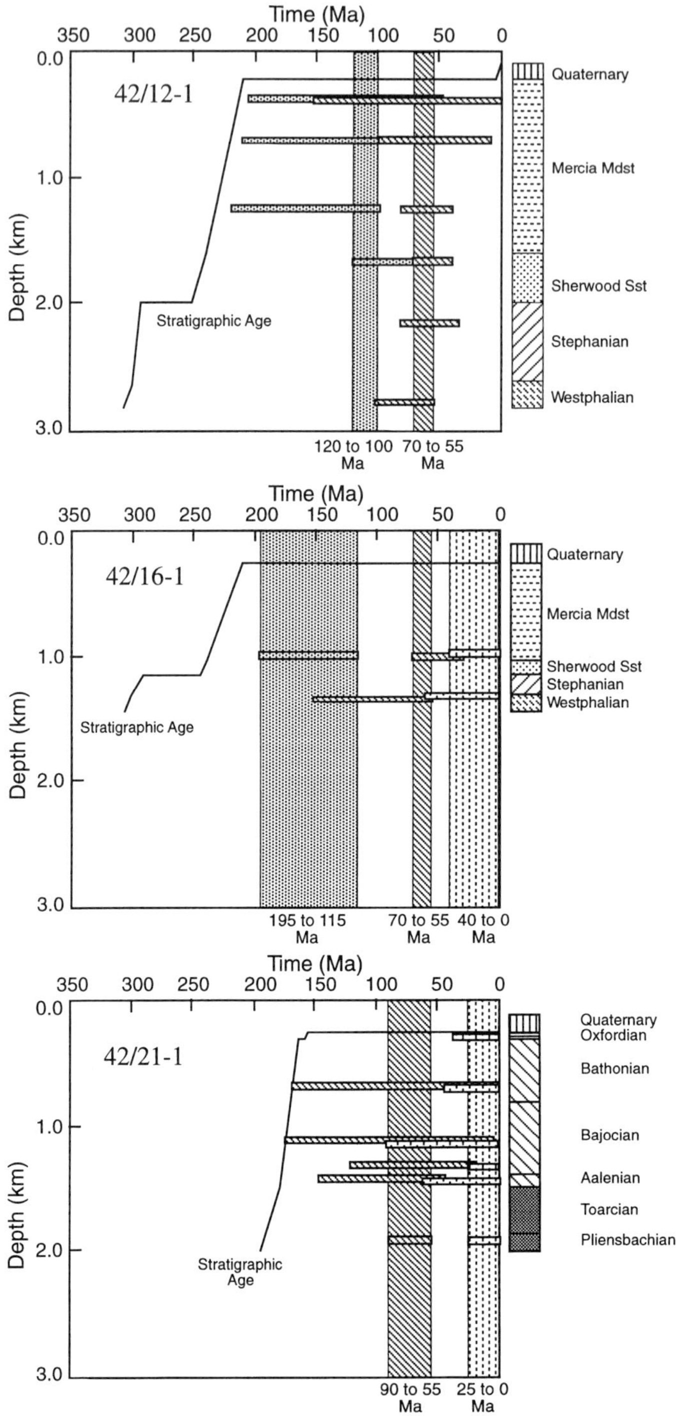

Fig. 2. Timing constraints derived from AFTA in individual samples from three CISB wells. Vertical shaded bars highlight the range of timings with which data from all samples for each well are consistent. Data from all three wells define three synchronous events, in which cooling began between 120 and 115 Ma, 70 and 55 Ma, and 25 and 0 Ma. It should be noted that AFTA data from the shallower sample in well 42/16-1 define three separate cooling episodes, whereas other samples define two or one episodes. This depends on the quality of the AFTA data, the magnitude of peak palaeotemperatures in individual episodes and the spread of chlorine contents in apatite grains from each sample.

where appropriate, former depths of burial can be estimated (non-linear profiles are interpreted as reflecting processes not directly related to depth of burial).

We again emphasize that it is the range of palaeo-gradients allowed by the palaeo-temperature constraints derived from the AFTA and VR data that is used to assess the likely nature of processes responsible for the observed palaeo-thermal effects, rather than relying on inference based on regional geological evidence, which may be amenable to a range of subjective interpretations.

Results from Central Irish Sea Basin wells

Identification of palaeo-thermal episodes

Figure 2 illustrates the range of timing for the onset of cooling derived from AFTA data in samples from three of the four CISB wells analysed in this study. In each well, this timing information is compared with the variation in stratigraphic age through the well. Two samples analysed from well 42/17-1 failed to yield any apatite. In most of the samples, the AFTA data require at least two episodes of cooling, as shown, whereas, as mentioned previously, the shallower sample analysed from well 42/16-1 shows very clear evidence of three distinct cooling episodes.

The vertical bars in each plot highlight the range of timing consistent with all samples from each well. Given the relatively close proximity of these three wells, it seems reasonable to assume that the palaeo-thermal effects recognized in each well represent synchronous events. On this basis, inspection of Fig. 2 shows that results from all three wells can be explained in terms of three palaeo-thermal episodes, with cooling beginning in the Early Cretaceous (between 120 and 115 Ma), Late Cretaceous–Early Tertiary (between 70 and 55 Ma) and Late Tertiary (between 25 and 0 Ma) times. It should be stressed that these quoted ranges refer to the time at which cooling began, and it is not implied either that all cooling in each episode occurred within each interval or that cooling necessarily encompassed the entire interval.

Quantification of palaeotemperatures in individual episodes

Figure 3 shows palaeotemperatures derived from AFTA and VR in individual samples from the four wells, plotted against sample depth (with respect to kelly bushing in each well). Also shown are present-day thermal gradients derived from corrected bottom hole temperature (BHT) data in each well. Results from the four wells show very similar features, as highlighted in the following discussion of each well.

Well 42/12-1. Results from this well were originally described by Duncan *et al.* (1998) and the results shown in Fig. 3 are as reported by those workers. In most of the AFTA samples, all tracks were totally annealed before the Early Cretaceous cooling episode, and provide only minimum estimates of the maximum palaeo-temperature in this episode. AFTA data also provide estimates of Late Cretaceous to Early Tertiary palaeotemperatures, as shown. VR data from the Carboniferous section in this well are between 1.77 and 1.95%, suggesting maximum palaeotemperatures in the range 160–200 °C. As AFTA data provide only minimum estimates, which are *c.* 40–80 °C less than the maximum values derived from VR, it is not immediately clear whether the maximum palaeotemperatures indicated by the VR data were attained during the Early Cretaceous episode revealed by AFTA or possibly during an earlier episode. Circum-stantial evidence supporting an Early Cretaceous maximum comes from the observation that a linear profile with a similar gradient to the present-day temperature profile can satisfy all the AFTA-based palaeotemperature constraints and also those from VR as illustrated in Fig. 3. Thus, if the VR data were to represent maximum temperatures reached in an earlier (pre-Cretaceous) episode, the Early Cretaceous palaeo-thermal episode would have to be described by a much lower palaeogeothermal gradient, for which no support is found in any of the results from other wells. As will be discussed, the consistency of results from this well with those from wells 42/16-1 and 42/17-1 strongly support the conclusion that the VR data in this well represent maximum palaeotemperatures reached during the Early Cretaceous episode, immediately before the onset of cooling between 120 and 115 Ma.

Well 42/16-1. New results from Triassic and Carboniferous units in this well reveal at least three palaeo-thermal episodes, which, on the basis of data from all four wells, are interpreted as representing the Early Cretaceous, Late Cretaceous–Early Tertiary and Late Tertiary episodes described above. Vitrinite reflectance values between 0.85 and *c.* 1.2% from the Carboniferous section define maximum palaeo-temperatures between 130 and 160 °C. AFTA samples from similar depths were totally

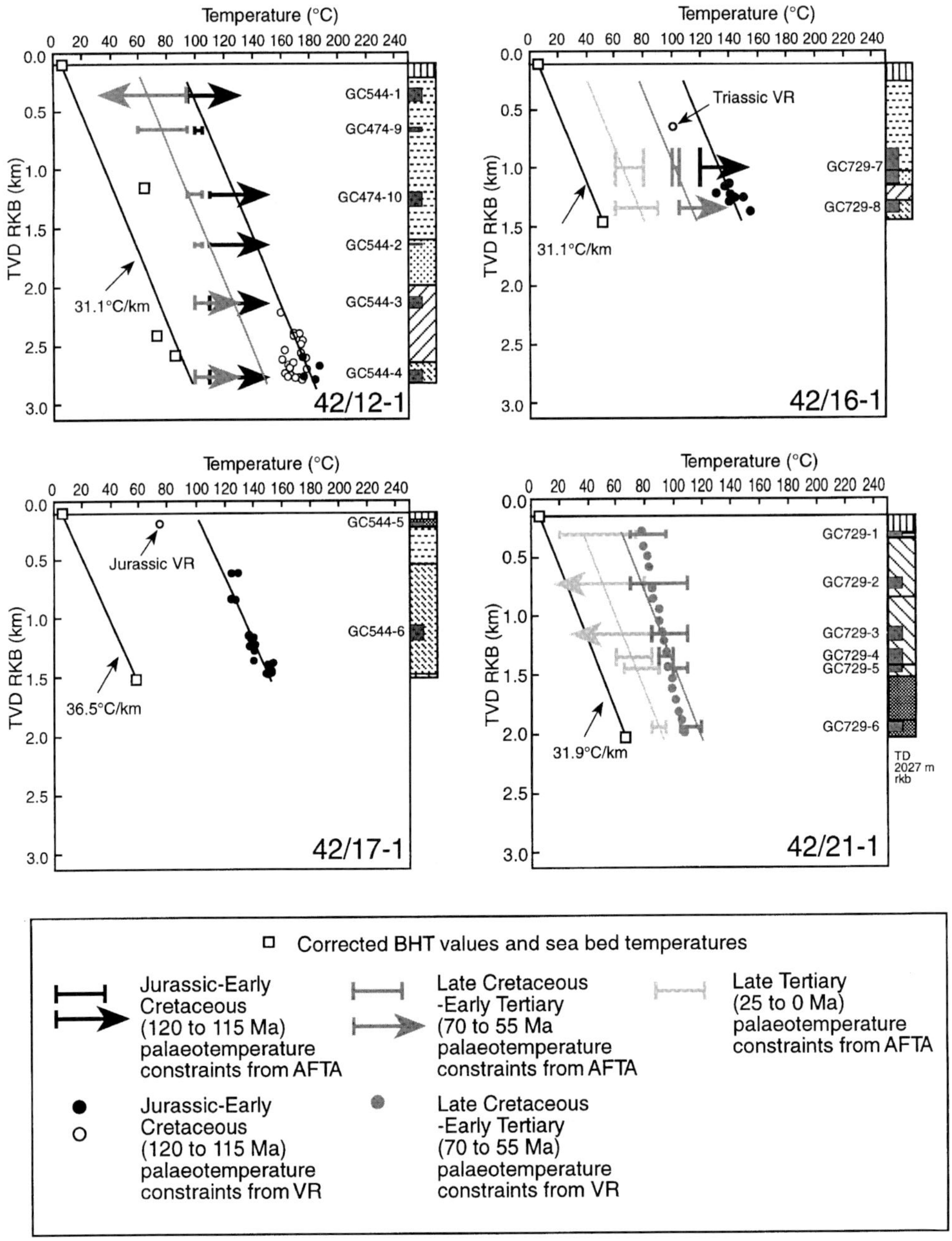

Fig. 3. Palaeotemperature constraints derived from AFTA and VR data in individual samples from each well, plotted against depth. Summary stratigraphic columns are also shown for comparison (details in Fig. 1). Palaeotemperature constraints are coded for discrete palaeo-thermal episodes as identified in Fig. 2. Present-day temperature profiles, together with corrected BHT data, are also shown for each well. Profiles parallel to the present-day temperature profiles are drawn through the palaeotemperatures characterizing individual episodes for each well, as an aid to discussion of the palaeotemperature interpretation in the text. Two sets of VR data are available from well 42/12-1, with more reliable data represented by filled symbols.

annealed before Late Cretaceous to Early Tertiary cooling, and provide a minimum estimate of 105 °C in that episode. AFTA data from the overlying Triassic section were totally annealed before Early Cretaceous cooling, and provide a minimum estimate of 120 °C in that episode, as well as discrete estimates of peak palaeotemperatures during two subsequent episodes. (Note that differences in apatite composition between samples result in different values for the minimum estimate of the maximum palaeotemperature.) In the deeper AFTA sample, effects of the Early Cretaceous episode were overprinted by the Late Cretaceous–Early Tertiary episode. Thus, as with well 42/12-1, the maximum palaeotemperatures indicated by VR are slightly higher than the Early Cretaceous values revealed by AFTA, and this raises the question of whether the VR data represent this or an earlier episode (as the values from the two systems differ by only *c.* 10 °C, interpretation in terms of a common event would seem to be justified, but this cannot be definitely confirmed from these data alone). Once again, comparison of data from other wells (discussed below) strongly supports an interpretation in which the VR data do indeed represent the Early Cretaceous episode. As illustrated in Fig. 3, palaeotemperature constraints derived from AFTA and VR data for the three episodes can be described by linear profiles with gradients similar to present-day values. The Early Cretaceous episode is the best constrained of the three, as constraints are available from AFTA and VR, whereas for the two most recent episodes, a wider range of palaeo-gradients would be allowed by the palaeotemperature constraints from the two AFTA samples alone (this is considered more quantitatively in a later section). A single VR value reported from the Triassic section falls below the profile drawn through the Early Cretaceous palaeotemperature constraints. If we are correct in attributing maximum palaeotemperatures throughout the Triassic and older section to the Early Cretaceous episode, then VR data from the Triassic section should also reflect that episode, in which case the palaeotemperature constraint from Triassic VR data should be collinear with the profile through the deeper samples for this episode. Experience in the region suggests that VR data from the Triassic section are prone to serious errors, as suitable lithologies for analysis are rare, and often represent contaminant material (from cavings or some other source). Therefore little significance is attributed to this value in the interpretation of these data favoured here.

Well 42/17-1. Results from this well were also described in detail by Duncan *et al.* (1998). As mentioned above, two AFTA samples collected from this well failed to yield any apatite. Thermal history information for this well is therefore dependent on VR data interpreted in a regional context and by comparison with data from neighbouring wells. VR values between 0.8 and *c.* 1.3% from Carboniferous units suggest maximum palaeotemperatures between 120 and 160 °C, whereas a single value from the Jurassic section gives a value of *c.* 80 °C. As with the Triassic VR value for well 42/16-1 (discussed above), this Jurassic value may be unreliable and, omitting this value, the data can be described by a linear profile with a slope similar to that of the present-day temperature profile. In the absence of AFTA data from this well, no direct indication is available of the timing of the maximum palaeotemperatures derived from the VR data. Comparison of values from this well with those from wells 42/12-1 and 42/16-1 (see later discussion) again strongly suggests that the VR data from this well represent the Early Cretaceous episode.

Well 42/21-1. New AFTA data from this well show a major difference from those for the three wells discussed so far, in that they clearly show that the thick Jurassic section intersected in this well began to cool from maximum palaeotemperatures in Late Cretaceous to Early Tertiary time (sometime between 70 and 55 Ma), and show no evidence of Early Cretaceous effects. Late Tertiary cooling is also detected from AFTA, as illustrated in Figs 2 and 3. Maximum palaeotemperatures derived from VR data for this well are highly consistent with those from AFTA, as shown in Fig. 3, confirming that the preserved section in this well cooled from maximum palaeotemperatures in Late Cretaceous to Early Tertiary time. However, one point to note from the results for this well shown in Fig. 3 is the lack of scatter in the VR values about the trend with depth. Typical VR datasets show appreciably more scatter about the general trend than seen in the results for this well, and we therefore view these data with some suspicion (although, as noted above, the reported values are highly consistent with the results from AFTA, and in broad terms these maturity levels are considered to be broadly correct).

Comparison of palaeotemperature profiles in different wells

Figure 4 shows a comparison of palaeotemperature profiles characterizing the three palaeothermal episodes in different wells. In the upper plot, palaeotemperature constraints interpreted as representing the Early Cretaceous episode emphasize the remarkable similarity in these values from wells 42/16-1 and 42/17-1, whereas values from well 42/12-1 appear to be slightly lower in magnitude although defining a similar overall trend. An interpretation of all these data in terms of a common episode offers the simplest explanation of all these results, and there seems no reason to invoke any earlier episodes to explain the VR data.

The central plot in Fig. 4 emphasizes the similar overall magnitude of Late Cretaceous–Early Tertiary palaeotemperatures in the three wells. Results from well 42/16-1 appear to be higher than those from the other two wells by *c.* 15 °C, but overall the similarity is most striking. The trend of the VR-derived values for well 42/21-1 is rather different from that defined by AFTA-derived constraints for that well and the other two wells, but as noted above, these VR data are regarded as being possibly unreliable, and AFTA data from well 42/21-1 suggest a trend more similar to the present-day temperature profile.

The lower plot in Fig. 4 emphasizes the similarity between Late Tertiary palaeotemperatures for wells 42/16-1 and 42/21-1. AFTA data from well 42/12-1 do not show any evidence of the Late Tertiary cooling episode, as those data are dominated by the earlier episodes. However, on the basis of the similarity of data from the other wells, it seems likely that Late Tertiary cooling also affected the section preserved in this well, as well as that in well 42/17-1 for which no

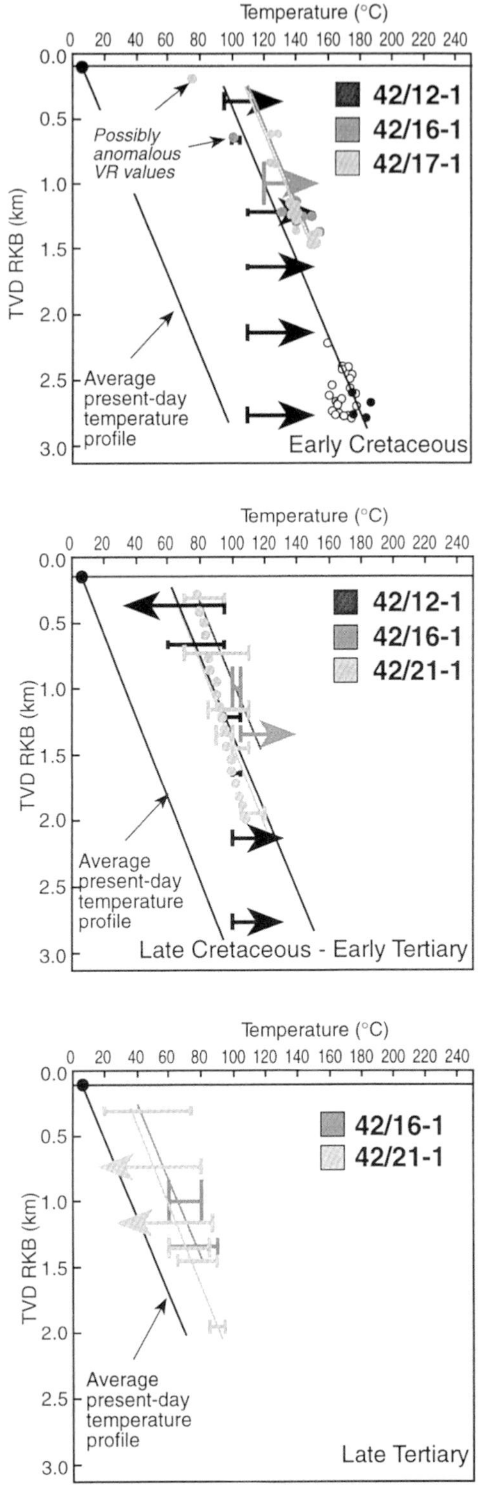

Fig. 4. Palaeotemperature constraints characterizing the Early Cretaceous (upper), Late Cretaceous–Early Tertiary (centre) and Late Tertiary (lower) palaeothermal episodes for each well (see Fig. 3) are plotted together on common axes, to facilitate comparison. For each palaeo-thermal episode, the magnitude of maximum or peak palaeotemperatures for different wells are remarkably similar. Again, profiles parallel to the present-day temperature profiles are drawn through the palaeotemperatures from each well. As shown in Fig. 5, results from all three episodes are generally consistent with constant palaeogeothermal gradients through time. It should be noted that VR data for well 42/21-1 are thought to be possibly unreliable, as they define a different trend from the majority of the data, although the overall values seem to be broadly correct.

AFTA data are available and thus no constraints are possible on Late Tertiary effects.

Characterizing mechanisms of heating and cooling

As illustrated in Figs 3 and 4, the variation of palaeotemperature with depth characterizing the three palaeo-thermal episodes in all four wells can be described by linear profiles, more or less parallel to the present-day profile. This can be assessed more quantitatively using the methods described by Bray *et al.* (1992) to define the range of palaeogeothermal gradient and removed section that are consistent (within 95% confidence limits) with the palaeotemperature constraints for each episode in each well. Figure 5 shows the results of this procedure. For each palaeo-thermal episode, the range of allowed values of palaeo-gradient and removed section are shown, with results from individual wells compared on a common scale. The range of present-day gradients in these wells is also highlighted in this plot, and it is clear that results for all three palaeo-thermal episodes, although allowing broad ranges of both parameters in most cases, are consistent with an overall interpretation in which palaeogeothermal gradients have remained close to present-day values since at least Early Cretaceous time. It should be noted that in performing these analyses, the Jurassic VR value from well 42/17-1 and the Triassic VR value in 42/16-1 were omitted. Inclusion of these data produces much less consistent interpretations. In particular, results from wells 42/16-1 and 42/17-1 would require much higher palaeogeothermal gradients, which are not allowed by results from well 42/12-1. Thus, different mechanisms of heating would need to be invoked for the different wells. Given the close proximity of all these wells, a consistent explanation of results from all three wells seems much more likely, and therefore VR data from Mesozoic units are excluded on the basis that they are unreliable.

Another point of note is that elevated palaeogeothermal gradients for the Late Cretaceous–Early Tertiary episode are clearly excluded by the palaeotemperature constraints provided by AFTA for wells 42/12-1 and 42/21-1, for which only values around (or less than) the lower end of the range of present-day values are allowed. For well 42/16-1, a broader range of palaeo-gradients is allowed, as a result of the relatively narrow depth range over which palaeotemperature constraints are available, and also because the deepest sample in that well provided only a minimum limit on the maximum palaeotemperature. Similar comments apply to the Late Tertiary results from this well and well 42/21-1.

In producing the allowed range shown in Fig. 5 for the Late Cretaceous–Early Tertiary episode for well 42/21-1, only the AFTA-based palaeotemperature constraints have been used. As already noted, the VR data show much less scatter than expected in a well-behaved dataset, which suggests that these data should be treated with some caution. The VR values as reported define a much lower palaeo-gradient (*c.* 10 °C km^{-1}), which would make results from this well inconsistent with those from other wells. Given the overall consistency of the palaeotemperature values characterizing this episode in Fig. 4, a similar interpretation in all wells is much more likely than local differences of this nature between wells.

In summary, AFTA and VR data from the four wells analysed for this study can be interpreted as representing the effects of three palaeo-thermal episodes, and the palaeotemperature constraints characterizing each episode can be explained in terms of linear profiles with palaeogeothermal gradients close to present-day thermal gradients in the region. This suggests, in turn, that the most likely explanation of these palaeo-thermal episodes is that heating was almost solely due to deeper burial, with little or no contribution from elevated basal heat flow, which would be manifested in these results by significantly higher palaeogeothermal gradients compared with present-day values. On this basis, Fig. 5 shows that for the Early Cretaceous episode, deeper burial by *c.* 3 km of additional section, subsequently removed by progressive uplift and erosion since Early Cretaceous time, is required to explain the observed palaeotemperatures. For the Late Cretaceous–Early Tertiary episode, *c.* 2 km of additional burial is required, whereas for the Late Tertiary episode the appropriate value is *c.* 1 km.

These values were derived assuming a constant palaeo-surface temperature through time, equal to the present-day temperature of 6 °C. If the palaeo-surface temperature was higher in the past, then the quoted values of removed section can be easily converted to apply to other values of palaeo-surface temperature by subtracting or adding the difference in depth equivalent to the difference between this value and the new palaeo-surface temperature, for the appropriate palaeogeothermal gradient. For instance, if the palaeogeothermal gradient was 50 °C km^{-1} and the palaeo-surface temperature was 10 °C higher than the value assumed in this

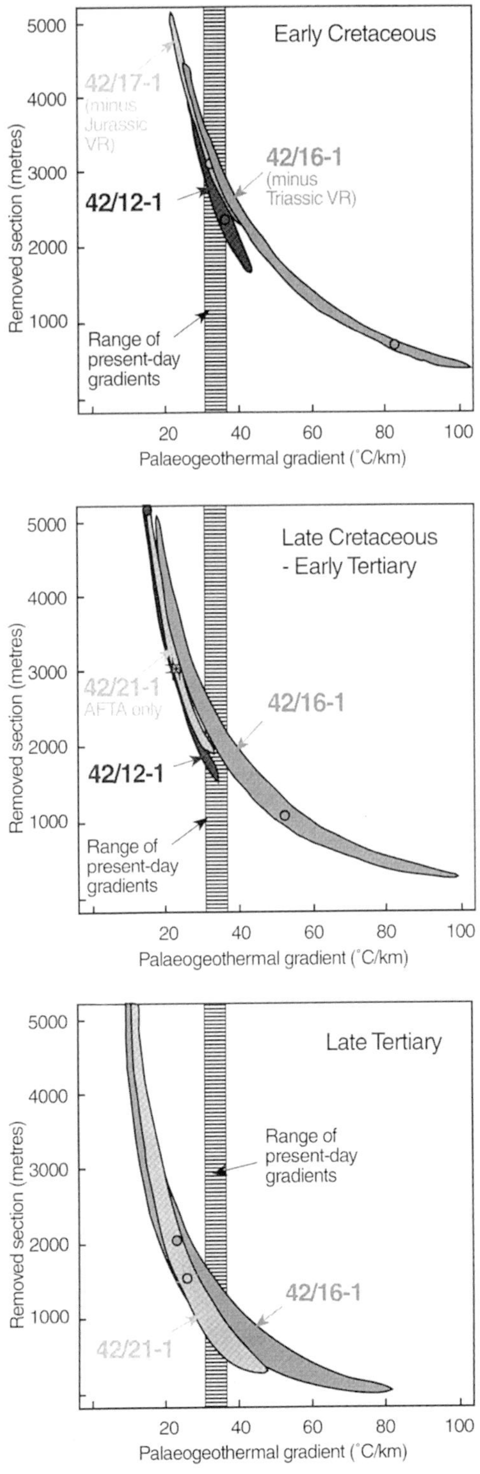

paper, the estimated eroded section should be reduced by 200 m.

It should be noted that, although results from wells 42/12-1, 42/16-1 and 42/17-1 show that at least 1 km of section must have been removed between the onset of the Early Cretaceous and the subsequent Late Cretaceous–Early Tertiary palaeotemperature peak, the total amount of section removed in this interval (and thus the amount of subsequent reburial during Late Cretaceous and Early Tertiary time) is not otherwise constrained. It is possible that all of the *c*. 3 km of additional Triassic to Early Cretaceous section responsible for producing the Early Cretaceous palaeotemperatures could have been removed, followed by deposition of *c*. 2 km of Late Cretaceous section. Alternatively, *c*. 2 km of section might have been removed during Early Cretaceous uplift and erosion, after which another *c*. 1 km of Late Cretaceous section was deposited to produce the required *c*. 2 km of additional burial during Late Cretaceous–Early Tertiary time. A wide variety of alternative scenarios are also possible, within the overall constraint of maximum burial depths required to explain the palaeotemperature data. Similar comments apply to the interval between the Late Cretaceous–Early Tertiary and Late Tertiary episodes in all four wells. Integration of the results of this study with regional geological information on thicknesses of overburden preserved in regions not affected by the events under

Fig. 5. Plots showing the range of values of palaeogeothermal gradient and removed section required to explain the palaeotemperature constraints characterizing the Early Cretaceous (upper), Late Cretaceous–Early Tertiary (centre) and Late Tertiary (lower) palaeo-thermal episodes for each well, for a palaeo-surface temperature of 6 °C. The effects of higher palaeo-surface temperatures can be allowed for as described in the text. The range of present-day thermal gradients in the four wells is indicated by the vertical band. Results from all four wells are consistent with a model in which palaeogeothermal gradients were close to present-day values since at least Early Cretaceous time, whereas significantly elevated palaeo-gradients appear to be ruled out for the Early Cretaceous episode in wells 42/12-1 and 42/17-1, and for the late Cretaceous–Early Tertiary episode in wells 42/12-1 and 42/21-1. Results for the Late Tertiary episode in wells 42/16-1 and 42/21-1 are consistent with a broad range of palaeo-gradients because of the narrow range of depths over which constraints are available (42/16-1) coupled with the rather broad range of palaeotemperatures allowed by the AFTA data in most cases (both wells).

discussion is required to provide better definition of this aspect of the geological evolution of the region, but this is beyond the scope of this paper.

Thermal history synthesis

Reconstructed thermal histories for units intersected in Central Irish Sea Basin wells 42/16-1 and 42/21-1, based on the results presented in preceding sections, are shown in Fig. 6. As the results of this study appear to rule out appreciably higher palaeogeothermal gradients (Fig. 5), these reconstructions employ a constant palaeo-gradient equal to the present-day values (30 °C km^{-1} in well 42/16-1, 31.9 °C km^{-1} in well 42/21-1), and the corresponding values of removed section from Fig. 5. Given the overall similarity in results from wells 42/12-1, 42/16-1 and 42/17-1, the reconstructions in wells 42/12-1 and 42/17-1 are likely to be very similar to that illustrated for well 42/16-1.

To produce the reconstructions shown in Fig. 6 for well 42/16-1 an additional 3475 m of post-Late Triassic sediment were deposited between 208 and 125 Ma, 2975 m of which were removed by uplift and erosion between 120 and 110 Ma; a further 2050 m were deposited between 110 and 65 Ma; 2250 m were removed by uplift and erosion between 65 and 60 Ma; a further 1000 m were deposited between 60 and 15 Ma; the remaining total of 1300 m of additional section were removed between 15 and 2 Ma.

For well 42/21-1, an additional 2000 m of post-Oxfordian sediment were deposited between 155 and 65 Ma, with 1500 m removed by uplift and erosion between 65 and 60 Ma, followed by a further 500 m deposited between 60 and 15 Ma, and the remaining total of 1000 m of additional section removed between 15 and 2 Ma.

It should be noted that although the onset of cooling in these reconstructions is shown as 120 Ma, 65 Ma and 15 Ma in the three episodes, any time between 120 and 115 Ma, between 70 and 55 Ma, and between 25 and 0 Ma, respectively, would be allowed by the AFTA data from these wells. As mentioned above, for well 42/16-1, the proportion of the total amount of additional section deposited in the earliest episode that was actually removed before the recommencement of burial in Late Cretaceous time is not defined precisely. In these reconstructions, we have selected one option out of many, simply to serve as an illustration of the overall nature of the history. Similar comments apply to the later episodes in both wells.

It should also be appreciated that although we have assumed here that heating was solely due to

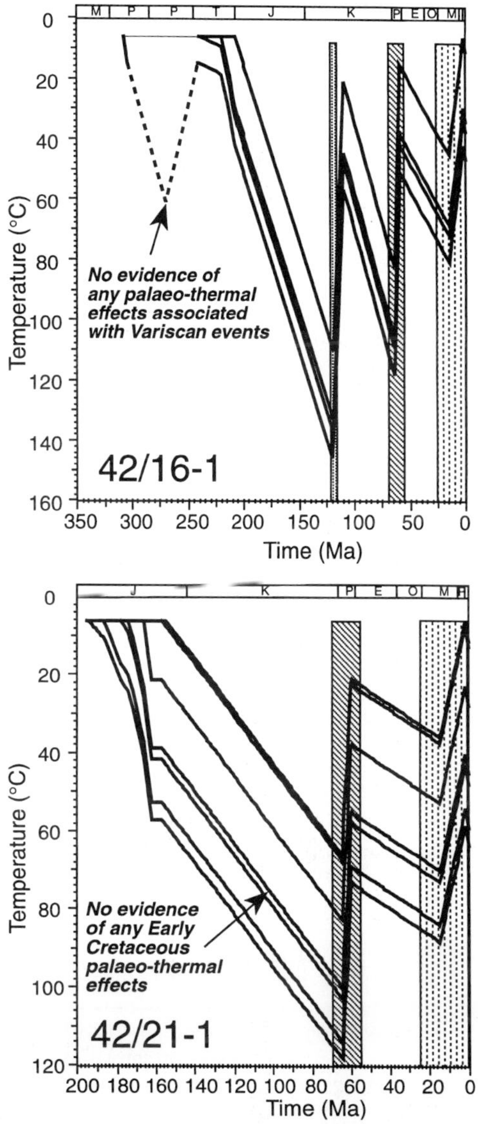

Fig. 6. Reconstructed thermal histories for units preserved in wells 42/16-1 and 42/21-1. Equivalent histories for wells 42/12-1 and 42/17-1 are thought to be very similar to that shown here for well 42/16-1. Although results from well 42/16-1 show no evidence of any latest Carboniferous (Variscan) palaeo-thermal effects, appreciable heating and cooling may have occurred at that time provided that peak palaeotemperatures were lower than the maximum values reached in the Early Cretaceous episode. Similar comments apply to the possibility of Early Cretaceous heating and cooling for well 42/21-1. Vertical shaded bands are similar to those described in Fig. 2.

deeper burial, alternative scenarios are possible and a variety of combinations of palaeo-gradient and removed section are capable of satisfying the palaeotemperature constraints from AFTA in these wells, as shown by the contoured regions in Fig. 5. However, all such combinations of palaeogeothermal gradient and removed section in each well result in reconstructed thermal histories for the preserved units that are very similar to those shown in Fig. 6, being tightly constrained by the AFTA and VR data presented in each well.

Factors such as possible non-linearity of the palaeotemperature profiles, particularly through the removed section where no constraints are available, complicate the estimation of removed section, and palaeo-surface temperatures may well have been higher than the values assumed here, as noted above. In addition, heating rates may have differed from those assumed in obtaining the palaeotemperature constraints from AFTA and VR in these wells. All these factors may introduce systematic errors into the estimation of burial depths from the palaeotemperature data, and thus exact reconstruction of burial histories in these wells is difficult. But, from the simple considerations outlined above, it is clear that these factors can account for only a few hundred metres of removed section, and higher palaeogeothermal gradients, which might reduce the amount of required burial, can be ruled out, as discussed above. Thus the palaeotemperature data clearly require removal of a total of *c.* 3 km in well 42/16-1 and *c.* 2 km in well 42/21-1, but attempts to determine more precisely the amounts of removed overburden using such approaches are not justified because of the various uncertainties involved.

However, we emphasize that despite these uncertainties in reconstructing former burial depths, the reconstructed thermal histories for units within the preserved section shown in Fig. 6 are not subject to any of these uncertainties. Thus, the main aspects of the reconstructed thermal histories shown in Fig. 6 (specifically the timing of cooling phases and the magnitude of maximum palaeotemperatures in each episode at specific horizons through each well) are well constrained by the AFTA and VR data from these wells, and can be used with confidence to predict patterns of hydrocarbon generation, etc.

Possible latest Carboniferous effects

As discussed above, on the basis of the integration of AFTA and VR data for wells 42/12-1, 42/16-1 and 42/17-1 (Fig. 3), and comparison of results from these three wells (Fig. 4), it seems clear that units within the Carboniferous section in wells 42/12-1, 42/16-1 and 42/17-1 reached their maximum palaeotemperatures during Early Cretaceous time. No significant palaeo-thermal effects have been identified that can be associated with the Late Carboniferous to Early Triassic unconformity in these wells. Given the regional occurrence of major heating and cooling associated with Variscan events represented by this unconformity across onshore Ireland and in many parts of the UK (see next section), some palaeo-thermal effects undoubtedly affected the Carboniferous and older section in the CISB, as shown schematically in the thermal history reconstruction illustrated for well 42/16-1 in Fig. 6. Nevertheless, it is worth noting that considerable additional burial and associated heating could have occurred during the time interval represented by the Variscan unconformity, provided that peak palaeotemperatures at this time did not exceed those reached in the Early Cretaceous episode.

Comparison with results from surrounding regions

One striking aspect of the results of this study is the similarity of the reconstructed thermal histories for three wells (42/12-1, 42/16-1 and 42/17-1), and for all four wells for post-Early Cretaceous time. The lack of detectable Early Cretaceous palaeo-thermal effects for well 42/21-1 stands in stark contrast to results from the other three wells.

Results from Central Irish Sea Basin wells 42/12-1, 42/16-1 and 42/17-1 are also highly consistent in most respects with those from neighbouring regions, where evidence for Early Cretaceous, Early Tertiary and Late Tertiary palaeo-thermal episodes is widespread. As illustrated in Fig. 7, previous thermal history studies from the East Irish Sea Basin (Green *et al.* 1997) and from onshore Ireland (Green *et al.* 2000) have provided evidence of Early Cretaceous, Early Tertiary and Late Tertiary cooling episodes that correlate closely with those identified in the Central Irish Sea in this study. Reconstructed thermal histories for post-Carboniferous times in these regions are almost identical to that illustrated for well 42/16-1 in Fig. 6. It seems reasonable to conclude that results from all these areas represent the same episodes, which thus appear to be truly regional in nature.

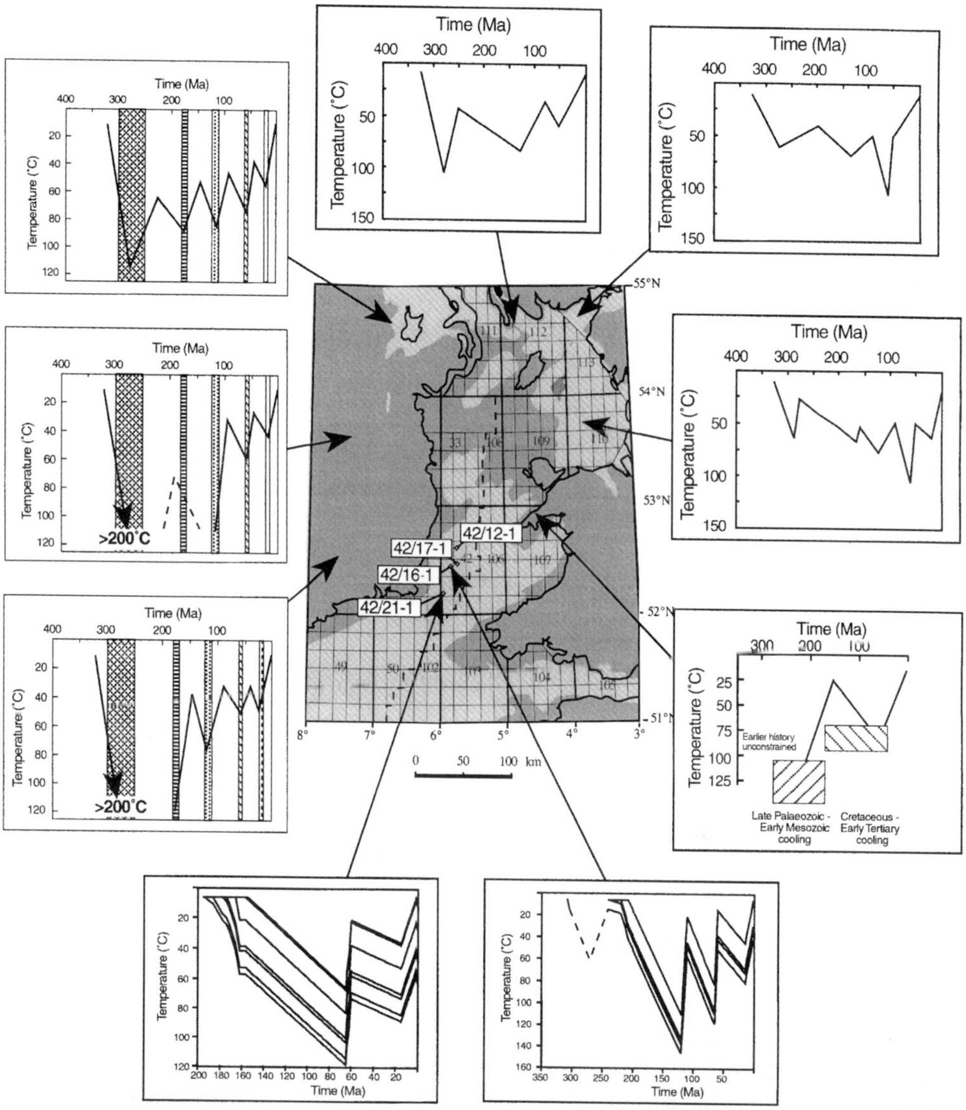

Fig. 7. Reconstructed thermal histories from regions surrounding the CISB are remarkably similar to those identified in this study, at least for post-Carboniferous time. Histories for samples from onshore Ireland are taken from Green *et al.* (1998), for the East Irish Sea Basin from Green *et al.* (1997) and for North Wales from Duncan *et al.* (1998). Histories for South Wales are essentially identical to those shown for onshore Ireland (Geotrack, unpublished results). The palaeo-thermal episodes identified in the four CISB wells in this study appear to be of truly regional extent.

The main difference in the results from the CISB wells, compared with adjacent regions, is the absence in the CISB of detectable latest Carboniferous effects, which dominate thermal histories of outcropping Carboniferous and older rocks from onshore Ireland (Green *et al.* 2000). The reasons for this are unknown, but as discussed by Duncan *et al.* (1998), the preservation of Stephanian sediments in the CISB is consistent with the lack of pronounced Variscan erosion in this region.

The effects of Early Cretaceous cooling are also seen across southern England (Bray *et al.* 1998) and SW England and SW Wales (unpublished Geotrack results), and Early Tertiary cooling is also recognized in these areas as well as northern, central and eastern England (Green 1989; Lewis *et al.* 1992; Green *et al.* 1993) and northern Scotland (Thompson *et al.* 1999), further emphasizing the regional extent of these episodes.

The dominance of individual episodes shows some variation, with the Early Tertiary episode, for example, being dominant in the East Irish Sea Basin (EISB), which has important implications for hydrocarbon prospectivity as discussed below. This is highlighted by the absence of detectable Early Cretaceous palaeo-thermal effects for well 42/21-1 in this study. This is thought to be due to the location of this well within a separate structural regime from the other three wells, aligned with the North Celtic Sea, St. George's Channel and Cardigan Bay basins. As reported by Murdoch *et al.* (1995), AFTA data from the North Celtic Sea Basin show that the Early Tertiary cooling episode is dominant in that region, and the results from well 42/21-1 provide similar conclusions. Appreciable Early Cretaceous cooling may have affected these basins, but with peak palaeotemperatures lower than subsequent maximum values reached in Early Tertiary time (analogous to the masking of postulated Variscan effects for well 42/16-1 in Fig. 6).

Implications for regional hydrocarbon prospectivity

In the EISB and adjacent areas, Green *et al.* (1997) showed that the area in which source rocks reached maximum maturity levels immediately before Early Tertiary inversion is restricted largely to the main EISB hydrocarbon province. In surrounding areas, by contrast, the main phase of hydrocarbon generation occurred during earlier episodes (from latest Carboniferous to Early Cretaceous time). The lack of hydrocarbon discoveries in these regions suggests either that the hydrocarbon generation pre-dated structure formation or that any hydrocarbons accumulated in earlier episodes were lost during subsequent uplift and/or tilting.

Results presented here indicate that potential source rocks within most of the CISB reached maximum maturity levels during Early Cretaceous time, which represents the termination of the main phase of hydrocarbon generation. Any hydrocarbons accumulated at that time are likely to have undergone phase changes and redistribution during at least three discrete phases of uplift and erosion, significantly decreasing the chances of commercial amounts surviving to the present day.

To the south, in the vicinity of well 42/21-1, the Early Tertiary episode appears to become dominant, raising the possibility that in this region conditions similar to those characterizing the main EISB hydrocarbon province may apply. However, because of the much larger thicknesses of preserved Jurassic section, any Carboniferous source rocks in most of this region are likely to have reached much higher maturity levels than in the EISB, whereas Jurassic source rocks are only marginally mature. These factors suggest much higher levels of exploration risk in this area, although the Dragon discovery in UK Quad 103 (Tanner 1999) shows that conditions suitable for operation of a viable petroleum system existed in this region at some stage.

Conclusions

Thermal history reconstruction of hydrocarbon exploration wells in the Central Irish Sea Basin typically reveals three major regional episodes of heating and cooling, related respectively to deep burial, and uplift and erosion. Maximum post-depositional palaeotemperatures generally occurred in Early Cretaceous time (120–115 Ma), with cooling from subsequent palaeotemperature peaks beginning in Late Cretaceous–Early Tertiary (70–55 Ma) and Late Tertiary (25–0 Ma) time. These cooling episodes coincide with similar episodes in surrounding regions.

It is clear that significant additional risk is associated with timing of hydrocarbon generation in the CISB. In future exploration in the Central Irish Sea Basin and adjacent regions, this risk can be much reduced through recognition of the major palaeo-thermal episodes that have affected the region, and the variation in the magnitude of their effects across the region, in order to identify regions where the main phase of hydrocarbon generation post-dated structuring. As with the EISB, definition of areas in which the main phase of hydrocarbon generation occurred during Early Tertiary time or later is likely to highlight the most prospective areas.

We are grateful to PAD, Dublin for provision of sample material from four Central Irish Sea Basin wells for this study. AFTA® is the registered trademark of Geotrack International.

References

ARMSTRONG, J.P., D'ELIA, V.A.A. & LØBERG, R. 1995. Holywell Shale: a potential source of hydrocarbons in the East Irish Sea. *In*: CROKER, P.F. & SHANNON, P.M. (eds) *The Petroleum*

Geology of Ireland's Offshore Basins. Geological Society, London, Special Publications, **93**, 37–38.

ARMSTRONG, J.P., SMITH, J., D'ELIA, V.A.A. & TRUEBLOOD, S.P. 1997. The occurrence and correlation of oils and Namurian source rocks in the Liverpool Bay–North Wales area. *In*: MEADOWS, N.S., TRUEBLOOD, S., HARDMAN, M. & COWAN, G. (eds) *Petroleum Geology of the Irish Sea and Adjacent Areas*. Geological Society, London, Special Publications, **124**, 195–211.

BRAY, R., GREEN, P.F. & DUDDY, I.R. 1992. Thermal history reconstruction using apatite fission track analysis and vitrinite reflectance: a case study from the UK East Midlands of England and the Southern North Sea. *In*: HARDMAN, R.P.F. (ed.) *Exploration Britain: Geological Insights for the Next Decade*. Geological Society, London, Special Publications, **67**, 3–25.

BRAY, R., GREEN, P.F. & DUDDY, I.R. 1998. Multiple heating episodes in the Wessex Basin: implications for geological evolution and hydrocarbon generation. *In*: UNDERHILL, J.R. (ed.) *Development, Evolution and Petroleum Geology of the Wessex Basin*. Geological Society, London, Special Publications, **133**, 199–213.

BURNHAM, A.K. & SWEENEY, J.J. 1989. A chemical kinetic model of vitrinite reflectance maturation. *Geochimica et Cosmochimica Acta*, **53**, 2649–2657.

COLTER, V.S, 1997. The East Irish Sea Basin—from caterpillar to butterfly, a thirty year metamorphosis. *In*: MEADOWS, N.S., TRUEBLOOD, S., HARDMAN, M. & COWAN, G. (eds) *Petroleum Geology of the Irish Sea and Adjacent Areas*. Geological Society, London, Special Publications, **124**, 1–9.

CORCORAN, D. & CLAYTON, G. 1999. Interpretation of vitrinite reflectance profiles in the Central Irish Sea area: implications for the timing of organic maturation. *Journal of Petroleum Geology*, **22**, 261–286.

DUDDY, I.R., GREEN, P.F., BRAY, R.J. & HEGARTY, K.A. 1994. Recognition of the thermal effects of fluid flow in sedimentary basins. *In*: PARNELL, J. (eds) *Geofluids: Origin, Migration and Evolution of Fluids in Sedimentary Basins*. Geological Society, London, Special Publications, **78**, 325–345.

DUDDY, I.R., GREEN, P.F., HEGARTY, K.A. & BRAY, R.J. 1991. Reconstruction of thermal history in basin modelling using apatite fission track analysis: what is really possible. *Proceedings of the First Offshore Australia Conference (Melbourne)*, **III**, 49–61.

DUDDY, I.R., GREEN, P.F. & LASLETT, G.M. 1988. Thermal annealing of fission tracks in apatite 3. Variable temperature behaviour. *Chemical Geology (Isotope Geoscience Section)*, **73**, 25–38.

DUNCAN, W.I., GREEN, P.F. & DUDDY, I.R. 1998. Source rock burial history and sea effectiveness: key facets to understanding hydrocarbon exploration potential in the East and Central Irish Sea Basins. *AAPG Bulletin*, **82**, 1401–1415.

FLOODPAGE, J., WHITE, J. & NEWMAN, P. 1999. The hydrocarbon prospectivity of the Irish Sea: insights from recent exploration of the Central Irish Sea, Peel and Solway Basins. *In*: CROKER, P.F. & O'LOUGHLIN, O. (eds) *The Petroleum Exploration of Ireland's Offshore Basins Conference, Dublin, 29–30 April 1999, Extended Abstracts*. Petroleum Affairs Division, Department of the Marine and Natural Resources, Dublin, 28–31.

GREEN, P.F. 1989. Thermal and tectonic history of the East Midlands shelf (onshore U.K.) and surrounding regions assessed by apatite fission track analysis. *Journal of the Geological Society, London*, **146**, 755–773.

GREEN, P.F., DUDDY, I.R. & BRAY, J.R. 1995. Applications of thermal history reconstruction in inverted basins. *In*: BUCHANAN, J.G. & BUCHANAN, P.G. (eds) *Basin Inversion*. Geological Society, London, Special Publications, **88**, 149–165.

GREEN, P.F., DUDDY, I.R. & BRAY, R.J. 1997. Variation in thermal history styles around the Irish Sea and adjacent areas: implications for hydrocarbon occurrence and tectonic evolution. *In*: MEADOWS, N.S., TRUEBLOOD, S., HARDMAN, M. & COWAN, G. (eds) *Petroleum Geology of the Irish Sea and Adjacent Areas*. Geological Society, London, Special Publications, **124**, 73–93.

GREEN, P.F., DUDDY, I.R., BRAY, R.J. & LEWIS, C.L.E. 1993. Elevated palaeotemperatures prior to early Tertiary cooling throughout the UK region: implications for hydrocarbon generation. *In*: PARKER, J.R. (eds) *Petroleum Geology of Northwest Europe: Proceedings of the 4th Conference*. Geological Society, London, 1067–1074.

GREEN, P.F., DUDDY, I.R., GLEADOW, A.J.W. & LOVERING, J.F. 1989*a*. Apatite fission track analysis as a palaeotemperature indicator for hydrocarbon exploration. *In*: NAESER, N.D. & MCCULLOH, T. (eds) *Thermal History of Sedimentary Basins—Methods and Case Histories*. Springer, New York, 181–195.

GREEN, P.F., DUDDY, I.R., GLEADOW, A.J.W., TINGATE, P.R. & LASLETT, G.M. 1986. Thermal annealing of fission tracks in apatite. 1. A qualitative description. *Chemical Geology (Isotope Geoscience Section)*, **59**, 237–253.

GREEN, P.F., DUDDY, I.R., HEGARTY, K.A., BRAY, R.J., SEVASTOPULO, G., CLAYTON, G. & JOHNSTON, D. 2000. The post-Carboniferous evolution of Ireland: evidence from Thermal History Reconstruction. *Proceedings of the Geologists Association*, **111**, 307–320.

GREEN, P.F., DUDDY, I.R., LASLETT, G.M., HEGARTY, K.A., GLEADOW, A.J.W. & LOVERING, J.F. 1989*b*. Thermal annealing of fission tracks in apatite. 4. Quantitative modelling techniques and extension to geological timescales. *Chemical Geology (Isotope Geoscience Section)*, **79**, 155–182.

GREEN, P.F., HEGARTY, K.A. & DUDDY, I.R. 1996. Compositional influences on fission track annealing in apatite and improvement in routine application of AFTA®. *American Association of Petroleum Geologists, San Diego, Abstracts with Program*, A56.

HARDMAN, M., BUCHANAN, J., HERRINGTON, P. & CARR, A. 1993. Geochemical modelling of the East Irish Sea Basin: its influence on predicting hydrocarbon type and quality. *In*: PARKER, J.R. (eds) *Petroleum Geology of Northwest Europe: Proceedings of the 4th Conference*. Geological Society, London, 809–821.

LASLETT, G.M., GREEN, P.F., DUDDY, I.R. & GLEADOW, A.J.W. 1987. Thermal annealing of fission tracks in apatite. 2. A quantitative analysis. *Chemical Geology (Isotope Geoscience Section)*, **65**, 1–13.

LEWIS, C.L.E., GREEN, P.F., CARTER, A. & HURFORD, A.J. 1992. Elevated late Cretaceous to Early Tertiary palaeotemperatures throughout Northwest England: three kilometres of Tertiary erosion? *Earth and Planetary Science Letters*, **112**, 131–145.

MADDOX, S.J., BLOW, R. & HARDMAN, M. 1995. Hydrocarbon prospectivity of the Central Irish Sea Basin,with reference to Block 42/12, offshore Ireland. *In*: CROKER, P.F. & SHANNON, P.M. (eds) *The Petroleum Geology of Ireland's Offshore Basins*. Geological Society, London, Special Publications, **93**, 59–77.

MURDOCH, L.M., MUSGRAVE, F.W. & PERRY, J.S. 1995. Tertiary uplift and inversion history in the North Celtic Sea Basin and its influence on source rock maturity. *In*: CROKER, P.F. & SHANNON, P.M. (eds) *The Petroleum Geology of Ireland's Offshore Basins*. Geological Society, London, Special Publications, **93**, 297–319.

SCOTCHMAN, I.C. & THOMAS, J.R.W. 1995. Maturity and hydrocarbon generation in the Slyne Trough, northwest Ireland. *In*: CROKER, P.F. & SHANNON, P.M. (eds) *The Petroleum Geology of Ireland's Offshore Basins*. Geological Society, London, Special Publications, **93**, 385–411.

STUART, I.A. & COWAN, G. 1991. The south Morecambe Field, blocks 110/2a, 110/3a, 110/8a, UK East Irish Sea. *In*: ABBOTS, I.L. (eds) *United Kingdom Oil and Gas Fields, 25 Years Commemorative Volume*. Geological Society, London, Memoir, **14**, 527–541.

SUMMER, N.S. & VEROSUB, K.L. 1989. A low temperature hydrothermal maturation mechanism for sedimentary basins associated with volcanic rocks. *In*: PRICE, P.A. (eds) *Origin and Evolution of Sedimentary Basins and their Economic Potential*. Geophysical Monograph, American Geophysical Union, **48**, 129–136.

SWEENEY, J.J. & BURNHAM, A.K. 1990. Evaluation of a simple model of vitrinite reflectance based on chemical kinetics. *AAPG Bulletin*, **74**, 1559–1570.

TANNER, H.C. 1999. The petroleum geology of the St George's Channel Basin. *In*: CROKER, P.F. & O'LOUGHLIN, O. (eds) *The Petroleum Exploration of Ireland's Offshore Basins Conference, Dublin, 29–30 April 1999, Extended Abstracts*. Petroleum Affairs Division, Department of the Marine and Natural Resources, Dublin, 44–45.

THOMPSON, K., UNDERHILL, J.R., GREEN, P.F., BRAY, R.J. & GIBSON, H.J. 1999. Evidence from apatite fission track analysis for the post-Devonian burial and exhumatuion history of the northern Highlands, Scotland. *Marine and Petroleum Geology*, **16**, 27–39.

ZIAGOS, J.P. & BLACKWELL, D.D. 1986. A model for the transient temperature effect of horizontal fluid flow in geothermal systems. *Journal of Volcanology and Geothermal Research*, **27**, 371–397.

The geology and geophysics of the SW Kinsale gas accumulation

JOHN M. O'SULLIVAN

Marathon International Petroleum Ireland Limited, Centre Park House, Centre Park Road, Cork, Ireland

Present address: Marine Institute, Parkmore Industrial Estate, Ballybrit, Galway, Ireland (e-mail: john.osullivan@marine.ie)

Abstract: The SW Kinsale gas accumulation is located in blocks 48/20 and 48/25 in the North Celtic Sea Basin, 50 km off the south coast of Ireland. The field lies in *c.* 100 m of water. The discovery well, 48/25-2, was drilled in 1971 and encountered gas-bearing fluviatile sandstones in the Lower Cretaceous Wealden sequence. In 1995, well 48/25-3 was drilled close to the discovery well on the southwestern limb of the Kinsale Head anticline, which is thought to have formed during Tertiary basin inversion. The well data indicate that this southwestern area is in pressure isolation from the main Kinsale Head Field, which lies in the central and eastern sectors of the structure. In 1997 a 3D seismic survey was acquired to assess the suitability of the field as a potential gas storage site. These data suggest that SW Kinsale is in structural isolation from the main Kinsale Head Field. Mapping of the 3D volume reveals SW Kinsale to comprise a relatively simple low-relief anticline. There is *c.* 160 m of closure from −810 m true vertical depth sub-sea (TVDSS) at its crest to −968 m TVDSS along a syncline to the north. The accumulation is thought to have a shared gas–water contact with the main Kinsale Head Field at −945 m TVDSS, beneath which lies a transition zone to −968 m TVDSS. The area within closure is close to 1200 ha. These data also suggest that the reservoir has undergone a minimal degree of structural compartmentalization. A major Wealden channel axis is interpreted to transect the field. Southwest Kinsale is thought to contain 1.1–1.4 BCM (billion cubic metres) gas initially in place with about 0.85 BCM recoverable. The field was recently developed as a single well sub-sea tieback to the Kinsale Bravo platform. First gas deliveries from the field took place in late 1999.

The SW Kinsale gas accumulation is located in blocks 48/20 and 48/25, *c.* 50 km off the south coast of Ireland. The area lies at a water depth of some 100 m in the North Celtic Sea Basin, SW of the main Kinsale Head Field and due south of the Ballycotton Field (Fig. 1). The field was discovered in 1971 by well 48/25-2, which encountered gas-bearing shelfal and fluviatile sandstones of Early Cretaceous Albian and Wealden age, respectively. Mapping of the available 2D seismic data suggested that the trapping geometry was a major mid-basin anticlinal structure, with caprocks provided by Lower Cretaceous shales (Colley *et al.* 1981). The dominant fault orientation that was mapped from these 2D data trends primarily in a NE–SW direction. Faults with both normal and reverse displacements were mapped throughout the area.

Further appraisal drilling revealed that the Cretaceous section of the North Celtic Sea Basin comprises one overall transgressive sequence (Fig. 2). The mixed alluvial–fluvial-dominated systems of the Wealden sequence give way progressively to the more continuous shelfal sands, the Greensand, of Albian age. The deeper water Upper Cretaceous mudstones and chalks cap this succession. The results of appraisal drilling indicated that the higher quality and more continuous Greensand reservoir was best developed in the central and eastern portions of the Kinsale Head area. It is for this reason that gas production to date has been focused on these sectors of the field complex.

Depth mapping of the 2D data over Kinsale Head revealed a potential structural saddle between the main field and the SW Kinsale area (Fig. 1). In addition, these data suggested that the area was relatively unfaulted, and that well 48/25-2 had penetrated SW Kinsale close to the local structural culmination at Lower Cretaceous level. The objective of the present paper is to describe the regional setting and exploration history of the SW Kinsale gas accumulation. The results of a recent high-resolution 3D seismic survey are described and discussed with a view to elucidating the size, structure and volumetric estimates of the field.

From: SHANNON, P.M., HAUGHTON, P.D.W. & CORCORAN, D.V. (eds). 2001. *The Petroleum Exploration of Ireland's Offshore Basins*. Geological Society, London, Special Publications, **188**, 189–199. 0305-8719/01/$15.00 © The Geological Society of London 2001.

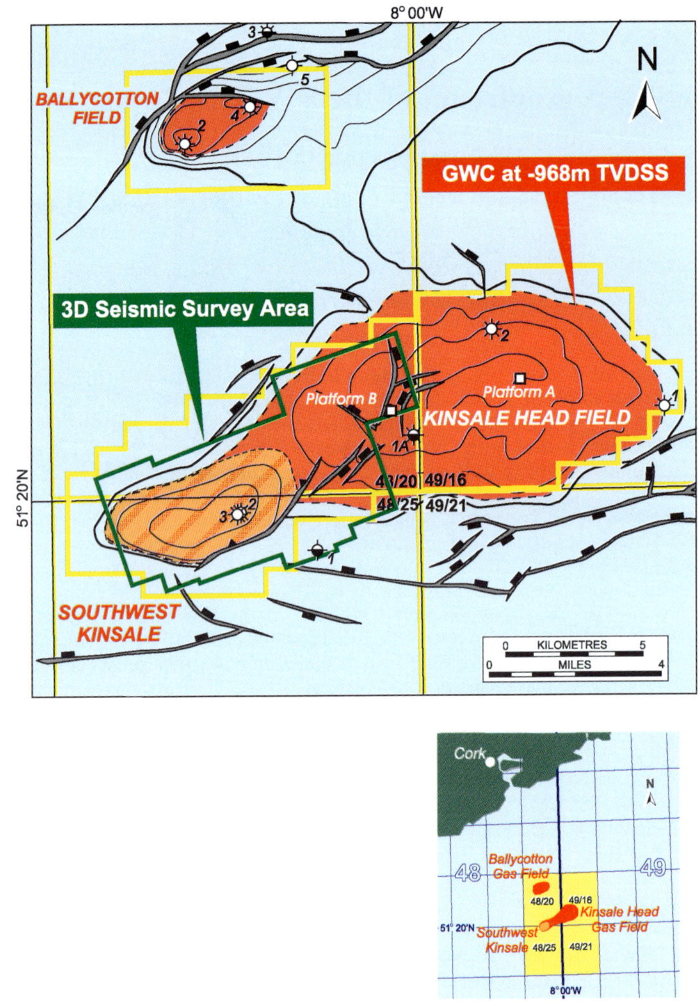

Fig. 1. Top Wealden formation depth structure of the SW Kinsale gas accumulation as mapped from 2D seismic data. The map shows the interpreted structural isolation of SW Kinsale, the gas–water contact (GWC) at −968 m TVDSS and the position of SW Kinsale 3D seismic survey. The yellow boxes denote defined field development areas for the Kinsale Head and Ballycotton fields. The inset map shows the general location of the fields.

Regional setting

The North Celtic Sea Basin is one of a number of parallel elongate NE–SW-trending basins that lie off the south coast of Ireland. Basement is considered to be of Devonian to Carboniferous age (Griffiths 1995), and although well control is limited, regional studies suggest that basin development was initiated during Triassic time (Musgrove *et al.* 1995). These early Mesozoic sub-rift systems are considered to be of limited areal extent, and are thought to have exploited earlier Charnian, Caledonian and Variscan lines of weakness. The Triassic succession of the North Celtic Sea Basin, as elsewhere in NW Europe, is interpreted to have been deposited in arid desert-type conditions. This led to the deposition of alluvial, fluvial and aeolian continental red beds with periodic marine incursions leading to the development of sabkha and playa-lake deposits. Reworking of Devonian and Carboniferous material from the Irish Massif to the north, as well as that of the Pembrokeshire Ridge to the south, is considered to have been the main sediment source at this time (Griffiths 1995; Musgrove *et al.* 1995).

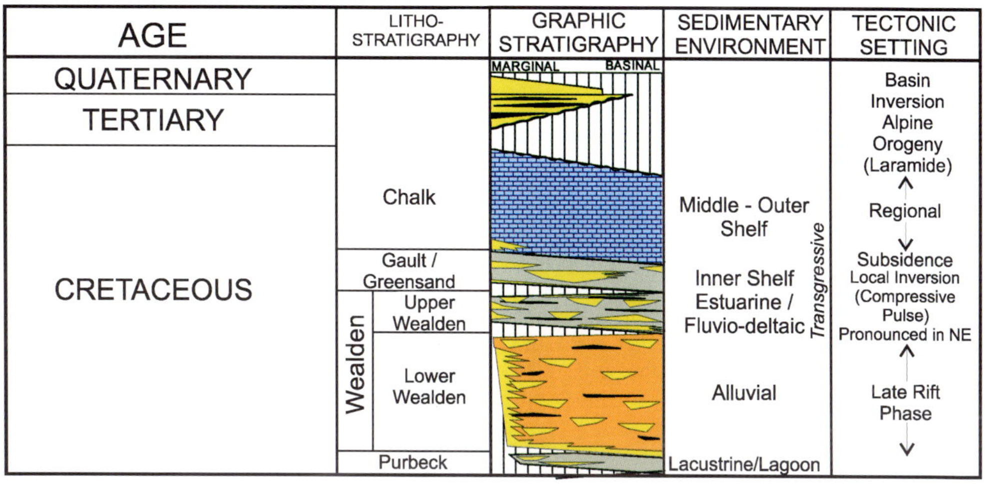

Fig. 2. Generalized Cretaceous–Recent lithostratigraphy, sedimentary environment and structural development of the North Celtic Sea Basin..

The main phase of rifting in the basin was initiated during the Jurassic period, with two overall basinwide transgressions recorded from the underlying red bed Permo-Triassic succession, to the Upper Cretaceous chalks (Rowell 1995). The Lower Jurassic succession is dominated by a shaly marine sequence, which is thought to be the main hydrocarbon source in the basin (Murphy *et al.* 1995). These shales were then succeeded by shallow-water marine shelfal limestones of Mid-Jurassic age. The main extensional phase of rifting occurred in Late Jurassic time with the development of significant hanging-wall synclinal depocentres (Rowell 1995). The associated footwall uplift of areas close to these depocentres resulted in the development of a facies architecture, which is thought to be related to contemporaneous basin structure. This resulted in rapid facies changes that are observed throughout the Upper Jurassic succession (Naylor & Shannon 1982). The latter is dominated by marine shales and limestones but also locally contains a considerable amount of terrestrial material.

The Jurassic–Cretaceous boundary in the North Celtic Sea Basin is characterized by the non-marine and lacustrine shales of the Purbeckian sequence. These shales are important as they signify the onset of the non-marine mixed fluvial–alluvial Wealden succession (Ewins & Shannon 1995), as well as being the source for a poor-quality waxy oil that is observed in wells throughout the basin (Murphy *et al.* 1995). The Wealden section provides the main reservoir interval for the SW Kinsale gas accumulation. Shelfal sandstones of the Albian Greensand, marking an episode of major marine transgression, progressively overlie the Wealden sequence. These sandstones are more continuous and better developed than those of the underlying Wealden sequence and form the main reservoir interval for the Kinsale Head Field. The Greensand itself was eventually overstepped by marine claystones of the Gault Clay as the basin continued to undergo post-rift thermal sag (Colley *et al.* 1981; Taber *et al.* 1995). Well data indicate that these claystones grade into a sandy facies toward the northern margin of the basin. The overlying Upper Cretaceous sequence comprises a thick succession of middle–outer shelfal chalks. The chalk section marks the culmination of the Cretaceous transgression, as recorded in the stratigraphy of the North Celtic Sea Basin (Fig. 2). Regional Tertiary basin inversion (Murdoch *et al.* 1995) has led to the removal of much of the younger section across the area. In addition, this compressive event has led to the formation of a number of mid-basin, inversion-induced anticlines with some associated strike-slip faulting.

Field history

The SW Kinsale gas accumulation was discovered in 1971 with the drilling of well 48/25-2. This well encountered gas in both the Greensand and Wealden (Lower Cretaceous) reservoirs and was the discovery well for the Kinsale Head Field. The well tested dry gas from the Wealden sequence at rates of up to 0.57 MMSCMD (million standard cubic metres per day). The field is thought to have been sourced from shales of Early Jurassic age (Taber *et al.* 1995).

Reservoir performance monitoring and simulation carried out by Marathon on the Kinsale Head Field in the early 1990s indicated that a potential volume of unswept gas might lie in the Wealden sandstones of the SW Kinsale area. This concept was supported by pressure differences, which had been observed between wells tested in the Wealden formation of the Kinsale Head Field. In addition, the complex architecture of the Wealden fluviatile systems, together with the potential structural isolation of the area, also supported the presence of an unswept accumulation.

The SW Kinsale area was further evaluated in 1995 with the drilling of appraisal well 48/25-3. This well, which twinned the original discovery well 48/25-2, was drilled close to the crest of the structure. As expected, the well encountered gas in the primary Wealden reservoir objective. The well was drillstem tested and flowed at rates of up to 0.66 MMSCMD on a 2.86 cm choke. The gross reservoir interval, averaged over the two wells, measured some 23 m with an average net to gross ratio of 0.35. The average porosity encountered was 22% and repeat formation test (RFT) data revealed that these Wealden sandstones were close to virgin pressure. A minor drawdown of some 3.45 bar was observed and it is unclear if this is due to minor pressure communication with the main Kinsale Head Field or to a calibration shift between the two measurements. Pressure data from the overlying Greensand interval exhibited a marked depletion, supporting the view that the platforms were effectively draining this reservoir unit.

Given the discovery of an undepleted gas volume close to the Kinsale Head Field, various development scenarios were investigated. Initial planning favoured an offshore gas storage site, which could be utilized in peak shaving and transmission support service. The results of geotechnical studies indicated that the Wealden sandstones had good potential to act as gas storage reservoirs. One area of concern remained regarding the accuracy of the subsurface reservoir model. To address this concern, a high-resolution 3D seismic survey was acquired over the SW Kinsale area.

High-resolution 3D seismic survey

The SW Kinsale 3D seismic survey was acquired over blocks 48/20 and 48/25 during the winter of 1997 (Fig. 1). The total survey area was some 53.5 km^2. The primary objective of the survey was to accurately image the Wealden reservoirs, which lie at *c.* 900 m true vertical depth sub-sea (TVDSS) (*c.* 600 ms two-way travel time (TWT)). Inlines were acquired at a spacing of 25 m with shots made at a 6.25 m interval. The sail-lines were orientated in a NW–SE direction to optimize the volume for structural imaging. The final bin size after processing was 12.5 m (inline) by 6.25 m (crossline).

The source used for the survey was a bolt sleeve gun array with a total capacity of 2.3 l. Three 600 m streamers towed at a depth of 3 m led to three common mid-point (CMP) subsurface lines being acquired for each sail-line pass. The data were sampled at a 1 ms interval over a total record length of 1.5 s. The recorded data were passed at a 12.5–308 Hz bandwidth with a final processed central frequency of 100 Hz at the reservoir level.

The main aim of the 3D survey was to provide a more refined depth model of the Wealden reservoir sandstones. It was hoped that the results of the analysis would address concerns regarding the structural isolation and reservoir compartmentalization of the SW Kinsale area. Given the relatively complex reservoir architecture of the Wealden fluviatile systems, it was also hoped that these data would provide a clearer insight into reservoir stacking patterns. Once fully integrated, the data could then be used in reservoir simulation, which in turn would lead to the definition of optimal drilling locations.

Although the SW Kinsale 3D data quality is very good, a number of artefacts remain in the data. Static shifts between sail-lines, which are caused by local tidal variations, were corrected for using a complex tidal correction algorithm. This routine used actual tidal gauge data recorded from the nearby Kinsale Head platforms as an input. Although this correction algorithm would have been sufficient for a conventional seismic survey, it proved inadequate for such a high-frequency dataset. To correct for these inline static shifts, a low-order polynomial surface was derived using the sea-bed pick as an input, and a residual was calculated. This was then applied to the subsurface picks to correct the data.

Vertical static shifts or 'banding' in the amplitude domain also proved problematic. These amplitude shifts tend to occur in the inline direction and are constant above, through and beneath the zone of interest. The origin of these artefacts is uncertain, although various studies have indicated a potential linkage with the sail-line static shifts, structure and/or hydrocarbon-induced velocity effects. These distortions have a considerable impact on amplitude-based attribute analysis. The simplest form of solution to correct for these effects is to normalize the extracted reservoir amplitudes against some nearby reflector of relatively constant impedance. The best such horizon is the base of the Upper Cretaceous Chalk, which was successfully used in correcting the data.

The survey area is well known for the significant amount of water-borne multiple energy recorded in reflection seismic data. The presence of a high-impedance chalk at the sea bed results in both waterborne as well as intra-Chalk 'pegleg' multiple energy, which pose a considerable technical challenge through processing. The 3D data are heavily contaminated with both types of multiple, and this makes any attempt at coherent event-based attribute analysis or seismic inversion extremely difficult. No post-stack multiple attenuation was attempted on the data.

Seismic interpretation

Interpretation of the 3D data volume was focused on the Lower Cretaceous interval. Seismic picks were made on the base Chalk, top Greensand and top Wealden reflections. An intra-Chalk pick, which corresponds to the top of a high-velocity chalk layer, was also carried out around the data volume to be used later in depth conversion.

Initial inspection of the data through the use of regular and flattened timeslice data reveals a significant amount of structural information. A timeslice taken at the Wealden reservoir level (Fig. 3) clearly illustrates a time closure associated with SW Kinsale. In addition, the bounding fault on the southern limb of the feature is also clearly imaged, as well as a number of important NW–SE-trending fault elements that occur within the structure itself. Review of an inline and crossline (Figs 4 and 5), over the crestal area, illustrates that the predominant fault orientation is in the inline direction (i.e. NW–SE). These lines also support the presence of a local time closure at Lower Cretaceous level, as well as exhibiting some of the significant coherent noise artefacts mentioned above. Mapping of the Lower Cretaceous events confirms the previous 2D mapping, which indicated that a separate time closure occurred in the SW Kinsale area (Fig. 6). These maps also support the view that wells 48/25-2 and 48/25-3 were drilled close to the crest of the feature.

Structural analysis of the 3D volume was carried out using the base Chalk seismic event. This event was considered to be most accurate because of the quality of the pick, with most faults seen at this level also cutting through the Wealden reservoirs. A number of map analyses such as dip and edge detection were carried out on the base Chalk time horizon data. These

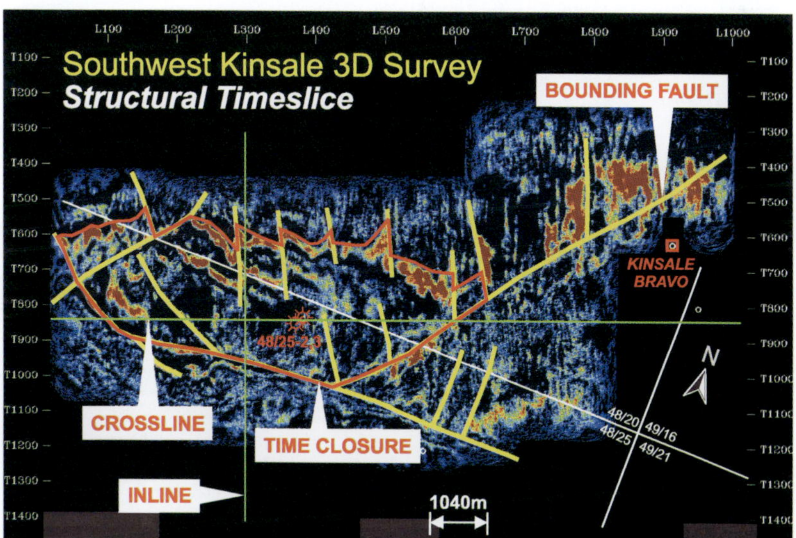

Fig. 3. Timeslice at the Wealden reservoir level through the SW Kinsale 3D seismic survey (TWT is 584 ms). Interpreted lineaments are shown in yellow and the area of interpreted time closure is shown in red. The seismic colour coding denotes the seismic reflection amplitude. The locations of Fig. 4 (inline) and Fig. 5 (crossline) are shown.

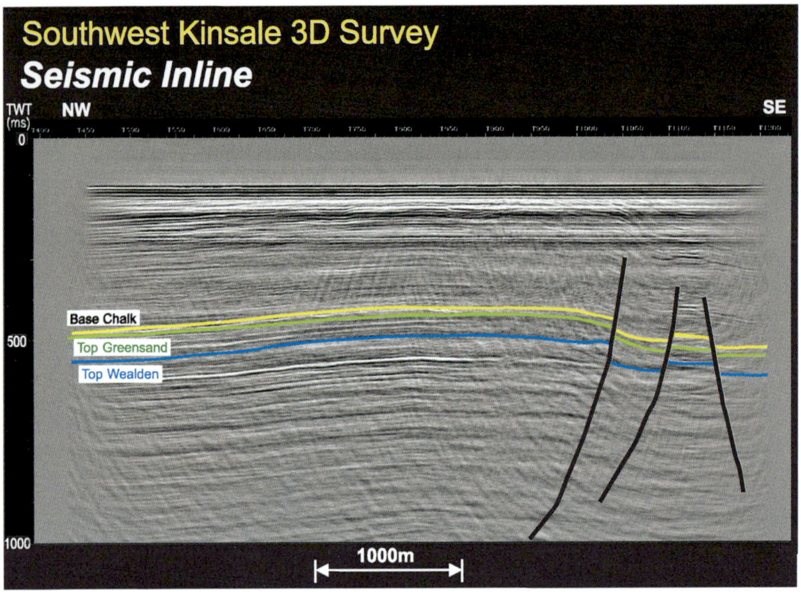

Fig. 4. Seismic inline (L300) over the crest of the SW Kinsale gas accumulation, North Celtic Sea Basin. The line shows picks on base Chalk (yellow), top Greensand (green) and top Wealden (blue) events. Significant faults are shown in black. (See Fig. 3 for location.)

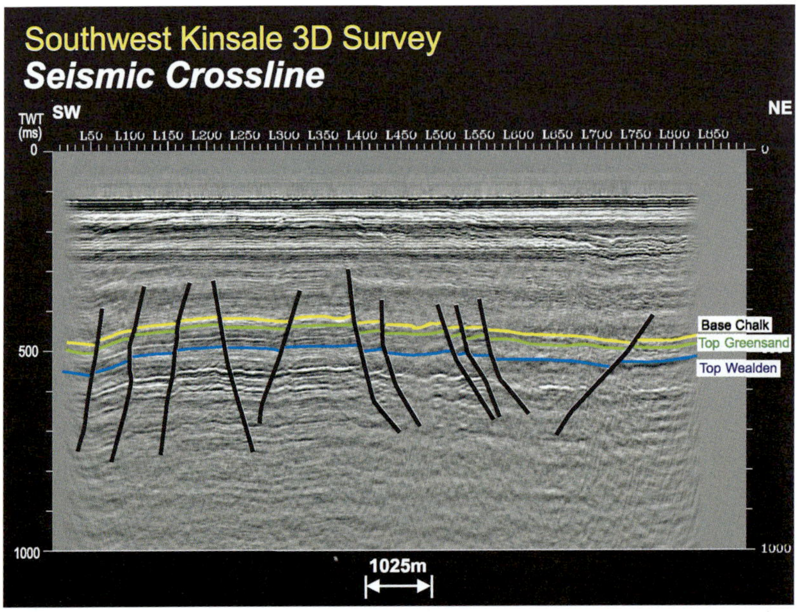

Fig. 5. Seismic crossline (T836) over the crest of the SW Kinsale gas accumulation, North Celtic Sea Basin. The line shows picks on base Chalk (yellow), top Greensand (green) and top Wealden (blue) events. Significant faults are shown in black. (See Fig. 3 for location.)

routines prove extremely useful in the detection of linear discontinuities (i.e. faults) within the data volume. The standard dip detection display picks out the major bounding fault on the southern limb of the field and confirms the timeslice interpretation (Fig. 3). The dip map also tends to suggest that the bounding fault comprises a number of linked and coalesced fault segments rather than being a single, continuous structure. In addition, a number of minor NW–SE-trending elements can be seen on the north and south flanks of the SW Kinsale feature. These are interpreted as Riedel shears, which are thought to have developed as a result of transfer motion along the bounding fault during structural inversion. They tend to throw to the west on the north flank and to the east on the south flank; however, they do not appear to link across the crest of the structure.

It is possible to go beyond conventional map analysis routines, given the considerable frequency bandwidth preserved in the 3D volume. Using a variety of workflows and map analysis processes, the base Chalk surface was interrogated to both second- and third-order degrees. The objective of these more detailed analyses was to assess potential fracture patterns from the volume, which might prove useful in reservoir simulation. A second-order map analysis of the base Chalk event (Fig. 7) clearly illustrates the amount of detail preserved in the dataset. The analysis clearly detects clusters of minor fractures within the SW Kinsale structure orientated with NE–SW, NW–SE and east–west trends.

Depth conversion method

A number of depth conversion routines were carried out on the SW Kinsale area. All of the velocity models were constructed utilizing log data from wells 48/25-2 and 48/25-3. Depth of burial related velocity functions were considered, but were thought to be inappropriate because of the presence of the chalk overburden as well as the previously mentioned late-stage structural inversion of the basin. Analysis of the time–depth data indicates that the Upper Cretaceous interval velocities 'on-structure' tend to be higher then those of deeper wells 'off-structure'. This is due to the stratified nature of the chalk overburden, which comprises two distinct velocity layers. The upper, lower velocity, layer tends to have been removed over the crests of structures during basin inversion, resulting in higher interval velocities than in locations away from the structural high.

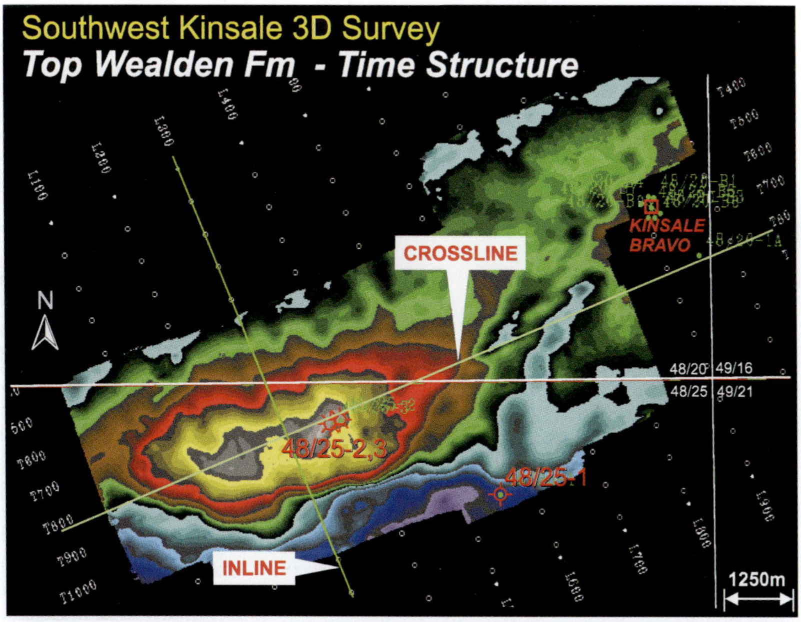

Fig. 6. Top Wealden time structure illustrating the isolation of SW Kinsale. The TWT contours are coloured filled with structural elevation shown in yellow or white. This display also shows the location of wells 48/25-2 and 48/25-3. The locations of Fig. 4 (inline) and Fig. 5 (crossline) are shown.

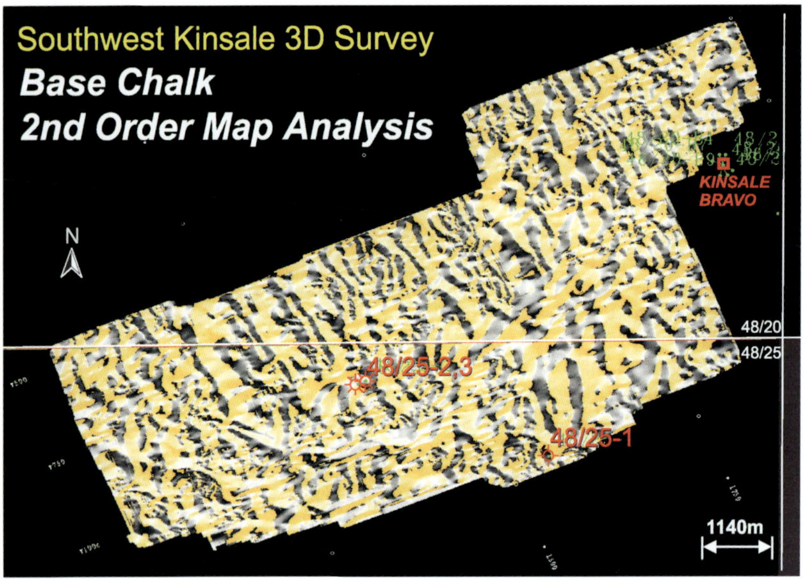

Fig. 7. Base Chalk time horizon second-order map analysis. This map is derived from the base Chalk time horizon data and it shows faulting and fracturing of the Lower Cretaceous section. Linear discontinuities (i.e. faults and fractures) are shown in black, and these have NE–SW, NW–SE and east–west trends. The map also shows the location of wells 48/25-2 and 48/25-3.

The velocity model considered to be most accurate comprises three layers, two within the Chalk and one from base Chalk to top Wealden level. The intra-Chalk event was constructed by isochroning a constant time (95 ms TWT, derived from well data) above the base Chalk event. This constant time isochron method was considered to be most accurate, as multiple contamination within the Chalk makes seismic picking extremely challenging. The wells were then used to calculate interval velocities, which were assigned to each layer. Error analysis from 'off-structure' wells (i.e. 48/25-1), indicates that this is a robust vertical depth migration technique for the Lower Cretaceous sequence of the North Celtic Sea Basin.

Depth structure mapping on the top Wealden event shows a strong correlation with the mapped time surface (Fig. 6). The SW Kinsale area appears to be in structural isolation from the main field, confirming the previous 2D mapping (Fig. 1). The mapped closure occurs at −968 m TVDSS; this corresponds to the deepest zone of relatively high gas saturations seen in the area. This closure is mapped as a relatively simple fault-bounded anticline, with *c.* 160 m of vertical relief from −81 m TVDSS at the crest to −96 m TVDSS at the spill. SW Kinsale is thought to have a shared gas–water contact (GWC) with the Wealden gas accumulation of the main Kinsale Head Field at about −9 m TVDSS. It is thought that a broad transition zone is then developed to −968 m TVDSS. The total area above the GWC is close to 1200 ha.

Seismic attribute analysis

A number of attribute analyses were carried out on the reservoir zones of the Upper Wealden sandstones within the 3D volume. As discussed above, these sandstones are thought to have been deposited in a predominantly fluviatile environment. In addition, core studies from the Wealden sequences in various wells in the Kinsale Head Field suggest that the primary palaeo-drainage direction was towards the SE. It was hoped that seismic attribute mapping might provide a more detailed insight into the depositional architecture of the Wealden sandstones.

Given the thin nature of the hydrocarbon-bearing intervals as well as their complex depositional architecture, extraction and mapping of r.m.s. amplitudes within envelopes was considered to be the best approach. A seismic window was created, by adding a constant time isochron to the top Wealden event. The window length was determined from reservoir interval thickness and velocities derived from well log data. The r.m.s. amplitudes were then extracted and mapped. The resultant map (Fig. 8), shows a

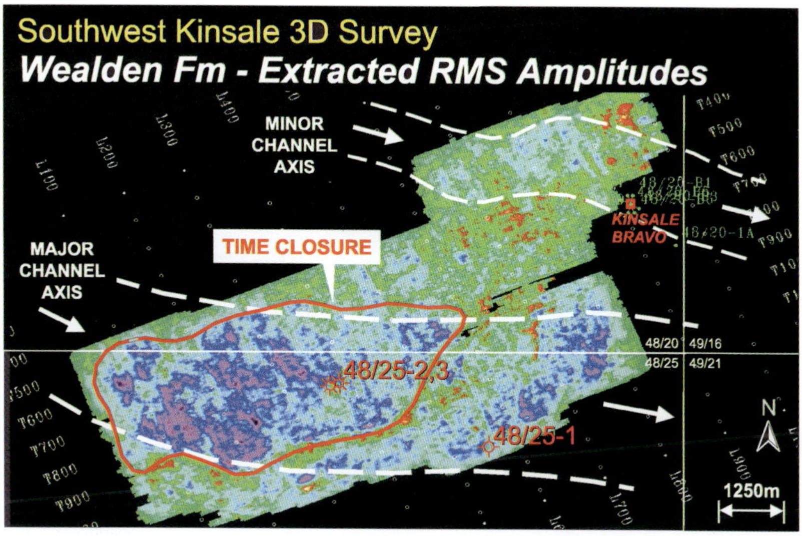

Fig. 8. Map of extracted Wealden formation r.m.s. amplitudes within a defined window beneath the top Wealden event. Higher amplitudes are shown in blue or purple and these are interpreted to correspond to Wealden sand depositional fairways. The mapped time closure associated with SW Kinsale is shown in red.

strong amplitude response trending ESE–WNW across the SW Kinsale area. These anomalous amplitudes correlate closely with the sandstones encountered in wells 48/25-1, 48/25-2 and 48/25-3, which are all located within the 3D survey area. A component of the seismic response over the crest of the structure may, in part, be associated with hydrocarbon charge. A weaker amplitude anomaly can be seen at the Kinsale Field Bravo platform area to the NE. This anomaly is interpreted to be a minor channel axis, which has been encountered by drilling in that area.

Field volumetric estimates and development plan

Following analysis of the geotechnical and reservoir engineering data, it was decided that the SW Kinsale reserves would be best utilized through a primary depletion plan. Given the relatively small closure (*c.* 1200 ha) in addition to the reservoir uncertainty, volumetric estimates tended to show significant variance with minor changes in input parameters. Original work on the Kinsale Head area had suggested that a regional GWC occurred at −968 m TVDSS. This depth tied closely with high gas saturations in the main field, as well as being the lowest closing contour in the Kinsale Head area at top Wealden level.

More recent reservoir engineering studies have, however, shown that the effective GWC is probably located close to −945 m TVDSS. Wells tend to encounter high residual gas saturations beneath this level, in addition to a waxy oil that has caused some production problems in the main Kinsale Head Field. The zone below −945 m TVDSS is now considered to be a broad transition zone of biodegraded oil and residual gas. Volumetric estimates of the Wealden reservoirs in SW Kinsale range between 1.1 and 1.4 BCM GIIP based on a GWC at −945 m TVDSS. The ultimate recovery predicted for these reservoirs is 60–70%, on the basis of production data from the main Kinsale Head Field.

Conclusions

The understanding of the SW Kinsale subsurface reservoir model has been greatly enhanced through the acquisition and interpretation of a high-resolution 3D seismic survey. These new data have significantly improved the confidence in the risked volumetric estimate, as well as in the ultimate recoverable reserves. The application of such state-of-the-art geophysical technology has proved to be a vital tool in the decision-making process, and has resulted in the economic development of an erstwhile dormant asset. The SW Kinsale development plan is very similar to that of the nearby Ballycotton Field

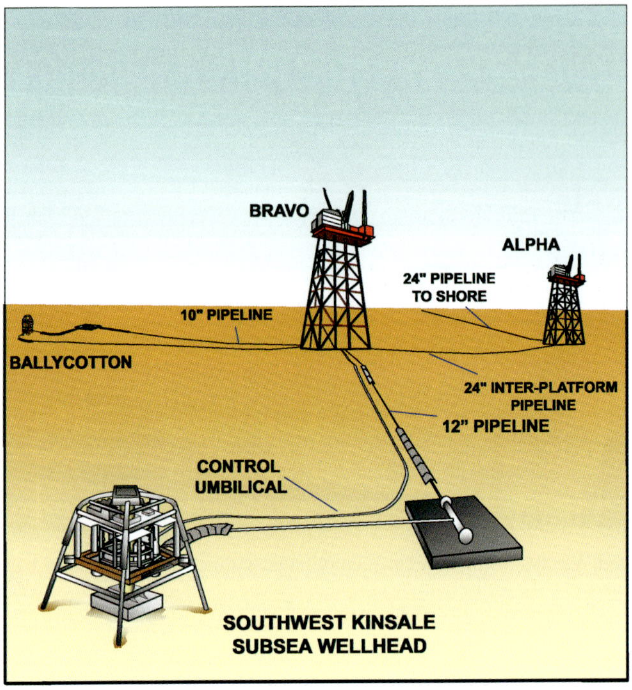

Fig. 9. The SW Kinsale field development concept. The field was developed through a single sub-sea wellhead. Well 48/25-3 was re-entered and completed in the Wealden sequence, and tied back to the Kinsale platforms via a sub-sea pipeline.

(Murray 1995). The field will be depleted via a single sub-sea wellhead located over well 48/25-3, which was re-entered and completed. The wellhead was linked to the Kinsale Bravo platform through a sub-sea pipeline and control umbilical (Fig. 9). The gas flows ashore to the Inch Gas Terminal via the Kinsale Alpha platform. First gas deliveries from SW Kinsale took place in late 1999.

I would like to thank Marathon International Petroleum Ireland Limited for permission to publish this paper. In particular I would like to thank A. Ring and J. Cockings of Marathon for their invaluable input and guidance. Many thanks go to the two Hilarys for technical support and to B. Golden, who did such a fine job on the draughting. Finally, I would like to acknowledge the input of the technical referees, whose comments greatly improved the structure and content of this paper. The SW Kinsale 3D seismic survey was funded in part by Bord Gais Eireann and in part by the EU sponsored *TENS* scheme.

References

COLLEY, M.G., MCWILLIAMS, A.S.F. & MYERS, R.C. 1981. Geology of the Kinsale Head Gas Field, Celtic Sea, Ireland. *In*: ILLING, L.V. & HOBSON, G.D. (eds) *Petroleum Geology of the Continental Shelf of North-West Europe*. Heyden, London, 504–510.

EWINS, N.P. & SHANNON, P.M. 1995. Sedimentology and diagenesis of the Jurassic and Cretaceous of the North Celtic Sea and Fastnet Basins. *In*: CROKER, P.F. & SHANNON, P.M. (eds) *The Petroleum Geology of Ireland's Offshore Basins*. Geological Society, London, Special Publications, **93**, 139–169.

GRIFFITHS, P.S. 1995. Predictive model for the development and distribution of Triassic reservoir sands offshore southeast Ireland based on seismic sequence geometries at the Variscan unconformity. *In*: CROKER, P.F. & SHANNON, P.M. (eds) *The Petroleum Geology of Ireland's Offshore Basins*. Geological Society, London, Special Publications, **93**, 79–80.

MURDOCH, L.M., MUSGROVE, F.W. & PERRY, J.S. 1995. Tertiary uplift and inversion history in the North Celtic Sea Basin and its influence on source rock maturity. *In*: CROKER, P.F. & SHANNON, P.M. (eds) *The Petroleum Geology of Ireland's Offshore Basins*. Geological Society, London, Special Publications, **93**, 297–319.

MURPHY, N.J., SAUER, M.J. & ARMSTRONG, J.P. 1995. Toarcian source rock potential in the North Celtic Sea Basin, offshore Ireland. *In*: CROKER, P.F. & SHANNON, P.M. (eds) *The Petroleum Geology of*

Ireland's Offshore Basins. Geological Society, London, Special Publications, **93**, 193–208.

MURRAY, M.V. 1995. Development of small gas fields in the Kinsale Head area. *In*: CROKER, P.F. & SHANNON, P.M. (eds) *The Petroleum Geology of Ireland's Offshore Basins*. Geological Society, London, Special Publications, **93**, 259–260.

MUSGROVE, F.W., MURDOCH, L.M. & LENEHAN, T. 1995. The Variscan fold–thrust belt southeast of Ireland and its control on early Mesozoic extension and deposition: a method to predict the Sherwood Sandstone. *In*: CROKER, P.F. & SHANNON, P.M. (eds) *The Petroleum Geology of Ireland's Offshore Basins*. Geological Society, London, Special Publications, **93**, 81–100.

NAYLOR, D., SHANNON, P.M. 1982. *The Geology of Offshore Ireland and West Britain*. Graham & Trotman, London.

ROWELL, P. 1995. Tectono-stratigraphy of the North Celtic Sea Basin. *In*: CROKER, P.F. & SHANNON, P.M. (eds) *The Petroleum Geology of Ireland's Offshore Basins*. Geological Society, London, Special Publications, **93**, 101–137.

TABER, D.R., VICKERS, M.K. & WINN, Jr, R.D. 1995. The definition of the Albian 'A' Sand reservoir fairway and aspects of associated gas accumulations in the North Celtic Sea Basin. *In*: CROKER, P.F. & SHANNON, P.M. (eds) *The Petroleum Geology of Ireland's Offshore Basins*. Geological Society, London, Special Publications, **93**, 227–244.

Late Tertiary faulting, footwall uplift and topography in western Ireland

MICHAEL E. BADLEY

Badley Earth Sciences, North Beck House, North Beck Lane, Hundleby PE23 5NB, UK
(e-mail: mike@badleys.co.uk)

Abstract: The Erriff Fault in south Mayo, Ireland, is a rare example within the British Isles and Ireland of a sizeable onshore fault (normal displacement >500 m) at which a correlatable stratigraphic horizon is preserved in both the footwall and the hanging wall. That this displacement probably accumulated during Late Tertiary time makes the Erriff Fault (and other similar faults in the area) all the more interesting. Structural forward modelling indicates that footwall uplift and hanging-wall subsidence accompanying faulting, followed by further isostatic uplift following glacial erosion during the Pleistocene glaciations, exercised a fundamental control on the overall form of present-day topography in western Ireland.

The uplifted landforms of the western fringe of the British Isles and Ireland have a long history of investigation. Typical are key papers in the 1960s (Dewey & McKerrow 1963; George 1966, 1967), which discussed landform evolution in western Ireland, Northern Ireland and Hebridean Scotland, respectively. George (1966, 1967) convincingly documented evidence that indicates a Neogene origin for landforms in this western fringe of the British Isles. More recent work (e.g. Japsen 1998) has documented the widespread extent of Neogene uplift in NW Europe. The cause of such regional uplift remains enigmatic but the potential for normal faulting to generate not only local subsidence but also local uplift and topography is well documented (e.g. King *et al.* 1988; Stein *et al.* 1988; Ellis *et al.* 1999). This paper discusses the possible contribution of recent normal faulting to landform evolution in western Ireland.

The sub-Carboniferous surface

Dewey & McKerrow (1963) first suggested that the high relief of the lands west of Lough Corrib could be due to uplift occurring as late as Mio-Pliocene time, rather than being the result of hard rocks resistant to weathering following earlier uplift (Fig. 1). The high relief consists of bevelled summits and wide bevelled tablelands deeply dissected by glacial erosion. Two small outliers, with up to 10 m of basal Carboniferous quartz breccias, conglomerates and sandstones lie unconformably on Ordovician rocks in the footwall of the Erriff Fault at an altitude of *c.* 650 m near Maumtrasna summit in the Partry Mountains (see arrow in Fig. 1). These rocks are identical to basal Carboniferous beds near Clonbur and on the northern shores of Clew Bay (at sea level). Dewey & McKerrow (1963) showed how the plateau surface beneath the Carboniferous outliers can be traced eastwards to continue directly below the basal Carboniferous sandstones north of Tourmakeady on the west side of Lough Mask, thereby demonstrating that the plateau surface, on a regional scale, approximates the exhumed base of the Carboniferous succession. Only in the immediate vicinity of the outliers does the plateau surface coincide exactly with the basal Carboniferous level. Elsewhere erosion has lowered the plateau surface to a few metres below, but essentially subparallel with, the basal Carboniferous level. The contours in Fig. 1 (after Dewey & McKerrow 1963, fig. 4) show the reconstructed sub-Carboniferous surface based on interpolation of the basal Carboniferous level from areas of outcrop, the bevelled plateau surfaces and, to the SW, the Dalradian mountain peaks in Connemara. The two profiles in Fig. 2 (after cross-sections 5 and 6 in fig. 2 of Dewey & McKerrow 1963) illustrate the interpreted relationship between present-day topography and the reconstructed sub-Carboniferous surface. Although significant glacial erosion has occurred, the intersections of the actual and projected sub-Carboniferous surface with the topographic profile represent uneroded remnants of the sub-Carboniferous surface.

From: SHANNON, P.M., HAUGHTON, P.D.W. & CORCORAN, D.V. (eds). 2001. *The Petroleum Exploration of Ireland's Offshore Basins*. Geological Society, London, Special Publications, **188**, 201–207. 0305-8719/01/$15.00

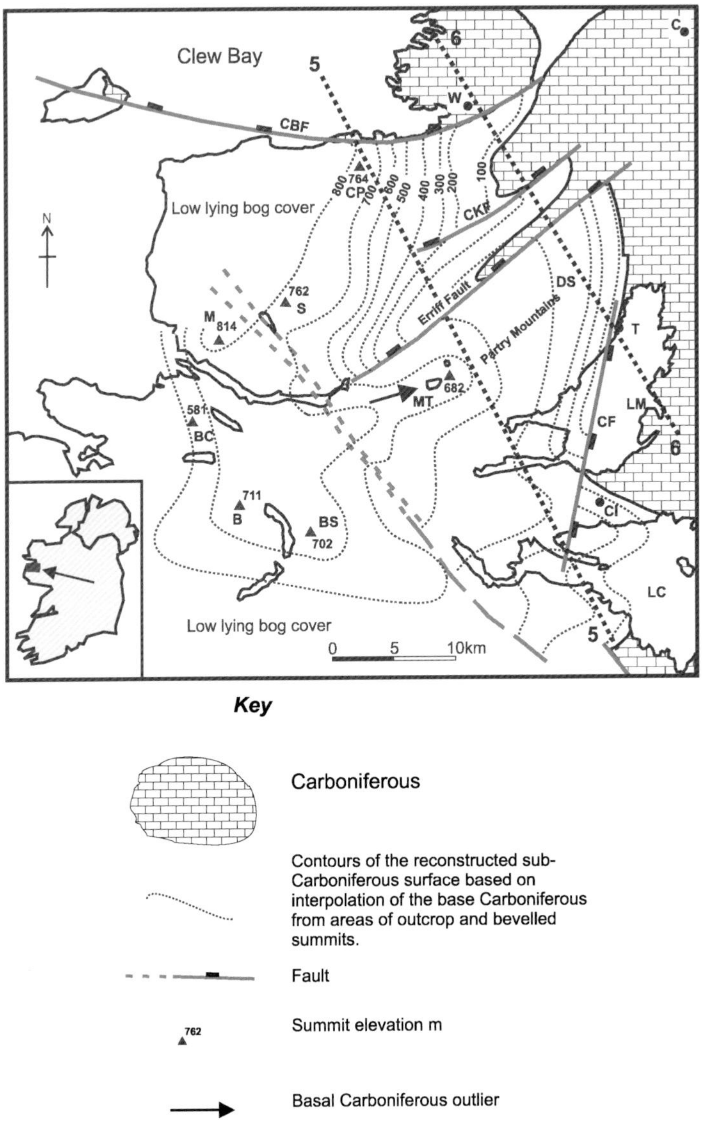

Fig. 1. Structure map of the reconstructed sub-Carboniferous surface in Connemara and south Mayo, western Ireland, after Dewey & McKerrow (1963). Dotted contours (in metres) represent isopleths on the sub-Carboniferous surface interpolated from the base of the Carboniferous outcrop (shaded grey) and bevelled summits. The locations of profiles 5 and 6 of Dewey & McKerrow (1963) are also indicated. B, Bencorr; BC, Benchoona; BS, Barrslievenaroy; C, Castlebar; CBF, Clew Bay Fault; CF, Clonbur Fault; CKF, Carrowkennedy Fault; Cl; Clonbur; CP, Crough Patrick; DS, Droimchogaidh Sill; LC, Lough Corrib; LM, Lough Mask; M, Mweelrea; MT, Maumtrasna; S, Sheeffry Hills; T, Tourmakeady; W, Westport.

Faults affecting the sub-Carboniferous surface

The map and cross-sections (Figs 1 and 2) show that significant normal faulting has affected the basal Carboniferous surface. In the north the Erriff, Carrowkennedy and Clew Bay faults successively downthrow the basal Carboniferous beds in a northward direction. A normal displacement of *c*. 520 m on the Erriff Fault can be estimated directly from the elevation difference between the basal Carboniferous beds in the footwall near Maumtrasna and in hanging-wall exposures in the valley to the NW. A minimum normal throw of >120 m at basal Carboniferous level is predicted on the Carrowkennedy Fault

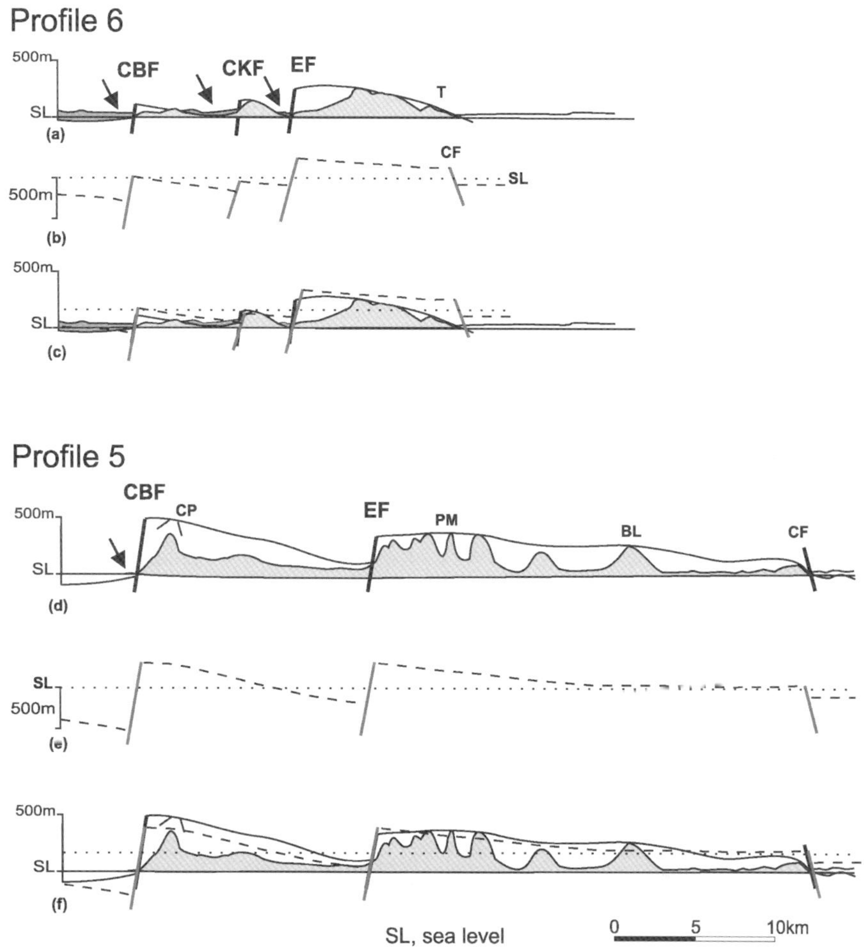

Fig. 2. Profiles 5 and 6 of Dewey & McKerrow (1963) showing the reconstructed sub-Carboniferous surface, forward modelled profiles using the 'flexural cantilever' method, and comparison of reconstructed and modelled profiles. The location of the profiles is shown in Fig. 1. Arrows indicate Carboniferous outcrop in the hanging walls of the Clew Bay, Carrowkennedy and Eriff faults. (**a**) Profile 6 of Dewey & McKerrow (1963) of the reconstructed sub-Carboniferous surface. The intersections of the reconstructed sub-Carboniferous surface with the topographic profile represent unconsumed remnants of the sub-Carboniferous surface. (**b**) Forward model of the sub-Carboniferous surface of Profile 6 using the flexural cantilever model. Throws on the faults in (**a**) have been converted to heaves assuming an original 60° fault dip. (**c**) Superimposed reconstructed and modelled profiles after the datum of the modelled profile has been shifted upwards by 208 m to compensate for long-wavelength isostatic uplift accompanying removal of material by post-faulting glacial erosion. (**d**) Profile 5 of Dewey & McKerrow (1963) of the reconstructed sub-Carboniferous surface. The intersections of the reconstructed sub-Carboniferous surface with the topographic profile represent unconsumed remnants of the sub-Carboniferous surface. (**e**) Forward model of the sub-Carboniferous surface of Profile 6 using the flexural cantilever model. Throws on the faults in (**a**) have been converted to heaves assuming an original 60° fault dip. (**f**) Superimposed reconstructed and modelled profiles after the datum of the modelled profile has been shifted upwards by 208 m to compensate for long-wavelength isostatic uplift accompanying removal of material by post-faulting glacial erosion. BL, Ben Levy; CBF, Clew Bay Fault; CF, Clonbur Fault; CKF, Carrowkennedy Fault; CP, Crough Patrick; PM, Party Mountains, T, Tourmakeady.

and a minimum throw of 850 m on the Clew Bay Fault. For these faults the minimum throws are estimated from outcrops of basal Carboniferous beds in their hanging walls and the extrapolated sub-Carboniferous surface based on summit levels. Evidence for the magnitude of offset on the Clonbur Fault, the easternmost fault in the cross-sections, is more difficult to determine.

Basal Carboniferous beds are present to the immediate NW of Lough Mask and Carboniferous beds within Lough Mask. However, the outcrop pattern is of no further help in defining the location of the putative Clonbur Fault. McManus (1967) documented small faults (of a few metres throw) on the eastern flank of the Partry Mountains with Carboniferous sediments in their hanging walls, but his observations fail to constrain the location of the Clonbur Fault.

No unequivocal evidence is available to date the timing of faulting or to determine whether the observed offsets are the cumulative effects of several episodes of faulting, but a Tertiary age for the main component of faulting is deduced from the following observations. Dewey & McKerrow (1963) based a Tertiary age for the faulting in the area on the circumstantial evidence of a few metres offset of a Tertiary teschenite dyke by a fault on Benchoona, south of Killary Harbour. They considered that the large NNE–SSW-trending faults are of the same, Tertiary, age. The Droimchogaidth Sill, dated at 55 $\pm$ 1 Ma, is exposed at the plateau surface in the footwall of the Erriff Fault. It was intruded at a depth of $>$100 m (Mohr 1982) and the plateau surface here must be younger than the sill. In the hanging wall of the Erriff Fault, near Castlebar, a Tertiary sill is exposed within the Lower Carboniferous succession. The rock surface here must be younger than the sill and older than the Quaternary deposits that rest on the bedrock surface. This suggests that the Erriff Fault displaces a post-55 Ma, but pre-Quaternary, surface. On a more regional basis there is also extensive evidence of both Tertiary age faulting and landform evolution. In Northern Ireland, 200 km to the NE, George (1967) presented unequivocal evidence of significant faulting extending into, at least, Oligocene time. The base of Paleocene basalts is downthrown to the SE by $>$3 km across the NE Lough Neagh Basin Fault. Oligocene sediments are also preserved in the hanging wall of the Tow Valley Fault in Antrim (Fyfe *et al.* 1993). George (1967) convincingly demonstrated a Neogene age origin for the landforms in Northern Ireland.

There is also evidence offshore of some late Tertiary faulting, although it appears to be fairly localized. For example, to the NW of south Mayo, 100 km offshore in the Slyne Basin, Dancer *et al.* (1999) showed seismic data with young faulting interpreted to offset a prominent truncational unconformity of Miocene age. Shannon *et al.* (1999, fig. 5) showed a Tertiary fault to the east of the Galway Graben, now named the Cillian Basin by Naylor *et al.* (1999), on the Porcupine High. It is likely that most of the extensive fault systems in the offshore basins, however, were too deeply buried to reactivate easily in late Tertiary time.

The observation that the landscape in the area is essentially of Neogene age (George 1967), coupled with the disposition of the basal Carboniferous bed in the footwall and hanging wall of the Erriff Fault and the preservation of the sub-Carboniferous surface (where not locally removed by recent glacial erosion), suggests strongly that faulting was relatively recent (but pre-glacial). The possibility that a component of the faulting is of Mesozoic age cannot be ruled out and would not invalidate the modelling results described below. Widespread mid-Mesozoic faulting is well documented both offshore of western Ireland (Shannon *et al.* 1999) and also onshore in Northern Ireland (George 1967). The role of Mesozoic faulting in landform evolution, however, is more problematic. The offset of the basal Carboniferous surface across the Erriff Fault gives an exact estimate of fault-related relief. It seems implausible to suggest that this topography could either have persisted since Mesozoic time or alternatively for erosion of an uplifted topography to coincidentally reveal the basal Carboniferous surface in both the footwall and hanging wall of the Erriff Fault. In the absence of clear evidence to the contrary, the data support the conclusions of George (1967) that Irish landscape is mainly of Neogene origin.

Faulting and topography

When the papers by Dewey & McKerrow (1963) and George (1967) were written, the association of footwall uplift and topography accompanying normal faulting had not been documented. Since then, the occurrence of both footwall uplift and hanging-wall subsidence as a consequence of normal faulting is well known and researched (see summaries by Yielding & Roberts 1992; Roberts & Yielding 1994). Quantitative modelling techniques, developed and applied widely to forward model normal faults, have proven to be reliable predictors of both magnitude and shape of topography associated with normal faults (King *et al.* 1988). The flexural cantilever method (Kusznir & Ziegler 1992; Kusznir *et al.* 1995) is especially suited to forward modelling the topography associated with large basement-involved faults. The approach calculates crustal-scale, faulting-induced strain through a coupled simple–pure shear model. The method incorporates isostatic loads and integrates the 'waveform-type' strain resulting from contemporaneous movement on sets of adjacent faults. Destructive or constructive interference of the finite motion on adjacent faults occurs where they are spaced closer than one-half wavelength (a function of crustal properties) and results in enhanced or reduced footwall uplift or hanging-wall subsidence. For example, forward models using the method explain well the occurrence of deeply eroded horsts. Predictions of the magnitude of emergent topography and the consequent spatial and temporal occurrence of erosional unconformities in the footwalls of fault blocks, within otherwise water-filled extensional basins, also match observations (e.g. Roberts *et al.* 1993; Berger & Roberts 1999).

This forward modelling approach has been applied to the profiles shown in Fig. 2 to investigate whether the topography revealed by the sub-Carboniferous datum surface can be explained as a consequence of footwall uplift accompanying normal faulting. The modelling sums the cumulative effects of many faulting episodes into a single step and would still be valid even if some of the fault displacement had occurred in mid-Mesozoic time. Figure 2a and d shows present-day topography and the reconstructed sub-Carboniferous surface of Dewey & McKerrow (1963). Figure 2b and c shows the forward modelled profiles using the flexural cantilever forward modelling approach. In the forward model the faults are assumed to dip at 60°. Observed throws have been converted to heaves accordingly. The models used a crustal thickness of 32 km and an effective elastic thickness of 3 km, a value that other modelling studies have indicated to be appropriate for continental crust of average thickness and normal temperature profile (e.g. Kusznir & Ziegler 1992; Hendrie *et al.* 1994). The original datum (sub-Carboniferous surface) was assumed to have been horizontal and at, or near, sea level before faulting. As expected following crustal extension, the hanging walls are below the original datum (the model has assumed that these hanging-wall lows are water filled and isostatically compensated). The profiles do not match present-day topography, being too low in elevation, but they do have a similar shape to the sub-Carboniferous surface shown in Fig. 2a and d. For example, the modelled Profile 5 (Fig. 2e) shows hanging walls of both the Clew Bay and Erriff faults below sea level, but their elevation difference is similar to that in Fig. 2d. The modelled predictions for Profile 6 show a similar correspondence of hanging-wall elevational differences between model and actual profile, although the elevations do not match present-day topography (Fig. 2a and b). In addition to creating local footwall islands, faulting also causes the submergence of large areas of low-lying topography in the hanging walls of faults. This occurs through a combination of local deformation around the faults and regional subsidence resulting from isostatic compensation accompanying the fault-related extension. Overall the area of land above sea level decreases, counteracting the increase in local topography at the footwall islands. Consequently, it is unlikely that local onshore faulting will have led to markedly greater rates of sediment supply in the offshore basins.

Glacial sea-level fall, erosion and isostatic response

It is presumed that faulting (mainly, but not necessarily exclusively, of Tertiary age) was the cause of differential topography that was present long before the onset of the Quaternary ice ages. All of the fault scarps show evidence of extensive modification by glacial erosion (Dewey & McKerrow 1963) and presumably the emergent footwalls also suffered erosion in late Tertiary time before the glaciations. Glacial erosion not only modified geomorphology but will also, through the removal of rock, have produced an isostatic uplift. Molnar & England (1990) discussed and quantified the link between increased erosion as a result of climate change and its isostatic effects on topography. These concepts have been used to investigate the isostatic uplift resulting from erosion accompanying glaciation. In considering the likely isostatic effect of glacial erosion on landscape evolution the change in base level accompanying ice age sea-level falls is a key consideration. Sea-level fall will expose previously submerged areas of both footwall and hanging wall, facilitating erosion and isostatic uplift. The Quaternary period has experienced several major glacial episodes, each accompanied by a sea-level fall of *c.* 120 m (Peltier 1998). Isostatic uplift calculations, however, must take into account the cumulative effects of erosion accompanying each sea-level fall. Isostatic uplift following the removal of material during the first ice age will have been followed by further sea-level fall, erosion and further isostatic uplift with each ice age. Lowering the base level increases the volume of rock vulnerable to erosion and so enhances the potential amount of isostatic uplift.

It is difficult to estimate the base-level adjustment needed to take into account the cumulative effects of successive ice ages each accompanied by a 120 m sea-level fall, as erosion has clearly not removed all emergent topography.

Isostatic calculations indicate that lowering base level by only 120 m would predict present-day submerged hanging walls and insufficient emergent topography. The cumulative base-level fall of 200 m used in the calculations is sufficient to make both footwall and hanging wall emergent and subaerial. A larger fall would not significantly increase the erosion potential of the landscape.

Profile 5 (Fig. 2d) of Dewey & McKerrow (1963), which shows present-day topography and the reconstructed sub-Carboniferous surface, has been used to estimate the magnitude of the isostatic uplift. In the 42 km of profile between the Clew Bay Fault and the Clonbur Fault it is estimated that around 10.3 km^2 of rock has been removed beneath the sub-Carboniferous surface. The lateral scale of the present-day topographic irregularities (Fig. 2) is very small compared with the tens of kilometre wavelengths over which isostatic rebound occurs and can be ignored in isostatic calculations. The 10.3 km^2 of eroded rock distributed evenly over the entire 42 km profile equates to a uniform layer of about 250 m and provides an estimate of the average amount of erosion along the profile. Removing 250 m of crust with a density of 2750 kg m^{-3}, and assuming a mantle density of 3300 kg m^{-3}, will result in a net isostatic uplift of 208 m (Molnar & England 1990). Ignoring the self-cancelling short-term effects of ice loading and then melting, the calculation above indicates that the profiles shown in Fig. 2b and e should be adjusted to a new datum *c.* 208 m shallower. Figure 2c and f shows the result of superimposing the modelled profiles, datum shifted by 208 m, onto profiles of present-day topography and the reconstructed sub-Carboniferous surface. The effect of this datum adjustment is to bring the modelled and reconstructed sub-Carboniferous surface into reasonable agreement, indicating that footwall uplift accompanying normal faulting is a plausible cause of the uplift evidenced by the sub-Carboniferous surface.

Conclusions

Both shape and topography (after adjustments for erosion-related isostatic effects) predicted by the forward models match well with the reconstructed sub-Carboniferous surface. This indicates that footwall uplift and hanging-wall subsidence accompanying (predominantly Late Tertiary) faulting could be the primary cause of the topography of the basal Carboniferous surface adjacent to the Erriff, Carrowkennedy and Clew Bay faults. It is unlikely that the faulting will have led to greater rates of sediment supply in the offshore basins. First, the footwalls appear to have experienced most erosion in Quaternary time, and, second, extensive areas of hanging wall will have been submerged by the regional subsidence resulting from isostatic compensation accompanying the fault-related extension.

The shorter-wavelength fault-related topography has been further modified by recent longer-wavelength (tens of kilometres) uplift associated with isostatic compensation following dissection of the sub-Carboniferous surface during Pleistocene glaciations. This uplift has enhanced topographic relief and elevated the Erriff and Carrowkennedy hanging walls above sea level.

I am very grateful to my colleagues K. Baxter and A. Roberts for valuable discussion and assistance, and to N. Kusznir for use of his forward modelling software STRETCH. The paper benefited greatly from constructive reviews and comments by D. Naylor and A. Phillips.

References

Berger, M. & Roberts, A.M. 1999. The Zeta Structure; a footwall degradation complex formed by gravity sliding on the western margin of the Tampen Spur, Northern North Sea. *In*: Fleet, A.J. & Boldy, S.A.R. (eds) *Petroleum Geology of Northwest Europe: Proceedings of the 5th Conference*. Geological Society, London, 107–116.

Dancer, P.N., Algar, S.T. & Wilson, I.R. 1999. Structural evolution of the Slyne Trough. *In*: Fleet, A.J. & Boldy, S.A.R. (eds) *Petroleum Geology of Northwest Europe: Proceedings of the 5th Conference*. Geological Society, London, 445–453.

Dewey, J.F. & McKerrow, W.S. 1963. An outline of the geomorphology of Murrisk and north-west Galway. *Geological Magazine*, **100**, 260–275.

Ellis, M.A., Densmore, A.L. & Anderson, R.S. 1999. Development of mountainous topography in the Basin Ranges, USA. *Basin Research*, **11**, 21–41.

Fyfe, J.A., Long, D., Evans, D. 1993. *United Kingdom offshore regional report: the geology of the Malin–Hebrides area*. British Geological Survey & HMSO, London.

George, T.N. 1966. Geomorphic evolution in Hebridean Scotland. *Scottish Journal of Geology*, **2**, 1–34.

George, T.N. 1967. Landforms and structure in Ulster. *Scottish Journal of Geology*, **3**, 413–448.

Hendrie, D.B., Kusznir, N.J., Morley, C.K. & Ebinger, C.J. 1994. Cenozoic extension in northern Kenya: a quantitative model of rift basin development in the Turkana region. *Tectonophysics*, **236**, 409–438.

Japsen, P. 1998. Regional velocity–depth analysis, North Sea Chalk: a record of overpressure and

Neogene uplift and erosion. *AAPG Bulletin*, **82**, 2031–2074.

KING, G.C.P., STEIN, R.S. & RUNDLE, J.B. 1988. The growth of geological structures by repeated earthquakes, 1. Conceptual framework. *Journal of Geophysical Research*, **93**, 13307–13318.

KUSZNIR, N.J. & ZIEGLER, P.A. 1992. The mechanics of continental extension and sedimentary basin formation: a simple-shear/pure-shear flexural cantilever model. *Tectonophysics*, **215**, 117–131.

KUSZNIR, N.J., ROBERTS, A.M. & MORLEY, C.K. 1995. Forward and reverse modelling of rift basin formation. *In*: LAMBIASE, J.J. (ed.) *Hydrocarbon Habitat in Rift Basins*. Geological Society, London, Special Publications, **80**, 33–56.

MCMANUS, J. 1967. Faulting of the sub-Carboniferous surface in eastern Murrisk, Co. Mayo. *Geological Magazine*, **104**, 228–231.

MOHR, P. 1982. Tertiary dolerite intrusions of West–Central Ireland. *Proceedings of the Royal Irish Academy, 82B*, 53–82.

MOLNAR, P. & ENGLAND, P. 1990. Late Cenozoic uplift of mountain ranges and global climate change: chicken or egg? *Nature*, **346**, 29–34.

NAYLOR, D., SHANNON, P., MURPHY, N. 1999. *Irish Rockall Basin region—a standard structural nomenclature system*. Petroleum Affairs Division, Special Publication, **1/99**.

PELTIER, W.R. 1998. Global glacial isostatic adjustment and coastal tectonics. *In*: STEWARD, I.S. & VITA FINZI, C. (eds) *Coastal Tectonics*. Geological Society, London, Special Publications, **146**, 1–29.

ROBERTS, A.M. & YIELDING, G. 1994. Continental extensional tectonics (a review chapter). *In*: HANCOCK, P.L. (ed.) *Continental Deformation*. Pergamon, Oxford, 223–250.

ROBERTS, A.M., YIELDING, G., KUSZNIR, N.J., WALKER, I. & DORN-LOPEZ, D. 1993. Mesozoic extension in the North Sea: constraints from flexural backstripping, forward modelling and fault populations. *In*: PARKER, J.R. (ed.) *Petroleum Geology of Northwest Europe: Proceedings of the 4th Conference*. Geological Society, London, 1123–1136.

SHANNON, P.M., JACOB, A.W.B., O'REILLY, B.M., HAUSER, F., READMAN, P.W. & MAKRIS, J. 1999. Structural setting, geological development and basin modelling in the Rockall Trough. *In*: FLEET, A.J. & BOLDY, S.A.R. (eds) *Petroleum Geology of Northwest Europe: Proceedings of the 5th Conference*. Geological Society, London, 421–431.

STEIN, R.S., KING, G.C.P. & RUNDLE, J.B. 1988. The growth of geological structures by repeated earthquakes, 2, Field examples of continental dip-slip faults. *Journal of Geophysical Research*, **93**, 13319–13331.

YIELDING, G. & ROBERTS, A. 1992. Footwall uplift during normal faulting—implications for structural geometries in the North Sea. *In*: LARSEN, R.M., BREKKE, H., LARSEN, B.T. & TALLERAAS, E. (eds) *Structural and Tectonic Modelling and its Application to Petroleum Geology*. Norwegian Petroleum Society (NPF) Special Publication, **1**, 289–304.

Exploring in the Slyne Basin: a geophysical challenge

P. NICK DANCER[1] & NICK W. PILLAR[2]

[1]*Enterprise Oil plc, 4th Floor Embassy House, Herbert Park Lane, Ballsbridge, Dublin 4, Ireland (e-mail: Nick.Dancer@DUBLIN.EntOil.com)*

[2]*Enterprise Oil plc, Grand Buildings, Trafalgar Square, London WC2N 5EJ, UK*

Abstract: The Slyne Basin lies *c.* 60 km offshore west of Ireland, in water depths of 200–500 m. It consists of three asymmetric half-graben that are separated by complex structural transfer zones. Sporadic exploration in the basin over the last 20 years has resulted in the drilling of four exploration wells, which have yielded one gas discovery. Well 18/20-1 (Corrib) successfully tested a faulted anticlinal structure and encountered gas in the Triassic Sherwood Sandstone Formation. Although a number of other potential hydrocarbon traps have been identified in the Slyne Basin, the poor quality of the seismic data, plus the presence of complex transfer zones, has generated considerable uncertainty with respect to the correlation of seismic markers. A primary control on the seismic data quality is the presence of near-sea-bed, high-velocity, Tertiary volcanic and Cretaceous chalk layers. These result in very strong and long multiple trains, energy scattering, mode conversion and attenuation. Studies suggest that improved signal penetration can be achieved when the seismic acquisition is focused on the low-frequency end of the spectrum. However, predictive multiple attenuation has proved ineffective because of the complex nature of the multiple generators. An approach based on detailed velocity analysis and the judicious parameterization of more than one pass of Radon demultiple has yielded good results. This approach, coupled with 3D acquisition and processing with its inherent increase in signal-to-noise ratio, has led to a dramatic improvement in the seismic data quality in the Corrib area.

Large tranches of the Atlantic margin, west of Ireland, are lightly explored. Although the first seismic data in the Slyne area were acquired in 1970, and subsequent seismic surveys were acquired throughout the 1970s, only one well had been drilled in the Slyne Basin before 1996. Well 27/13-1 was drilled by Elf in 1981 and reached a total depth (TD) of 2725 m MDBRT (measured depth below rotary table) in Rhaetian sediments (Fig. 1). Although this well encountered oil shows in good quality sandstone reservoirs of Mid-Jurassic age, and excellent quality Lower Jurassic source rocks (Scotchman & Thomas 1995), no further drilling occurred in the area for 15 years.

In 1996, Enterprise Oil and partners, Santa Fe and Statoil, drilled two exploration wells in the Slyne Basin. The first well, 27/5-1, encountered significant oil shows in Middle Jurassic sandstones, and penetrated a Triassic interval, equivalent to the Sherwood Sandstone Group (SSG), a Zechstein Halite equivalent, and a sequence of sandstones, coals and shales of Westphalian B age. The second well, 18/20-1, which was planned as a deviated well, again encountered oil shows in the Middle Jurassic sequence, but also penetrated a significant gas column in the SSG. Well 18/20-1 was plugged and abandoned as a gas discovery (named Corrib). Encouraged by this success, a 660 km^2 3D seismic survey (E97IE11) was acquired and processed during 1997 (Fig. 1). The interpretation of this survey led to the drilling of the Corrib appraisal well 18/20-2 in 1998. This well was sidetracked for operational reasons, the sidetrack (18/20-2z) reaching a TD of 3730 m MDBRT. The gas column encountered in the SSG was successfully tested at a stabilized rate of 63 MMSCFD (million standard cubic feet per day) through a 2 inch choke. Subsequent to this well, a subset of the Corrib 3D dataset was reprocessed to produce a post-stack depth migration (PostSDM) volume. Utilizing this improved dataset, a second appraisal well, 18/25-1, was drilled on the Corrib discovery during 1999. This well reached a TD of 3741 m MDBRT and tested gas from the SSG at a rate of 64 MMSCFD through a 11/4 inch choke.

The Corrib 3D dataset also covered a separate tilted fault-block structure (the Shannon prospect), located *c.* 12 km to the south of Corrib. Exploration well 18/25-2 was drilled on this

From: SHANNON, P.M., HAUGHTON, P.D.W. & CORCORAN, D.V. (eds). 2001. *The Petroleum Exploration of Ireland's Offshore Basins*. Geological Society, London, Special Publications, **188**, 209–222. 0305-8719/01/$15.00

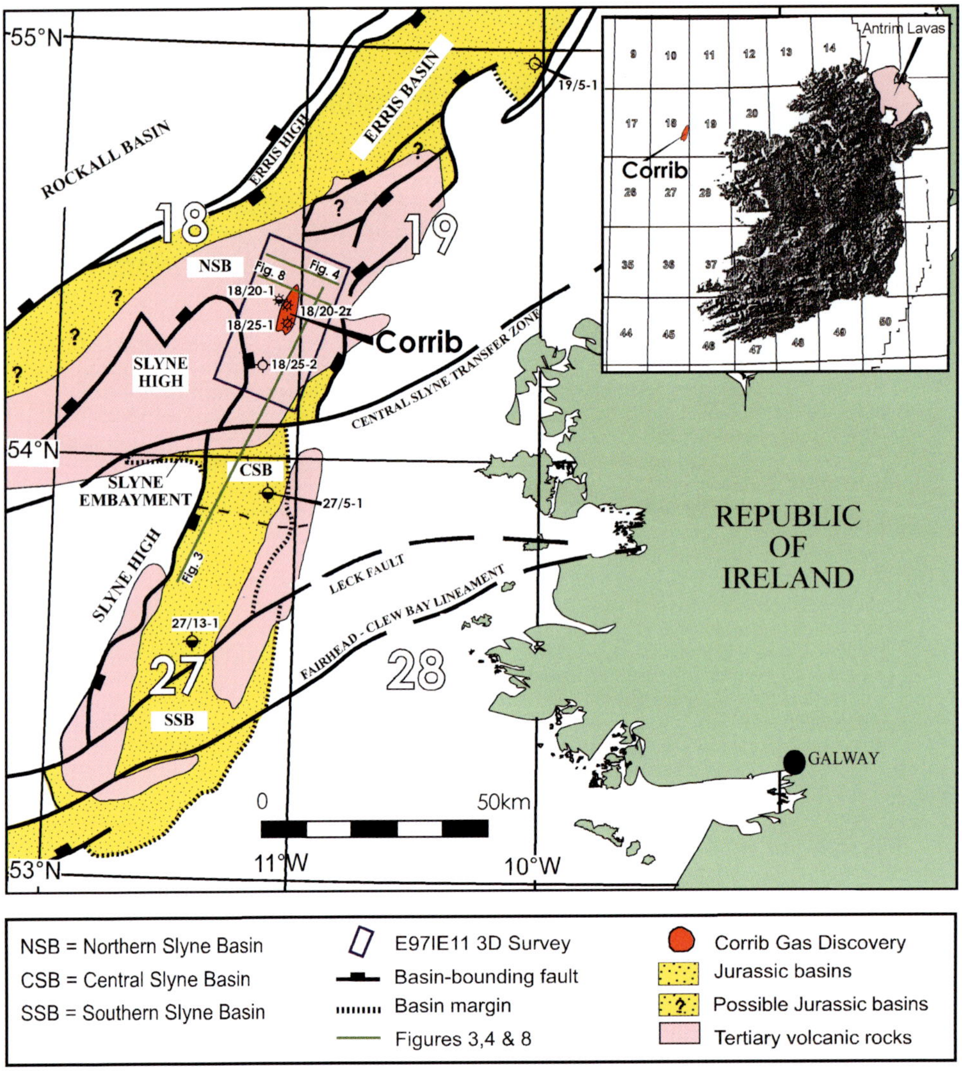

Fig. 1. Simplified tectonic elements map of the Slyne Basin. The distribution of near-surface Tertiary volcanic rocks is shown (based on aeromagnetic and well data) with respect to the location of the nearest outcrop analogues, the Antrim lavas. Also shown are the locations of the seismic profiles referred to in the text.

structure in 1999 but the well was plugged and abandoned as a dry hole.

Constraints on exploration in the Slyne Basin

A number of geological and logistical factors have hindered exploration in the Slyne Basin. First, the quality of the seismic data is extremely poor throughout parts of the basin. Second, the Slyne Basin is transected by a number of strike-slip faults and complex structural transfer zones (Trueblood & Morton 1991), which generate considerable uncertainty with respect to the correlation of seismic markers. Third, the water depth in the area, ranging from 200 to 500 m, has until recently been considered too deep for oil and gas developments in the harsh physical environment of the NE Atlantic margin. Fourth, the prevailing climatic conditions of the Atlantic seaboard constrain both seismic and drilling activity to a weather window of April–September. The time constraints imposed by this weather window present a significant challenge for the planning, budgeting and

Fig. 2. Escarpment at Garron Point, County Antrim, illustrating an erosional chalk topography infilled by basalt. Intra-basalt layering is also observed as a result of the development of thin palaeosols between the lava flows. (Telegraph pole in foreground for scale.)

execution of exploration programmes. Seismic data acquired during the summer months may have to be processed and interpreted in parallel with the decision-making and well-planning efforts for the following drilling season.

Since 1970, some 28 2D seismic surveys and one 3D seismic survey (E97IE11) have been acquired in the Slyne area. Most of the data can be described as fair to poor in quality. The primary reason for the poor quality of the seismic data is a series of geological conditions that contribute to a degradation of the seismic signal. In addition, the water depth range in the Slyne Basin results in multiples that have a period of 400–500 ms throughout the seismic record.

The near-sea-bed geology is a significant control on the quality of the seismic data in the Slyne area. Underlying a 20–100 m interval of soft sea-bed sediments (of Late Miocene to Recent age) lies an interval of stacked volcanic deposits, which have been locally dated to Eocene time (Dancer *et al.* 1999). Interpretation of aeromagnetic data suggests that this volcanic unit is widespread, covering an area of $>2500\,km^2$ in the northern Slyne and southern Erris basins (Fig. 1). The upper surface of this volcanic unit generates strong multiples as well as refractions and mode-converted energy. In addition, the upper surface of the volcanic rocks is likely to be highly rugose as a result of subaerial exposure and weathering. Several distinct periods of eruption are inferred from the magnetic phase reversals recorded by these lavas. Discrete lava flows are likely to be separated by interbedded weathered layers. By analogy with outcrops from County Antrim, individual lava flows are heterogeneous and contain massive jointed sections and vesicular flow bands (Wilson & Manning 1978). This heterogeneity, combined with the surface and internal rugosity of the flows, leads to increased scattering and absorption of the seismic energy near the sea bed.

The base of the volcanic unit is also an irregular surface. In County Antrim basaltic lavas are observed to infill a palaeo-topography of eroded chalk (Fig. 2). Intra-basalt layering is also observed as a result of the development of thin palaeosols between intermittent lava flows. This irregular top chalk surface, and the weathered layers, are also detrimental to the propagation of coherent seismic energy. Furthermore, the chalk encountered by drilling in the Slyne Basin differs from the contemporaneous chalk of the North Sea. Chalks from County Antrim and the Slyne Basin are characterized by an exceptional hardness, which results from diagenetic recrystallization of calcite in the original matrix pore space and a relatively high flint content (Wilson & Manning 1978). In the Slyne Basin, the high sonic velocity and density of the chalk creates a significant impedance contrast with the underlying, lower impedance, stratigraphic units (Lower Cretaceous–Middle Jurassic sandstones and shales). This impedance inversion generates another series of multiples in the seismic data.

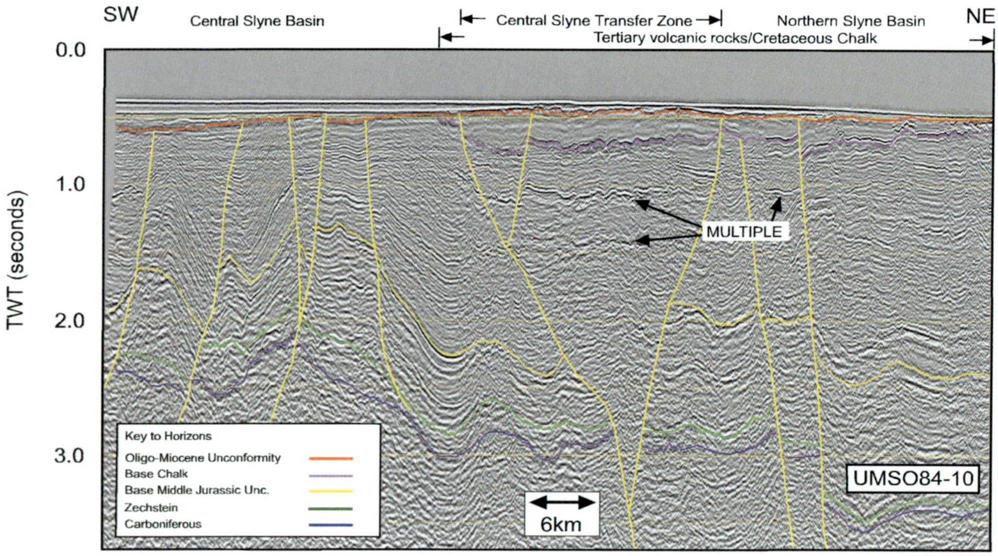

Fig. 3. A 2D seismic profile, UMSO84-10, along the strike of the Slyne Basin. The degradation of data quality beneath the Tertiary volcanic rocks should be noted, particularly in the area of the Central Slyne Transfer Zone. Water bottom multiples generated from the base of the chalk layer are prominent in the area of the transfer zone, between 1.0 and 1.5 s. Significant post-Oligocene inversion of the Central Slyne Transfer Zone should also be noted.

These near-surface volcanic and chalk layers are a particular problem in the Northern Slyne Basin, although near-surface volcanic rocks also occur in the Southern Slyne Basin (Fig. 1). However, in the Central and Southern Slyne basins, zones of poor-quality seismic data are predominantly caused by the presence of structurally complex transfer zones (Fig. 3). These zones are characterized by high noise content and a lack of coherent reflectors, which results in uncertain correlation of seismic markers across these zones. In addition, gas chimneys may be localized along some of these transfer zones, thus contributing to a degradation of the seismic data. Finally, the presence of igneous sills in the Middle Jurassic section masks the deeper reflectors that are used to map Triassic (SSG) prospects.

Understanding the acquisition and processing challenges

Improved seismic data quality was recognized as an essential prerequisite for the definition of drillable prospects in the Slyne Basin. Selective reprocessing of older vintages of seismic data led to some improvement in data quality, although in the very poor data areas, little improvement was observed. This suggested that a revised acquisition strategy would be necessary to improve the data. In 1993, a trial 2D programme was undertaken (in a fair-quality data area, without near-surface volcanic rocks), with a variety of acquisition parameters, to optimize acquisition for future larger-scale surveys. This testing confirmed that an increase in fold of coverage significantly improved the signal-to-noise ratio, and that fold of coverage was the most significant acquisition parameter. In 1994 a high-fold acquisition programme was undertaken in the area. Results indicated that the 2D lines that were acquired through near-surface basalts were poor in quality relative to the results subsequently achieved via 3D acquisition and processing (Fig. 4a and b). This 1994 survey was acquired with a 18.75 m shot interval (12.5 m group interval, 240 channels, 3000 m streamer), as the contracted vessel could not acquire the desired record length using a 12.5 m shot interval (which had been used in the 1993 trial programme). This shot interval, after alternate trace drop (common processing practice) maintained a common mid-point (CMP) spacing of 6.25 m but reduced the fold to 40 for each CMP. Subsequent reprocessing with all traces (6.25 m CMP spacing, 80 fold) showed a

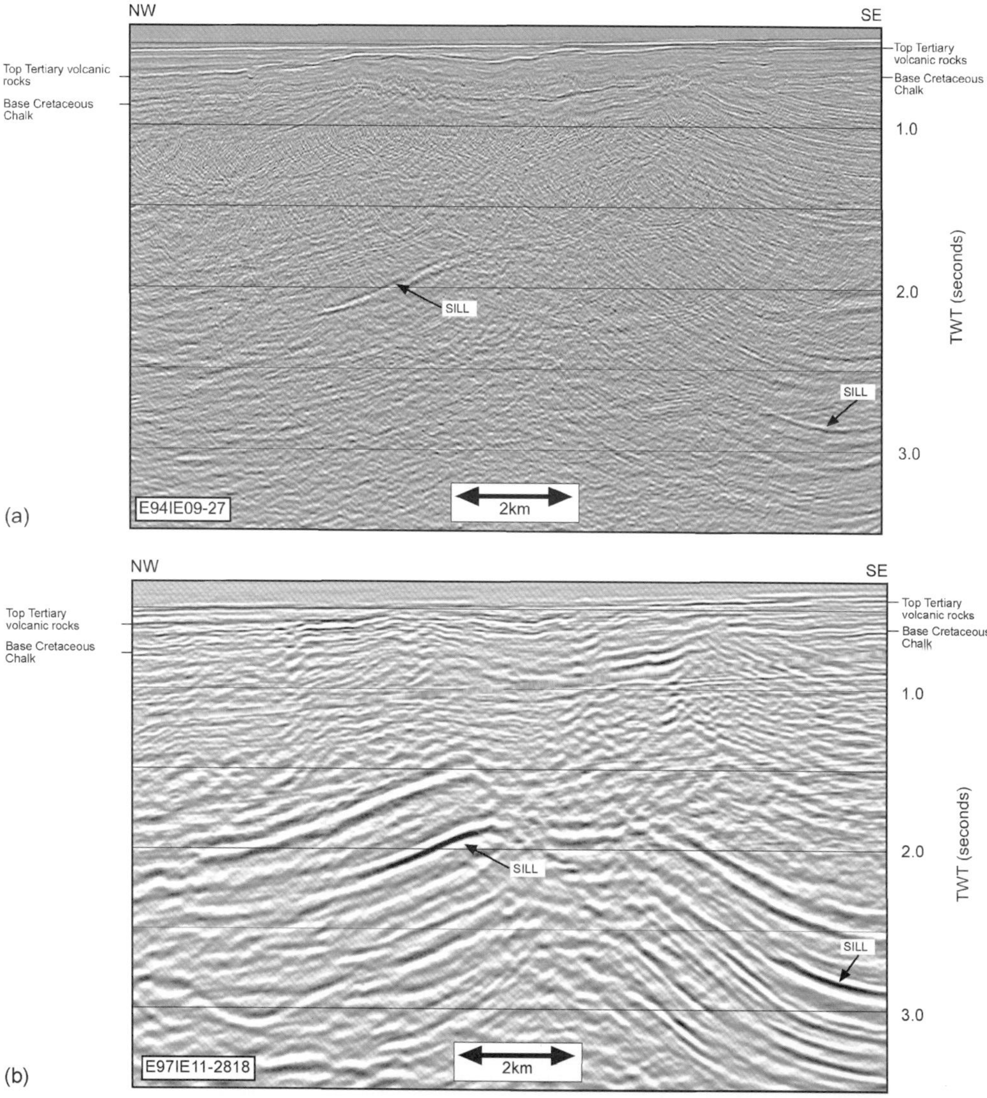

Fig. 4. Comparison of 2D and 3D seismic data across the Corrib structure. (**a**) A 2D seismic profile, E94IE09-27 (see Fig. 1 for location). The low signal-to-noise ratio and the pervasive migration 'smiles', which are a product of the over-migration of residual multiples not removed by the multiple attenuation routine, should be noted. Reliable events deeper in the section are dominated by reflections from discontinuous igneous sills. (**b**) A 3D seismic profile, Inline 2818, coincident with the 2D profile. The dominant frequency is lower than in the 2D section but there is a dramatic uplift in the signal-to-noise ratio below 1.5 s. Although some multiple energy remains in the 3D section, residual noise has been minimized where the multiple attenuation has been successful.

significant improvement over the original processed sections.

Subsequent to the drilling of the Corrib discovery well, 18/20-1, which was drilled in a window of fair-quality 2D seismic data, a series of processing tests were performed. These tests included 'full elastic' wavefield and ray-trace modelling in an attempt to establish the factors that give rise to the deterioration of the seismic data and to identify possible solutions to these problems. A 3D survey design and evaluation exercise was also carried out to determine the optimum acquisition parameters required for such a seismic programme.

Table 1. V_p (compressional wave velocity), Q (quality factor) and ρ (density) values for the plane-layered Corrib Earth model used in the 2D ray-tracing exercise

Stratigraphic interval	V_p (m s^{-1})	Q	ρ (g cm^{-3})
Sea surface to sea bed	1468		1
Sea bed to top volcanic rocks	1579	62000	1.7
Top volcanic rocks to Base Cretaceous Unconformity	2970	80	2.6
Base Cretaceous Unconformity to Bajocian Limestone	V0 = 2440 k = 0.5	400	2
Bajocian Limestone to Broadford Beds	V0 = 2750 k = 0.25	300	2.4
Broadford Beds to Top Mercia Group	V0 = 3550 k = 0.15	250	2.6
Top Mercia Group to Top SSG	4390	350	2.5

V_0 and k functions (V_0 is velocity at zero depth; k is compaction factor) were used to define the interval velocity for some layers. Q is a measure of the attenuation characteristics of the layers; high values represent low absorption and low values (such as the Top Volcanic rocks–Base Cretaceous Unconformity layer) indicate high absorption.

Modelling

A 2D ray-tracing exercise was performed on a geological model derived from a representative 2D seismic line across the Corrib structure. CMP gathers were simulated by ray tracing through this plane-layered Earth model. The model parameters are summarized in Table 1. The synthetic CMP gathers generated by this modelling were then compared with the actual CMP gathers, from the same relative position, along the 2D line (Fig. 5a and b). This ray-tracing model presents a 2D simulation of the primaries, simple multiples and peglegs of the water bottom, and volcanic and chalk layers, generated for all horizons in the model. 'Random noise' was then added to the record to better facilitate a comparison with the real records. However, mode conversion and out-of-plane effects were not incorporated in the ray-tracing model.

Marked differences are observed between the real and synthetic CMP gathers (Fig. 5a and b). On the synthetic CMP panel, primary energy as well as simple and pegleg multiples are discernible. In particular, there is a good primary reflector at 2.5 s, and although multiples are present, they do not dominate the CMP record (Fig. 5a). In contrast, on the real CMP records, the multiple energy is far more pervasive, dominating the record throughout, with a series of multiple events that reverberate with a period of 400 ms (Fig. 5b). Stronger refractions and multiples of these refractions are also present in the real CMP records. There is very little evidence for primary reflections and the strong event at 2.5 s on the synthetic CMP is not observed on the equivalent real CMP gather.

These results suggest that the model does not capture the complexity of the real subsurface geology. Although the macro-scale model is accurate (constrained by the seismic and well data) the micro-scale of the real Earth is not captured by the synthetic model. The internal heterogeneity and spatial variation of both the volcanic and chalk layers is a likely reason for this difference between model and real data.

From the real CMP gathers, it can be seen that it is the reverberations created by the shallow layers (volcanic rocks and chalk) that dominate the seismic record. However, the strength of the reverberation varies spatially: some of the CMPs show well-defined multiple events, whereas others are more noise dominated with the multiple periodicity less obvious. This lateral variation is a product of the near-sea-bed geology, reflecting variations in thickness and distribution of the lava flows. In addition, lithological and topographical variations will generate significant scattered noise in both the inline and crossline directions. Another factor affecting the strength of the multiples (and of the primary energy and the overall signal-to-noise ratio) is the amount of refraction. On the real CMP gathers, considerable variation in the amplitude of refractions is observed (Fig. 5b). Where refractions are strong, less energy is propagating through the near-surface layers and more energy is being trapped in the near surface.

Apart from the generation of significant multiple and refracted energy, these high-velocity near-surface layers are important with respect to mode conversion. To understand wavefield propagation, and in particular associated mode conversion through the near-surface

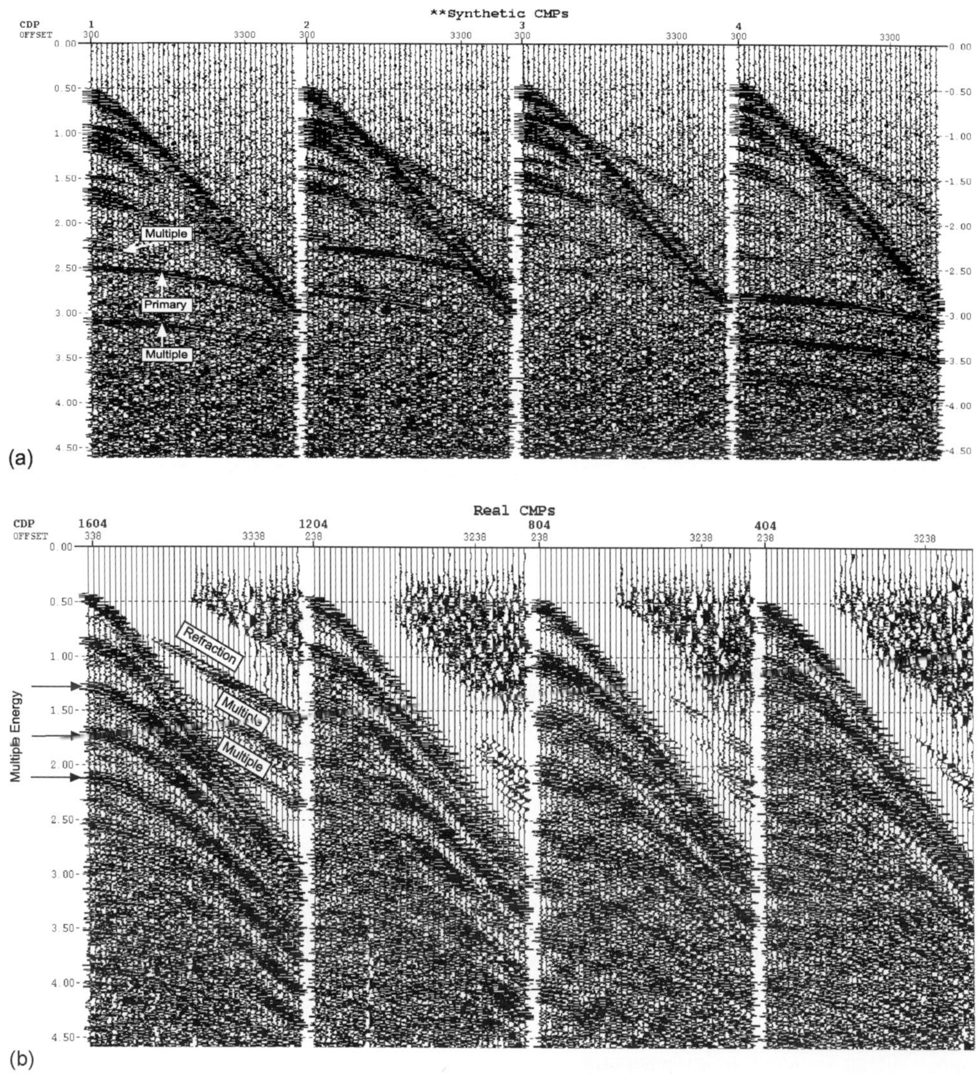

Fig. 5. Synthetic vs real CMP gathers, Corrib area. (**a**) Synthetic CMP gathers, without normal move-out (NMO) correction, showing regular multiples as well as primary reflections. These synthetic CMPs were generated via 2D ray-tracing performed on a plane-layered Earth model developed from 2D seismic interpretation. (**b**) Real CMP gathers, without NMO correction, showing heavy multiple contamination, no primary reflectors and significant variation in amplitude and coherence of multiple events between CMPs. These real CMPs are located in the same relative position along the 2D profile as the synthetic CMPs.

volcanic and chalk layers, a 2D model using a plane-layered Earth was constructed for the purpose of a full elastic wavefield simulation (using PGS Seres software). This model indicated that there is significant mode conversion at both the top and base of the volcanic layer and from the base of the chalk layer. In addition, the model suggested that there was extremely low-energy propagation through this section to the deeper horizons, of both compressional (P) and shear (S) waves. Significantly, the model also indicated that low-frequency energy propagated better than that at higher frequencies.

2D seismic processing and reprocessing

In the Slyne Basin the traditional approach to 2D processing has been to attack noise and multiple contamination at an early stage in the processing sequence. The general solution has been to focus

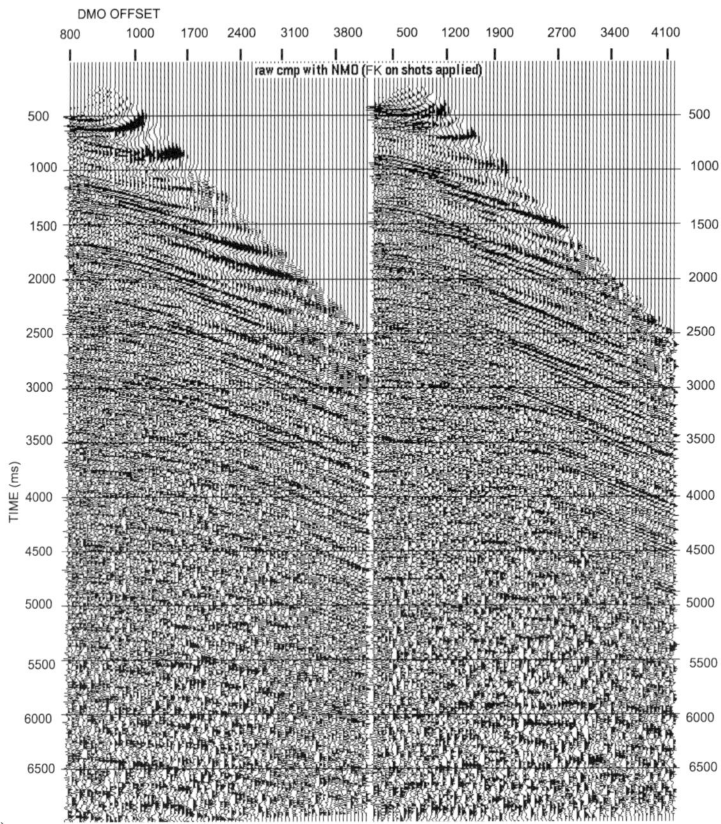

(a)

Fig. 6. Corrib 3D dataset: CMPs with and without Radon demultiple. (**a**) Raw CMPs, without Radon demultiple, but with final NMO applied. The primary signal is poorly resolved because of the dominance of multiple contamination (pervasive dipping events). (**b**) Processed CMPs with Radon demultiple and final NMO applied. After three passes of Radon, most of the multiple energy has been removed and the primary signal (flat events) can be observed.

on noise attenuation and multiple suppression in the shot and receiver domain with two or three passes of demultiple routines commonly applied. However, a number of predictive multiple attenuation techniques, such as wave-equation demultiple, have proved ineffective because of variation in the complexity and intensity of the multiples on a shot-by-shot basis. The inability of these processes to predict the true amplitude created by the spatially varying interference pattern of the multiples underlies their ineffectiveness. In addition, there is a significant amount of refracted multiple energy with non-hyperbolic move-out, which is difficult to remove from the CMP records. In many cases, the result of this traditional approach has been to remove both the multiple data and the primary signal, leaving significant amounts of residual noise. This residual noise (residual refracted multiples) degrades the section, especially via the migration process, which organizes this random noise into coherent 'smiles' (Fig. 4a). The limited transmitted energy is dominated by the lower frequencies.

Trial processing indicated that the identification of the correct velocity field is also a critical factor for successful multiple attenuation. This has proved to be a difficult task in the Slyne area, especially where the seismic data contain limited primary signal. Heavy multiple attenuation early in the processing sequence can result in a degradation of the data and a reduced ability to determine an appropriate velocity function. In addition, before the drilling of wells 27/5-1, 18/20-1 and 18/20-2z, limited well data were available to help constrain velocity interpretation in the Slyne Basin. Higher velocities prevail in the shallow part of the section because of the

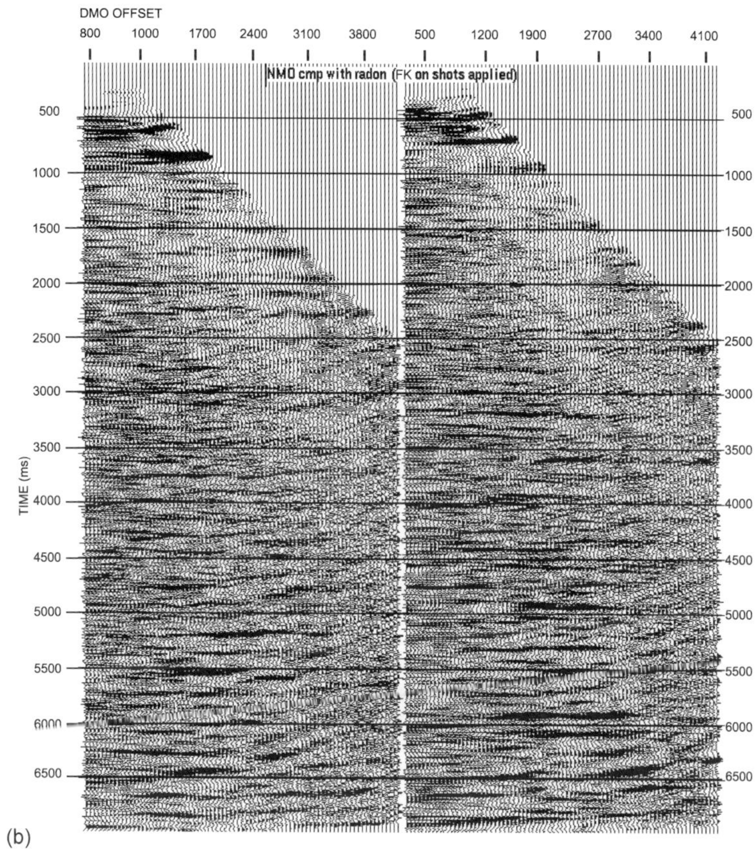

Fig. 6. Continued.

presence of volcanic rocks and chalk close to the sea bed. However, lower-velocity sediments of Early Cretaceous to Mid-Jurassic age underlie the Cretaceous chalk. On velocity gathers, this configuration makes it difficult to distinguish between primary reflectors from Middle Jurassic sediments and fast multiples and refracted multiples generated by the overlying stratigraphy.

During 1996, a number of further reprocessing trials were performed on selected 2D lines in the Corrib area. These trials focused on the issues of velocity analysis (now with the benefit of well data) and multiple attenuation. Tests indicated that a significant improvement in the deeper imaging could be achieved by filtering out the higher end of the frequency spectrum and by utilizing iterative velocity analysis and interpretation. It was also noted that improved seismic sections were obtained where a limited multiple attenuation scheme was utilized. Application of this approach to the subsequent 3D seismic survey resulted in a dramatic improvement in the data quality.

3D acquisition and processing

During 1998, Enterprise Oil acquired a 660 km^2 survey (E97IE11) over the Corrib discovery and the Shannon prospect (Fig. 1). The survey was acquired in a dip orientation (NW–SE) for optimum velocity interpretation. The survey was designed to enhance the low-frequency end of the amplitude spectrum by towing source and streamers at 10 m below sea surface. The analysis of a number of older vintage 2D datasets revealed that peak frequencies of 25–30 Hz prevailed on shot records at the target depth. All previous 2D acquisition programmes had utilized minimum or maximum sub-sea depths of 5–7 m for source and 6–8 m for streamer. This configuration resulted in higher frequencies and improved resolution of the near-surface layers and locally the Middle Jurassic section, but resulted in very limited imaging of the deeper reflectors.

Previous experience with processing 2D datasets in the area had indicated that good results achieved on individual test lines were not being replicated through the batch processing of an entire dataset. To avoid a similar problem with the 3D dataset, the processing contractor was based in the Operator's office for the duration of the project. This facilitated frequent daily interaction between the processing and interpretation geophysicists and helped to expedite and improve the testing and decision-making process. The more extensive testing of the data, before the production processing, resulted in a 3D dataset that is consistently better than the 2D data throughout the area. However, there is variation in the quality of the 3D data and some deterioration in the quality of the 3D data is still observed where the Tertiary volcanic rocks approach the sea bed. This suggests that a revised acquisition strategy may be appropriate, for future surveys, in this particular geological setting.

The greatest contributions to the improved data are derived from 3D dip move-out corrections (DMO) and migration. These processes are effective in improving the signal-to-noise ratio, in particular through the attenuation of crossline noise. In addition, 3D velocity analysis offers consistency through the tight spatial concentration of analyses and this in turn leads to improved reliability of the picks. After the success of well 18/20-2z it was decided to reprocess a subset of the 3D survey over the Corrib structure. The revised processing sequence, which incorporated Radon demultiple together with the prior experience in velocity picking, yielded a significant improvement in the imaging of the sub-volcanic/chalk section (Fig. 4b)

Key processing steps: Radon demultiple and velocity analysis

A range of demultiple techniques, including wave equation demultiple, frequency wavenumber (FK) demultiple and surface multiple attenuation, have been tested and evaluated on datasets from the Slyne Basin. Application of these techniques has proved to be of limited value. For example, the wave equation technique could not accommodate the amplitude variations generated by the constructive–destructive interference of the various multiples. To date, Radon demultiple has proved to be the most effective demultiple technique in the area.

The Radon demultiple technique utilizes the transformation of data into offset vs velocity slowness in the seismic domain, to discriminate between the multiple and primary data (Durrani & Bisset 1984). A three-pass approach has proved to be most beneficial in the Slyne Basin. The nature of the shallow geology and the resultant near-surface velocity structure creates refracted multiples that dominate the middle to far offset ranges. By using an initial pass of the linear Radon application the refracted multiples can be eliminated successfully. A second pass of the parabolic Radon application removes much of the remaining simple and pegleg multiples. A final modelled output, of the primaries only, further reduces the residual noise. The velocity analysis benefits greatly from the second pass parabolic Radon and the third pass modelled primary output (Fig. 6a and b).

The first water bottom–volcanic multiple remains problematic, as there is much diffracted high-frequency and high-amplitude energy, which is difficult to remove. However, examination of the amplitude spectrum indicates that some improvement in the CMP gather is observed after the application of Radon demultiple. Before Radon demultiple the source and receiver notch at 75 Hz (10 m source and streamer) is filled with noise, whereas after demultiple the notch is clearly observed (Fig. 7a and b).

Although the application of Radon demultiple has been a key to improving the data quality, the most significant factor has been the improved velocity control offered by 3D acquisition. The data redundancy of a 3D grid and the ability to sum gathers crossline for input to velocity analysis has dramatically improved the signal-to-noise ratio of the velocity gathers and greatly assisted the interpretation of the initial velocity function (Fig. 8a and b). This in turn has allowed a more accurate targeting of the demultiple process that limits damage to the primary energy (Fig. 9a and b). Inaccurate picking of the velocity function will result in a lower signal-to-noise ratio on the stacked section, as a result of the presence of residual multiple energy. In spite of these improvements the data still contain some residual multiples that have similar move-out to the primary energy. These multiples are impossible to remove with conventional, velocity discriminant, demultiple techniques.

Future work in the Corrib area will focus on improving the sub-basalt/chalk imaging through pre- and post-stack 3D depth migration, which has resulted in a significant improvement in the imaging of the SSG reservoir. Additional modelling studies are also planned to better

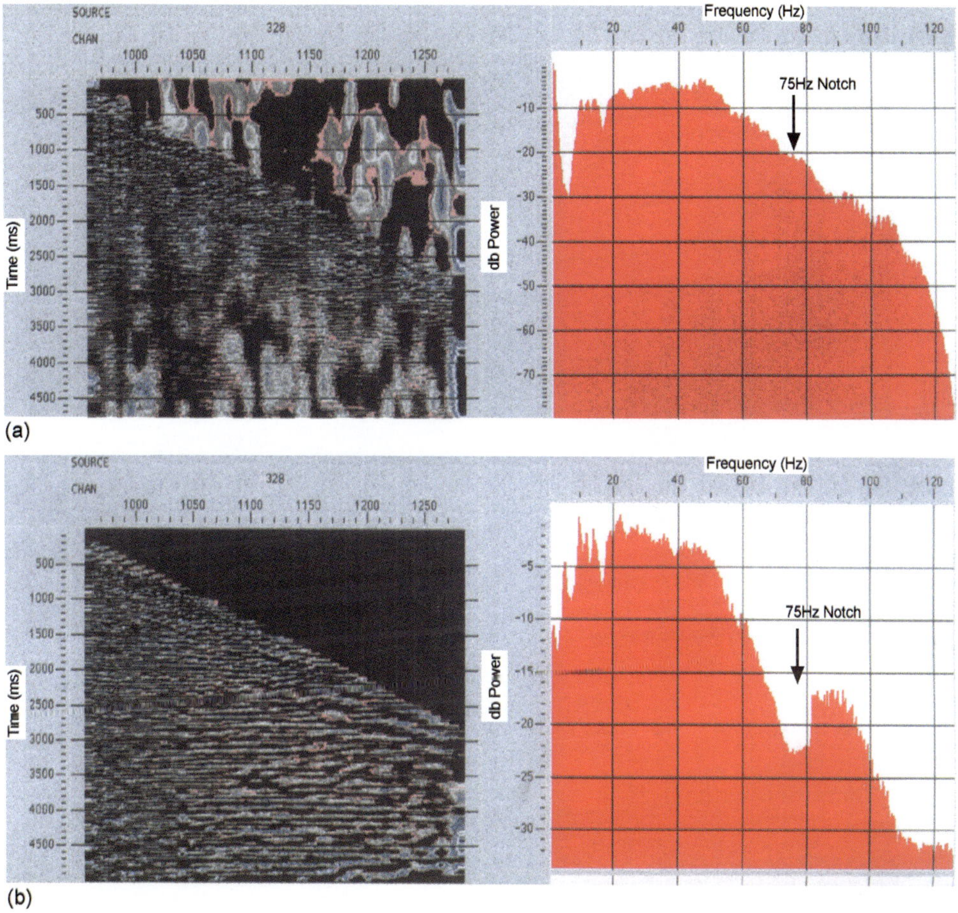

Fig. 7. Corrib 3D dataset: noise attenuation via Radon demultiple. (**a**) CMP gather and amplitude spectrum, pre-Radon, with the source and receiver notch (75 Hz) not visible because of noise contamination. (**b**) CMP gather and amplitude spectrum, post-Radon, with source and receiver notch (75 Hz) visible after multiple and noise attenuation.

understand wavefield propagation through this section. It is important to emphasize that difficult data require significant amounts of time and skill to undertake the iterative process required for optimal results. Combined insights from the operations, processing and interpretation geophysicists are essential, to develop and parameterize the individual acquisition and processing steps and optimize the seismic image.

Conclusions

Exploration in the Slyne Basin has, until recently, been hindered by poor-quality seismic data. Although potential hydrocarbon traps have been recognized, lack of confidence in the seismic data has deterred exploration drilling. The seismic data are poor because of the confluence of a number of factors that conspire to produce seismic data that are rich in multiples and noise, but limited in signal. These factors include moderate water depths, hard and seismically fast near-sea-bed stratigraphy, significant velocity inversions, rugose interfaces and lateral velocity variation. However, the application of low-frequency 3D acquisition techniques and careful 3D seismic processing, with particular attention to velocity picking, has led to a significant improvement in the data. This improved seismic dataset has been an important factor in the successful appraisal of the Corrib discovery.

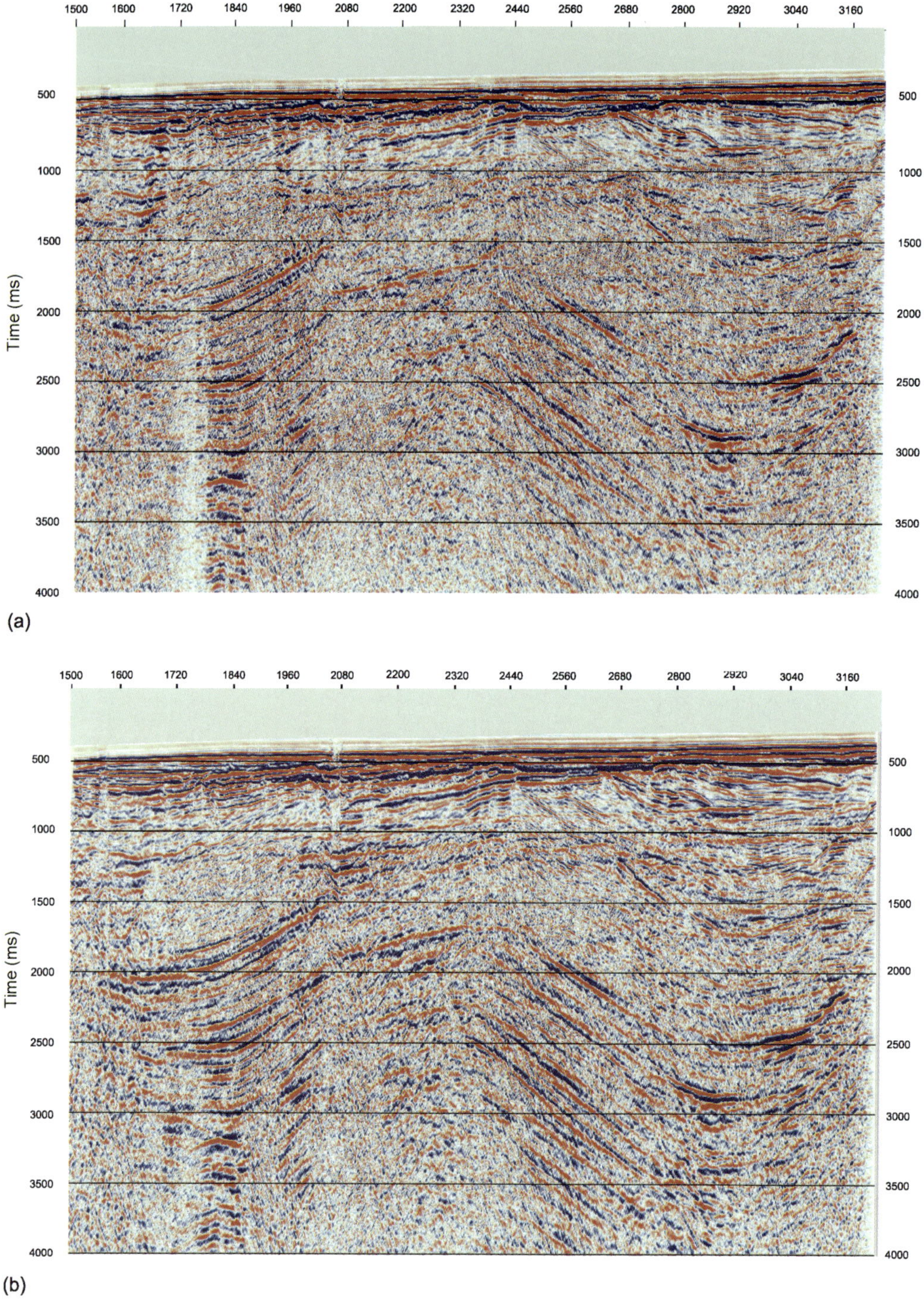

Fig. 8. Corrib 3D dataset: improved signal-to-noise ratio from crossline summation of CMP gathers. (**a**) Stack of 3D seismic profile, Inline-3060, using one crossline CMP, which manifests both primary as well as multiple reflections and considerable dipping noise. (**b**) Stack of 3D seismic profile, Inline-3060, with summation of three crossline CMPs, demonstrating improved signal-to-noise ratio with clearer primary reflections, which aids velocity interpretation.

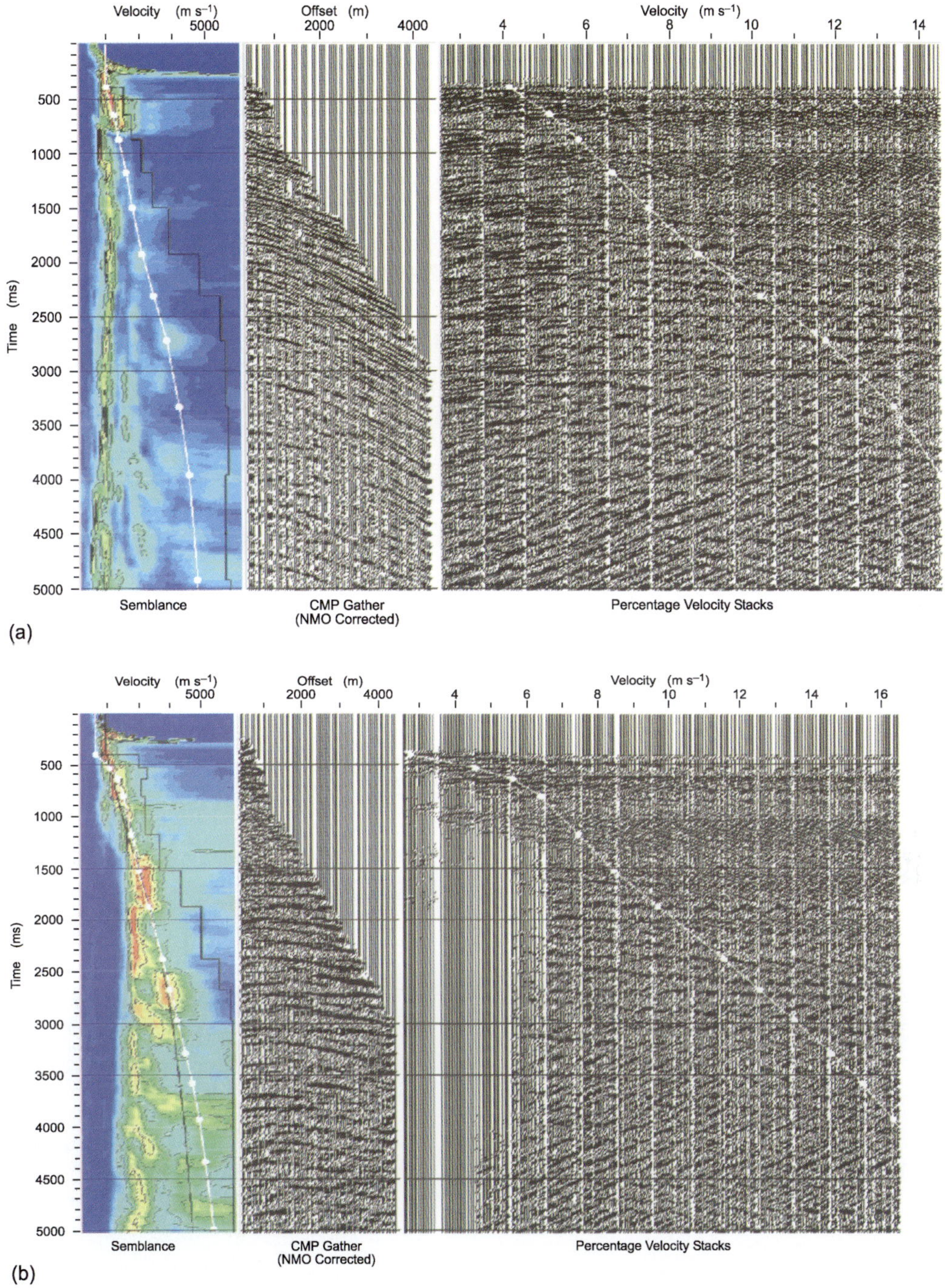

Fig. 9. Corrib 3D dataset: velocity interpretation pre- and post-Radon. (**a**) Velocity interpretation panel with NMO-corrected CMP gather, before Radon. On the semblance panel the multiple trend is clear, with very little energy observed close to the inferred velocity function (white line). (**b**) Velocity interpretation panel with NMO-corrected CMP gather, post-Radon. Successful multiple attenuation via Radon enhances the observed primary energy on both the semblance panel and the CMP gather (flat events).

We gratefully acknowledge permission to publish this work from Enterprise Oil plc and our partners in the 2/93 and 3/94 Licences, Statoil Exploration (Ireland) Ltd., and Marathon Oil (Hibernia) Ltd. The authors would like to thank the many people who have been involved in the seismic acquisition and processing of data in the area: N. Turton (Geco); C. Warner and N. Oliver (PGS); P. Harrison and J. Zimmerman (Enterprise Oil). In particular, special thanks are due to S. T. Sampanthan and Xiaobin Bob Ge (Just Geo, Inc.) for their work on the seismic reprocessing and parameterization of the Radon Transform multiple attenuation technique. We are also grateful to Just Geo for providing some of the figures, and to S. Coffey (Enterprise Oil) for the preparation of the figures. We would like to thank the referees (C. Bean and T. Chapman) for their constructive comments. The opinions expressed herein are those of the authors.

References

DANCER, P.N., ALGER, S.T. & WILSON, I.R. 1999. Structural evolution of the Slyne Trough. *In*: FLEET, A.J. & BOLDY, S.A.R. (eds) *Petroleum Geology of Northwest Europe: Proceedings of the 5th Conference*. Geological Society, London, 445–453.

DURRANI, T.S. & BISSET, D. 1984. The Radon transform and its properties. *Geophysics*, **49**, 1180–1187.

SCOTCHMAN, I.C. & THOMAS, J.R.W. 1995. Maturity and hydrocarbon generation in the Slyne Trough, northwest Ireland. *In*: CROKER, P.F. & SHANNON, P.M. (eds) *The Petroleum Geology of Ireland's Offshore Basins*. Geological Society, London, Special Publications, **93**, 385–411.

TRUEBLOOD, S. & MORTON, N. 1991. Comparative sequence stratigraphy and structural styles of the Slyne Trough and Hebrides Basin. *Journal of the Geological Society, London*, **148**, 197–201.

WILSON, H. E. & MANNING, P. I. 1978. *Geology of the Causeway Coast*. Memoir of the Geological Survey of Northern Ireland, Sheet 7.

Sub-basalt imaging using converted waves: numerical modelling

F. MARTINI[1], C. LAFOND[2], S. KACULINI[2] & C. J. BEAN[1]

[1]*Department of Geology, University College Dublin, Belfield, Dublin 4, Ireland*
(e-mail: francesca.martini@ucd.ie)

[2]*ELF-GRC, 30 Buckingham Gate, London SW1E 6NN, UK*

Abstract: Both ray-tracing and frequency–wavenumber integration modelling have been used to investigate the behaviour of P waves and P to S converted waves in simple geological models, which involve a basaltic layer above a lower-velocity layer of sediments. Model parameters were adjusted to study the behaviour of these waves, under conditions of changing water depth, basalt layer thickness and a number of permutations with respect to basalt–sediment stratification. Synthetic shot gathers were generated and then analysed and processed as real data. The general approach was to document changes in reflection curves with offset, changes in the stacking velocity, variations in the imaging of deeper reflectors under different model scenarios and to understand which offsets are useful for converted wave processing. Although this work is preliminary, some general principles have been identified, which may be of use in an operational context. For example, the modelling indicates that S waves have larger amplitudes than P waves at long offsets and that P to S converted waves arrive before the sea bottom arrival at the far offset. Consequently, the recording and identification of converted waves is favoured by long offset (c. 10 km) acquisition arrays. However, the modelling also indicates that the presence of sediment–basalt interlayering makes the identification of converted waves more difficult.

In many regions of the world, the presence of high-velocity layers poses a significant challenge to the seismic imaging of deeper reflectors. These high-velocity layers commonly result from the presence of evaporites, carbonates or igneous rocks within the sedimentary column. Many of the basins of the NE Atlantic margin are characterized by extensive igneous activity of Early Tertiary age (White & McKenzie 1989). This activity has resulted in the widespread emplacement of significant quantities of basalt as lava flows or sills and dykes. In some of these basins, such as the Slyne Basin to the northwest of Ireland, these lava flows are preserved at or close to the sea bed (Dancer & Pillar 2001). Elsewhere, they commonly occur at 2–3 km below the sea bed as a result of post-extrusion sedimentation and subsidence. The thickness of these high-velocity basalt layers varies from a few metres to kilometres, but in most cases they create a significant barrier to the imaging of the pre-Tertiary structure using conventional seismic acquisition methods.

One of the problems encountered, in the presence of basaltic rocks, is the relatively limited penetration of compressional (P) waves that generally occurs when a high-velocity basaltic layer rests upon a layer of significantly lower velocity. In addition, the P-wave energy may be subject to attenuation by wave absorption or internal scattering, which results from the presence of intra-basalt weathering surfaces and rugosity (Purnell 1992). Published examples (e.g. Tatham & Goolsbee 1984) suggest that high-velocity layer boundary effects may also be important, notwithstanding the magnitude of internal absorption or scattering, and that these effects are magnified as the velocity contrast between the basalt layer and the country rock increases. Moreover, basalt can be strongly variable in morphology and seismic velocity (Samson *et al.* 1995) and can also be highly heterogeneous and anisotropic (Kiorboe & Petersen 1995). In relative terms, converted waves show good penetration of these high-velocity layers (Li *et al.* 1998). The converted waves include P waves and shear (S) waves that experience one or more conversions along the path from source to receiver.

This paper addresses a number of questions regarding the use of converted waves for sub-basalt imaging. Under what conditions are conversions to S waves observed? How do they compare with P-wave arrivals? At which offsets are they recorded? How can stacking be

From: SHANNON, P.M., HAUGHTON, P.D.W. & CORCORAN, D.V. (eds). 2001. *The Petroleum Exploration of Ireland's Offshore Basins*. Geological Society, London, Special Publications, **188**, 223–235. 0305-8719/01/$15.00

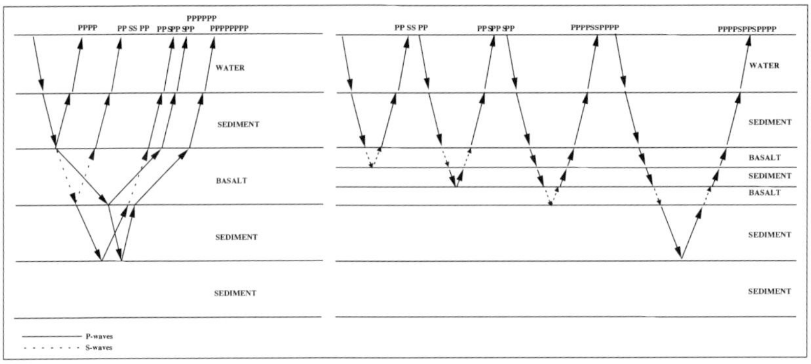

Fig. 1. Wave conversion and nomenclature for converted waves. This modelling study considers only those converted waves that contain S-wave segments within the basalt layer.

optimized to best use the S-wave information? The study concentrates on the wave penetration problem, in particular a comparison of PP and PS converted waves. Some findings in the literature, related to the use of converted waves for sub-basalt imaging, are verified.

Modelling techniques

This study has utilized both ray-tracing and frequency–wavenumber integration (F–K) modelling techniques. Initial analysis involved data generated from ray-tracing by considering only P waves and converted waves in the basalt layer (see Fig. 1 for converted wave nomenclature). This permitted a superficial analysis of the data, by focusing attention on the primary reflections and the converted waves only. Multiple reflections and other converted waves were not considered at this stage. A second phase of analysis involved data produced from F–K modelling, which incorporated most kinds of wave propagation energy (i.e. converted waves, refracted waves and multiples), except for environmental noise.

Models

A number of simple 2D models were generated as templates for analysis (Fig. 2; Table 1). The entire media were modelled as isotropic and elastic (i.e. with no intrinsic attenuation). All the models used for the F–K modelling are 11 km long, whereas the models used for ray-tracing are 15 km in length (Table 2). Long offset synthetic shot gathers were generated for each model but the gather sizes were constrained by program limitations.

The basic model (Model 1, Fig. 2a) consists of a basaltic layer of 500 m thickness overlain by a 500 m layer of sediments and 500 m of water. Two further layers of sediments, each 500 m thick, are present beneath the basalt (Fig. 2a; Table 1 for general model lithology). In Layer 2, a velocity gradient is modelled with velocity

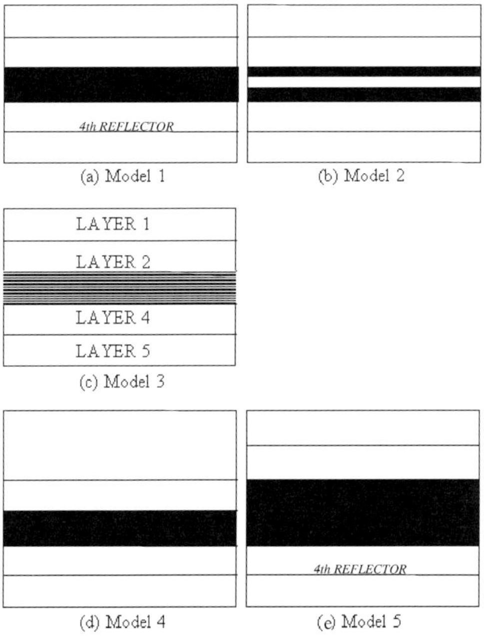

Fig. 2. Simple 2D models used as templates for analysis. All media are assumed to be isotropic and elastic, with no intrinsic attenuation. (**a**) Model 1 consists of a basalt layer of 500 m thickness, overlain by a 500 m layer of sediments and 500 m of water. (**b**) In Model 2 the single basalt layer is replaced with two basalt layers *c.* 160 m thick separated by a 160 m layer of sediments. (**c**) In Model 3 the basalt layer is replaced by a series of 50 interbedded layers of basalt and sediments, each of 10 m thickness. (**d**) In Model 4 the water depth is increased from 500 m to 1000 m. (**e**) In Model 5 the basalt thickness is doubled from 500 m to 1000 m.

Table 1. *General parameters for five-layer 2D model, with composition, P-wave velocity (V_p), S-wave velocity (V_s) and density (ρ) values for each layer*

Layer	Composition	V_P (m s^{-1})	V_S (m s^{-1})	Density (g cm^{-3})
1	Water	1500	0	1.0
2	Sediments	1800–3000	3	1.5
3	Basalt	5800	2900	2.9
4	Sediments	2900	1550	2.5
5	Sediments	3500	1850	2.6

gradually changing from 1800 m s^{-1} at the upper boundary to 3000 m s^{-1} at the lower one. This offers a more realistic simulation of increasing velocity with depth as a result of sediment compaction.

Model 2 is developed by replacing a single 500 m basalt layer with two basaltic layers *c.* 160 m thick, separated by a layer of sediments (Fig. 2b). In Model 3, a series of 50 interbedded layers of basalt and sediment, each of 10 m thickness, is introduced (Fig. 2c). In Model 4 the water depth is increased from 500 m to 1000 m (Fig. 2d), and in Model 5 the basalt layer thickness is increased to 1000 m (Fig. 2e). Identical P- and S-wave velocity and density values, for each layer, are utilized in all five models.

As a result of modelling conditions and limitations on the value of the parameters, imposed by the computer code, different acquisition parameters were used to run the ray-tracing and discrete wavenumber modelling (Table 2). For example, it was not possible to employ the same long offsets with the F–K modelling as the ray-tracing data. However, the ray-traced data suggest that little converted wave information is available at offsets longer than 10 km. The maximum offset at which converted waves were recorded is 9100 m. Ray-tracing was not possible for Model 3 (Fig. 2c) because the thickness of the basalt layers (10 m) is smaller than the modelled signal wavelength (minimum wavelength is *c.* 20 m).

Previous workers (Purnell *et al.* 1990; Purnell 1992; Hanßen 1998) considered waves that convert from P to S on the downward journey and from S to P on the upward journey. White & Stephen (1980) demonstrated that these waves display almost perfect conversion efficiency when the S-wave velocity in the overlying sediments is very low (as for unconsolidated sediments) and the basalt S-wave velocity equals the P-wave velocity of the underlying sediments. For higher S-wave velocity in the sediments, the conversion efficiency decreases proportionally, but the V_P(underlying sediment) $\approx$ V_S(basalt) scenario still gives a high conversion efficiency coefficient. This situation also leads to efficient coupling between S waves within the basalt layer and P waves outside. The coincidence of 2900 m s^{-1} for the S-wave velocity in the basalt and P-wave velocity in the underlying sediments is one of the simplifying assumptions in the models analysed.

Table 2. *Model acquisition parameters for ray-trace modelling and F–K modelling; models used for F–K modelling are 11 km long, whereas models used for ray-trace modelling are 15 km in length*

Parameter	Ray-trace modelling	F–K modelling
Number of receivers	298	100
Receiver interval	50 m	100 m
Minimum offset	100 m	220 m
Maximum offset	14950 m	10120 m
Number of shot points	100	100
Source interval	100 m	100 m
Central frequency of source signal	20 Hz	20 Hz
Total number of samples per trace	2048	3001
Sampling interval	2 ms, resampled to 4 ms	2 ms, resampled to 4 ms

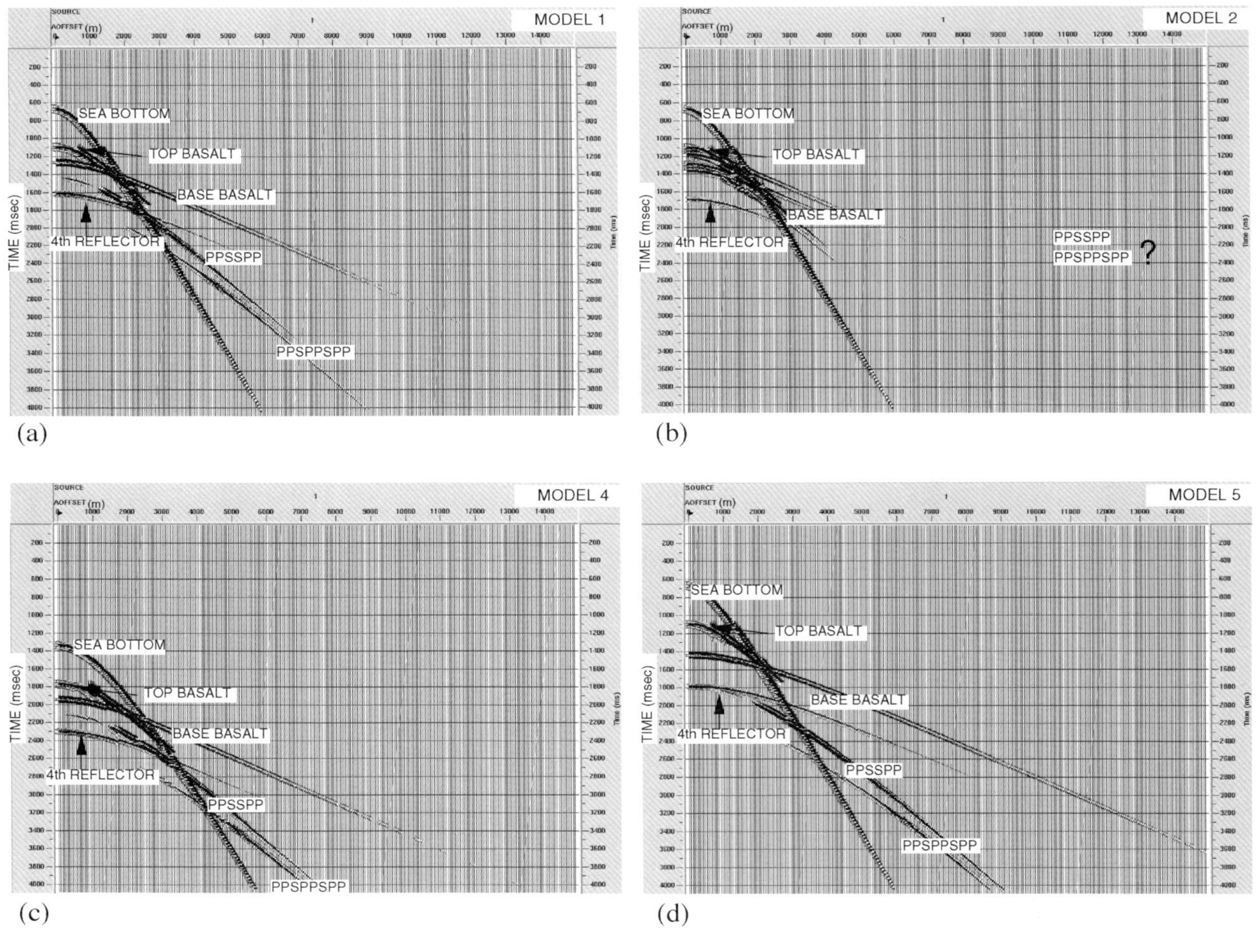

Fig. 3. Shot gathers obtained from ray-trace modelling with key P waves and converted waves labelled. There is no gain applied to the shot gathers and all model outputs are truncated at 4000 ms. (**a**) Model 1, converted waves are readily identified; (**b**) Model 2, converted waves are not easily recognized; (**c**) Model 4, increased water depth moves the converted waves closer to the sea-bed reflection and makes the processing of these converted waves more difficult; (d) Model 5, converted waves are strong and distinct from the sea-bottom arrival. Ray-tracing modelling has not been possible for Model 3 because the bed thickness of the intercalated basalt–sediment units within Layer 3 is smaller than the signal wavelength.

Results from ray-tracing

The shot gathers obtained from the ray-tracing simulations are shown in Fig. 3. Results from the basic model (Model 1, Fig. 3a) indicate that the sea bottom is represented by a very strong arrival and is recorded over all offsets. The top basalt reflection is recorded out to 2700 m, where a turning wave effect is introduced because of the velocity gradient in the second layer. These turning waves, also called diving waves, arise in the presence of a strong velocity gradient, which induces reversal of the downward component of the seismic rays and bends the rays back to the surface before reflection from the interface is achieved. The arrival at *c.* 1260 ms is the reflection from the base basalt, which is recorded up to the far offset. At 1600 ms the fourth reflector is observed.

Analysis of the gather for Model 2 (Fig. 3b) reveals a more complex result than observed for Model 1. In this case there is interference between the arrivals from the top and the base of the two basalt layers. Also, the converted waves PPSSPP, PPSPPSPP, PPPPSSPPPP and PPPPSPPSPPPP cannot easily be identified. It should be noted that Model 2 contains two basalt layers in contrast to Models 1, 4 and 5, which have only one basalt layer (Fig. 2). In this study, only wavetrains that have an S-wave segment in the basalt layer are considered, i.e. in Models 1, 4 and 5 only PPSSPP and PPSPSSPP waves are considered, and for Model 2 only PPSSPP and PPSPPSPP waves (from the uppermost basalt

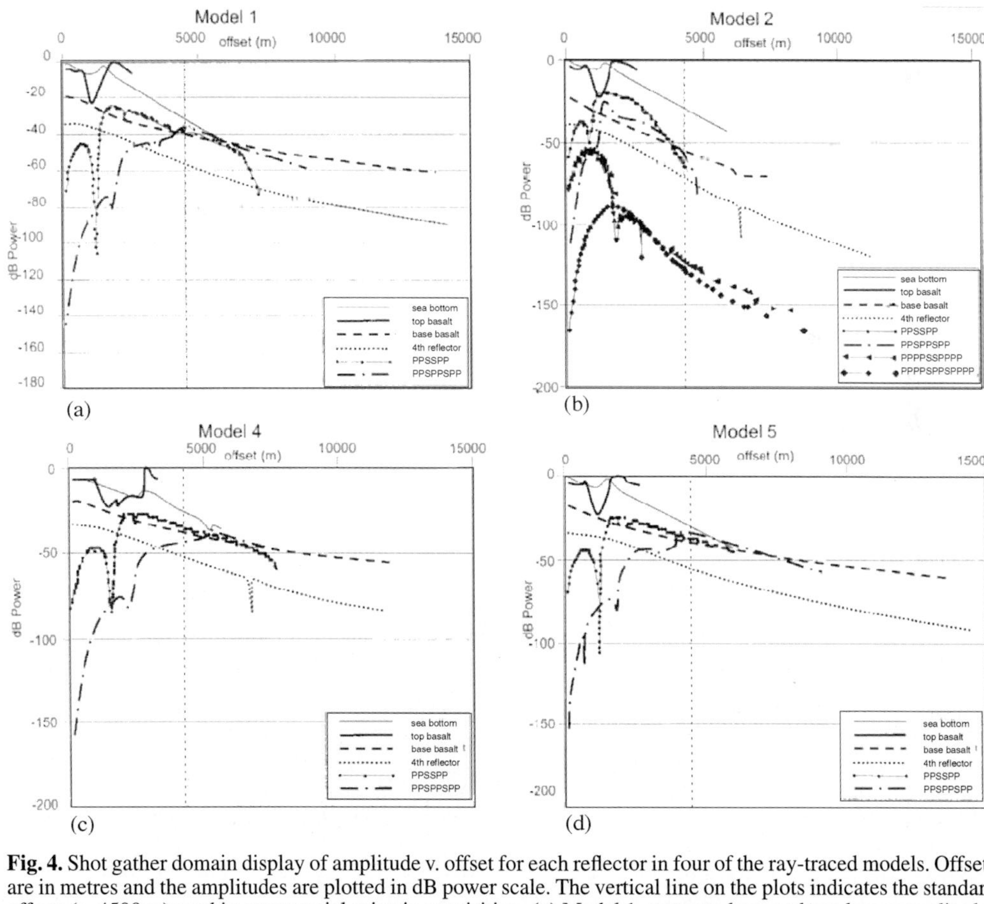

Fig. 4. Shot gather domain display of amplitude v. offset for each reflector in four of the ray-traced models. Offsets are in metres and the amplitudes are plotted in dB power scale. The vertical line on the plots indicates the standard offsets (*c.* 4500 m) used in commercial seismic acquisition. (**a**) Model 1, converted waves have larger amplitudes than P waves for long offsets; (**b**) Model 2, at long offsets larger amplitudes are observed for the PPSSPP and PPSPPSPP waves than for P waves; (**c**) Model 4, PPSSPP and PPSPPSPP waves have amplitudes greater than the fourth reflector at long offsets and have amplitudes similar to the base basalt reflector at offsets in excess of 4500 m; (**d**) Model 5, PPSSPP wave has larger amplitudes than the P-wave arrival for both the base basalt and the fourth reflector, at offsets greater than 1450 m.

layer), and PPPPSSPPPP and PPPPSPPSPPPP waves (from the lower basalt layer) are analysed.

Model 4 (Fig. 3c) indicates that progression to deeper water moves the converted arrivals closer to the sea-bed reflection, and this makes the converted wave processing more difficult. In the case of thicker basalt (Model 5, Fig. 3d) the converted waves are fairly strong, and distinct from the sea-bottom arrival. The P-wave arrivals manifest good resolution and are easily distinguished on the shot gathers.

Ray-tracing; amplitude analysis

One of the aims of this study is to determine the offsets at which converted waves are recorded and to compare them with the P-wave arrivals that are utilized during conventional processing. To achieve this comparison, an amplitude analysis for all the arrivals in the shot gathers was performed for Models 1, 2, 4 and 5.

Model 1 (Fig. 4a). For reflections from below the top of the basalt, all amplitudes are very strongly attenuated as expected. However, below the basalt, converted waves (PPSPPSPP waves) have relatively larger amplitudes than P waves for long offsets. This may be important for real data, as all arrivals from beneath the basalt will have attenuated amplitudes that can be difficult to identify in presence of multiples and noise. However, far-offset arrivals are free from multiple contamination and arrive before the sea-bottom reflection, and so they can be clearly identified on the shot gathers.

Model 2 (Fig. 4b). In common with Model 1, the amplitudes of all reflections from below the top basalt are very strongly attenuated. However, at long offsets relatively larger amplitudes are observed for the PPSSPP and PPSPPSPP waves than for P waves. Amplitude values for the PPPPSSPPPP and PPPPSPPSPPPP waves are measurable for offsets out to 8150 m and 8750 m, respectively, but the low magnitude of these values would be difficult to utilize in real data that contain noise. Also of note is the fact that the PPSSPP wave has a larger amplitude than the base basalt reflection for offsets longer than 1250 m and a larger amplitude than the reflector below the basalt (fourth reflector) for offsets longer than 1100 m. In addition, the PPSPPSPP wave manifests amplitudes greater than the base basalt reflection between 1100 and 4100 m offset.

Model 4 (Fig. 4c). In this model, the converted waves (PPSSPP and PPSPPSPP waves) have amplitudes greater than the fourth reflector at long offsets and have amplitudes similar to the base basalt reflector at offsets in excess of 4500 m.

Model 5 (Fig. 4d). In this case, the PPSSPP wave has larger amplitudes than the P-wave arrival for both the base basalt reflector and the fourth reflector beyond an offset of 1450 m. At an offset of 2850 m, the PPSPPSPP wave has larger amplitudes than the fourth reflector. Also, the PPSPPSPP wave has amplitudes larger than the base basalt reflector at offsets in excess of 4500 m.

Comparison of P-wave and S-wave model amplitudes

Four models are compared with a view to determining some of the critical influences on sub-basalt imaging. In particular, this assessment attempts to identify the conditions most favourable for the recording of converted waves of 'useful' amplitude. By comparing the amplitudes of the same reflector from different models, we can obtain some insight into how the different model scenarios change the imaging at depth.

P waves

Analysis of the deep reflectors has been effected in both the shot gather and the stacked section domains. The P-wave amplitudes of the base basalt reflector, in the shot gather domain, are plotted versus offset in Fig. 5 (Models 1, 2, 4 and 5). For conventional offsets (up to *c.* 4500 m), the largest amplitudes are shown for Model 1 (massive basalt layer 500 m thick). In Model 5 (basalt thickness doubled), the amplitudes decrease to about 60% of those in Model 1. In Model 2 (two basalt layers interbedded with a sediment layer), the amplitudes are further decreased, with values less than half those of Model 1. The worst case scenario is shown by Model 4 (the water depth doubled to 1 km), where the zero offset amplitude values are *c.* 40% of those observed in Model 1. However, the amplitude curve for Model 4 has the same shape as for Model 1, but is shifted down on the amplitude scale, as observed in Fig. 5.

In all cases amplitude decay is at least −15 dB at 4500 m offset. This is partly due to spherical divergence (the decrease in amplitudes of a wavefront because of geometric spreading),

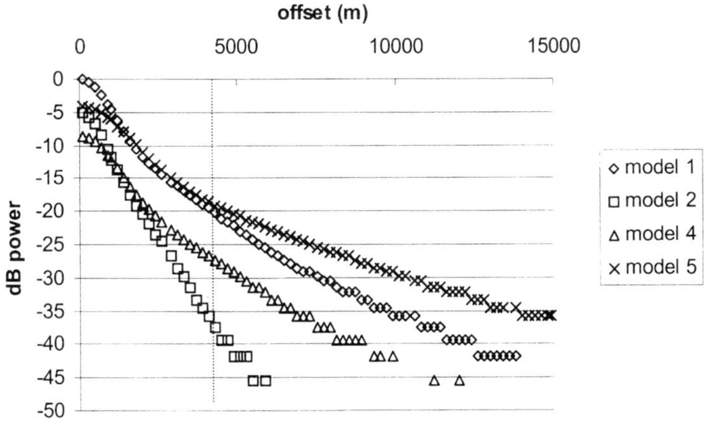

Fig. 5. Shot gather domain display of P-wave amplitude v. offset plotted for the base basalt reflector in Models 1, 2, 4 and 5. Offsets are in metres and the amplitudes are plotted in dB power scale. The dotted line indicates standard commercial seismic acquisition offsets (*c.* 4500 m).

which is present in the synthetic amplitudes. For a homogeneous medium without attenuation wave amplitudes decay as $1/\sqrt{r}$ (2D case), where r is the radius of a spherical wavefront. In the case of a layered Earth, amplitude decay can be described approximately by $1/[v^2(t)t]$ (Newman 1973), where t is the two-way travel time and $v(t)$ is the r.m.s. velocity of the primary reflections (those reflected only once) averaged over a survey area (Yilmaz 1987).

For longer offsets, the largest P-wave amplitudes pertain to Model 5 (thick basalt) in spite of the longer raypaths encountered. These P-wave amplitudes are considerably larger than the P-wave amplitudes encountered in the deep-water case (Model 4, Fig. 5). This results from the fact that, for a given offset, the incidence angle of a ray at the top basalt is smaller for shallower thick basalt compared with a deeper one, and the transmission coefficient is therefore higher. Consequently, a slower decay with offset of P-wave amplitudes is observed in the case of a thick, shallow basalt layer. The greatest attenuation occurs in Model 2, where the basalt contains interbedded sediment layers, which further attenuate wave transmission (Fig. 5).

The P-wave amplitudes of the reflector below the basalt (fourth reflector) are plotted versus offset in Fig. 6 (Models 1, 2, 4 and 5). The P-wave amplitudes of this reflector (fourth reflector) in the shot gather domain manifest similar behaviour with offset to that of the base basalt reflector (Figs 5 and 6).

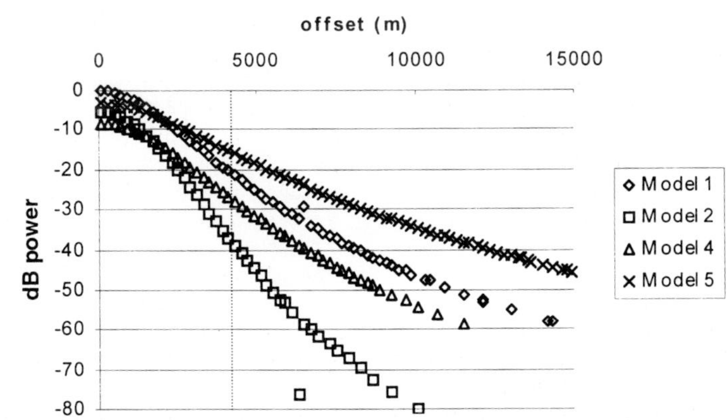

Fig. 6. Shot gather domain display of P-wave amplitude v. offset plotted for the reflector below the basalt (fourth reflector) in Models 1, 2, 4 and 5. Offsets are in metres and the amplitudes are plotted in dB power scale. The dotted line indicates standard commercial seismic acquisition offsets (*c.* 4500 m).

S-waves

A shot gather domain plot of amplitude v. offset, for waves converted in the basalt, is presented for three models in Fig. 7 (Models 1, 4 and 5). Model 2 arrivals have been excluded from this plot as the PPSSPP and PPSPSPP waves have very high amplitude (these reflections are actually from the uppermost basalt layer) and are not comparable with the other models. It is observed that waves that have been converted to S waves at the top basalt and reconverted to P waves below the base of the basalt (PPSPPSPP waves) are recorded, with significant amplitudes, at large offsets (up to 10 km) for all models. It is also noted that Model 5 (thicker basalt) yields the strongest arrivals at long offsets and in general offers better continuity of the relatively high-amplitude arrivals over the offset range.

Stacking

The shot gather results from these simulations have been repeated laterally to give a complete dataset for analysis. These data were processed as real data using ProMAX™. Shot–receiver geometries were designed to yield Common Depth Points (CDPs) with a maximum fold of coverage of 75. Two different approaches to velocity analyses were evaluated (Fig. 8a and b).

The initial velocity analysis identifies the primary P-wave reflections, as is usually done for a conventional stack, by picking velocity maxima. In this case four velocity maxima are recognized in the velocity spectrum and these picks correspond to the four main reflectors in the model (Fig. 8a). The stacking velocity function obtained from this analysis was used to generate the stack section displayed in Fig. 9a. However, the base basalt reflector and the fourth reflector are not well resolved on this stacked section as a result of the presence of a high ambient noise level at a distance of *c.* 1 km from the origin of the model.

A second velocity analysis was performed, to pick the converted waves in the basalt layer (Fig. 8b). The appropriate stacking velocities for events down to the top of the basalt are readily identified on the velocity spectra. However, problems usually arise when attempting to pick stacking velocities beneath the basalt. It is noted from the simple data in Fig. 8b that the velocity maxima for the PP events from the top of the basalt, and for the PP and PPSSPP events from the base of the basalt layer, form a triangular shape on the contoured velocity spectrum. Li *et al.* (1998) demonstrated that the separation of these three points depends on the V_p/V_s ratio of the basalt. The stacked section obtained from this Normal Move Out (NMO) velocity function is shown in Fig. 9b. A significant improvement in the imaging of the fourth reflector is observed. This image is clearer and most of the noise below the bottom basalt, clearly visible on the conventional stack section, has disappeared permitting increased resolution at *c.* 1600 ms.

Examination of an NMO-corrected CDP gather (corrected with the two different velocity functions identified in Fig. 8a and b) helps to illuminate the improvement obtained by picking the converted waves on the velocity spectra (Fig. 10a and b). On the one hand, correcting the

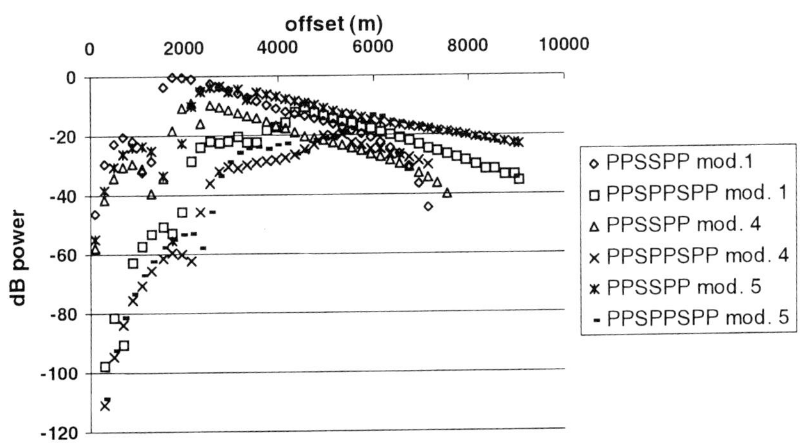

Fig. 7. Shot gather domain display of S-wave (converted in the basalt) amplitude v. offset plotted for Models 1, 4 and 5. Offsets are in metres and the amplitudes are plotted in dB power scale. Model 5 yields the the strongest arrivals at long offsets, although the PPSPPSPP waves are recorded with significant amplitudes, at long offsets for all three models.

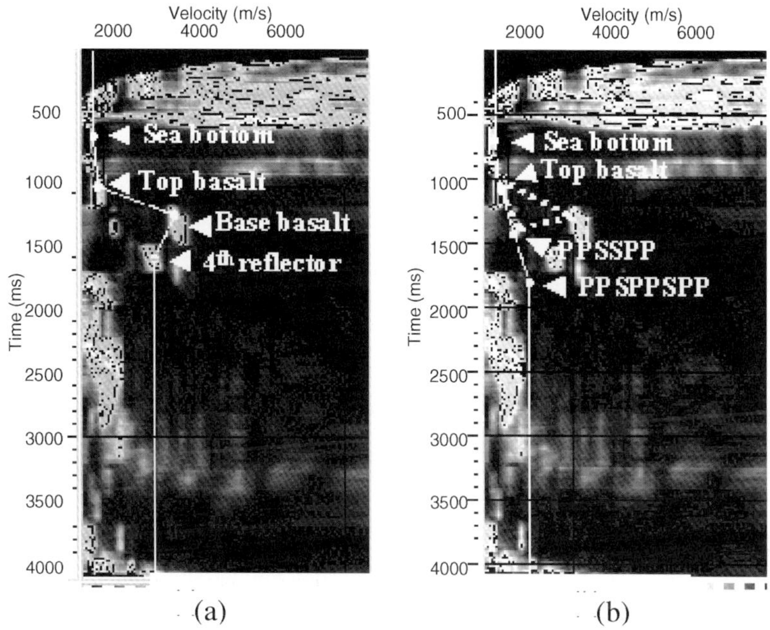

Fig. 8. Contoured velocity spectrum for CDP from Model 1 with key events labelled. (**a**) The velocity function picked for a conventional P-wave stack; (**b**) the velocity function picked for converted wave processing. The white dotted line shows the triangular shape created by the PP wave from the top basalt, the PP wave from the base basalt and the PPSSPP wave from the base basalt.

data for the P-wave reflections (Fig. 8a) (by picking the velocity maxima associated with the base basalt reflector and the fourth reflector), results in the velocity being too high to correct all the arrivals visible under the fourth reflector. In particular, the sea-bottom arrival is problematic below 1260 ms, where it crosses the top basalt reflector (Fig. 10a). The appropriate NMO velocity for the base basalt reflector is too high to correct the sea-bottom tail.

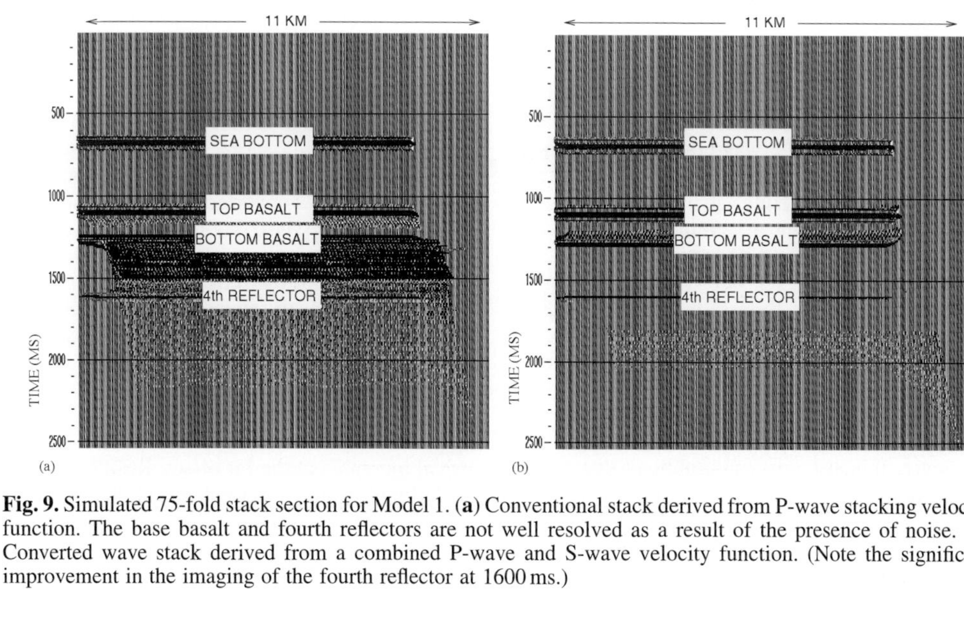

Fig. 9. Simulated 75-fold stack section for Model 1. (**a**) Conventional stack derived from P-wave stacking velocity function. The base basalt and fourth reflectors are not well resolved as a result of the presence of noise. (**b**) Converted wave stack derived from a combined P-wave and S-wave velocity function. (Note the significant improvement in the imaging of the fourth reflector at 1600 ms.)

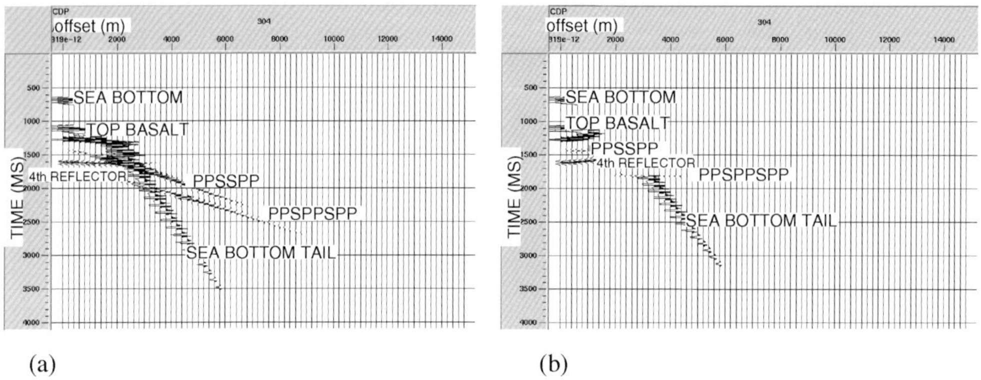

Fig. 10. NMO-corrected CDP gather from Model 1. (**a**) NMO corrected with P-wave velocity function. The appropriate NMO velocity is too high to correct the sea-bottom tail, which interferes with the base basalt and fourth reflectors. (**b**) NMO corrected with combined P-wave and S-wave velocity function, which is sufficiently slow to correct the sea-bottom arrival and facilitate its partial removal via application of a stretch mute and NMO correction.

On the other hand, by picking from the velocity spectrum a trend that incorporates the converted wave maxima (Fig. 8b), the NMO velocity is sufficiently low to correct the sea-bottom arrival between 1260 and 1600 ms (and it is subsequently removed by the stretch mute applied together with the NMO correction). At times after the PPSPPSPP-wave arrival, the sea-bottom tail still remains, but it is not in phase with other arrivals. Furthermore, the amplitudes are very small at those offsets and give a weak noise signal between 1800 and 2000 ms on the stack section, which does not affect the imaging of the fourth reflector (Fig. 9b). Even without the optimal NMO correction velocity, the power of stacking results in a good stacked signal for the base basalt and the fourth reflectors.

The problem of the sea-bottom tail is not usually encountered when working with conventional data, as a mute is normally applied. In this case, however, a mute was not applied, because of the goal of investigating the converted waves, which, on the far offsets, arrive before the sea-bottom reflection. Ideally, removal of the sea-bottom reflection could be achieved via spectral techniques (e.g. F–K) but this has not been attempted as part of this study. In this case, the sea-bottom arrival is so strong that its presence affects the velocity analysis. For example, a strong maximum observed on the velocity spectrum at 1100 ms and 1550 m s^{-1} apparently corresponds to the top basalt reflector (Fig. 11a). However, this velocity maximum is due to the contribution of two arrivals: the top basalt arrival on the near offset and the sea-bottom tail on the far offset (Fig. 11b). This interference introduces an error in the estimation of the NMO velocity for the top basalt reflector and consequently affects the estimate of the velocity of the later arrivals.

Results from F–K modelling

Results of the F–K modelling with respect to Model 1 are shown in Fig. 12a. Although the model is simple, the shot gather is difficult to interpret as all the waves are modelled, including multiples. As a result, the identification of primary P-wave and converted S-wave reflections is difficult and these reflections are highlighted here courtesy of ray-trace modelling (Fig. 12a). Some of the stronger multiple events and other converted waves, which camouflage the primary arrivals and converted waves of interest, are labelled in Fig. 12b.

Results of F–K modelling with respect to Model 2 are shown in Fig. 12c. In this case, the shot gather is even more difficult to interpret. Without any processing, the reflector above the basalt and the converted waves cannot be identified. The shot gather from Model 3 (Fig. 10d) looks clearer, but neither the base basalt reflector nor the reflector below the basalt (equivalent to fourth reflector) can be identified. Model 4 (water depth 1000 m) allows the identification of the P-wave reflections and the converted waves (Fig. 12e). The converted wave arrivals move down the gather as the water depth increases and interference with the sea-bottom arrival increases at the far offset. Results of F–K modelling for Model 5 (thick basalt) are shown in Fig. 12f, with events again highlighted courtesy of ray-tracing. All the waves are recognized, even though the converted waves have very small

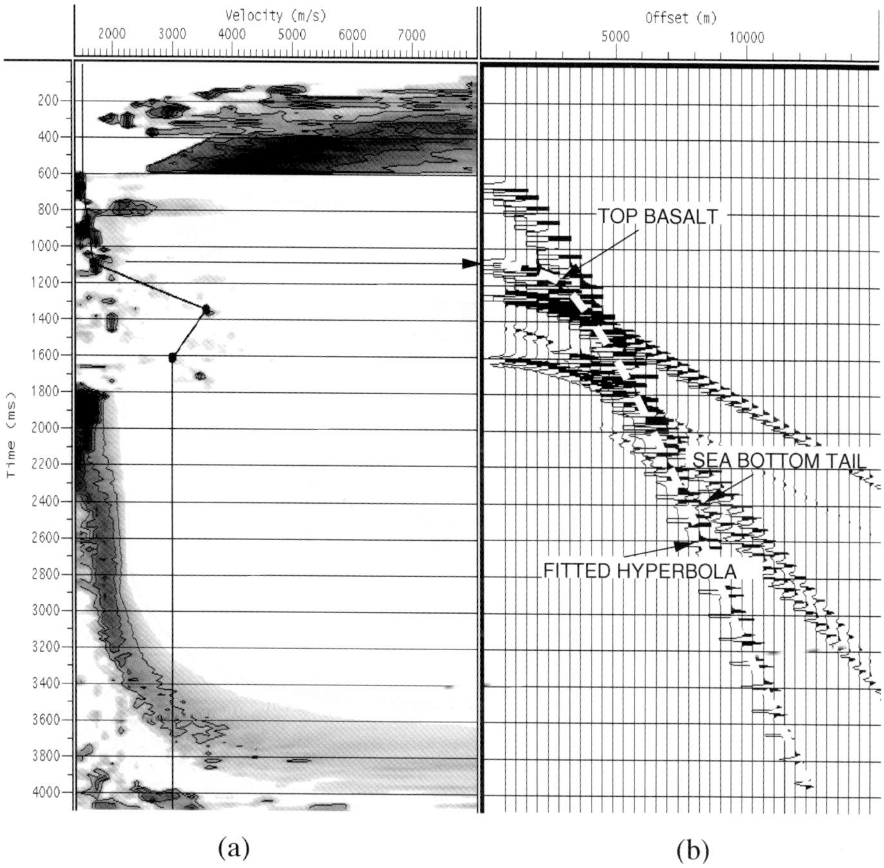

Fig. 11. Contoured velocity spectrum for CDP from Model 1 with maximum display offset of 5 km. (**a**) Velocity spectrum showing the function obtained by picking the P-wave velocity maxima. (**b**) The appropriate NMO hyperbola fits the top of basalt on the near offsets and the sea-bottom tail on the far offset This constructive interference results in the strong maximum observed on the velocity spectrum at 1100 ms and 1550 m s^{-1}, but also yields erroneous interval velocities.

amplitudes. In general, converted waves are more readily recognized with thicker basalt layers and increasing the basalt thickness increases the separation between the converted arrivals and the primary PP-wave reflections at mid-to-far offsets.

The results from F–K modelling show how difficult S-wave identification and processing can be in full wavefield synthetic data. This suggests that, in spite of the insights provided by ray-trace modelling, imaging via converted waves may be very difficult in the case of real data that contain additional noise components.

Conclusions

A preliminary evaluation of some ray-tracing and F–K modelling techniques helps to illuminate some of the critical factors with respect to sub-basalt seismic imaging. Some general principles have been identified, which could be useful for the acquisition and processing of seismic data, in areas containing basalt in the overburden:

(1) P-waves are useful at standard acquisition offsets only in the case of a shallow thick basalt layer.

(2) For long offsets, the largest P-wave amplitudes for the base basalt reflection and later arrivals are recorded for thick shallow basalt layers, in spite of the longer raypaths. This is because, for a given offset, the incidence angle is smaller for thicker basalt and the transmission coefficient is therefore larger.

(3) Converted S waves have relatively larger amplitudes than P waves for large offsets. This is important for real data, as all the P-wave arrivals

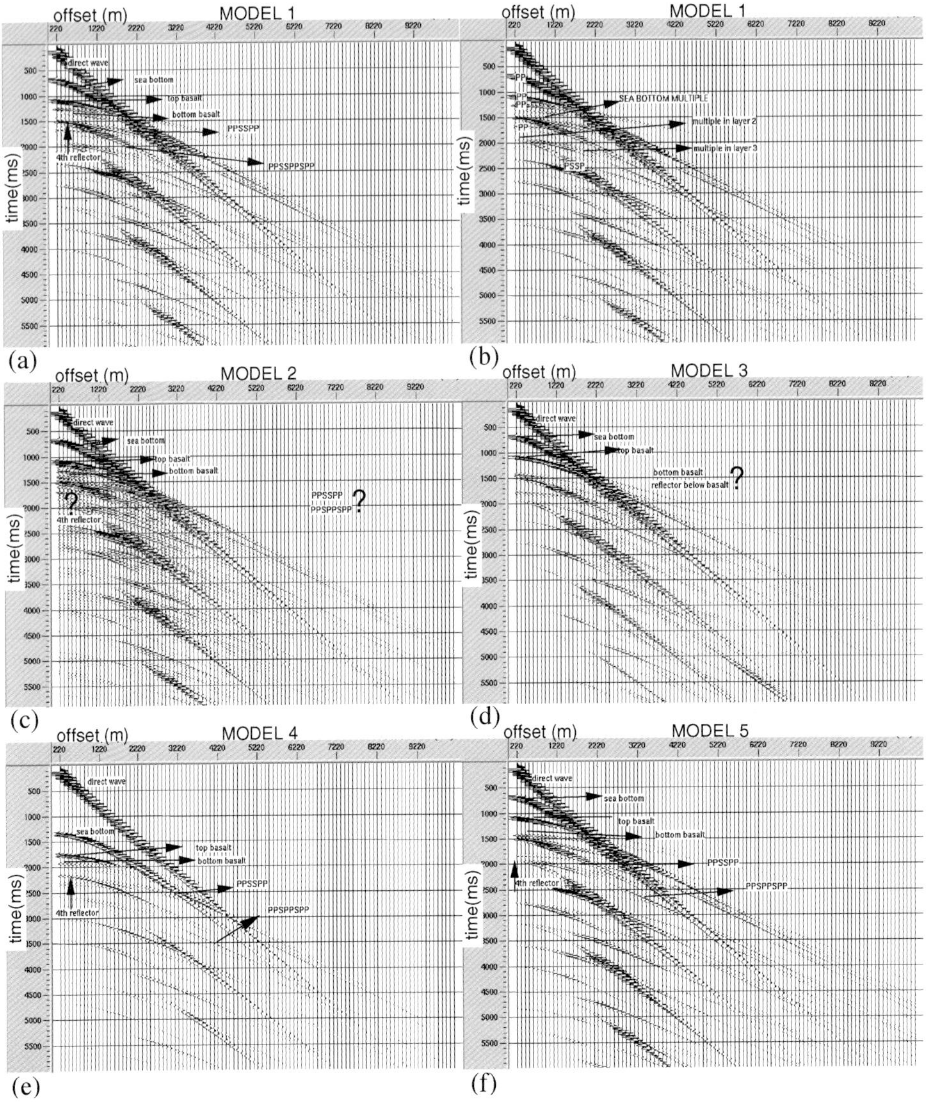

Fig. 12. F–K modelling results. Shot gather domain display of all waves, including multiples, for five models. All the events are highlighted courtesy of ray-trace modelling. (**a**) Model 1, P-wave arrivals and converted waves of interest in this study (S-wave segment in the basalt only) are highlighted. (**b**) Model 1, multiple reflections and other converted waves are highlighted, to show what makes the P-wave arrivals, and converted waves of interest, difficult to be identify. (**c**) Model 2, reflector below the basalt (fourth reflector) and the converted waves cannot be identified. (**d**) Model 3, shot gather looks clearer, but it is not possible to identify the base basal, the fourth reflector below or the converted waves. (**e**) Model 4, identification of P-wave arrivals and converted waves is possible. (**f**) Model 5, all the waves are recognized but the converted waves have small amplitude.

from under the basalt have small amplitudes and can be difficult to identify in presence of multiples and noise.

(4) Converted S waves are recorded at offsets of up to *c.*10 km. Long offset acquisition is required, as standard acquisition offsets are usually restricted to *c.* 4500 m.

(5) Travel time curves are non-hyperbolic for large offset acquisition (the small spread approximation is no longer valid). As a result, a three-term analysis may be required for NMO velocity optimization.

(6) In the case of thick basalt layers, converted S waves have strong amplitudes and manifest

continuity of the relatively high-amplitude arrivals over a large offset range.

(7) In the case of interbedded basalt and sediment layers, only converted S waves in the uppermost basalt layer have amplitudes that may be useful.

(8) Velocity maxima for PP events from the top of the basalt and PP and PS events from the base basalt form a triangular shape on contoured velocity spectra. This attribute may facilitate the identification of PS-wave arrivals.

The authors wish to acknowledge D.V. Corcoran, X. Y. Li and an anonymous reviewer, who helped to improve the manuscript. F.M. is sponsored by PIP Project 97/21.

References

DANCER, P.N. & PILLAR, N.W. 2001. Exploring in the Slyne Basin: a geophysical challenge. *In*: SHANNON, P.M., HAUGHTON, P.D.W. & CORCORAN, D.V. (eds) *The Petroleum Exploration of Ireland's Offshore Basins*. Geological Society, London, Special Publications, **188**, 209–222.

HANßEN, P. 1998. *Optimum conversion of shear waves for sub basalt imaging*. Edinburgh Anisotropy Project Sponsors Report.

KIORBOE, L. & PETERSEN, S.A. 1995. Seismic investigation of the Faeroe basalts and their substratum. *In*: SCRUTTON, R.A., STOKER, M.S., SHIMMIELD, G.B. & TUDHOPE, A.W. (eds) *The Tectonics, Sedimentation and Palaeoceanography of the North Atlantic Region*. Geological Society, London, Special Publications, **90**, 111–122.

LI, X.Y., MACBETH, C. & HITCHEN, K. 1998. *Using converted shear-waves for imaging beneath basalt in deep water plays*. Edinburgh Anisotropy Project Report.

NEWMAN, P. 1973. Divergence effects in a lateral earth. *Geophysics*, **38**, 481–488.

PURNELL, G.W. 1992. Imaging beneath a high-velocity layer using converted waves. *Geophysics*, **57**, 1444–1452.

PURNELL, G.W., MCDONALD, J.A., SEKHARAN, K.K. & GARDINER, G.H.F. 1990. Imaging beneath a high velocity layer using converted waves. *Expanded Abstracts of the 60th Annual International Meeting, Society of Exploration Geophysicists*, 752–755.

SAMSON, C., BARTON, P.J. & KARWATOWSKI, J. 1995. Imaging beneath an opaque basaltic layer using densely sampled wide-angle OBS data. *Geophysical Prospecting*, **43**, 509–527.

TATHAM, R.H. & GOOLSBEE, D.V. 1984. Separation of S-wave and P-wave reflections offshore western Florida. *Geophysics*, **49**, 493–508.

WHITE, R.S. & MCKENZIE, D.P. 1989. Magmatism at rift zones: the generation of volcanic continental margins and flood basalts. *Journal of Geophysical Research*, **94**, 7685–7729.

WHITE, R.S. & STEPHEN, R.A. 1980. Compressional to shear wave conversion in oceanic crust. *Geophysical Journal of the Royal Astronomical Society*, **63**, 547–565.

YILMAZ, O. 1987. Seismic data processing. *In*: DOHERTY, S.M. (ed.) *Investigations in Geophysics, 2*. Society of Exploration Geophysicists, Tulsa, OK.

The Mesozoic evolution of the southern North Atlantic region and its relationship to basin development in the south Porcupine Basin, offshore Ireland

SARAH JOHNSTON[1], ANTHONY G. DORÉ[2] & ANTHONY M. SPENCER[3]

[1]*Statoil Exploration (Ireland) Ltd, Statoil House, 6 George's Dock, IFSC, Dublin 1, Ireland (e-mail: sarah.johnston@statoil.com)*

[2]*Statoil (UK) Ltd, 11a Regent Street, London SW1Y 4ST, UK*

[3]*Statoil, Forushagen, 4035 Stavanger, Norway*

Abstract: The Mesozoic history of a number of Atlantic borderland sedimentary basins can be related to the early opening history of the southern North Atlantic Ocean. Regional tectonic controls such as plate motion vectors and the pre-existing tectonic grain had an important role in basin development and are expressed as local tectonostratigraphic events. The evolving palaeogeographies for the region are demonstrated in a series of maps based on computer-generated plate reconstructions. The Porcupine Basin, centrally located in the study area, lay close to the intersection of three plate boundaries that separated Eurasia from North America and controlled opening of the Bay of Biscay. The south Porcupine Basin, where there is relatively poor data control, is considered in the context of broader plate-tectonic controls, which were also responsible for the development of contiguous and better understood basins during Mesozoic time. This approach provides new insight into the structural evolution and likely facies development in the south Porcupine Basin, allowing broad inferences for petroleum prospectivity to be made. Initial Permo-Triassic fault-controlled extension led to continental deposition, which, if associated with aeolian and/or fluvial reservoir rocks, will mostly be too deep to be prospective. Thermal subsidence during Early Jurassic time was associated with flooding and fine-grained clastic deposition with anticipated moderate source rock potential. Regional uplift of the northern Porcupine area during Mid-Jurassic time forced shorelines and shelves southwards and the south Porcupine Basin could contain good reservoir quality sandstones and possible waxy deltaic-type source rocks of this age. In Late Jurassic time, major crustal extension took place with potential for reservoir and source rocks in locally expanded footwall successions. Further extensional faulting occurred in earliest Cretaceous (Neocomian) time with further synrift plays possible at this level. Growth of the Porcupine Median Volcanic Ridge is attributed to Barremian–Aptian time and related to continuing extension associated with a northwesterly arm of a triple junction positioned to the south of the Porcupine area. Strong subsidence of the basin centre during this time will have a significant impact on source rock maturation and flank trap geometries in the south Porcupine Basin.

The Porcupine Basin is a large, underexplored area located on the continental shelf, 200 km west of Ireland (Figs 1 and 2). It can be geographically subdivided into the northern and the southern sectors based on a bathymetric divide that equates with the limit of exploration drilling to date (Fig. 2). The south Porcupine Basin has water depths >1500 m with a northern limit at latitude 51°40′. This area was the focus of the 1998 south Porcupine frontier licensing round. The geological history of this southern area has been described previously by Masson & Miles (1986), Tate & Dobson (1988) and Tate (1992), largely on the basis of seismic mapping.

The North Porcupine Basin trends north–south and contains up to 9 km of Mesozoic and Tertiary section (Moore & Shannon 1995) displaying a typical 'steer's head' profile (Croker & Shannon 1987; Croker & Klemperer 1989. It is penetrated by 26 hydrocarbon exploration wells and is extensively covered by regional 2D seismic data. In this paper the term 'North Porcupine Basin' encompasses both the North Porcupine and Porcupine basins of Naylor *et al.* (1999). Although this area provides an excellent database, which can be used to obtain a better understanding of the frontier south Porcupine Basin, some fundamental differences

From: SHANNON, P.M., HAUGHTON, P.D.W. & CORCORAN, D.V. (eds). 2001. *The Petroleum Exploration of Ireland's Offshore Basins*. Geological Society, London, Special Publications, **188**, 237–263. 0305-8719/01/$15.00

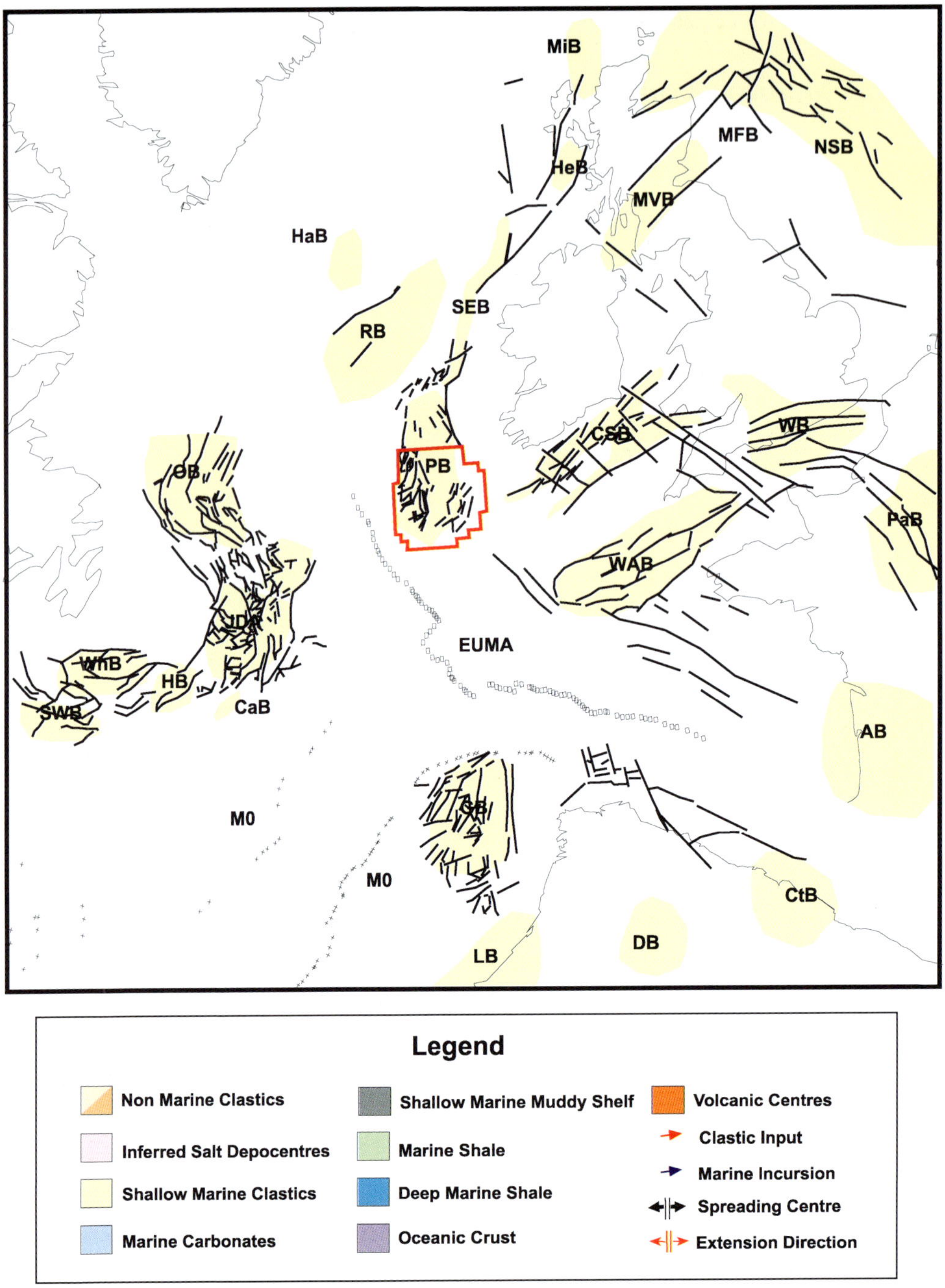
MiB
MFB
NSB
HeB
MVB
HaB
SEB
RB
OB
PB
CSB
WB
PaB
WAB
IDA
EUMA
WhB
HB
SWB
CaB
AB
M0
M0
GB
CtB
DB
LB
Legend
Non Marine Clastics
Inferred Salt Depocentres
Shallow Marine Clastics
Marine Carbonates
Shallow Marine Muddy Shelf
Marine Shale
Deep Marine Shale
Oceanic Crust
Volcanic Centres
Clastic Input
Marine Incursion
Spreading Centre
Extension Direction

are apparent: the Porcupine Median Volcanic Ridge (Tate & Dobson 1988) is restricted to the south Porcupine Basin, and the Cretaceous section is considerably thinner in the northern Porcupine area (Fig. 3a and b).

Notwithstanding these differences, the Mesozoic development of the south Porcupine Basin shares characteristics with the North Porcupine and also with other pre-Atlantic opening, contiguous basins such as the Jeanne d'Arc Basin, Lusitanian Basin and Celtic Sea basins (Fig. 4). The aim of this paper is to develop a better understanding of the evolution of the south Porcupine Basin by inferring the likely Mesozoic tectonics, depositional regime and prospectivity over the southern North Atlantic region as a whole, and in turn, relate these to this frontier basin ahead of drilling.

The Porcupine Basin is one of a series of Atlantic borderland basins that developed within a tectonostratigraphic framework related to the protracted opening history of the North Atlantic Ocean. Basin development took place under complex kinematic influences associated with the evolution of a nearby triple junction (Figs 5 and 6; Sibuet & Collette 1991). Permo-Triassic, Late Jurassic and Early–Mid-Cretaceous rifting events are recorded in the south Porcupine Basin, and these can be related to wider plate-tectonic controls. Although some of these episodes were marked by a particular tectonic grain (sometimes over a wide area) we have found no evidence for, nor need to evoke, plate-wide extension vectors. On the other hand, there is reasonable circumstantial evidence for inheritance of basement trends, and good evidence of multiple fault reactivation in the region as a whole.

Exploration history

Historically, exploration activity in the Porcupine Basin has concentrated on the northern part of the basin, where water depths vary between 50 and 1000 m. Most of the 27 exploration wells drilled to date have targeted Mesozoic tilted fault-block structures. These wells were located on the basin margins because of the limitations of drilling in deep water during the 1970s, which saw the most active period of exploration. Significant oil shows, and hydrocarbons capable of flowing to the surface, have been encountered but a commercially significant discovery has yet to be made.

An area covering over 27 000 km^2 was declared open to exploration with the announcement of the South Porcupine frontier licensing round by the Irish licensing authorities in 1998. This area is covered by over 10 000 km of speculative 2D seismic data (Fig. 2). Water depths range between 1000 and 2500 m. A single well penetration (43/13-1) drilled by BP in 1988 to test a Mesozoic tilted fault-block play is located on the NW margin of the licencing area (Fig. 2). Southeast of the licencing area, well 62/7-1 (Fig. 2), drilled by Esso in 1982, also tested a Mesozoic tilted fault-block structure. Therefore, whereas the North Porcupine Basin remains underexplored, the south Porcupine Basin can be described as a truly frontier area. In such a frontier setting, where there is a lack of direct information, Atlantic borderland basins (Fig. 1) that developed in a comparable tectonostratigraphic setting are of use in providing additional data, constraining petroleum exploration risk and developing play

Fig. 1. Plate-tectonic reconstruction of the southern North Atlantic region at 110 Ma (Aptian time) generated using PLATES (see text) showing the location of Mesozoic sedimentary basins in the area and observed fault patterns (after Chadwick *et al.* 1989; Petrie *et al.* 1989; Welsink *et al.* 1989; Wernicke & Tilke 1989; Wilson *et al.* 1989; Keen & Williams 1990; Ziegler 1990). Outline of the south Porcupine Basin frontier licensing round area is shown in red. Abbreviations in this and subsequent figures are as follows: AB, Aquitaine Basin; AD, Avalon Dyke, AM, Armorican Massif; AML, Avalon Meguma Lineament; BBRZ, Bay of Biscay Rift Zone; BH, Bank High; BTH, Beothuk Basin; BV, Barra Volcanic Ridge System; CaB, Carson–Bonnition Basin; CB, Cheshire Basin; CGFZ, Charlie–Gibbs Fracture Zone; CL, Clare Lineament; CM, Cornubian Massif; CSB, Celtic Sea basins; CtB, Cantabrian Basin; DB, Duero Basin; DT, Dominion Transfer; EUMA, European Magnetic Edge Anomaly; FC, Flemish Cap; GB, Galicia Bank, GH, Grampian High; HaB, Hatton Basin; HB, Horseshoe Basin; HeB, Hebrides Basin; HH, Hebrides High; HCM/NEA, Hatton Continental Margin and proto-NE Atlantic; IBM, Iberian Meseta; IM, Irish Massif; JDB, Jeanne d'Arc Basin; KB, Kish Basin; KCT, Kingscourt Basin, LB, Lusitanian Basin; LBM, London–Brabant Massif; LBR, Labrador Sea–Baffin Bay Rift; MB, Malin Basin; MFB, Moray Firth Basin; MiB, Minch Basin; MVB, Midland Valley Basin; NFL, Newfoundland; NSB, North Sea Basin; OB, Orpheus Basin; OK, Orphan Knoll; PaB, Paris Basin; PB, Porcupine Basin; PH, Porcupine High; PL, Peel Basin; PMVR, Porcupine Median Volcanic Ridge; PS, Porto Seamount; RB, Rockall Basin; RH, Rockall High; SB, Slyne Basin; SEB, Slyne and Erris basins; SM, Scottish Massif; SP, Shetland Platform; SPB, South Porcupine Basin; SWB, South Whale Basin; TAP, Tagus Abyssal Plain; TBH, Tail of the Bank High; TL, Tagus Lineament; UB, Ulster Basin; WAB, Western Approaches Basin; WB, Worcester Basin; WH, Welsh High; WhB, Whale Basin; WHP, West Hebrides Platform.

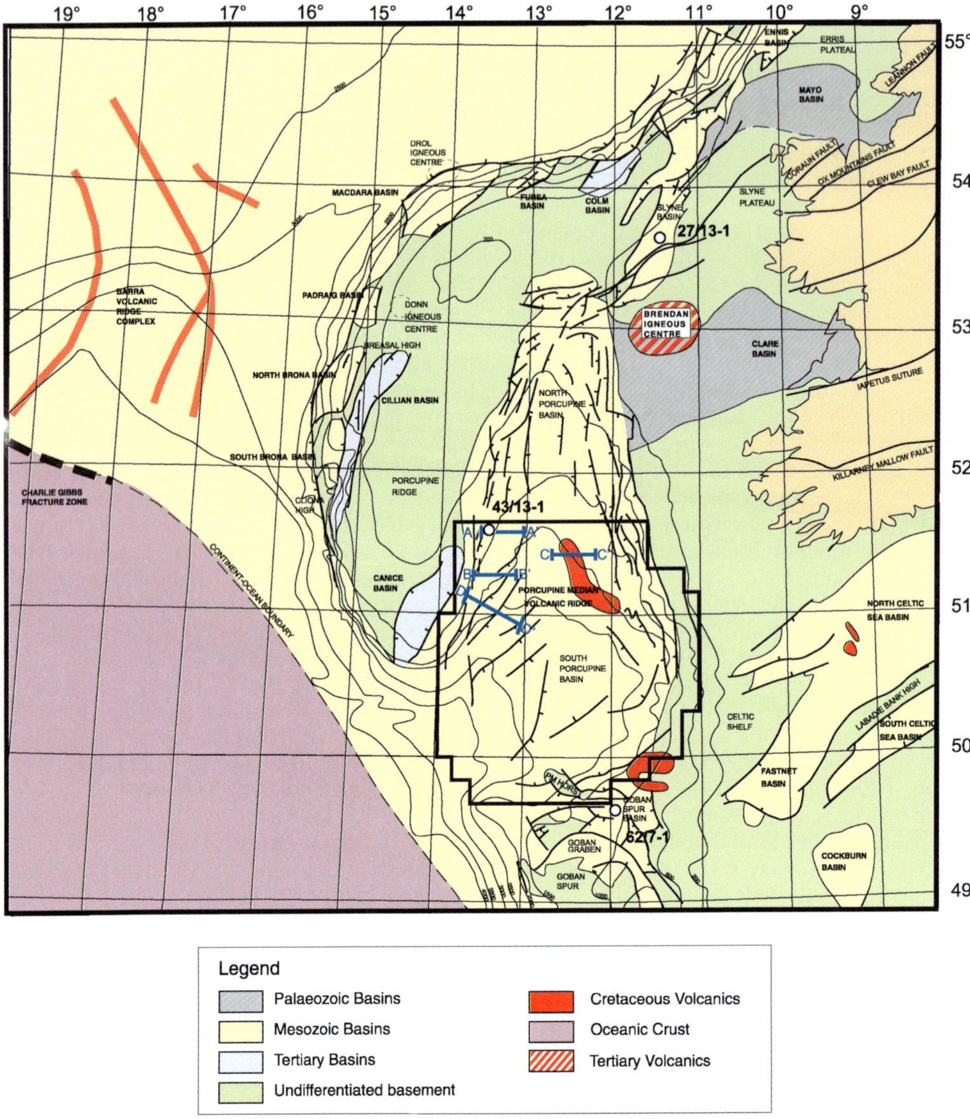

Fig. 2. Tectonic elements of the North and South Porcupine Basins in relation to surrounding Palaeozoic, Mesozoic and Tertiary basins, Cretaceous and Tertiary volcanic centres, basement blocks, and oceanic crust (modified from Naylor *et al.* 1999). Present-day bathymetric contours are in metres. The location of the South Porcupine frontier licensing round area is outlined in black. The location of Figures 13–16 is annotated in blue.

concepts. In this study, we consider a broad area incorporating the Grand Banks, offshore Newfoundland and the NW European margin. We refer specifically to the Jeanne d'Arc Basin, Celtic Sea basins and the Lusitanian Basin (Fig. 1). We have assembled this information in a series of kinematic and palaeogeographical maps, incorporating plate-tectonic reconstructions and regional tectonostratigraphy.

Plate-tectonic and palaeogeographical reconstructions

The plate-tectonic base maps cover an area stretching from the central Atlantic to the southern part of the NE Atlantic (Fig. 5). They were made using PLATES, an interactive plate modelling program developed at the University of Texas. Plate motion models (using finite difference poles of rotation) are based on updated

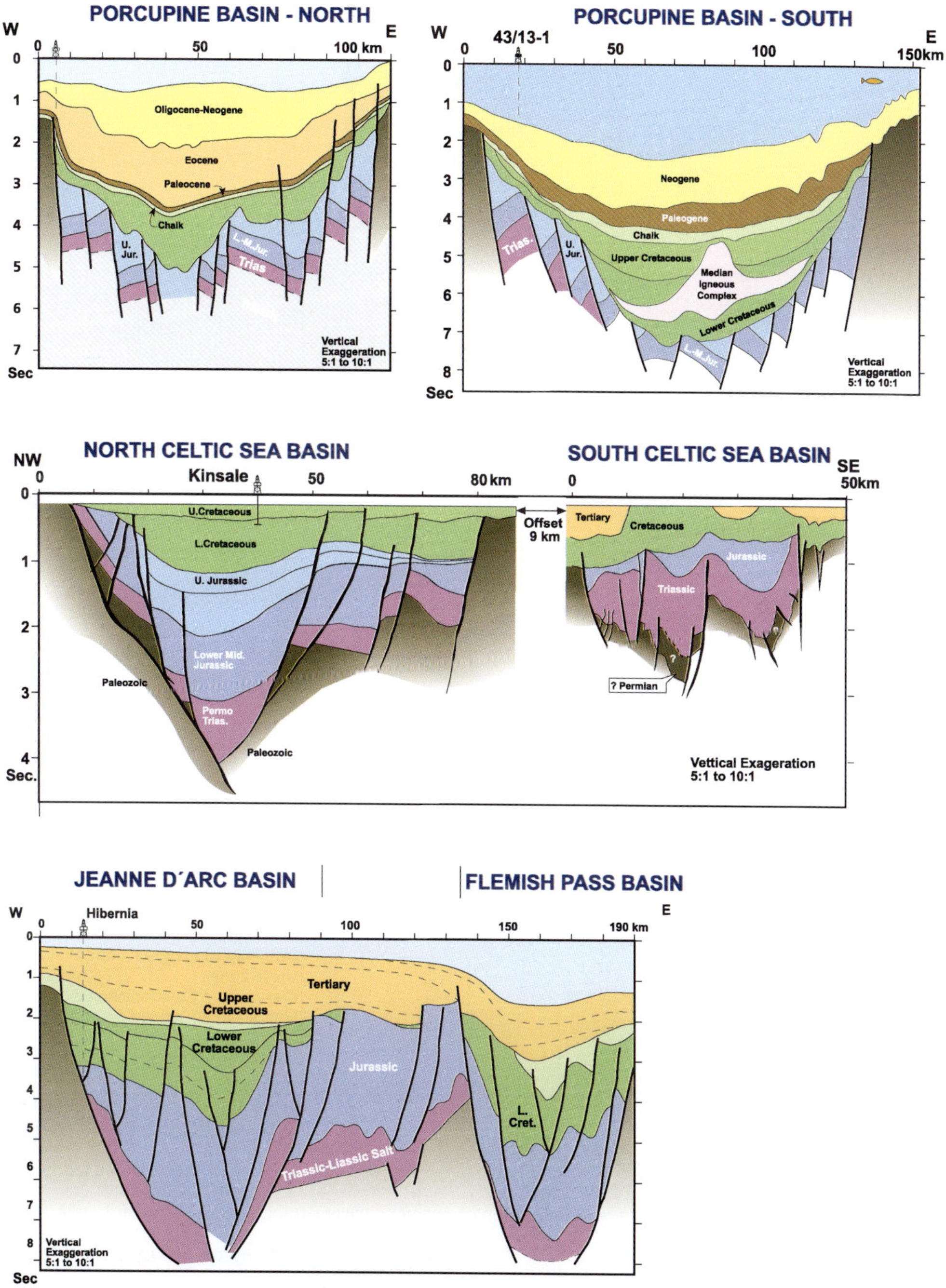

Fig. 3. Comparative geoseismic cross-sections through (**a**) the North and (**b**) south Porcupine Basins, (**c**) the North and South Celtic Sea basins and (**d**) the Jeanne d'Arc and Flemish Pass basins.

palaeomagnetic and hotspot track data and also draw from published academic geological and geophysical studies. PLATES uses text data files for plate characterization (outlines, identities, rotations, etc.), which can be manipulated, allowing customization of both plate outlines (allowing intra-plate deformation to be incorporated) and plate rotations. This provides a

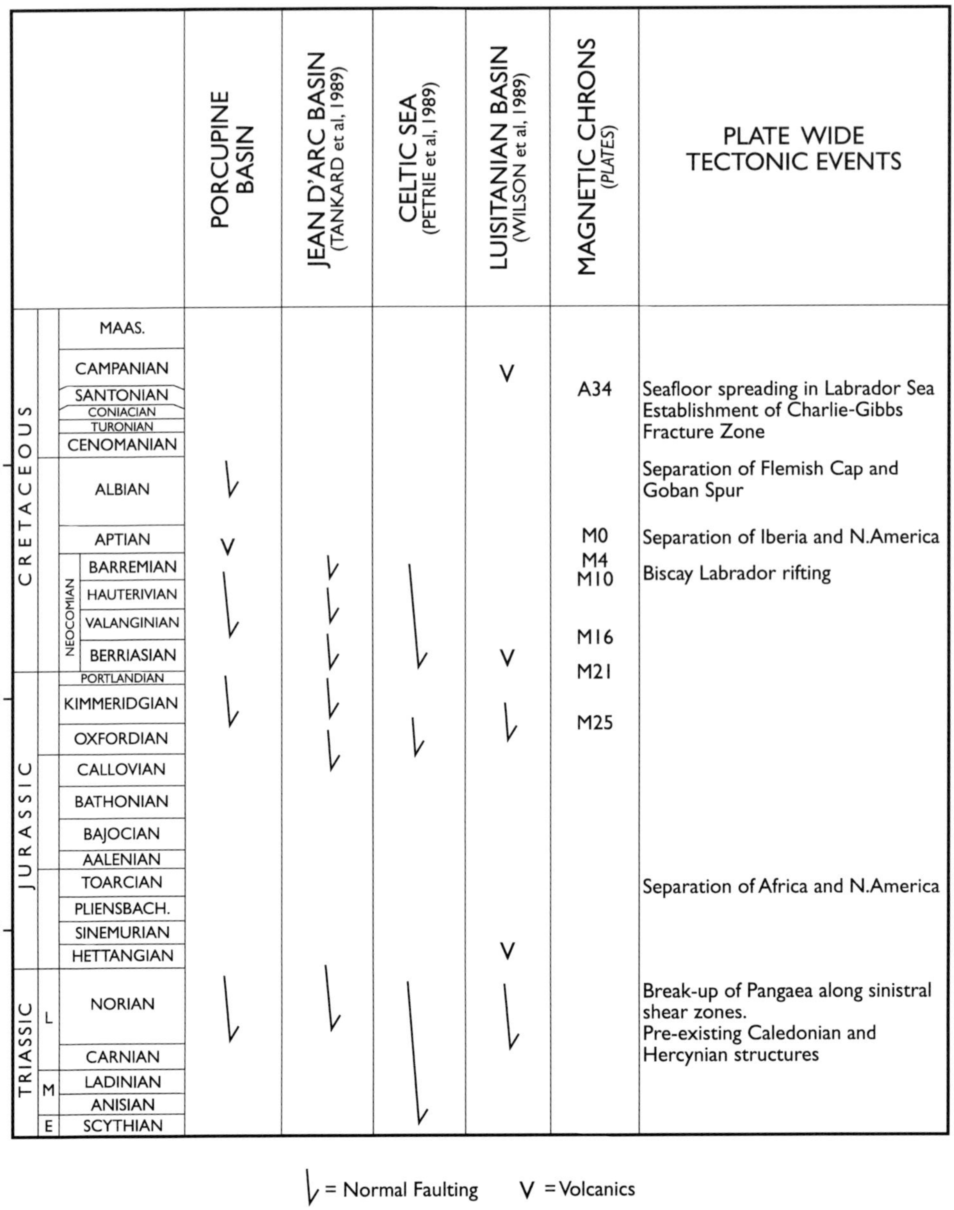

Fig. 4. Comparative tectonic histories of southern North Atlantic margin basins showing major extensional events (arrows) and episodes of volcanism (v), and the timing of magnetic chrons (modified after Tankard *et al.* 1989).

versatile package in which discrepancies in plate reconstructions can be isolated and modified.

The palaeogeographical maps presented in Figs 7-11 (Permo-Triassic to Hauterivian time) are drawn on a reconstruction made at 131 Ma, showing the continental plates of North America, Greenland, Rockall, Eurasia and Iberia in their positions before North Atlantic sea-floor spreading. Figures 1 and 12 (Barremian to Albian time) are based on a reconstruction at 110 Ma, when separation occurred between North America and Iberia, and opening of the Bay of Biscay took place. Magnetic chron M0 (Fig. 1) clearly shows plate separation and marks the continent–ocean boundary at the North American and Iberian plates margins. The European Magnetic Edge

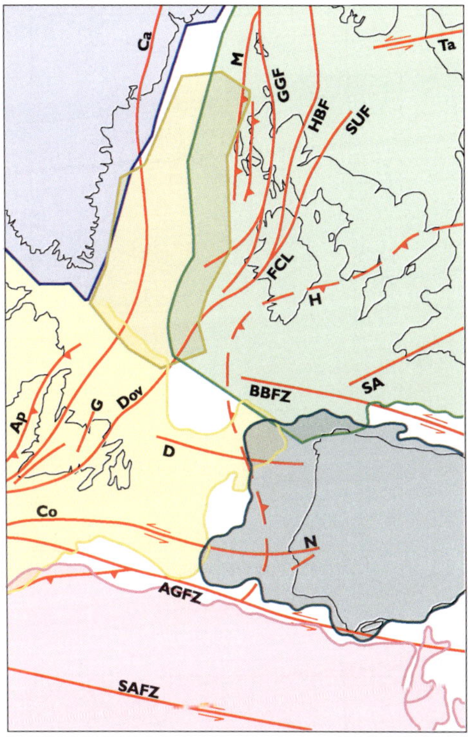

Fig. 5. Plate-tectonic reconstruction generated at 210 Ma (Late Triassic time) using PLATES (see text) showing the location of the Greenland plate (blue), Rockall plate (brown), Eurasian plate (green), North American plate (yellow), Iberian plate (blue) and African plate (pink) in their pre-Atlantic-opening positions. Crustal extension accounts for the overlap between plate boundaries. Observed fracture zones, shear zones and large-scale crustal lineaments (after (Tankard & Welsink 1989; Tankard *et al.* 1989; Verhoef & Srivastava 1989; Welsink *et al.* 1989; Ziegler 1990) which may, in part, account for the break-up of Pangaea (Ziegler 1990) are illustrated in red. Abbreviations for the lineaments and fracture zones are as follows: AGFZ, Azores Fracture Zone; Ap, Appalachian Front; BBFZ, Bay of Biscay Fracture Zone; Ca, Caledonian Front; Co, Collector; Dov, Dover; FCL, Fair Head–Clew Bay Line; G, Gander; GGF, Great Glen Fault; H, Hercynian Front; HBF, Highland Boundary Fault; M, Moine Thrust; N, Nazare Line; SA, South American; SAFZ, South Atlas Fracture Zone; SUF, Southern Uplands Fault; Ta, Tagus Lineament; To, Tornquist line.

Anomaly (Fig. 1) marks the northernmost extent of oceanic sea-floor spreading in the immediate area surrounding the Porcupine Basin.

The Mesozoic basins in the southern North Atlantic region are represented in their reconstructed positions on the palaeogeographical maps. In these reconstructions (Figs 7–12) we suggest links between basins and structural lineaments now separated by the Atlantic Ocean. These connections are, of course, conjectural and require confirmation by more detailed basin-to-basin comparisons. The palaeogeography shown on each map uses available stratigraphic information extrapolated between the basins. Present-day coastlines are added to indicate the position of palaeogeographical features relative to known landmarks. The plate edges are taken to be at the present-day continent–ocean boundaries, which for consistency are taken to be at the present-day 2000 m bathymetric contour. It should be noted that the latitudes and longitudes shown on the maps do not represent palaeo-geographical reference points. They are correct for present-day Ireland and the UK, and are intended as a reference frame to allow comparison between the series of maps.

Before Atlantic opening, the study area was located in a central position on the supercontinent of Pangaea, which existed between Carboniferous and Late Triassic times (Fig. 5; Ziegler 1990). This supercontinent appears to have been inherently unstable, with the result that assembly and the beginnings of continental break-up were virtually simultaneous (Doré *et al.* 1999). An early phase of fragmentation took place in Early Permian time (Coward 1995), including oceanic spreading in Tethys, which gave rise to the development of Gondwanaland in the south and Laurasia in the north. Further break-up of Pangaea may have been associated with movement along a series of deep-seated, ENE–WSW-trending, crustal shear zones with sinistral displacement (Ziegler 1990). However, it is not possible to infer a single extensional vector that could account for fragmentation of the supercontinent (Doré *et al.* 1999). Crustal fabrics that were established during and before the early break-up of Pangaea had a major influence on the later development of Atlantic borderland rift basins. Reactivation of these Variscan and earlier Caledonian structures gave rise to many basin-defining structures throughout the southern North Atlantic region (Fig. 5; Coward 1990; Bartholomew *et al.* 1993; Doré *et al.* 1999). The episodic northward propagation of the Atlantic Ocean is associated with the subsequent break-up of Laurasia, which included the continental plates of North America, Laurentia–Greenland, Iberia and Eurasia. The south Porcupine Basin lies on the Eurasian plate adjacent to the intersection of these four plate margins.

In the ensuing discussion, we have subdivided the Mesozoic evolution of the area into six

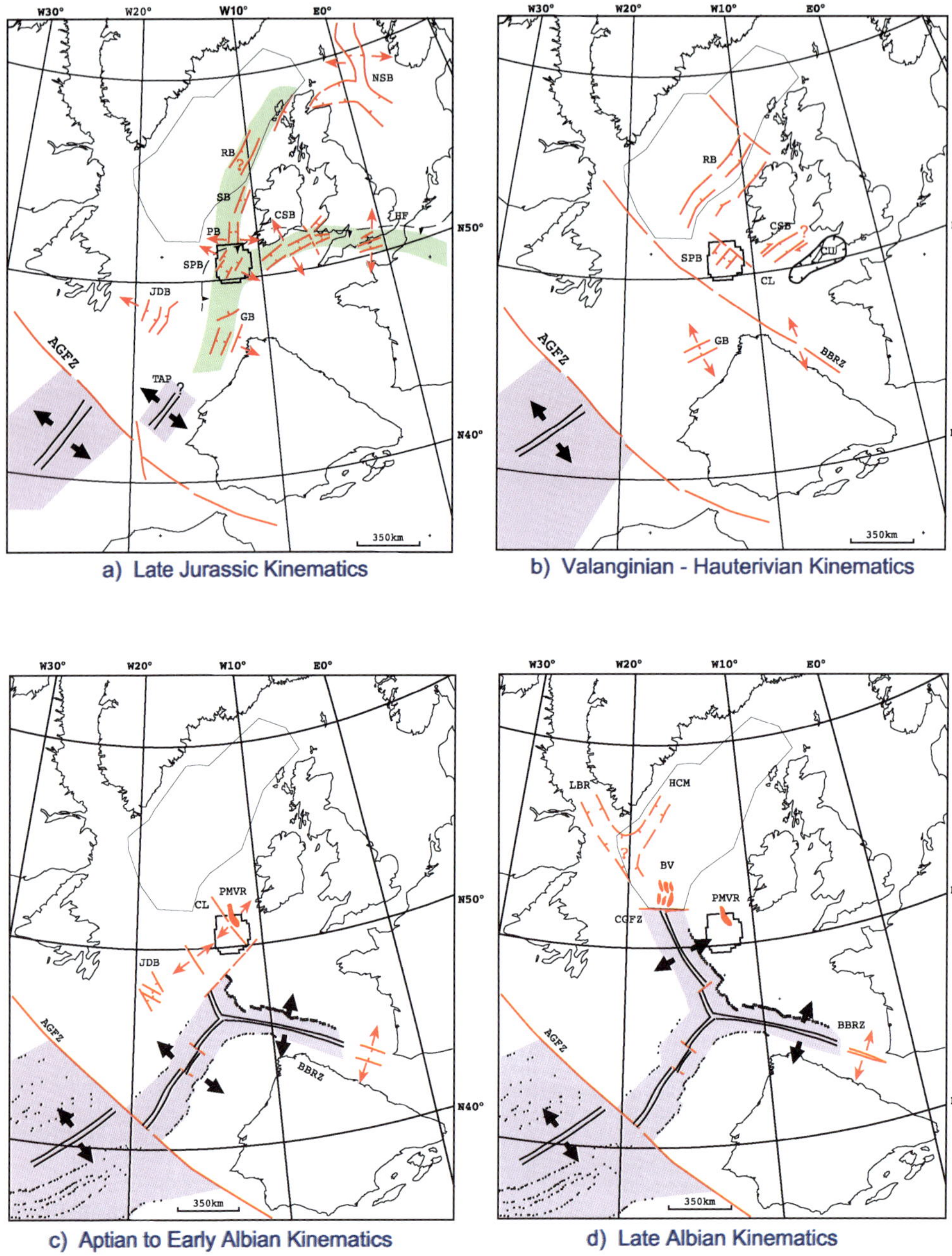

Fig. 6. Summary of the key kinematic events in the southern North Atlantic region which gave rise to Mesozoic basin development in the Porcupine area during (**a**) Late Jurassic, (**b**) Valanginian to Hauterivian, (**c**) Aptian to early Albian, (**d**) Late Albian times (see text). Plate-tectonic reconstructions created using PLATES are generated at 131 Ma (**a**, **b**), when plate separation took place between Africa and North America and 110 Ma (**c**, **d**) when plate separation took place between Iberia and North America. Abbreviations used are as in Figs. 1 and 5.

phases: the Permo-Triassic, Early Jurassic, Mid-Jurassic, Late Jurassic, Early Cretaceous (Valanginian–Hauterivian) and Early Cretaceous (Barremian–Albian) phases. These represent key phases in the inferred development of the south Porcupine Basin. The tectonic framework, palaeogeography and prospectivity of each phase are discussed in turn below.

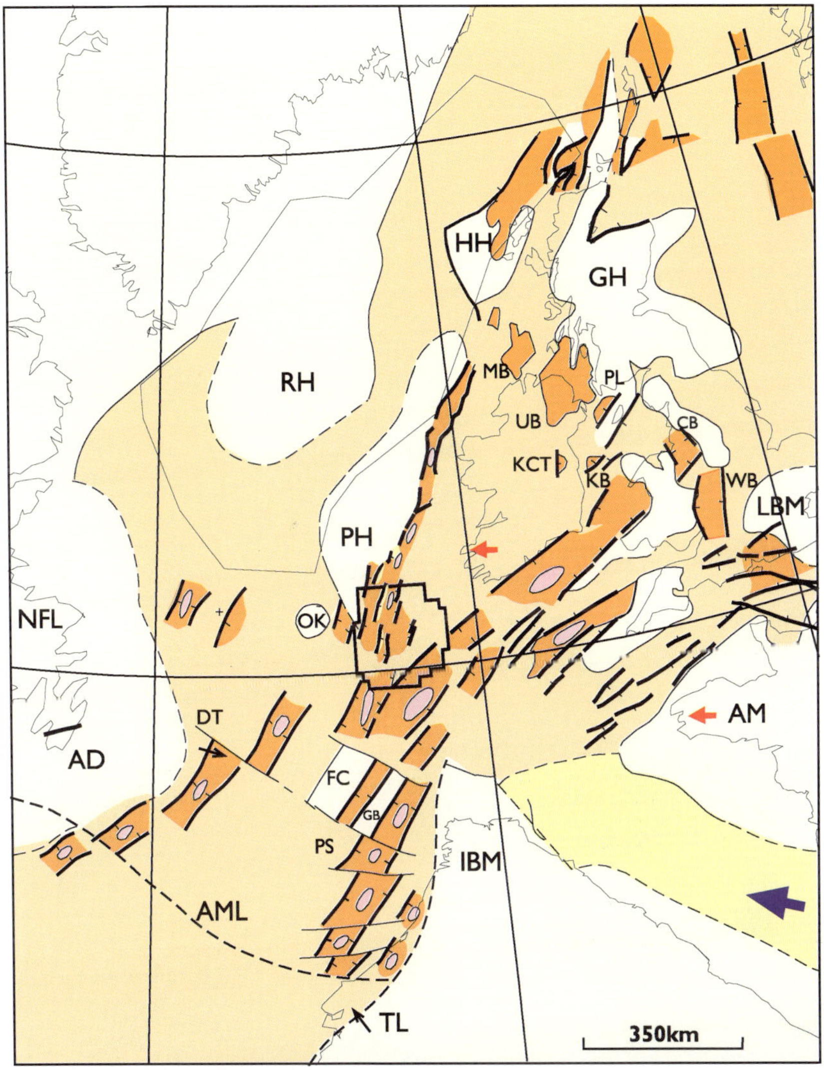

Fig. 7. Palaeogeographical map summarizing the gross depositional environments inferred for the Permian to Late Triassic interval based on a plate-tectonic reconstruction generated at 131 Ma using PLATES. The positions of inferred active faults and basins containing known Triassic strata are indicated (abbreviations are as for Fig. 1).

Permo-Triassic evolution: break-up of Pangaea

Tectonic setting

Permo-Triassic extension in the Porcupine Basin is interpreted to have created a predominantly NE–SW fault trend as a result of reactivation of both Hercynian and Caledonian structural lineaments (Shannon 1991). The south Porcupine Basin lay south of the inferred 'Hercynian Front', which is extrapolated from its position onshore (Gardiner & Sheridan 1981). Relaxation of NE–SW-trending Variscan thrusts is believed by some workers to have produced strongly asymmetrical, alternating half-grabens with successive down to the NW and down to the SE faults controlling graben geometry, for example in the Celtic Sea (Gardiner & Sheridan 1981). Inferred Permo-Triassic fault development is illustrated in Figs 5 and 7. The faults shown in Fig. 3 are conjectural for the south Porcupine Basin because earlier faulting is largely obscured by later Jurassic and Cretaceous rift events.

A similar extensional setting is anticipated for structural development of the Slyne and Erris

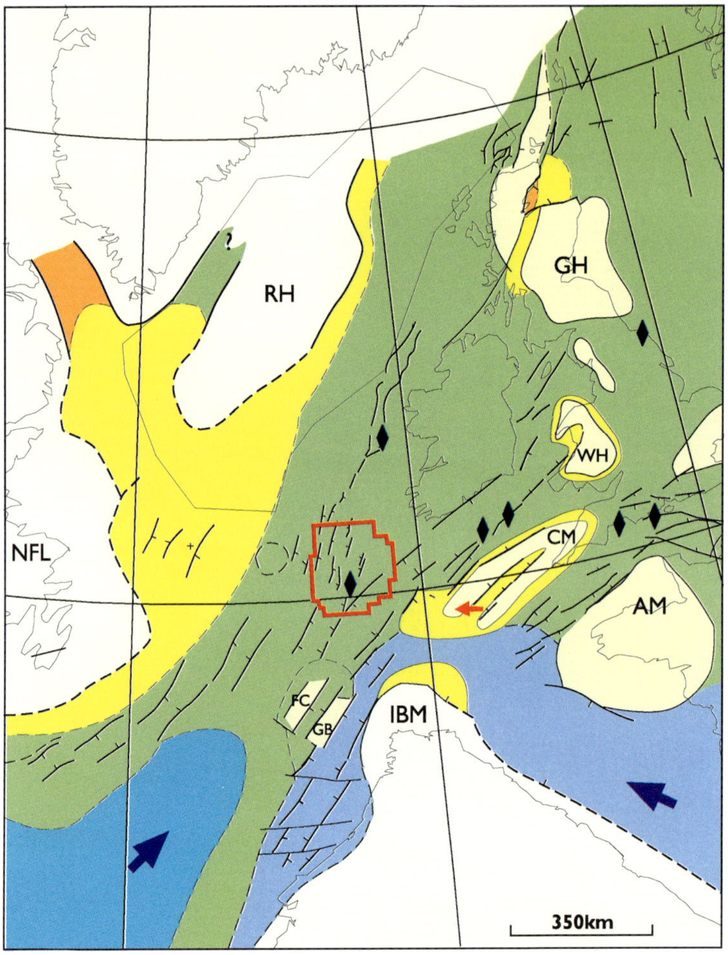

Fig. 8. Palaeogeographical map summarizing the gross depositional environments inferred for the stratigraphic interval from Sinemurian to Toarcian (Early Jurassic) time based on a plate-tectonic reconstruction generated at 131 Ma using PLATES (abbreviations are as in Fig. 1).

basins, where relaxation of Caledonian thrusts north of the Hercynian Front may have given rise to a more NNE–SSW basin trend (Tate & Dobson 1988; Chapman *et al.* 1999). Further to the south, crustal extension and basin formation in western Iberia may have been linked to crustal detachments following discontinuities in the Hercynian basement (Wilson *et al.* 1989). Extension west of Iberia was linked to a westward-dipping crustal detachment that controlled basin development (Wilson *et al.* 1989).

West of the Grand Banks, the Fundy and the Orpheus pull-apart basins have been interpreted by Tankard *et al.* (1989) to have developed in response to right-lateral movement along the Avalon Meguma Suture. Similar movement along the Bay of Biscay Fracture Zone (BBFZ, Fig. 6) may also have resulted in basin formation in the Orphan Basin, Flemish Pass Basin and Orphan Knoll (Fig. 5) areas, and may possibly have been responsible for structural complexity in the south Porcupine Basin during this rift event.

Palaeogeography

The inferred Permo-Triassic palaeogeography and gross depositional environments are shown in Fig. 7. Western and Central Europe drifted northwards into the trade wind belt during Permian time, where continental-style sedimentation took place in a series of intermontane

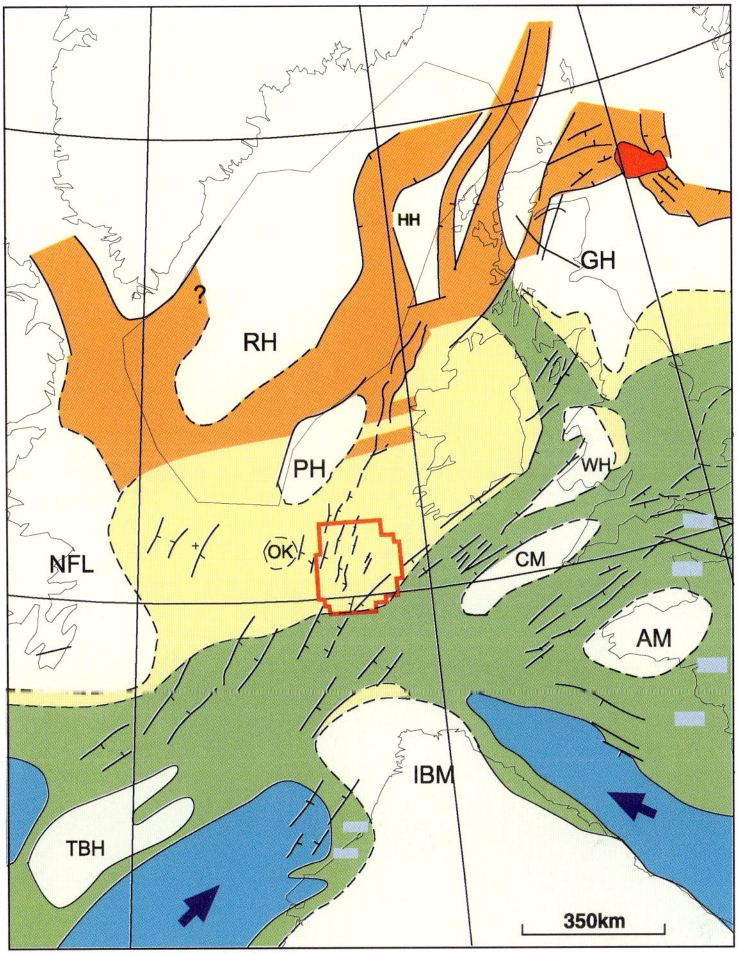

Fig. 9. Palaeogeographical map summarizing the gross depositional environments inferred for the stratigraphic interval from Bajocian to Bathonian (Mid-Jurassic) time based on a plate-tectonic reconstruction generated at 131 Ma using PLATES (abbreviations are as in Fig. 1).

peripheral collapse basins (Uchupi 1988). Well data and onshore outcrop throughout the area from the North Sea (Coward 1995) to the Lusitanian Basin (Wilson *et al.* 1989) support this interpretation of a wholly non-marine sedimentary succession. Cyclic climatic conditions with wetter, more humid intervals gave rise to increased runoff, increased alluvial or fluvial sedimentation and the development of ephemeral lakes (Uchupi 1988). During drier cycles, a reduction in runoff resulted in reworking of existing alluvial–fluvial sand bodies by aeolian processes. The prevailing westerly-directed trade winds (Jackson *et al.* 1995) may have preferentially accumulated aeolian sediments against the margins of the actively subsiding easterly dipping half-grabens in this rift system. The precipitation of evaporites may have taken place towards the axes of the half-grabens as ephemeral lakes dried up.

A similar depositional setting is interpreted for the Slyne and Erris basins to the north of the Porcupine Basin, where there are thick Permo-Triassic successions of non-marine mixed clastic deposits and evaporites (Chapman *et al.* 1999). East of the Porcupine area, the inferred Late Permian deposits in the Celtic Sea basin (Fig. 1) were dominated by alluvial-fan coarse-grained clastic sediments that were deposited along the downthrown side of syndepositional normal faults (Petrie *et al.* 1989). These clastic deposits grade up into evaporitic mudstones and massive halites in the Celtic Sea basins, Bristol Channel Basin, St. George's Channel Basin and Cardigan

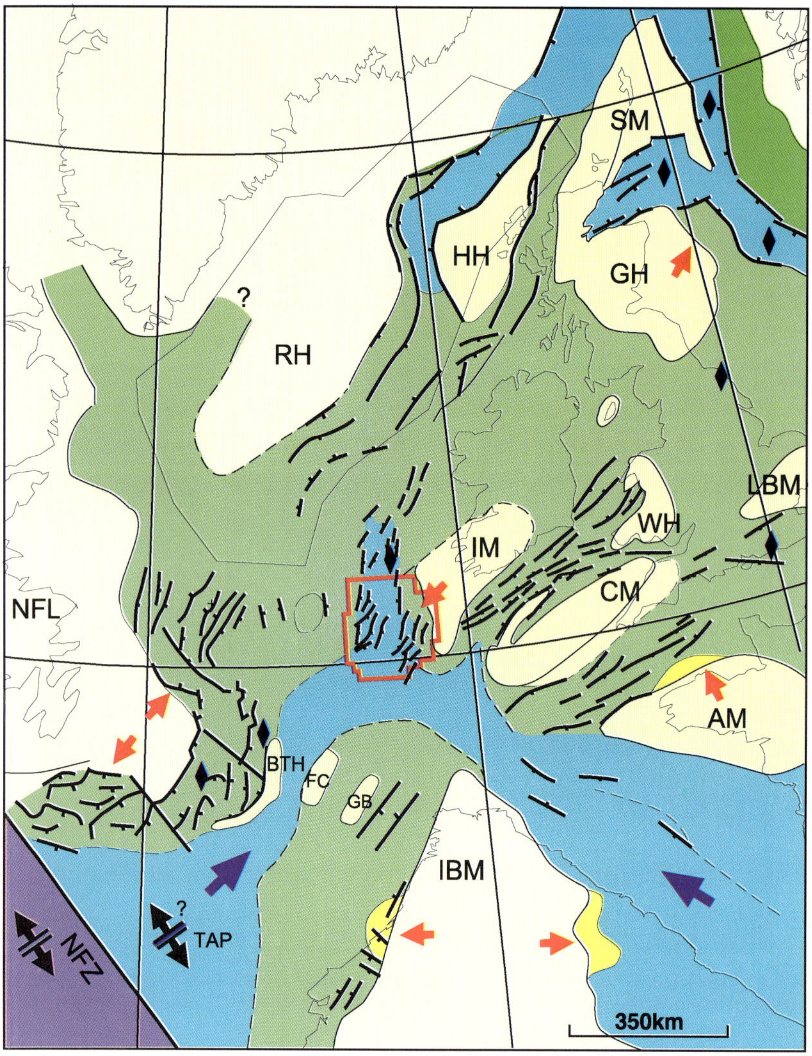

Fig. 10. Palaeogeographical map summarizing the gross depositional environments inferred for the stratigraphic interval from Kimmeridgian to Portlandian (Late Jurassic) time based on a plate-tectonic reconstruction generated at 131 Ma using PLATES (abbreviations are as in Fig. 1).

Bay Basin (Petrie *et al.* 1989). Strata of Triassic age were encountered in well 26/21-1 in the North Porcupine Basin, and an intra-Kimmeridgian salt section has been drilled in well 35/19-1, where it may have been injected upwards across the basin-bounding fault along a salt wall from possible undrilled Triassic strata present deeper in the basin.

To the south of the general Porcupine area, a thick succession (60–388 m) of Permo-Triassic non-marine clastic sediments, halites and gypsum is preserved in western Iberia (Wilson *et al.* 1989). Permo-Triassic sedimentation is well characterized in the Lusitanian Basin, where a combination of good onshore exposure combined with data from 12 offshore wells provides detailed lithostratigraphic information. The Upper Triassic Silves Formation in the Lusitanian Basin has been described as a clastic succession that was shed westwards and southwards into a series of tilted half-grabens (Wilson *et al.* 1989). Good reservoir-quality sandstones deposited by braided streams are anticipated on the eastern flanks of the sub-basins. A thin shale

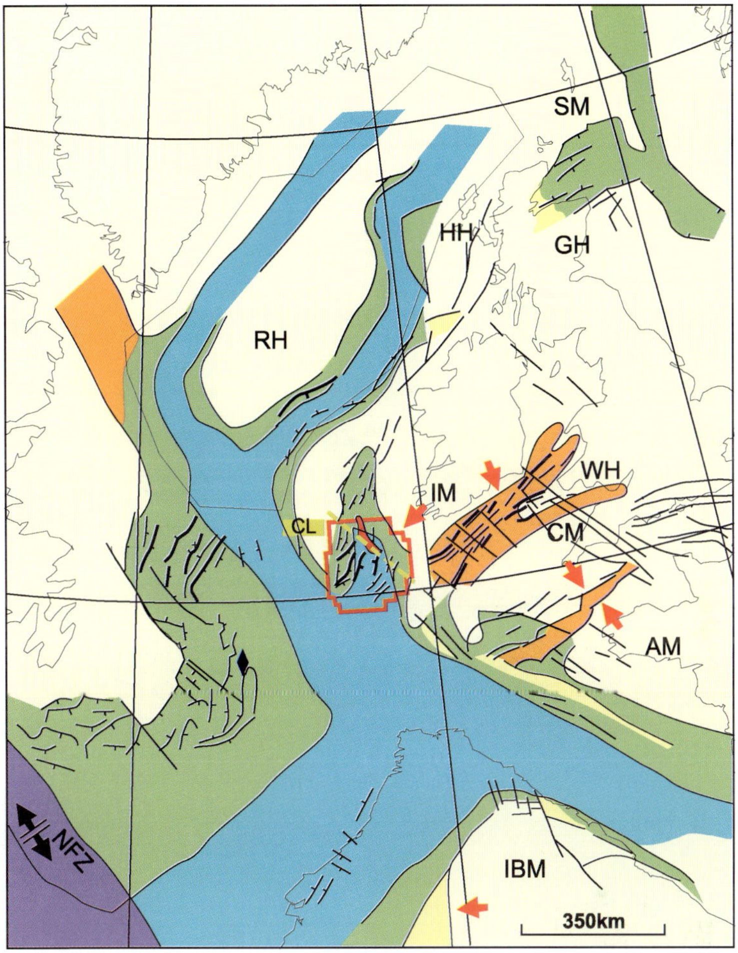

Fig. 11. Palaeogeographical map summarizing the gross depositional environments inferred for the stratigraphic interval from Valanginian to Hauterivian (Early Cretaceous) time based on a plate-tectonic reconstruction generated at 131 Ma using PLATES (abbreviations are as in Fig. 1).

at the top of the succession represents a marine incursion from the Tethyan Ocean through the Bay of Biscay Fracture Zone. Similarly, the Late Triassic succession of the Jeanne d'Arc Basin exhibits non-marine clastic deposition of the Eurydice Formation (Tankard *et al.* 1989). This is succeeded by thick Triassic–Liassic evaporites of the Argo Formation.

Thus, in common with Permo-Triassic continental deposits elsewhere in NW Europe, sedimentation is likely to have been climatically and fault controlled in the south Porcupine Basin. However, a Permo-Triassic succession has not been penetrated by wells in the south Porcupine Basin and its presence is therefore entirely conjectural, based on the regional observations described above.

Implications for prospectivity in the south Porcupine Basin

The Permo-Triassic interval has the potential to develop a mixed succession of non-marine sandstones and clastic deposits, shales and evaporites in the south Porcupine Basin. The best reservoir potential is likely to occur within the inferred aeolian sandstones, which, if present, will probably to be areally restricted to the western margins of Permo-Triassic sub-basins because of the prevailing westerly wind directions and development of accommodation space. Possible bedded evaporites or sabka deposits may provide semi-regional seals. However, the hydrocarbon charging potential for such successions is

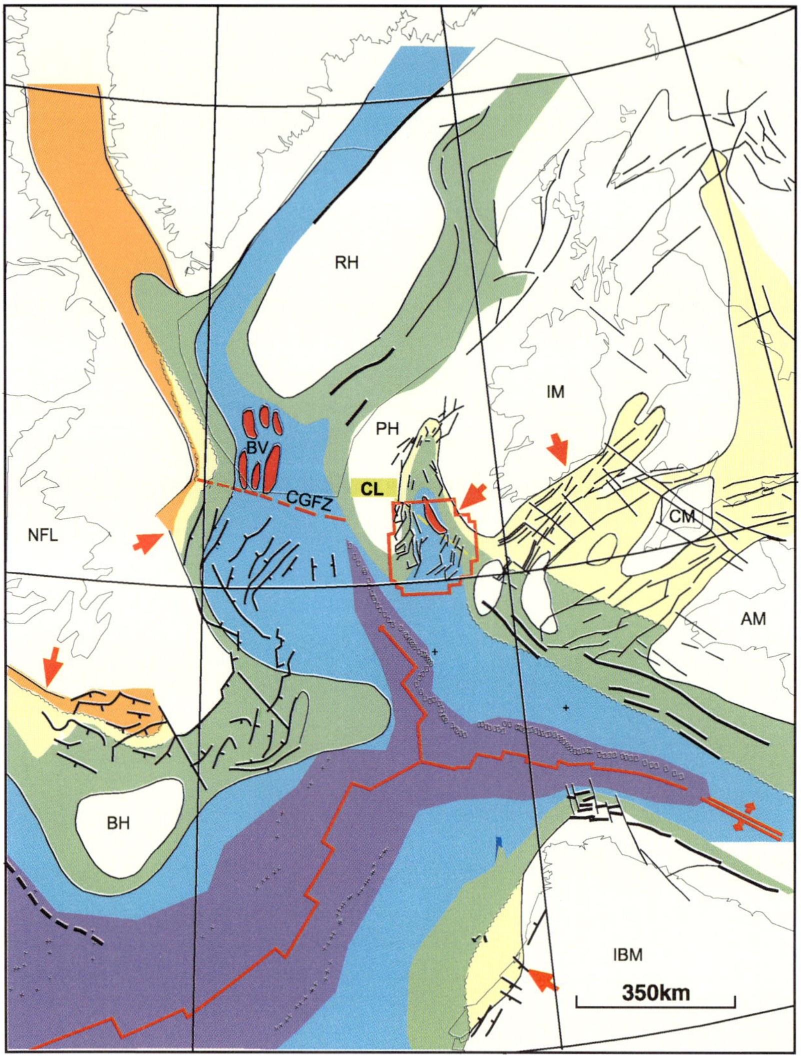

Fig. 12. Palaeogeographical map summarizing the gross depositional environments inferred for the stratigraphic interval from Barremian to Albian (Early Cretaceous) time based on a plate-tectonic reconstruction generated at 110 Ma using PLATES (abbreviations are as in Fig. 1). Sea-floor spreading took place between Iberia and North America at Chron M0 time (118 Ma, Aptian time) and oceanic crust was formed south of the South Porcupine Basin by Late Albian time.

unknown, and identifying traps will be difficult because of complex (multiply rifted) structure and poor seismic resolution at this level.

Early Jurassic phase

Tectonic setting

Early Jurassic extensional faulting, which may be related to incipient sea-floor spreading in the central Atlantic, is recorded in the Lusitanian Basin, offshore Nova Scotia and Newfoundland (e.g. Roberts *et al.* 1999). Closer to the south Porcupine Basin, in the North Celtic Sea Basin and the Hebridean Basin, minor Early Jurassic extensional faulting has been reported (e.g. Morton 1989; Roberts *et al.* 1999). In general, however, this area was tectonically quiescent and was dominated by a thermal subsidence regime following the phase of Permo-Triassic rifting.

Palaeogeography

During post-rift thermal subsidence, a marine transgression took place as far north as the Barents Sea (Roberts *et al.* 1999). The old Triassic rift system became flooded from the south across the Tagus Abyssal Plain and from the east through the Biscay Rift Zone (Figs 6 and 8). A mononotonous series of mudstones, shales and carbonates were deposited in the Grand Banks area (the Murre Formation of the Jeanne d'Arc Basin) and these have poor source-rock potential (Tankard *et al.* 1989). Meanwhile, in the Lusitanian Basin, limestones of the Coimbra Formation are overlain by the deeper-water shaly carbonates of the Toarcian Brehna Formation, the latter forming a source rock in this area (Wilson *et al.* 1989). Restricted anoxic shales deposited in the deep half-graben setting of the old Triassic rift system may account for the pervasive development of Liassic source rocks across NW Europe in Yorkshire, the Wessex Basin and the contiguous Paris Basin (Scotchman 2001).

Well data indicate that the richness of the Toarcian source rock developed in the North Celtic Sea Basin deteriorates westwards towards the Fastnet Basin (Murphy *et al.* 1995). A thick Liassic shale section was encountered in the Goban Spur well 62/7-1 to the south of the Porcupine Basin area (Fig. 2). This exhibits residual total organic carbon (TOC) values (Cook 1987) and has low potential as a hydrocarbon source rock. Meanwhile, in the Slyne Basin to the north, well 27/13-1 penetrated a rich oil-prone source rock of Toarcian age (Scotchman 2001).

The presence of a sandy Sinemurian shelf is inferred in the North Celtic Sea Basin and in the Fastnet Basin where an overall coarsening-upwards sequence is interpreted from well-log information (Kessler & Sachs 1995). These inferred clastic deposits were probably derived locally from the Cornubian Massif, which lay to the east and may have been one of several shallow-marine sequences on shelves that fringed palaeohighs (Fig. 8) that developed at this time. The Sinemurian shelf sandstones, which form an important reservoir unit in the North Celtic Sea Basin (Kessler & Sachs 1995), probably did not extend as far west as the south Porcupine Basin (Fig. 8). To the far west, on the North American plate, a probable prograding shallow-marine clastic shelf is inferred in this study, and may have resulted from uplift of a hinterland of unknown affinity (Roberts *et al.* 1999).

Lower Jurassic units have not been encountered in wells in the Porcupine Basin. However, by inference from surrounding basins, a marine shale is predicted to have been deposited throughout the basin area and the Irish Massif is interpreted to have been totally or partially submerged during this period of marine transgression (Fig. 6).

Implications for prospectivity in the south Porcupine Basin

The comparison with nearby areas suggests that Lower Jurassic source rocks may have been deposited in the south Porcupine Basin. However, the low TOC values present in the nearest well (62/7-1), and the deterioration in quality of the Toarcian source rock of the Celtic Sea basins westwards, point to limited source rock potential. Comparisons with the Celtic Sea basins area also suggest that there is low potential to develop good reservoir rock during this stratigraphic interval. However, Lower Jurassic shales may provide good seals to the underlying Upper Triassic plays.

Mid-Jurassic phase

Tectonic setting

The Mid-Jurassic period saw the initiation of sea-floor spreading in the central Atlantic in Bajocian–Bathonian time, at *c.* 175–180 Ma (see discussion by Driscoll *et al.* (1995)). Farther north, post-rift thermal subsidence with only minor faulting prevailed along much of the proto-Atlantic seaboard. The generally quiescent tectonic setting was interrupted, however, by a period of uplift or restriction in the North Atlantic which appears to have closed the Mesozoic seaway (Doré *et al.* 1999). This broad uplift, of late Bajocian to Bathonian age, included the northern part of the Porcupine Basin and the subsequent forced regression may have caused southward clastic progradation (Fig. 9).

Palaeogeography

Regional evidence supports the presence of an extensive shallow-marine progradational siliciclastic shelf system (Fig. 9), which may have been sourced from an uplifted landmass to the NW of the Porcupine area (Roberts *et al.* 1999). Such deposits were developed in the Jeanne d'Arc Basin during Bajocian–Bathonian time when mobilization of salt in the underlying Lower Jurassic Argo Formation took place as a result of sediment loading (Tankard *et al.* 1989) (Fig. 3). Bathonian to Callovian sandstones are reported in the Golconda well in the Flemish Pass Basin (Foster & Robinson 1993). The Deep Sea Drilling Project (DSDP) well at Site 111 (Ocean Drilling Program Leg 149) located in the Orphan Knoll, which lay to the west of the Porcupine Basin at that time, encountered a sandstone section of Bajocian age (Tate & Dobson 1988). Meanwhile, more open-marine conditions persisted further south. In the southern Grand Banks area, a monotonous series of calcareous shales and limestones were deposited in a broad epeiric basin setting that was undergoing uniform subsidence (Welsink *et al.* 1989), and open-marine carbonates are reported in the Middle Jurassic sequence of the Lusitanian Basin (Wilson *et al.* 1989).

This clastic progradational system is seen to the north of the Porcupine Basin in the Slyne and Erris basins, where a paralic setting with occasional marine incursions is interpreted from the Upper Bajocian–Bathonian sequence encountered in well 27/13-1 (Fig. 2; Chapman *et al.* 1999). A similar paralic setting is known to extend northwards into the Hebrides Basin, where Tethyan fauna have been reported (Morton 1989, 1992), whereas deposition of the fluvio-deltaic Brent province took place in the North Sea as a result of the development of the North Sea Dome (Underhill & Partington 1993).

The southerly extent of this inferred shallow-marine progradational system can be reasonably well established from well data in the North Celtic Sea Basin, where a monotonous series of shelf mudstones and siltstones form a complete stratigraphic sequence deposited between Sinemurian and Bathonian times (Murphy *et al.* 1995). Progradation and shallowing within this unit is inferred from a very gradual coarsening-upward log signature for the Bajocian–Bathonian sequence. The extent of maximum progradation is difficult to establish but a shoreface section interpreted in the North Celtic Sea Basin (Kessler & Sachs 1995) provides a southerly palaeogeographical limit. A thin shoreface sandstone of Bathonian age is encountered in well 62/7-1 (Fig. 2; Cook 1987), which also helps to constrain the southernmost limit of maximum progradation of this shallow-marine clastic succession.

In the North Porcupine Basin in block 26/28, Middle Jurassic fluvio-deltaic sandstones of poor reservoir quality are reported to be derived from the NW (MacDonald *et al.* 1987). A reinterpretation of biostratigraphic data in wells in the North Porcupine Basin (proprietary unpublished report) suggests, however, that sediments of Bajocian–Bathonian age may be missing in this area. It is this observation that leads us to suggest that the northern part of the Porcupine Basin, in common with other parts of the Atlantic seaway in the vicinity of the UK and Ireland, became uplifted or restricted during Mid-Jurassic time. By comparison with surrounding basins, a Bajocian to Bathonian shallow-marine clastic system is predicted in the south Porcupine Basin, representing a forced regression associated with the uplift of the northern part of the basin (Fig. 9).

Implications for prospectivity in the south Porcupine Basin

The Middle Jurassic Bajocian to Bathonian succession in the south Porcupine Basin is interpreted to represent an overall north-to-south prograding marine siliciclastic system (Fig. 9) with the potential to develop reservoir quality sandstones in an upper shoreface setting. Revised biostratigraphy (proprietary unpublished data) means that these sandstones are untested in the Porcupine Basin. Middle Jurassic source rocks have been encountered in wells 26/28-1, 34/15-1 and 35/6-1. These are oil-prone waxy lacustrine organic-rich shales with good oil-source potential (Butterworth *et al.* 1999), which may have been deposited in a delta top setting. Similar source rocks may extend southwards into the southern Porcupine area.

Late Jurassic phase

Tectonic setting

During Late Jurassic time, plate separation between America and Africa continued, and sea-floor spreading was decoupled from the area to the north by the Grand Banks–Azores–Gibraltar Fracture Zone (e.g. Srivastava *et al.* 1990; Fig. 6). A possible spreading axis may have existed immediately north of this in what is now the Tagus Abyssal Plain. This interpretation, however, remains controversial (compare Boillot & Malod (1988) with Srivastava *et al.* (1990)). Farther north, a very widespread episode of crustal stretching and extension took place. This has been recognized along the entire NW European margin from offshore Iberia to the Barents Sea. This extensional episode may reflect fragmentation of the plate in response to both central Atlantic and Tethyan speading (e.g. Doré *et al.* 1999).

In the Porcupine Basin this phase of extension created the most significant and largest fault offsets observed in the basin (Fig. 13) together with considerable synrift sedimentary expansion (McCann *et al.* 1995; Shannon *et al.* 1995). The dominant fault trend is north–south, as is the overall trend of the Jurassic basin. Our observations in the South Porcupine exploration area indicate major normal offsets and sedimentary expansion against both north–south and NE–SW faults, with no discernible evidence of oblique slip on either fault set (see Figs 14–16, below). On the basis of this evidence, we believe that the local extension vector was E–W or WNW–ESE, as also suggested by Shannon *et al.* (1995). These workers described a similar extensional direction for the Jeanne d'Arc Basin, where major movement also occurred on north–south and NE–SW faults. On Galicia Bank significant normal displacements also occurred on north–south-trending faults (Boillot *et al.* 1989). These faults would rotate to a more NNE trend relative to a fixed Eurasian plate, taking into account later rotation of Iberia. Nevertheless, there is still good consistency with the extension direction in the Jeanne d'Arc Basin, which lay almost adjacent to Galicia Bank before plate separation (Fig. 1).

Galicia Bank, the Jeanne d'Arc Basin, the Porcupine Basin and the Slyne Basin seem to define a general northerly-trending Late Jurassic system of rifts, perhaps extending northwards into the north–south section of the Rockall Basin west of the Hebrides and southwards into the Tagus Abyssal Plain (Fig. 6a). This approximately east–west, Late Jurassic extension is also characteristic of the northern North Sea, the Mid-Norwegian shelf and East Greenland. However, the evidence for a consistent plate-wide extension vector is sharply contradicted by the North Celtic Sea Basin, which probably underwent NW–SE orthogonal extension in Late Jurassic time (Petrie *et al.* 1989; Rowell 1995). Farther east, the Jurassic basin system trends more east–west in the Bristol Channel Basin, and in the Dorset region the extension vector appears to be north–south. It seems that south of the Variscan deformation front the basement grain (particularly that of Variscan origin) played a particularly important role in intraplate extension, as has been suggested by numerous workers (e.g. Shannon 1991; Rowell 1995), allowing the extension vector to change from east–west or WNW–ESE (South Porcupine) to NW–SE (North Celtic Sea) to north–south (Dorset). We suggest that this configuration results from stress variations within an anisotropic crust, and does not require some or most of the basins to have undergone oblique slip.

The timing of this extensional episode appears to vary, even given possible uncertainty in the dating. Shannon *et al.* (1995) assigned the principal faulting in the Jeanne d'Arc and Porcupine basins to Tithonian–early Valanginian time, and a similar timing has been suggested for Galicia Bank (Boillot & Malod 1988). The North Celtic Sea episode has been assigned to Oxfordian to Tithonian time (Petrie *et al.* 1989; Rowell 1995). In the south Porcupine Basin, we suggest Kimmeridgian to Portlandian extension. It is probable therefore that rifting was multiphase, with basins experiencing their acme of extension at different times.

Palaeogeography

The Late Jurassic depositional system is characterized by a rapidly deepening, bifurcating marine rift system with marine connections to the south, across the Tagus Abyssal Plain and through the Biscay Fault Zone to the east (Fig. 10). This deep marine system was prone to the development of anoxia and hence to the deposition of source rocks (Doré *et al.* 1999). Widespread deposition of the Kimmeridgian Clay Formation took place in the North Sea and equivalent world-class Upper Jurassic source rocks developed offshore from NW Europe (Butterworth *et al.* 1999).

In the Grand Banks area, uplift of the Avalon Terrace during Tithonian time gave rise to a broad basement arch, which became an important provenance area for the derivation of clastic sediments (Tankard *et al.* 1989). The Upper Jurassic Egret Member of the Jeanne d'Arc Basin is the main reservoir in the Hibernia oil field, where simultaneous deposition of calcareous organic-rich shales provided the prolific Egret Member source rock (Sinclair *et al.* 1999). The Avalon uplift may have resulted from continent–continent translation along the Newfoundland Fracture Zone in association with central Atlantic and early North Atlantic sea-floor spreading (Fig. 6a). Simultaneous uplift of the Iberian Meseta with associated volcanism gave rise to a progradational wedge of Kimmeridgian siliciclastic deposits (the Grès Superiores Formation), which was shed westwards across the underlying carbonate platform (Wilson *et al.* 1989). This uplift, which extended into Hauterivian time, may also have been caused by translation across the Azores–Gibraltar Fracture Zone (Fig. 6a).

In Late Jurassic time, a non-marine to marginal marine, muddy shelf system with major clastic input along the northern margin from the Irish Massif is interpreted from well data in the Celtic Sea basins (Petrie *et al.* 1989). At this time, the Irish Massif may also have acted

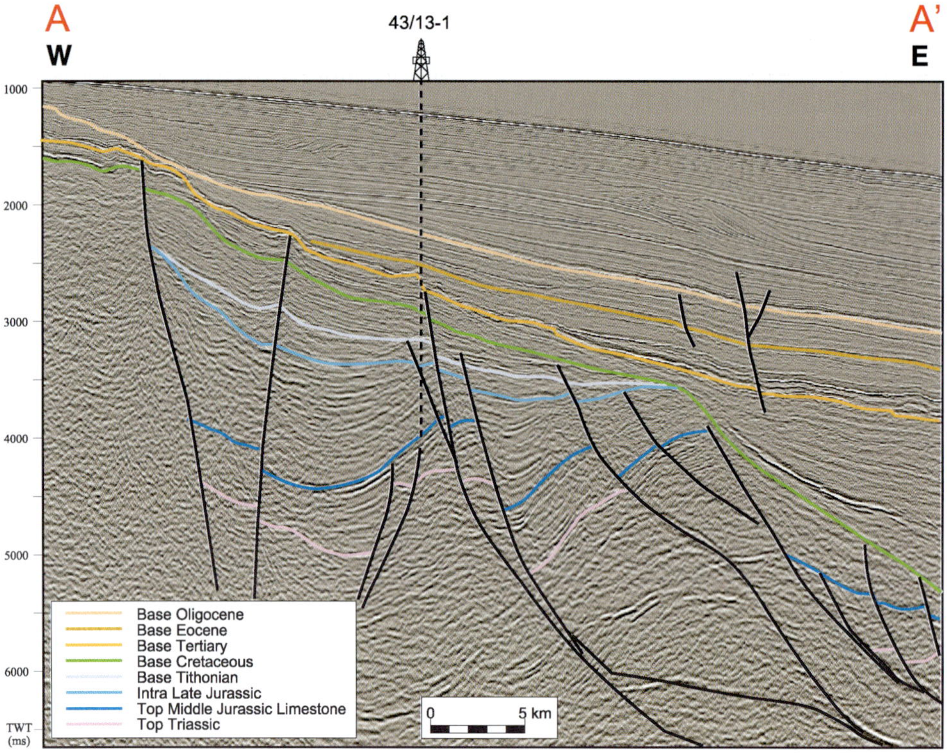

Fig. 13. Seismic line SPB97-113 (A–A' in Fig. 2) on the western flank of the south Porcupine Basin showing location of well 43/13-1 and Late Jurassic synrift expansion against the north–south-trending basin-bounding fault in this area.

as a localized source of clastic input westwards from a Devonian–Carboniferous provenance area.

The Upper Jurassic section has been penetrated by a number of wells in the North Porcupine Basin, where fully marine conditions are interpreted to have been established in the south (Butterworth *et al.* 1999). An open-marine connection to the SW can be interpreted from the regional palaeogeography (Fig. 10). However, a complete synrift section is not encountered, as most wells were located on footwall crests of tilted fault blocks, to test the underlying pre-rift section. Well 43/13-1, in the south Porcupine Basin, penetrated a dominantly muddy Kimmeridgian to Tithonian section in a footwall crestal position (Fig. 13). Uplifted footwall blocks associated with active extensional faulting may have given rise to localized alluvial fan and turbiditic sandstone deposits on the hanging-wall dip slopes away from the fault-block crests, and may provide good reservoir potential.

Implications for prospectivity in the south Porcupine Basin

The main outline of the North Porcupine Basin was established after Late Jurassic extension. The Late Jurassic depositional setting in the south Porcupine Basin is based on an analogy between it and well data to the north and data from surrounding basins. The pervasive anoxic deep-marine setting, one of the factors that gave rise to the development of a world-class source rock throughout parts of NW Europe, is also thought to have persisted in the south Porcupine Basin. Although immature in well 43/13-1, the Upper Jurassic sequence is thought to become more mature towards the centre of the south Porcupine Basin. Possible sandstones, supplied from localized, uplifted footwall blocks during this period of active rifting, may provide reservoirs with the Base Cretaceous unconformity providing the overlying seal. However, this play remains untested in the North Porcupine Basin.

Early Cretaceous (Valanginian–Hauterivian) phase

Tectonic setting

The plate motions discussed herein are mainly from Srivastava *et al.* (1990) with some modifications based on subsequent work by Sibuet & Collette (1991); García-Mondéjar (1996). Sea-floor spreading continued south of the Azores–Gibraltar Fracture Zone, creating a NW–SE divergence between the African and American plates. North of the fracture zone, Srivastava *et al.* (1990) indicated a NW–SE extension vector west of Iberia and across the proto-Bay of Biscay. This represents a rotation of the extensional direction from the Late Jurassic east–west extension described for Galicia Bank (Boillot & Malod 1988) and the Jeanne d'Arc Basin (Shannon *et al.* 1995).

In the south Porcupine Basin, extensional reactivation and expansion of the marine sedimentary sequence against some of the larger faults is seen in the seismic data (Figs 12 and 13). Although there is some uncertainity in the seismic ties, the section is probably of Neocomian age. Of the two main fault sets (north–south and NE–SW) the NE–SW faults are most clearly reactivated (Fig. 13), suggesting a consistency in extension direction with that described by Srivastava *et al.* (1990) to the south. On the basis of this consistency, we suggest that the relative importance of the NE–SW faults in south Porcupine Basin, and the swing of the basin into the NE–SW Porcupine Seabight Basin, is attributable to Neocomian extension. Using the plate kinematics, together with observations from the UK sector (e.g. Doré *et al.* 1999), we infer that the main outlines of the Rockall Basin also formed at this time. The North Porcupine Basin remained relatively inactive tectonically, probably as a result of the focusing of extension on the Rockall Basin and South Porcupine rift segments (Fig. 6b).

Eastwards, in the North Celtic Sea Basin, some significant differences appear to be present in both tectonic regime and sedimentary style. Extension took place in a mainly non-marine setting, the result of a significant intra-plate uplift extending eastwards to the Weald and centred on the Cornubian Peninsula (McMahon & Underhill 1995). Petrie *et al.* (1989) assigned extension to the Berriasian–Aptian interval, but according to Rowell (1995) the faulting is principally of Valanginian age. Both sets of workers believe that this faulting was either caused by, or accompanied by, right-lateral movements on the bounding faults of the basin, attributed to plate motions along the proto-Biscay structure to the south. However, if the faulting is of Valanginian age, it is difficult to see how the postulated strike-slip relates to the relative plate motion (see, e.g. Srivastava *et al.* 1990, fig. 18b). There are also significant differences in the interpreted internal faulting of the North Celtic Sea Basin between Petrie *et al.* (1989), who suggested dip-slip on NW–SE 'transfer' faults (NE–SW extension), and Rowell (1995), who suggested north–south extension with synsedimentary dip-slip on east–west faults.

The apparent discrepancy between the kinematics of the Porcupine Basin and the North Celtic Sea Basin may be attributable to a number of factors (see also the Late Jurassic kinematics). First, the inverted and exhumed nature of the North Celtic Sea Basin (Murdoch *et al.* 1995) may add a degree of difficulty to resolving the earlier extensional history. Second, the contrasting trends of the two Mesozoic basins suggest a different crustal structure: the importance of control by pre-existing (Caledonian and Variscan) basement fractures has been widely stressed by Celtic Sea workers (e.g. Petrie *et al.* 1989; Shannon 1991; Rowell 1995) but is less evident in the Porcupine Basin. Problems in reconciling the two basins could therefore relate to the difficulty in distinguishing orthogonal extension from oblique slip along basement trends in the North Celtic Sea. Finally, it is also possible that stress in Late Jurassic and Early Cretaceous time was simply not evenly distributed or unidirectional between the two basins.

Palaeogeography

Active extension during Valanginian–Hauterivian time created accommodation space for the deposition of the Hibernia Formation sandstone in the Jeanne d'Arc Basin (Sinclair *et al.* 1999) and the terrigenous facies of the Torres Vedras Formation in the Lusitanian Basin (Wilson *et al.* 1989). This stratigraphic interval also coincides with the main rift event in Flemish Pass Basin and the development of a massive sandstone unit of 50 m thickness penetrated in the Baccalieu well on the western margin of the basin, together with an oil-prone marine source rock (Foster & Robinson 1993).

In contrast, the palaeogeographical setting of the Celtic Sea basins is dominantly non-marine at this time (Figs 11 and 13) with up to 1000 m of lacustrine and lagoonal mudstones deposited in the basin axes whereas thin clastic deposits are reported on the margins of the basins, particularly to the north (Petrie *et al.* 1989). The Valanginian to Hauterivian succession was

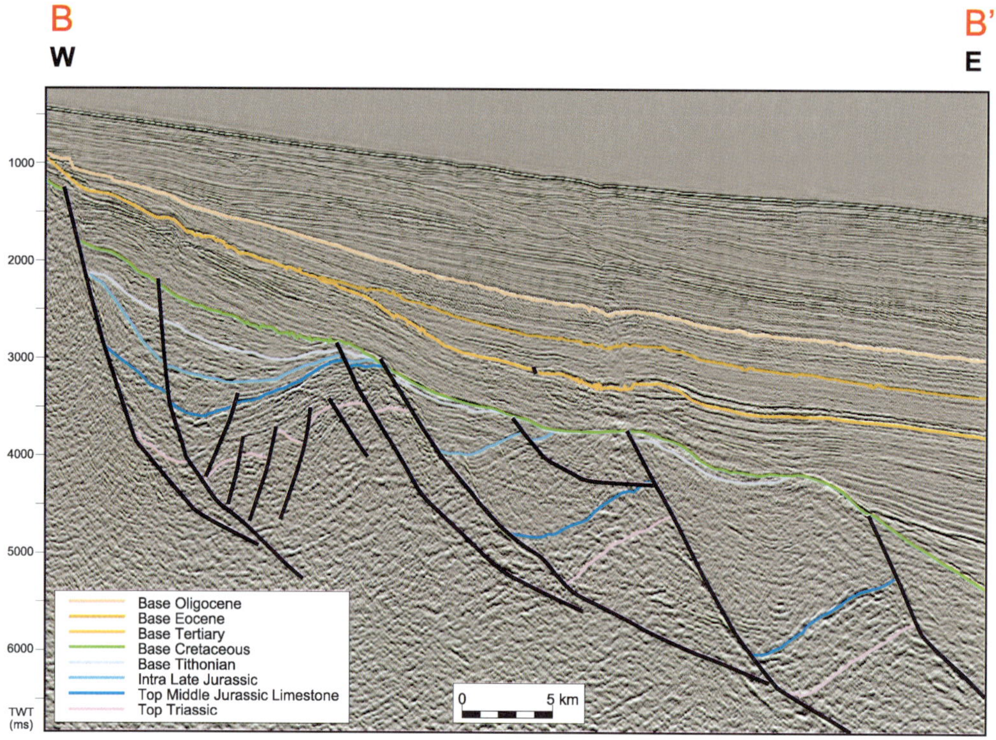

Fig. 14. Seismic line GSP97-019 (B–B′ in Fig. 2) showing Mesozoic tilted fault blocks on the western flank of the south Porcupine Basin with Late Jurassic synrift extension and Neocomian reactivation against a NE–SW-trending basin-bounding fault.

penetrated by a number of wells in the North Porcupine Basin where the predominant sedimentation pattern consists of marine shelfal mudstones, siltstones and thin sandstones (Moore & Shannon 1995). They may also have prograded southwards into deeper water. The Valanginian to Hauterivian sedimentary package forms the lower part of sequence PK1 described by Moore & Shannon (1995).

In the south Porcupine Basin, early Cretaceous synrift wedges are seen banked against the dominant NE–SW faults. We suspect that these wedges contain a similar facies assemblage to the PK1 sequence of Moore & Shannon (1995), possibly with the development of scarp-derived sandstones in proximal locations. In the centre of the basin, the inception of the Porcupine Median Volcanic Ridge (PMVR) (e.g. Masson & Miles 1986; Tate & Dobson 1988) has been ascribed to earliest Cretaceous time. However, our examination of the most recent seismic data suggests that Neocomian strata may extend beneath the feature (Fig. 15). This observation, and consideration of the plate-tectonic constraints (see Barremian–Aptian section below) lead us to suggest a younger age for the median volcanic ridge.

Implications for prospectivity of the south Porcupine Basin

Reactivation along the NE–SW-trending faults that dominate the south Porcupine Basin may have given rise to siliciclastic reservoir units derived from the uplifted Porcupine Bank to the west and the Irish Massif to the east. The lesser scale of the Valanginian–Hauterivian faulting compared with that of the Late Jurassic faulting suggests that such reservoirs will be very localized. Further potential for reservoir development may exist within the thin turbiditic sandstone units located towards the basin centre. This period of rapid deepening and the accumulation of a thick sequence of basinal shales, siltstones and mudstones is important to the burial history and thermal maturation of the underlying Jurassic source rocks.

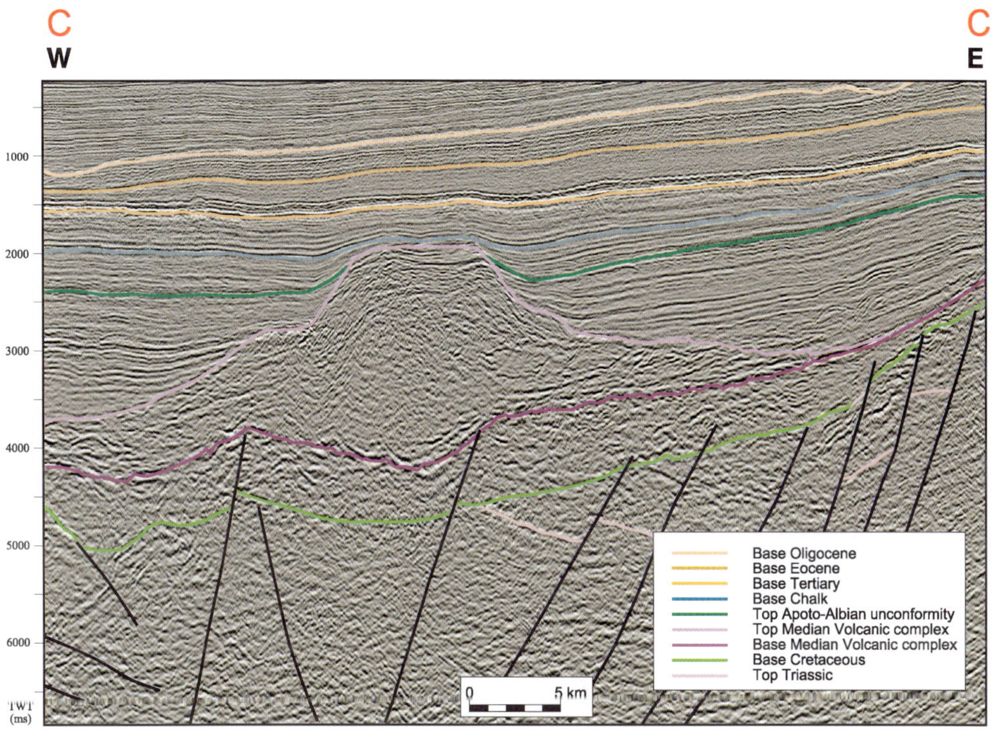

Fig. 15. Seismic line GSP97-019 (C–C′ in Fig. 2) showing the Porcupine Median Volcanic Ridge and underlying sediments of possible Neocomian age. Tilted fault-block structures of Late Triassic or Late Jurassic age are poorly imaged beneath the ridge.

Early Cretaceous (Barremian–Albian) phase

Tectonic setting

At about Chron M0 time (118 Ma, Aptian time), a major change in plate kinematics took place. Sea-floor spreading began between Iberia and the Grand Banks, and shortly after M0 NNE–SSW plate divergence began across the Bay of Biscay, activity that continued until Chron 34 time (84 Ma, Santonian time; Srivastava *et al.* 1990). At some stage, probably in Aptian time, a spreading centre started to propagate northwestwards from a triple junction west of Galicia Bank. By late Early Albian time, oceanic crust was developed off the Goban Spur (Sibuet & Collette 1991) and probably reached the Charlie–Gibbs Fracture Zone by Santonian time. In the Jeanne d'Arc Basin, mid-Aptian to Albian extensional faulting took place along WNW–ESE and NW–SE lines (Shannon *et al.* 1995). These trends match very well with those of the Biscay and northerly arm of the Atlantic triple junction, and it follows logically that the faults represent extension propagating ahead of the spreading ridges (Fig. 6c).

Likewise, in the south Porcupine Basin, the major NW–SE-trending PMVR, thought to consist of a series of coalesced, partly subaqueously extruded volcanic cones (Masson & Miles 1986; Tate & Dobson 1988), overlies thinned crystalline crust in the basin centre (White *et al.* 1992). This igneous feature also suggests NE–SW extension, and fits with the idea of distributed extension in advance of the northerly spreading ridge (Fig. 6c). The PMVR is difficult to date precisely on seismic data because of the difficulty of correlating seismic reflectors basinward into the basin centre. However, we believe that Lower Cretaceous strata probably extend beneath the feature, suggesting that it was initiated within or after Neocomian time. On the basis of the plate kinematics described above, its most likely age is (?)Barremian to Aptian time. Basin modelling suggests that the intrusion was probably one manifestation of more widespread extension, resulting in rapid subsidence of the basin centre (Baxter *et al.* 2001). However, such extension was not manifested as brittle faulting of the upper crust, leading to the hypothesis of depth-dependent extension described elsewhere in this volume (Baxter *et al.* 2001).

The PMVR became inactive and was onlapped during Albian time, probably as a result of bypassing of the Porcupine Basin by the northwestward-propagating spreading ridge (Fig. 6d). Spreading was inhibited south of the Charlie–Gibbs Fracture Zone until Chron 27 time (early Paleocene time), when it extended into the Labrador Sea (Chalmers *et al.* 1993), and Chron 24b time, when it propagated into the NE Atlantic. However, extension probably again propagated ahead of the spreading ridge in Cretaceous times. For example, the Barra Volcanic Ridge System, a series of elongate igneous bodies in the axial part of the SW Rockall Basin and just north of the Charlie–Gibbs Fracture Zone (Scrutton & Bentley 1988) represents intrusion into highly thinned crust. A mode of origin similar to that of the PMVR is suggested. The Barra Volcanic Complex is imprecisely dated but, by inference and the model shown in Figure 6d, we suspect that it is of Albian age or younger.

Palaeogeography

A widespread period of uplift and associated volcanic activity is interpreted throughout the area from Barremian to Albian times (Fig. 12), which is thought to have been associated with the plate-wide reorganizations described above. A second phase of uplift is interpreted for the Avalon Platform, which shed clastic deposits into the Jeanne d'Arc Basin and gave rise to the Avalon and Ben Nevis Formation clastic reservoir intervals (Keen & Williams 1990; Sinclair *et al.* 1999). Continued rejuvenation of the Iberian Meseta during Barremian to Albian time gave rise to the deposition of the Tores Vedras Formation clastic sequence in the Lusitanian Basin, which eventually was drowned during Cenomanian time as indicated by the succeeding deeper-marine carbonates of the Cacaem Formation (Dumestre & Carvalho 1987).

Marine transgression from the SW, which was initiated in Hauterivian time, continued northeastwards into the Celtic Sea basins, reaching its maximum extent around the area of Quadrant 48 early in Aptian time (Petrie *et al.* 1989). Elsewhere in the basins, non-marine deposition continued from Hauterivian into Aptian time in response to uplift of the basin margins (Petrie *et al.* 1989). On the Goban Spur, open-marine shallow-shelf carbonates, in part reefal, were deposited (Cook 1987). Carbonates encountered at DSDP Site 111 (ODP Leg 149 Shipboard Scientific Party 1993) indicate that the Orphan Knoll was also uplifted although not fully emergent at this time. We believe that the relative uplift experienced by the Goban Spur and Orphan Knoll areas at this time may be associated with a more widespread phase of uplift as this region became the footwall margin to the Biscay Labrador Rift. Thus the Porcupine Bank may have become a provenance area for clastic input into the south Porcupine Basin during Barremian to Aptian time.

In the North Porcupine Basin, deposition of marine shelfal mudstones, siltstones and thin sandstones continued until Late Aptian time, when the M0 break-up unconformity is found in wells and on seismic data (Moore & Shannon 1995). This unconformity, described as the boundary between sequence PK2 (of Albian–Cenomanian age) and the underlying sequence PK1 (of Valanginian–Late Aptian age), is conformable towards the basin centre and is marked by erosional truncation towards the basin margins (Moore & Shannon 1995). The PK2 sequence is dominated by marine shelf mudstone and sandstone deposits, with deeper-water basinal facies evident towards the southernmost part of the basin (Moore & Shannon 1995).

In the south Porcupine Basin, rapid subsidence and accumulation of deep-marine sediments continued as mid-Cretacous extension provided accommodation space in the form of a deep 'sag' basin surrounding the median volcanic ridge (Baxter *et al.* 2001). The south Porcupine Basin gradually accumulated a thick sequence of Barremian to Aptian age sediments with restriction of accommodation space occurring in Late Aptian time. On seismic data we recognize a sedimentary progradational package, which can be dated to Albian time in wells to the north. The PMVR was onlapped by this Albian prograding sequence, which marks the cessation of volcanic and tectonic activity in the south Porcupine Basin (Fig. 16). Regional subsidence, with only minor reactivation of the basin-bounding faults, typified the following Late Cretaceous interval, with the onset of pelagic chalk deposition in Cenomanian time.

Implications for prospectivity in the south Porcupine Basin

Aptian shales and mudstones that were deposited under restricted anoxic conditions have been recorded in the North Porcupine Basin, where a possible Ryazanian to Aptian oil-prone shale source (TOC $<2.7\%$) has been reported (Butterworth *et al.* 1999). This interval may

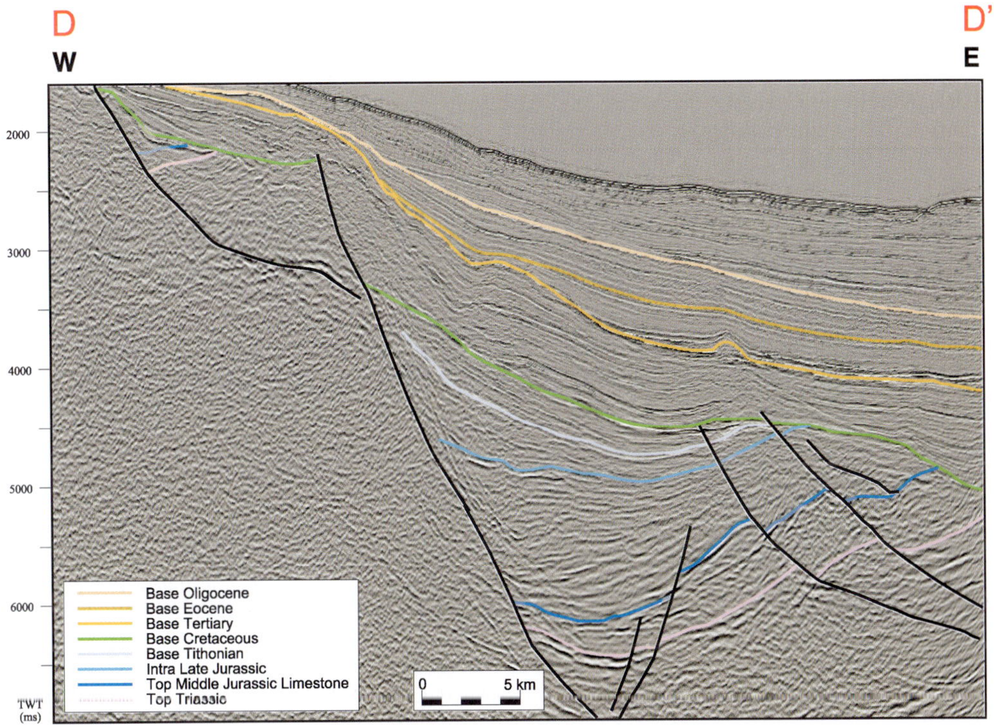

Fig. 16. Seismic line SPB97-023 (D–D′ in Fig. 2) showing Late Jurassic synrift expansion and Neocomian reactivation of NE–SW-trending basin margin fault.

provide an oil-prone source rock in the south Porcupine Basin. Locally derived fault-bounded clastic fans, deltas and associated facies, turbidite or mass-flow deposits, channel sands and shallow-marine shelf sands have been reported in the North Porcupine Basin (Moore & Shannon 1995) and may also be present in the south Porcupine Basin as a reservoir sequence.

Conclusions

Plate-tectonic reconstructions, regional data and local seismic observations have been used to constrain a new interpretation of the Mesozoic history of the south Porcupine Basin. Development of the basin is considered to have been strongly affected by the northward propagation of Atlantic spreading, which included complex changes in plate vectors and the development of a nearby triple junction. The likely sequence of events was as follows:

(1) Permo-Triassic extensional basins, representing fragmentation of Pangaea along basement weaknesses, are typical of the North Atlantic margin and are anticipated (but not directly observed) in the south Porcupine Basin. A NE–SW structural trend is probable, and continental sedimentation is anticipated. Aeolian or fluvial sandstones may be present, but will mostly be too deep to be prospective.

(2) Early Jurassic time saw the onset of sea-floor spreading in the central Atlantic, but in the proto-North Atlantic the tectonic regime was of post-rift subsidence and/or minor extensional faulting. The key event was the flooding of the old Permo-Triassic rift basins, which created links between Tethys, the central Atlantic and the Boreal (Arctic) sea. Fine-grained clastic lithologies with moderate source-rock potential and good seal potential are expected in the south Porcupine Basin.

(3) Central Atlantic sea-floor spreading continued in Mid-Jurassic time and post-rift sedimentation without large-scale faulting took place over much of the proto-North Atlantic. However, a significant fall of base level, probably associated with regional uplift, created major marine restrictions to the north of the south Porcupine Basin. Prograding deltaic to shallow-marine deposits emanating from this northerly hinterland could contain reservoir-quality sandstones and, possibly, waxy deltaic-type source rocks.

(4) In Late Jurassic time further sea-floor spreading took place in the central Atlantic, and much of the proto-North Atlantic and its borderlands were subjected to major crustal extension. The main outlines of the Porcupine Basin were formed by this episode. As with much of the proto-North Atlantic, the south Porcupine Basin appears to have undergone net east–west extension, but this vector was by no means uniform across the Eurasian plate. Significant stratigraphic expansion against north–south and NE–SW faults and observed half-graben geometries on seismic profiles in the south Porcupine Basin indicate potential for both source rock presence and locally derived clastic reservoirs.

(5) In earliest Cretaceous (Neocomian) time, changes in plate motions north of the Atlantic spreading centre occurred as a precursor to plate separation. These events were accompanied by further movement on extensional faults in the south Porcupine Basin, particularly those of a NE–SW trend. We suggest that the South Porcupine and Rockall basins may also have been initiated at this time. Continued marine sedimentation is inferred in the south Porcupine Basin, in contrast to the terrestrial deposits of the North Celtic Sea Basin. Synrift wedges, less voluminous than those of Jurassic time, could contain reservoir lithologies.

(6) In approximately Aptian times, plate separation began between Iberia and the Grand Banks, and across the Biscay plate boundary. A triple junction existed to the south of the south Porcupine Basin, and extension took place ahead of a northwesterly propagating arm of the spreading system. We suggest that the Porcupine Median Volcanic Ridge (PMVR), believed from seismic observation to be of Barremian–Aptian age, is linked to this extension. Strong subsidence of the basin centre at this time may also reflect extension, albeit without significant brittle faulting of the upper crust. This subsidence will have been a significant factor in source-rock maturation and flank trap geometry in the south Porcupine Basin.

(7) The PMVR became inactive and was onlapped in Albian time, probably as a result of bypass of the south Porcupine Basin by the spreading ridge as it propagated northwestwards towards the location of the proto-Charlie–Gibbs Fracture Zone, which developed by Santonian times. Fine-grained deposits are thought to have predominated, with carbonates forming on the shelves and intrabasinal highs and mudstones dominating in the subsiding basinal sags. Regional subsidence with only minor fault reactivation continued for the remainder of Cretaceous time, with the onset of widespread pelagic chalk deposition in Cenomanian time.

We acknowledge the technical contribution by workers from Shell and BP–Amoco (in particular T. Primmer) to the study and for their permission to publish this work, and also D. Corcoran for his contribution to the ideas presented in this paper. We gratefully acknowledge the permission granted by Fugro–Geoteam AS to publish part sections from their 1997 2D regional speculative survey, and Geco-Prakla to publish part sections from their 1997 2D GSP regional survey. We thank E. Kinsella, J. Kipps and D. Geraghty for assistance in drafting the figures for this paper. Statoil is thanked for permission to publish this work.

References

BARTHOLOMEW, I.D., PETERS, J.M. & POWELL, C.M. 1993. Regional structural evolution of the North Sea: oblique slip and the reactivation of basement lineaments. *In*: PARKER, J.R. (ed) *Petroleum Geology of Northwest Europe: Proceedings of the 4th Conference*. Geological Society, London, 1109–1122.

BAXTER, K., BUDDIN, T., CORCORAN, D. & SMITH, S. 2001. Structural modelling of the south Porcupine Basin, offshore Ireland: implications for the timing, magnitude and style of crustal extension. *In*: SHANNON, P.M., HAUGHTON, P.D.W. & CORCORAN, D.V. (eds) *The Petroleum Exploration of Ireland's Offshore Basins*. Geological Society, London, Special Publications, **188**, 275–290.

BOILLOT, G. & MALOD, J. 1988. The north and north-west Spanish continental margin: a review. *Revista de la Sociedad Geolœgica de España*, **1**, 3–4.

BOILLOT, G., MOUGENOT, D., GIRARDEAU, J. & WINTERER, E.L. 1989. Rifting processes on the West Galicia margin, Spain. *In*: TANKARD, A.J. & BALKWILL, H.R. (eds) *Extensional Tectonics and Stratigraphy of the North Atlantic Margins*. American Association of Petroleum Geologists, Memoir, **46**, 363–377.

BUTTERWORTH, P., HOLBA, A., HERTIG, S., HUGHES, W. & ATKINSON, C. 1999. Jurassic non-marine source rocks and oils of the Porcupine Basin and other North Atlantic Margin basins. *In*: FLEET, A.J. & BOLDY, S.A.R. (eds) *Petroleum Geology of Northwest Europe: Proceedings of the 5th Conference*. Geological Society, London, 471–486.

CHADWICK, R.A., LIVERMORE, R.A. & PENN, I.E. 1989. Continental extension in southern Britain and surrounding areas and its relationship to the opening of the North Atlantic Ocean. *In*: TANKARD, A.J. & BALKWILL, H.R. (eds) *Extensional Tectonics and Stratigraphy of the North Atlantic Margins*. American Association of Petroleum Geologists, Memoirs, **46**, 411–424.

CHALMERS, J.A., PULVERTAFT, T.C.R., CHRISTIANSEN, F.G., LARSEN, H.C., LAURSEN, K.H. & OTTERSEN, T.G. 1993. The southern West Greenland continental margin: rifting history, basin development, and petroleum potential. *In*: PARKER, J.R. (ed.) *Petroleum Geology of Northwest Europe: Proceedings of the 4th Conference*. Geological Society, London, 915–931.

CHAPMAN, T.J., BROKS, T.M., CORCORAN, D.V., DUNCAN, L.A. & DANCER, P.N. 1999. The structural evolution of the Erris Trough, offshore northwest Ireland, and implications for hydrocarbon generation. *In*: FLEET, A.J. & BOLDY, S.A.R. (eds) *Petroleum Geology of Northwest Europe: Proceedings of the 5th Conference*. Geological Society, London, 455–469.

COOK, D.R. 1987. The Goban Spur—exploration in a deep water frontier basin. *In*: BROOKS, J. & GLENNIE, K.W. (eds) *Petroleum Geology of North West Europe*. Graham & Trotman, London, 623–632.

COWARD, M.P. 1990. The Precambrian, Caledonian and Variscan framework to NW Europe. *In*: HARDMAN, R.F.P. & BROOKS, J. (eds) *Tectonic Events Responsible for Britain's Oil and Gas Reserves*. Geological Society, London, Special Publications, **55**, 1–33.

COWARD, M.P. 1995. Structural and tectonic setting of the Permo-Triassic basins of northwest Europe. *In*: BOLDY, S.A.R. (ed.) *Permian and Triassic Rifting in Northwest Europe*. Geological Society, London, Special Publications, **91**, 7–39.

CROKER, P.F. & KLEMPERER, S.L. 1989. Structure and stratigraphy of the Porcupine Basin: relationship to deep crustal structure and the opening of the North Atlantic. *In*: TANKARD, A.J. & BALKWILL, H.R. (eds) *Extensional Tectonics and Stratigraphy of the North Atlantic Margins*. American Association of Petroleum Geologists, Memoirs, **46**, 445–459.

CROKER, P.F. & SHANNON, P.M. 1987. The evolution and hydrocarbon prospectivity of the Porcupine Basin, offshore Ireland. *In*: BROOKS, J. & GLENNIE, K.W. (eds) *Petroleum Geology of North West Europe*. Graham & Trotman, London, 633–642.

DORÉ, A.G., LUNDIN, E.R., JENSEN, L.N., BIRKELAND, Ø., ELIASSEN, P.E. & FICHLER, C. 1999. Principal tectonic events in the evolution of the northwest European Atlantic margin. *In*: FLEET, A.J. & BOLDY, S.A.R. (eds) *Petroleum Geology of Northwest Europe: Proceedings of the 5th Conference*. Geological Society, London, 41–61.

DRISCOLL, N.W., HOGG, J.R., CHRISTIE-BLICK, N. & KARNER, G.D. 1995. Extensional tectonics in the Jeanne d'Arc Basin, offshore Newfoundland: implications for the timing of break-up between the Grand Banks and Iberia. *In*: SCRUTTON, R.A., STOKER, M.S., SHIMMIELD, G.B. & TUDHOPE, A.W. (eds) *The Tectonics, Sedimentation and Palaeoceanography of the North Atlantic Region*. Geological Society, London, Special Publications, **90**, 1–28.

DUMESTRE, M.A. & CARVALHO, F.F. 1987. Petroleum geology of the Lusitanian basin. *Oil and Gas Journal*, **85**, 54–58.

FOSTER, D.G. & ROBINSON, A.G. 1993. Geological history of the Flemish Pass Basin, offshore Newfoundland. *AAPG Bulletin*, **77**, 588–609.

GARCÍA-MONDÕJAR, J. 1996. Plate reconstruction of the Bay of Biscay. *Geology*, **24**, 635–638.

GARDINER, P.R.R. & SHERIDAN, D.J.R. 1981. Tectonic framework of the Celtic Sea and adjacent areas with special reference to the location of the Variscan Front. *Journal of Structural Geology*, **3**, 317–331.

JACKSON, D.I., JACKSON, A.A., EVANS, D., WINGFIELD, R.T.R., BARNES, R.P. & ARTHUR, M.J. 1995. *The Geology of the Irish Sea*. United Kingdom Offshore Regional Report, British Geological Survey.

KEEN, M.J. & WILLIAMS, G.L. (eds). *Geology of the Continental Margin of Eastern Canada. Geology of Canada, Vols 1 and 2*. Geological Survey of Canada.

KESSLER, L.G. II & SACHS, S.D. II 1995. Depositional setting and sequence stratigraphic implications of the Upper Sinemurian (Lower Jurassic) sandstone interval, North Celtic Sea/St George's Channel Basins, offshore Ireland. *In*: CROKER, P.F. & SHANNON, P.M. (eds) *The Petroleum Geology of Ireland's Offshore Basins*. Geological Society, London, Special Publications, **93**, 171–191.

MACDONALD, H., ALLAN, P.M. & LOVELL, J.P.B. 1987. Geology of oil accumulation in block 26/28. Porcupine Basin, offshore Ireland. *In*: BROOKS, J. & GLENNIE, K.W. (eds) *Petroleum Geology of North West Europe*. Graham & Trotman, London, 643–651.

MASSON, D.G. & MILES, P.R. 1984. Mesozoic seafloor spreading between Iberia, Europe and North America. *Marine Geology*, **56**, 279–287.

MASSON, D.G. & MILES, P.R. 1986. Structure and development of Porcupine Seabight sedimentary basin, offshore southwest Ireland. *AAPG Bulletin*, **70**, 536–548.

MCCANN, T., SHANNON, P.M. & MOORE, J.G. 1995. Fault styles in the Porcupine Basin, offshore Ireland: tectonic and sedimentary controls. *In*: CROKER, P.F. & SHANNON, P.M. (eds) *The Petroleum Geology of Ireland's Offshore Basins*. Geological Society, London, Special Publications, **93**, 371–383.

MCMAHON, N.A. & UNDERHILL, J.R. 1995. The regional stratigraphy of the southwest United Kingdom and adjacent offshore areas with particular reference to the major intra-Cretaceous unconformity. *In*: CROKER, P.F. & SHANNON, P.M. (eds) *The Petroleum Geology of Ireland's Offshore Basins*. Geological Society, London, Special Publications, **93**, 323–326.

MOORE, J.G. & SHANNON, P.M. 1995. The Cretaceous succession in the Porcupine Basin, offshore Ireland: facies distribution and hydrocarbon potential. *In*: CROKER, P.F. & SHANNON, P.M. (eds) *The Petroleum Geology of Ireland's Offshore Basins*. Geological Society, London, Special Publications, **93**, 345–370.

MORTON, N. 1989. Jurassic sequence stratigraphy in the Hebrides Basin, NW Scotland. *Marine and Petroleum Geology*, **6**, 243–260.

MORTON, N. 1992. Dynamic stratigraphy of the Triassic and Jurassic of the Hebrides Basin, NW

Scotland. *In*: PARNELL, J. (ed.) *Basins on the Atlantic Seaboard: Petroleum Geology, Sedimentology and Basin Evolution*. Geological Society, London, Special Publications, **62**, 97–110.

MURDOCH, L.M., MUSGROVE, F.W. & PERRY, J.S. 1995. Tertiary uplift and inversion history in the North Celtic Sea Basin and its influence on source rock maturity. *In*: CROKER, P.F. & SHANNON, P.M. (eds) *The Petroleum Geology of Ireland's Offshore Basins*. Geological Society, London, Special Publications, **93**, 297–319.

MURPHY, N.J., SAUER, M.J. & ARMSTRONG, J.P. 1995. Toarcian source rock potential in the North Celtic Sea Basin, offshore Ireland. *In*: CROKER, P.F. & SHANNON, P.M. (eds) *The Petroleum Geology of Ireland's Offshore Basins*. Geological Society, London, Special Publications, **93**, 193–207.

NAYLOR, D., SHANNON, P. & MURPHY, N. 1999. *Irish Rockall Basin Region—a Standard Structural Nomenclature System*. Petroleum Affairs Division, Special Publication **1/99**.

ODP Leg 149 Shipboard Scientific Party, 1993. ODP drills the west Iberia rifted margin. *EOS Transactions, American Geophysical Union*, **74**, 454–455.

PETRIE, S.H., BROWN, J.R., GRANGER, P.J. & LOVELL, J.P. 1989. Mesozoic history of the Celtic Sea Basins. *In*: TANKARD, A.J. & BALKWILL, H.R. (eds) *Extensional Tectonics and Stratigraphy of the North Atlantic Margins*. American Association of Petroleum Geologists, Memoirs, **46**, 433–444.

ROBERTS, D.G., THOMPSON, M., MICHENER, B., HOSSACK, J., CARMICHAEL, S. & BØRNSETH, H.-M. 1999. Palaeozoic to Tertiary rift and basin dynamics: mid-Norway to the Bay of Biscay—a new context for hydrocarbon prospectivity in the deep water frontier. *In*: FLEET, A.J. & BOLDY, S.A.R. (eds) *Petroleum Geology of Northwest Europe: Proceedings of the 5th Conference*. Geological Society, London, 7–40.

ROWELL, P. 1995. Tectono-stratigraphy of the North Celtic Sea Basin. *In*: CROKER, P.F. & SHANNON, P.M. (eds) *The Petroleum Geology of Ireland's Offshore Basins*. Geological Society, London, Special Publications, **93**, 101–137.

SCOTCHMAN, I.C. 2001. Petroleum geochemistry of the Lower and Middle Jurassic in Atlantic margin basins of Ireland and the UK. *In*: SHANNON, P.M., HAUGHTON, P.D.W. & CORCORAN, D.V. (eds) *The Petroleum Exploration of Ireland's Offshore Basins*. Geological Society, London, Special Publications, **188**, 31–60.

SCRUTTON, R.A. & BENTLEY, P.A.D. 1988. Major Cretaceous volcanic province in the southern Rockall Trough. *Earth and Planetary Science Letters*, **91**, 198–204.

SHANNON, P.M. 1991. The development of Irish offshore sedimentary basins. *Journal of the Geological Society, London*, **148**, 181–189.

SHANNON, P.M., WILLIAMS, B.P.J. & SINCLAIR, I.K. 1995. Tectonic controls on Upper Jurassic to Lower Cretaceous reservoir architecture in the Jeanne d'Arc Basin with some comparisons from the Porcupine and Moray Firth Basins. *In*: CROKER, P.F. & SHANNON, P.M. (eds) *The Petroleum Geology of Ireland's Offshore Basins*. Geological Society, London, Special Publications, **93**, 467–490.

SIBUET, J.-C. & COLLETTE, B.J. 1991. Triple junctions of the Bay of Biscay and North Atlantic: new constraints on the kinematic evolution. *Geology*, **19**, 522–525.

SINCLAIR, I.K., EVANS, J.E., ALBRECHTSONS, E.A. & SYDORA, L.J. 1999. The Hibernia Oilfield—effects of episodic tectonism on structural character and reservoir compartmentalisation. *In*: FLEET, A.J. & BOLDY, S.A.R. (eds) *Petroleum Geology of Northwest Europe: Proceedings of the 5th Conference*. Geological Society, London, 517–528.

SRIVASTAVA, S.P., ROEST, W.R., KOVACS, L.C., OAKLEY, G., LEVESQUE, S., VERHOEF, J. & MACNAB, R. 1990. Motion of Iberia since the Late Jurassic: results from detailed aeromagnetic measurements in the Newfoundland Basin. *Tectonophysics*, **184**, 229–260.

TANKARD, A.J. & WELSINK, H.J. 1989. Mesozoic extension and styles of basin formation in Atlantic Canada. *In*: TANKARD, A.J. & BALKWILL, H.R. (eds) *Extensional Tectonics and Stratigraphy of the North Atlantic Margins*. American Association of Petroleum Geologists, Memoirs, **46**, 175–195.

TANKARD, A.J., WELSINK, H.J. & JENKINS, W.A.M. 1989. Structural styles and stratigraphy of the Jeanne d'Arc basin, Grand Banks of Newfoundland. *In*: TANKARD, A.J. & BALKWILL, H.R. (eds) *Extensional Tectonics and Stratigraphy of the North Atlantic Margins*. American Association of Petroleum Geologists, Memoirs, **46**, 265–282.

TATE, M.P. 1992. The Clare Lineament: a relic transform fault west of Ireland. *In*: PARNELL, J. (ed.) *Basins on the Atlantic Seaboard: Petroleum Geology, Sedimentology and Basin Evolution*. Geological Society, London, Special Publications, **62**, 375–384.

TATE, M.P. & DOBSON, M.R. 1988. Syn- and post-rift igneous activity in the Porcupine Seabight Basin and adjacent continental margin W of Ireland. *In*: MORTON, A.C. & PARSON, L.M. (eds) *Early Tertiary Volcanism and the Opening of the NE Atlantic*. Geological Society, London, Special Publications, **39**, 309–334.

UCHUPI, E. 1988. The Mesozoic–Cenozoic geologic evolution of Iberia, a tectonic link between Africa and Europe. *Revista de la Sociedad de Geolœgica de España*, **1**, 257–294.

UNDERHILL, J.R. & PARTINGTON, M.A. 1993. Jurassic thermal doming and deflation in the North Sea: implications of sequence stratigraphic evidence. *In*: PARKER, J.R. (ed.) *Petroleum Geology of Northwest Europe: Proceedings of the 4th Conference*. Geological Society, London, 337–345.

VERHOEF, J. & SRIVASTAVA, S.P. 1989. Correlation of sedimentary basins across the North Atlantic as obtained from gravity and magnetic data, and its relations to the early evolution of the North Atlantic. *In*: TANKARD, A.J. & BALKWILL, H.R. (eds) *Extensional Tectonics and Stratigraphy of the*

North Atlantic Margins. American Association of Petroleum Geologists, Memoirs, **46**, 131–148.

WELSINK, H.J., DWYER, J.D. & KNIGHT, R.J. 1989. Tectono-stratigraphy of the passive margin off Nova Scotia. *In*: TANKARD, A.J. & BALKWILL, H.R. (eds) *Extensional Tectonics and Stratigraphy of the North Atlantic Margins*. American Association of Petroleum Geologists, Memoirs, **46**, 215–232.

WERNICKE, B. & TILKE, P.G. 1989. Extensional tectonics framework of the U.S. Central Atlantic passive margin. *In*: TANKARD, A.J. & BALKWILL, H.R. (eds) *Extensional Tectonics and Stratigraphy of the North Atlantic Margins*. American Association of Petroleum Geologists, Memoirs, **46**, 23–40.

WHITE, N., TATE, M. & CONROY, J.-J. 1992. Lithospheric stretching in the Porcupine Basin, west of Ireland. *In*: PARNELL, J. (ed.) *Basins on the Atlantic Seaboard: Petroleum Geology, Sedimentology and Basin Evolution*. Geological Society, London, Special Publications, **62**, 327–349.

WILSON, R.C.L., HISCOTT, R.N., WILLIS, M.G. & GRADSTEIN, F.M. 1989. The Lusitanian Basin of West–Central Portugal: Mesozoic and Tertiary tectonic, stratigraphic and subsidence history. *In*: TANKARD, A.J. & BALKWILL, H.R. (eds) *Extensional Tectonics and Stratigraphy of the North Atlantic Margins*. American Association of Petroleum Geologists, Memoirs, **46**, 341–361.

ZIEGLER, P.A. 1990. *Geological Atlas of Western and Central Europe*. Shell Internationale Petroleum Maatschappij B.V. The Hague.

Aspects of the structure of the Porcupine and Porcupine Seabight basins as revealed from gravity modelling of regional seismic transects

H. JOHNSON[1], J. D. RITCHIE[1], R. W. GATLIFF[1], J. P. WILLIAMSON[2], J. CAVILL[1] & J. BULAT[1]

[1]*British Geological Survey, Murchison House, West Mains Road, Edinburgh EH9 2LF, UK (e-mail: h.johnson@bgs.ac.uk)*

[2]*British Geological Survey, Kingsley Dunham Centre, Keyworth, Nottingham NG12 5GG, UK*

Abstract: The Porcupine Basin is characterized by a large central free air gravity anomaly high (+55 mGal) flanked by local lows. In contrast, the Porcupine Seabight Basin has low-amplitude anomalies in its centre, flanked by edge anomalies. Two transects, one in each of these basins, have been modelled using satellite gravity data; the upper parts of the transects are constrained by interpretation of recent commercial seismic reflection data and two wells. Results from the modelling suggest that the Porcupine Basin is not in isostatic equilibrium. In contrast, the essentially zero free air anomaly over the centre of the Porcupine Seabight Basin suggests that this basin is isostatically compensated. The difference in isostatic compensation between the two basins may reflect a fundamental contrast between the strength of the crust; the crust underlying the Porcupine Basin possesses the greater strength. The Clare Lineament may represent a fundamental boundary within the 'Avalonian Terrane' that juxtaposes basement blocks of differing rheologies.

The North Porcupine Basin (NPB), Porcupine Basin (PB) and Porcupine Seabight Basin (PSB) form a large north–south-trending structure *c.* 150 km to the west and SW of Ireland (Fig. 1). The NPB and PB are separated by the east–west-trending North Porcupine Wrench Fault System (e.g. Tate 1993). The boundary between the PB and the PSB is placed at *c.* 51°N (Naylor *et al.* 1999). The bathymetric trough associated with the Porcupine basins is *c.* 250 km in length and broadens in width from 65 km in the north to 100 km towards its SW limit. Water depths range from only 200 m in the north of the trough to 3000 m in the south. The PB and PSB are separated from the Rockall Basin to the west and north by a combination of the Porcupine High and the Slyne–Erris basins. Towards the south, the PSB passes laterally to Cretaceous oceanic crust of the Porcupine Abyssal Plain. The eastern flanks of the PB and PSB are defined by a combination of the Late Palaeozoic Clare Basin in the north and basement (including Palaeozoic) highs with or without thin Cenozoic cover further south (Naylor *et al.* 1999).

Data

This transect modelling study is based on >5000 km of recently acquired commercial seismic data that straddle the PB and PSB (Fig. 2) and public domain free air gravity anomaly data (Fig. 3) derived from satellite altimetry (Sandwell & Smith 1995). The gravity data were used to model two geoseismic transects for investigation of the crustal structure beneath the basins (Figs 2 and 3). It should be noted that Sandwell & Smith (1997) showed that the shortest gravity anomaly wavelength that is resolved by the Geosat/ERS-1 data is 20 km. Comparison of these data with shipborne data in the Gulf of Mexico by Yale & Sandwell (1999) supports this observation. The seismic data were tied to the only two wells that intersected them, namely Shell 34/19-1 (completed in 1978) and BP 43/13-1 (completed in 1988) on the western margin of the PB (Fig. 2). The wells terminated within (?)lower Permian silty anhydritic mudstones at 3208 m (Tate & Dobson 1989) and middle–upper Jurassic mudstones, siltstones and sandstones at 5128 m, respectively.

From: SHANNON, P.M., HAUGHTON, P.D.W. & CORCORAN, D.V. (eds). 2001. *The Petroleum Exploration of Ireland's Offshore Basins*. Geological Society, London, Special Publications, **188**, 265–274. 0305-8719/01/$15.00

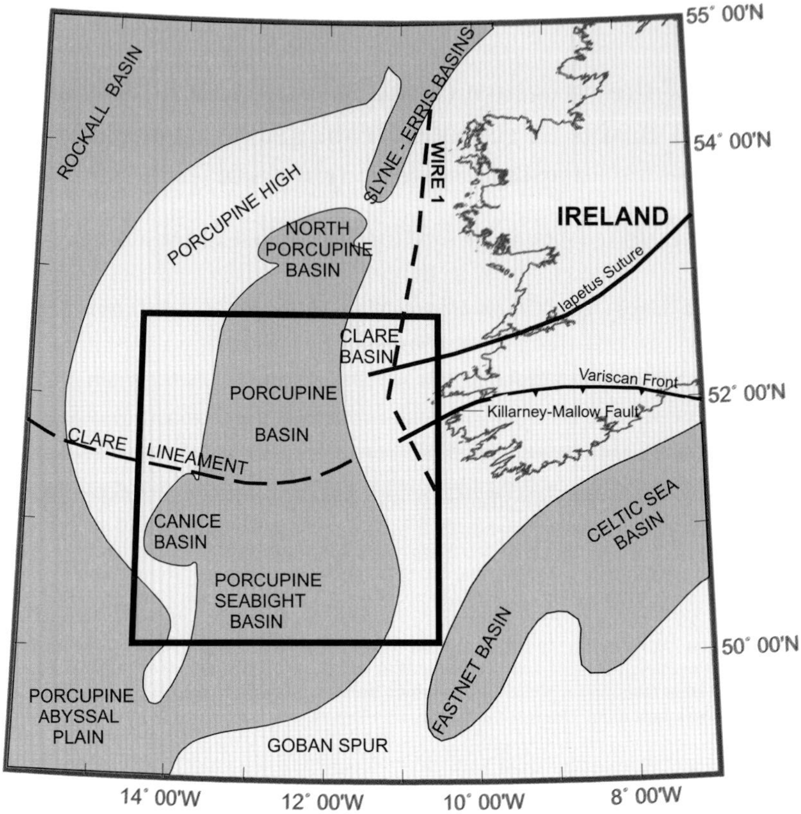

Fig. 1. Generalized structural elements of the Porcupine and Porcupine Seabight basins and adjacent areas. The area within the highlighted box is illustrated in Figure 2.

Previous work

The deep crustal structure of the PB and PSB is poorly understood. For example, west of the WIRE 1 line the continuations of fundamental basement structures such as the Iapetus Suture and the Variscan Front are unconstrained (Fig. 1). Similarly poorly constrained is the eastward continuation of the Clare Lineament, which may represent a transform fault that formed a precursor to the Charlie–Gibbs Fracture Zone (Tate 1992). The nature of the basement beneath the PB and PSB has been the subject of some debate. On the basis of magnetic data, Young & Bailey (1974) and Lefort & Max (1984) considered the crust, at least in the PSB, to be of oceanic character. Lefort & Max (1984) preferred a mid-Jurassic age for the oceanic crust, juxtaposed with Cretaceous oceanic crust of the Porcupine Abyssal Plain to the south. However, Masson & Miles (1986) considered that the amplitudes of the magnetic anomalies are too small to be characteristic of oceanic crust. Croker & Shannon (1987) and Conroy & Brock (1989) suggested that the crust within the PSB is of thinned continental character. This model is similar to that proposed for the adjacent Rockall Basin by Joppen & White (1990).

The PB and PSB have been the focus of a number of deep seismic normal incidence and refraction wide-angle reflection experiments (e.g. Whitmarsh *et al.* 1974; Makris *et al.* 1988; Croker & Klemperer 1989) and potential field studies (e.g. Young & Bailey 1974; Masson *et al.* 1985; Conroy & Brock 1989; Tate *et al.* 1993). Crustal thicknesses below mainland Ireland attain typical continental values of 30 km or so (e.g. Jacob *et al.* 1985). In contrast, results from the BIRPS WIRE experiment indicate that the Moho lies at about 35 km on the continental shelf to the east of the PB and the PSB (Klemperer & Hobbs, 1991; Klemperer *et al.* 1991). Results from the COOLE refraction/wide-angle reflection profiles 3A and 3B (Makris *et al.* 1988) (Fig. 2) suggest that the crust is 25–28 km thick on the eastern flank of the PB–PSB and thins

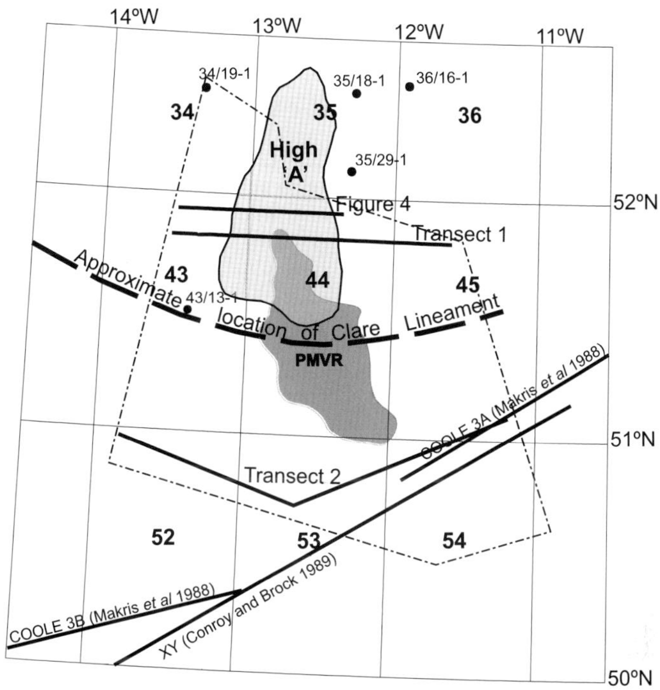

Fig. 2. Location of transects 1 and 2, and deep seismic refraction–normal incidence and wide-angle reflection experiments and potential field profile. PMVR, Porcupine Median Volcanic Ridge; High 'A', positive free air gravity anomaly (see Fig. 3) in commercial seismic reflection profiles. The area within the polygon highlighted in a dot–dash line indicates the approximate extent of the commercial seismic reflection profiles interpreted in this study.

gradually towards the SW. Potential field modelling by Conroy & Brock (1989) on a line that bridges the gap between the COOLE profiles (Fig. 2, line X–Y), suggests that the crystalline crust thins from 28 km at the eastern margin of the PB–PSB to only 8 km within the centre of the PSB. A two-layer velocity model of the crystalline basement was proposed, comprising an upper layer of velocity 5.9–6.1 km s^{-1} and a lower layer of 6.4–6.6 km s^{-1}, values typical of upper and lower continental crust (Makris *et al.* 1988). This model is similar to that proposed for the adjacent Rockall Basin, although the RAPIDS line suggests that basement on the southern flank of the Mayo Basin immediately to the north of the NPB has a three-layer configuration (O'Reilly *et al.* 1995). On the COOLE lines, it should be noted that no anomalous layers with velocity of 7.2–7.4 km s^{-1} are reported to occur between the base of the crystalline crust (6.6 km s^{-1}) and the upper mantle (8.0 km s^{-1}) (Makris *et al.* 1988). Such anomalous velocity layers are commonly thought to represent underplated basic magma emplaced, for example, during phases of rifting. However, underplating of the PB and PSB has been invoked by Tate *et al.* (1993) to explain the discrepancy between their large stretching factors (β ranging up to more than six in the centre of the PSB), calculated from observed amounts of post-rift subsidence within the PB and PSB, and the much smaller values (β approximately four in the PSB) derived from the variation in crystalline crustal thickness determined by the COOLE experiment (Makris *et al.* 1988). One possible explanation for this discrepancy could be that the underplated material is dispersed throughout the lower crust and is not readily recognizable as a distinct body with a characteristic interval velocity structure. Notably, Croker & Shannon (1987) and Conroy & Brock (1989) suggested that the crust within the PB and PSB incorporates considerable quantities of intruded igneous material. Clearly, the presence of any unrecognized underplated material within the crust will result in an underestimate of crustal stretching from seismic experiments.

There are numerous tectonostratigraphic models for the development of the PB and PSB. Croker & Shannon (1987) and Croker & Klemperer (1989) suggested a three-fold subdivision of the sedimentary succession with Devonian–Permian pre-rift, Triassic–Jurassic synrift and Cretaceous–Cenozoic post-rift successions. Other workers have modified these tectonostratigraphic models in detail (e.g. Shannon 1991, 1993; Tate 1993; McCann *et al.* 1995a, b; Williams *et al.* 1999; Johnston *et al.* 2001). Within the post-rift Lower Cretaceous succession of the Porcupine region, a notably large buried mound characterized by chaotic seismic facies has been termed the Porcupine Median Volcanic Ridge (PMVR) (e.g. Tate 1992, 1993). There is no significant gravity anomaly associated with the PMVR and it does not feature on the transects modelled here (Fig. 2).

Transects

Two transects, one in the PB and one in the PSB, have been modelled using a 2D gravity program. The aim of the work was to investigate the gross crustal structure beneath the basins. The free air gravity image of the Porcupine area, together with the location of the transects, is shown in Fig. 3. It is noticeable that the PB is characterized by a large central gravity anomaly high (+55 mGal), termed High 'A' (Figs 2 and 3), flanked by local lows. In contrast, the PSB has low-amplitude anomalies in its centre, flanked by edge anomalies similar in form to the edge anomalies observed at continental boundaries. The cause of these anomalies, in the broadest sense, is the interaction between the gravity effects of the sea-floor topography and the Moho relief. These typical signatures may be modified by the presence of thick sedimentary accumulations (Karner & Watts 1982).

The upper parts of the transects are constrained by regional interpretation of the seismic reflection and well data (see Fig. 4). The fill of the PB and PSB is divided into essentially pre-rift, synrift and post-rift packages, which are characterized by tilt-block, wedge-shaped and onlapping saucer-shaped geometries (Fig. 5a and b). We interpret the pre-rift succession to comprise (?)Carboniferous (or possibly even Devono-Carboniferous) to middle Jurassic rocks. It seems likely that this essentially pre-rift succession includes deposits formed during both rift and thermal sag phases. The major synrift succession imaged in the commercial seismic data is considered to comprise middle to upper Jurassic strata (although possibly extending in age to Mid-Ryazanian time). The post-rift succession is believed to comprise Cretaceous and Cenozoic strata and includes the PMVR. Consequently, the timing of vulcanicity is at variance with that predicted by the McKenzie (1978) model of basin formation. According to Baxter *et al.* (1999), depth-dependent stretching of the lower crust may have continued until Aptian times and is assumed to be associated with the Early Cretaceous eruption of the PMVR.

Depth conversion

Accurate depth conversion of the sedimentary successions on the transects is difficult to achieve, because of the limited well control on interval velocities. Regional seismic interpretation studies have indicated that considerable lateral facies variation occurs within, for example, the Cenozoic and Cretaceous successions of the PB. Within the Cretaceous sequence, local developments of deltaic facies pass basinwards into distal mudstones and associated clastic fans (Shannon *et al.* 1993; Moore & Shannon 1995). Clearly, the velocity structure of stratigraphic units within the Porcupine region will reflect these lateral facies changes. However, because of a paucity of well data it is currently not possible to develop a sophisticated velocity model that would accurately reflect these lateral facies changes. Consequently, a simple layer-cake structure model was constructed on velocity data derived from BP well 43/13-1, and assumed the following interval velocities: water 1.48 km s^{-1}, Cenozoic units 2.00 km s^{-1}, Cretaceous units 3.80 km s^{-1}, upper–middle Jurassic units 3.60 km s^{-1}, middle Jurassic–(?)Carboniferous units 4.50 km s^{-1}.

Densities

The densities used for modelling sedimentary successions on the transects are based partly on those given by Conroy & Brock (1989). However, they assigned a density of 2.74 Mg m^{-3} for the Jurassic to Triassic sequence. This value appears excessively high, particularly in the less deeply buried successions, and we have preferred a lower value of 2.40 Mg m^{-3} for the upper–middle Jurassic synrift and 2.50 Mg m^{-3} for the pre-rift succession. A two-layer model of crystalline upper and lower crust has been assumed, based on regional seismic refraction evidence (Makris *et al.* 1988). The densities used for the modelling are as follows: sea water 1.03 Mg m^{-3}, Cenozoic and Cretaceous units 2.20–2.40 Mg m^{-3}, upper–middle Jurassic units 2.40 Mg m^{-3}, middle

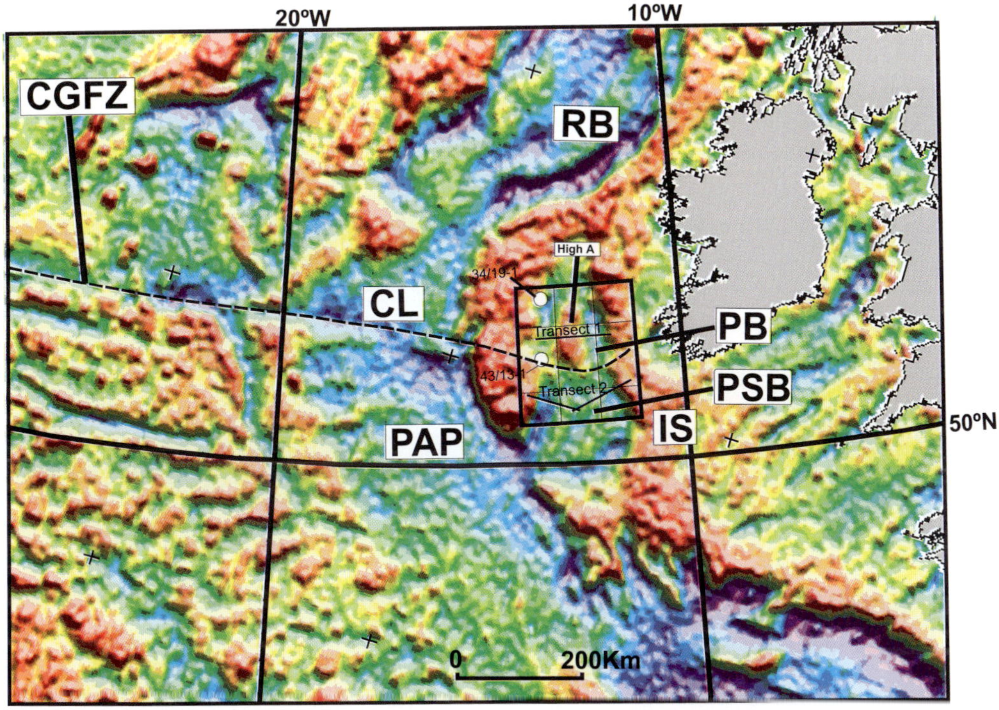

Fig. 3. Free air gravity image of the Porcupine and Porcupine Seabight basins and adjacent areas derived from satellite altimetry (Sandwell & Smith 1995). Red and blue indicate positive and negative gravity anomalies, respectively; illumination is from the north. The highlighted box includes transects 1 and 2, and High A, and is illustrated in Fig. 2. CGFZ, Charlie–Gibbs Fracture Zone; RT, Rockall Trough; CL, Clare Lineament; PB, Porcupine Basin; PSB, Porcupine Seabight Basin; IS, Irish Shelf; PAP, Porcupine Abyssal Plain.

Jurassic– (?)Carboniferous units 2.50 Mg m^{-3}, crystalline upper crust 2.75 Mg m^{-3}, crystalline lower crust 2.85 Mg m^{-3}, upper mantle 3.30 Mg m^{-3}.

Transect 1

The depth-converted pre-, syn- and post-rift successions within the southern part of the PB are constrained by interpretation of a coincident seismic reflection line shot to 9 s two-way time (Fig. 5a). To produce a major positive gravity anomaly over the basin (High A, Figs 2 and 3), a significant amount of high-density material must underlie the sedimentary rocks. The prominent convex-upward reflector (Figs 4 and 5a) that is observed towards the western margin of the basin on Transect 1 is interpreted to represent the top of the crystalline upper crust (i.e. base (?)Carboniferous). As noted above, a two-layer density model comprising crystalline upper and lower crust is assumed and a reasonable fit obtained between the observed and calculated gravity fields across the basin as a whole (Fig. 5a). Crystalline crust thins from *c.* 24 km on the basin flanks of the PB to *c.* 7.5 km towards the basin axis, with the Moho rising from *c.* 30 km on the flanks to *c.* 16 km in the basin centre, where sediment thickness reaches *c.* 8 km. An alternative model for Transect 1 could include the presence of a major axial dyke within the lower and upper crystalline crust of the PB (Tate *et al.* 1993; McGrane *et al.* 1999). The presence of such a dyke within the crystalline crust cannot be precluded, because it is likely to possess densities in the range 2.74–2.85 Mg m^{-3} and would therefore be indistinguishable on gravity modelling from the surrounding crystalline crust. However, it should be noted that evidence for a dyke within the crystalline crust is not observed on the seismic reflection data; neither is a dyke imaged within the overlying sedimentary succession. Notably, in other regions of the world where axial dykes have been described, their size and development appears to be directly proportional to the degree of crustal extension associated with rift development (Gunn 1997). This relationship would therefore imply that a greater degree of crustal stretching should have occurred in the PB compared with the PSB, but published

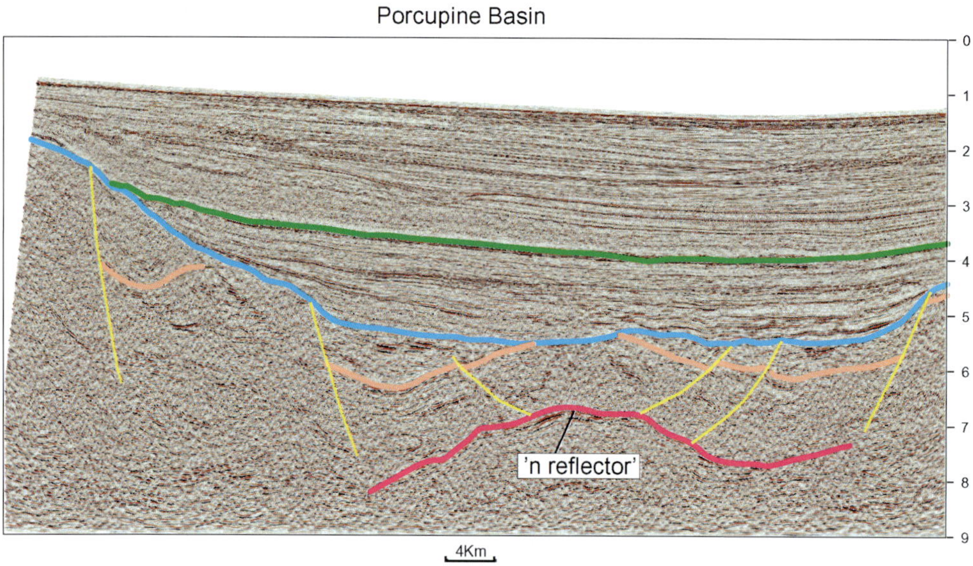

Fig. 4. Seismic profile within the Porcupine Basin displaying the prominent upbowed n reflector (red); orange reflector, base synrift succession (middle–upper Jurassic units); blue reflector, base post-rift (Cretaceous–Cenozoic units); green reflector, base Cenozoic succession; major faults shown in yellow. (For location, see Fig. 2.)

estimates of crustal stretching within the PB–PSB suggest that this is unlikely (see fig. 6 of Tate *et al.* 1993).

On initial inspection, there is some similarity between the very bright and unbroken deep reflection within the PB (Fig. 4) and the high-amplitude and continuous 'S reflector' described from the Galicia margin (Reston *et al.* 1995). Before the opening of the North Atlantic, the Porcupine and Galicia areas were approximately juxtaposed along-strike (Tankard & Balkwill 1989). The S reflector was considered by Reston *et al.* (1995) to represent a lithospheric detachment fault cutting down to the west. However, the geometry of the strongly upbowed deep reflection within the PB, which we call the 'n reflector' (Fig. 4), contrasts with that of the S reflector of Reston *et al.* (1995) (Fig. 6a). Notably, a bowed-up mid-crustal reflector, similar to the n reflector, has been described beneath the Gjallar Ridge on the outer Vøring Margin, and is interpreted to represent a mylonite zone separating the sedimentary succession from the underlying plutonic or metamorphic rocks (Lundin & Doré 1997; Doré *et al.* 1999). Doré *et al.* (1999) postulated an association between underplating and the updomed mid-crustal reflector beneath the Gjallar Ridge, and interpreted the Ridge as a core complex. Following Doré *et al.* (1999), we have modelled the n reflector as the top of the crystalline crust, although an association of underplating, plutonism and core complex formation within the Porcupine Basin remains uncertain. It should be noted that an alternative interpretation of the n reflector may be that it represents the brittle–ductile transition. A model of a brittle–ductile transition formed by essentially pure shear stretching of a basin is illustrated in Fig. 6b and shows an upbowed geometry similar to that of the n reflector.

Transect 2

The depth-converted pre-, syn- and post-rift successions within the northern part of the PSB on Transect 2 (Fig. 5b) are similarly constrained by our seismic mapping. A reasonable fit was obtained between the observed and calculated gravity fields across Transect 2. In contrast to Transect 1, there is no large gravity anomaly over the centre of the PSB and no equivalent of the n reflector is imaged on the seismic reflection data. Crystalline crust thins from 26 km on the flanks of the PSB to 6.5 km around the basin axis, with the Moho rising from 27 km on the flanks to 18 km in the basin centre, where sediment thickness reaches a maximum of 9 km.

Discussion

The difference in free air gravity between the PB and the PSB is puzzling. Modelling suggests that

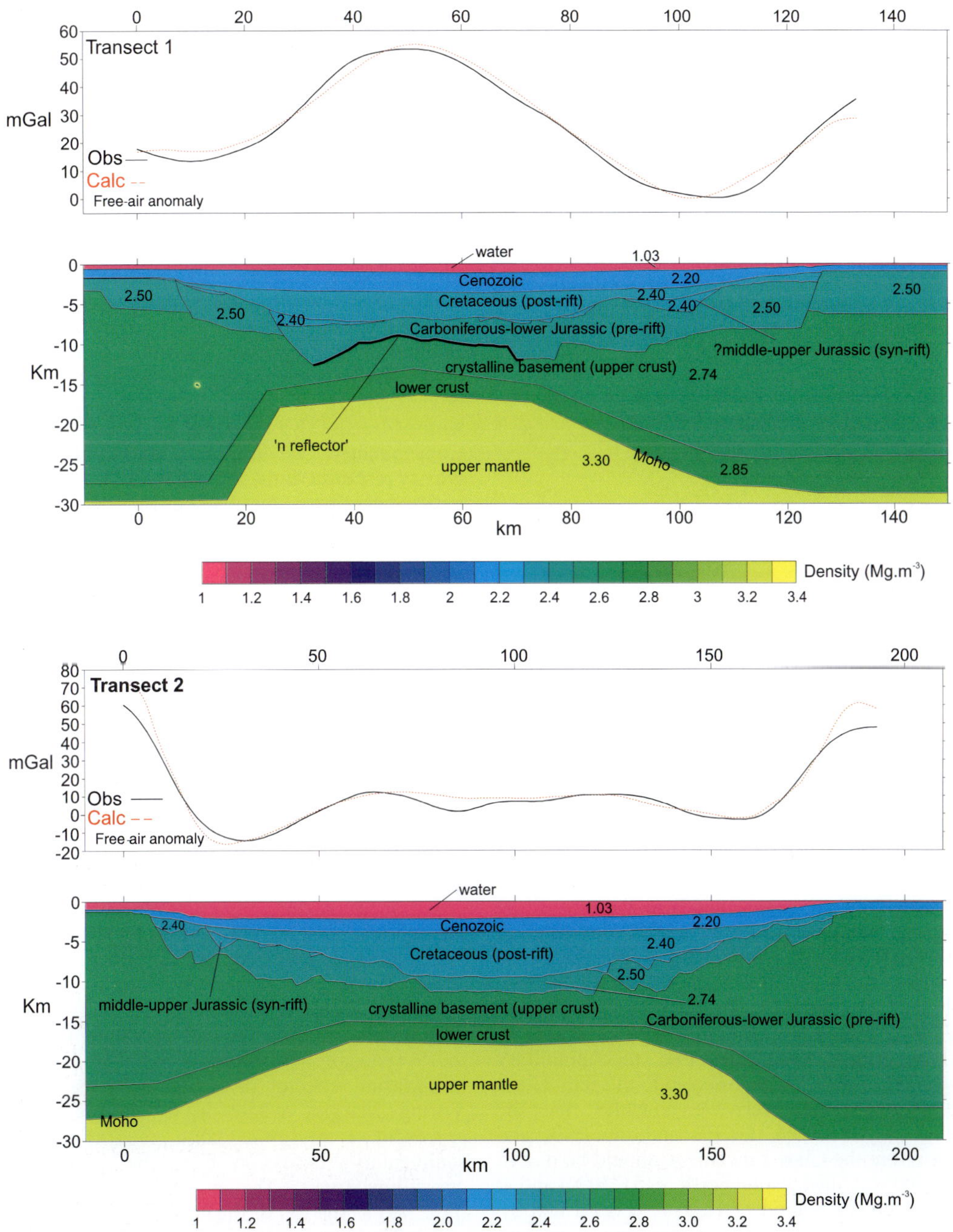

Fig. 5. (**a**) Transect 1; 2D gravity model across the Porcupine Basin. (**b**) Transect 2; 2D gravity model across the Porcupine Seabight Basin. Location of transects is shown in Fig. 2.

both basins have a broadly similar thickness of crystalline crust and sedimentary rocks, but the depth of water within the PSB is significantly greater than that in the PB. Assuming isostatic compensation of present-day water depths, we would therefore expect the depth to the Moho beneath the PB to be greater than that below the PSB, whereas our gravity modelling suggests that the Moho lies at a similar level beneath both basins. We explain this by postulating that the crust underlying the PB has the greater strength, so that the sedimentary load of the basin is at

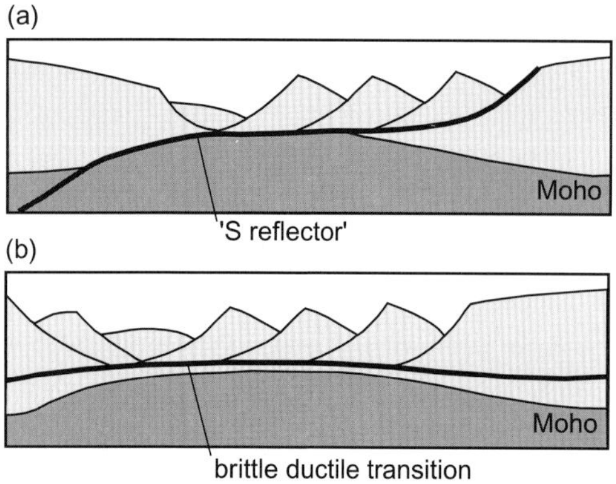

Fig. 6. Schematic illustration of alternative models for the evolution of the Galicia margin and its conjugate margin (modified after Reston *et al.* 1995). (**a**) West-dipping simple shear model of the S reflector favoured by Reston *et al.* (1995); (**b**) pure shear model of brittle–ductile transition zone with geometry similar in style to the n reflector in the Porcupine Basin.

least partially supported by the crust, giving rise to a positive gravity anomaly (High A, Fig. 2). Within the PSB, weaker crust has allowed Airy isostatic compensation to occur. However, it should be noted that the gravity and seismic modelling presented in Transect 1 is not unique. An alternative model for Transect 1 could include the presence of a major axial dyke intruded into the crystalline crust, although there is no supporting evidence for such a dyke on the seismic data.

Notwithstanding the non-uniqueness of the gravity modelling presented here, the contrasting gravity field over the PB compared with that over the PSB suggests a significant difference in the nature of the crystalline crust beneath the two basins. The boundary between the PB and the PSB is placed at *c.* 51°N (Naylor *et al.* 1999). Tate (1992) postulated that this boundary corresponds to the location of the Clare Lineament. However, we consider that the boundary between the PB and the PSB, and therefore the Clare Lineament, should be placed a little way towards the north (*c.* 51°30′N) and at the southern margin of High A (Figs 2 and 3). Speculatively, the Clare Lineament may represent a fundamental boundary that separates basement blocks of differing rheologies, which underlie each of the basins. As noted above, Tate (1992) proposed that the Clare Lineament represents a transform fault that formed a precursor to the Charlie–Gibbs Fracture Zone (CGFZ). The CGFZ probably originated during, or slightly after, Santonian time and is the largest transform fault in the North Atlantic region. Megson (1987, fig. 6) illustrated a North Atlantic refit and proposed that the Clare Lineament–Clare Trend was continuous with the Dover Fault in Newfoundland and that these major lineaments correspond to a westerly continuation of the Iapetus Suture as defined in the UK and Ireland. However, Tate (1992) considered that the Clare Lineament does not represent a westerly continuation of the Iapetus Suture and that the two structures are unrelated, because of their apparent offset and different trends. Although we prefer a more northerly location for the Clare Lineament (Figs 2 and 3) than that of Tate (1992, fig. 1), this does not significantly reduce the offset with the known trace of the Iapetus Suture as defined on the WIRE 1 line west of Ireland (Klemperer *et al.* 1991) (Fig. 1). Consequently, we interpret the contrasting gravity response of the PB and the PSB to reflect a difference in crustal characteristics within the Avalon Composite Terrane (Pharaoh *et al.* 1996). Speculatively, the Clare Lineament may represent the site of a fundamental boundary within the poorly understood Avalonian Composite Terrane west of Ireland.

Conclusions

The Porcupine Basin is characterized by a large central free air gravity anomaly high (+55 mGal), in marked contrast to the Porcupine Seabight Basin, where low-amplitude anomalies are observed. Two transects, one in each of these basins, have been modelled using satellite gravity data; the upper parts of these transects are constrained by interpretation of recent commercial seismic reflection profiles and two wells. The findings of this study are as follows.

(1) Towards the western margin of the Porcupine Basin, a prominent, convex-upward deep reflector is observed and informally termed the ‘n reflector’. This appears to be similar to a mid-crustal reflector described from the outer Vøring Margin, where it is interpreted to represent a mylonite zone separating the sedimentary succession from the underlying crystalline rocks. The n reflector has been modelled as the top of the crystalline crust.

(2) The preferred model for the Porcupine Basin transect indicates that crystalline crust thins from *c.* 24 km on the basin flanks to *c.* 7.5 km towards the basin axis, with the Moho rising from 30 km on the flanks to 16 km in the basin centre, where sediment thickness reaches 8 km. An alternative model could include the presence of a major axial dyke, but there is little unequivocal seismic evidence to support such a model. Evidence from other parts of the world

suggests that a dyke of this type would be more likely to occur in the Porcupine Seabight Basin, where crustal stretching is considered to be greater.

(3) Modelling of the Porcupine Seabight Basin transect suggests that crystalline crust thins from 26 km on the basin flanks to 6.5 km near the basin axis, with the Moho rising from 27 km on the flanks to 18 km in the basin centre, where sediment reaches a maximum thickness of 9 km.

(4) Although the depth of water within the Porcupine Seabight Basin is significantly greater than that in the Porcupine Basin, gravity modelling suggests that the Moho lies at a similar level beneath both of the basins, and consequently that the Porcupine Basin is not isostatically compensated.

(5) It is postulated that the crust underlying the Porcupine Basin is stronger than that beneath the Porcubine Seabight Basin, so that the sedimentary load of the Porcupine Basin is at least partially supported by the crust, giving rise to a large positive gravity anomaly.

(6) The boundary between the Porcupine Basin and the Porcupine Seabight Basin corresponds to the Clare Lineament and is interpreted to represent the site of a fundamental boundary within the Avalonian Composite Terrane.

We would like to thank Fugro-Geoteam AS and especially P. Broad for permission to illustrate the seismic profile. Petroleum Affairs Division is gratefully acknowledged for allowing access to the well information. We also thank P. M. Shannon, J. J. Conroy and an anonymous referee for constructive reviews and comments and S. M. Jones for preparing the figures. This paper is published with the approval of the Director of the British Geological Survey (NERC).

References

BAXTER, K., BUDDIN, T., CORCORAN, D. & SMITH, S. 1999. Structural modelling of the south Porcupine Basin, offshore Ireland: implications for the timing, magnitude and style of crustal extension. *In*: CROKER, P.F. & O'LOUGHLIN, O. (eds) *The Petroleum Exploration of Ireland's Offshore Basins, Dublin, 29–30 April 1999. Extended Abstracts*. Petroleum Affairs Division, Department of the Marine and Natural Resources, Dublin, 72–75.

CONROY, J.J. & BROCK, A. 1989. Gravity and magnetic studies of crustal structure across the Porcupine basin west of Ireland. *Earth and Planetary Science Letters*, **93**, 371–376.

CROKER, P.F. & KLEMPERER, S.L. 1989. Structure and stratigraphy of the Porcupine Basin: relationships to deep crustal structure and the opening of the North Atlantic. *In*: TANKARD, A.J. & BALKWILL, H.R. (eds) *Extensional Tectonics and Stratigraphy of the North Atlantic Margins*. American Association of Petroleum Geologists, Memoirs, **46**, 445–459.

CROKER, P.F. & SHANNON, P.M. 1987. The evolution and hydrocarbon prospectivity of the Porcupine Basin, offshore Ireland. *In*: BROOKS, J. & GLENNIE, K.W. (eds) *Petroleum Geology of North West Europe*. Graham & Trotman, London, 633–642.

DORÉ, A.G., LUNDIN, E.R., JENSEN, L.N., BIRKELAND, Ø., ELIASSEN, P.E. & FICHLER, C. 1999. Principal tectonic events in the evolution of the northwest European Atlantic margin. *In*: FLEET, A.J. & BOLDY, S.A.R. (eds) *Petroleum Geology of Northwest Europe: Proceedings of the 5th Conference*. Geological Society, London, 41–61.

GUNN, P.J. 1997. Regional magnetic and gravity responses of extensional sedimentary basins. *AGSO Journal of Australian Geology and Geophysics*, **17** (2), 115–131.

JACOB, A.W.B., KAMINSKI, W., MURPHY, T., PHILLIPS, W.E.A. & PRODEHL, C. 1985. A crustal model for a northeast–southwest profile through Ireland. *Tectonophysics*, **113**, 75–103.

JOHNSTON, S., DORÉ, A.G. & SPENCER, A.M. 2001. The Mesozoic evolution of the southern North Atlantic and its relationship to basin development in the south Porcupine Basin, offshore Ireland. *In*: SHANNON, P.M., HAUGHTON, P.D.W. & CORCORAN, D.V. (eds) *The Petroleum Exploration of Ireland's Offshore Basins*. Geological Society, London, Special Publications, **188**, 237–263.

JOPPEN, M. & WHITE, R.S. 1990. The structure and subsidence of Rockall Trough from two-ship seismic experiments. *Journal of Geophysical Research*, **95**, 19821–19837.

KARNER, G.D. & WATTS, A.B. 1982. On isostasy at Atlantic-type continental margins. *Journal of Geophysical Research*, **87**, 2923–2948.

KLEMPERER, S.L. & HOBBS, R. 1991. *The BIRPS Atlas: Deep Seismic Reflection Profiles around the British Isles*. Cambridge University Press, Cambridge.

KLEMPERER, S.L., RYAN, P.D. & SNYDER, D.B. 1991. A deep seismic reflection transect across the Irish Caledonides. *Journal of the Geological Society, London*, **148**, 149–164.

LEFORT, J.P. & MAX, M.D. 1984. Development of the Porcupine Seabight: use of magnetic data to show the direct relationship between early oceanic and continental structures. *Journal of the Geological Society, London*, **141**, 663–674.

LUNDIN, E.R. & DORÉ, A.G. 1997. A tectonic model for the Norwegian passive margin with implications for the NE Atlantic: Early Cretaceous to break-up. *Journal of the Geological Society, London*, **154**, 545–550.

MAKRIS, J., EGLOFF, R., JACOB, A.W.B., MOHR, P., MURPHY, T. & RYAN, P. 1988. Continental crust under the southern Porcupine Seabight west of Ireland. *Earth and Planetary Science Letters*, **89**, 387–397.

MASSON, D.G. & MILES, P.R. 1986. Structure and development of Porcupine Seabight sedimentary basin, offshore southwest Ireland. *AAPG Bulletin*, **70**, 536–548.

MASSON, D.G., MONTADERT, L., SCRUTTON, R.A., et al. 1985. Regional geology of the Goban Spur continental margin. *In*: de Graciansky, P.C. & Poag, C.W. (eds) *Initial Reports of the Deep Sea Drilling Project 80*. US Government Printing Office, Printing Office, Washington, DC, 1115–1139.

McCann, T., Shannon, P.M. & Moore, J.G. 1995*a*. Fault patterns in the Cretaceous and Tertiary (end syn-rift, thermal subsidence) successions of the Porcupine Basin, offshore Ireland. *Journal of Structural Geology*, **17** (2), 201–214.

McCann, T., Shannon, P.M. & Moore, J.G. 1995*b*. Fault styles in the Porcupine Basin, offshore Ireland: tectonic and sedimentary controls. *In*: Croker, P.F. & Shannon, P.M. (eds) *The Petroleum Geology of Ireland's Offshore Basins*. Geological Society, London, Special Publications, **93**, 371–383.

McGrane, K., Readman, P.W., O'Reilly, B.M. & Shannon, P.M. 1999. Gravity fabrics, basin structuring and morphology in the Rockall Trough and Porcupine Basin. *In*: Croker, P.F. & O'Loughlin, O. (eds) *The Petroleum Exploration of Ireland's Offshore Basins, Dublin, 29–30 April 1999. Extended Abstracts*. Petroleum Affairs Division, Department of the Marine and Natural Resources, Dublin, 162–163.

McKenzie, D. 1978. Some remarks on the development of sedimentary basins. *Earth and Planetary Science Letters*, **40**, 25–32.

Megson, J.B. 1987. The evolution of the Rockall Trough and implications for the Faeroe–Shetland Trough. *In*: Brooks, J. & Glennie, K.W. (eds) *Petroleum Geology of North West Europe*. Graham & Trotman, London, 653–665.

Moore, J.G. & Shannon, P.M. 1995. The Cretaceous succession in the Porcupine Basin, offshore Ireland: facies distribution and hydrocarbon potential. *In*: Croker, P.F. & Shannon, P.M. (eds) *The Petroleum Geology of Ireland's Offshore Basins*. Geological Society, London, Special Publications, **93**, 345–370.

Naylor, D., Shannon, P. & Murphy, N. 1999. *Irish Rockall Basin Region—a Standard Structural Nomenclature System*. Petroleum Affairs Division, Special Publication **1/99**.

O'Reilly, B.M., Hauser, F., Jacob, A.W.B., Shannon, P.M., Makris, J. & Vogt, U. 1995. The Erris and eastern Rockall troughs: structural and sedimentological development. *In*: Croker, P.F. & Shannon, P.M. (eds) *The Petroleum Geology of Ireland's Offshore Basins*. Geological Society, London, Special Publications, **93**, 413–421.

Pharaoh, T.C., Morris, J.H., Long, C.B. & Ryan, P.D. (compilers) 1996. *Tectonic Map of Britain, Ireland and Adjacent Areas*, 1:1 500 000. Nottingham: British Geological Survey, Keyworth.

Reston, T.J., Krawczyk, C.M. & Hoffman, H.J. 1995. Detachment tectonics during Atlantic rifting: analysis and interpretation of the S reflection, the west Galicia margin. *In*: Scrutton, R.A., Stoker, M.S., Shimmield, G.B. & Tudhope, A.W. (eds) *The Tectonics, Sedimentation and Palaeoceanography of the North Atlantic Region*. Geological Society, London, Special Publications, **90**, 93–109.

Sandwell, D.T. & Smith, W.H.F. 1995. Marine gravity from satellite altimetry (online). Available by anonymous ftp: <ftp://topex.ucsd.edu/pub/global_grav_2min/world_grav.img.7.2.

Sandwell, D.T. & Smith, W.H.F. 1997. Marine gravity anomaly from Geosat and ERS-1 satellite altimetry. *Journal of Geophysical Research*, **102**, 10039–10054.

Shannon, P.M. 1991. Irish offshore basins: geological development and petroleum plays. *In*: Spencer, A.M. (ed.) *Generation, Accumulation, and Production of Europe's Hydrocarbons*. Special Publication, European Association of Petroleum Geoscientists, **1**, 99–109.

Shannon, P.M. 1993. Oil and gas in Ireland: exploration, production and research. *First Break*, **11**, 429–433.

Shannon, P.M., Moore, J.G., Jacob, A.W.B. & Makris, J. 1993. Cretaceous and Tertiary basin development west of Ireland. *In*: Parker, J.R. (ed.) *Petroleum Geology of Northwest Europe: Proceedings of the 4th Conference*. Geological Society, London, 1057–1066.

Tankard, A.J. & Balkwill, H.R. 1989. Extensional tectonics and stratigraphy of the North Atlantic margins: introduction. *In*: Tankard, A.J. & Balkwill, H.R. (eds) *Extensional Tectonics and Stratigraphy of the North Atlantic Margins*. American Association of Petroleum Geologists, Memoirs, **46**, 1–6.

Tate, M.P. 1992. The Clare Lineament: a relic transform fault west of Ireland. *In*: Parnell, J. (ed.) *Basins on the Atlantic Seaboard: Petroleum Geology, Sedimentology and Basin Evolution*. Geological Society, London, Special Publications, **62**, 333–349.

Tate, M.P. 1993. Structural framework and tectono-stratigraphic evolution of the Porcupine Seabight Basin, offshore western Ireland. *Marine and Petroleum Geology*, **10**, 95–123.

Tate, M.P. & Dobson, M.R. 1989. Pre-Mesozoic geology of the western and north-western Irish continental shelf. *Journal of the Geological Society, London*, **146**, 229–240.

Tate, M., White, N. & Conroy, J.-J. 1993. Lithospheric extension and magmatism in the Porcupine Basin West of Ireland. *Journal of Geophysical Research*, **98** (B8), 13905–13923.

Whitmarsh, R.B., Langford, J.J., Buckley, J.S., Bailey, R.J. & Blundell, D.J. 1974. The crustal structure beneath Porcupine Ridge as determined by explosion seismology. *Earth and Planetary Science Letters*, **22**, 197–204.

Williams, B.P.J., Shannon, P.M. & Sinclair, I.K. 1999. Comparative Jurassic and Cretaceous tectonostratigraphy and reservoir development in the Jeanne d'Arc and Porcupine Basins. *In*: Fleet, A.J. & Boldy, S.A.R. (eds) *Petroleum Geology of Northwest Europe: Proceedings of the 5th Conference*. Geological Society, London, 487–499.

Yale, M.M. & Sandwell, D.T. 1999. Stacked global gravity profiles. Society for Exploration Geophysicists, Geophysics Online, published electronically at: http://www.geo-online.org/manuscript/yale98115.pdf.

Young, D.G.G. & Bailey, R.J. 1974. An interpretation of some magnetic data off the west coast of Ireland. *Geological Journal*, **9**, 137–146.

Structural modelling of the south Porcupine Basin, offshore Ireland: implications for the timing, magnitude and style of crustal extension

K. BAXTER[1], T. BUDDIN[2], D. V. CORCORAN[3] & S. SMITH[4]

[1]*Badley Earth Sciences Ltd, North Beck House, North Beck Lane, Hundleby, Spilsby PE23 5NB, UK (e-mail: ken@badleys.co.uk)*

[2]*BP Amoco Exploration, Chertsey Road, Sunbury-on-Thames TW16 7LN, UK*

[3]*Statoil Exploration (Ireland) Ltd, Statoil House, 6 George's Dock, IFSC, Dublin 1, Ireland*

[4]*Shell UK Exploration and Production Ltd, Shell Mex House, Strand, London WC2R 0DX, UK*

Abstract: Regional structural synthesis together with 2D forward and reverse flexural isostatic basin modelling techniques have been used to investigate the extensional and subsidence history of the southern part of the Porcupine Basin. Two structural interpretations of seismic line GSP97-19 have been considered: (1) a Mid–Late Jurassic rift basin based upon seismic interpretation of well-defined tilted fault blocks, with subsidence modelling of the thick overlying sediment section predicting high lithosphere stretching factors of up to $\beta = 6$; (2) a Mid-Jurassic–Early Cretaceous rift responsible for a thick Barremian–Aptian synrift sequence within the basin centre resulting in reduced maximum lithospheric stretching factors of $\beta = 2.3$. The variance in published estimates of crustal thickness beneath the basin cannot distinguish between these scenarios. A comparison between stretching factors and the amount of observable upper-crustal faulting suggests that depth dependent lithospheric stretching may be a feature of the basin, as in other sedimentary basins along the Atlantic margin, and is directly associated with the onset of Cretaceous plate break-up in the Atlantic.

The Porcupine Basin is a large north–south-trending sedimentary basin, located on the continental shelf *c.* 200 km west of Ireland (Fig. 1). A number of rift events are documented to have affected the basin. Pre-Jurassic rift events are poorly resolved on seismic data, although episodes of crustal extension during Permo-Triassic to earliest Jurassic times have been suggested (Croker & Shannon 1987; Naylor & Anstey 1987; Shannon *et al.* 1995). In contrast, a series of well-defined Mid–Late Jurassic tilted fault blocks and synrift sediment packages resulting from NW–SE extension are well documented from seismic and well data (e.g. Masson & Miles 1986; Croker & Shannon 1987; Naylor & Anstey 1987; Tate & Dobson 1989; Tate *et al.* 1993; Shannon *et al.* 1995; Williams *et al.* 1999).

Previous proposed ages for the onset of Jurassic rifting are varied between Bajocian (180 Ma) (Tate *et al.* 1993) and earliest Tithonian time (Williams *et al.* 1999). The post-rift unconformity marking the transition from active rifting to thermal subsidence has been suggested to be located at the Jurassic–Cretaceous boundary by Tate *et al.* (1993), and assigned to Early–Middle Valanginian time by Williams *et al.* (1999). A thick, largely symmetrical Cretaceous–Tertiary succession overlies the Jurassic rift, and this has previously been interpreted as post-rift (Shannon 1992; Tate *et al.* 1993).

Geophysical studies in the Porcupine Basin have shown that the central part of the basin is characterized by a shallowing of the crust–mantle boundary and have produced estimates of minimum crustal thicknesses of 5–10 km (Masson & Miles 1986; Conroy & Brock 1989; Needham *et al.* 1999). Numerous previous workers cited above have suggested that this represents a high level of crustal attenuation beneath the Porcupine Basin, probably related to the onset of Cretaceous sea-floor spreading in the Atlantic Ocean.

This study concentrates on the southern part of the Porcupine Basin, close to the maximum thickness of post-Jurassic sediment using interpretations of structure and stratigraphy from seismic line GSP97-19, tied to well 43/13-1, and integrating these with 2D basin modelling techniques. The location of the line and well are shown in Fig. 1. The objectives of

From: SHANNON, P.M., HAUGHTON, P.D.W. & CORCORAN, D.V. (eds) 2001. *The Petroleum Exploration of Ireland's Offshore Basins*. Geological Society, London, Special Publications, **188**, 275–290.
0305-8719/01/$15.00

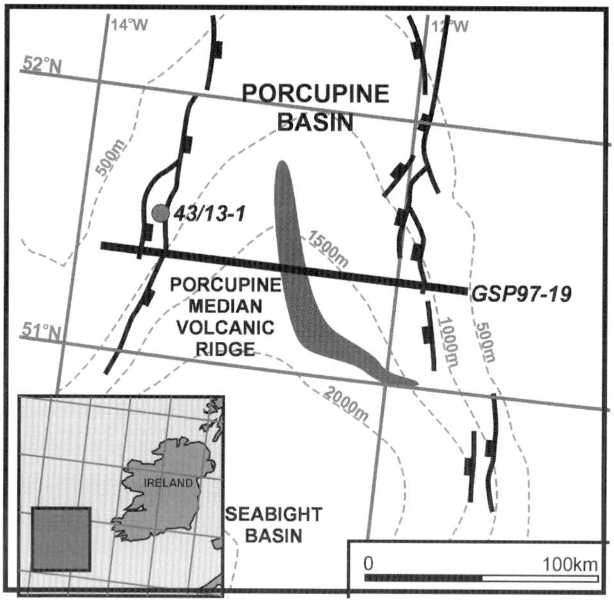

Fig. 1. Location of the Porcupine Basin showing the position of seismic line GSP97-19 and well 43/13-1. Also shown are the major structural features and the location of the Porcupine Median Volcanic Ridge (PMVR). Dashed lines represent present-day water depths contours at 500 m intervals.

the study were to: (1) evaluate the timing and magnitude of upper-crustal faulting; (2) evaluate the magnitude of lithospheric stretching (β) responsible for the thick Cretaceous–Tertiary post-rift sediment section; (3) consider these results in the light of previous work in the area, particularly subsidence models and geophysical estimates of crustal thickness. Assessing the lithospheric stretching history was particularly important in a hydrocarbon exploration context, to evaluate the heat flow history of the basin for maturation studies. Previous subsidence modelling of the basin exists (Tate *et al.* 1993) based primarily on 1D modelling of well data. This approach is continued in this paper, using 2D forward and reverse basin modelling techniques. The primary differences between these approaches and results are also discussed.

Review of modelling techniques

Flexural isostatic modelling of continental lithosphere extension allows the mechanisms for the evolution of basin accommodation space to be explored. This study uses the combined reverse and forward modelling techniques described by Roberts *et al.* (1993) and Kusznir *et al.* (1995).

Reverse modelling methods involve the removal of stratigraphic units and associated decompaction of underlying sediments to 'rewind' the basin history through time towards a restoration of the synrift basin geometry (Fig. 2a). The procedure is a modified form of flexural backstripping (Watts *et al.* 1982) that includes reverse thermal modelling to account for the amount of post-rift subsidence generated during each stratigraphic interval (Roberts *et al.* 1993). The reverse thermal modelling procedure estimates the rate of thermal subsidence from the lithospheric stretching (β) factor. The β factors used in the model can be determined by forward modelling, by estimating cumulative fault heave across the section, by estimating changes in crustal thickness, or by a trial-and-error method until the model gives a best-fit restoration. This technique differs from previous subsidence modelling in the Porcupine Basin by Tate *et al.* (1993) by using a 2D model rather than their 1D approach. A 2D model allows the definition of lithosphere flexural strength that allows the regional distributions of lithosphere loads by flexure rather than local (Airy) isostatic compensation as used in the 1D approach. As outlined by Tate *et al.* (1993), the differences between flexural and local isostatic models during backstripping may be minimal if the flexural strength is low or if the associated loads are long wavelength. However, restorations in areas of more pronounced stratigraphic topography (e.g. restorations across fault blocks) may

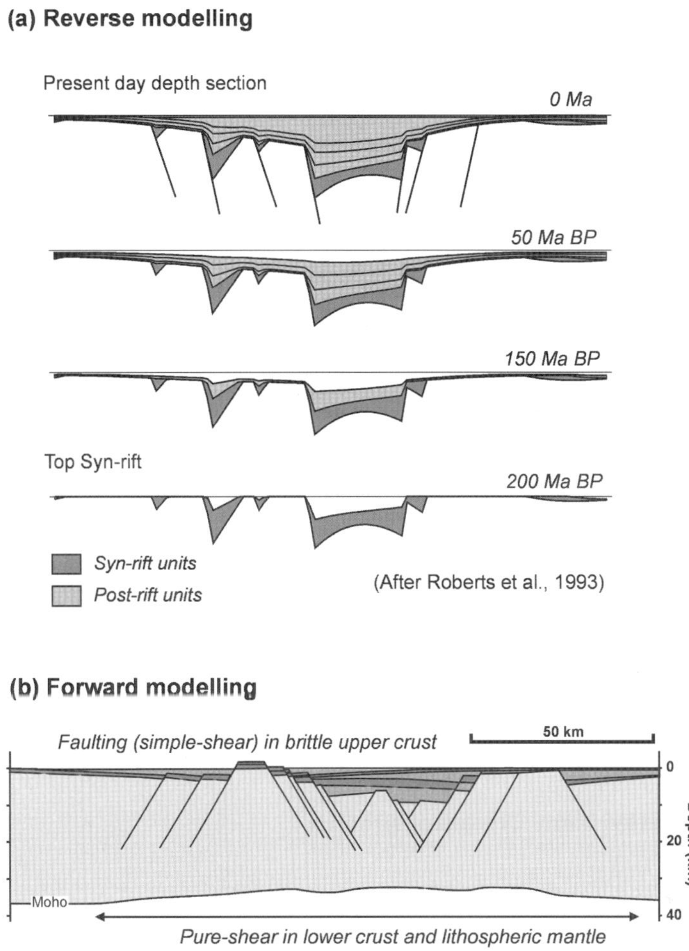

Fig. 2. Overview of the modelling techniques. (**a**) Reverse modelling allows the backstripping, sediment decompaction and reverse thermal modelling of post-rift sediments. A successfully reverse modelled section generates a reconstruction of the end synrift basin constrained by palaeobathymetric indicators and erosional unconformities. This allows the amount of thermal perturbation generated during rifting (quantified by the β factor) to be determined. (**b**) Forward modelling allows the response of the upper crust to extension to be quantified by modelling the evolution of basin geometry. Deformation by faulting in the brittle upper crust and pure shear in the lower crust and lithospheric mantle generates changes in the distribution of mass during deformation and thermal perturbation, which are compensated by flexural isostasy.

result in an underestimation of palaeobathymetry using a 1D approach, which in turn may lead to an incorrect restoration and errors in the prediction of β. It is in these areas where 2D reverse modelling has immediate benefit. The 2D approach also allows the rapid restoration of complete cross-sections (which would be extremely time-consuming using a 1D approach at intervals across the section) including the restoration of topographic surfaces and unconformities, for example across fault blocks. By restoring complete horizons the user can evaluate the restorations alongside the structural interpretation for the section: this is not as straightforward in a 1D well-based model. The structural interpretation can then be evaluated by forward modelling with the backstripped restoration as a constraint. This approach is described below.

Forward modelling methods are based upon the flexural cantilever model of continental extension (Kusznir & Egan 1989; Kusznir *et al.* 1991) combined with the McKenzie (1978) thermal subsidence and sediment compaction model to produce a 2D theoretical model for basin evolution. The model begins with an

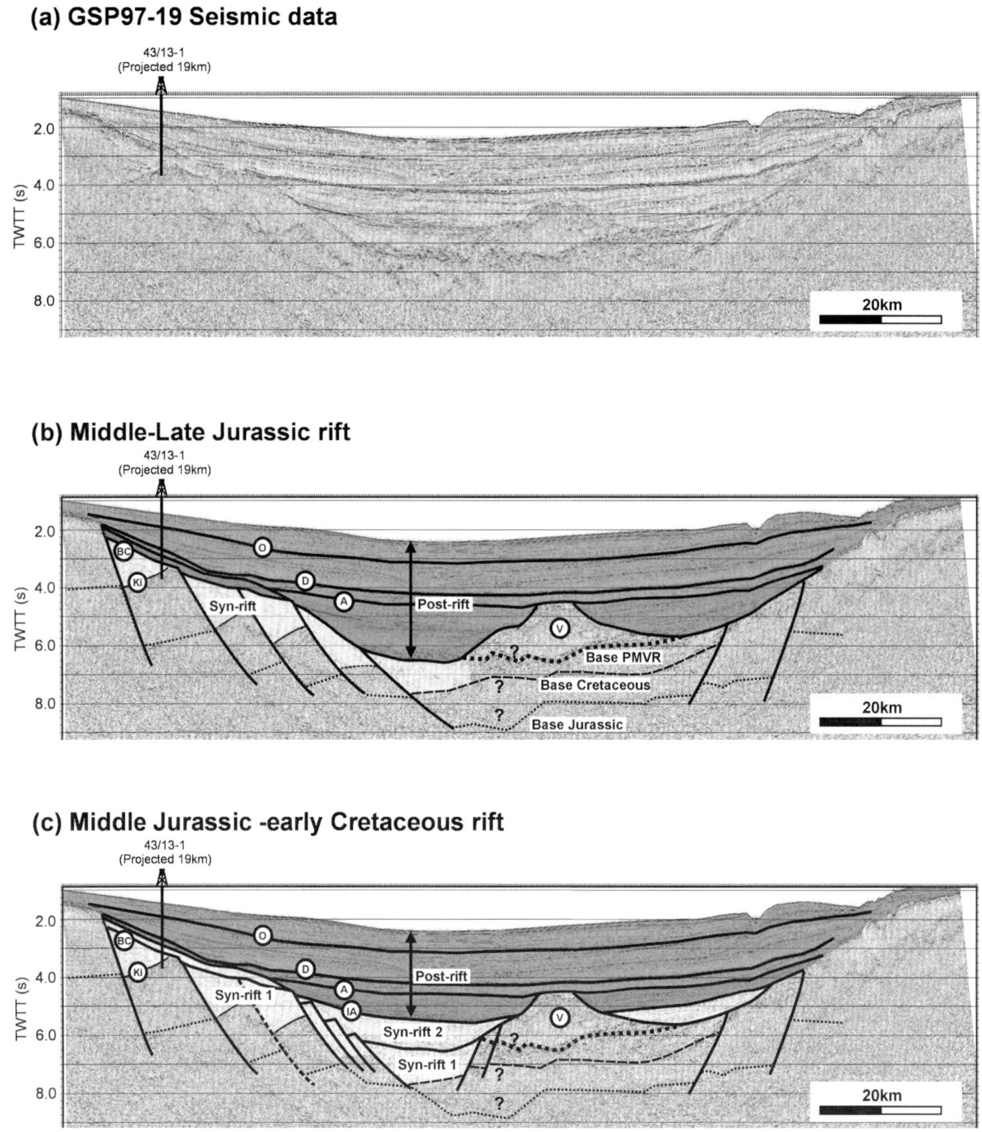

Fig. 3. Seismic line GSP97-19 and alternative structural interpretations. Key horizons: O, Top Oligocene; D, Top Danian Chalk; A, Top Aptian; IA, intra-Aptian; BC, Base Cretaceous; Ki, intra-Oxfordian; V, Porcupine Median Volcanic Ridge (PMVR). (**a**) Seismic line GSP97-19 showing location of well 43/13-1. (**b**) 'Conventional' interpretation of the basin showing rifting in Mid–Late Jurassic time with faulting on the western flank generating a series of rotated fault blocks apparent beneath the Base Cretaceous reflector. Although most faulting is shown on the western flank, faults have also been interpreted from reflector geometries to underlie the eastern flank of the basin. (**c**) Interpretation of the basin with rifting continuing into Early Cretaceous time. Similar to the interpretation in (**b**), Mid–Late Jurassic faulting produced fault-block rotation and synrift deposition (denoted 'synrift 1'), which is well imaged on the western flank. Continued rifting is interpreted into Early Cretaceous time (denoted 'synrift 2') with the slope separating the western flank of the basin from the deep Barremian–Aptian depocentre representing an Early Cretaceous fault scarp. Active Early Cretaceous faults are shown as continuous lines, inactive (Jurassic) faults are shown dashed.

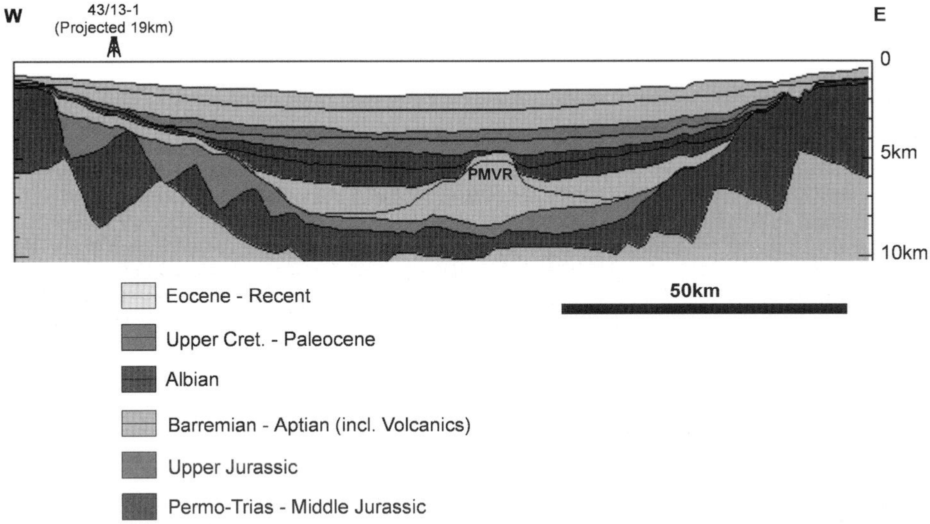

Fig. 4. GSP97-19: depth conversion and interpreted stratigraphy used for reverse modelling. The base of the Permo-Triassic interval is not imaged on seismic data; however, the interval is included in the stratigraphy to allow for decompaction during backstripping.

undeformed section of lithosphere and extends it by faulting (simple shear) in the brittle upper crust, and distributed plastic deformation (pure shear) in the lower crust and lithospheric mantle to produce a sedimentary basin (Fig. 2b). The effect of this extension is to mechanically thin the crust and lithospheric mantle, thus generating a change in the distribution of mass within the lithosphere. Pure-shear deformation within the crust and lithospheric mantle is quantified by a spatially varying β factor, which is used to compute the synrift thermal perturbation of the lithosphere temperature field and the subsequent post-rift thermal re-equilibration (McKenzie 1978). Changes in the distribution of mass as a result of mechanical thinning of the lithosphere, syn- and post-rift thermal effects and additional loads (e.g. erosion and sediment fill) are compensated by flexural isostasy. Sediment compaction effects are also included.

Combining the reverse and forward modelling techniques (Kusznir *et al.* 1995) allows constraints from one modelling technique to be passed to the other, thus promoting convergence of model iterations. A typical approach would initially reverse model a section to the onset of the post-rift phase to produce a restoration of the end synrift palaeobathymetry and topography. Further constraint on crust and lithosphere extension (parameterized by β) and the effective elastic thickness is then obtained by forward modelling, using fault positions and extensions observed on the section, with the reverse modelled template used as a constraint on the resulting basin geometry. The improved laterally varying β factor, and further constrained elastic thickness would then be input back into the reverse model and the process repeated until the section could be successfully reverse and forward modelled to give an acceptable match between basin reconstructions generated in the two modelling techniques.

Seismic interpretation and modelling approach

A number of seismic lines were interpreted as part of an industry study of the southern part of the Porcupine Basin. This paper concentrates on the interpretation and modelling of seismic line GSP97-19. Key interpreted horizons are shown in Fig. 3. Interpretation of GSP97-19 is constrained to the west by well 43/13-1, which provides palaeobathymetric control on the Jurassic and Cretaceous succession on the western basin flank. A series of widely spaced faults separated by tilted Jurassic fault blocks are clearly apparent on the western flank of the basin. This fault-block morphology is truncated close to the Base Cretaceous Unconformity (BCU) with some local erosion of footwall blocks. Faulting on the eastern flank of the basin is less clearly defined from seismic data and is characterized by more closely spaced faults with reduced throw. The Porcupine Median Volcanic Ridge (PMVR)

Table 1. *Lithologies and interval ages based upon well 43/13-1 (Ewan 1998)*

Interval	Base age (Ma)	Description	Model lithology
Oligocene–Recent	23	Marine basinal mudstones	Mixed sand–shale
Eocene–Oligocene	42	Shelf sandstones–basinal mudstones	Mixed sand–shale
Paleocene–Oligocene	60.5	Shelf sandstones–basinal mudstones	Mudstone–shale
Albian–Danian	97	Chalk	Chalk
Albian	108	Shelf sandstones–basinal mudstones	Mixed sand–shale
Aptian	112	Shelf sandstones–basinal mudstones	Mixed sand–shale
Barremian	124	Shelf–basinal mudstones	Mudstones–siltstones
Early Cretaceous	143	Volcaniclastic + clastic deposits	Clastic deposits
Late Jurassic	160	Shelf sandstones–mudstones	Mixed sand–shale

is clearly apparent in the centre of the basin onlapped by sediments interpreted as being of at least Aptian age. Seismic resolution in the deeper part of the basin is poor beneath the BCU and beneath the PMVR. An interpretation has been made of the base of the PMVR, the BCU and the Jurassic section in the deeper part of the basin (shown in Fig. 3) based upon prominent reflections in the seismic data, although these interpretations are considered speculative.

Two structural interpretations have been considered for this seismic line, producing different ages for the end of rifting in this area of the Porcupine Basin: (1) a Mid–Late Jurassic rift that terminated at the BCU (Fig. 3b); (2) a Mid-Jurassic rift that continued into Early Cretaceous time with the onset of full post-rift thermal subsidence at the end of Aptian time (Fig. 3c).

The models were applied by initially reverse modelling the section to estimate the lithospheric stretching factor necessary to correctly restore the section to the onset of the post-rift phase. The reverse model was based upon the depth-converted interpretation of GSP97-19, shown in Fig. 4. Depth conversion of the section was undertaken using interval velocities based upon stacking velocities for the deeper section, and well-derived velocities for the shallower section (BP 1998). The interpreted fault geometries and throws were then input into the forward model to investigate the synrift evolution of the basin. The forward models use the reverse model restorations as a constraint on basin geometry. It should be noted that only limited lithological data are available in the area from which decompaction parameters can be deduced. In this study these have been taken primarily from well 43/13-1 and have been simplified into dominant lithologies (Table 1). Decompaction parameters used in the reverse models (Table 2) are based upon lithological properties from Sclater & Christie (1980), with averaged uniform parameters applied to each interval. The interpretation and modelling results from the two interpretations described above are discussed in separate sections below followed by a discussion of the results.

Mid–Late Jurassic rift

Seismic interpretation

This interpretation shows a Mid–Late Jurassic rift with the onset of post-rift thermal subsidence at the BCU (Fig. 3b), and is similar to the

Table 2. *Reverse model interval parameters derived from Sclater & Christie (1980)*

Lithology	Density ($g\ cm^{-3}$)	Porosity (%)	Decay constant
Mixed sand–shale	2.68	56	0.39
Mudstone–shale	2.72	63	0.51
Chalk	2.71	70	0.71
Clastic deposits	2.65	49	0.27
Crustal basement	2.8	–	–

structural models described by previous workers cited above. The dominant structuring is in Mid–Late Jurassic time on large normal faults with fault-block tilting and footwall erosion beneath the western flank of the basin clearly imaged on seismic data. The onlapping Cretaceous and Tertiary intervals are interpreted to represent passive sediment fill during the post-rift thermal subsidence from this rift event.

Model results

This model uses the interpretation of lithosphere extension as proposed by previous workers (Tate *et al.* 1993) and applied to the modelled line. The main period of lithospheric extension was in Mid-Jurassic time, starting at 180 Ma and lasting 30–40 Ma, based upon backstripping of subsidence profiles, the observation of Jurassic synrift sediments from well data and the interpretation of Jurassic fault blocks and related stratigraphic sequences. The onlapping 'steer's head' geometry of the overlying Lower Cretaceous to Recent stratigraphy was proposed to result from the subsequent post-rift thermal re-equilibration.

The magnitude and timing of lithosphere extension as proposed by Tate *et al.* (1993) for

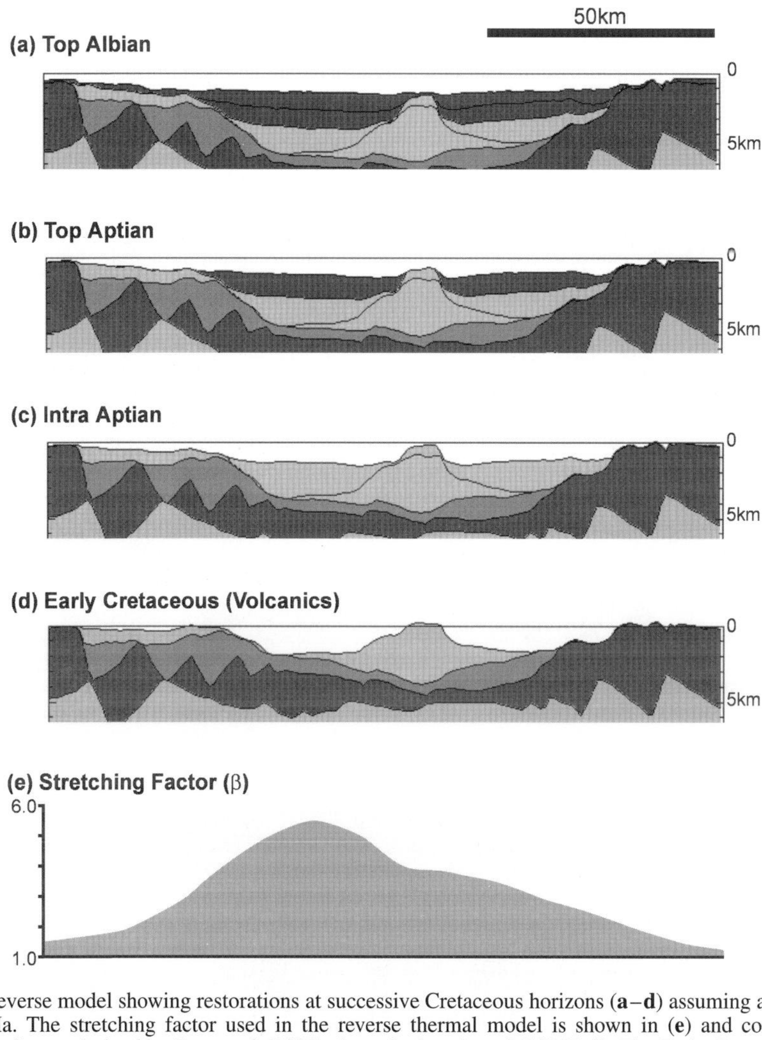

Fig. 5. Reverse model showing restorations at successive Cretaceous horizons (**a**–**d**) assuming a Jurassic rift age of 160 Ma. The stretching factor used in the reverse thermal model is shown in (**e**) and corresponds to the stretching factors derived by Tate *et al.* (1993) along the location of GSP97-19. The Early Cretaceous restoration in (**d**) represents Top Barremian–Hauterivian time, correlated from 43/13-1. The PMVR has not been backstripped in this restoration and can only be dated from the onlap of Aptian sediments.

the southern part of the Porcupine Basin was initially tested within the reverse model by applying a β profile taken directly from the Tate *et al.* (1993) study (see their fig. 6). Thermal perturbation within the reverse model was assigned an age of 160 Ma representing an average for previously proposed age ranges. Restoration of GSP97-19 is shown in Fig. 5 at key seismic horizons using a Jurassic stretching factor of up to *c.* $\beta = 5.5$. This model gives a consistent fit with palaeobathymetry estimates from the Early Cretaceous section in well 43/13-1 (Ewan 1998). Restoring the section before the Early Cretaceous restoration (Fig. 5d) involves removing the loading effects of the PMVR. A speculative interpretation of the base of the PMVR has been made from prominent reflections within the seismic data (Fig. 3b and c) and the unloading effects of removing the ridge have been calculated. Clastic lithology properties have been assumed for the unloading of the volcaniclastic sequence, with an associated decompaction and flexural restoration of the underlying Jurassic section. The unloading of the PMVR been attempted, to generate a template for forward modelling the Jurassic rift (as shown in Fig. 6b), but it is noted that uncertainty in the restoration is likely in the central part of the basin, because of uncertainties in the interpretation of the base of the PMVR and the thickness of the underlying Jurassic section.

The forward model for Jurassic rifting is shown in Fig. 6. Fault locations and heaves were taken directly from the seismic interpretation, with the reverse modelled restoration at the BCU used as a constraint on the basin and fault-block geometries. The model concentrated on the Jurassic structures, sediment thickness and erosional unconformities on the western flank of the basin, as these are well constrained by seismic and biostratigraphic data. The model gives a good fit to the observed sediment

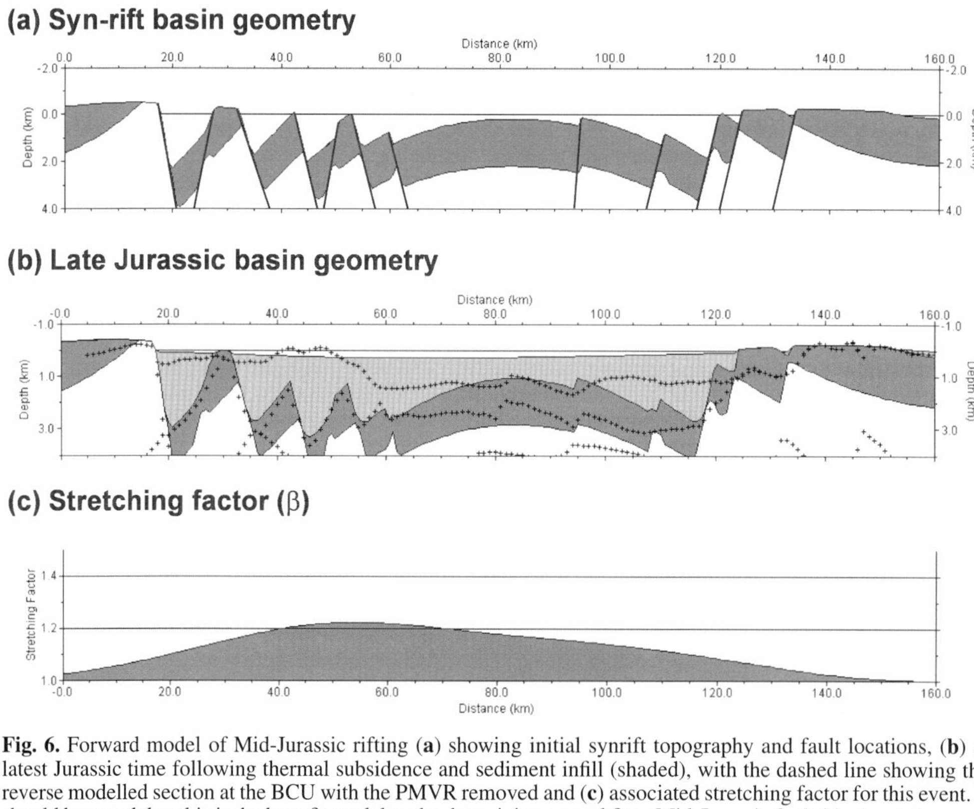

Fig. 6. Forward model of Mid-Jurassic rifting (**a**) showing initial synrift topography and fault locations, (**b**) at latest Jurassic time following thermal subsidence and sediment infill (shaded), with the dashed line showing the reverse modelled section at the BCU with the PMVR removed and (**c**) associated stretching factor for this event. It should be noted that this is the best-fit model to the data giving a good fit to Mid-Jurassic fault-block rotations and synrift sediment thickness on the western flank of the basin. However, the model fails to generate the slope between the western flank and the main depocentre. If fault throw were increased on the edge of the western flank to generate this accommodation space, a significant reduction in the fit to the observed fault geometries would result on the western basin flank. Upper-crustal faulting in this model represents 20% extension.

thickness, degree of erosion of footwall crests and fault-block rotations. The latter are particularly sensitive to the effective elastic thickness (T_e), with a fairly low best-fit T_e of 5 km. The misfit between the model and the restoration in the centre of the basin beneath the PMVR is compounded by the interpretation problems discussed above and is not considered further here. The main misfit between the model and the restoration is at the significant slope that separates the shelf area on the western flank and the deeper basin area further to the east. This slope is clearly apparent in the Early Cretaceous restoration in Fig. 5d, but is not generated in the forward model using the observed faults. If this slope were considered to represent a remnant Jurassic fault, the necessary basin offset would result in significant footwall uplift and degrade the fit between the model and the restored section on the basin western flank. The downstepping of the BCU by at least 2 km suggests that this is a structural rather than a stratigraphic feature that developed in post-BCU time. The short-wavelength nature of this feature suggests that this is not the result of flexure resulting from additional crustal loads; the template shown in Fig. 6b includes the removal of the PMVR during backstripping and clearly shows the slope still preserved.

Figure 6c shows the amount of extension in the forward model, assuming uniform extension with depth. This shows that normal fault heaves represent only *c.* 22% extension, significantly lower than the lithosphere extension of up to 450% ($\beta = 5.5$) necessary in the reverse model to drive post-rift thermal subsidence. The possible reasons for this discrepancy are discussed below.

Mid-Jurassic–Early Cretaceous rift

Seismic interpretation

As discussed above, the main area of misfit between the interpretation of a wholly Jurassic rift and the associated forward model is the significant scarp feature present in Early Cretaceous time immediately adjacent to the western basin flank. These observations have led to a second interpretation (Fig. 3c) in which the scarp is considered to be a structural feature. The steep basin slope between the western flank and the main Early Cretaceous depocentre is characterized by incoherent reflections, which have been interpreted to represent closely spaced (resulting in degraded seismic data quality) normal faults. These vertically offset the BCU by up to 2 km between the western basin shelf and the main depocentre. This interpretation suggests that extension on normal faults in Jurassic time continued into Early Cretaceous time, generating the basin accommodation space for thick Lower Cretaceous (BCU–Top Aptian) sediments in the centre of the basin. Small fault offsets at the BCU, mostly on reactivated Jurassic faults, are also interpreted on the western flank of the basin, suggesting that post-BCU faulting was not isolated to the edge of the basin flank. Other workers have also noted post-BCU faulting, particularly to the north and around the flanks of the Porcupine Basin (Needham *et al.* 1999). The sediment package overlying the BCU on the western flank suggests deeper palaeobathymetries in Hauterivian time, with upward shallowing during Barremian time possibly reflecting renewed faulting on the basin flank. Erosional truncation of the Top Barremian units on the basin flank is apparent from well and seismic data, increasing towards the main basin bounding fault, suggesting fault activity into Early Aptian time.

Model results

To reverse model the section assuming continued rifting into Early Cretaceous time, the timing for rifting must be defined to determine the thermal perturbation history. Because the models used in this study are instantaneous rift models, average rift ages are generally assigned rather than start or end rift ages. However, a rift event that was active from Mid-Jurassic to Aptian time would have been active for up to *c.* 80 Ma, during which interval early rift thermal perturbations will have significantly re-equilibrated by the onset of the post-rift phase. Determining a thermal age for the models is therefore very difficult. In this study we have assigned an Early Cretaceous age (Barremian) to the thermal models, to assess what effects a significantly younger thermal perturbation would have on the resulting post-rift subsidence history and predicted β factors.

Figure 7 shows the reverse model of the section. The stretching factor in Fig. 7e represents the best-fit β profile to restore Early Cretaceous palaeobathymetries on the western basin flank, and has a maximum of $\beta = 2.3$ beneath the PMVR, decreasing beneath the basin flanks.

Forward modelling used the restoration at the intra-Aptian seismic horizon (Fig. 7c) as a template to constrain a forward model of Early Cretaceous rifting. In particular, this model attempted to generate the palaeobathymetries across the western flank and the adjacent thick Lower Cretaceous sediment section. The model

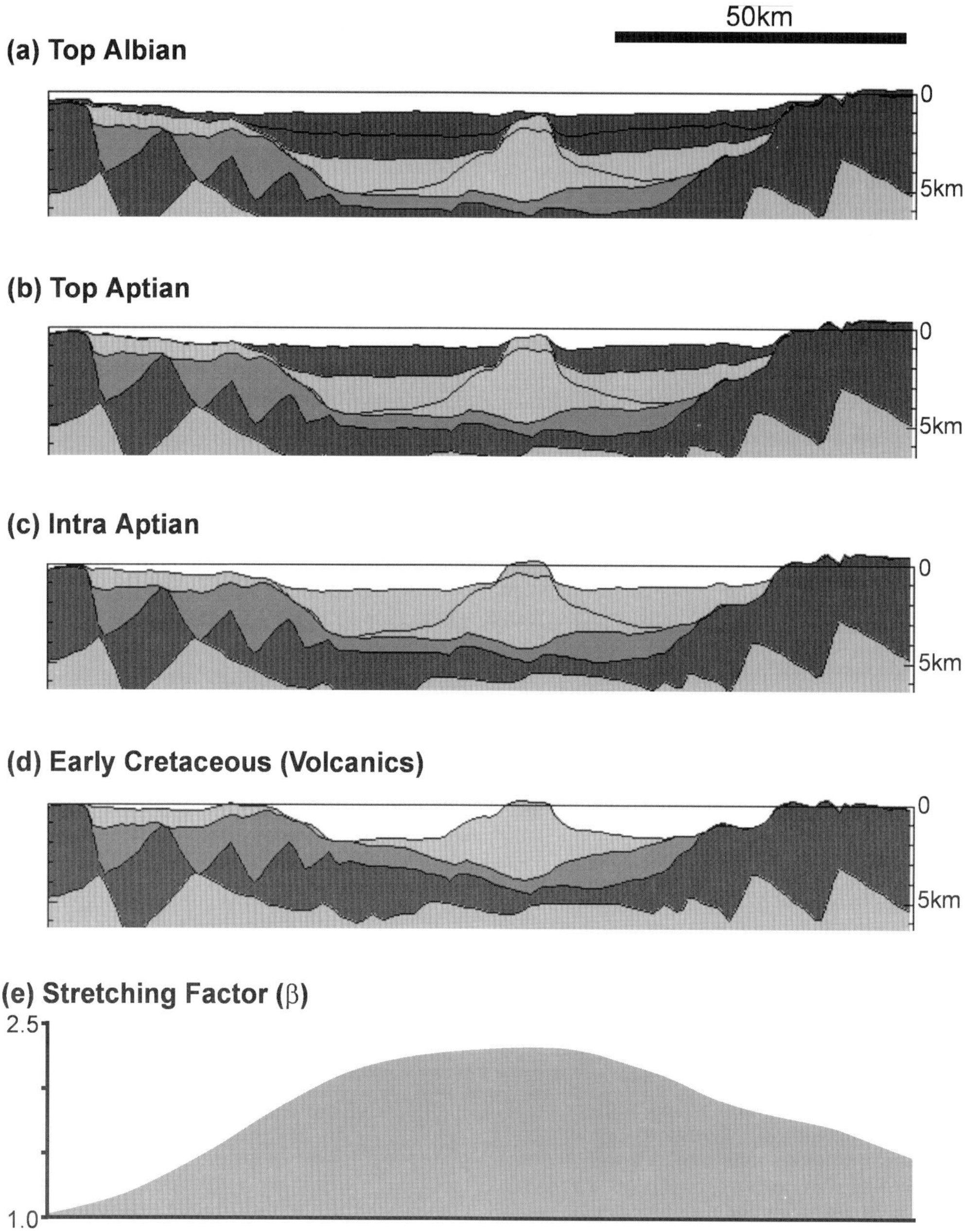

Fig. 7. Reverse model of a Mid-Jurassic to Early Cretaceous rift showing successive restorations at Cretaceous horizons (**a**–**d**). The model assumes an Early Cretaceous (Barremian–Early Aptian) lithospheric stretching event at 120 Ma. The stretching factor used in the reverse thermal model is shown in (**e**) and represents the best fit to restore the western basin flank to the palaeobathymetries predicted from well 43/13-1.

also includes increased extension in the lower crust and lithospheric mantle to generate the thermal perturbation at rifting, and includes associated Barremian to intra-Aptian thermal subsidence. The best-fit forward model is shown in Fig. 8. Faults located to the east of the PMVR are difficult to identify on seismic data although a number of reflections suggest fault-block rotation immediately beneath the Aptian section. Within the model, faults in this area have been

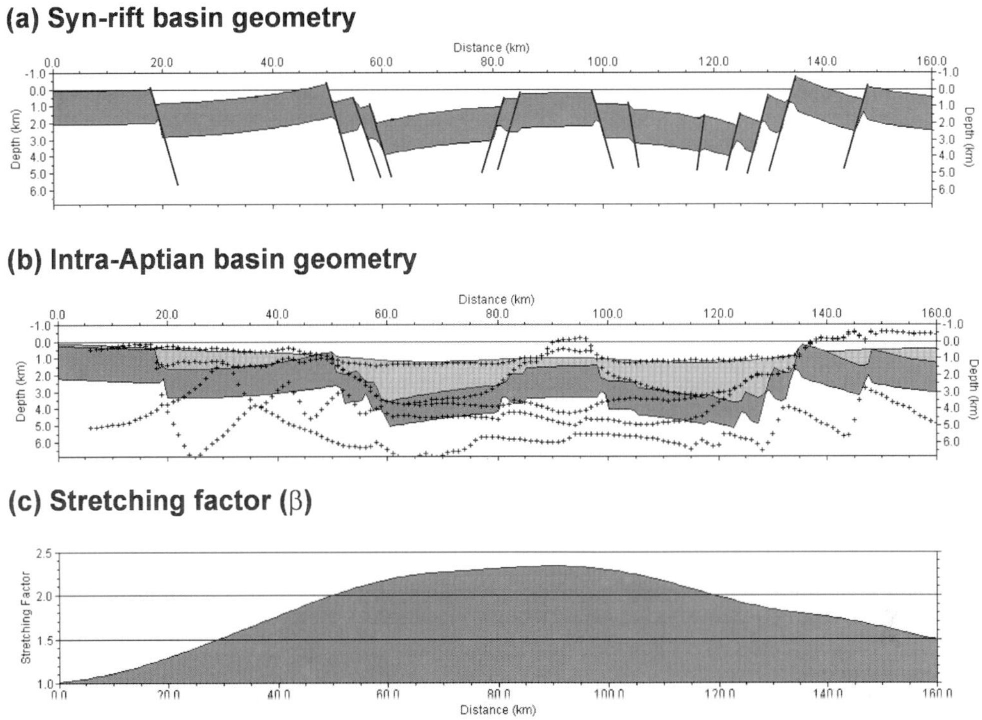

Fig. 8. Forward model of Early Cretaceous (Barremian–Aptian) rifting (**a**) showing initial synrift topography and fault locations, and (**b**) following early post-rift and sediment fill (shaded) at intra-Aptian time. Dashed lines in (**b**) represent the intra-Aptian restoration from Fig. 7c. It should be noted that modelling the Aptian depocentre as controlled by faulting on the eastern edge of the western flank gives a good fit to the observed basin geometry and Early Cretaceous sediment thickness in the depocentre and on the western flank. Upper-crustal faulting in this model generates 10% extension. The depocentre to the east of the PMVR shows a good fit to the basin thickness and may suggest similar faulting on the eastern margin of the basin at this time. (**c**) β factor for this event includes a component of lower-crustal thinning to match the stretching factor predicted from reverse modelling.

included to generate the observed basin geometry and sediment thickness in this part of the basin. The location and throw of faults west of the PMVR were based upon fault geometries from the seismic interpretation. In this part of the model a good fit is obtained between the reverse modelled template for the orientation and thickness of sediments on the western basin flank, and the observed sediment thickness within the main depocentre. Extension on upper-crustal faults in this model is *c.* 10%. The interpretation of the main Aptian depocentre as the result of continued faulting into Early Cretaceous time appears to be a viable mechanism for the observed Early Cretaceous basin geometry. It should be noted, however, that upper-crustal extension on normal faults is still significantly lower than the maximum lithospheric stretching of $\beta = 2.3$ predicted from the reverse model.

Discussion

Modelling results

The modelling described above has investigated two structural and thermal models for the evolution of the southern part of the south Porcupine Basin: a wholly Jurassic event with β factors similar to those described by earlier workers, and a longer-lived rift event that continued into in Early Cretaceous time. These models give very different results for maximum lithospheric stretching with β factors of 5.5 and 2.3, respectively. However, the considerable time period of the synrift phase means that a detailed thermal history cannot be constrained from seismic data alone, hence the second of our models was biased towards an Early Cretaceous rift age to give a younger 'end member' model. Reverse modelling using the two distinct rift ages and interpretations generates very similar restorations of Early Cretaceous palaeobathymetries on the western basin flank, and it is therefore expected that variations in β and rift age between the two models could give similar restorations. From this study it is therefore clear that reverse modelling alone cannot determine a unique model for the stretching history beneath the basin.

Forward modelling of the rift basin gives a good fit to the Jurassic fault blocks and sediments on the western basin flank. However, assuming the onset of post-rift subsidence at the BCU does not suggest a mechanism for the deep bathymetry in the basin axis, and it is difficult to generate a model in the deeper part of the basin using the observed normal faults. Although an alternative (non-structural) mechanism for generating the deep Early Cretaceous basin cannot be discounted, the model of Jurassic extension continuing into Early Cretaceous time gives an improved fit to the observed basin geometry and appears viable from the seismic interpretation.

Crustal structure

For the two models described above, the maximum amount of lithospheric stretching is significantly different. The modelling results from the wholly Jurassic rift concur with those of Tate *et al.* (1993) with predicted lithosphere stretching of up to $\beta = 5.5$. However, if Jurassic rifting continued into Early Cretaceous time with a dominant thermal perturbation in Cretaceous time (as in our second model), lithospheric stretching is reduced to a maximum of $\beta = 2.3$. Applying additional observations to distinguish between these models is problematic. Crustal thickness measurements have been used by a number of previous workers, and are discussed in the light of our results.

Various geophysical techniques have been used to measure crustal thickness beneath the Porcupine Basin, with minimum crustal thickness measurements of 7.5–10 km (Masson & Miles 1986), 8 km (Conroy & Brock 1989) and 5 km (Needham *et al.* 1999). These have been routinely used to predict crustal β factors that can then be compared with those predicted from subsidence modelling. Assuming unthinned crust at the flanks of the basin of *c.* 30 km thickness (Tate *et al.* 1993) would predict maximum β factors within the basin of between three and six given the variance of these crustal thickness measurements. These values are similar to the variation in maximum lithosphere stretching described by our two models, and do not immediately distinguish a preferred model.

A direct comparison between β factors derived from crustal thicknesses and those predicted from subsidence modelling should be carefully considered. β factors derived from thermal subsidence modelling (McKenzie 1978) are an estimation of the thermal perturbations generated by particular tectonic events that are then related to lithosphere thinning. However, as the present-day crustal thickness beneath the Porcupine Basin reflects the entire deformation history of the basin (rather than a particular tectonic event), it includes any pre-Jurassic phases of crustal attenuation. Therefore estimates of β factors derived from crustal thickness measurements beneath the Porcupine Basin can only define a maximum estimate for stretching resulting from a Jurassic event. The estimates of crustal thinning β factors from beneath the Porcupine Basin would therefore appear to support both of our models. However, stretching factors of >6 as described by Tate *et al.* (1993) in the southernmost part of the Porcupine Basin would appear to be less favourable.

Lithosphere stretching models

In both model scenarios the amount of extension on upper-crustal normal faults (as observed from seismic data and used in the forward models) is significantly lower than the amount of lithospheric stretching (predicted by the β factors in the reverse model) necessary to drive post-rift thermal subsidence. For the wholly Jurassic event, lithospheric stretching of up to *c.* $\beta = 5.5$ was coincident with crustal extension on normal faults of *c.* 20% ($\beta = 1.2$). Assuming Jurassic rifting continued into Cretaceous time, as in our second model, predicts a reduced lithosphere stretching factor of up to *c.* $\beta = 2.3$ associated with 20% (β *c.* 1.2) Jurassic faulting and 10% ($\beta = 1.1$) Early Cretaceous faulting. It is apparent from these results and those of Tate *et al.* (1993) that there is a significant discrepancy between the magnitudes of subsidence-derived lithospheric extension and extension in the upper crust that is observed as normal faults. These observed discrepancies might be manifested in a number of ways. Walsh *et al.* (1991) proposed that up to 40% of regional extension could be missed by summing fault offsets on seismic profiles, as a result of faulting below seismic resolution. However, this is still far short of the observed discrepancies within the southern part of the Porcupine Basin. Alternatively, unresolved Jurassic or Early Cretaceous structures may exist within the poor quality seismic data beneath the PMVR, although on seismic lines immediately to the north of the PMVR, where lithosphere extension is still predicted to be high, additional large extensional structures are not apparent. Two possible interpretations of these discrepancies are made here and can be applied to both model scenarios: (1) extreme upper-crustal extension exists but is not apparent from seismic data; (2) extreme upper-crustal extension did not occur. These two scenarios are discussed below.

Whole-crustal thinning. An important feature of the β profiles for both reverse models described above is that the largest β factors are beneath the centre of the basin, which is routinely poorly resolved on seismic data. Tate *et al.* (1993) noted that at stretching higher than $\beta = 2$, stretching processes within the crust are likely to become complex, with considerable internal deformation resulting in poor seismic imaging of associated structures. Therefore, if the upper crust beneath the centre of the basin has been attenuated by the β factors of between 2.3 and 5.5 described in our basin models, seismic evidence is likely to be limited. These limits fit within the ranges apparent from crustal thickness measurements and therefore extreme upper-crustal thinning beneath the centre of the basin appears to be viable.

Depth-dependent crustal thinning. An alternative interpretation of these discrepancies in upper crust v. lithosphere stretching is that they reflect high-magnitude lithosphere extension at depth that was partitioned from significantly lower-magnitude extension on normal faults within the brittle upper crust. Previous workers (e.g. Baxter *et al.* 1997; Roberts *et al.* 1997; Davis 1999) have proposed depth-dependent lithosphere extension in other offshore areas, based upon observed discrepancies between upper-crustal and lithospheric extension. An important feature of these previous studies is that high-magnitude lithosphere extension occurred beneath continental margins immediately before continental break-up. Our second model of continued rifting into Early Cretaceous time, rather than a wholly Jurassic rift, may concur better with this. A younger rift age to Aptian time would suggest associated lithosphere stretching beneath the Porcupine Basin immediately before the onset of Aptian–Albian break-up further to the west in the Atlantic (Masson & Miles 1984; Johnston *et al.* 1999). An alternative depth-dependent model for the Rockall Trough area has been proposed by Shannon *et al.* (1999), in which crustal attenuation is characterized by high stretching within the upper crust (based upon observations of highly attenuated upper crust from seismic and gravity studies) and significantly reduced stretching within the lower crust and lithospheric mantle (derived from reverse modelling). However, estimates of minimum crustal thickness of 5–10 km beneath the south Porcupine Basin (Masson & Miles 1986; Conroy & Brock 1989; Needham *et al.* 1999) are broadly consistent with the range in stretching factors defined in this and other modelling studies (e.g. Tate *et al.* 1993). Stretching models similar to that described for the Rockall Trough are yet to be proposed for the south Porcupine Basin.

Regional tectonic setting and Atlantic break-up

Considering the interpretation of a Mid-Jurassic lithosphere extension event raises the question of how this fits into the regional tectonic evolution of the NE Atlantic region. Mid-Jurassic plate reconstructions (e.g. Ziegler 1988; Johnston *et al.* 1999) show that the Porcupine Basin area was undergoing intra-continental extension, before eventual plate break-up along the incipient proto-Atlantic spreading centre in Aptian–Albian time (Masson & Miles 1984). However, it is not immediately apparent why high lithosphere extension of up to 500% was occurring beneath the Porcupine Basin during Mid–Late Jurassic time, at least 50 Ma before plate break-up. Our second model of continuing rifting into Early Cretaceous time allows high lithospheric extension beneath the Porcupine Basin (reduced to a maximum of 130%) to be active immediately before Atlantic break-up. In this scenario the Mid-Jurassic rift is interpreted as an early phase of intra-continental rifting (of *c.* 20% extension), similar to that evident over much of the NW Atlantic region at this time (Ziegler 1988; Johnston *et al.* 1999).

A model of a relatively low-magnitude Jurassic rift that was overprinted by Early Cretaceous rifting associated with increased magnitude lithosphere extension at depth may give an improved correlation between the evolution of the Porcupine Basin and other sedimentary basins further north along the Atlantic margin. A major rifting event has been suggested in the Rockall Trough during Early–Mid-Cretaceous time with the development of normal faults and a significant sediment depocentre (Musgrove & Mitchener 1996; England & Hobbs 1997). This rift event has also been inferred to show a similar discrepancy between extension observed on upper-crustal faults and the deeper lithosphere extension necessary to

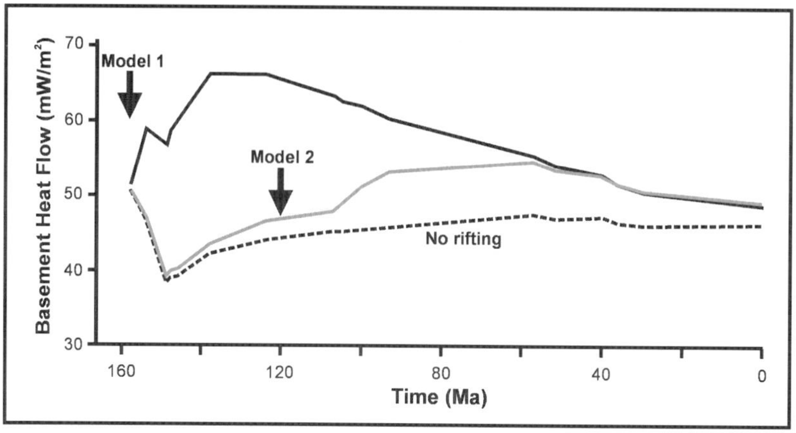

Fig. 9. Heat flow models for the western basin flank using the two tectonic models discussed. Model 1 assumes a Mid–Late Jurassic stretching factor of $\beta = 4$ and predicts maximum heat flow in Early Cretaceous time of *c.* 68 mW m^{-2}. Model 2 assumes a two-phase rift with a Mid–Late Jurassic event of $\beta = 1.2$ followed by Barremian–Aptian stretching of $\beta = 1.5$ and predicts maximum heat flow in latest Cretaceous time of 55 mW m^{-2}. Heat flow resulting from no applied rifting is also shown.

drive post-rift thermal subsidence (Nadin *et al.* 1999). Rifting in the Rockall Basin was active slightly later than in the Porcupine Basin, with fault activity extending to Santonian time (Nadin *et al.* 1999). The similarities in lithosphere extension between these two basins, and the northward younging of Cretaceous rifting from the Porcupine Basin to the Rockall Basin, are consistent with a model of northward development of Early Cretaceous break-up, and anomalous deeper lithosphere extension beneath the adjacent rift basins immediately before break-up.

Implications for heat flow models

One of the aims of this study was to assess alternative structural models for the Porcupine Basin so as to model the heat flow history of the basin for direct input to maturation studies. The 1D heat flow model applied the stretching factors described above to predict basement heat flows. The model incorporates sediment properties (defined from 43/13-1), sediment burial history and additional crust and sediment radiogenic heat sources. Thermal evolution of the lithosphere within the model allows thermal perturbation and re-equilibration from either whole-lithosphere extension (similar to McKenzie 1978) or depth-dependent lithosphere extension. A number of models have been run for the western flank of the Porcupine Basin using the β factors predicted from reverse modelling.

The results of the model are shown in Fig. 9. β factors used in the models represent the stretching factors predicted from the reverse models in the vicinity of well 43/13-1. Model 1 represents a Mid-Jurassic stretching event with $\beta = 4.0$. Model 2 represents a Mid-Jurassic to Early Cretacous rift and has been modelled as a two-phase rift event with Mid-Jurassic extension of $\beta = 1.2$ followed by Early Cretaceous extension of $\beta = 1.5$. The increased thermal perturbation generated by the Mid-Jurassic rift (Model 1) gives higher basement heat flows compared with that from the lower-magnitude Mid-Jurassic–Early Cretaceous events (Model 2), and these heat flows remain higher until Early Tertiary time, after which the heat flow curves are similar. A feature of these heat flow curves is the delay in maximum heat flow relative to the timing of lithosphere extension. This is the result of the initial thermal perturbation originating at depth, with a delay of up to 30 Ma before maximum heat flow is apparent in the sedimentary section. This differs from models that consider uniform extension of the lithosphere with depth (e.g. McKenzie 1978; Jarvis & McKenzie 1980) in which maximum heat flow is concurrent with the end of rifting. The magnitude of maximum heat flow is also significantly reduced to that predicted from whole-lithospheric stretching (McKenzie 1978). Finite rifting models (e.g. Jarvis & McKenzie 1980) show a decrease in maximum heat flow, but this is the result of thermal re-equilibration during rifting, rather than the partitioning of stretching within the lithosphere.

Summary and conclusions

(1) Quantitative 2D structural modelling of synrift and post-rift responses to continental extension has been applied to the interpretation of seismic line GSP97-19 in the southern part of the Porcupine Basin, correlated with well 43/13-1 on the western flank of the basin.

(2) Two alternative scenarios have been tested using a combined forward and reverse modelling approach: (a) a dominant Mid–Late Jurassic rift with observed upper-crustal faulting of *c.* 20% extension and associated lithosphere extension up to $\beta = 5.5$, similar to models proposed by previous workers; (b) Mid-Jurassic rifting (*c.* 20% extension) that continued into Early Cretaceous time with additional upper-crustal faulting of *c.* 10% extension and deeper lithosphere extension up to $\beta = 2.3$ (130% extension).

(3) Both of these models show a distinct discrepancy between predicted lithosphere stretching and the magnitude of extension observed on upper-crustal faults. A model has been proposed that reflects an element of depth-dependent lithosphere extension. However, seismic data quality in the centre of the basin renders this inconclusive at present. Both of the above models fit within the bounds of crustal thickness measurements of previous workers and therefore a preferential model is not obvious. Forward modelling of the Early Cretaceous basin geometry gives a preferred model of rifting continuing into Early Cretaceous time, although alternative non-structural mechanisms cannot be ruled out. The interpretation of extension continuing into Early Cretaceous time suggests a regional tectonic model in which rifting and associated lithospheric extension was directly related to the progressive northward development of Early Cretaceous break-up in the Proto-Atlantic. This is also reflected in a similar style of Early Cretaceous extension in the Rockall Trough, which evolved slightly later as break-up propagated northwards.

(4) Heat flow on the western flank of the basin has been modelled for input to maturation studies, with the results used to evaluate the impact of different structural models on predicted source peak maturation. The key element in this modelling is the use of a depth-dependent thermal perturbation, which results in a delay of up to 30 Ma between rifting and maximum basement heat flow.

The authors would like to acknowledge coworkers whose work contributed to the development of the ideas in this paper, and to state that the opinions expressed here are those of the authors and do not necessarily represent those of the companies involved. P. Shannon, M. Tate and D. Praeg are thanked for constructive comments on the manuscript. Schlumberger Geco-Prakla is acknowledged for permission to publish interpretations of seismic data.

References

BAXTER, K., COOPER, G.T., O'BRIEN, G.W., HILL, K.C. & STURROCK, S. 1997. Flexural isostatic modelling as a constraint on basin evolution, the development of sediment systems, and palaeo-heat flow: application to the Vulcan Sub-basin, Timor Sea. *APPEA Journal*, **37**, 137–153.

BP 1998. *Velocity Analysis and Depth Conversion for the South Porcupine Licence Round. Paradigm Geophysical.* BP internal report.

CONROY, J.J. & BROCK, A. 1989. Gravity and magnetic studies of crustal structure across the Porcupine Basin west of Ireland. *Earth and Planetary Science Letters*, **93**, 371–376.

CROKER, P.F. & SHANNON, P.M. 1987. The evolution and hydrocarbon prospectivity of the Porcupine Basin, offshore Ireland. *In*: BROOKS, J. & GLENNIE, K.W. (eds) *Petroleum Geology of North West Europe*. Graham & Trotman, London, 633–642.

DAVIS, M. 1999. *Lithospheric stretching at rifted continental margins.* PhD thesis, University of Liverpool.

ENGLAND, R.W. & HOBBS, R.W. 1997. The structure of the Rockall Trough imaged by deep seismic reflection profiling. *Journal of the Geological Society, London*, **154**, 497–502.

EWAN, D. 1998. *Summary Chronostratigraphy—Porcupine Basin West Flank.* BP internal report.

JARVIS, G.T. & MCKENZIE, D.P. 1980. Sedimentary basin formation with finite extension rates. *Earth and Planetary Science Letters*, **48**, 42–52.

JOHNSTON, S., DORÉ, A.G. & SPENCER, A.M. 1999. Plate tectonic setting of basin development in the southern Porcupine Basin area. *In*: CROKER, P.F. & O'LOUGHLIN, O. (eds) *The Petroleum Exploration of Ireland's Offshore Basins, Dublin, 29–30 April 1999. Extended Abstracts.* Petroleum Affairs Division, Department of the Marine and Natural Resources, Dublin, 2–4.

KUSZNIR, N.J. & EGAN, S. 1989. Simple-shear and pure-shear models of extensional sedimentary basin formation: application to the Jeanne d'Arc basin, Grand Banks of Newfoundland. *In*: TANKARD, A.J. & BALKWILL, H.R. (eds) *Extensional Tectonics and Stratigraphy of the North Atlantic Margins.* American Association of Petroleum Geologists, Memoirs, **46**, 305–322.

KUSZNIR, N.J., MARSDEN, G. & EGAN, S. 1991. A flexural cantilever simple-shear/pure-shear model of continental lithosphere extension: applications to the Jeanne d'Arc Basin, Grand Banks & Viking Graben, North Sea. *In*: ROBERTS, A.M., YIELDING, G. & FREEMAN, B. (eds) *The Geometry of Normal*

Faults. Geological Society, London, Special Publications, **56**, 41–61.

KUSZNIR, N.J., ROBERTS, A.M. & MORLEY, C.K. 1995. Forward and reverse modelling of rift basin formation. *In*: LAMBIASE, J.J. (ed.) *Hydrocarbon Habitat in Rift Basins*. Geological Society, London, Special Publications, **80**, 33–56.

MASSON, D.G. & MILES, P.R. 1984. Mesozoic seafloor spreading between Iberia, Europe and North America. *Marine Geology*, **56**, 279–287.

MASSON, D.G. & MILES, P.R. 1986. Structure and development of Porcupine Seabight sedimentary basin, offshore southwest Ireland. *American Association of Petroleum Geologists Bulletin*, **70**, 536–548.

MCKENZIE, D. 1978. Some remarks on the development of sedimentary basins. *Earth and Planetary Science Letters*, **40**, 25–32.

MUSGROVE, F.W. & MITCHENER, B. 1996. Analysis of the pre-Tertiary rifting history of the Rockall Trough. *Petroleum Geoscience*, **2**, 353–360.

NADIN, P.A., HOUCHEN, M. & KUSZNIR, N.J. 1999. Evidence for pre-Cretaceous rifting in the Rockall Trough: an analysis using quantitative 2D structural/stratigraphic modelling. *In*: FLEET, A.J. & BOLDY, S.A.R. (eds) *Petroleum Geology of Northwest Europe: Proceedings of the 5th Conference*. Geological Society, London, 371–378.

NAYLOR, D. & ANSTEY, N.A. 1987. A reflection seismic study of the Porcupine Basin, offshore west Ireland. *Irish Journal of Earth Sciences*, **8**, 187–210.

NEEDHAM, C.E.J., CHILOVI, C., BOCCA, P. & TILTMAN, C. 199. A proposed model for the structural development of the south Porcupine Basin. *In*: CROKER, P.F. & O'LOUGHLIN, O. (eds) *The Petroleum Exploration of Ireland's Offshore Basins, Dublin, 29–30 April 1999. Extended Abstracts*. Petroleum Affairs Division, Department of the Marine and Natural Resources, Dublin, 68–71.

ROBERTS, A.M., LUNDIN, E.R. & KUSZNIR, N.J. 1997. Subsidence of the Vøring Basin and the influence of the Atlantic Continental Margin. *Journal of the Geological Society, London*, **154**, 551.

ROBERTS, A.M., YIELDING, G., KUSZNIR, N.J. & DORN-LOPEZ, D. 1993. Mesozoic extension in the North Sea: constraints from flexural backstripping, forward modelling and fault populations. *In*: PARKER, J.R. (ed.) *Petroleum Geology of Northwest Europe: Proceedings of the 4th Conference*. Geological Society, London, 1123–1136.

SCLATER, J.G. & CHRISTIE, P.A.F. 1980. Continental stretching: an explanation of the post-mid-Cretaceous subsidence of the Central North Sea Basin. *Journal of Geophysical Research*, **85**, 3711–3739.

SHANNON, P.M. 1991. The development of Irish offshore sedimentary basins. *Journal of the Geological Society, London*, **148**, 181–189.

SHANNON, P.M., JACOB, A.W.B., O'REILLY, B.M., HAUSER, F., READMAN, P.W. & MAKRIS, J. 1999. Structural setting, geological development and basin modelling in the Rockall Trough. *In*: FLEET, A.J. & BOLDY, S.A.R. (eds) *Petroleum Geology of Northwest Europe: Proceedings of the 5th Conference*. Geological Society, London, 421–431.

SHANNON, P.M., WILLIAMS, B.P.J. & SINCLAIR, I.K. 1995. Tectonic controls on Upper Jurassic to Lower Cretaceous reservoir architecture in the Jeanne d'Arc Basin, with some comparisons from the Porcupine and Moray Firth Basins. *In*: CROKER, P.F. & SHANNON, P.M. (eds) *The Petroleum Geology of Ireland's Offshore Basins*. Geological Society, London, Special Publications, **93**, 467–490.

TATE, M.P. & DOBSON, M.R. 1989. Pre-Mesozoic geology of the western and northwestern Irish continental shelf. *Journal of the Geological Society, London*, **146**, 229–240.

TATE, M., WHITE, N. & CONROY, J.-J. 1993. Lithospheric extension and magmatism in the Porcupine Basin West of Ireland. *Journal of Geophysical Research*, **98**, 13905–13923.

WALSH, J., WATTERSON, J. & YIELDING, G. 1991. The importance of small-scale faulting in regional extension. *Nature*, **351**, 391–393.

WATTS, A.B., KARNER, G.D. & STECKLER, M.S. 1982. Lithosphere flexure and the evolution of sedimentary basins. *Philosophical Transactions of the Royal Society, London, Series A*, **305**, 249–281.

WILLIAMS, B.P.J., SHANNON, P.M. & SINCLAIR, I.K. 1999. Comparative Jurassic and Cretaceous tectono-stratigraphy and reservoir development in the Jeanne d'Arc and Porcupine basins. *In*: FLEET, A.J. & BOLDY, S.A.R. (eds) *Petroleum Geology of Northwest Europe: Proceedings of the 5th Conference*. Geological Society, London, 487–499.

ZIEGLER, P.A. (ed.) 1988. *Evolution of the Arctic–North Atlantic and the Western Tethys*. American Association of Petroleum Geologists, Memoir, **43**.

Provenance implications of reworked palynomorphs in Mesozoic successions of the Porcupine and North Porcupine basins, offshore Ireland

J. SMITH & K. T. HIGGS

Department of Geology, University College Cork, Cork, Ireland

(e-mail: j.smith@ucc.ie)

Abstract: Reworked Late Carboniferous palynomorphs are recorded from mudrocks within Upper Jurassic and Lower Cretaceous sandstone-prone horizons from five wells (26/28-A1, 26/28-A1Z, 35/19-1, 35/8-1 and 35/8-2) in the Porcupine and North Porcupine Basin area, offshore western Ireland. Reworked palynomorphs are recognized by their anomalous biostratigraphic age; differences in colour and preservation are less significant. The reworked palynomorphs are remarkably well preserved and are thermally relatively immature in comparison with *in situ* onshore palynomorphs of a similar age. This indicates that the source of the reworked palynomorphs was not the onshore Upper Carboniferous successions of the Irish mainland. They were probably derived, along with the associated sand, from the immediate basin margin to the east, as this is the only area where there is suitably aged material with an appropriate thermal history. Reworking of rift flank Late Carboniferous material was thus an important factor in the provenance of Mesozoic sandstone horizons throughout the Porcupine and North Porcupine basins and surrounding areas. The study demonstrates that the investigation of reworked palynomorphs can be an important aspect of provenance studies in general.

The aim of this paper is to illustrate the application of reworked palynomorphs to provenance studies of potential Mesozoic reservoir sandstones in the Porcupine Basin, offshore western Ireland, and to highlight the general utility of this technique in basin analysis. Although palynologists often report the presence of reworked palynomorphs in preparations, they are seldom considered further (Streel & Bless 1980). Notable exceptions to this include the studies by Bless & Streel (1976); Streel & Bless (1980); Truswell (1982); Truswell 1983a, 1983b), Gaupp & Batten (1983); Truswell & Drewry (1984); Guy-Ohlson *et al.* (1987); Eshet *et al.* (1988); Van de Laar & Fermont (1989); Batten (1991); Tyson (1995). The work by Truswell (1982; 1983a, 1983b) and Truswell & Drewry (1984) concentrated on the effects of ice movement and melt-out and may have little relevance to non-glacial environments (Batten 1991; Tyson 1995). The distribution of reworked palynomorphs within depositional environments was reviewed by Tyson (1995), but their use in the investigation of sedimentary provenance was not discussed. The role of reworked palynomorphs is thus a much neglected tool in sedimentary provenance investigations.

Geological setting

The Porcupine and North Porcupine basins (Fig. 1) are north–south-orientated Mesozoic to Cenozoic sedimentary basins on the continental shelf off the west coast of Ireland (Naylor & Shannon 1982; structural nomenclature from Naylor *et al.* 1999). They are bounded to the north by the Slyne High, to the west by the Porcupine High, to the east by the Irish Mainland Shelf and merge with the Goban Spur to the south (Moore 1992). The basins contain up to 10 km thickness of post-Palaeozoic sediments (Shannon 1992), and up to 200 m of Upper Carboniferous deltaic sandstone-prone successions have been drilled in the North Porcupine and Porcupine Basin area (Croker & Shannon 1987). However, the detailed distribution and lithology of the Carboniferous–Lower Palaeozoic successions in and around the North Porcupine and Porcupine Basin area are sparsely documented and, as such, poorly understood. In this study, the palynological sampling has focused on the Upper Jurassic and Lower Cretaceous sandstone-prone horizons from five wells (26/28-A1, 26/28-A1Z, 35/19-1, 35/8-1 and 35/8-2), details of which are given below. The environment of deposition of these sandstone-prone horizons ranges from shallow

From: SHANNON, P.M., HAUGHTON, P.D.W. & CORCORAN, D.V. (eds) 2001. *The Petroleum Exploration of Ireland's Offshore Basins*. Geological Society, London, Special Publications, **188**, 291–300.

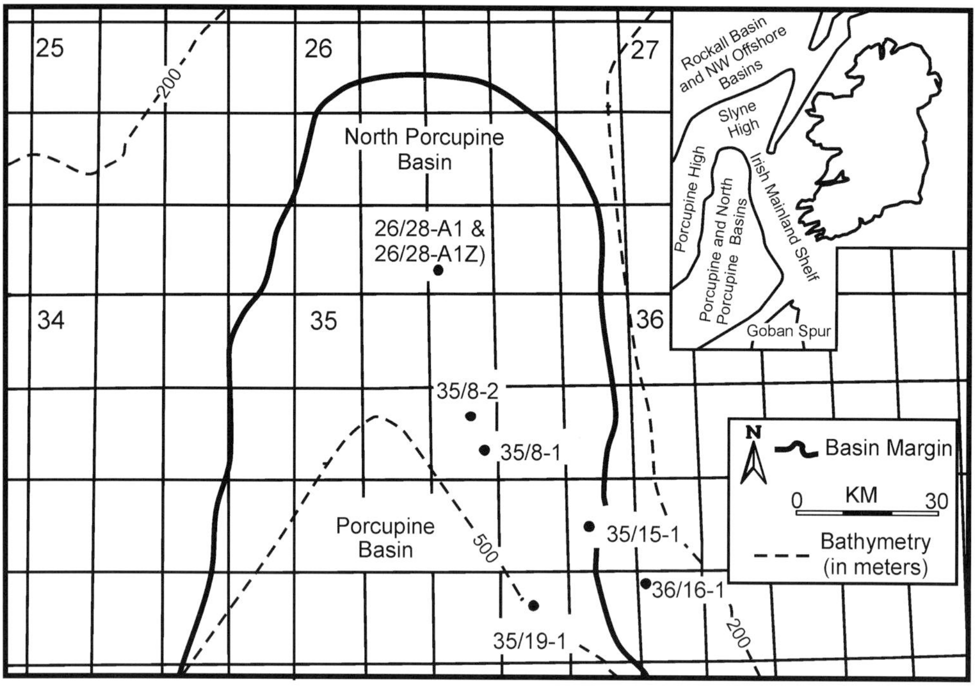

Fig. 1. Location of the study area and the wells referred to in the text. Regional geology after Shannon & Naylor (1998) and structural nomenclature from Naylor *et al.* (1999).

marine–alluvial fan sandstones in 26/28-A1 and 26/28-A1Z area, to deep (below storm wave base) submarine fan systems in 35/8-2 (Robinson & Canham 2001). In addition, completion logs and well reports from two wells (35/15-1 and 36/16-1) from the eastern margin of the Porcupine Basin (see Fig. 1) and previously published material were studied to give an indication of the local Palaeozoic successions present on the immediate basin margin.

Material and methods

Guy-Ohlson *et al.* (1987) found that the greatest concentrations of reworked Carboniferous miospores in the Swedish Jurassic sections they investigated correlated with texturally immature coarse-grained siltstones. This is probably due to the fact that the reworked Carboniferous spores are significantly larger than the *in situ* Jurassic palynomorphs (Guy-Ohlson *et al.* 1987). Therefore for this study, where possible, samples were collected from this type of lithology. In addition, sampling in three of the wells (35/8-2, 35/8-1 and 35/19-1) was deliberately biased towards horizons within cores that were associated with waning flow conditions immediately after sand deposition, as it was assumed by one of the authors (J.S.) that reworked palynomorph material is likely to come out of suspension under these conditions.

The samples were processed using standard palynological preparation techniques (Traverse 1988; Van Bergan *et al.* 1990) to remove all mineral components after sieving through a 20 μm mesh. With the exception of the samples from 26/28-A1 and 26/28-A1Z (which had to be bleached before acid digestion), the samples were not oxidized. Standard palynological counting was not carried out, as the number of reworked grains identified in each sample was too low (<300 grains per sample) to allow any meaningful information to be derived. Instead, the distribution of identified reworked material within the various sections was recorded in a 'present-or-absent' format for each of the samples in the section.

Recognizing reworked palynomorphs

Of all fossil groups, palynomorphs are the most likely to survive a sedimentary reworking event. This is due to a number of factors including the durability of their sporopollenin walls, their small size and their relative abundance in certain environments and stratigraphic levels (Batten

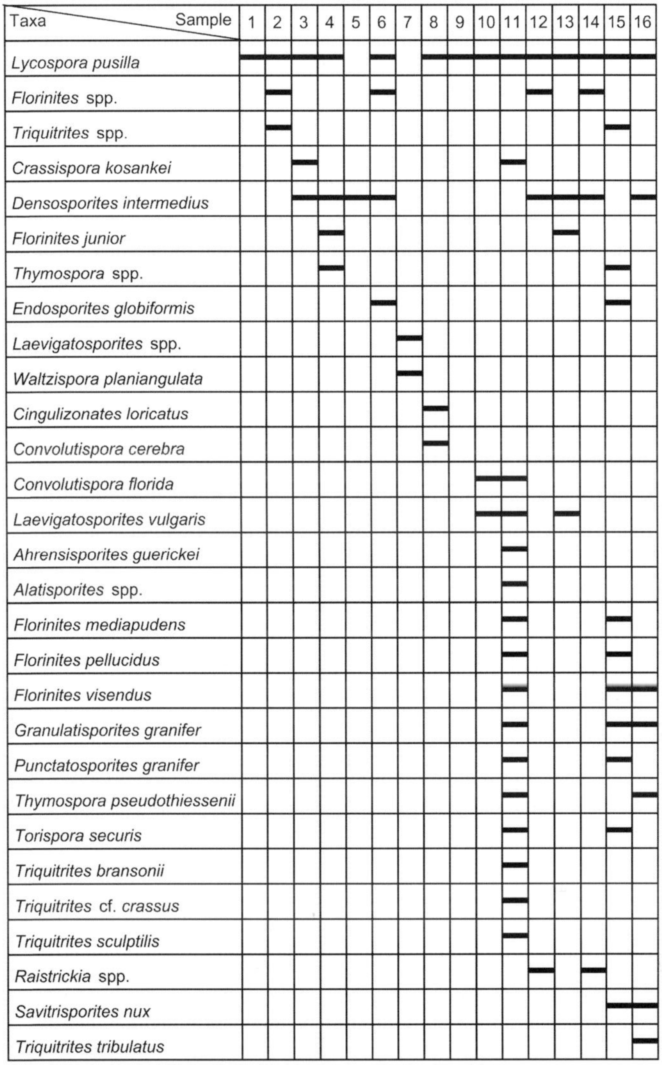

Fig. 2. Distribution of reworked Carboniferous palynomorphs recovered from well 26/28-A1. The sample numbers increase with increasing depth in the well and the taxa are ordered in terms of first downhole appearance.

1991). The criteria for recognizing the presence of reworked palynomorphs within a palynological assemblage were discussed at length by Batten (1991) and therefore will be only briefly summarized here. The most reliable indication that a palynomorph has been reworked is when it is found in association with significantly younger palynomorphs. Where the stratigraphic age difference is large this type of reworking is obvious. However, as the age difference decreases, it becomes harder to recognize reworking, and when taxa from intervals above and below are morphologically identical (such as with Hauterivian and Barremian spores) it may be impossible to recognize reworking (Batten 1991).

In addition to an anomalous biostratigraphic age, reworked palynomorphs may be recognized by differing from the *in situ* palynomorphs as follows: differences in colour (palynomorphs darken in colour with increasing thermal maturity); differences in preservation state (i.e. the condition of the palynomorph); the presence or absence of pyrite or damage associated with pyrite in the palynomorph wall; differences in autofluorescence. Of the above list, the difference in colour is the most reliable indicator. However, care must be taken to ensure that the colour difference is not due to a variation in the relative thickness of the palynomorph walls, as thicker-walled palynomorphs will appear darker in transmitted light.

Results

The results of the investigation of reworked palynomorphs are outlined below and represented graphically in Figs 1–6. A selection of the reworked material is illustrated in Fig. 7. As the depths of the individual samples could not be released, the samples are numbered in increasing order of depth within the well in question in Figs 1–6. The taxa are arranged in order of first downhole occurrence.

Well 26/28-A1

A total of 27 core and 15 cuttings samples were analysed from a series of Upper Jurassic sandstone-prone horizons in this well, 16 of which contained reworked material. The distribution of the reworked palynomorph taxa identified within the sandstone-prone horizons is detailed in Fig. 2.

The reworked miospores recovered are indicative of a latest Westphalian C to Westphalian D age. The reworked palynomorphs are well preserved with little sign of abrasion, fungal attack or pyrite damage. As this material (and the material from 26/28-A1Z) had to be oxidized during preparation, no conclusions may be drawn from the colour of the reworked material.

Well 26/28-A1Z

A total of 76 cuttings samples were analysed from a series of Upper Jurassic sandstone-prone horizons in this well, 13 of which contained reworked material. The reworked miospores recovered are indicative of a latest Westphalian C to Westphalian D age and their distribution is detailed in Fig. 3. Again, the reworked miospores are well preserved.

Well 35/8-2

A total of 24 core samples, selected from horizons associated with waning flow conditions immediately after sand deposition, were analysed from a series of Kimmeridgian–Volgian sandstone-prone horizons (turbidite beds), 16 of which contained reworked material. The reworked miospores recovered are indicative of a Westphalian age (no older than late Westphalian A) and their distribution is detailed in Fig. 4.

Although not as well preserved as the reworked miospores present in wells 26/28-A1

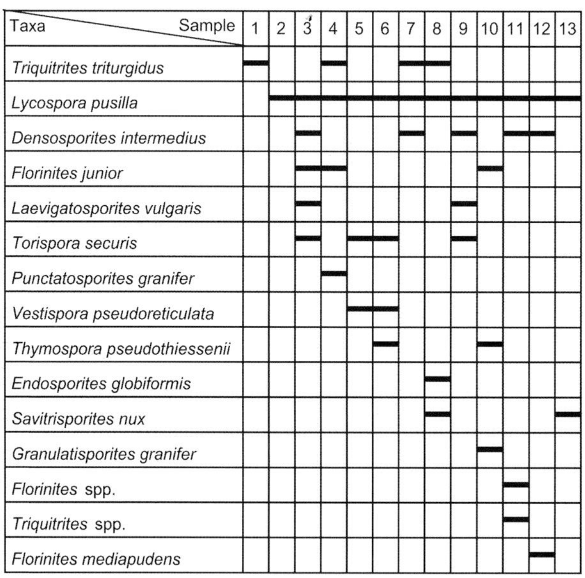

Fig. 3. Distribution of reworked Carboniferous palynomorphs recovered from well 26/28-A1Z. The sample numbers increase with increasing depth in the well and the taxa are ordered in terms of first downhole appearance.

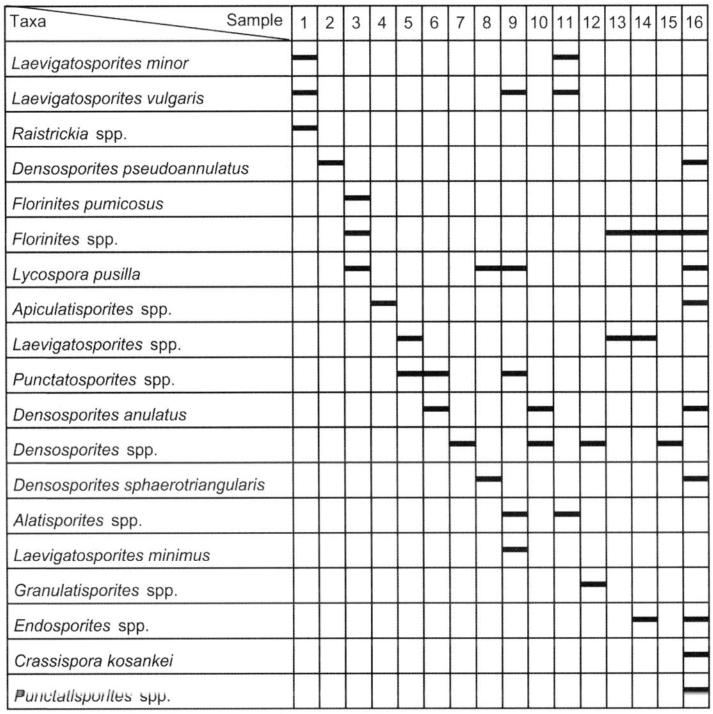

Fig. 4. Distribution of reworked Carboniferous palynomorphs recovered from well 35/8-2. The sample numbers increase with increasing depth in the well and the taxa are ordered in terms of first downhole appearance.

and 26/28-A1Z, the reworked miospores present show only minor damage and are relatively well preserved. The reworked material is orange in colour, indicating a Spore Colour Index (SCI) of six, which equates with a Thermal Alteration Index (TAI) of 2.4–2.6 (Marshall 1990), whereas the *in situ* material has an SCI of four (TAI of 2.2–2.4; Marshall 1990).

Well 35/8-1

A total of 16 cutting samples were analysed from a series of Barremian sandstone-prone horizons, 12 of which contained reworked palynomorphs. Most of the reworked material is indicative of a latest Westphalian C to Westphalian D age and its distribution is detailed in Fig. 5. The preservation

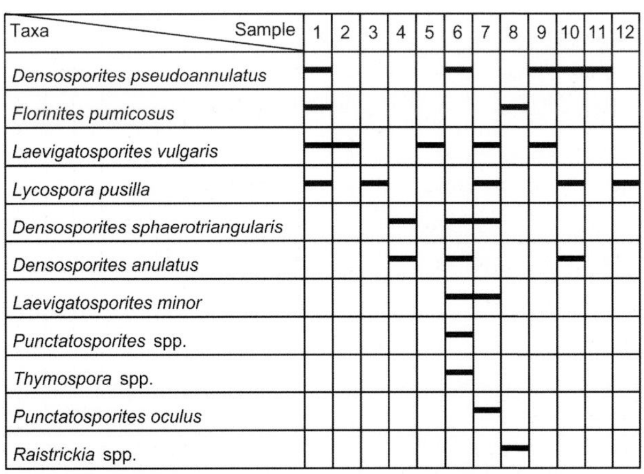

Fig. 5. Distribution of reworked Carboniferous palynomorphs recovered from well 35/8-1. The sample numbers increase with increasing depth in the well and the taxa are ordered in terms of first downhole appearance.

Taxa \ Sample	1	2	3	4	5	6	7	8	9
Densosporites pseudoannulatus	—					—			
Lycospora pusilla	—			—			—	—	—
Densosporites anulatus		—	—	—				—	—
Densosporites spp.		—							
Laevigatosporites minor		—	—		—	—		—	
Crassispora kosankei					—				—
Laevigatosporites vulgaris					—		—		
Punctatosporites spp.					—				
Florinites pumicosus									—

Fig. 6. Distribution of reworked Carboniferous palynomorphs recovered from well 35/19-1. The sample numbers increase with increasing depth in the well and the taxa are ordered in terms of first downhole appearance.

style and coloration of these Westphalian palynomorphs is similar to that of the reworked material in well 35/8-2, as is the SCI of the *in situ* material. In addition to reworked Westphalian palynomorphs, three of the samples, samples 6, 7 and 10, contained rare examples of the acritarch *Multiplicisphaeridium*. This long-ranging Early Palaeozoic acritarch genus was probably originally reworked into the Carboniferous sediments and subsequently reworked along with the Carboniferous material into the sandstones of Barremian age. That the occurrence is due to direct input by marine Early Palaeozoic or Devonian material is considered unlikely primarily because of the distance from the nearest marine Devonian or older sources, although the possibility of as yet undocumented suitably aged material being present on the basin flanks cannot be ruled out. The fact that the acritarchs are all much darker than the rest of the reworked palynomorphs (SCI nine, TAI 3.5; Marshall 1990) suggests that they have had a different thermal history, supporting the multiple reworking hypothesis. Reworked acritarchs have also been reported in Lower Carboniferous successions in southern Ireland (Clayton *et al.* 1980) and the Upper Carboniferous succession of NW Ireland (Smith 1995).

Well 35/19-1

A total of 12 core samples, selected from silty horizons above sandstone units (assumed to be associated with waning flow conditions immediately after sand deposition) were analysed from a series of Berriasian sandstone-prone horizons, nine of which contained reworked material. The reworked material is indicative of a Westphalian age, no older than late Westphalian A, and its distribution is detailed in Fig. 6. The preservation of both the *in situ* and reworked material in this well is rather poor, with pyrite development common in both sets of palynomorphs. The reworked and *in situ* material are both somewhat darker than the reworked material from 35/8-2 and 35/8-1, having an SCI of seven (TAI of 2.6–2.8; Marshall 1990).

Discussion

All of the Upper Jurassic and Lower Cretaceous sandstone-prone horizons from the five wells (26/28-A1, 26/28-A1Z, 35/19-1, 35/8-1 and 35/8-2) investigated contained reworked palynomorphs of Westphalian age. However, the number of samples containing reworked palynomorphs was much higher in wells where the sampling was biased towards environments associated with waning flow conditions, i.e. wells 35/8-2 (66% of samples), 35/8-1 (75% of samples) and 35/19-1 (75% of samples), than in wells where the sampling was more random, i.e. wells 26/28-A1 (38% of samples) and 26/28-A1Z (17% of samples).

The precision to which the age of the reworked material present in each of the intervals can be dated proved variable. Although the assemblages recovered from each of the intervals were typically latest Westphalian C to Westphalian D in character, this was definitive only in wells 26/28-A1, 26/28-A1Z and 35/8-1. It is therefore considered likely (accepting that the finer-grained lithologies are part of the same dispersal systems) that the sandstones in all five of these

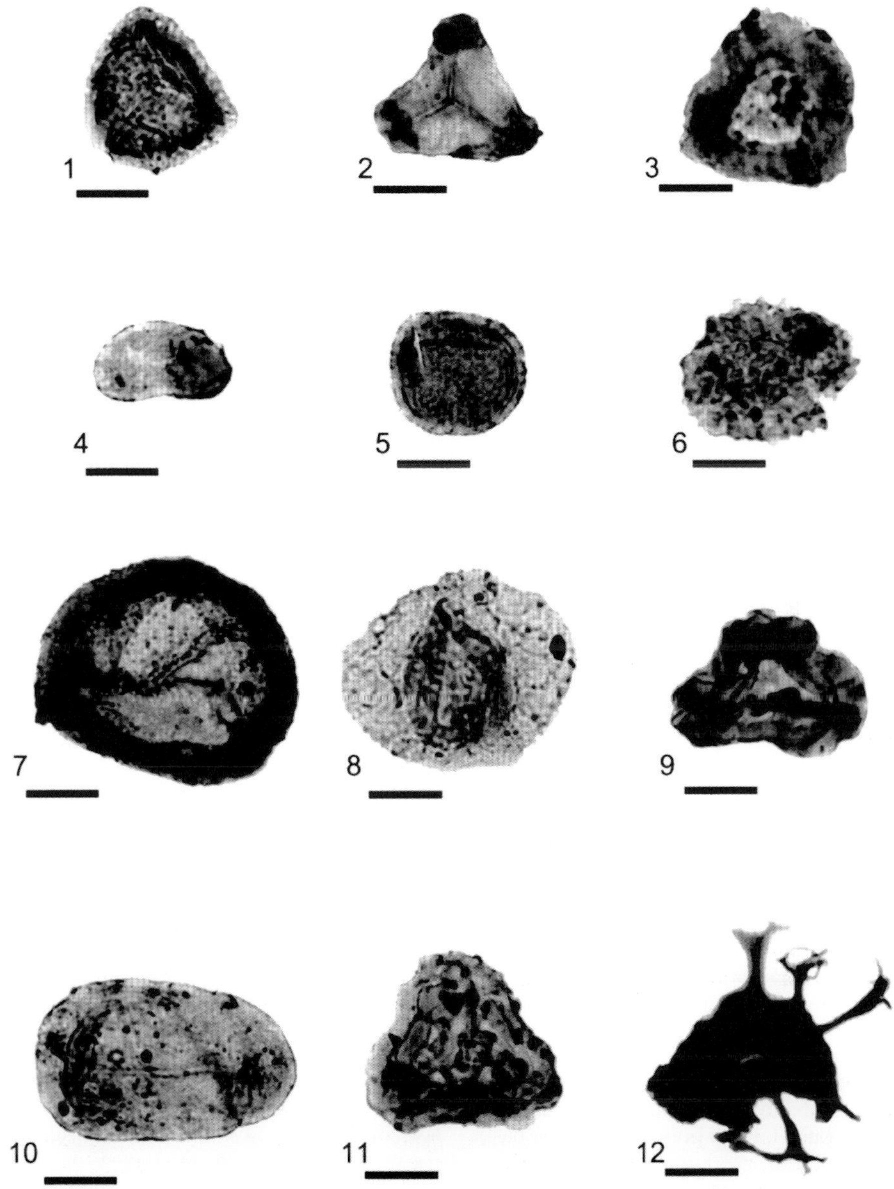

Fig. 7. Examples of the reworked palynomorphs identified in wells from the Porcupine and North Porcupine basins, offshore Ireland. Scale bar represents 20 μm. 1, *Lycospora pusilla*. 2, *Triquitrites bransoni*. 3, *Densosporites anulatus*. 4, *Torispora securis*. 5, *Punctatosporites granifer*. 6, *Thymospora theissenii*. 7, *Crassispora kosankei*. 8, *Florinites junior*. 9, *Triquitrites sculptilis*. 10, *Laevigatosporites vulgaris*. 11, *Savitrisporites nux*. 12, *Multiplicisphaeridium* sp.

potential reservoir intervals were derived, at least partly, from material of latest Westphalian C to Westphalian D age.

The reworked palynomorphs in all of the wells, with the exception of well 35/19-1, are remarkably well preserved with little evidence of mechanical or chemical damage. This is particularly true of the reworked material from wells 26/28-A1 and 26/28-A1Z. In all of the wells, the style and state of preservation is similar to that of the *in situ* material, differing only in having a slightly higher SCI. Remarkably well-preserved reworked palynomorphs are considered to be a result of both the proximity of the source area to the area of redeposition and a negligible influence of oxidation during weathering, transport and redeposition (Van de Laar & Fermont 1989). This assumes that the palynomorphs were not reworked enclosed in clasts or pellets of the source lithology, as this could protect the palynomorphs from both mechanical and chemical damage during transportation (Clayton, pers. comm.). However, although this possibility cannot be discounted, it is thought unlikely on account of the uniform preservation. At least some of the reworked palynomorphs would be released from the enclosing matrix during sediment recycling and hence be subject to enhanced alteration. The uniform nature of the preservation thus suggests that the material under investigation was released from its enclosing matrix and transported as discrete particles.

Vitrinite reflectance (R_m) data indicate that the maximum burial of the Carboniferous material in the Porcupine Basin area occurred before Mid-Jurassic times (Robeson *et al.* 1988). Therefore, the reworked Westphalian palynomorphs, if they were derived from the Porcupine Basin area, must have been thermally mature before being reworked into the Upper Jurassic and Lower Cretaceous sandstone-prone horizons. Because thermally mature palynomorphs tend to be more rigid and therefore more prone to mechanical damage than thermally immature palynomorphs (Guy-Ohlson *et al.* 1988; Batten 1991), the preservation state of the reworked Westphalian palynomorphs under investigation may indicate a relatively short transportation distance from the source area to the area of redeposition. The source area of the reworked palynomorphs and therefore the source of the host Upper Jurassic and Lower Cretaceous sandstone-prone horizons was thus probably the Upper Carboniferous successions known to be present along the basin margin to the east.

Wells along the eastern margin of the Porcupine Basin demonstrate that although the age of the Upper Palaeozoic strata immediately below the Mesozoic unconformity is variable (Robeson *et al.* 1988), the majority of the strata are Late Westphalian in age and have a similar thermal history to that of the reworked material (Robeson *et al.* 1988). Approximately 1170 m of (?)Lower Carboniferous to Lower Namurian strata are present in well 35/15-1 (Robeson *et al.* 1988; Tate & Dobson 1989; see Fig. 1 for location). The section comprises predominantly calcareous sandstones and siltstones, with fossiliferous limestones becoming more common in the upper half of the section, and it is unconformably overlain by a Lower Cretaceous succession (Robeson *et al.* 1988; Tate & Dobson 1989). The Carboniferous section is *c.* 1420 m thick in well 36/16-1 (Robeson *et al.* 1988; Tate & Dobson 1989; see Fig. 1 for location). This section is relatively complete and ranges in age from Late Namurian to Westphalian D time. Lithologically, the section comprises a series of Upper Namurian to Westphalian B fine-grained sandstones and siltstones interpreted as a prograding prodelta to delta-top abandonment facies and a series of Westphalian B to Westphalian D carbonaceous claystones, siltstones, sandstones, coals and rare argillaceous limestones interpreted as a delta-top facies (Tate & Dobson 1989).

The relative proximity of this source area would reduce the amount of mechanical damage during transport and redeposition, and the area is thought to have been emergent during the period of redeposition (Robeson *et al.* 1988), with the erosion culminating during mid-Cimmerian uplift (Tate & Dobson 1989).

An onshore source area is ruled out for the following reasons. With the exception of an isolated occurrence of Westphalian D aged material in County Wexford, all of the preserved Silesian rocks are of Westphalian A age (Clayton *et al.* 1986) and therefore significantly older than the reworked material. The possibility that the material was derived from an onshore Late Westphalian source that has been subsequently completely eroded away is considered unlikely as the preserved onshore Westphalian rocks have a much higher thermal maturity and therefore different Palaeozoic thermal history (Robeson *et al.* 1988; Clayton *et al.* 1989).

The possibility that the Westphalian miospores were reworked into Jurassic-aged sediments and then subsequently reworked into the Upper Jurassic and Lower Cretaceous sandstone-prone horizons under investigation has also been considered but can be discounted, as the well-preserved nature of the reworked material suggests a single reworking event. The reworked Jurassic forms reported in well 35/8-2 by Robinson & Canham (2001) are considered to be derived from a thin cover of Jurassic-aged material resting unconformably on the Carboniferous successions that was caught up in the reworking of the Carboniferous material.

It is interesting to note the complete absence of any reworked terrestrial Devonian material in any of the samples. It could be argued that the regional Devonian successions of southern Ireland, being typically red sandstones, are unlikely to produce many reworked palynomorphs, as they have a very low content of palynomorphs themselves.

The preferred interpretation is therefore that the Westphalian successions on the margin of the Porcupine Basin were recycled (in a single step) to supply at least part of the sediment budget to the Late Jurassic and Early Cretaceous Porcupine basin fill. The regional terrestrial Devonian and onshore Westphalian successions can be discounted in any provenance reconstructions. The presence of the Palaeozoic acritarch genus *Multiplicisphaeridium* in several samples from well 35/8-1 is considered to represent an earlier phase of reworking associated with the deposition of the Westphalian strata in the source area and has no real significance to the provenance of the Upper Jurassic–Lower Cretaceous sandstone-prone horizons.

Conclusions

Reworked Carboniferous miospores are pervasive in the Mesozoic sandstone-prone horizons of the Porcupine Basin and adjacent areas, offshore western Ireland (Robeson *et al.* 1988). In wells 26/28-A1, 26/28-A1Z, 35/19-1, 35/8-1 and 35/8-2 in the Porcupine Basin (see Fig. 1) the reworked miospores are of Westphalian (probable latest Westphalian C to Westphalian D) age. The source of these reworked miospores, and hence the provenance of the Mesozoic sandstone-prone horizons, is the Carboniferous succession present on the eastern margin of the Porcupine Basin. A possible onshore source that has subsequently been completely eroded away is discounted as the reworked and *in situ* Westphalian miospores of the Porcupine Basin area have a different pre-Jurassic thermal history from that of the onshore Westphalian miospores.

The deliberate biasing of the sampling strategy towards mudrock lithologies stratigraphically above sandstones (assumed to represent waning flow conditions) resulted in a marked increase in the number of samples containing reworked palynomorphs, and this approach is recommended for future provenance studies. However, useful information concerning the age of possible source areas may also be derived from samples taken as part of a standard biostratigraphic investigation. The ability to date the source area of at least some of the basin fill using reworked palynomorphs is a powerful, if underutilized, technique and its use complements the more 'traditional' techniques used in provenance studies.

This work was carried out while one of the authors (J.S.) was a research assistant at the Department of Geology, UCC, and funded by Marathon International Petroleum Hibernia Ltd and FORBAIRT. Phillips Petroleum Company United Kingdom Ltd, AGIP (UK) Ltd, Marathon International Petroleum Hibernia Ltd and Statoil Exploration (Ireland) Limited are thanked for permission to publish the results. The manuscript was greatly improved both by the reviewers, J. Marshall and M. Stephenson, and by P. Haughton.

References

BATTEN, D.J. 1991. Reworking of plant microfossils and sedimentary provenance. *In*: MORTON, A.C., TODD, S.P. & HAUGHTON, P.D.W. (eds) *Developments in Sedimentary Provenance Studies*. Geological Society, London, Special Publications, **57**, 79–90.

BLESS, M.J.M. & STREEL, M. 1976. The occurrence of reworked miospores in a Westphalian C microflora from South Limburg (the Netherlands) and its bearing on paleogeography. *Mededelingen Rijks Geologische Dienst*, **27**, 1–39.

CLAYTON, G., HAUGHEY, N., SEVASTOPULO, G.D. & BURNETT, R. 1989. *Thermal Maturation Levels in the Devonian and Carboniferous Rocks in Ireland*. Geological Survey of Ireland, Dublin.

CLAYTON, G., JOHNSTON, I.S., SEVASTOPULO, G.D. & SMITH, D.G. 1980. Micropalaeontology of a Courceyan (Carboniferous) borehole section from Ballyvergin, County Clare, Ireland. *Journal of Earth Sciences, Royal Dublin Society*, **3**, 81–100.

CLAYTON, G., SEVASTOPULO, G.D. & SLEEMAN, A.G. 1986. Carboniferous (Dinantian and Silesian) and Permo-Triassic rocks in south County Wexford, Ireland. *Geological Journal*, **21**, 355–374.

CROKER, P.F. & SHANNON, P.M. 1987. The evolution and hydrocarbon prospectivity of the Porcupine Basin, offshore Ireland. *In*: BROOKS, J. & GLENNIE, K.W. (eds) *Petroleum Geology of the Continental Shelf of North West Europe*. Graham & Trotman, London, 633–642.

ESHET, Y., DRUCKMAN, Y., COUSMINER, H.L., HABIB, D. & DRUGG, W.S. 1988. Reworked palynomorphs and their use in the determination of sedimentary cycles. *Geology*, **16**, 662–665.

GAUPP, R. & BATTEN, D.J. 1983. Depositional setting of Middle to Upper Cretaceous sediments in the Northern Calcareous Alps from palynological evidence. *Neues Jahrbuch für Geologie und Paläontologie, Monatshefte*, **1983/10**, 585–600.

GUY-OHLSON, D., LINDQVIST, B. & NORLING, E. 1987. Reworked Carboniferous spores in Swedish Mesozoic sediments. *Geologiska Föreningens i Stockholm Förhandlinger*, **109**, 295–306.

GUY-OHLSON, D., OHLSON, N.G. & LINDQVIST, B. 1988. Fossil palynomorph deformation and its relationship to sedimentary deposition. *Geologiska Föreningens i Stockholm Förhandlinger*, **110**, 111–119.

MARSHALL, J.E.A. 1990. Determination of thermal maturity. *In*: BRIGGS, D.E. & CROWTHER, P.R.

(eds) *Palaeobiology: a Synthesis*. Blackwell, Oxford, 511–515.

MOORE, J.G. 1992. A syn-rift to post-rift transition sequence in the main Porcupine Basin, offshore western Ireland. *In*: PARNELL, J. (ed.) *Basins on the Atlantic Seaboard: Petroleum Geology, Sedimentology and Basin Evolution*. Geological Society, London, Special Publications, **62**, 333–349.

NAYLOR, D. & SHANNON, P.M. 1982. *The Geology of Offshore Ireland and West Britain*. Graham & Trotman, London.

NAYLOR, D., SHANNON, P.M. & MURPHY, N. 1999. *Irish Rockall Basin Region—a Standard Structural Nomenclature System*. Petroleum Affairs Division, Special Publication, **1/99**.

ROBESON, D., BURNETT, R.D. & CLAYTON, G. 1988. The Upper Palaeozoic geology of the Porcupine, Erris and Donegal Basins, offshore Ireland. *Irish Journal of Earth Sciences*, **9**, 153–175.

ROBINSON, A.J. & CANHAM, A.C. 2001. Reservoir characteristics of the Upper Jurassic sequence in the 35/8-2 discovery, Porcupine Basin. *In*: SHANNON, P.M., HAUGHTON, P.D.W. & CORCORAN, D.V. (eds) *The Petroleum Exploration of Ireland's Offshore Basins*. Geological Society, London, Special Publication, **188**, 301–321.

SHANNON, P.M. 1992. Early Tertiary submarine fan deposits in the Porcupine Basin, offshore Ireland. *In*: PARNELL, J. (ed.) *Basins on the Atlantic Seaboard: Petroleum Geology, Sedimentology and Basin Evolution*. Geological Society, London, Special Publications, **62**, 351–373.

SHANNON, P.M. & NAYLOR, D. 1998. An assessment of Irish offshore basins and petroleum plays. *Journal of Petroleum Geology*, **21**, 125–152.

SMITH, J. 1995. *A palynofacies analysis of the Carboniferous Leitrim Group in the Lough Allen Basin, northwest Ireland*. PhD thesis, National University of Ireland.

STREEL, M. & BLESS, M.J.M. 1980. Occurrence and significance of reworked palynomorphs. *In*: BLESS, M.J.M., BOUCKAERT, J. & PAPROTH, E. (eds) *Pre-Permian around the Brabant Massif in Belgium, The Netherlands and Germany*. Mededelingen Rijks Geologische Dienst, **32**, 69–80.

TATE, M.P. & DOBSON, M.R. 1989. Pre-Mesozoic geology of the western and north-western Irish continental shelf. *Journal of the Geological Society, London*, **146**, 229–240.

TRAVERSE, A. 1988. *Paleopalynology*. Unwin Hyman, Boston, MA.

TYSON, R. 1995. *Sedimentary Organic Matter: Organic Facies and Palynofacies*. Chapman and Hall, London.

TRUSWELL, E.M. 1982. Palynology of seafloor samples collected by the 1911–14 Australasian Antarctic Expedition: implications for the geology of coastal east Antarctica. *Journal of the Geological Society of Australia*, **29**, 343–356.

TRUSWELL, E.M. 1983*a*. Recycled Cretaceous and Tertiary pollen and spores in Antarctic marine sediments: a catalogue. *Palaeontographica B*, **186**, 121–174.

TRUSWELL, E.M. 1983*b*. Geological implications of recycled palynomorphs in continental shelf sediments around Antarctica. *In*: OLIVER, R.L., JAMES, P.R. & JAEP, J.B. (eds) *Antarctic Earth Science*. Australian Academy of Science, Canberra, A.C.T., 394–399.

TRUSWELL, E.M. & DREWRY, D.J. 1984. Distribution and provenance of recycled palynomorphs in surficial sediments of the Ross Sea, Antarctica. *Marine Geology*, **59**, 187–214.

VAN BERGAN, P.F., JANSSEN, N.M.M., ALFERINK, M. & KERP, J.H.F. 1990. Recognition of organic matter types in standard palynological slides. *In*: FERMONT, W.J.J. & WEEGINK, J.W. (eds) *Proceedings of the International Symposium on Organic Petrology, Zeist, January 1990*. Mededelingen Rijks Geologische Dienst, **45**, 9–21.

VAN DE LAAR, J.G.M. & FERMONT, W.J.J. 1989. On-shore Carboniferous palynology of the Netherlands. *Mededelingen Rijks Geologische Dienst*, **43**, 35–73.

Reservoir characteristics of the Upper Jurassic sequence in the 35/8-2 discovery, Porcupine Basin

A. J. ROBINSON[1] & A. C. CANHAM[2]

[1]*Chevron Europe Ltd, 43–45 Portman Square, London W1H 0AN, UK*

Present address: Chevron Overseas Petroleum, Inc., 6001 Bollinger Canyon Road, San Ramon, CA 94583, USA (e-mail: roaj@chevron.com)

[2]*Geochem Group Ltd, Chester Street, Saltney, Chester CH4 8RD, UK*

Present address: Integrated Reservoir Solutions Ltd, Norian House, Bridges Way, Ellesmere Port CH65 4LB, UK

Abstract: The 35/8-2 well, drilled by Phillips Petroleum in 1981, tested oil and gas from a Late Jurassic reservoir within a plunging, tilted fault block. The reservoir is overpressured by 5000 psi and, despite having good porosity and high net to gross ratio, exhibits very low permeability. Three hydrocarbon-bearing sandstone intervals were encountered in the well, and more than 120 m of core was taken from these reservoir sections. The sedimentological structures identified in core are clearly indicative of a deep-water (below storm wave base) submarine fan system, comprising a wide spectrum of turbidite and debris-flow, channel and lobe deposits. Detailed petrographic analysis of the reservoir was undertaken to identify the key factors afffecting reservoir quality. The sandstones are chemically immature lithic arenites, containing abundant quartz, feldspar and rock fragments. Reservoir quality is generally poor, largely as a result of the chemical instability of the sandstones. Diagenetic overprinting is the main agent of downgrading the sandstones, with kaolinites and carbonate cements blocking pore throats and reducing permeability. Facies type, sand to mud ratio and ultimately the clastic provenance of the sandstones are also critical factors affecting the reservoir quality in these sandstones. Hydrocarbon flow was predominantly from a single sandstone facies (B1) and was confined to the 'A' sand interval in well 35/8-2, but it is inferred that this facies may occur away from the wellbore in all three sand intervals. Uncertainties in the results of the drillstem test and pressure data demonstrate that it is not possible to confirm whether the hydrocarbons present in the reservoir are gas condensate or volatile oil. Uncertainty in the extent of the productive reservoir facies, hydrocarbon columns and hydrocarbon type are challenges that must be overcome before the reserves contained within the 35/8-2 structure can be developed.

The Upper Jurassic interval in the Porcupine Basin encompasses sandstone reservoirs deposited in a range of sedimentary environments, from fluvial through deltaic and shallow marine, to deep marine. Hydrocarbon shows are common in wells penetrating the Upper Jurassic interval (Croker & Shannon 1987), and oil and gas have flowed on test from non-marine and marginal marine reservoirs in the Connemara Field in Block 26/28 (MacDonald *et al.* 1987). Of the 29 wells drilled to date in the Porcupine Basin, only one well, 35/8-2, has tested significant hydrocarbons from an Upper Jurassic deep marine sandstone reservoir. This paper examines the results of a detailed evaluation of this well and the implications for Upper Jurassic prospectivity in deep marine turbidite sandstones in the Porcupine Basin.

The Porcupine Basin (Fig. 1) is a Mesozoic rift basin, bounded on three sides by relatively shallow shelf platforms (the Irish Mainland Shelf and Porcupine High). It is probable that these shallow features are composed of Precambrian and Caledonian metamorphic rocks and deformed Upper Palaeozoic strata, with only a thin Mesozoic and Tertiary sediment cover (Naylor & Shannon 1982). These platform areas underwent long periods of uplift and erosion throughout Mesozoic time, and are likely to have provided a significant source for clastic sediment deposition into the Porcupine Basin.

The structural setting and depositional history of the Porcupine Basin (Fig. 2) has been discussed by several workers (Croker & Shannon 1987; Tate 1993; Sinclair *et al.* 1994; Williams *et al.* 1999) and reflects periods of extension

From: SHANNON, P.M., HAUGHTON, P.D.W. & CORCORAN, D.V. (eds). 2001. *The Petroleum Exploration of Ireland's Offshore Basins*. Geological Society, London, Special Publications, **188**, 301–321. 0305-8719/01/$15.00

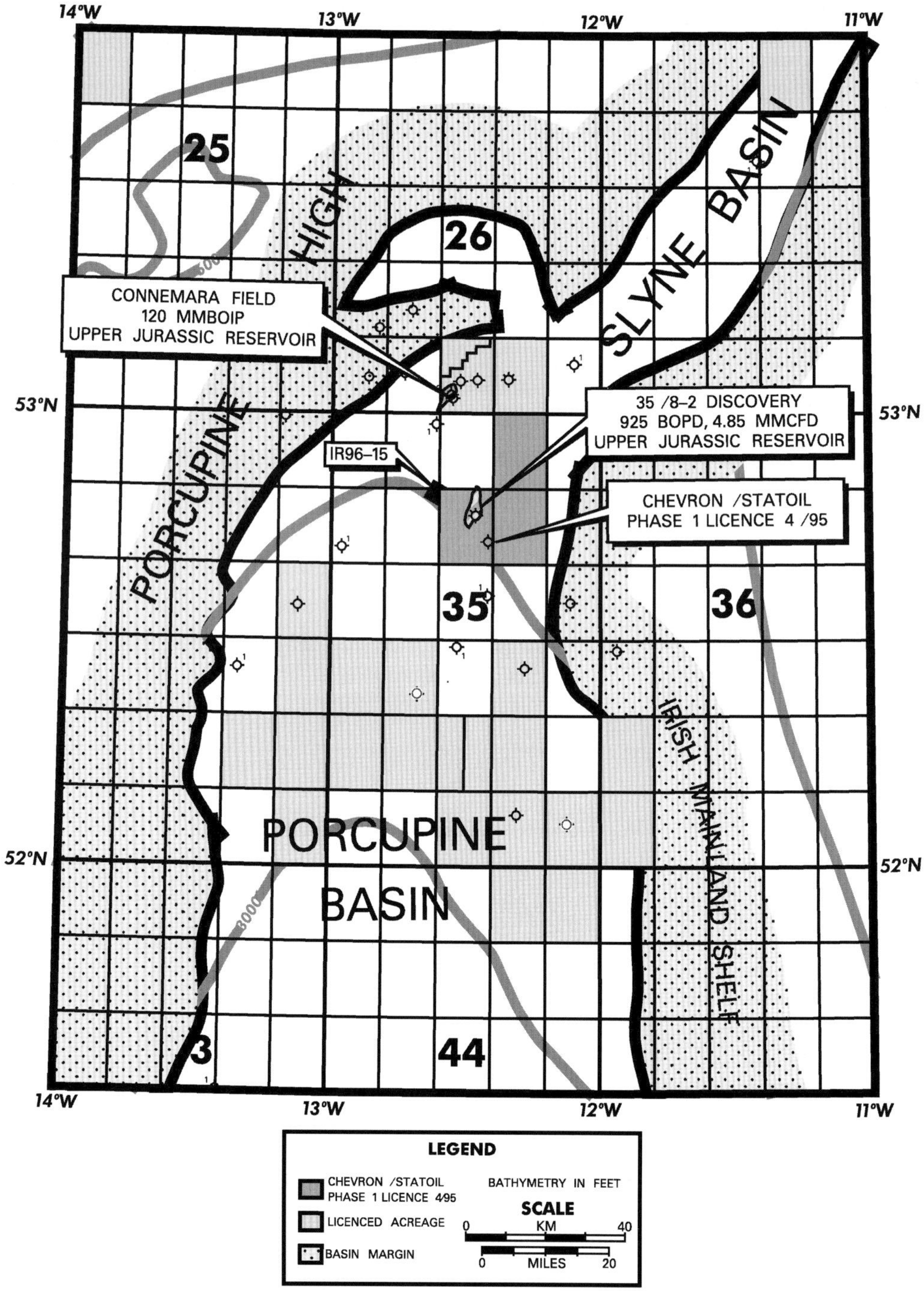

Fig. 1. Porcupine Basin location map showing basin margins, bathymetry and well locations. Frontier Licence 4/95 is highlighted and Jurassic discoveries labelled. Location of seismic line IR96-15 is also shown (see Fig. 4).

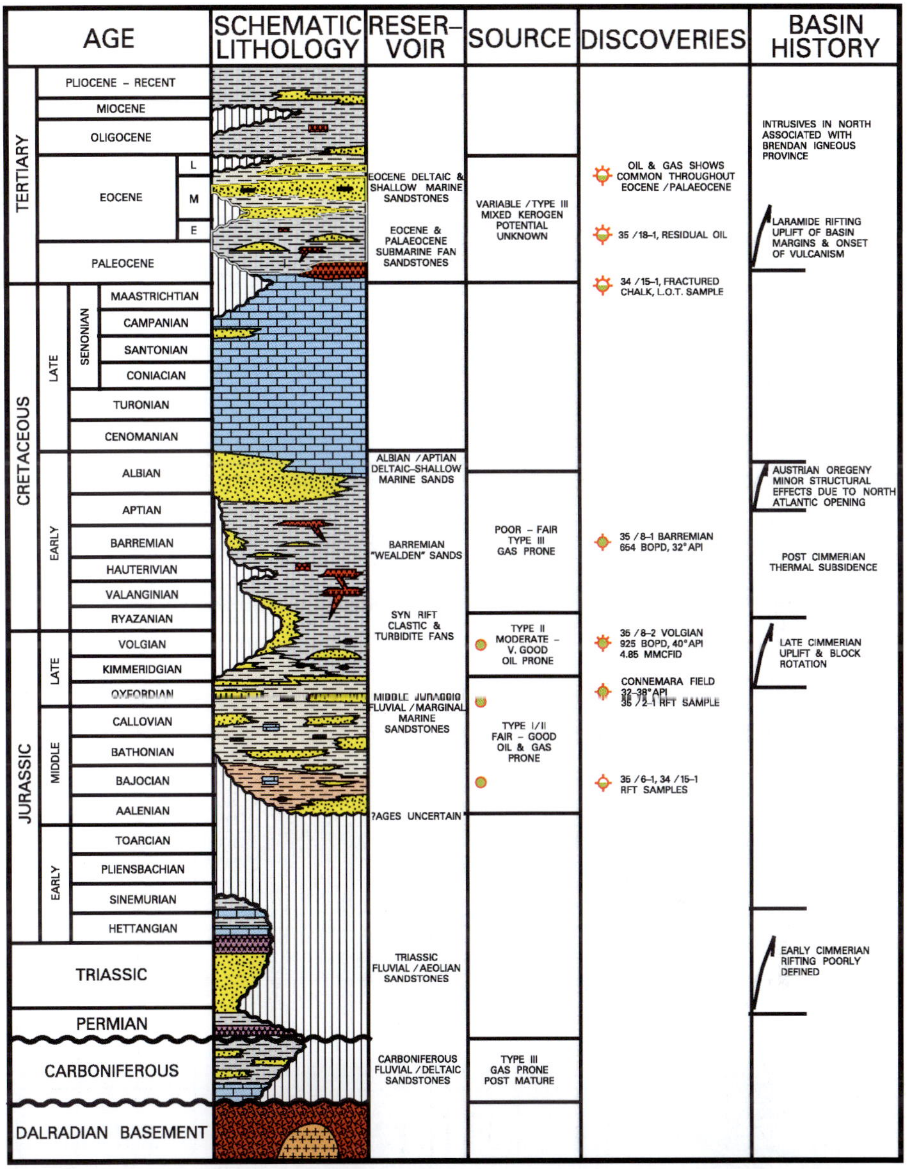

Fig. 2. Generalized stratigraphy, lithology and tectonic history of the Porcupine Basin, indicating major unconformities, reservoir and source rock intervals and hydrocarbon discoveries to date. Based on a variety of sources including Croker & Shannon (1987) and Sinclair *et al.* (1994).

related to the many phases of opening of the North Atlantic (Masson & Miles 1986; Doré *et al.* 1997). The dominant extensional episode affecting the basin is the Late Cimmerian event (Fig. 2), with an east–west extensional direction as reflected in the north–south strike of the Porcupine Basin. In a detailed review, Sinclair *et al.* (1994) discussed evidence for the timing of rifting commencing in latest Oxfordian to Early Kimmeridgian time, with the 'onset warp' reflecting a change from the dominantly continental basin to a rapidly subsiding and expanding marine basin.

The stratigraphy of the Porcupine Basin (Fig. 3) reflects the tectonic and sedimentary effects of the rifting episodes outlined above. Wells drilled on the basin margins have encountered Lower Palaeozoic sandstones and

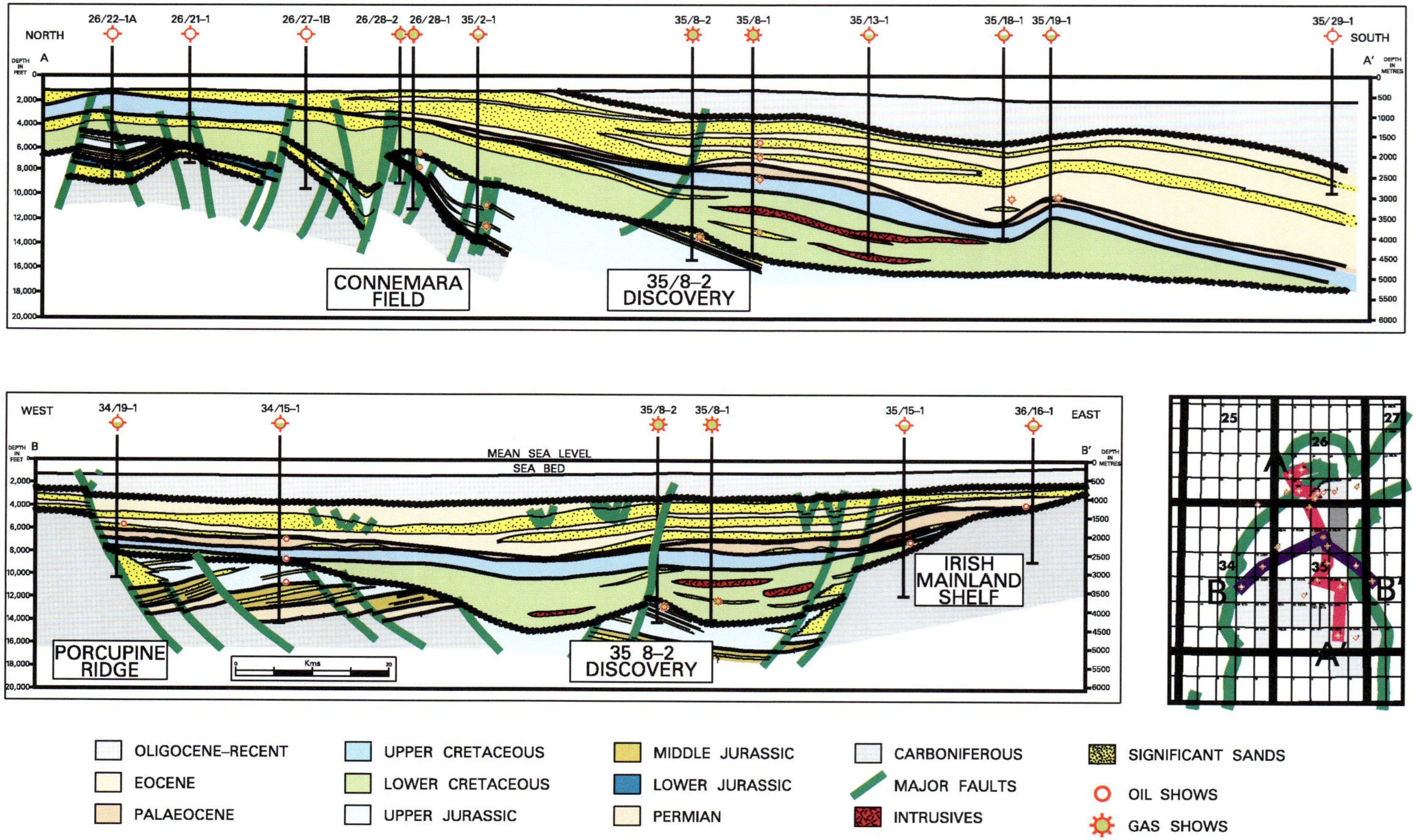

Fig. 3. Schematic cross-sections north–south (top) and east–west (refer to inset location map for lines of section) through the northern part of the Porcupine Basin. The following should be noted: (1) relative positions of wells 35/8-2 and 26/28-1 and 2 (Connemara Field); (2) location of well 35/8-2 on an isolated tilted fault block near the axis of the basin; (3) the proximity of the basin margins and narrow basin morphology.

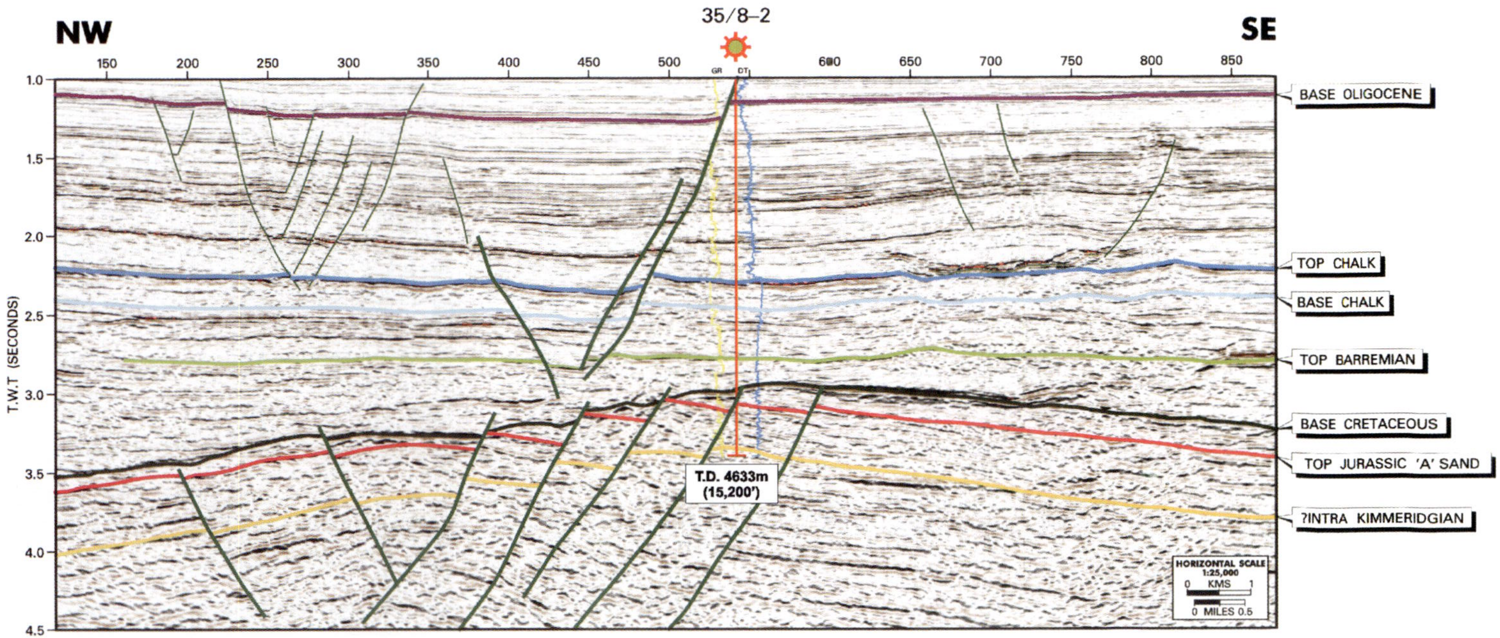

Fig. 4. Interpreted seismic line IR96-15 through the 35/8-2 discovery well. Location is shown in Figs. 1 and 14. Well logs displayed adjacent to wellbore: gamma ray (left, yellow) and sonic (right, blue).

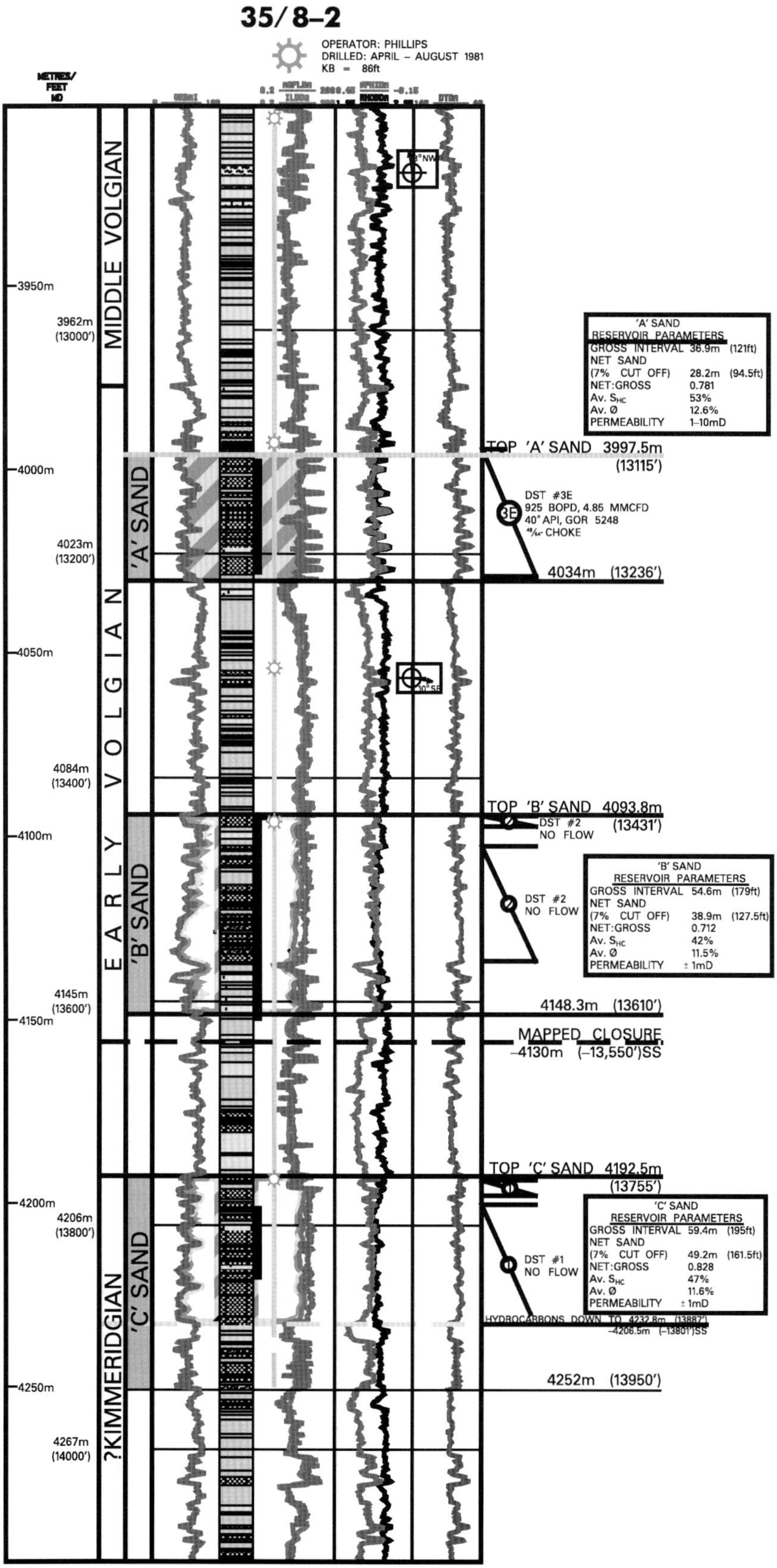
35/8–2
OPERATOR: PHILLIPS
DRILLED: APRIL – AUGUST 1981
KB = 86ft
METRES/ FEET MD
MIDDLE VOLGIAN
EARLY VOLGIAN
?KIMMERIDGIAN
'A' SAND
'B' SAND
'C' SAND
3950m
3962m (13000')
4000m
4023m (13200')
4050m
4084m (13400')
4100m
4145m (13600')
4150m
4200m
4206m (13800')
4250m
4267m (14000')
'A' SAND RESERVOIR PARAMETERS
GROSS INTERVAL 36.9m (121ft)
NET SAND (7% CUT OFF) 28.2m (94.5ft)
NET:GROSS 0.781
Av. S$_{HC}$ 53%
Av. Ø 12.6%
PERMEABILITY 1–10mD
TOP 'A' SAND 3997.5m (13115')
DST #3E
925 BOPD, 4.85 MMCFD
40° API, GOR 5248
CHOKE
4034m (13236')
TOP 'B' SAND 4093.8m (13431')
DST #2 NO FLOW
'B' SAND RESERVOIR PARAMETERS
GROSS INTERVAL 54.6m (179ft)
NET SAND (7% CUT OFF) 38.9m (127.5ft)
NET:GROSS 0.712
Av. S$_{HC}$ 42%
Av. Ø 11.5%
PERMEABILITY ± 1mD
DST #2 NO FLOW
4148.3m (13610')
MAPPED CLOSURE –4130m (–13,550')SS
TOP 'C' SAND 4192.5m (13755')
'C' SAND RESERVOIR PARAMETERS
GROSS INTERVAL 59.4m (195ft)
NET SAND (7% CUT OFF) 49.2m (161.5ft)
NET:GROSS 0.828
Av. S$_{HC}$ 47%
Av. Ø 11.6%
PERMEABILITY ± 1mD
DST #1 NO FLOW
HYDROCARBONS DOWN TO 4232.8m (13887') –4206.5m (–13801')SS
4252m (13950')

shales, and basement schist and gneiss (Croker & Shannon 1987). The identification of Carboniferous and Permian sections on the western terraces of the Porcupine Basin in wells 34/19-1 and 34/15-1 (original operator's well reports and composite logs) is considered tenuous in the light of new proprietary biostratigraphic data. These non-marine sections are frequently dominated by reworking, which renders the sparse *in situ* microfauna difficult to recognize and thus reduces confidence in derived ages. This problem persists throughout the Jurassic non-marine sequence of fluvial and lacustrine facies, which have been assigned ages ranging from Bajocian to Early Kimmeridgian time (Croker & Shannon 1987; MacDonald *et al.* 1987). Recent multi-well proprietary studies, however, have been unable to firmly identify *in situ* microfauna older than Callovian time in these well sections.

The continental and marginal marine Jurassic sediments are succeeded by marine sandstones and shales of Kimmeridgian to Volgian age, reflecting the overall relative rise in sea level throughout Kimmeridgian time. This relative rise in sea level is attributed to both eustatic rise and increasing rates of tectonic subsidence (Sinclair *et al.* 1994). Shallow marine bioclastic sandstones, identified in the Connemara Field as being of Late Oxfordian age by Sinclair *et al.* (1994), may be as young as Early Volgian time, and are inferred to represent a northward-transgressing marine shoreface. The overlying shale section has a significantly lower sandstone content, indicating further retreat of clastic source areas. Much of this section is described as bituminous, and representative of a restricted, dysaerobic or anoxic marine environment, with considerable source rock potential.

The Upper Jurassic section penetrated in well 35/8-2 comprises hemipelagic claystones and bituminous shales with development of three significant sandstone intervals, 37-60 m (121–195 ft) thick. The sandstones are believed to represent sudden pulses of clastic input to the narrow, rapidly subsiding Porcupine Basin during the height of the Late Cimmerian rifting event.

Well 35/8-2

Well 35/8-2 was drilled by Phillips Petroleum in 1981 to test a south-plunging, easterly tilted Jurassic fault block on the eastern margin of the Porcupine Basin. The well was located 200 km west of Ireland, in a water depth of 422 m (1386 ft), and 30 km south of the hydrocarbon discovery in Block 26/28, which was subsequently named the Connemara Field. Seismic line IR96-15 (Fig. 4) illustrates the structure targeted by well 35/8-2. The well drilled a thick Tertiary and Upper Cretaceous section, and encountered increasing formation pressure with gas shows through the Lower Cretaceous section. The Top Jurassic level was reached at 3856 m (12 650 ft) measured depth (MD), and three hydrocarbon-bearing sandstone intervals were encountered (Fig. 5). These sandstones are 37–60 m (121–195 ft) thick and are separated by claystones up to 61 m (200 ft) thick. Cores taken in each of the sandstones had oil shows throughout. High mud weights (15.5 ppg (pounds per gallon)) used to control the well pressures masked hydrocarbon shows in the well during drilling, but subsequent petrophysical analysis confirmed the likelihood of movable hydrocarbons and the well underwent an extensive testing programme.

The lower two hydrocarbon-bearing sandstones ('B' and 'C' sands) failed to flow on test, whereas the upper 'A' sand produced 925 BOPD (barrels of oil per day) of 40° API (American Petroleum Institute) oil and 4.85 MMSCFD (million standard cubic feet per day) of gas. Plans to stimulate the 'B' and 'C' sands by fracturing were abandoned because off deteriorating weather conditions and mechanical difficulties encountered during the well test programme. The well was plugged and abandoned as an oil and gas discovery and, after subsequent evaluation, no further wells were drilled on the structure and Phillips surrendered the acreage in 1985.

New evaluation of the 35/8-2 discovery

Chevron undertook a thorough investigation of this discovery as part of the Phase I evaluation of Frontier Licence 4/95. Seismic mapping of the new 2D data (IR96 survey) acquired over the structure was combined with a detailed sedimentological and petrographical analysis of over 120 m of core, petrophysical analysis and a review of the pressure and drillstem test (DST) data. These studies were designed to address questions regarding: (1) the depositional environment and sedimentary and diagenetic

Fig. 5. Wireline logs and interpreted lithology over the reservoir section in well 35/8-2, showing cored intervals (black bars) and summary of DSTs and reservoir parameters.

Table 1. *Facies names and codes used in the description of core from well 35/8-2 and associated interpretation of depositional environments.*

Facies code	Facies name	Description	Depositional environment
A1	Massive muddy gravel	Matrix-supported massive conglomerates with high mudstone content (10–50%), with intra- and extraformational clasts	Highly concentrated sediment gravity flow, probably cohesive, 'freezing' in place
A2	Conglomerates and pebbly sandstone	Gravel-rich intra- and extraformational clasts in a coarse quartz-rich sand matrix; typically fining upwards with low-angle cross-stratification	Highly concentrated sediment gravity flow, probably turbiditic; deposition probably within a submarine channel complex
B1	Massive sandstone	Ungraded sandstone facies with local grain-size trends, rare coarse clasts and mudstone clasts, common fluid escape and injection structures	Turbidity current deposition by rapid fallout from suspension, in a broad submarine channel or lobe environment
C1	Thick-bedded sandstone and mudstone	Sandstone beds 0.3–1.0 m, erosive base, fine to coarse grained, overlain by planar and trough cross-laminations, some ripples with mudstone drapes	High-density turbidite currents with erosive base overlain by graded bed features, later reworked by lower-density currents
C2	Medium-bedded sandstone and mudstone	As above, with sandstone beds 0.1–0.3 m, very fine to fine grained, subparallel laminations dominant	Classic 'Bouma' style turbidity current deposits
C3	Thin-bedded sandstone and mudstone	As above, sandstone bed thickness <0.1 m, very fine to fine grained, subparallel laminations and rippled tops dominant	Low-density turbidity currents, probably in a distal submarine lobe setting
D1–D2	Thick- to thin-bedded, irregularly laminated siltstones and silty claystones	Rarely encountered in core, interbedded, laminated dark grey mudstone and fine sandstone	Very dilute, low-energy turbidity currents
D3	Bioturbated siltstones and silty claystones	Burrow fills are tentatively identified, could also represent minor load structures in facies D3	As above
E2	Laminated claystones and silty claystones	Subfissile to fissile shales, highly carbonaceous and pyritic	Predominantly suspension fallout

characteristics of the sandstones; (2) the reservoir quality and reasons for the very low permeabilities observed in the sandstones; (3) the failure of the B and C sands to flow on test; (4) how the DSTs were conducted and what the results reveal about the nature of the hydrocarbons in the reservoir; (5) the range of reserves estimates and potential development options.

The Upper Jurassic sandstones in well 35/8-2 were reported as ranging from Kimmeridgian to Early Volgian age from the operator's original biostratigraphy. However, a recent proprietary biostratigraphy study, utilizing new and existing palynological and micropalaeontological slides, suggests that the section is of Early to Mid-Volgian age, but is dominated by reworking of older (Carboniferous, Mid-Jurassic and Kimmeridgian) microfauna. The section is clearly the Late Cimmerian synrift section but the difficulty in precise age dating has led to the naming convention of A, B and C sands to avoid confusion.

Sedimentology and petrography

A sedimentological and petrographical examination of the cored intervals was undertaken using both slabs of the core. Twelve sandstone samples were selected for analysis and 120 m of core was logged in detail. The sandstone samples were prepared into conventional thin sections for description and point counting. A subset of four samples was also selected for scanning electron microscopy (SEM) and X-ray diffraction (XRD) analysis to aid in the identification of clay minerals and factors influencing reservoir quality. After detailed examination of the core, a subset of three samples was selected for cathodoluminescence (CL) microscopy, and four samples were also selected for stable isotopic analysis of carbonate and pyrite cements, to better characterize the diagenetic history.

A facies scheme (Table 1) was devised to subdivide the core and assist in the interpretation of depositional environments. Examples of several of these facies types are shown in Fig. 6. The cored sandstone intervals (Fig. 7) can be interpreted as the products of distinct and different depositional mechanisms within a submarine fan environment. The C sand interval at 4195.0–4213.6 m (13 763–13 824 ft) MD comprises a sequence of sandstones in which each individual unit fines upwards. The individual fining-upward sandstone units are stacked in a series of units that overall coarsen upwards.

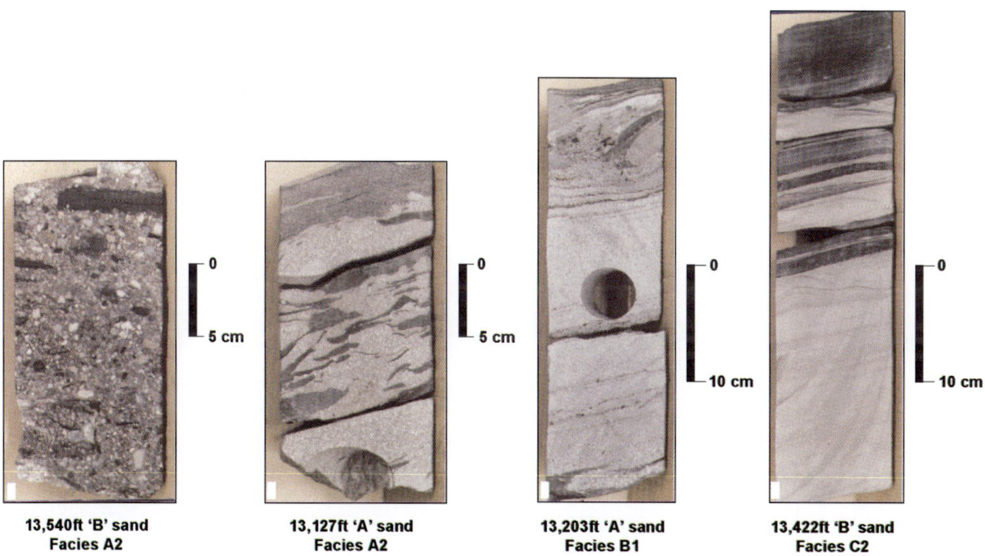

Fig. 6. Examples of facies from the 35/8-2 cored intervals: (**a**) poorly sorted pebble-grade conglomerate with coarse sand matrix including a variety of clastic source material including igneous and metamorphic quartz, feldspar and lithic fragments, claystone intraclasts and abundant carbonaceous material; (**b**) massive coarse-grained sandstone with entrained claystone rip-up clasts; (**c**) massive sands with coarsening-upward profile, coarse-grained erosive base, sediment loading and dewatering features, and discrete laminations of carbonaceous material; (**d**) fine-grained thick-bedded sandstones with sub-parallel and cross-laminations, interbedded with dark grey claystone.

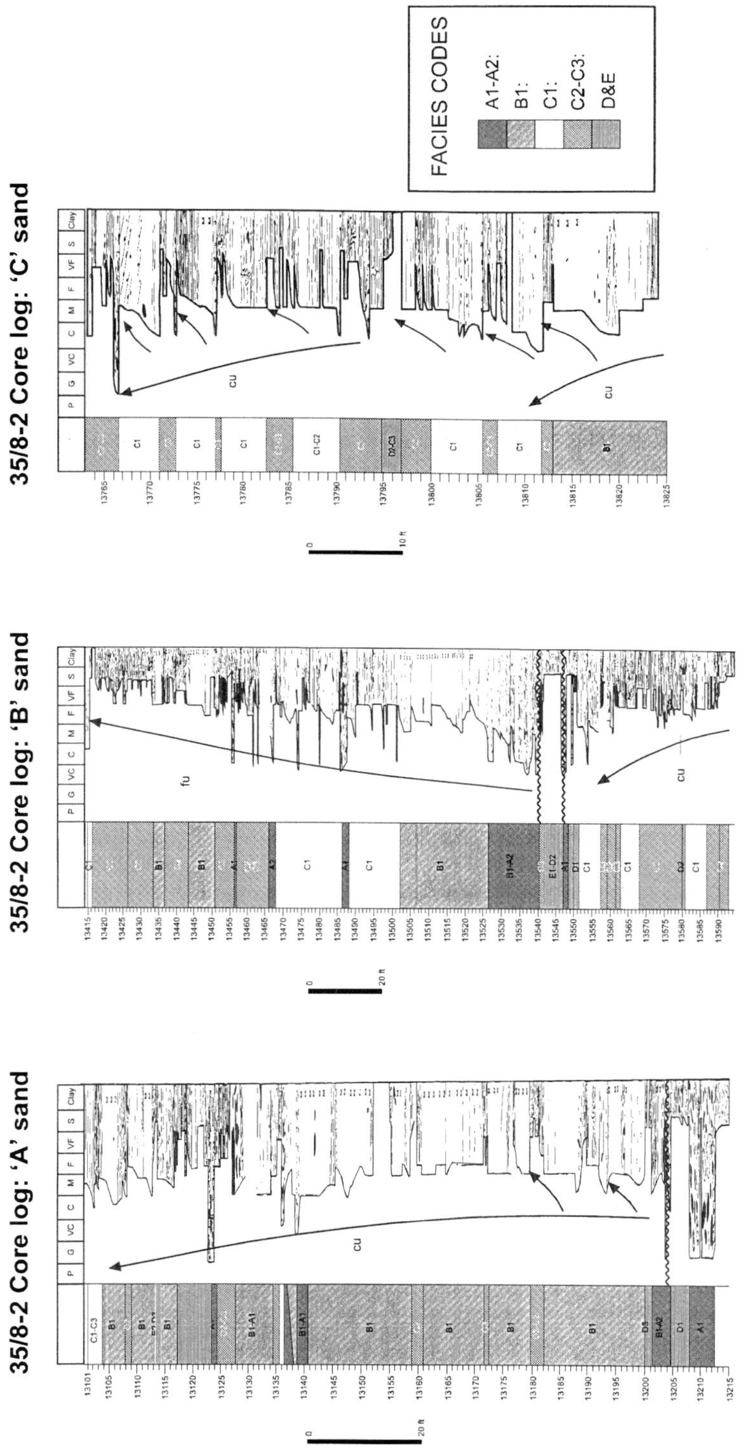

Fig. 7. Simplified core logs showing grain-size variations and facies stacking patterns. fu, fining-upward profile; cu, coarsening-upward profile.

Within each coarsening-upward section there is a trend from mud-dominated intervals with thinly bedded turbidites (Facies C2–C3) to thickly bedded more massive sandstones with markedly erosive bases (Facies C1–B1). The overall C sand interval could be interpreted as a prograding submarine fan lobe sequence developed in a lower to middle fan setting (Shanmugam & Moiola 1991).

The B sand cored interval at 4088.6–4143.8 m (13 414–13 595 ft) MD shows a number of distinctive trends. The basal 14 m (45 ft) of the interval is similar to that encountered in the C sand, with an overall coarsening-upward sequence comprising claystones and thinly bedded turbidites grading up into thicker, sand-dominated turbidites. This is capped by a 3 m (10 ft) interval of laminated claystones, deposited by suspension fallout, but also containing mud-dominated debris-flow deposits. Above this, a sandy sequence fines upwards over a 42 m (137 ft) interval. Thickly bedded, sand- and gravel-dominated, high-density turbidity current deposits (Facies A2–B1) at the base are interpreted to represent channelized sandstones and these grade upwards into more 'classical' sand-dominated turbidites (Facies B1–C3), which become progressively thinner and finer grained. The overall fining-upward sequence is interpreted to represent a gradation from upper to lower fan, indicating retrogradation induced by local tectonic subsidence or a rise in relative sea level.

The A sand interval between 3993.2 and 4027.9 m (13 101 and 13 215 ft) MD comprises a basal 3 m (10 ft) interval of claystones deposited by suspension fallout and distal silty turbidites associated with thick mud-supported debris-flow intraformational conglomerates (Facies A1). Above this, a composite sequence of stacked, overall coarsening-upward sandy units is present. Each coarsening-upward unit comprises individual beds that fine upwards. The interval is dominated by massive sandstones of Facies B1, which show abundant dewatering structures. The overall sequence is interpreted to represent fan lobe progradation in a lower to middle fan environment.

Depositional setting

A deep marine (below storm wave base) depositional environment is identified, comprising deposition by debris flows, high-density, 'classical' and low-density turbidites, and hemipelagic fallout of fines from suspension. On the basis of the scale of sedimentary sequences present and the geographical and geological setting, the depositional environment is inferred to represent a submarine fan or slope apron setting, developed adjacent to an actively faulted basin margin (Fig. 8). The system comprises upper fan high-density turbidite channel fills, grading laterally to slumped slope deposits, lower to middle fan lobe deposits, and hemipelagic claystones in distal settings. Each of the

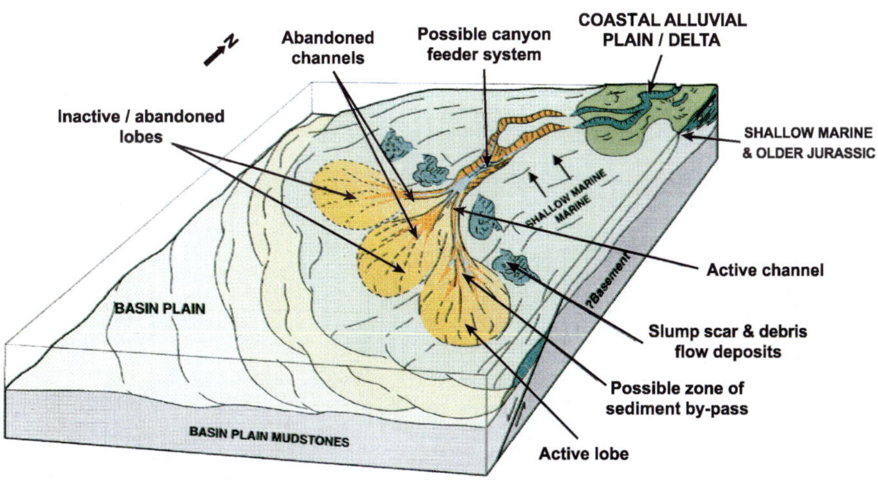

Fig. 8. Schematic depositional model for the sand intervals encountered in well 35/8-2. The inferred proximity of the shoreline with a narrow shallow-marine shelf affording little opportunity for reworking of sediment derived from a major fluvial–deltaic clastic source area to the north and east should be noted. Sediments are interpreted to have been deposited by several different flow mechanisms.

major sandstone bodies represents an episode of sediment progradation into the area, and intervening claystones record abandonment and a return to quiet basinal conditions.

Seismic data quality does not permit meaningful estimation of sandbody geometry and, with a single well penetration, the depositional environment cannot be determined precisely. Broadly speaking, comparison of slope apron and submarine fan depositional systems suggests that fan systems tend to have a higher degree of internal organization with well-defined trends in bed thickness, sediment type and grain size. Careful examination of the cores from well 35/8-2 revealed the trends discussed above, and it is therefore suggested that the overall sequence was deposited in a submarine fan setting. Sediment available was sand dominated, with a significant component of gravel, sourced both extra- and intra-formationally. High ratios of sand to mud, relatively high depositional gradients (indicated by significant debris-flow and high-density turbidity current deposits), the scale of sequence development and the limited basin size all suggest an immature, actively faulted basin margin setting (North Sea type of Shanmugam & Moiola 1991). The abundance of shallow marine organisms, combined with prolific amounts of terrigenous plant and wood debris, suggests a narrow shelf, with sand and gravel transported directly from fluvial sources and resedimented as a submarine fan system.

Detrital mineralogy

The 35/8-2 sandstones are classified as chemically immature lithic arenites, containing abundant quartz, feldspar and rock fragments. Some of the samples are more feldspathic, and can be classified as feldspathic arenites, often modified by calcareous and dolomitic cementation. Figure 9 shows the average detrital composition for the 12 samples analysed. Quartz (27–50.5%) is the most abundant detrital mineral, predominantly monocrystalline with subordinate polycrystalline grains. Under CL, quartz grains are generally non-luminescent, but locally, quartz displays two luminescent characteristics, first a pink coloration suggesting a volcanogenic origin and second a faint blue colour indicating a metamorphic source. Feldspar (9–19%) is an important constituent within all three sandstone intervals. Alkali feldspar is dominant, typically untwinned orthoclase with occasional twinned microcline. There is a comparative paucity of plagioclase, which ranges in appearance from relatively fresh (dominant) to dusty. Rock fragments (7–34.5%) are particularly abundant, often forming a principal portion of the composition. Clasts of igneous and metamorphic origin are dominant. Igneous clasts included quartz feldspar aggregates, mica-rich quartz and inclusion-rich quartz. Metamorphic clasts comprise quartz mica-schists, strained quartz, feldspar with mica inclusions, seriticized feldspars

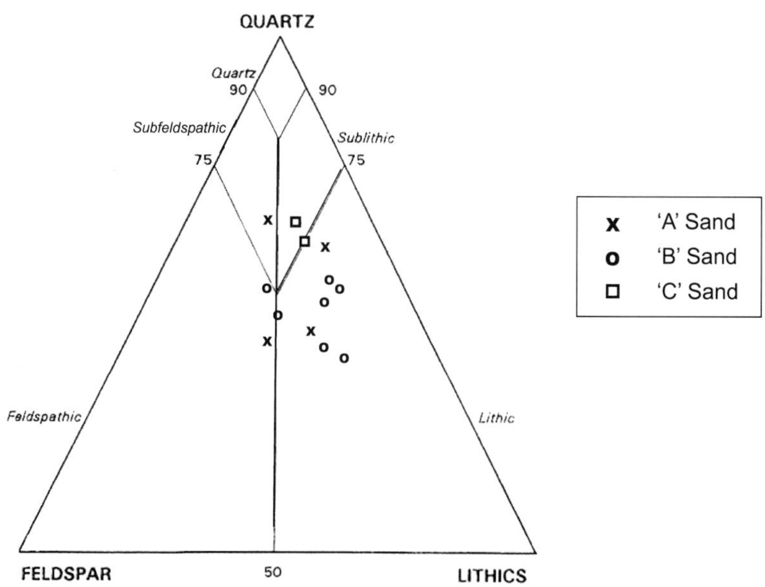

Fig. 9. Sandstone classification diagram for the cored Upper Jurassic sandstones in well 35/8-2, coded by sand body.

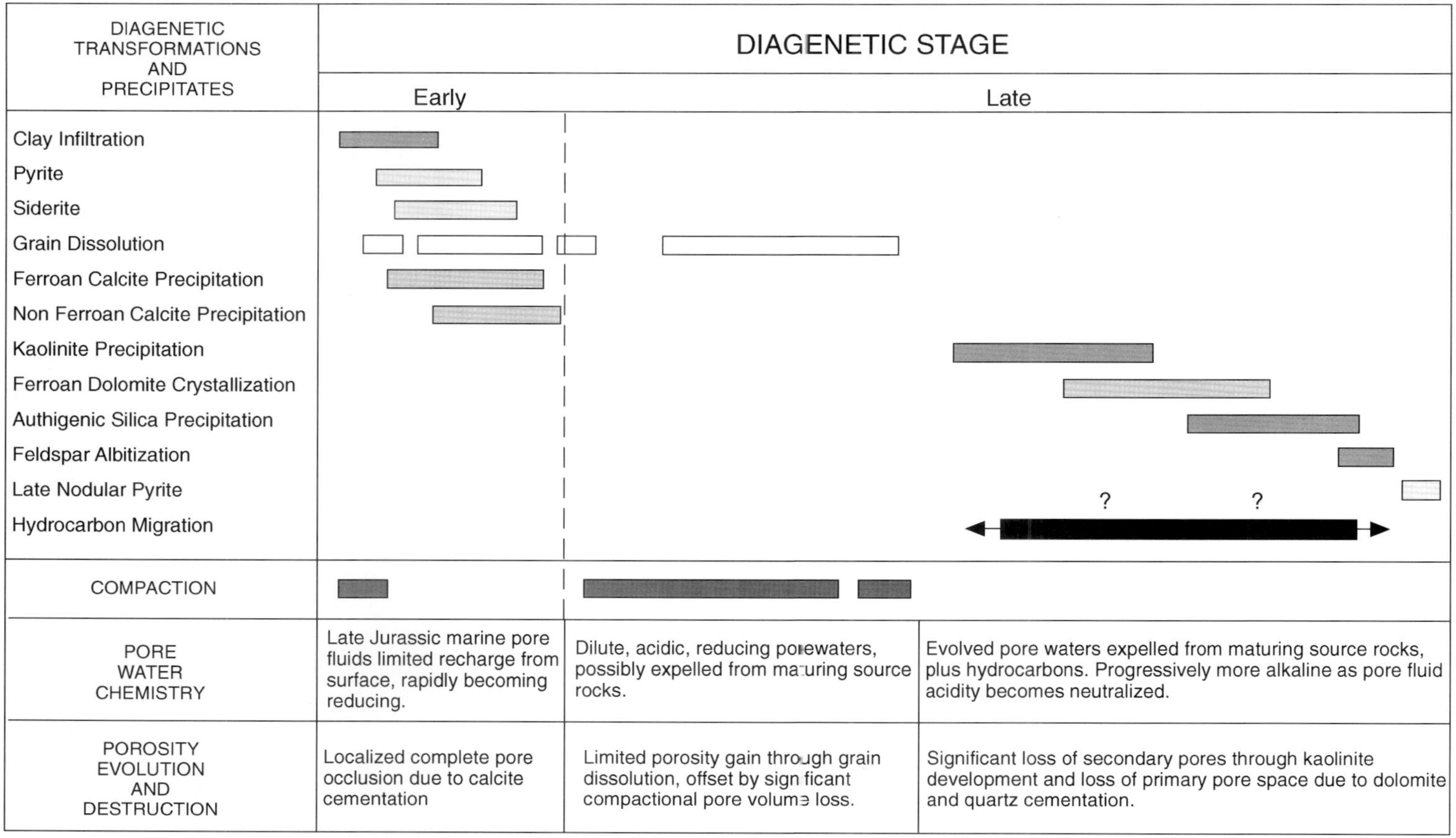

Fig. 10. Diagenetic sequence for the sandstone reservoirs in well 35/8-2. Early diagenesis is inferred to represent burial depths of up to 1 km (temperatures of 35–47 °C); late diagenesis occurred at burial depths of 2–4 km (75–103 °C). Temperatures are estimated from oxygen isotope data.

and occasional metasedimentary clasts. Mica is notable by its limited abundance and diversity, only muscovite being present. Heavy minerals (0–2%) are dominated by subrounded zircon and apatite, with occasional tourmaline. Optically non-resolvable clays (0–7.5%) occur as amorphous aggregates in pore peripheral situations, and as occasional grain rims. Organic fragments (0–6%) are important in a single sample, occurring as both dispersed flakes and concentrated along bedding planes. Bioclasts (0–1%) occur as occasional disarticulated, broken and abraded skeletal fragments.

The clastic source area probably comprised mixed Palaeozoic sediments and a Precambrian regional metamorphic and plutonic igneous crystalline basement terrain. Recycling of older sediments is also confirmed by the abundance of reworked Carboniferous microfauna contained within the sandstone intervals. Clasts, which in the conglomerate facies reach up to large pebble size, are typically subangular to subrounded.

Diagenetic history

The sandstones have undergone considerable diagenetic modifications, which, combined with the effects of compaction, have led to the poor reservoir characteristics observed. A summary of the diagenetic transformations and precipitates is shown in Fig. 10, with photomicrograph examples in Fig. 11. Five forms of authigenic

Fig. 11. Photomicrographs illustrating diagenetic precipitates. In all cases, field of view is 1.5 mm × 1 mm; plane-polarized light. (**a**) Typical sandstone illustrating poorly to moderately sorted, very fine sand grade lithic arenite, displaying moderate compaction. The porosity (stained blue) is randomly developed and poorly connected. Noteworthy features are bored skeletal fragment (top left) and the presence of pore-occluding subhedral to euhedral ferroan dolomite (pale brown), variably developed discontinuous quartz overgrowths and grain-coating tangentially oriented illite. $Ø_{he}$ 13.9%, $Ø_{M}$ 9.5%, K_{AH} 4.0 mD. (**b**) Moderately to poorly sorted, weakly compacted calcareous feldspathic arenite. Early-formed pervasive ferroan calcite cement (stained blue) causes near complete porosity removal and supports the oval shell fragment against compaction. (**c**) A large secondary dissolution pore (stained blue) showing euhedral ferroan dolomite rhombs that line the edge, and pseudo-vermicular kaolinite in the centre. Dolomite is interpreted to replace and grow from earlier ferroan calcite cement, and grows by displacing kaolinite crystals. (**d**) Ferroan dolomite crystals (small euhedral rhombs stained blue) both enclosed by and enclosing authigenic quartz overgrowths forming an interlocking pore-occluding mosaic.

carbonate precipitates have been identified, using carbon and oxygen isotope geochemistry and CL microscopy.

Early shallow burial diagenesis, under temperatures and pressures common to the environment of deposition, began with filtrational introduction of clay into the interstitial pore spaces. Water escape and sandstone injection driven by rapid deposition and sediment–water density imbalances may have assisted introduction of clay minerals. Bacterial reduction of seawater sulphate allowed finely disseminated pyrite to develop, and indicates an anoxic, sulphidic environment. As burial progressed, changes in the pore fluid regime resulted in disequilibrium conditions and dissolution of feldspars and biogenic carbonate material, providing enhanced pore fluid bicarbonate. Localized ferroan and non-ferroan calcite cements in the form of 'doggers' developed at low temperatures (35–47 °C) at *c.* 1 km burial depth. These early calcite cements have both marine carbonate and oxidized organic debris as carbon sources.

Later deep burial diagenesis led to further grain dissolution (mainly of feldspars and rock fragments), with localized kaolinite precipitation in grain dissolution pores. Later, magnesium-rich fluids allowed dolomitization of the early calcite cements and dolomite precipitation at temperatures of 75–103 °C (2–3 km burial depth). Late diagenesis is completed by the precipitation of authigenic silica as syntaxial quartz overgrowths. There is some overlap in the diagenetic sequence between quartz and dolomite precipitation. Figure 11d shows an example of this, where ferroan dolomite crystals are both enclosed by, and are enclosing, authigenic quartz overgrowths. Albitization of alkali feldspars is noted and isotopically heavy nodular pyrites are recorded from core as late diagenetic cements ending the diagenetic sequence. Hydrocarbons stain authigenic kaolinite, suggesting that oil emplacement post-dated grain dissolution and kaolinite development. Basin modelling suggests that the adjacent Upper Jurassic source rocks began to generate hydrocarbons in the axis of the Porcupine Basin during Late Cretaceous time.

Reservoir potential

Reservoir quality in the 35/8-2 sandstones is generally poor, largely because of the chemical instability of the clastic components (high proportion of feldspars and labile rock fragments), which, through burial diagenesis, has led to the low permeabilities encountered. Diagenetic overprinting is the main agent of downgrading of reservoir potential, particularly in the form of kaolinite blocking pore throats. Compactional pore volume loss is significant and has led to quartz overgrowths that bridge pore throats and significantly reduce connectivity.

The porosity and permeability cross plots (Fig. 11) show that higher permeabilities (2–10 mD range) are encountered in the A sand interval, as compared with an average value of 1 mD in the B and C sand intervals. Coding the porosity–permeability values by facies type shows that the higher-permeability points recorded from the A sand are predominantly from Facies B1 (high-density turbidites). This significant variation separates the Facies B1 sandstones in the A sand interval from other examples of Facies B1 and the remaining facies types. Although Facies B1 sandstones are also encountered in the B and C sands, they are typically thinner and less abundant. The Facies B1 sandstones in the A sand show good development of early-formed water escape structures (dish and pillar structures). These are believed to have led to the repacking of grain fabrics and more extensive expulsion of detrital clays than seen elsewhere, resulting in a corresponding enhancement of permeabilities. Other factors that control the reservoir potential include bed thickness and relative abundance of interbedded claystones, as well as the detrital clay content.

Prediction of reservoir quality

The three sandstone bodies all contain a similar variety of facies (Fig. 12) indicative of deposition in a submarine fan setting. The intervals include evidence of fan lobe and channel fill deposition, along with debris-flow and hemipelagic deposition. Therefore the occurrence of better quality Facies B1, with associated water escape structures that may enhance permeability, is unlikely to be restricted to the A sand. Laterally across the depositional system it is inferred that the B and C sand intervals may have better reservoir facies developed than were encountered by the 35/8-2 wellbore. The morphology and internal geometry of the sandstone intervals cannot, however, be defined using currently available seismic data.

Petrophysical analysis

A detailed petrophysical analysis using modern multi-mineral composition techniques was undertaken to assess the three hydrocarbon-bearing sandstones in well 35/8-2. Porosity calculations were calibrated against the core measurements, and a porosity transform used to

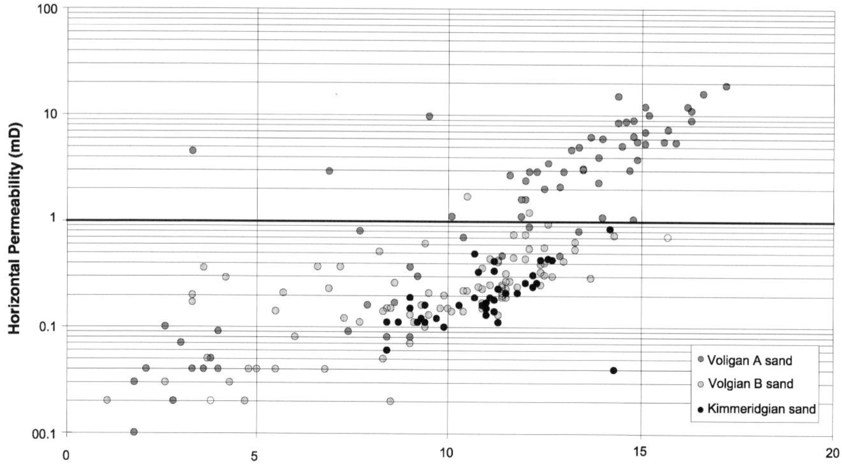

(a)

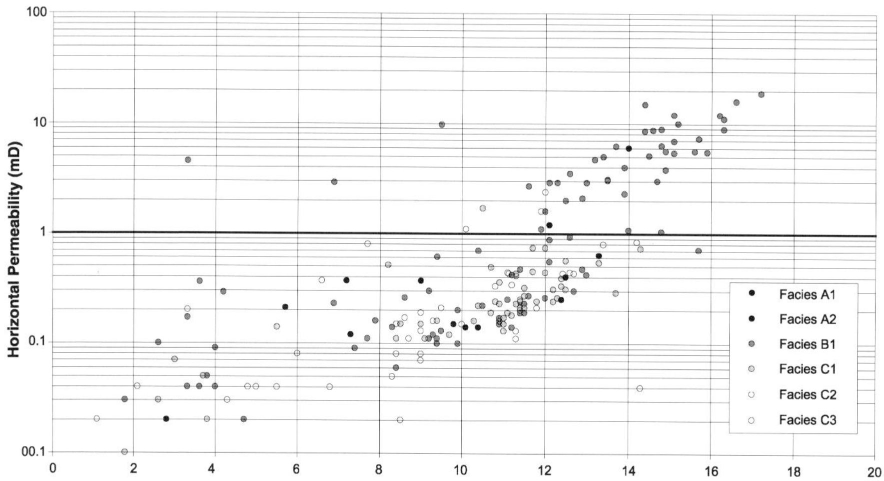

(b)

Fig. 12. Helium porosity v. horizontal permeability cross plots for cored intervals, coded by (**a**) reservoir interval and (**b**) facies type. The 1 mD permeability cut-off is highlighted.

estimate permeability. Net sand was defined on the basis of 1 mD permeability and 7% porosity cut-offs. Two values of R_w (water resistivity) are used to calculate the hydrocarbon saturation; a value of 0.45 ohm m was recorded from a repeat formation test (RFT) sample taken at 4014.5 m (13 171 ft) MD. This value is considered to be rather fresh for this environment, and the water sample (derived from within the oil leg) was probably contaminated. Using information from the Connemara Field, a value of 0.15 ohm m has been used to provide an alternative calculation, resulting in higher hydrocarbon saturation values. Petrophysical calculations for the A, B and C sand intervals are tabulated in Table 2. A clear increase in calculated S_w (water saturation) values is encountered at 4232.8 m (13 887 ft) MD and this is taken as the lowest known hydrocarbons in the well for calculation of reserves estimates, at −4206.5 m (−13 801 ft) TVDSS (true vertical depth sub-sea).

Table 2 *Summary of petrophysical parameters by reservoir interval*

Interval	Gross sand, G	Net sand, N	N:G	Average S_w[1] (%)	Average S_w[2] (%)	Average Ø (%)	Permeability (mD)
Facies A 3997.5–4034.3 m MD (13 115–13 236 ft MD)	36.9 m (121. ft)	28.8 m (94.5 ft)	0.781	71	47	12.6	2–20
Facies B 4093.8–4148.3 m MD (13 431–13 610 ft MD)	54.6 m (179 ft)	38.9 m (127.5 ft)	0.712	84	58	11.5	1–2
Facies C 4192.5–4252.0 m MD (13 755–13 950 ft MD)	59.4 m (195. ft)	49.2 m (161.5 ft)	0.828	73	53	11.6	1–2

[1] Calculated using an $\boldsymbol{R}_{\mathrm{w}}$ value of 0.45 ohm m.
[2] Calculated using an $\boldsymbol{R}_{\mathrm{w}}$ value of 0.15 ohm m.

DST, fluid and pressure analysis

The successful drillstem test (DST 3E, 3995.9–4029.5 m (13 110–13 220 ft) MD) from the A sand interval tested 925 BOPD and 4.85 MMSCFD with a drawdown pressure of 8800 psig. The test had a high skin factor of +7, indicating the possibility of 'condensate banking' or formation damage (the test was conducted over 6 weeks after the well had reached total depth). The surface-recombined gas and condensate samples characterized the reservoir fluid as a retrograde gas condensate with a dew-point of 8520 psig. However, the bottomhole flowing pressure (1788 psig) was significantly below the dew-point and therefore condensate drop-out will have occurred in the reservoir. The effect of this would be to give incorrect gas–oil ratio (GOR) and fluid API measurements at the surface (too high), and unrepresentative surface-recombined fluid samples. Given the inconclusive nature of the data available, it is not possible to confirm whether the hydrocarbons present are gas condensate or volatile oil. It is certain, however, that a considerable quantity of gas is present in the reservoir.

The RFT data recorded in the well were reviewed to assess the quality of the measurements, fluid gradients and the likelihood of pressure communication between the three sandstone intervals. A limited number (nine out of 53) of the pre-tests were found to be of sufficient quality to determine that the elevated pressures noted in this well are real. Most of these tests are from the A sand interval and give a calculated fluid gradient of 0.123 psi ft^{-1}, suggesting a fairly dry gas. The produced fluids suggest a gradient of 0.179 psi ft^{-1}, which is indicative of a condensate gas. The discrepancy between these gradients is attributed to the RFT strain gauge measurements that have poor accuracy and low resolution. The low permeabilities observed in the core undoubtedly result in the failure of the majority of the pressure tests in the B and C sands. It therefore remains impossible to confirm if there is pressure communication between the three sandstones, and it is possible that independent hydrocarbon columns may exist in individual sandstones within the structure. If future appraisal drilling on the 35/8-2 accumulation is to be attempted, the depletion strategy for the reservoir must take into account the high-pressure, low-permeability reservoir, and wells would almost certainly require fracture stimulation to induce more promising flow rates.

Mapping and trap configuration

The depth structure map to the top of the hydrocarbon-bearing Jurassic A sand interval is shown in Fig. 13. Resolution of this mapping horizon becomes difficult in the fault terraces to the NW of well 35/8-2, and these terraces have considerable structural complexity. The deeper 'Intra Kimmeridgian' marker (see Fig. 4) has been used to assist the interpretation across these fault terraces, and is used to imply that the interval between these two mapping horizons retains a relatively uniform thickness across the area. This interpretation, however, does not allow for potential syndepositional thickness changes across these fault terraces, which may have been actively subsiding concurrent with sandstone deposition. Erosion at the Base Cretaceous–Late Cimmerian unconformity is interpreted to affect the section younger in age than the A sand interval, but the possibility of erosional truncation of these sands is not discounted. The depth map on the Top Jurassic A sand shows the crest of the structure to be at −3871 m (−12 700 ft) TVDSS, and the lowest closing contour to the northeasterly spill point is −4130 m (−13 550 ft) TVDSS. This mapped closure lies just beneath the B sand interval in the well (see Fig. 5); however, as discussed above, the estimated lowest hydrocarbons in the well are at −4206.5 m (−13 801 ft) TVDSS. Consequently, a variety of possible trap configurations and hydrocarbon columns may be proposed to explain the well results in the context of the overall mapped accumulation (Figs 14 and 15). A three-point reserves spreadsheet (Otis & Schneidermann 1997), which takes gross rock volumes and ranges of key reservoir parameters as input, was used to calculate reserves estimates for four possible trap configurations. In all cases, a gas recovery factor of 64%, condensate yield of 190 bbls per MMCF and condensate recovery factor of 22% is used.

In the minimum case, Case 1, only those hydrocarbons proven by DST 3E in the A sand interval are deemed effective. No volumes are attributed to the A sand outside the fault block penetrated by the well. The estimated mean case proven hydrocarbon reserves for the structure are 33 BCF (billion cubic feet) and 2 MMBC (million barrels of condensate). Case 2 assumes that only the A sand is an effective reservoir, but across the entire mapped structure, and has associated mean reserves of 355 BCF and 23 MMBC. Case 3 allows all three sand intervals to be effective across the structure, and has estimated mean reserves of 532 BCF and 35 MMBC. The upside (Case 4), assumes that independent columns are present, and the resulting mean recoverable reserves are 764 BCF and 50 MMBC.

Further Jurassic potential

The 35/8-2 structure is an isolated, plunging, faulted anticline. Adjacent fault terraces along the eastern margin of the Porcupine Basin may offer potential for hydrocarbon entrapment in similar Upper Jurassic deep-marine sandstones.

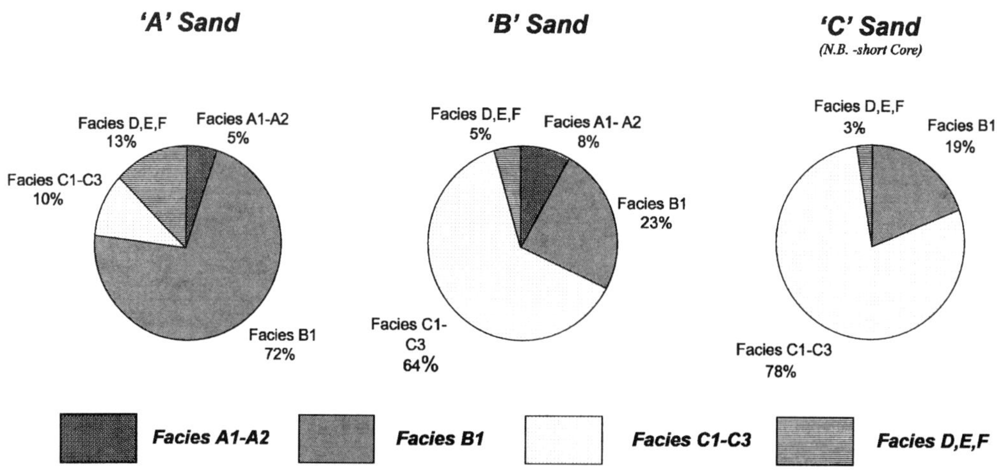

Fig. 13. Proportions of major facies groups by sand interval. Dominance of Facies B1 in the A sand interval, with greater proportions of interbedded sandstones and claystones (facies C1–C3) in the B and C sand intervals, should be noted.

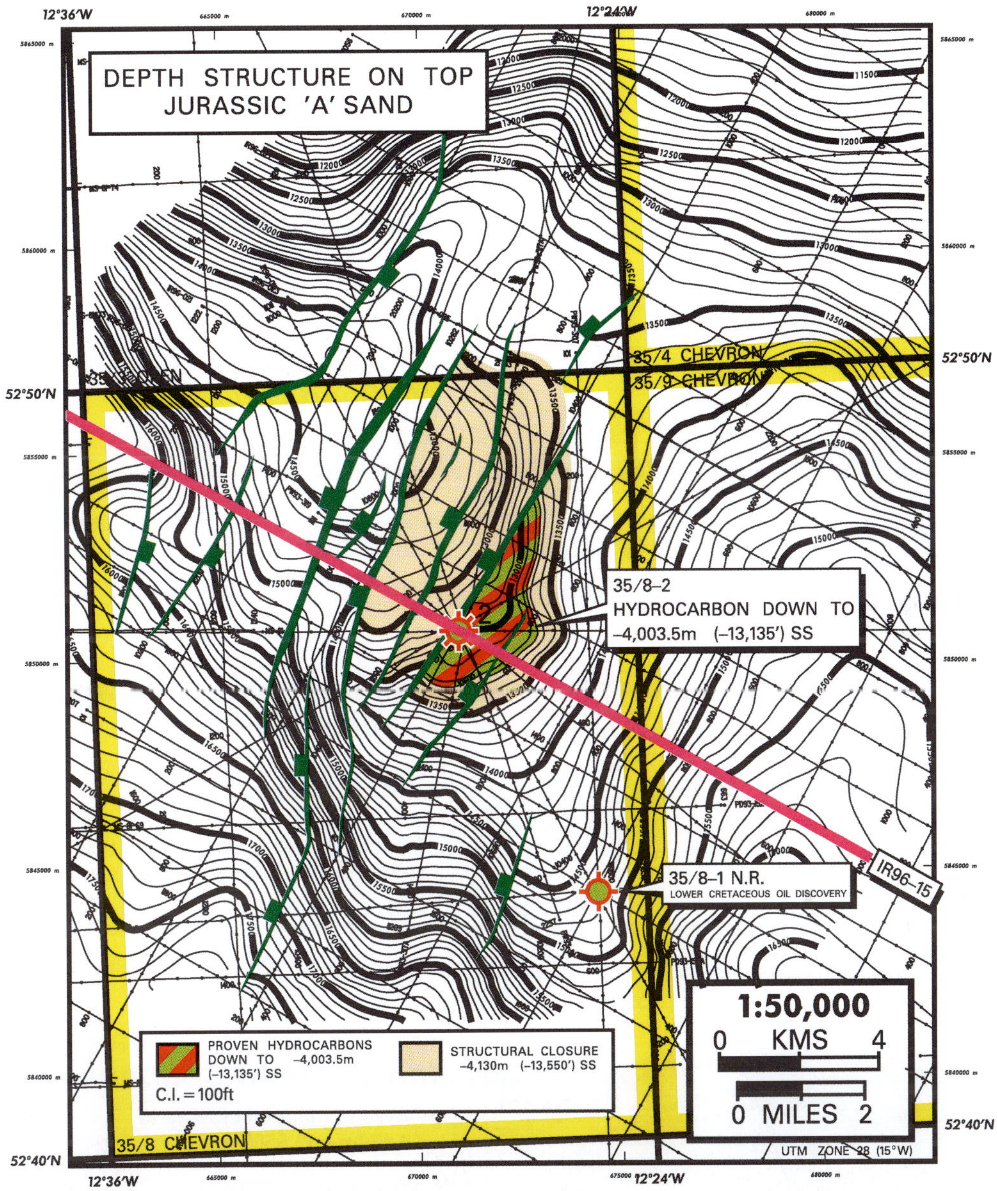

Fig. 14. Depth structure map on the top Jurassic A sand reflector. Noteworthy features are: (1) location of seismic line IR96-15 (Fig. 4); (2) mapped proven hydrocarbons to −4003.5 m (−13 135 ft) sub-sea (SS) and structural closure at −4130 m (−13550 ft) SS. Adjacent well 35/8-1 did not penetrate the Upper Jurassic A sand interval. N.R., not reached.

However, with closer proximity to the basin margin the risk of sediment bypass and lack of sandstone deposition increases. Potential for truncation of the 35/8-2 submarine fan sandstones may exist to the south and west of the discovery, where the Late Cimmerian unconformity erodes deeply into the Jurassic section, and an overlying seal would be provided by Lower Cretaceous claystones. The possibility of stratigraphic hydrocarbon traps within the 35/8-2 sandstones and other equivalent deep marine deposits is not discounted, but with the current well control and seismic datasets, this play is not resolvable.

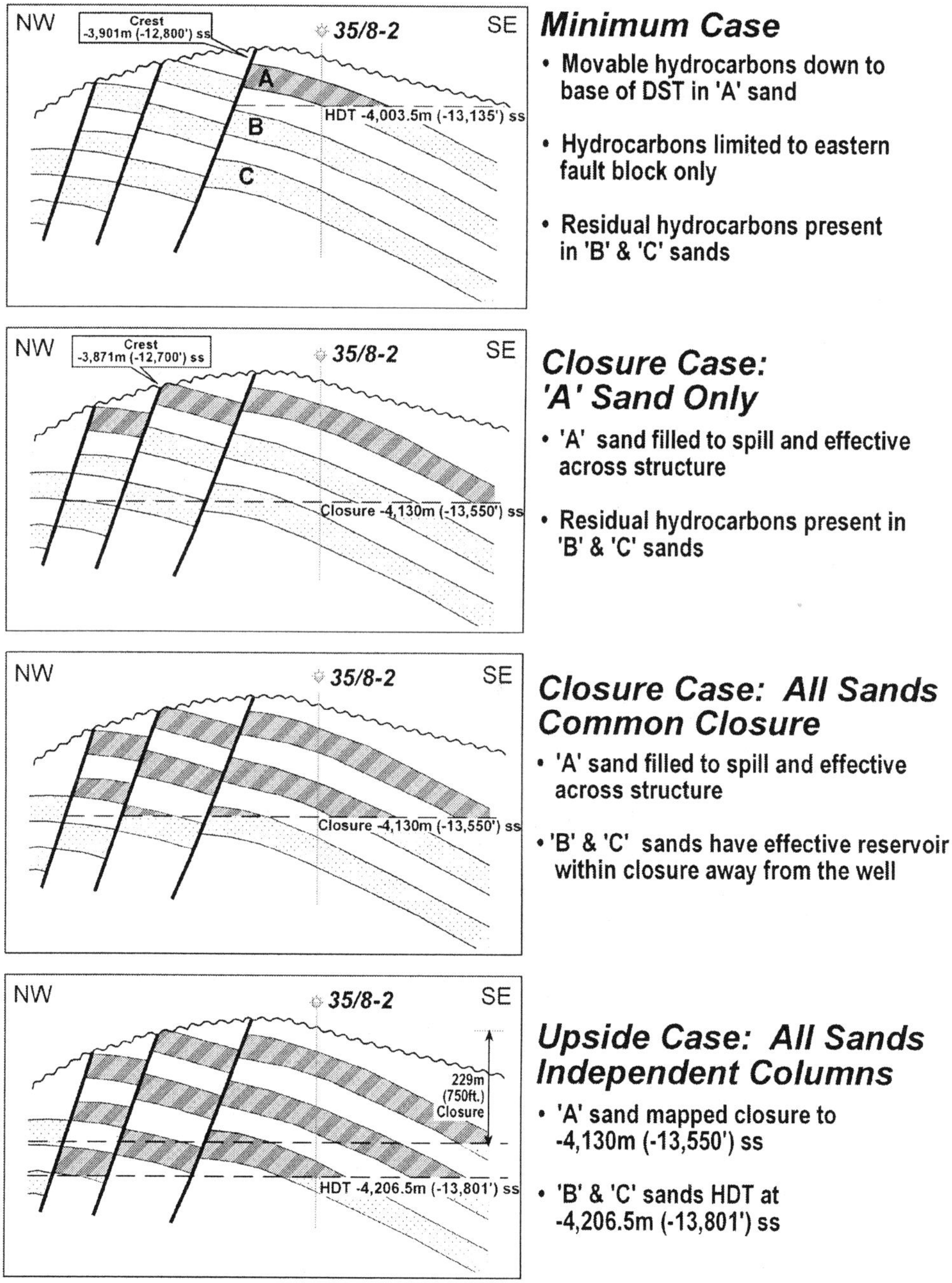

Fig. 15. Potential trap configurations used as inputs for reserves calculations in the 35/8-2 structure.

Future Jurassic exploration in the Porcupine Basin must consider the detrimental effects of diagenesis on deeply buried reservoir sandstones. At shallow depths (e.g. Connemara Field), the reservoir permeability is not significantly affected (MacDonald *et al.* 1987). The potentially attractive turbidite reservoirs encountered in well 35/8-2 are, however, more deeply buried and consequently suffer from severe diagenetic overprinting. This is partly due to the chemical instability of the sandstones, derived from a mixed metamorphic and igneous basement terrain. Provenance studies may provide an indication of where a more mature clastic source area may be developed adjacent to the basin.

Conclusions

The hydrocarbon-bearing Upper Jurassic sandstones encountered in the 35/8-2 well are unique to date within the Porcupine Basin. Despite good porosity and high net/gross ratios, these reservoirs exhibit very low permeability. The initial clastic composition of the sediment, derived from the north and east, included high proportions of rock fragments and feldspars, containing relatively unstable minerals that were particularly susceptible to the effects of burial diagenesis. The subrounded detrital quartz grains and abundance of terrigenous material indicate limited sediment transport distance, affording little opportunity for winnowing under fluvio-deltaic or shallow marine conditions before being rapidly resedimented into the deep marine environment.

In locations where a more mature clastic source (e.g. Lower Palaeozoic or Triassic sandstones) adjacent to the Porcupine Basin is eroded, and the Late Jurassic shallow marine shelf is broader and more effective in removing labile constituents, the deep marine Upper Jurassic sandstones may represent a viable hydrocarbon play within the Porcupine Basin.

The uncertainties surrounding hydrocarbon type, trap configuration and extent of the productive facies, combined with water depth, distance from market and costs in producing from a high-pressure reservoir, are all challenges that must be overcome before the reserves in the 35/8-2 structure can be economically developed.

The authors would like to thank the directors of Chevron Europe Ltd and Statoil (Ireland) Ltd for permission to publish this paper. The Petroleum Affairs Division of the Department of the Marine and Natural Resources, and Phillips Petroleum Company Ltd allowed access to the core material for logging and sampling purposes. We would like to thank our coworkers at Chevron (D. Lewis, J. Lockett, N. Bremner, F. Hayes, D. Way, P. Nesom, G. Rorrison and S. Stephens), Geochem (E. Smith and J.Cole) and Statoil (G. Haarr, B. MacTiernan and J. Conroy) for their contributions to this study. I. Sinclair (Hibernia Management and Development Company Ltd, Newfoundland) and P. Shannon (University College, Dublin) provided valuable revisions to the manuscript. The opinions expressed in this paper are those of the authors alone and are not necessarily shared by Chevron or Statoil.

References

CROKER, P.F. & SHANNON, P.M. 1987. The evolution and hydrocarbon prospectivity of the Porcupine Basin, offshore Ireland. *In*: BROOKS, J. & GLENNIE, K.W. (eds) *Petroleum Geology of North West Europe*. Graham & Trotman, London, 633–642.

DORÉ, A.G., LUNDIN, E.R., BIRKELAND, Ø., ELIASSEN, P.E. & JENSEN, L.N. 1997. The NE Atlantic Margin: implications of late Mesozoic and Cenozoic events for hydrocarbon prospectivity. *Petroleum Geoscience*, **3**, 117–131.

MACDONALD, H., ALLAN, P.M. & LOVELL, J.P.B. 1987. Geology of oil accumulation in Block 26/28, Porcupine Basin, offshore Ireland. *In*: BROOKS, J. & GLENNIE, K.W. (eds) *Petroleum Geology of North West Europe*. Graham & Trotman, London, 643–651.

MASSON, D.G. & MILES, P.R. 1986. Development and hydrocarbon potential of Mesozoic sedimentary basins around margins of the North Atlantic. *AAPG Bulletin*, **70**, 721–729.

NAYLOR, D. & SHANNON, P.M. 1982. *The Geology of Offshore Ireland and West Britain*. Graham & Trotman, London.

OTIS, R.M. & SCHNEIDERMANN, N. 1997. A process for evaluating exploration prospects. *AAPG Bulletin*, **81**, 1087–1109.

SHANMUGAM, G. & MOIOLA, R.J. 1991. Types of submarine fan lobes: models and implications. *AAPG Bulletin*, **75**, 156–179.

SINCLAIR, I.K., SHANNON, P.M., WILLIAMS, B.P.J., HARKER, S.D. & MOORE, J.G. 1994. Tectonic control on sedimentary evolution of three North Atlantic borderland Mesozoic basins. *Basin Research*, **6**, 193–217.

TATE, M.P. 1993. Structural framework and tectonostratigraphic evolution of the Porcupine Seabight Basin, offshore western Ireland. *Marine and Petroleum Geology*, **10**, 95–123.

WILLIAMS, B.P.J., SHANNON, P.M. & SINCLAIR, I.K. 1999. Comparative Jurassic and Cretaceous tectono-stratigraphy and reservoir development in the Jeanne d'Arc and Porcupine basins. *In*: FLEET, A.J. & BOLDY, S.A.R. (eds) *Petroleum Geology of Northwest Europe: Proceedings of the 5th Conference*. Geological Society, London, 487–499.

Comparative Tertiary stratigraphic evolution of the Porcupine and Rockall basins

A. McDONNELL & P. M. SHANNON

Department of Geology, University College Dublin, Belfield, Dublin 4, Ireland
(e-mail: angela.mcdonnell@ucd.ie)

Abstract: The Tertiary development of the Porcupine and Rockall basins is compared in terms of stratigraphy and sedimentation. Four main unconformities are correlated between the basins and these are interpreted as being of Paleocene (C40), latest Eocene–Early Oligocene (C30), latest Early Miocene (C20) and Early Pliocene (C10) age. Seismic stratigraphic analysis of both basins suggests a greater similarity in post-Eocene deposition than in the Paleocene to Eocene stratigraphy. During Early Tertiary time a regressive succession, punctuated by minor transgressions, marks a major interruption in the general post-rift thermal subsidence pattern of the region. This regression, possibly triggered by lithospheric thermal effects and/or ridge-push stresses, resulted in deltaic and submarine fan deposition in the Porcupine Basin, with submarine channel trends indicating that sediment was sourced mainly from the Porcupine High to the north and west. Sand deposition in the Porcupine Basin occurred principally during Mid- to Late Eocene times. In contrast, Early Eocene sand input is postulated in the Rockall Basin, whereas deposition during the Mid- to Late Eocene times was more mud-prone. In Oligocene and Mio-Pliocene times, sediment build-ups, interpreted as contourites, developed towards the margins of both basins, with sedimentation principally influenced by oceanographic circulation patterns at this time. During Neogene to Recent times limited marginal sediment influx occurred in the Porcupine Basin whereas sediment input continued locally, during Neogene time, in the Hebrides region of the Rockall Basin.

The Porcupine and Rockall basins are the two largest frontier basins west of Ireland (Fig. 1). They are both lightly explored. Twenty-nine exploration and appraisal wells have been drilled in the northern part of the Porcupine Basin, whereas the southern, deep-water, area of the basin remains undrilled. No exploration wells have been drilled to date (early 2001) in the Irish sector of the Rockall Basin. Consequently, many aspects of the stratigraphic evolution of these basins remain poorly constrained.

In common with most basins in the Atlantic margin region, it has been suggested that the Porcupine and Rockall basins developed in response to episodic crustal extension during Permo-Triassic to Early Cretaceous times (Croker & Shannon 1987; Tate 1993; Sinclair *et al.* 1994; Shannon *et al.* 1995; Walsh *et al.* 1999). The most significant phase of extension in both basins is thought to have occurred during Late Jurassic to Early Cretaceous time as Atlantic sea-floor spreading propagated northwards from the central Atlantic region (Johnston *et al.* 2001). The subsequent lithospheric response to this rifting resulted in a phase dominated by passive thermal subsidence, which created the accommodation space for the Late Cretaceous and Tertiary sedimentary fill in these basins.

The Porcupine and Rockall basins display similar cross-sectional profiles within the Upper Cretaceous and Tertiary successions (Fig. 2). In particular, the Tertiary succession onlaps the margins of both basins in a broadly symmetrical fashion. A summary of the uppermost Cretaceous and Tertiary lithostratigraphy for the Porcupine and Rockall basins is illustrated in Fig. 3. Basin flank uplift occurred in both basins in Aptian–Albian times, possibly in response to regional extension and sea-floor spreading between the Bay of Biscay and Newfoundland (Knott *et al.* 1993). Clastic sediment influx occurred at this time in the Porcupine and Rockall basins where synrift accommodation space was generated through selective fault reactivation (Corfield *et al.* 1999; Johnston *et al.* 2001). A transition to passive thermal subsidence followed in both basins. During Campanian to Maastrichtian time, sea-floor spreading began in the Labrador Sea, and the widespread sea-level highstands at this time resulted in extensive chalk

From: SHANNON, P.M., HAUGHTON, P.D.W. & CORCORAN, D.V. (eds). 2001. *The Petroleum Exploration of Ireland's Offshore Basins*. Geological Society, London, Special Publications, **188**, 323–344. 0305-8719/01/$15.00

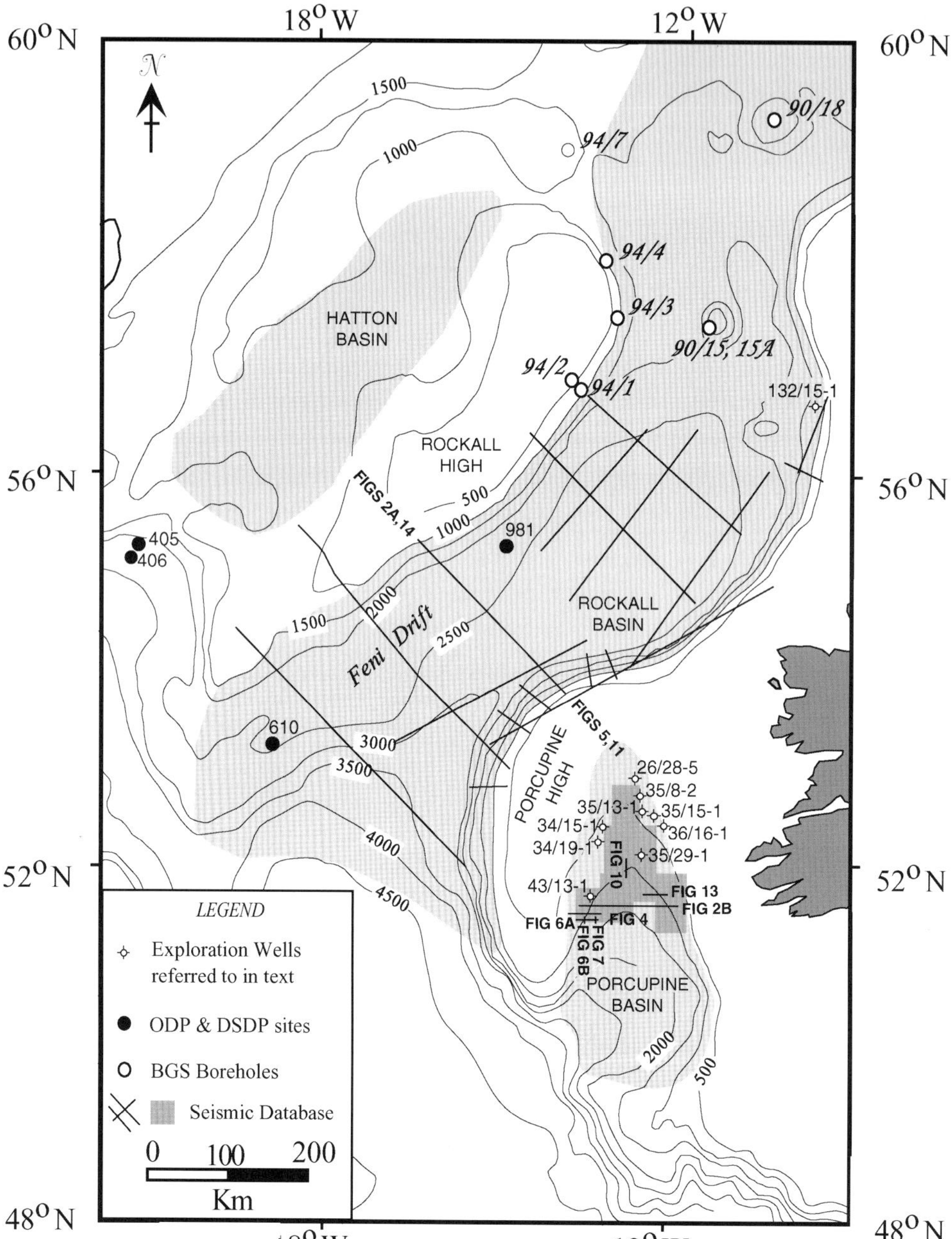

Fig. 1. Location map of the Rockall and Porcupine basins showing bathymetry and the seismic data grid used in this study. Locations of wells and seismic profile figures referred to in the text are also highlighted.

deposition in both basins (Moore & Shannon 1995; Corfield *et al.* 1999; Johnston *et al.* 2001). Sea-floor spreading in the Labrador Sea, succeeded by the initiation of spreading in the Norwegian–Greenland Sea during Early Eocene time (Knott *et al.* 1993), culminated in the separation of North America and Greenland followed by the break-up of Greenland and Eurasia (Roberts *et al.* 1999). In addition to the effects of these plate motions, postulated influences on the tectonic evolution of the Porcupine and Rockall region, during the

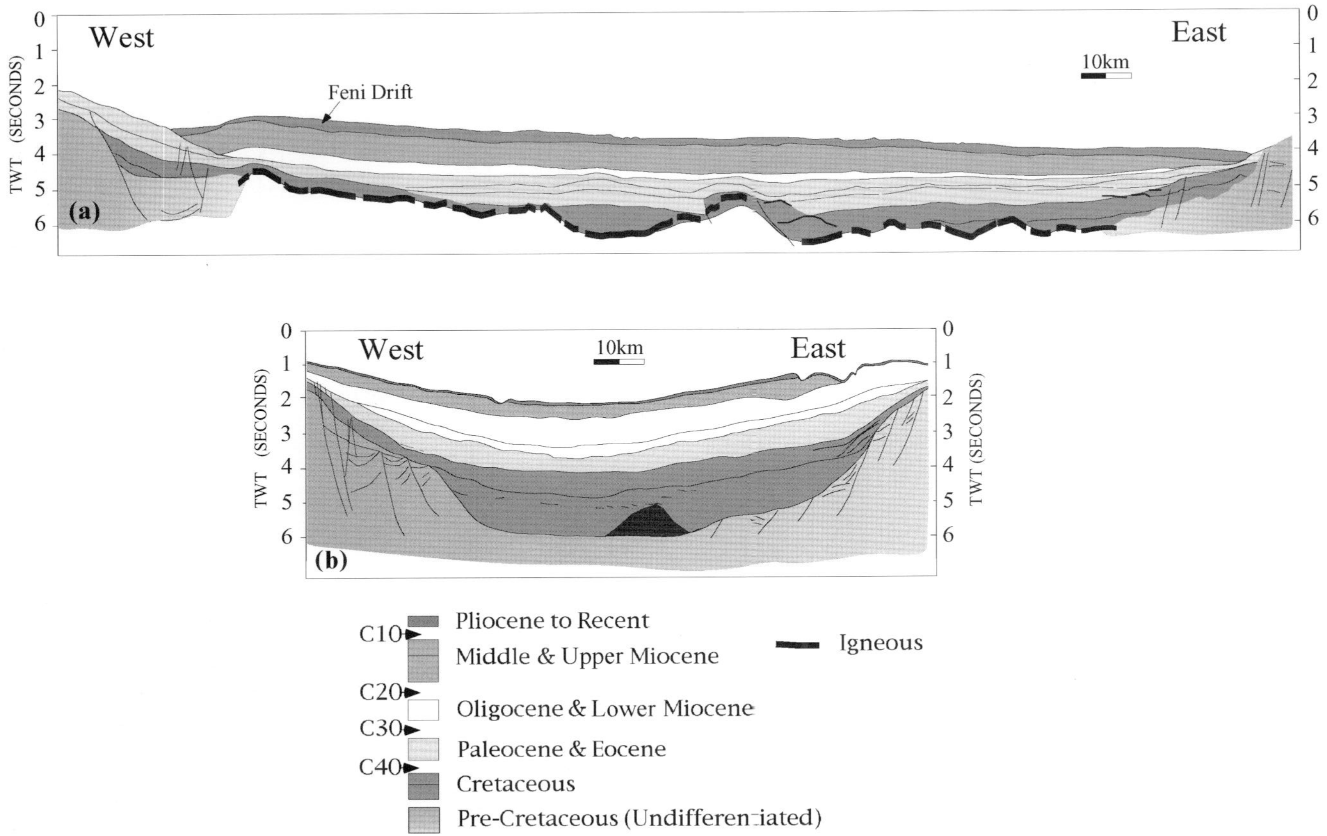

Fig. 2. Geoseismic cross-sections of (**a**) the Rockall Basin and (**b**) the Porcupine Basin, illustrating gross morphology of Cretaceous and Tertiary sections in each basin (see Figs. 1, 8, 9 and 12 for locations of geoseismic sections). The Tertiary succession onlaps the margins of both basins in a symmetrical fashion. It should be noted that the Porcupine Basin cross-section is from the south of the basin, therefore it does not illustrate the maximum thickness of accumulated Tertiary sediments. The deeper (pre-Cretaceous) section in the central parts of the Rockall Basin is masked by igneous sills.

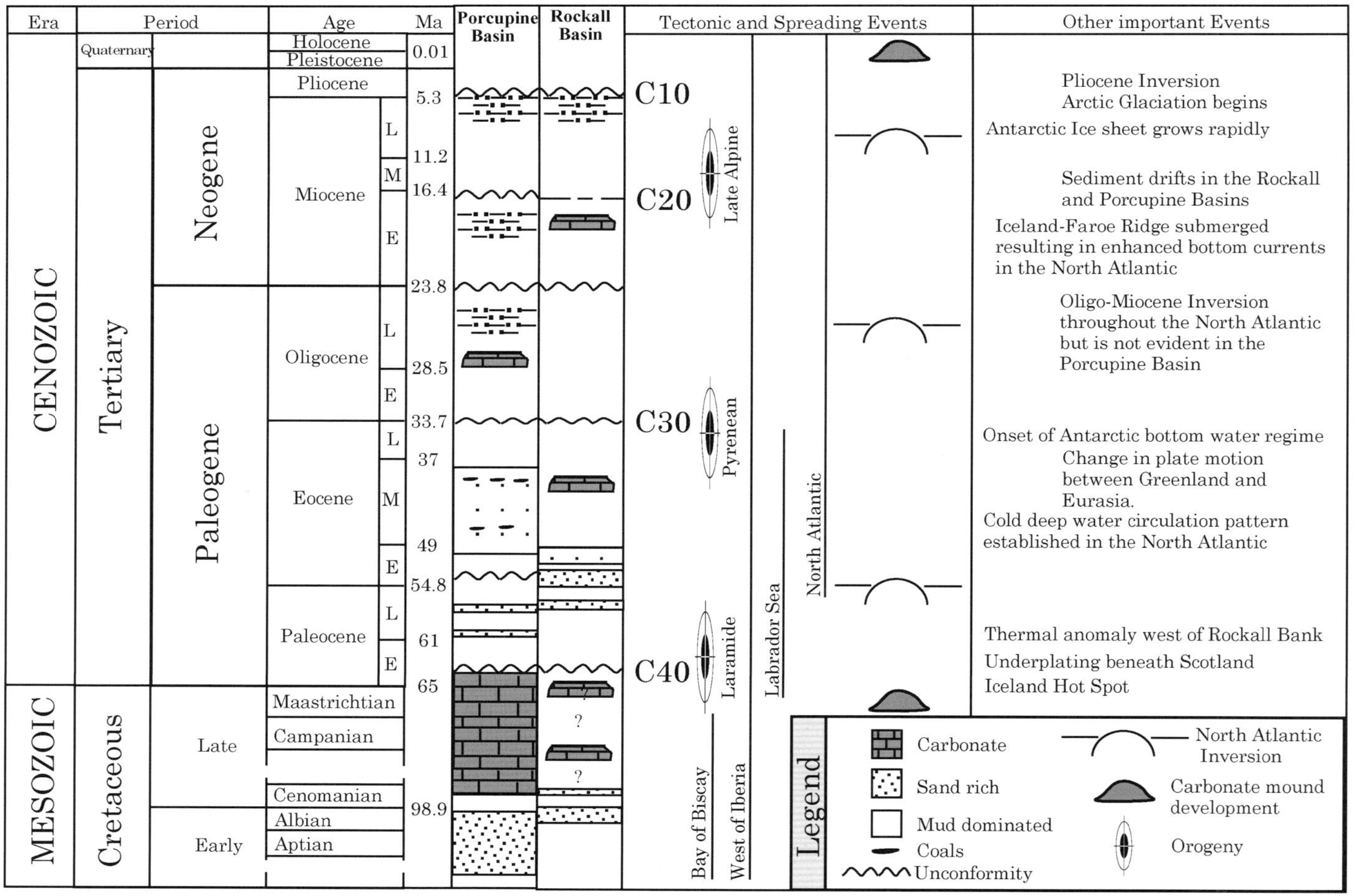

Fig. 3. Abbreviated lithostratigraphic column for the Tertiary successions in the Porcupine and Rockall basins. The proposed sequence stratigraphic scheme of Stoker *et al.* (2001) has been adopted for both basins in this study.

Tertiary period, include magmatic underplating beneath the NE Atlantic margin related to the Icelandic plume (Joppen & White 1990; Brodie & White 1995) and ridge-push forces induced by sea-floor spreading west of the Hatton Continental Margin (Shannon *et al.* 1993).

The northern part of the Porcupine Basin contains up to 4 km of Tertiary fill (Tate 1993). In contrast, the Rockall Basin is interpreted to contain a thinner Tertiary succession, with *c.* 2 km of strata postulated by Shannon *et al.* (1993). Aspects of the Tertiary stratigraphic evolution of these basins have been previously described by a number of workers (e.g. Moore & Shannon 1991; Shannon 1992; Tate 1993; Moore & Shannon 1995). Shannon *et al.* (1993) defined a number of seismic sequences in the Porcupine and Rockall basins. They proposed a general stratigraphic correlation and speculated upon the likely ages of some of the undrilled sequences in the Rockall Basin. Evidence for Palaeogene, coarse clastic, deltaic and submarine fan deposition in the Porcupine Basin has been presented by Moore & Shannon (1992); Shannon (1992). In contrast, most workers interpret the Neogene succession as representing a period of generally low sediment input into both basins, resulting in shale-dominated, deep marine sedimentation (Stoker 1997; Stoker *et al.* 2001). Slope failure features have been documented from both basins (Moore & Shannon 1991; Unnithan *et al.* 2001), and deposition of contourite drifts on the basin margins is interpreted as a response to changing oceanographic circulation patterns and is especially well developed in the Rockall Basin (Stoker *et al.* 2001).

The main objectives of this paper are to compare the seismic character of the interpreted Upper Cretaceous and Tertiary successions in the Porcupine and Rockall basins and to speculate on the likely controls on the Cenozoic stratigraphic evolution of these basins. Various seismic reflection notation schemes exist for the Tertiary stratigraphy of the Porcupine and Rockall basins (Roberts 1975; Miller & Tucholke 1983; Masson & Kidd 1986; Shannon *et al.* 1993). However, Stoker *et al.* (2001) have proposed a comprehensive scheme for the Rockall Basin, based on extensive seismic and shallow borehole data. This proposed nomenclature (C10, C20, C30, etc.) has been adopted for both basins in the present work.

Database and methods

An extensive seismic database exists for the Porcupine Basin in addition to the 29 exploration and appraisal wells drilled to date. Most of these wells penetrate to pre-Tertiary levels so that the stratigraphic age control for Paleogene sediments is good in the northern part of the Porcupine Basin, but the Neogene is poorly constrained. The database employed by the authors included information from 16 of these wells together with *c.* 5500 km of good-quality 2D seismic data shot in 1981 and augmented by some recent infill lines. Dip line spacing was *c.* 3–5 km with strike line spacing varying from 5 to 10 km (Fig. 1).

In contrast, a limited database was available for the Rockall Basin. With the exception of well 132/15-1 (drilled in UK waters), stratigraphic age control in the Rockall Basin is restricted to Neogene sediments, which can be dated using data from Deep Sea Drilling Project (DSDP) wells 405, 406 and 610 (Masson & Kidd 1986; Bull & Masson 1996), Ocean Drilling Program (ODP) well 981 (Stoker *et al.* 2001) and a number of British Geological Survey (BGS) shallow boreholes (Stoker *et al.* 2001). Seismic coverage for the Irish sector of the Rockall Basin comprised a number of widely spaced regional lines of varying vintages (1977–1996). This seismic coverage is largely confined to the eastern margin of the basin (Fig. 1). Approximately 4700 km of such data were available to the authors. In comparison with the Porcupine Basin (106 seismic profiles examined), interpretation of the Rockall Basin succession is largely speculative (16 seismic profiles examined).

A sequence stratigraphic approach, employing the techniques of Vail *et al.* (1977), was adopted to interpret the seismic data in each basin. Sequence boundaries were defined, based on reflector termination criteria, and seismic facies analysis was effected for a number of sequences. Correlation between the basins is largely based on unconformity relationships, seismic character jump correlations and the study of sequence boundary characteristics. However, the lack of significant Paleogene age constraints in the Rockall Basin and Neogene age constraints in the Porcupine Basin means that detailed inter-basin correlation of the seismic sequences, particularly those in the pre-Neogene successions, is extremely speculative.

Seismic units

The Upper Cretaceous to Recent successions in the Porcupine and Rockall basins have been subdivided into three major informal units comprising (1) uppermost Cretaceous, (2) Paleocene to Eocene and (3) Oligocene to Recent sequences. In each case the successions in the Porcupine Basin and the Rockall Basin are described and then compared. The main unconformities and correlatable surfaces identified in this study can, in general, be tied with those defined in the Rockall Basin by Stoker *et al.* (2001). Four major Tertiary correlation surfaces are recognized in both basins (Fig. 3): a Paleocene regional unconformity (C40), a latest Eocene–Early Oligocene unconformity surface (C30), a latest Early Miocene marker (C20) and an Early Pliocene unconformity event (C10). The C20 surface has chronostratigraphic significance in both the Porcupine and Rockall basins. It is unconformable in both basins and in the Rockall Basin it is diagenetically-enhanced (Dolan 1986).

Uppermost Cretaceous sequences

Porcupine Basin

Well data in the Porcupine Basin (e.g. 43/13-1, 35/15-1) record a chalk-dominated Campanian and Maastrichtian succession. Up to 600 m of chalk and chalky limestone was deposited in the centre and northern part of the basin (Croker & Shannon 1987). On seismic profiles the succession generally appears as a unit with moderate to relatively weak reflections capped by a very strong, high-amplitude and continuous reflection (C40) (Fig. 4). The unit shows higher-amplitude, more continuous reflections near the basin margin. This is interpreted to represent a lateral facies change where platform carbonates may have developed updip from the more basinal chalk that dominates most of the basin. In well 34/15-1 the chalk displays an upward-cleaning pattern, with less clastic material in the uppermost Maastrichtian to Danian interval. This may indicate a progressive flooding of the shelf. Further evidence for this transgression is provided by the onlap of uppermost Cretaceous reflections onto the basin flanks.

Rockall Basin

In the Rockall Basin a series of high-amplitude, continuous reflections observed along the eastern flank of the basin change basinwards into more discontinuous, moderate-amplitude reflections (Fig. 5). Well 132/15-1, on the eastern flank of the basin, penetrated a mud-dominated Upper Cretaceous interval with abundant limestone stringers. Here, the seismic facies of the Upper Cretaceous sequence consists of moderate-amplitude reflections of moderate continuity, topped by a very high-amplitude, continuous reflection similar to that observed in the Porcupine Basin.

Comparison

A close similarity is observed in the seismic character of the interpreted latest Cretaceous seismic unit in the Porcupine and Rockall basins (Figs 4 and 5). Age constraint on this succession is provided by the wells in the Porcupine Basin, most of which penetrated an Upper Cretaceous succession (Croker & Shannon 1987), and by extrapolation from well 132/15-1, which encountered Upper Cretaceous strata on the eastern margin of the Rockall Basin (Musgrove & Mitchener 1996). This seismic unit in both basins is characterized by parallel planar, moderate- to high-amplitude reflections that become weaker in a basinward direction. It is capped in both basins by a pronounced, continuous, high-amplitude reflection, which is a significant unconformity and is herein defined as the C40 event.

The general similarity in seismic facies of the Upper Cretaceous sequences in the Rockall and Porcupine basins suggests that an open marine depositional environment was common to both basins at that time. The limited presence of chalk in well 132/15-1 suggests a greater terrigenous influence on the eastern margin of the Rockall Basin than in the Porcupine Basin.

Paleocene to Eocene sequences

Porcupine Basin

Well data (e.g. 26/28-5, 34/19-1, 35/13-1, 35/15-1) indicate that the Paleocene succession in the Porcupine Basin is mud dominated whereas the main coarse clastic input occurred in Mid-Eocene to earliest Late Eocene times. The lower boundary of this Paleocene to Eocene unit is defined by a high-amplitude, continuous reflection (C40), which marks the top of the Cenomanian to Danian chalk sequence in the Porcupine Basin and represents the change from carbonate to clastic deposition.

A number of wedge-shaped seismic packages are recognized along the western margin of the Porcupine Basin (Fig. 6). These wedges extend for up to 8 km in the dip direction. The lower boundary is defined by the high-amplitude (C40) reflection marking the base of the Paleocene to Eocene section and this reflection is onlapped marginwards by internal reflections of the wedge. The upper boundary is a moderate-amplitude event onlapped by younger reflections. Internal reflections within these wedge-shaped deposits are continuous, of moderate to low amplitude, and display an aggradational stacking pattern. Amplitude typically increases towards the edges of these packages, particularly the basinward edge, and this increase generally corresponds to an increase in frequency. The structural setting of these packages, together with their seismic

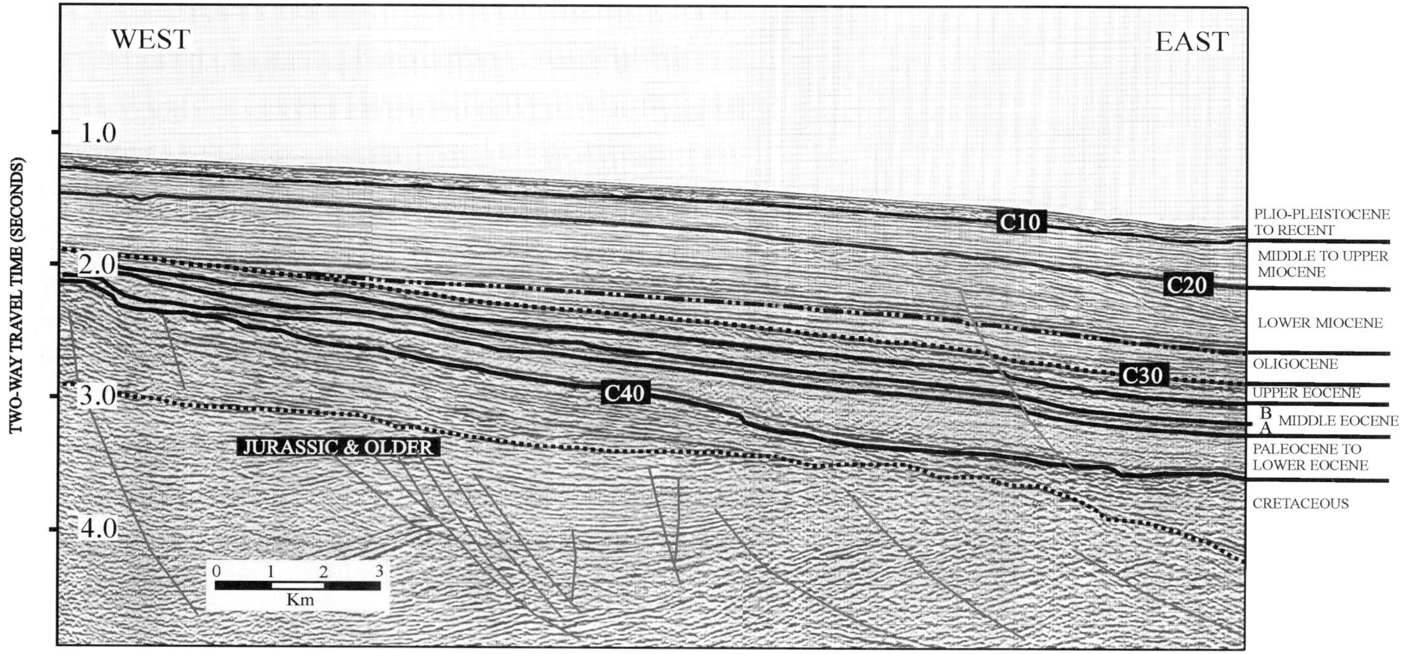

Fig. 4. Seismic profile from the western margin of the Porcupine Basin (see Figs. 1, 8, 9 and 12 for location). Readers should note the high-amplitude, continuous reflection associated with the C40 regional event, and the low-amplitude character of the Paleocene–Lower Eocene sequences, interpreted as a mud-prone section, in contrast to the higher-amplitude, more continuous reflections that distinguish the Middle to Upper Eocene section. Pronounced onlap of the western margin by the Oligocene succession is also observed. The stratigraphy is based on well ties in the north of the basin. Regional isochore maps of Middle Eocene seismic unit A and seismic unit B are presented in Figs. 8 and 9, respectively.

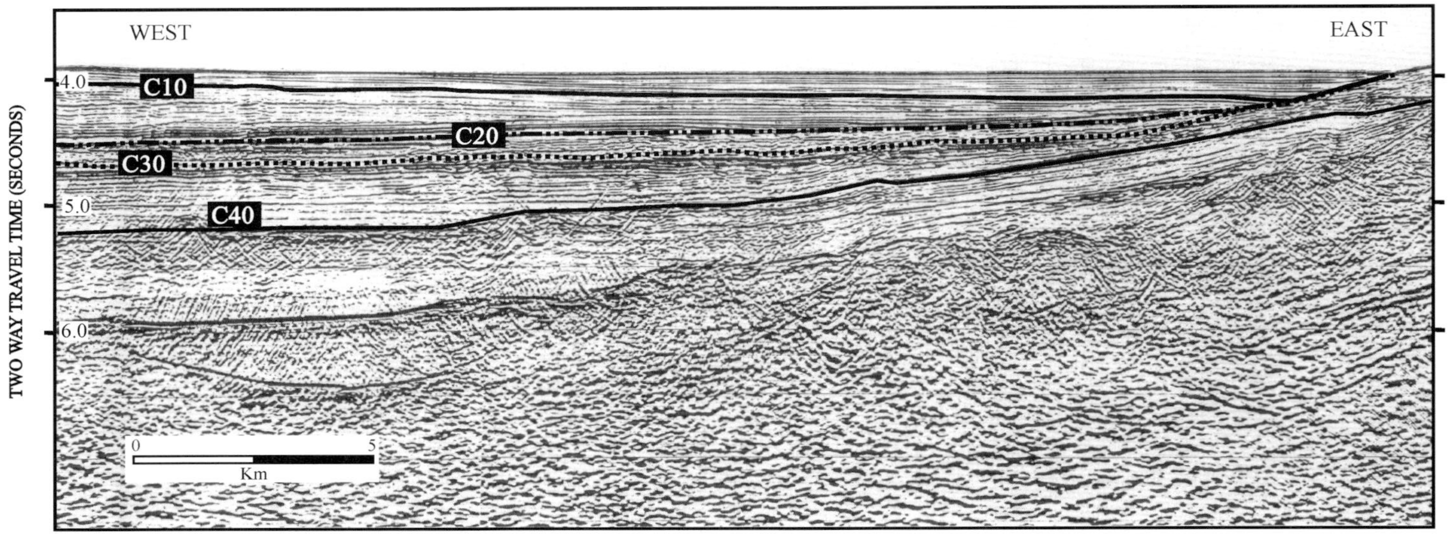

Fig. 5. Seismic profile from the eastern margin of the Rockall Basin illustrating the interpreted Tertiary succession (see Fig. 1 for location). Age interpretation is based primarily upon DSDP and BGS shallow boreholes and correlation with well 132/15-1. The very continuous, moderate- to high-amplitude Upper Cretaceous package, which is topped by a very high-amplitude, continuous, reflection (C40) similar to that observed in the Porcupine Basin, should be noted. The Paleocene–Eocene section locally onlaps the Upper Cretaceous section, particularly towards the base of the slope defined by the C40 boundary.

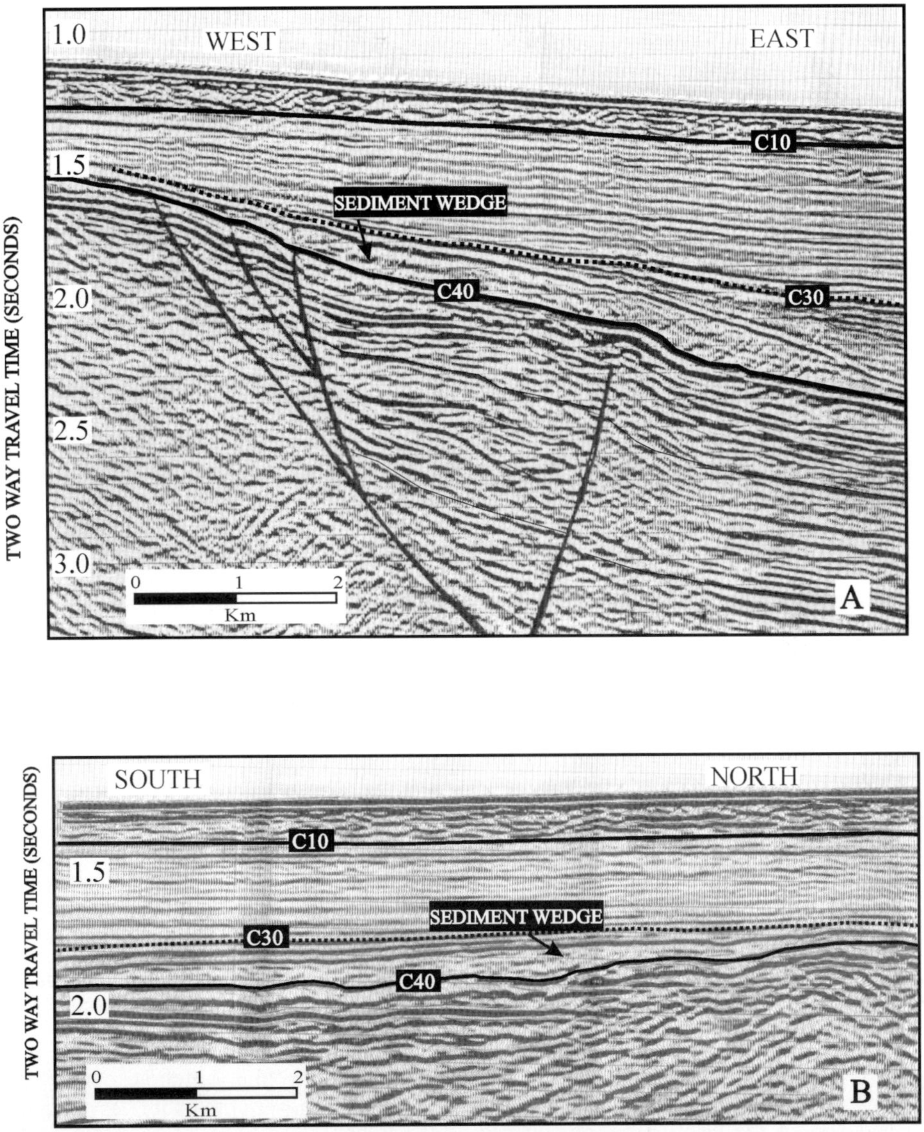

Fig. 6. Paleocene to lowermost Eocene wedge-shaped seismic package, present along the SW flank of the Porcupine Basin (see Figs. 1, 8, 9 and 12 for location. (**a**) West–east dip-section, illustrating the low-amplitude character of the package onlapping and downlapping the basin margin. (**b**) North–south section through the same feature illustrating onlap of upper boundary by younger reflections. The structural and stratigraphic setting of this wedge-shaped seismic package suggests the presence of an alluvial fan deposit on the margin of the basin.

configuration, is compatible with the development of a series of marginal alluvial fans. These wedges represent the earliest Tertiary sediment input into the basin with sediment sourced from the Porcupine High.

In the northern part of the basin a series of large-scale, southward-prograding clinoforms correspond to Late Paleocene and Eocene deltaic deposits (Moore & Shannon 1992). These deposits are coeval with localized basin floor submarine fan deposition in the deeper parts of the Porcupine Basin (Shannon *et al.* 1993).

On seismic profiles a mounded seismic facies, interpreted as being of Eocene age, is evident in the SW of the basin (Fig. 7). These mounds manifest an isochore thickness of 0.2–0.3 s TWT corresponding to *c.* 200–300 m in vertical relief. They have an approximate extent of 5–20 km in the strike dimension and a down-dip extent of some 40–50 km into the basin. They are characterized by convex-upward reflections which display bi-directional downlap at the mound edges. Internally the mounds comprise moderate- to high-amplitude, laterally continuous, reflections of moderate frequency. A decrease in seismic amplitude toward the mound fringes is occasionally observed. A depositional origin is inferred from thickening of individual drapes in the vicinity of the mound crest and depositional thinning and downlapping towards the mound fringes. The mound edges often show evidence of later incision and erosional modification.

These mounds were interpreted by Shannon (1992) as submarine fan deposits with the lateral seismic character change towards the mound fringes interpreted to represent a shale-prone fan fringe to a more sandy, channelized inner fan zone. High-amplitude, discontinuous, reflections onlap the erosive mound edges (Fig. 7). Mapping of this infill package in the present study reveals a linear, NW–SE, orientation in the west of the basin. It is interpreted as the deposit of a subsequent fan sequence, and the seismic facies of this infill package suggests that it represents a more proximal part of a younger high-energy complex. This is compatible with continued regression, causing later pulsed fan progradation across the older deposits, with the fan channels utilizing the underlying topographic lows generated by the prior fan deposition.

The Middle Eocene to Upper Eocene succession in the Porcupine Basin can be subdivided into a number of units (Fig. 4). Figure 8 is an isochore map of the Middle Eocene seismic unit A identified in Figs 4 and 7. A thick development of unit A occurs in the north and NW of the basin. Less pronounced depocentres are developed in the SW and SE of the basin, where the sediment lobes manifest an irregular geometry. In the SW of the basin a pronounced NW–SE isochore thickness represents the presence of a series of coalesced channel fill and basin floor lobes. Isochore thicknesses show a preferred NNE–SSW orientation in the north of the basin, north of 52°N, and a general NE–SW orientation along the SE margin. These are interpreted from seismic profiles to represent areas of incision and channelling caused by sediment transport from the basin margins. Both NW–SE and NE–SW orientations combine to create an axial drainage pattern in the south of the basin.

The overlying Middle to Upper Eocene sequence (incorporating Middle Eocene unit B and Upper Eocene seismic unit identified in Figs 4 and 7) reveals a more distinct lobate pattern on the basin flanks (Fig. 9). Inferred channel orientation, on the SW margin of the basin, varies from east–west to NW–SE. The isochore map (Fig. 9) indicates that the main depocentre was in the north of the basin at this time. Sediment in this part of the basin was sourced from the north and east. In contrast, in the south of the basin it is evident from submarine channel trends that the principal sediment sources were from the Porcupine High to the west, with a less pronounced clastic input from the east.

The Upper Eocene sequence of the Porcupine Basin (Figs 4 and 7) is dominated by moderate- to high-amplitude continuous reflections, which drape the mounds and onlap the southerly-prograding deltaic clinoforms in the north of the basin (Fig. 10). Wells 35/8-2 and 35/29-1 record a mud-prone Upper Eocene succession. Northerly flooding of the Eocene deltaic sequences at the end of Eocene time is interpreted to signify an end to significant coarse clastic input into the basin. A major basinwide unconformity developed in latest Eocene to Early Oligocene time (Figs 4 and 10) and this event has been correlated with the designated C30 event in the Rockall Basin of Stoker *et al.* (2001).

Rockall Basin

The lower boundary of the interpreted Paleocene to Eocene sequence in the Rockall Basin is marked by a basinwide, high-amplitude reflection (C40), which is locally onlapped by lower-amplitude, less continuous reflections (Fig. 5). The upper boundary of this sequence (C30) is identified as a high-amplitude event that is overlain by lower-amplitude, less continuous reflections.

The interpreted Paleocene to Eocene sequence thickens into the basin centre. Internally, this sequence is characterized by moderate- to high-amplitude reflections with reflection frequency decreasing basinwards. Seismic reflection amplitude strength tends to increase towards the top of the sequence, similar to that observed in the Paleocene to Eocene section of the Porcupine Basin, although reflection continuity is poorer in the Rockall Basin. At the eastern margin of the basin a wedge of low- to moderate-amplitude seismic reflections occurs near the base of this sequence (Figs 1 and 11). This wedge indicates that the Porcupine High may have acted as a provenance for the southern Rockall Basin

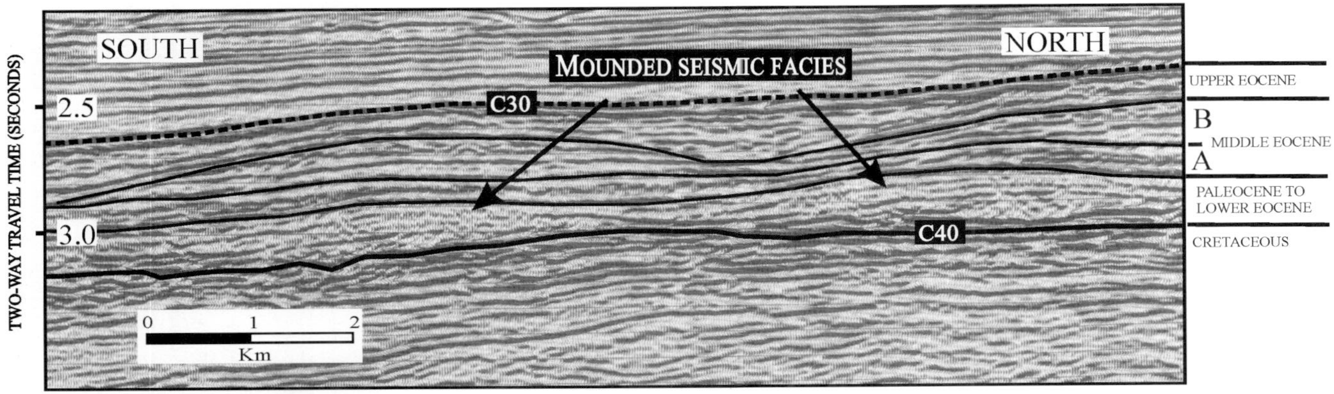

Fig. 7. Eocene mounds, interpreted as submarine fans by Shannon (1992). The mounds are characterized by convex-upward reflections that display bi-directional downlap at the mound edges. A depositional origin is inferred from thickening of individual drapes in the vicinity of the mound crest. Regional isochore maps of Middle Eocene seismic unit A and seismic unit B are presented in Figs. 8 and 9, respectively.

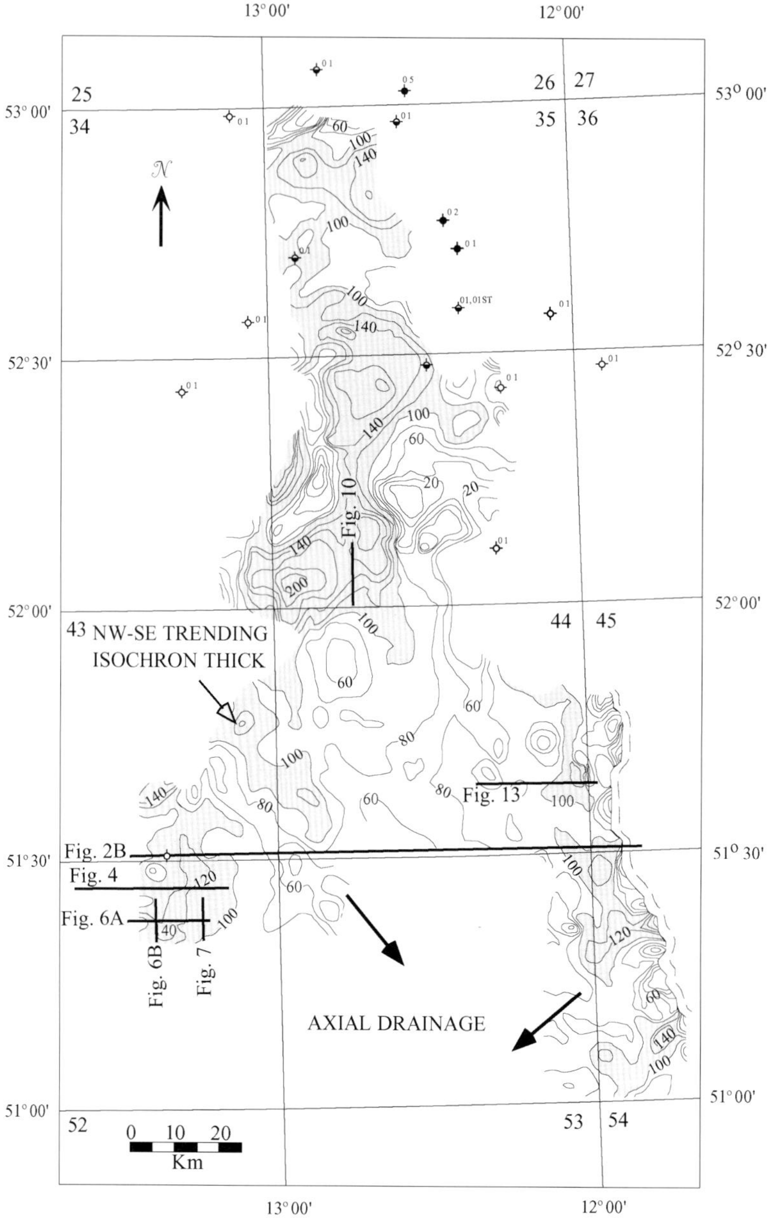

Fig. 8. Regional isochore map of Middle Eocene seismic unit A from the Porcupine Basin. Mapped interval is indicated in Figs. 4, 7, 10 and 13. Contour interval 20 ms TWT and shaded areas indicate isochore thicknesses. Principal sediment accumulation is in the north and west of the basin, with some smaller, isolated, lobate accumulations along the eastern margin. The pronounced NW–SE elongate trend in the SW of the basin, which is interpreted to represent canyon development from the western basin flank, should be noted. The dashed line represents the updip onlap of seismic unit A.

during Paleocene times. This wedge is onlapped by a seismic package of inferred Mid-Eocene age. Well 132/15-1, on the eastern flank of the Rockall Basin, recorded a highly tuffaceous, mud-dominated Paleocene succession with thin sandstones. The main sandy interval encountered was of Early Eocene (Ypresian) age, notably earlier than in the Porcupine Basin (Fig. 3).

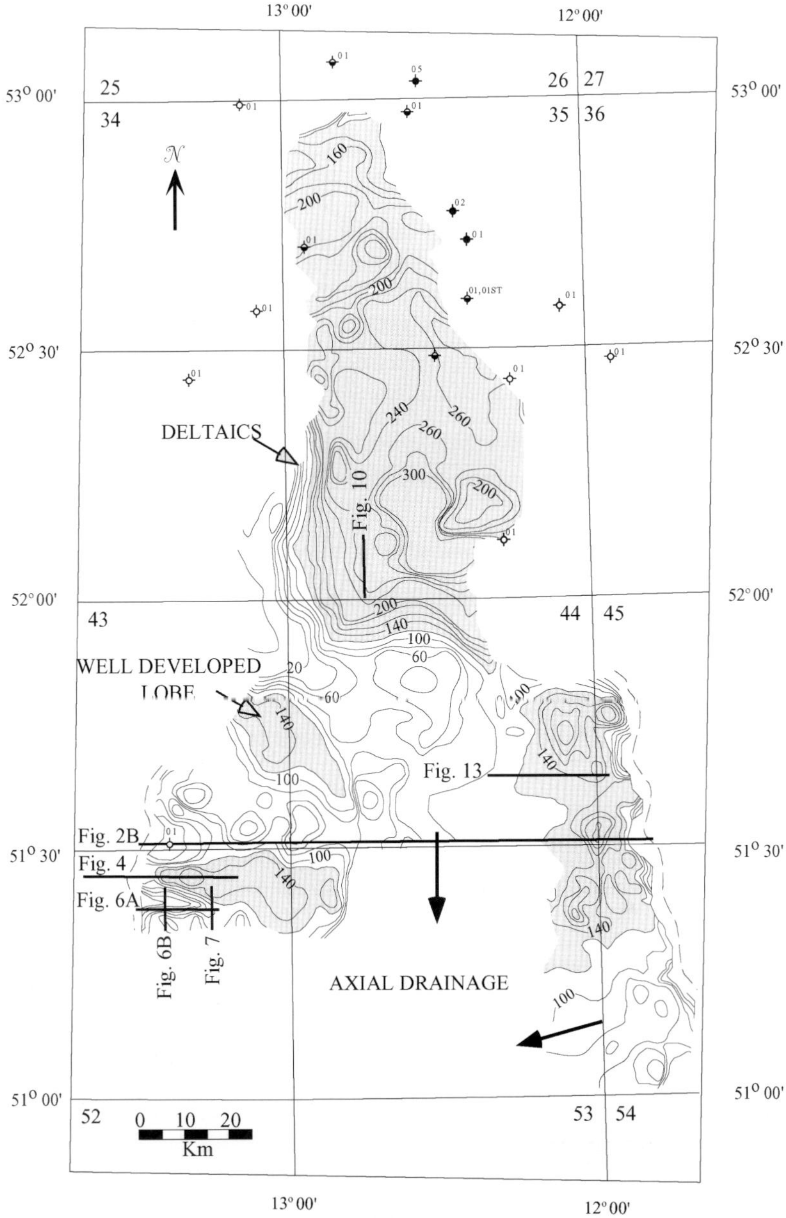

Fig. 9. Regional isochore map of the uppermost Middle Eocene to lower Upper Eocene seismic unit B in the Porcupine Basin. Mapped interval is indicated in Figs. 4, 7, 10 and 13. Contour interval is 20 ms TWT and shaded areas indicate isochore thicknesses. Principal sediment accumulation is in the north and east of the basin, with subsidiary, isolated, lobate depocentres in the SW and on the SE margin. Well-developed NW–SE and west–east isochore thicknesses are observed in the SW of the basin, indicating sediment transport from the Porcupine High to the west.

Comparison

Paleocene and Eocene sequences in both the Porcupine and Rockall basins are generally characterized by lower-amplitude, less continuous, reflections than the underlying Upper Cretaceous sequence. Some similarities in the stratal geometries of the Paleocene to Eocene

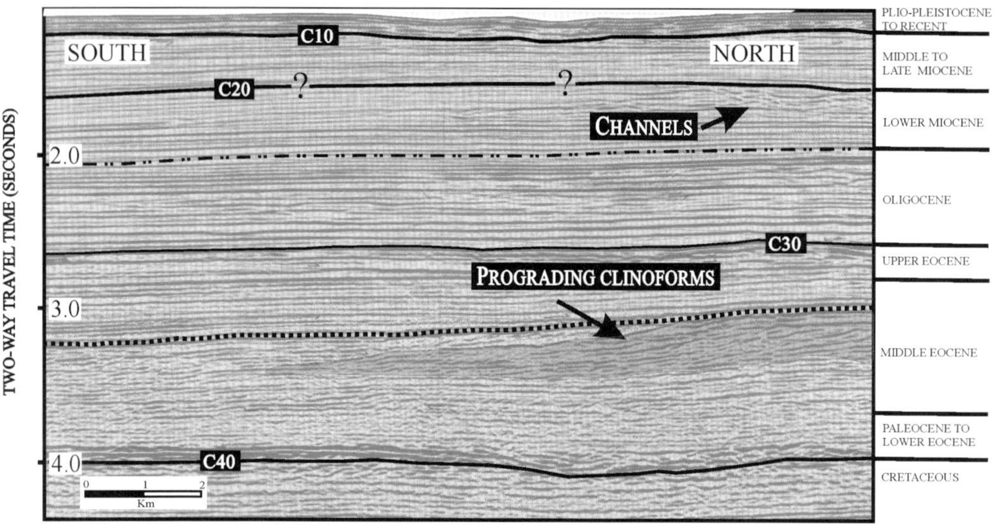

Fig. 10. Lower to Middle Eocene southward-prograding deltaic clinoforms, onlapped by Middle and Upper Eocene strata. Northerly-flooding of the Eocene deltaic sequences at the end of Eocene time is interpreted to signify an end to coarse clastic influx to the basin. The erosional truncation at the Early Pliocene (C10) unconformity should also be noted.

sequences are observed between the two basins. In particular, the onlapping relationship of the lowest Tertiary strata onto the high-amplitude basal reflection marking the top of the Cretaceous Chalk succession is markedly similar. Wedge-shaped basal units are recorded in both basins whereas the overlying seismic packages in both basins show a general upwards increase in reflection amplitude. From the seismic lines examined there is no evidence for the development of large-scale progradational clinoforms in the Eocene succession of the Rockall Basin.

Oligocene to Recent sequences

Porcupine Basin

The transition from Eocene to Oligocene strata in the Porcupine Basin is marked by a regional unconformity (C30), which is defined on seismic data by the onlap of continuous, moderate- to high-amplitude events (Fig. 4). A pronounced regional unconformity (C20) is also observed within the Miocene succession. A number of seismic sequences are contained in the interval bounded by the C30 and C20 events. One such

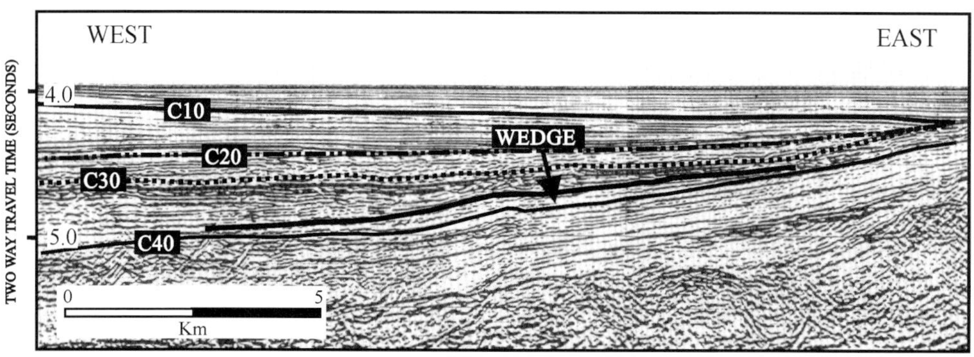

Fig. 11. Eastern flank of Rockall Basin, illustrating marginal clastic wedge of interpreted Paleocene to Eocene age.

sequence, the uppermost unit bounded by C20, is characterized by large-scale, mud-prone, slide and slump deposits (Fig. 4), which have previously been described by Moore & Shannon (1991). Well data (e.g. 43/13-1, 35/29-1) indicate that the overall Oligocene to Recent succession in the Porcupine Basin is mud dominated with occasional limestone stringers. This succession is manifest on seismic sections as consisting of very continuous, moderate- to high-amplitude seismic reflections. The isochore map for the Oligocene to Recent interval indicates that the greatest sediment accumulation occurred in the basin centre (Fig. 12).

The post-C20 stratigraphy of the basin is dominated by parallel reflections, of generally

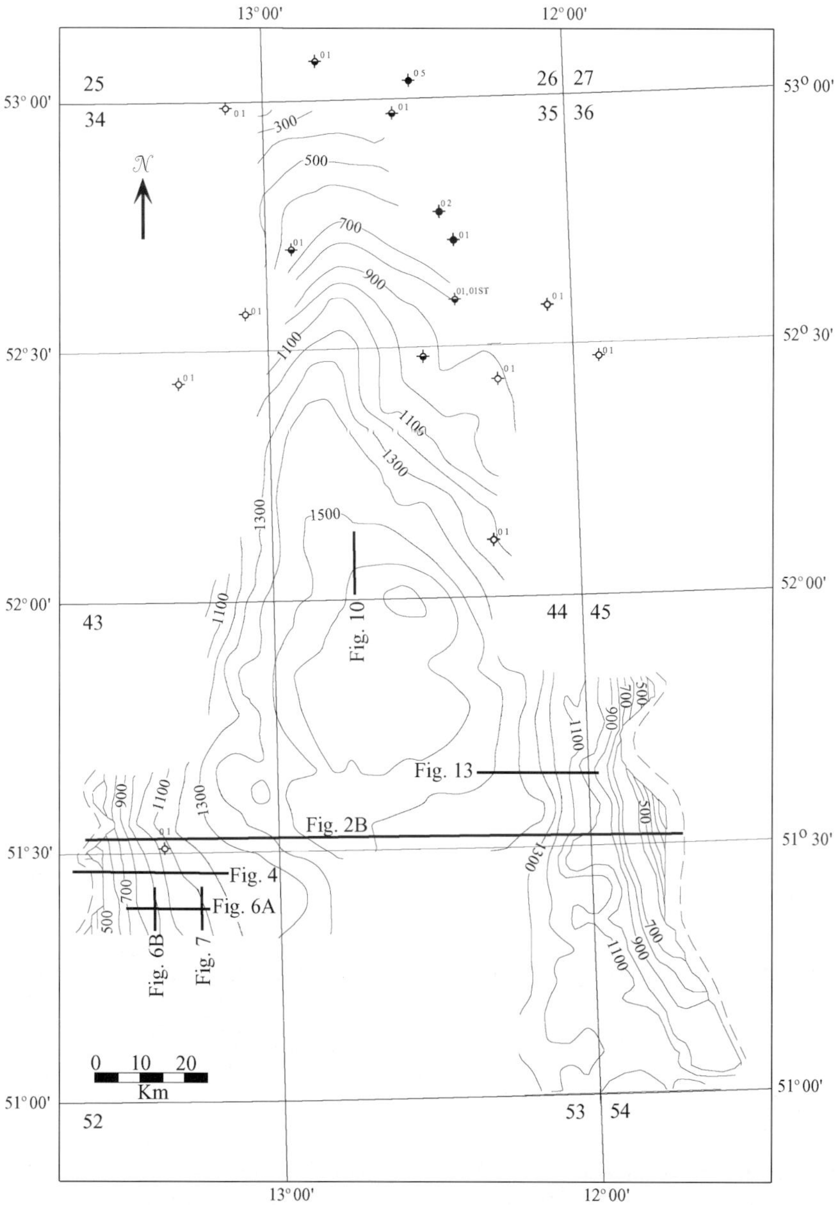

Fig. 12. Regional isochore map of the lowermost Oligocene to sea-bed interval, Porcupine Basin. Greatest sediment accumulation is in the basin centre, suggesting that the Porcupine Basin is not sediment starved at this time. Contour interval is 100 ms TWT.

moderate amplitude and sheeted geometry, which onlap the basin margin. The entire sequence thickens southwards towards the basin centre. A shale-dominated lithology is interpreted for this interval. Onlap of reflections onto the basin margin suggests a relative sea-level rise through Miocene to Recent times. In the SE of the basin the reflections display an aggradational trend and form a post-Middle Miocene, elongate sedimentary build-up (Fig. 13). This onlaps the basin margin and gives way basinwards to a more sheeted geometry. The build-up manifests both erosional and depositional features. The aggradational reflections are interpreted to represent contourite deposits, accreting onto the eastern flank of the basin, which developed as a response to changing current circulation conditions in the basin. The latest Early Miocene (C20) unconformity, which corresponds to a relative sea-level fall in the Haq *et al.* (1987) global sea-level curve, is noticeably erosive in the south of the basin. This erosional scouring may be related to intensifying bottom current activity at this time as noted elsewhere in the North Atlantic (Stoker 1997).

The youngest unconformity mapped in the Porcupine Basin is correlated with the designated C10 boundary of Stoker *et al.* (2001) in the Rockall Basin. The C10 to sea-bed seismic sequence has a largely unconformable base, and internally records high-amplitude reflections of varying continuity (Fig. 4). In addition, a number of buried carbonate mounds have been identified in this stratigraphic interval, and these mounds may have served as nucleation sites for the present-day carbonate mud mounds and bioherms observed at the sea bed in the Porcupine Basin (Henriet *et al.* 1999).

Rockall Basin

The transition to the interpreted Oligocene succession in the Rockall Basin is also marked by an unconformity surface (C30). It is present on the western margin of the southern Rockall Basin as a distinctive angular unconformity (Fig. 14). The Oligocene to Lower Miocene sequences in the Rockall Basin are characterized by moderate-amplitude reflections, which are more discontinuous than observed in the Porcupine Basin as a result of the presence of extensive synsedimentary faulting in this section.

A major sequence boundary is defined by the base of a package of high-amplitude reflections, which can be mapped throughout the southern Rockall Basin (Fig. 14). This high-amplitude event is dated to late Early Miocene time, from DSDP site 610, and is correlated with the C20 seismic marker of Stoker *et al.* (2001). The overlying Miocene sequence is dominated by high-amplitude, continuous reflections of basin-wide extent. In the west of the Rockall Basin the external geometry of the Lower–Middle Miocene to Recent succession displays a distinct mounded, elongate shape, which is known as the Feni Drift (Jones *et al.* 1970). The internal reflections display an aggradational pattern and onlap the western margin (Fig. 14). The main development of this drift occurred during the latest Early Miocene (C20) to Early Pliocene (C10) interval. This is approximately coeval with the relatively low-relief drift feature lying on the SE margin of the Porcupine Basin. The youngest unit recognized in the Rockall Basin comprises moderate-amplitude, continuous events with a wavy reflection configuration. The base of this unit is erosional, is dated as an intra-Early Pliocene unconformity, and correlates with the C10 reflection of Stoker *et al.* (2001). The wavy configuration has been interpreted to represent large sediment waves, which formed as a result of northerly-derived deep-water currents (Stoker 1997). There are no sediment waves of this scale associated with the sediment drift package in the Porcupine Basin, although sidescan sonar data, from the SW Porcupine Basin, records smaller-scale deep-water sediment waves (Kenyon *et al.* 1998).

Comparison

The seismic characters of the Oligocene to Recent successions in the Porcupine and Rockall basins display greater similarities than the observed stratal geometries for the Paleocene to Eocene strata. In general, the Oligocene to Recent sequences are characterized by seismic reflections displaying higher amplitude and greater continuity than observed in the Paleocene to Eocene sections. In both the Porcupine and Rockall basins the Oligocene to Recent succession generally thickens into the basin centre and displays onlap onto the basin flanks. Large-scale sediment drift features are recorded on the margins of both basins, capped by deep-water sediment waves, which are typically smaller in scale in the Porcupine Basin.

Discussion

A number of possible explanations are considered, to account for the observed Palaeogene stratigraphic record in the Porcupine and Rockall basins. Within both basins the well and seismic evidence points to relative sea-level falls during Late Paleocene (C40 event) and Late Eocene to Early Oligocene time (C30 event). Sea-level

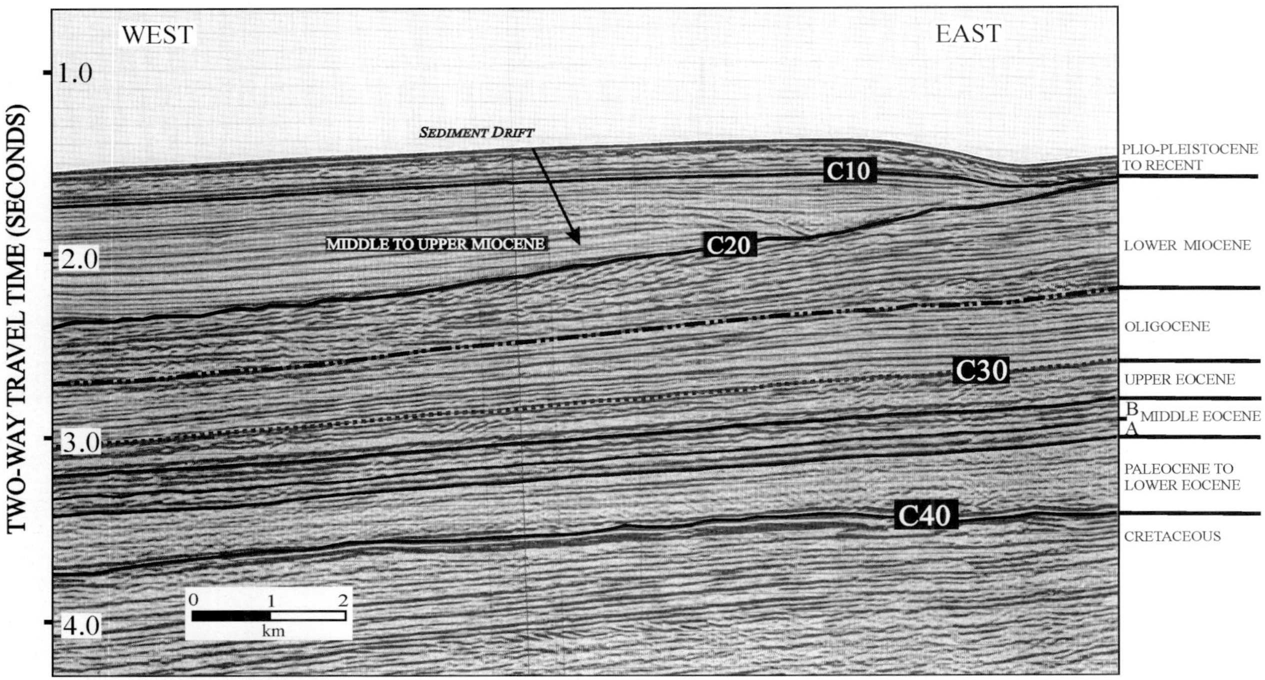

Fig. 13. Seismic profile from the eastern flank of the Porcupine Basin (see Figs. 1, 8, 9 and 12 for location). The strong aggradation of the Middle Miocene to Recent sediments onto the underlying unconformity surface (C20) should be noted. This is interpreted as a sediment drift feature accreting onto the eastern flank of the Porcupine Basin.

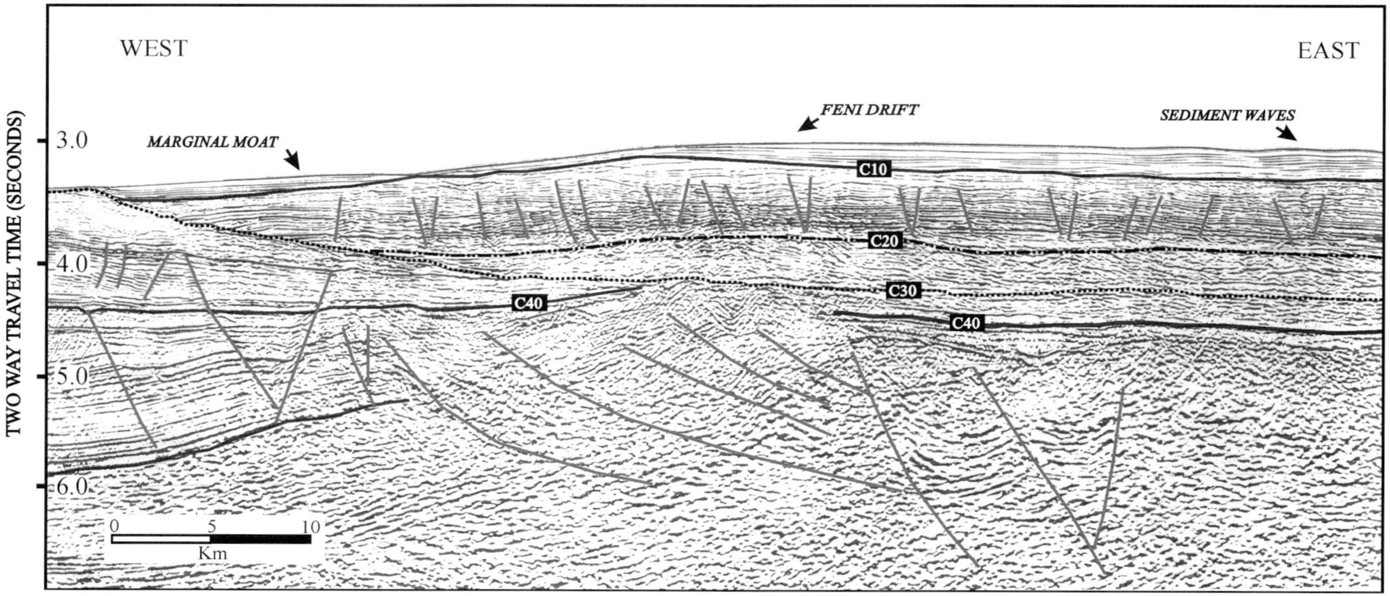

Fig. 14. Western flank of the Rockall Basin, illustrating build-up of the Feni Drift, and also highlighting the four regionally correlatable seismic markers (C40, C30, C20 and C10), identified in the basin (see Fig. 1 for location). Also to be noted is the abundance of synsedimentary faulting in the C20–C10 interval.

curves for Paleocene and Eocene time suggest that a number of eustatic sea-level falls occurred during this interval, including a major eustatic fall at *c.* 60 Ma (Vail *et al.* 1977; Haq *et al.* 1987). Masson *et al.* (1985) have described a locally developed Late Paleocene hiatus in the Goban Spur Basin and have correlated this hiatus with a eustatic controlled regressive event. However, the timing of this eustatic event precedes the onset of major coarse clastic input into the Porcupine and Rockall basins. It is therefore unlikely to be able to account *per se* for the observed Palaeogene stratigraphic record in the Porcupine and Rockall basins.

Lithospheric-scale thermal processes may have influenced the Early Tertiary tectonosedimentary evolution of these basins. Initiation of the Iceland plume in Paleocene times created over 1.5 km of uplift in the Faeroes region (Joppen & White 1990). This regional uplift resulted in the development of Early Tertiary deltaic and submarine fan complexes in the Faeroe–Shetland Basin (Ebdon *et al.* 1995). Although this may have influenced sedimentation in the north of the Rockall Basin (UK sector), its effects have been argued by Shannon *et al.* (1999) to have been significantly less in the Irish sector of Rockall Basin and also probably in the Porcupine Basin.

Substantial volumes of melt were generated by the Iceland plume (White 1988). Trapping of this melt in the lithosphere is suggested to have led to underplating beneath Scotland, which resulted in an estimated 1–2 km of uplift and denudation of the Scottish landmass (Brodie & White 1995). This led to the shedding of significant thicknesses of Paleocene sediments into the North Sea, with sediments also shed into the offshore basins to the west of Britain. However, the Paleocene succession of the Porcupine Basin is thin and relatively sand poor, and a thin, relatively sand-poor, Paleocene succession is recorded in well 132/15-1 of the Rockall Basin. Consequently, Paleocene regional uplift associated with the Iceland plume is thought to have contributed little to the generation of coarse clastic input into the basins of the western Irish margin.

O'Reilly *et al.* (1998) suggested the presence of a more localized Palaeogene thermal anomaly west of the Rockall High, close to the continental–oceanic margin, on the basis of the analysis of tectonic subsidence patterns and the regional gravity field. Those workers proposed that the Palaeogene submarine fan sedimentation in the Porcupine Basin was related to this anomaly, which was caused by magmatism and heating of the lithosphere. The influence of this anomaly waned by Mid- to Late Eocene times, resulting in significant subsidence throughout the region. O'Reilly *et al.* concluded that rapid subsidence at the end of the Eocene led to northerly flooding of the deltaic sequences in the Porcupine Basin, and resulted in the cessation of coarse clastic input into the basin at that time. White & Lovell (1997) suggested a causal link between pulses of clastic sedimentation in the North Sea and the activity of mantle plumes. Jones *et al.* (2001) proposed a model of plume-influenced transient uplift to explain the rapid increase in sediment flux into the Porcupine Basin during Eocene time.

Shannon *et al.* (1993) suggested that the sandy succession in the Eocene sequence of the Porcupine Basin may have resulted from ridge-push effects caused by sea-floor spreading and oceanic crustal development to the west of the Rockall region. The ridge-push forces are suggested to have induced intra-plate compressional stresses that resulted in uplift of the thick continental crust beneath the Porcupine High, thereby creating a potential provenance source area for clastic input to the Porcupine Basin. The age relationships between the sea-floor spreading to the west of the Rockall region and the timing of the regression are compatible. In addition, the rate of sea-floor spreading varies with time, resulting in consequential variations in the ridge-push stresses (Jones *et al.* 2001). This may account for the successive pulses of deltaic progradation and fan deposition in the basins. However, relatively little information is available on the precise intra-plate responses to sea-floor spreading and consequently it is difficult to quantify the likely effects of any ridge-push effects in the area. In summary, there are a number of potential influences on sediment deposition in the Porcupine and Rockall region at this time, but it would appear that North Atlantic sea-floor spreading plays the most influential role in the Palaeogene stratigraphic evolution of these basins.

The transition to Oligocene times is marked by a major unconformity (C30) in both basins. The change in seismic character at the base of the interpreted Oligocene sequence occurs in response to changing depositional styles in the basins. The sediments in the basin are interpreted to have been reworked and deposited in response to the onset of bottom current activity (Stoker 1997). The Porcupine and Rockall basins show enhanced subsidence rates at this time, which may be related to a change in the spreading rates in the North Atlantic as the opening of the Atlantic continued northwards (Doré *et al.* 1999). Also at this time a major climatic cooling took place, probably related to the formation of the

Antarctic Sea ice sheet, which resulted in a change in oceanic circulation patterns (Kennett *et al.* 1975).

The main accretion of the Feni Drift occurred from Early Miocene to Recent times (Stoker *et al.* 2001), when the drift migrated upslope onto the western flank of the basin. Marginal moats are also developed where stronger currents have scoured sediment from the base of the slope. From Mid-Miocene times the margins of the Rockall Basin were progressively onlapped as subsidence outpaced sedimentation in the region. Further north along the Hebridean margin, there was significant sediment input into the Rockall Basin during this Mid-Miocene to Holocene period, leading to the deposition of a number of major sediment fans such as the Barra Fan (Stoker 1997). The difference in drift morphology between the Rockall and Porcupine basins may be accounted for by the open nature of the Rockall Basin and the lack of topographic barriers to North Atlantic circulation. In contrast, the Porcupine Basin was a more enclosed oceanic system at that time, which may have resulted in weaker prevailing circulating water currents than in the Rockall Basin. Consequently, contourite build-ups are likely to be less well developed in the Porcupine Basin.

The C20 unconformity, identified in both basins, is also interpreted as being related to changing ocean current patterns. Stow & Holbrook (1984) argued that by Mid-Miocene times the Wyville-Thomson Ridge had subsided sufficiently to allow overflow of Arctic water into the Rockall Basin, resulting in enhanced bottom current activity leading to development of a basinwide unconformity at this time. In the Rockall Basin, however, the most easily mapped event of Early Miocene times is a diagenetic horizon of very high amplitude, related to the deposition of reworked volcanic ash in the Rockall Basin, which was subsequently altered to a smectite-rich horizon (Dolan 1986). Such a horizon is not evident in the Porcupine Basin at this time. Numerous wells in the Porcupine Basin, including well 36/16-1, encountered a number of glauconite-rich horizons in the Upper Eocene to Lower Miocene interval. Volcanogenic glauconite has been recorded in clays of Early and Late Cretaceous age in southern England and Northern Ireland, by Jeans *et al.* (1982). Those workers demonstrated the development of glauconite from volcanic debris of mafic composition, in contrast to smectite-rich clay layers, which developed from volcanic debris of acid or alkaline composition. Dolan (1986) attributed the smectite-rich horizon, recorded in the Rockall Basin, to volcanic ash derived from volcanic eruption in the Norwegian Basin. In contrast, it is suggested that the glauconite-rich formations in the Porcupine Basin are derived from volcanic debris of mafic origin, indicating a different source provenance to the altered (acidic or alkaline) volcanic glass of the Rockall Basin. The Early Pliocene C10 unconformity is interpreted as being related to the exhumation of Britain and Ireland (Japsen 1997), and the subsequent effects of glacial abrasion and unloading.

Conclusions

(1) Four main unconformities are correlated between the Porcupine and Rockall basins. These boundaries are interpreted as being of Paleocene (C40), latest Eocene–Early Oligocene (C30), latest Early Miocene (C20) and Early Pliocene (C10) age.

(2) Well data indicate that coarse clastic deposition occurred during Late Paleocene to Early Eocene times, in both the Porcupine and Rockall basins, with sand deposition continuing into Mid-Eocene time in the Porcupine Basin. Large-scale progradational clinoforms are not observed in the Rockall Basin as in the Porcupine Basin. The main Eocene sediment transport direction for the Porcupine Basin was from the north and west, with clastic input from the east in Rockall Basin.

(3) The Porcupine High served as a provenance area for sediment input into the Porcupine Basin during Eocene times. Regional lithospheric thermal effects and/or ridge-push stresses were probable contributing factors to the uplift of the Porcupine High at this time.

(4) The onset of Oligocene times was marked by a cessation of major clastic sediment supply to these basins, with sediment dispersal becoming more influenced by oceanographic circulation patterns. Oligocene and younger sequences in both basins are interpreted as representing open to deep marine deposition, with contourite drifts developed in response to the changing ocean currents.

(5) An unconformity of interpreted Plio-Pleistocene age (C10) is identified in both basins. This unconformity may represent the stratigraphic response to the latest Cenozoic exhumation of Britain and Ireland.

The support of Saga Petroleum ASA for Ph.D. studentship funding is acknowledged. The seismic sections illustrating the Rockall Basin are selected from a variety of non-exclusive proprietary surveys owned by Schlumberger Geco-Prakla. Permission to use these data is gratefully acknowledged. The authors also thank the referees for their helpful comments and suggestions.

References

BRODIE, J. & WHITE, N. 1995. The link between sedimentary basin inversion and igneous underplating. *In*: BUCHANAN, J.G. & BUCHANAN, P.G. (eds) *Basin Inversion*. Geological Society, London, Special Publications, **88**, 21–38.

BULL, J.M. & MASSON, D.G. 1996. The southern margin of the Rockall Plateau: stratigraphy, Tertiary volcanism and plate tectonic evolution. *Journal of the Geological Society, London*, **153**, 601–612.

CORFIELD, S., MURPHY, N. & PARKER, S. 1999. The structural and stratigraphic framework of the Irish Rockall Trough. *In*: FLEET, A.J. & BOLDY, S.A.R. (eds) *Petroleum Geology of Northwest Europe: Proceedings of the 5th Conference*. Geological Society, London, 407–420.

CROKER, P.F. & SHANNON, P.M. 1987. The evolution and hydrocarbon prospectivity of the Porcupine basin, offshore Ireland. *In*: BROOKS, J. & GLENNIE, K.W. (eds) *Petroleum Geology of North West Europe*. Graham & Trotman, London, 633–642.

DOLAN, J.F. 1986. The relationship between the 'R2' seismic reflector and a zone of abundant detrital and authigenic smectites, Deep Sea Drilling Project Hole 610, Rockall Plateau region, North Atlantic. *In*. RUDDIMAN, W.F., KIDD, R.B., THOMAS, E. *et al.* (eds) *Initial Reports of the Deep Sea Drilling Project, 94*. US Government Printing Office, Washington, DC, 1105–1109.

DORÕ, A.G., LUNDIN, E.R., JENSEN, L.N., BIRKELAND, Ä., ELIASSEN, P.E. & FICHLER, C. 1999. Principal tectonic events in the evolution of the northwest European Atlantic margin. *In*: FLEET, A.J. & BOLDY, S.A.R. (eds) *Petroleum Geology of Northwest Europe: Proceedings of the 5th Conference*. Geological Society, London, 41–61.

EBDON, C.C., GRANGER, P.J., JOHNSON, H.D. & EVANS, A.M. 1995. Early Tertiary evolution and sequence stratigraphy of the Faeroe–Shetland Basin: implications for hydrocarbon prospectivity. *In*: SCRUTTON, R.A., STOKER, M.S., SHIMMIELD, G.B. & TUDHOPE, A.W. (eds) *The Tectonics. Sedimentation and Palaeoceanography of the North Atlantic Region*. Geological Society, London, Special Publications, **90**, 51–69.

HAQ, B.U., HARDENBOL, J. & VAIL, P.R. 1987. Chronology of fluctuating sea levels since the Triassic. *Science*, **235**, 1156–1167.

HENRIET, J.P., DE MOL, B. & THE PORCUPINE–BELGICA '97 & '98 SHIPBOARD PARTIES 1999. Carbonate mounds, ring bioherms and past slope failures in the Porcupine Basin: prologue to a far-reaching story? *In*: CROKER, P.F. & O'LOUGHLIN, O. (eds) *The Petroleum Exploration of Ireland's Offshore Basins Conference, Dublin 29–30 April 1999, Extended Abstracts*. Petroleum Affairs Division, Department of the Marine and Natural Resources, Dublin, 90–92.

JAPSEN, P. 1997. Regional Neogene exhumation of Britain and the western North Sea. *Journal of the Geological Society, London*, **154**, 239–247.

JEANS, C.V., MERRIMAN, R.J., MITCHELL, J.G. & BLAND, D.J. 1982. Volcanic clays in the Cretaceous of Southern England and Northern Ireland. *Clay Minerals*, **17**, 105–156.

JOHNSTON, S., DORÉ, A.G. & SPENCER, A.M. 2001. The Mesozoic evolution of the southern North Atlantic region and its relationship to basin development in the south Porcupine Basin, offshore Ireland. *In*: SHANNON, P.M., HAUGHTON, P.D.W. & CORCORAN, D.V. (eds) *The Petroleum Exploration of Ireland's Offshore Basins*. Geological Society, London, Special Publications, **188**, 237–263.

JONES, E.J.W., EWING, J.I. & EITTREIM, S.L. 1970. Influences of Norwegian Sea Overflow Water on sedimentation in the northern North Atlantic and Labrador Sea. *Journal of Geophysical Research*, **75**, 1655–1680.

JONES, S.M., WHITE, N. & LOVELL, B. 2001. Cenozoic and Cretaceous transient uplift in the Porcupine Basin and its relationship to a mantle plume. *In*: SHANNON, P.M., HAUGHTON, P.D.W. & CORCORAN, D.V. (eds) *The Petroleum Exploration of Ireland's Offshore Basins*. Geological Society, London, Special Publications, **188**, 345–360.

JOPPEN, M. & WHITE, R.S. 1990. The structure and subsidence of Rockall Trough from two-ship seismic experiments. *Journal of Geophysical Research*, **95**, 19821 19837.

KENNETT, J.P., HOUTZ, R.E., ANDREWS, P.B. & 8 OTHERS 1975. Cenozoic palaeoceanography in the south west Pacific Ocean, Antarctic glaciation and the development of the circum Antarctic current. *In*: KENNETT, J.P., HOUTZ, R.E. *et al.* (eds) *Initial Reports of the Deep Sea Drilling Project, 29*. Washington, DC, US Government Printing Office, 1155–1169.

KENYON, N.H., IVANOV, M.K. & AKHMETZHANOV, A.M. 1998. *Cold Water Carbonate Mounds and Sediment Transport on the Northeast Atlantic Margin*. Intergovernmental Oceanographic Commission Technical Series, **52**.

KNOTT, S.D., BURCHELL, M.T., JOLLEY, E.J. & FRASER, A.J. 1993. Mesozoic to Cenozoic plate reconstructions of the North Atlantic and hydrocarbon plays of the Atlantic margins. *In*: PARKER, J.R. (ed.) *Petroleum Geology of Northwest Europe: Proceedings of the 4th Conference*. Geological Society, London, 953–974.

MASSON, D.G. & KIDD, R.B. 1986. Revised Tertiary seismic stratigraphy of the southern Rockall Trough. *In*: RUDDIMAN, W.F., KIDD, R.B., THOMAS, E. *et al.* (eds) *Initial Reports of the Deep Sea Drilling Project, 94*. US Government Printing Office, Washington, DC, 1117–1126.

MASSON, D.G., MONTADERT, L. & SCRUTTON, R.A. 1985. Regional geology of the Goban Spur continental margin. *In*: DE GRACIANSKY, P.C. & POAG, C.W. (eds) *Initial Reports of the Deep Sea Drilling Project, 80*. US Government Printing Office, Washington, DC, 1115–1139.

MILLER, K.G. & TUCHOLKE, B.E. 1983. Development of Cenozoic abyssal circulation south of the Greenland–Scotland Ridge. *In*: BOTT, M.H.P.,

SAXOV, S., TALWANI, M. & THEIDE, J. (eds) *Structure and Development of the Greenland–Scotland Ridge*. NATO Conference Series 4: Marine Sciences, **8**, 549–589.

MOORE, J.G. & SHANNON, P.M. 1991. Late Tertiary slump structures in the Porcupine Basin, offshore Ireland. *Marine and Petroleum Geology*, **8**, 184–197.

MOORE, J.G. & SHANNON, P.M. 1992. Palaeocene–Eocene deltaic sedimentation, Porcupine Basin, offshore Ireland: a sequence stratigraphic approach. *First Break*, **10**, 461–469.

MOORE, J.G. & SHANNON, P.M. 1995. The Cretaceous succession in the Porcupine Basin, offshore Ireland: facies distribution and hydrocarbon potential. *In*: CROKER, P.F. & SHANNON, P.M. (eds) *The Petroleum Geology of Ireland's Offshore Basins*. Geological Society, London, Special Publications, **93**, 345–370.

MUSGROVE, F.W. & MITCHENER, B. 1996. Analysis of the pre-Tertiary rifting history of the Rockall Trough, west of Britain. *Petroleum Geoscience*, **2**, 353–360.

O'REILLY, B.M., READMAN, P.W. & HAUSER, F. 1998. Lithospheric structure across the western Eurasian plate from a wide-angle seismic and gravity study: evidence for a regional thermal anomaly. *Earth and Planetary Science Letters*, **156**, 275–280.

ROBERTS, D.G. 1975. Marine geology of the Rockall Plateau and Trough. *Philosophical Transactions of the Royal Society of London, Series A*, **278**, 447–509.

ROBERTS, D.G., THOMPSON, M., MICHENER, B., HOSSACK, J., CARMICHAEL, S. & BJØRNSETH, H.-M. 1999. Palaeozoic to Tertiary rift and basin dynamics: mid-Norway to the Bay of Biscay—a new context for hydrocarbon prospectivity in the deep water frontier. *In*: FLEET, A.J. & BOLDY, S.A.R. (eds) *Petroleum Geology of Northwest Europe: Proceedings of the 5th Conference*. Geological Society, London, 7–40.

SHANNON, P.M. 1992. Early Tertiary submarine fan deposits in the Porcupine Basin, offshore Ireland. *In*: PARNELL, J. (ed.) *Basins on the Atlantic Seaboard: Petroleum Geology, Sedimentology and Basin Evolution*. Geological Society, London, Special Publications, **62**, 351–373.

SHANNON, P.M., JACOB, A.W.B., O'REILLY, B.M., HAUSER, F., READMAN, P.W. & MAKRIS, J. 1999. Structural setting, geological development and basin modelling in the Rockall Trough. *In*: FLEET, A.J. & BOLDY, S.A.R. (eds) *Petroleum Geology of Northwest Europe: Proceedings of the 5th Conference*. Geological Society, London, 421–431.

SHANNON, P.M., MOORE, J.G., JACOB, A.W.B. & MAKRIS, J. 1993. Cretaceous and Tertiary basin development west of Ireland. *In*: PARKER, J.R. (ed.) *Petroleum Geology of Northwest Europe: Proceedings of the 4th Conference*. Geological Society, London, 1057–1066.

SHANNON, P.M., WILLIAMS, B.P.J. & SINCLAIR, I.K. 1995. Tectonic controls on Upper Jurassic to Lower Cretaceous reservoir architecture in the Jeanne d'Arc Basin, with some comparisons from the Porcupine and Moray Firth Basins. *In*: CROKER, P.F. & SHANNON, P.M. (eds) *The Petroleum Geology of Ireland's Offshore Basins*. Geological Society, London, Special Publications, **93**, 467–490.

SINCLAIR, I.K., SHANNON, P.M., WILLIAMS, B.P.J., HARKER, S.D. & MOORE, J.G. 1994. Tectonic controls on sedimentary evolution of three North Atlantic borderland Mesozoic basins. *Basin Research*, **6**, 193–217.

STOKER, M.S. 1997. Mid–Late Cenozoic sedimentation on the continental margin off NW Britain. *Journal of the Geological Society, London*, **154**, 509–515.

STOKER, M.S., VAN WEERING, T.C.E. & SVAERDBORG, T. 2001. A Mid- to Late Cenozoic tectonostratigraphic framework for the Rockall Trough. *In*: SHANNON, P.M., HAUGHTON, P.D.W. & CORCORAN, D.V. (eds) *The Petroleum Exploration of Ireland's Offshore Basins*. Geological Society, London, Special Publications, **188**, 411–438.

STOW, D.A.V. & HOLBROOK, J.A. 1984. Hatton Drift contourites, northeast Atlantic. *In*: ROBERTS, D.G., SCHNITKER, D. *et al.* (eds) *Initial Reports of the Deep Sea Drilling Project,* 81. US Government Printing Office, Washington, DC, 1115–1139.

TATE, M.P. 1993. Structural framework and tectonostratigraphic evolution of the Porcupine Seabight Basin, offshore western Ireland. *Marine and Petroleum Geology*, **10**, 95–123.

UNNITHAN, V., SHANNON, P.M., MCGRANE, K., READMAN, P.W., JACOB, A.W.B., KEARY, R. & KENYON, N.H. 2001. Slope instability and sediment redistribution in the Rockall Trough: constraints from GLORIA. *In*: SHANNON, P.M., HAUGHTON, P.D.W. & CORCORAN, D.V. (eds) *The Petroleum Exploration of Ireland's Offshore Basins*. Geological Society, London, Special Publications, **188**, 439–454.

VAIL, P.R., MITCHUM, R.M. & THOMPSON, S. 1977. Seismic stratigraphy and global changes of sea-level, part 4: Global cycles of relative changes of sea-level. *In*: PAYTON, C.E. (ed.) *Seismic Stratigraphy: Applications to Hydrocarbon Exploration*. American Association of Petroleum Geologists, Memoirs, **26**, 83–97.

WALSH, A., KNAG, G., MORRIS, M., QUINQUIS, H., TRICKER, P., BIRD, C. & BOWER, S. 1999. Petroleum geology of the Irish Rockall Trough—a frontier challenge. *In*: FLEET, A.J. & BOLDY, S.A.R. (eds) *Petroleum Geology of Northwest Europe: Proceedings of the 5th Conference*. Geological Society, London, 433–444.

WHITE, R.S. 1988. A hot-spot model for early Tertiary Volcanism in the N. Atlantic. *In*: MORTON, A.C. & PARSON, L.M. (eds) *Early Tertiary Volcanism and the Opening of the NE Atlantic*. Geological Society, London, Special Publications, **39**, 3–13.

WHITE, N. & LOVELL, B. 1997. Measuring the pulse of a plume with the sedimentary record. *Nature*, **387**, 888–891.

Cenozoic and Cretaceous transient uplift in the Porcupine Basin and its relationship to a mantle plume

STEPHEN M. JONES, NICKY WHITE & BRYAN LOVELL

Bullard Laboratories, Madingley Rise, Madingley Road, Cambridge CB3 0EZ, UK (e-mail: jones@esc.cam.ac.uk)

Abstract: The Mesozoic and Cenozoic history of the Porcupine Basin may be broadly summarized as a Jurassic synrift phase, followed by Cretaceous and Cenozoic post-rift subsidence. Two periods, Early Cretaceous and Early Eocene times, do not fit the simple pattern of post-rift subsidence and are characterized by increased sedimentation. We recognize distinctive sedimentological responses to the basin flanks being either exposed or submerged, and infer that transient regional uplift caused the Early Eocene event. Modelling subsidence histories of wells and of the Porcupine Bank allows quantification of the magnitude and timing of anomalous uplift and subsidence. Transient uplift of 300–600 m occurred at the Paleocene–Eocene boundary, followed by subsidence of 500–800 m after Early Eocene time, over a period with a minimum length of 25 Ma and a maximum of 55 Ma. Renewed rifting is unlikely to be responsible for the Paleogene subsidence because it cannot account for the preceding uplift, and significant normal faults of Paleogene age are absent. A Paleogene uplift–subsidence cycle has also been noted in the basins surrounding Scotland and along Hatton continental margin. One way to explain regional subsidence between Eocene time and the present is that the European plate moved off the topographic swell above the Iceland plume following continental separation between Greenland and Europe in Early Eocene time. Another possibility is that an anomalously hot layer *c*. 50 km thick was emplaced beneath the entire region just before the onset of sea-floor spreading in Early Eocene time and was then dissipated by convection following continental separation. A Cretaceous transient uplift–subsidence cycle that shares many similarities with the Paleogene cycle is also recognized. Immediately following Late Jurassic rifting, 200–700 m transient uplift occurred in Early Cretaceous time, followed by 0–500 m subsidence coeval with the onset of sea-floor spreading at the Goban Spur margin. The Cretaceous uplift–subsidence cycle might also be caused by anomalously hot mantle.

It is generally accepted that a large region surrounding and including Britain and Ireland experienced uplift and exhumation during Paleogene time. This area is part of a larger region, characterized by voluminous Paleogene igneous activity, that extends westwards to western Greenland and northwards to the Lofoten margin, off the coast of Norway (White & McKenzie 1989). It is generally agreed that uplift and volcanism were related to activity of the Iceland plume but the shape and size of the plume and the relative significance of the mechanisms that caused the uplift remain unclear. The principal purpose of this study is to present evidence that constrains the magnitude and timing of Paleogene epeirogenic uplift and subsidence in and around the Porcupine Basin, offshore west of Ireland (Fig. 1), and discuss possible causes of this uplift.

First, evidence for transient and permanent uplift around Britain and Ireland is reviewed. Structural and sedimentological evidence is then described, which suggests that the Paleogene development of the Porcupine Basin cannot be explained by a combination of lithospheric stretching and global sea-level change. The geometry of the Porcupine Basin promoted deltaic sedimentation when the basin margins were emergent, allowing recognition of relative sea-level lowstands. Deltaic sedimentary rocks also provide good control in subsidence modelling. Hence, the Porcupine Basin is a particularly favourable location for research into possible Paleogene transient epeirogenic uplift. Subsidence analysis techniques that allow us to isolate the transient epeirogenic subsidence history are described and applied. Finally, some mechanisms that can explain the epeirogenic events revealed by our analysis are proposed.

From: SHANNON, P.M., HAUGHTON, P.D.W. & CORCORAN, D.V. (eds) 2001. *The Petroleum Exploration of Ireland's Offshore Basins*. Geological Society, London, Special Publications, **188**, 345–360.

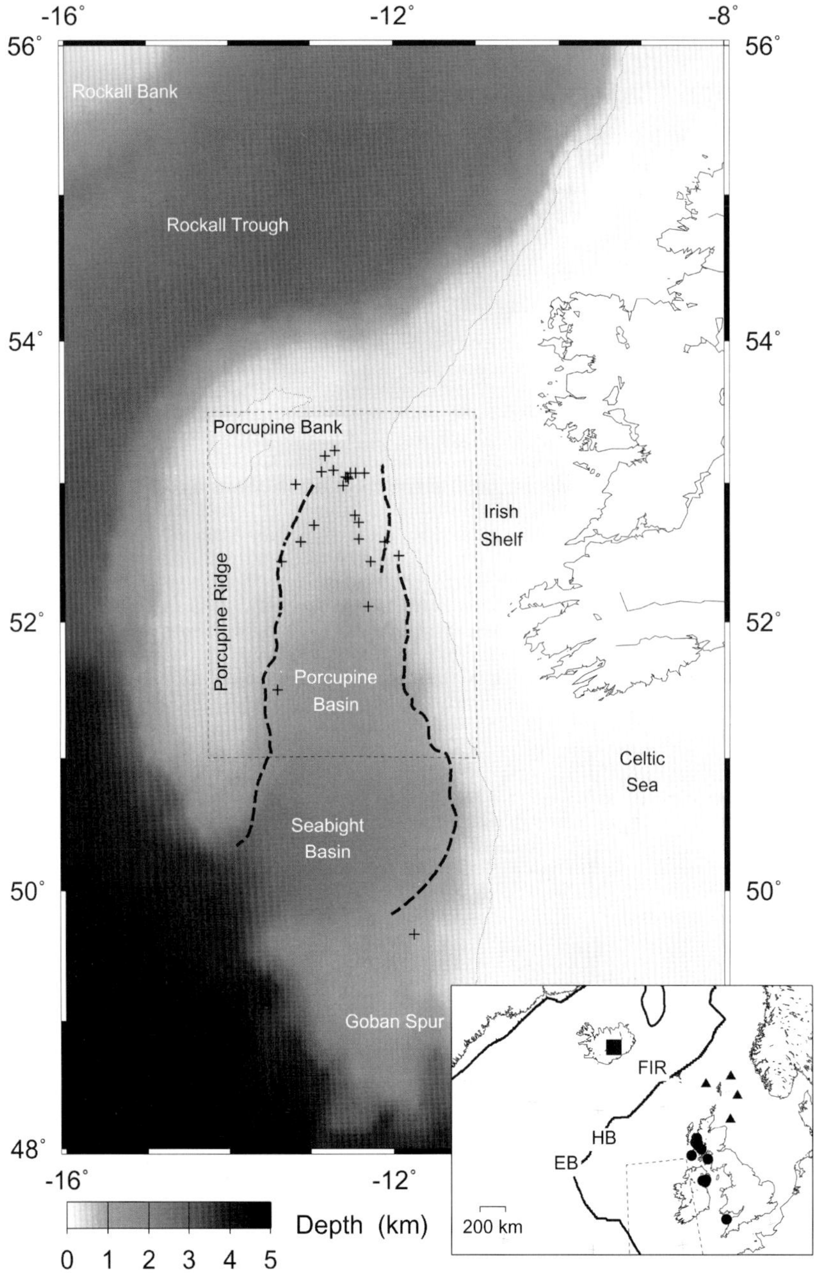

Fig. 1. Location map showing the physical features discussed in the text. The 200 m bathymetric contour is shown. Bold dashed lines delimit the Porcupine Basin. Dashed box shows the extent of the isopachs in Figure 4. +, wells. Inset map shows the regional setting of Porcupine Basin, together with specific sites with evidence for Paleocene–Eocene uplift. Evidence for transient uplift: ▲, basins where Late Paleocene–Early Eocene uplift has been quantified using subsidence analysis (Nadin *et al.* 1997). Evidence for permanent uplift: ●, central igneous complexes with evidence for Paleocene igneous underplating of the crust; EB and HB, Edoras Bank and Hatton Bank underplated continental margins; FIR, Faroe–Iceland Ridge. Bold lines mark continent–ocean boundaries; ■, present centre of the Iceland plume.

Paleogene transient and permanent uplift around Britain and Ireland

As yet, little effort has been made to quantify epeirogenic uplift in the Porcupine Basin. However, many studies that quantify Paleogene epeirogenic uplift related to the Iceland plume elsewhere across the NE Atlantic continental margin provide indirect evidence that the Porcupine Basin was affected by Paleogene events. In this section, published evidence for Paleogene transient and permanent uplift around Britain, Ireland and the adjacent continental margin is discussed, so that events in the Porcupine Basin described later may be seen in their regional context. We also point out how these regional events might be represented on a typical subsidence plot, to facilitate interpretation of the Porcupine Basin dataset presented later (Fig. 2).

Two types of epeirogenic uplift are expected when anomalously hot mantle underlies the lithosphere. Uplift results when the convecting mantle causes normal stresses to act on the base of the lithosphere, and when heat is added to lithosphere overlying hot asthenosphere by conduction; such uplift is transient and disappears when the convection wanes and the thermal anomaly dissipates. Permanent uplift can also occur if adiabatic decompression of hot asthenosphere causes melting and the igneous material is injected into the crust. There is good evidence that both permanent and transient uplift have affected Britain, Ireland and the adjacent continental shelf during Cenozoic time.

Nadin *et al.* (1997) argued that Cretaceous–Cenozoic sedimentation patterns in the northern North Sea and Faroe–Shetland basins can be explained by transient uplift that initiated in Paleocene time and decayed through the remainder of the Paleogene period (such uplift might appear on a subsidence plot as ABC, Fig. 2a). They assumed that the timing of rifting in each basin is known, so post-rift subsidence can be forward modelled using the observed synrift subsidence history. Post-rift marker horizons with sedimentologically well-constrained water depths are then compared with the anticipated post-rift subsidence curve to measure the magnitude of transient uplift through time (compare ABC and AC, Fig. 2a). The most important marker horizons are delta-top coals of Balder age (Paleocene–Eocene boundary). An Early Eocene delta system with lignites is well known from the Porcupine Basin (Moore & Shannon 1992), making this an important area for estimating transient uplift.

Joy (1992) and Hall & White (1994) documented a phase of anomalous Paleogene subsidence in the North Sea and Porcupine basins. Their phase of anomalous subsidence in the North Sea Basin corresponds to decay of the transient uplift event of Nadin *et al.* (1997) (e.g. BC, Fig. 2a). Anomalous Paleogene subsidence

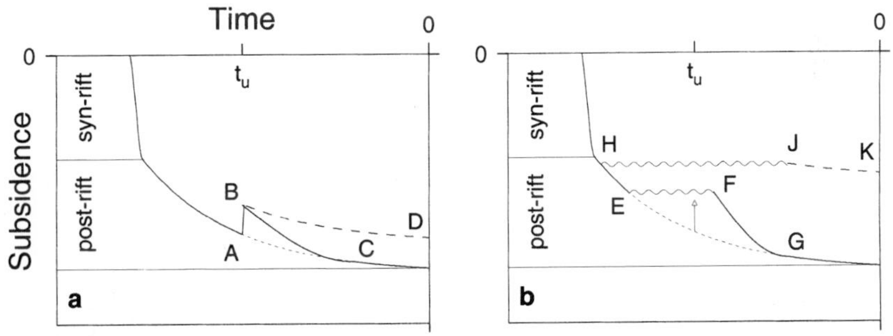

Fig. 2. Schematic illustrations of the appearance on a subsidence plot of an extensional sedimentary basin affected by transient or permanent uplift. Horizontal axes are time with the present at the right edge, vertical axes are water-loaded basement subsidence. A single, instantaneous phase of anomalous uplift occurs at time t_u in both examples; the path that basement subsidence would have followed in the absence of uplift is marked by a fine dashed line. In (**a**) sea depth just before uplift is greater than the amount of uplift AB, so after uplift the sediment surface remains below sea level and no erosion takes place. If the uplift is transient then it decays through time until it vanishes at C; the subsidence curve has the form ABC. If the uplift is permanent then it persists to the present and the subsidence curve has the form ABD. In (**b**) initial uplift shown by the fine arrow raises the sediment surface above sea level and erosion occurs. If the uplift is transient then the erosion surface eventually sinks below sea level at F after the sedimentary column has been eroded back to E, leaving an unconformity EF; the final subsidence curve is EFG. If uplift is permanent then the erosion surface remains above sea level for longer; the final subsidence curve is HJK.

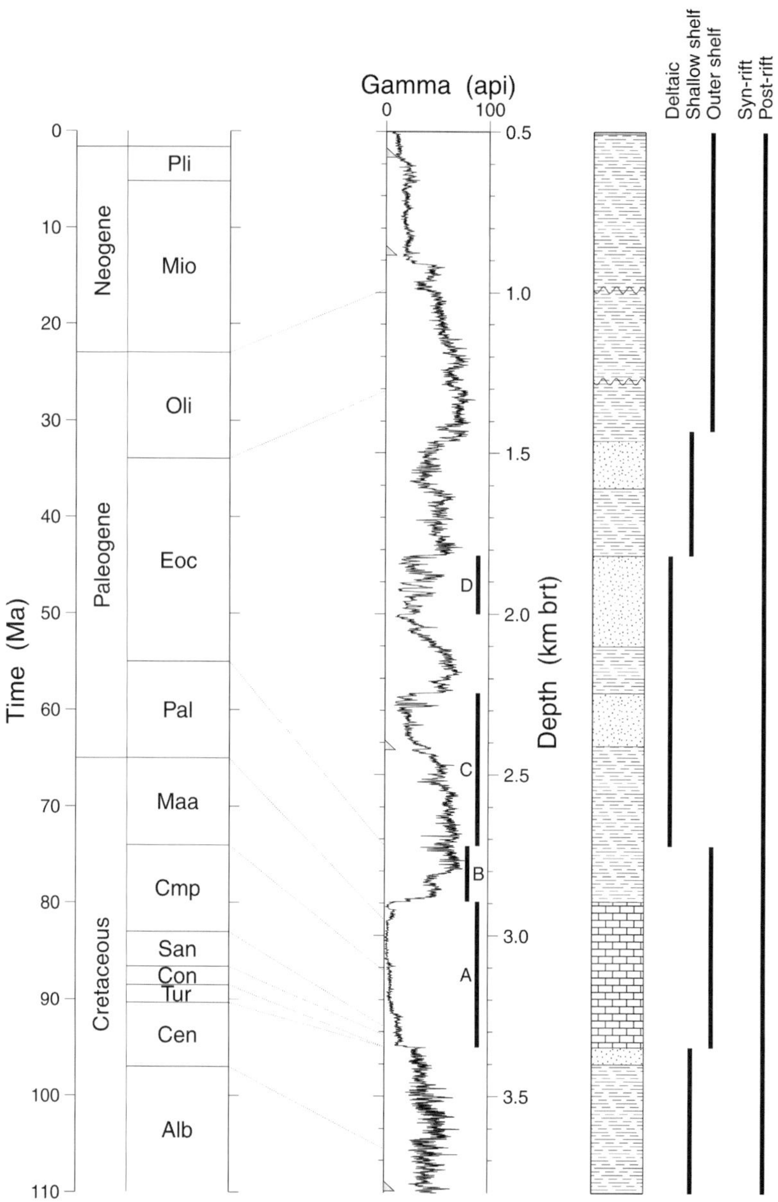

Fig. 3. Chronostratigraphy and lithostratigraphy of Porcupine Basin during part of the post-rift phase, displayed using a representative well (Shell 35/13-1). Black bars next to the gamma log show the sediment sequences plotted as isopachs in Figure 4. Tie-lines between chronostratigraphy and lithostratigraphy illustrate changes in sedimentation rate. The dramatic increase in sedimentation rate in Eocene time should be noted. Wavy lines on the stratigraphic column mark unconformities in deep-water Oligocene and Neogene sediments.

in the Porcupine Basin might be explained in the same manner.

The absence of the full complement of post-rift sedimentary rocks from Mesozoic extensional basins across Britain and Ireland suggests substantial exhumation. Jurassic synrift and older basement rocks are still close to sea level today, suggesting that the uplift that triggered the exhumation is permanent (e.g. HJK, Fig. 2b). Igneous underplating of the crust related to Paleogene igneous activity is thought to be an important cause of this uplift (Brodie & White

1995). Evidence for underplating beneath the Cenozoic igneous centres of Britain and Ireland is principally petrological. Many lavas have undergone fractional crystallization at a pressure equivalent to Moho depth, so a residue must remain near the base of the crust (Thompson 1974). The thickness of melt required to generate the observed trace element geochemistry is greater than the thickness of igneous rock present at the surface today (Brodie & White 1995).

There is also evidence for permanent uplift at the continental margin to the west and north of the Porcupine Basin (Fig. 1). Wide-angle seismic profiles indicate a pod of sub-crustal high-velocity material at Hatton Bank and Edoras Bank (Barton & White 1997). This material is interpreted as an igneous underplated body that formed during continental break-up above asthenosphere 150–200 °C hotter than normal. Seismic data have also been used to image underplated continental crust adjacent to the Faroe–Iceland Ridge, which is thought to have formed above the hottest part of the developing Iceland plume (Smallwood *et al.* 1999).

Although direct evidence for both the existence and timing of emplacement of igneous underplating is areally restricted at present, intrusive and extrusive Paleogene volcanism is widespread across the entire continental shelf to the west of Britain and Ireland, including the Porcupine Basin (Tate & Dobson 1988; Ritchie & Hitchen 1996). Addition of igneous material to the crust may have caused permanent uplift of the Porcupine Basin and adjacent areas. If permanent uplift affected a highly stretched region such as the Porcupine Basin, a subsidence history like ABD (Fig. 2a) would be expected.

The sedimentary record in a basin contains an indirect, yet quantifiable, record of regional uplift because sediment flux through time can be related to uplift and denudation of the sediment source regions, where their extent and location are known. Basins surrounding Scotland experienced a rapid and episodic influx of coarse, clastic sediments during Paleogene time in response to the regional uplift and denudation (White & Lovell 1997). The Porcupine Basin is an important element in understanding the link between uplift and erosion of a sediment source and the resulting offshore sediment accumulation because a sediment source area to the west of the basin can be delimited and clearly related to sediment fans that prograde into the basin.

Studies of apatite fission-track data from Ireland and East Greenland suggest that exhumation of North Atlantic continental margins may have continued from Eocene time to the present day (Clift *et al.* 1998; Green *et al.* 2001). Unconformities exist in the Oligocene and Neogene successions of the Porcupine Basin but these seem more directly linked to the scouring action of bottom currents (Stoker 1997; Fig. 3). The Cenozoic denudation history of Ireland may be a protracted one but the effects of possible Oligocene and Neogene events are relatively poorly understood and the mechanisms that might have caused them are even less clear. Here we focus on quantifying those Paleocene and Eocene events that can be most clearly associated with igneous activity and the Iceland plume.

In summary, several Paleogene uplift and exhumation events affecting Britain, Ireland and the adjacent continental shelf have been proposed. The acme of onshore igneous activity of the British Tertiary Igneous Province occurred during chron 26r (61–58 Ma), presumably coeval with permanent uplift resulting from igneous underplating of the crust (Saunders *et al.* 1997; Hamilton *et al.* 1998). Sedimentation rates in the North Sea Basin reached a maximum at roughly the same time, i.e. 59 Ma (White & Lovell 1997). Continental break-up between Greenland and Europe occurred slightly later, during chron 24r at *c.* 55 Ma, with associated igneous intrusive and extrusive activity and uplift of the continental margins (Barton & White 1997). Maximum transient uplift in the northern North Sea and Faroe–Shetland basins is held to occur at this time (Nadin *et al.* 1997). Anomalously rapid subsidence then ensued during Eocene and Oligocene times (Joy 1992; Hall & White 1994; Nadin *et al.* 1997).

Structural and sedimentological evidence for transient uplift

The Porcupine Basin is a north–south-trending, extensional sedimentary basin *c.* 400 km long and *c.* 150 km wide (Fig. 1). It lies off the western coast of Ireland, separated from the Rockall Trough and its underlying sedimentary basin by the basement-cored bathymetric features of the Porcupine Ridge to the west and the Porcupine Bank to the north. Well and seismic data indicate that several episodes of rifting occurred through Triassic and Jurassic time, culminating in a major rift event during Late Jurassic time (Croker & Shannon 1987; Tate 1993). West–east seismic lines show rotated fault blocks with associated Mesozoic synrift sediments covered by up to 7 km of Cretaceous and Cenozoic post-rift sediments. These are essentially unfaulted and onlap the basin margins to form a classic ‘steer’s head’ profile (Croker & Klemperer

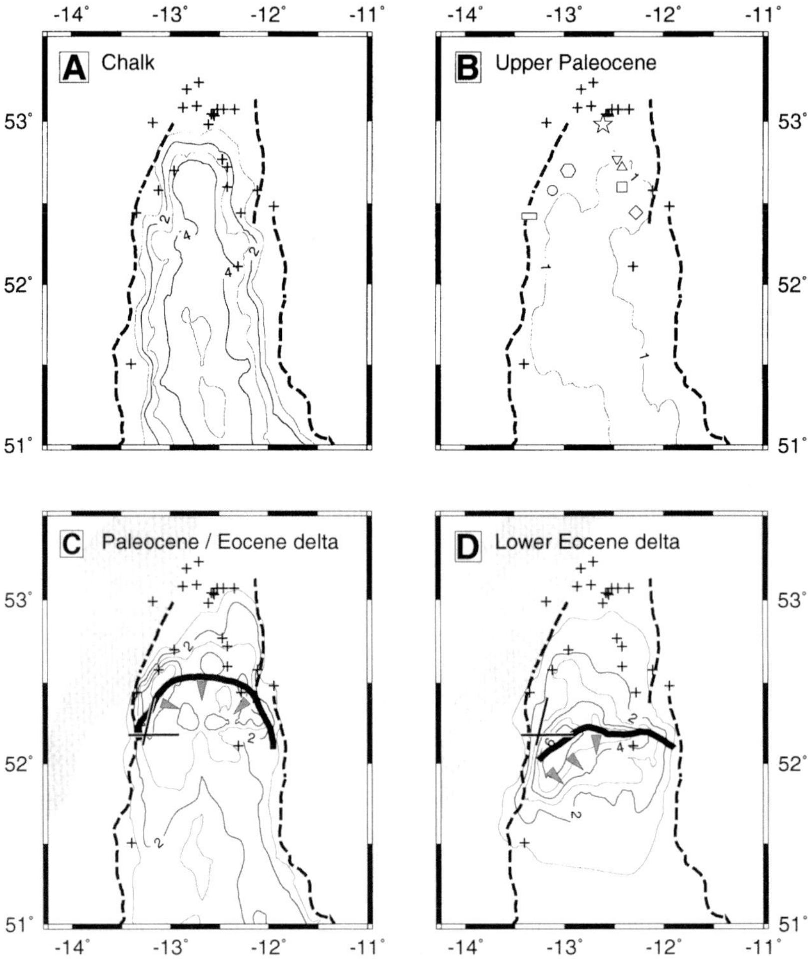

Fig. 4. Solid-grain sediment isopachs, contour interval 100 m. The isopachs were constructed using a grid of seismic data that encompasses the entire area of the maps in this figure. Porosity was removed using the standard porosity–depth relationship (Sclater & Christie 1980) so that sediment thicknesses, volumes and sedimentation rates for sequences of various ages can be compared realistically. (**a**) Chalk (Late Cretaceous–Early Paleocene time); (**b**) Late Paleocene; (**c**) first deltaic unit (<1 Ma at Paleocene–Eocene boundary); (**d**) final deltaic unit (Early Eocene time, *c.* 51–49 Ma). Ages of deltaic sequences are from our own analysis of the ranges of planktonic Foraminifera and calcareous nannoplankton in six wells, correlated with the Cenozoic time scale of Berggren *et al.* (1995). Bold black line shows the break in slope between delta topsets and foresets seen on seismic data; this is taken to indicate the position of the shoreline. Grey shading delimits the Porcupine Bank sediment source area, based on present-day bathymetry. Grey arrows indicate direction of foreset progradation, derived from seismic data. Two fine, straight lines on the western margin mark the seismic line drawings in Figure 5. +, wells. Symbols on Isopach B denote the wells used for subsidence modelling in Figure 7.

1989). Subsidence modelling of well and seismic data is consistent with a principal rifting event in Mid–Late Jurassic time (180–145 Ma) with stretching factors of $\beta \approx 1.2$ in the north increasing to $\beta > 6$ in the south (Tate *et al.* 1993).

The present-day bathymetric expression of the Porcupine Basin is a southward-dipping trough, enclosed to the west, north and east by basement ridges (Fig. 1). The bathymetry mimics the basement structure resulting from the history of rifting described above. This basement geometry was established in Cretaceous time, following the most recent major stretching event. Given this basic structure, two distinct sedimentological regimes may be recognized in the post-rift stratigraphic record, controlled directly by changes of relative sea level.

One of the two sedimentological regimes prevailed during Late Cretaceous and Paleocene time and then again from Late Eocene time to the present day. It was characterized by relatively low rates of sediment supply, deep-water deposition and no shoreline within the basin (Tate 1993). The Upper Cretaceous and Paleocene successions consist of deep-water limestones and marls with very low sedimentation rates (Fig. 3; Moore & Shannon 1995). These sedimentary sequences are laterally persistent, each sequence isopach mirroring the basin geometry (Fig. 4a and b). The Oligocene to Recent succession consists of deep-water mudstones (Shannon *et al.* 1993). This deep-water, sediment-starved regime has dominated for over half the post-rift period and in consequence the basin is not full of sediment at present.

The other distinctive sedimentological regime occurred during Early Cretaceous and Early Eocene times and was characterized by the development of deltas, which are easily recognized on seismic sections. Early Cretaceous deltas developed when the present basin geometry had only just formed (Moore & Shannon 1995). The Eocene delta complex is of central importance to the arguments outlined in this study. It can be divided into at least three deltaic sequences, described by several workers (e.g. Moore & Shannon 1992; Shannon *et al.* 1993; Tate 1993). The largest Eocene deltas formed in the north of the basin and prograded southwards down the axis, but there were also some smaller sediment fans, which apparently prograded into the basin from the west and the east (Fig. 4). The progradation direction of sediment fans on the western side of the basin during the initiation of the Early Eocene deltaic system suggests that these fans were supplied directly from Porcupine Bank and Porcupine Ridge (Fig. 4c). Seismic interpretation confirms that north to south progradation of these fans parallel to the basin margin was negligible (see McDonnell & Shannon 2001). It is unlikely that they were sourced by sediment arriving from the north that had somehow bypassed to the west of the main deltaic complex in the centre of the basin (Fig. 5).

The maximum area of the Porcupine Bank sediment source region is *c.* 1×10^4 km^2 (Fig. 4c).

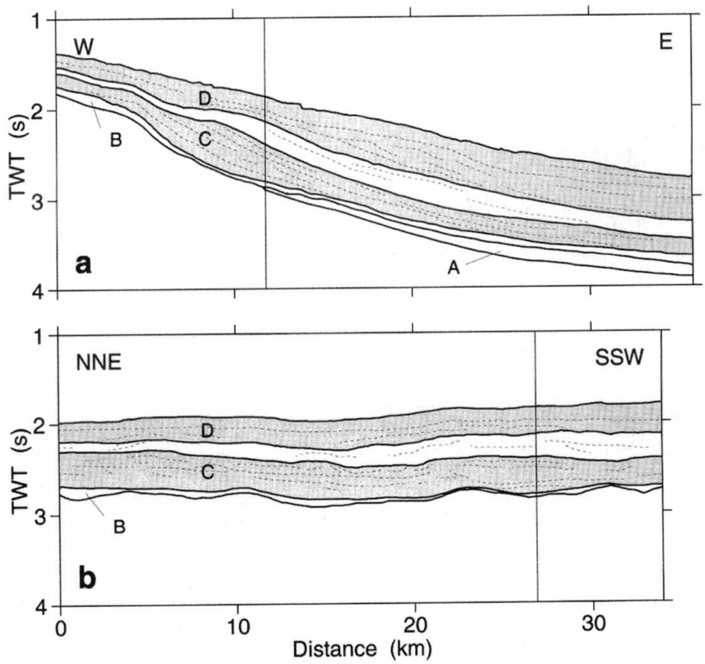

Fig. 5. Seismic line drawings to show the geometry of Eocene sediment fans prograding eastwards from Porcupine Ridge. Location of lines is shown in Figure 4. Letters refer to sequences displayed in Figures 3 and 4; the two deltaic sequences C and D are shaded. (**a**) Line perpendicular to Porcupine Ridge; internal reflectors in sequences C and D are sigmoidal and downlap onto the lower sequence boundary. (**b**) Line parallel to Porcupine Ridge; internal reflectors in sequences C and D are generally parallel to the sequence boundaries, suggesting that little sediment transport occurred parallel to the Ridge.

Integration of solid grain sediment isopachs, some of which are shown in Figure 4, indicates that the maximum volume of solid grain sediment likely to have been sourced from Porcupine Bank is 6400 km^3 during Early Eocene time, a period of 6 Ma. Data from modern rivers indicate that a source region of such a size can produce comparable sediment yields when uplifted by between 100 m and 1 km (Milliman & Syvitski 1992). This estimate is in agreement with the independent results of subsidence analysis presented later.

Porcupine Bank is of particular significance in this study because it represents a small, well-defined sediment source area directly related to sediment fans that prograded into the basin. Only when the basin flanks immediately adjacent to the enclosed northern end of the basin became emergent did sediment supply outpace tectonic subsidence. The sediment surface was then maintained near sea level over a wide area, forming a stable delta plain suitable for lignite development.

Thus, from Late Cretaceous to Paleocene time, and from Late Eocene time to the present, the Porcupine Basin and its surrounding margins were entirely below sea level. In contrast, during Early Eocene time, Porcupine Bank and the Irish Shelf were above sea level and a shoreline crossed the basin from west to east. Porcupine Bank and the Irish Shelf are basement blocks that are stable compared with the stretched Porcupine Basin. Both have a crustal thickness of *c.* 28 km (Whitmarsh *et al.* 1974; Makris *et al.* 1988) and both expose Variscan or older rocks at the sea bed (Robeson *et al.* 1988; Masson *et al.* 1989). These basin flanks suffered negligible tectonic subsidence during Mesozoic and Cenozoic time.

Given that the basin flanks are tectonically stable, what mechanism caused sea level to change as it did across a large region west of Ireland? Mechanisms that require topography to be supported by intra-plate stresses or to result from flexure are not tenable because the elastic thickness of the continental shelf around Britain and Ireland is too low. The relationship between gravity and topography indicates that the effective elastic thickness is <5 km beneath the North Sea (Barton & Wood 1984) and Scotland (Barton 1992). Subsidence modelling of stratigraphic profiles in two dimensions suggests similar effective elastic thicknesses (e.g. Marsden *et al.* 1990). Thus, relative sea-level variation recorded by the post-rift sedimentary succession must have been caused by either a eustatic regression–transgression cycle or transient epeirogenic uplift. Published eustatic sea-level curves indicate long-term transgression across the Paleocene–Eocene boundary followed by long-term regression to the present day (e.g. Haq *et al.* 1987). Global sea-level change is the opposite to the relative sea-level change recorded by the Cenozoic sedimentary record in the Porcupine Basin. Hence transient epeirogenic uplift is most likely to have caused relative sea-level changes in the Porcupine Basin.

Duration and magnitude of uplift

The Porcupine Basin is an extensional sedimentary basin and its subsidence history can be predicted using the finite-duration lithospheric stretching model (Jarvis & McKenzie 1980). Anomalous subsidence and uplift not accounted for by lithospheric stretching can then be isolated by comparing modelled subsidence curves with backstripped well sections. The most important source of error when backstripping well sections is uncertainty in depositional water depth. Eocene deltas are particularly significant in this respect because the lignite-bearing topsets were deposited at or near sea level. Unfortunately, Cretaceous water depths are poorly constrained by sedimentological and palaeontological evidence. The minimum depth is 200 m but the maximum water depth cannot be constrained more accurately than 'upper slope' (Dobson *et al.* 1991).

The subsidence analysis technique used to isolate anomalous subsidence is illustrated in Figure 6. The start and duration of the synrift period are assumed *a priori* from the geological history of the basin. Figure 6a illustrates the result of modelling the observed data with a single Jurassic stretching event, suggested by the geological history described above. A theoretical subsidence curve fitted to the observed synrift section A produces theoretical post-rift subsidence C, which is less than the observed post-rift subsidence B; this subsidence deficit occurs for all the Porcupine Basin wells that contain useful Paleogene data. The post-rift subsidence deficit arises because the observed synrift section is an incomplete record of the stretching history, either because the synrift is bound by unconformities (Fig. 6a) or because the exploration target did not require drilling to basement (Fig. 7b and d-f). It should be noted that Jurassic stretching produces modelled subsidence in excess of observed subsidence during Early Cretaceous and Early Eocene time. Lithospheric stretching of the magnitude observed here does not produce uplift, so another mechanism to explain this uplift is required.

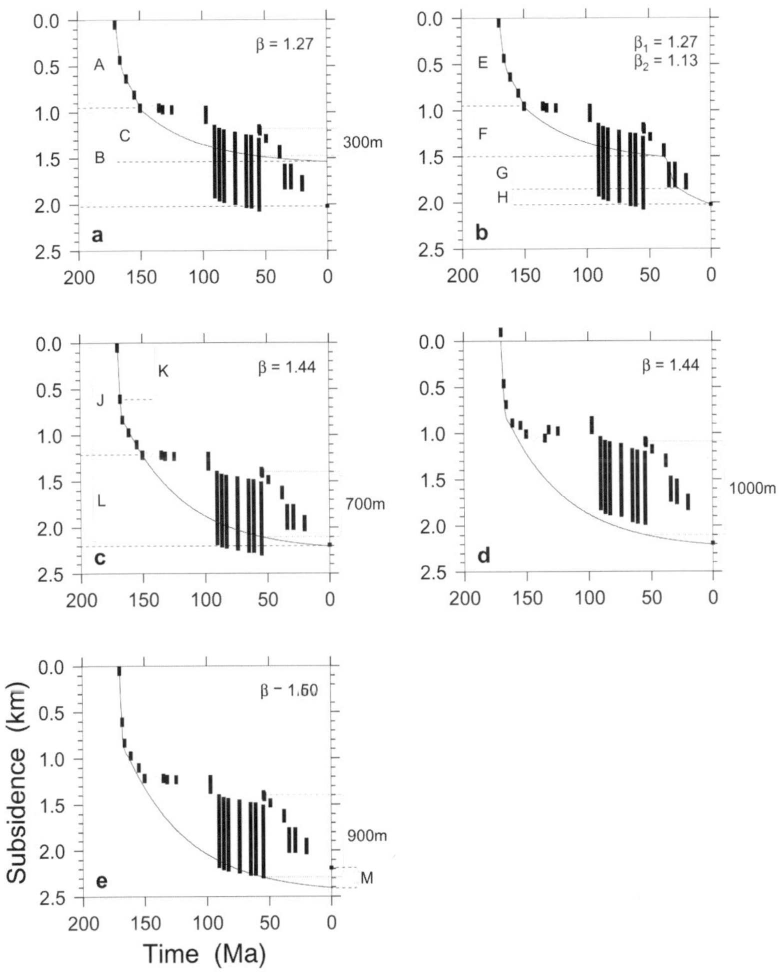

Fig. 6. Comparison of subsidence modelling techniques using a single well (Deminex 34/15-1). Backstripped well data plotted as bars, indicating uncertainty in depositional water depth. Wavy lines mark unconformities within the well. Curves are theoretical models assuming finite duration lithospheric stretching. (See text for full explanation of each model.) (**a**) Modelling of observed synrift section A predicts post-rift subsidence C, less than observed post-rift subsidence B. (**b**) Total observed subsidence can be modelled with two phases of rifting (in Jurassic and Oligocene time). (**c**) Total observed subsidence is better modelled by adding an extra layer K to represent eroded synrift section. (**d**) Theoretical model (**c**) with observed data incorporating the global sea-level curve of Haq *et al.* (1987). (**e**) Observed data as (**c**), modelled assuming present-day epeirogeneic uplift of M.

Hall & White (1994) argued that the simplest way to achieve the observed magnitude of post-rift subsidence is to invoke a period of Eocene–Oligocene stretching, illustrated in Figure 6b. Jurassic stretching produces synrift subsidence E and post-rift subsidence F. Then Eocene stretching produces further synrift subsidence G and post-rift subsidence H to match the present-day basement depth. However, this model is not a satisfactory explanation of the data because there is no evidence for the required amount of Eocene–Oligocene stretching in the Porcupine Basin (Hall & White 1994). Although small-scale Cenozoic faulting has been noted in the Porcupine Basin, these faults are layer-bound within the Eocene and Oligocene successions in most places and represent minimal extension (Naylor & Anstey 1987; Tate 1993; Hall & White 1994). The faults may instead represent volumetric contraction during compaction (Cartwright & Lonergan 1996). Even if Cenozoic stretching was admissible, a mechanism to generate uplift during Early Cretaceous and Early Eocene time would still be required.

The simplest way both to recreate the magnitude of post-rift subsidence observed in wells and to satisfy constraints on the stretching history imposed by faulting observed on seismic sections is illustrated in Figure 6c. The incomplete synrift section is augmented by a layer K so that a model of the resulting synrift section J predicts the observed post-rift subsidence magnitude L. The position of the synthetic synrift layer within the observed synrift section is not important for our purposes because the shape of the post-rift subsidence curve is relatively insensitive to details of the strain-rate history (White 1994). Here, the synthetic layer is always inserted into the unconformity between synrift and basement, or at the base of the drilled synrift section.

The method illustrated in Figure 6c can be used to fit model subsidence curves to all the Porcupine Basin wells with a well-dated Paleogene section (Fig. 7). However, two further issues must be considered before differences between observed and modelled subsidence are interpreted further. First, Figure 6d illustrates the effect of including variation in sea level in the backstripping calculations. Anomalous transient uplift during Cretaceous and Cenozoic time is increased, implying that the estimates of anomalous uplift presented in the following section are minima. Second, implicit in the model shown in Figure 6c is that transient epeirogenic uplift is negligible at the present day. If this is not the case then the data should be modelled as illustrated in Figure 6e, where M is

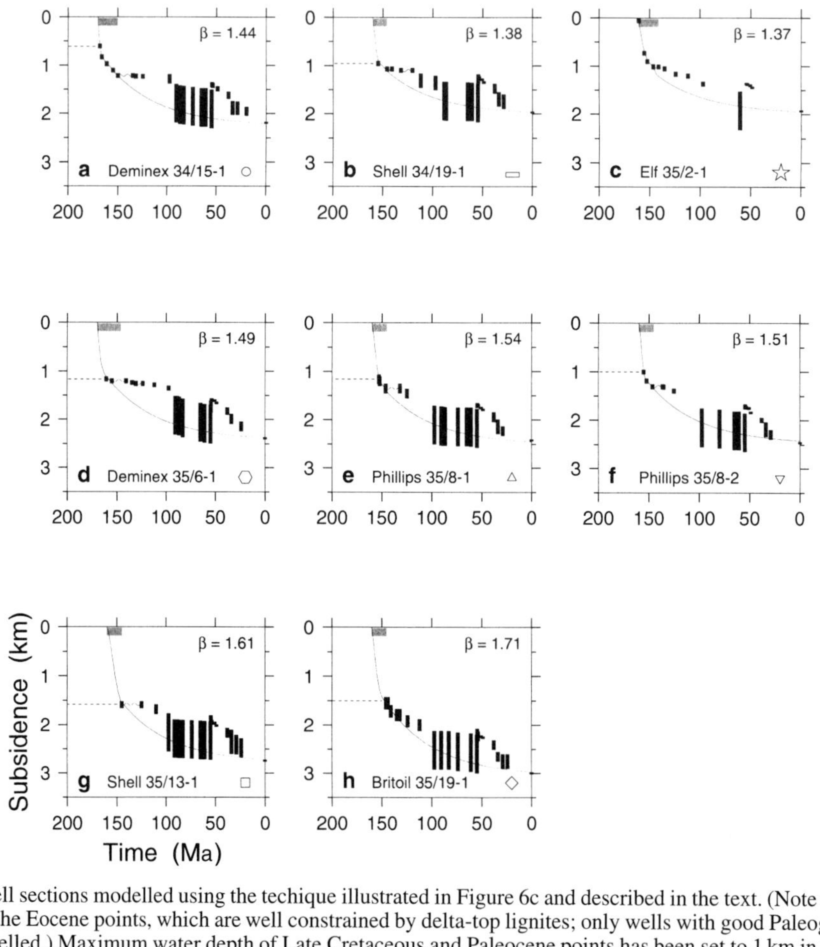

Fig. 7. Well sections modelled using the techique illustrated in Figure 6c and described in the text. (Note the small errors on the Eocene points, which are well constrained by delta-top lignites; only wells with good Paleogene data were modelled.) Maximum water depth of Late Cretaceous and Paleocene points has been set to 1 km in all cases. Symbols in the lower right corners are used to locate each well in Figure 4b. Dashed lines mark total depth (TD). Grey bars mark duration of synrift phase.

the present-day epeirogenic uplift. Present-day epeirogenic uplift can be easily measured for oceanic crust of known thickness. Just south of the Porcupine Basin, off Goban Spur, a deep seismic reflection (WAM) and a seismic refraction profile have indicated a non-volcanic continental margin and oceanic crust 5.4 km thick (Horsefield *et al.* 1994). Near the western end of WAM, the sea bed is 4.80–4.81 km deep, equivalent to 5.35–5.60 km after correcting for the effect of unusually thin crust and a sediment layer of 1.5–2.0 km thickness. The oceanic crust is aged 80–85 Ma and hence should lie at a depth of 5.50–5.57 km (Parsons & Sclater 1977). Thus, there is no clear evidence of transient uplift just south of the Porcupine Basin today. If present-day transient uplift does affect the Porcupine Basin itself, once again, our estimates of anomalous uplift should be regarded as minima.

Although present-day transient uplift has been accounted for, it is possible that permanent uplift of the Porcupine Basin occurred during Cretaceous and Cenozoic time. In our subsidence analysis scheme, permanent uplift occurring after the Jurassic stretching would also be included as extra subsidence M at the present day in Figure 6e. As a complete stretching history is unavailable in the Porcupine Basin, we argue that present-day permanent uplift cannot be determined by comparing observed data and theoretical models, although other studies have attempted to demonstrate permanent uplift across this region (Clift & Turner 1998).

The residual subsidence history of the Porcupine Basin since Jurassic time is summarized in Fig. 8. It should be noted that the term 'residual subsidence history' used here refers to the net subsidence resulting from both basin subsidence and basin uplift. The most prominent feature of the residual subsidence history is a phase of 500–800 m subsidence occurring after Early Eocene time. This subsidence took place over a minimum period of 25 Ma through Eocene and Oligocene time, and a maximum period of 55 Ma between Eocene time and the present (Fig. 8). Residual subsidence during Late Cretaceous time, and hence any transient uplift at the Paleocene–Eocene boundary, cannot be constrained satisfactorily using well data because of the uncertainty in determining depositional water depths during Late Cretaceous and Paleocene time. Indeed, using well data alone it would be possible to suggest that transient uplift was initiated during Early Cretaceous time and did not begin to decay until Eocene time. Fortunately, the history of submergence and exposure of Porcupine Bank provides a reliable minimum estimate of the magnitude of Paleocene–Eocene uplift.

The height of Porcupine Bank relative to sea level can be affected only by regional or eustatic sea-level changes because it is a stable basement block. The depth of the crest of Porcupine Bank today is 200 m: its equilibrium or zero residual depth, assuming negligible epeirogenic uplift at present (Fig. 1). If a minimum of 500 m transient subsidence has occurred since the start of the Eocene period (Fig. 8) then at that time the crest would have been at least 300 m above sea level (this estimate assumes a water load: actual air-loaded topography would have been 200 m). However, during Late Cretaceous and Paleocene time the sedimentary history indicates that the basin flanks were submerged, so the minimum water-loaded uplift of Porcupine Bank at the Paleocene–Eocene boundary was 300 m (Fig. 8). Eustatic sea-level variation between Eocene time and the present day does not alter the result of this calculation because it has the same effect on both the magnitude of transient subsidence and the equilibrium depth of Porcupine Bank.

The Cenozoic transient uplift–subsidence cycle was preceded by a separate transient

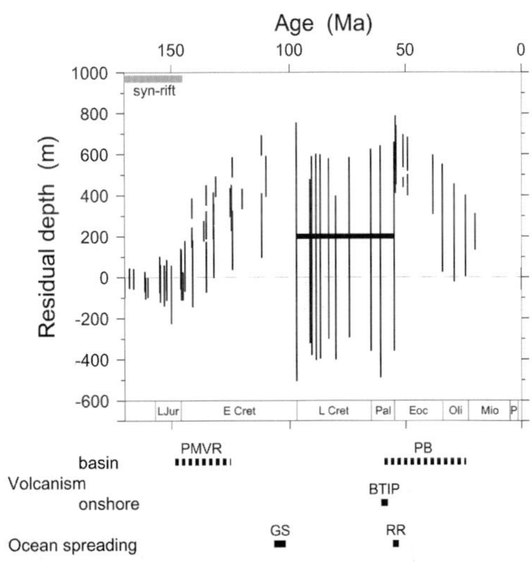

Fig. 8. Residual subsidence history calculated from the data presented in Figure 7 by subtracting the backstripped data from the modelled subsidence curve for each well. Bold black line is the maximum residual depth of Porcupine Bank, discussed in detail in the text. Contemporary igneous and tectonic events are represented by the following abbreviations: BTIP, British Tertiary Igneous Province; GS, Goban Spur margin (south of Porcupine Basin); PB, relatively minor magmatism in Porcupine Basin itself; PMVR, Porcupine Median Volcanic Ridge; RR, Reykjanes Ridge (west of Porcupine Basin).

uplift–subsidence cycle during Cretaceous time. Immediately following rifting, 200–700 m transient uplift occurred in Early Cretaceous time, followed by 0–500 m subsidence continuing into Late Cretaceous time.

Causes of anomalous uplift and subsidence

The primary purpose of this paper is to consider the possible effect of the Iceland plume on the Porcupine Basin, so most attention is devoted to explaining the Cenozoic uplift–subsidence cycle, rather than the earlier Cretaceous cycle. What was the regional extent of the Early Eocene transient uplift event? Transient uplift of this age has been demonstrated in the Faroe–Shetland Basin and the northern North Sea Basin (900 m and 375–525 m, respectively; Nadin *et al.* 1997) (Fig. 1). Apatite fission-track data suggest it occurred throughout Ireland (Clift *et al.* 1998; Green *et al.* 2001). A Paleocene–Early Eocene uplift event affected the Celtic Sea basins (south of Ireland), where middle Eocene to Oligocene shallow marine or lacustrine sediments lie unconformably upon Upper Cretaceous Chalk (Murdoch *et al.* 1995). This unconformity is also present in the St. George's Channel basin (east of Ireland). Early Eocene uplift affected an area extending at least 1000 km inboard of the continental margin between Edoras Bank and the Faroe Islands (Fig. 1 inset).

What caused the rapid post-Early Eocene subsidence? One possibility is that the basin moved off the topographic swell formed above the plume following the initiation of sea-floor spreading between Greenland and Europe. Europe is moving away from a plume centred on the Reykjanes Ridge at a velocity given by the half spreading rate, which has been *c.* 10 km Ma^{-1} since Eocene time (Srivastava & Tapscott 1986). In the Porcupine Basin, transient subsidence of 500–800 m occurred over a period of minimum 25 Ma and maximum 55 Ma (Fig. 8); in this time Europe has moved 250–550 km away from the plume. Thus the average subsidence gradient (ratio of vertical movement to horizontal movement) experienced by the Porcupine Basin was between 3.2×10^{-3} and 0.9×10^{-3}. White *et al.* (1995) isolated present-day convective plume support on a 1400 km transect along the Reykjanes Ridge by removing the effect of variable crustal thickness; it has a subsidence gradient of 1.4×10^{-3}. Assuming that the plume is radially symmetrical and also that it has maintained a similar shape similar since Eocene time, it is then possible to account for the transient subsidence in the Porcupine Basin simply by movement away from the plume. Both of these assumptions may be incorrect in detail. The present plume head may be elongated along the spreading centre (Jones *et al.* in press). Variation in crustal thickness and structure across the oceanic basins south of Iceland indicates that the shape and size of the plume head have varied through time (White 1997). Yet a plume head of different size and shape could still account for the rapid post-Early Eocene subsidence. For example, if the present-day plume is elongated by a factor of two, giving a subsidence gradient of 2.8×10^{-3} for movement perpendicular to the spreading centre, then transient subsidence in the Porcupine Basin can still be accounted for by movement away from the plume swell.

A second possible way that the Iceland plume could have influenced the subsidence history of the Porcupine Basin requires emplacement of a thermal anomaly beneath the continental shelf. The entire continental margin between Iceland and Edoras Bank (Fig. 1) was underlain by asthenosphere 150–200 °C hotter than normal at the time of continental break-up (Barton & White 1997). Variations in oceanic crustal thickness and the subsidence history of the margin show that the temperature beneath the spreading centre decayed to only *c.* 50 °C hotter than normal in 5–10 Ma following the start of sea-floor spreading. Such a decrease in temperature beneath a spreading centre can be explained if the anomalously hot asthenosphere was confined to a layer a few tens of kilometres thick, which was then rapidly cooled after decompression melting beneath the new spreading centre (Barton & White 1997). If such a hot layer was injected laterally beneath continental lithosphere before continental break-up it would cause transient uplift. Following continental separation, mantle flow would be dominated by upwelling, decompressional melting and cooling beneath the spreading centre, and the hot layer beneath the adjacent continent would not be maintained. Subsidence would result from dissipation of the thermal anomaly. A sub-lithospheric thermal anomaly would dissipate principally by convection. This model is attractive because it can explain sudden transient uplift during Early Eocene time as well as coeval onset of sea-floor spreading and subsidence of the continental margins. Assuming Airy isostasy, uplift resulting from the injection of a hot layer beneath the lithosphere is given by

$$U = D\rho_m \alpha \Delta T / (\rho_w - \rho_a)$$

where D is the layer thickness, ΔT is the anomalous temperature and the other variables

Table 1. *Symbols and values of variables used in text*

Symbol	Definition	Value
α	Thermal expansion coefficient	$3.28 \times 10^{-5}\,°C^{-1}$
ρ_a	Asthenosphere density (at 1333 °C)	$3.18\,Mg\,m^{-3}$
ρ_m	Mantle density (at 0 °C)	$3.33\,Mg\,m^{-3}$
ρ_w	Sea-water density	$1.03\,Mg\,m^{-3}$

are given in Table 1. A 50 km layer 200 °C hotter than normal would produce uplift of 500 m, a value that compares well with the residual subsidence history (Fig. 8).

The Early Cretaceous transient uplift event seen in the Porcupine Basin caused a regional unconformity between the synrift and post-rift succession (Fig. 7; Tate 1993). It occurred just after the end of rifting and well before the start of sea-floor spreading in the North Atlantic adjacent to the Goban Spur (Early to Mid-Albian time; Masson *et al.* 1981) but is coeval with the extrusion of the Porcupine Median Volcanic Ridge (Tate & Dobson 1988; Tate 1993; Fig. 8). A detailed discussion of the Cretaceous uplift–subsidence cycle in the Porcupine Basin is largely outside the scope of this paper, yet there are parallels with the Paleogene uplift–subsidence cycle. The Early Cretaceous unconformity in the Porcupine Basin can be correlated with an unconformity in the Celtic Sea Basins (Petrie *et al.* 1989). A coeval unconformity has been documented in other basins surrounding the Cornubian platform and assigned a Berriasian age (McMahon & Turner 1998). In the absence of any other suitable mechanism, McMahon & Turner (1998) tentatively proposed a thermal origin for this unconformity. The contemporary igneous activity demonstrated by the Porcupine Median Volcanic Ridge lends support to a mechanism involving anomalously hot mantle. Following the onset of sea-floor spreading off Goban Spur, transient subsidence ensued in Late Cretaceous time (Fig. 8). Delta development and subsequent drowning immediately after maximum transient uplift was similar to the Eocene deltaic cycle (Moore & Shannon 1995).

Conclusions

Evidence is presented that constrains the magnitude and timing of Paleogene epeirogenic uplift and subsidence in and around the Porcupine Basin, offshore west of Ireland. The possible causes of this uplift are discussed and our principal results are as follows:

(1) We recognize distinctive sedimentological responses to the basin flanks being either exposed or submerged. From Late Cretaceous to Paleocene times, and from Late Eocene time to the present, the Porcupine Basin and its surrounding margins were entirely below sea level. In contrast, during Early Eocene times, the Porcupine Bank and the Irish Shelf were above sea level, and a shoreline crossed the basin from west to east.

(2) The Mesozoic and Cenozoic history of the Porcupine Basin may be summarized as a Jurassic synrift phase, followed by Cretaceous and Cenozoic post-rift subsidence. The Early Eocene period of uplift and increased sedimentation is not predicted by this simple pattern of post-rift subsidence. Neither eustatic sea-level change nor variation in flexural support can account for the amplitude of the Eocene uplift–subsidence cycle. Transient epeirogenic uplift is most likely to have caused the Early Eocene relative sea-level changes noted in the Porcupine Basin.

(3) Modelling subsidence histories of wells and of Porcupine Bank allows quantification of the magnitude and timing of epeirogenic uplift and subsidence. Transient uplift of 300–600 m occurred at the Paleocene–Eocene boundary, followed by subsidence of 500–800 m after Early Eocene time, over a period with a minimum length of 25 Ma and a maximum of 55 Ma. A coeval uplift–subsidence cycle has also been noted in the Celtic Sea, in the basins surrounding Scotland and along the NW European continental margin.

(4) One way to explain regional subsidence between Eocene time and the present is that the European plate moved off the topographic swell above the Iceland plume following continental separation between Greenland and Europe in Early Eocene time. Another possibility is that an anomalously hot layer, 50 km thick, was emplaced beneath the entire region just before the onset of sea-floor spreading in Early Eocene time and was then dissipated by convection following continental separation.

(5) A Cretaceous transient uplift–subsidence cycle that shares many similarities with the Paleogene cycle is also recognized. Immediately following Late Jurassic rifting, 200–700 m transient uplift occurred in Early Cretaceous time, followed by subsidence of 0–500 m coeval with the onset of sea-floor spreading at the Goban Spur margin. The Cretaceous uplift–subsidence cycle might also be caused by anomalously hot mantle.

We thank B. Mitchener and J. Perry of BP-Amoco for providing seismic data. P. Bellingham, B. Clarke, J. Maclennan, F. Nimmo and the two referees provided helpful comments. S.M.J. was supported by a NERC studentship. This paper is Department of Earth Sciences Contribution 6100.

References

BARTON, A.J. & WHITE, R.S. 1997. Crustal structure of Edoras Bank continental margin and mantle thermal anomalies beneath the North Atlantic. *Journal of Geophysical Research*, **102**, 3109–3129.

BARTON, P.J. 1992. LISPB revisited: a new look under the Caledonides of northern Britain. *Geophysical Journal International*, **110**, 371–391.

BARTON, P.J. & WOOD, R.J. 1984. Tectonic evolution of the North Sea basin: crustal stretching and subsidence. *Geophysical Journal of the Royal Astronomical Society*, **79**, 987–1022.

BERGGREN, W.A., KENT, D.V., SWISHER, C.C. & AUBREY, M.-P. 1995. A revised Cenozoic geochronology and chronostratigraphy. *In*: BERGGREN, W.A., KENT, D.V., AUBREY, M.-P. & HARDENBOL, J. (eds) *Geochronology, Time Scales and Global Stratigraphic Correlation*. SEPM (Society for Sedimentary Geology) Special Publications, **54**, 129–212.

BRODIE, J. & WHITE, N. 1995. The link between sedimentary basin inversion and igneous underplating. *In*: BUCHANAN, J.G. & BUCHANAN, P.G. (eds) *Basin Inversion*. Geological Society, London, Special Publications, **88**, 21–38.

CARTWRIGHT, J.A. & LONERGAN, L. 1996. Volumetric contraction during the compaction of mudrocks: a mechanism for the development of regional-scale polygonal fault systems. *Basin Research*, **8**, 183–193.

CLIFT, P.D. & TURNER, J. 1998. Paleogene igneous underplating and subsidence anomalies in the Rockall–Faroe–Shetland area. *Marine and Petroleum Geology*, **15**, 223–243.

CLIFT, P., CARTER, A. & HURFORD, A. 1998. The erosional and uplift history of NE Atlantic passive margins: constraints on a passing plume. *Journal of the Geological Society, London*, **155**, 787–800.

CROKER, P.F. & KLEMPERER, S. 1989. Structure and stratigraphy of the Porcupine Basin: relationships to deep crustal structure and the opening of the North Atlantic. *In*: TANKARD, A.J. & BALKWILL, H.R. (eds) *Extensional Tectonics and Stratigraphy of the North Atlantic Margins*. American Association of Petroleum Geologists, Memoirs, **46**, 445–459.

CROKER, P.F. & SHANNON, P.M. 1987. The evolution and hydrocarbon prospectivity of the Porcupine Basin, offshore Ireland. *In*: BROOKS, J. & GLENNIE, K.W. (eds) *Petroleum Geology of Northwest Europe*. Graham & Trotman, London, 633–642.

DOBSON, M.R., HAYNES, J.R., BANNISTER, A.D., LEVENE, D.G., PETRIE, H.S. & WOODBRIDGE, R.A. 1991. Early Tertiary palaeoenvironments and sedimentation in the NE Main Porcupine Basin (well 35/13-1), offshore western Ireland—evidence for global change in the Tertiary. *Basin Research*, **3**, 99–117.

GREEN, P.F., DUDDY, I.R., BRAY, R.J., DUNCAN, W.I. & CORCORAN, D.V. 2001. The influence of thermal history on hydrocarbon prospectivity in the Central Irish Sea Basin. *In*: SHANNON, P.M., HAUGHTON, P.D.W. & CORCORAN, D.V. (eds) *The Petroleum Exploration of Ireland's Offshore Basins*. Geological Society, London, Special Publications, **188**, 171–188.

HALL, B.D. & WHITE, N. 1994. Origin of anomalous Tertiary subsidence adjacent to North Atlantic continental margins. *Marine and Petroleum Geology*, **11**, 702–714.

HAMILTON, M.A., PEARSON, D.G., THOMPSON, R.N., KELLY, S.P. & EMELEUS, H. 1998. Rapid eruption of Skye lavas inferred from precise U–Pb and Ar–Ar dating of the Rum and Cuillin plutonic complexes. *Nature*, **394**, 260–263.

HAQ, B.U., HARDENBOL, J. & VAIL, P.R. 1987. Chronology of fluctuating sea-levels since the Triassic. *Science*, **235**, 1156–1167.

HORSEFIELD, S.J., WHITMARSH, R.B., WHITE, R.S. & SIBUET, J.-C. 1994. Crustal structure of the Goban Spur rifted continental margin, NE Atlantic. *Geophysical Journal International*, **119**, 1–19.

JARVIS, G.T. & MCKENZIE, D.P. 1980. Sedimentary basin formation with finite extension rates. *Earth and Planetary Science Letters*, **48**, 42–52.

JONES, S.M., WHITE, N. & CLARKE, B. in press. Present and past influences of the Iceland plume on sedimentation. *In*: DORÉ, A.G., WHITE, N., STOKER, M.S., CARTWRIGHT, J.A. & TURNER, J.P. (eds) *Exhumation of the North Atlantic Margin: Timing, Mechanisms and Implications for Petroleum*. Geological Society, London, Special Publications.

JOY, A.M. 1992. Right place, wrong time: anomalous post-rift subsidence in sedimentary basins around the North Atlantic Ocean. *In*: STOREY, B.C., ALABASTER, T. & PANKHURST, R.J. (eds) *Magmatism and the Causes of Continental Break-up*. Geological Society, London, Special Publications, **68**, 387–393.

MAKRIS, J., EGLOFF, R., JACOB, A.W.B., MOHR, P., MURPHY, T. & RYAN, P. 1988. Continental crust under the southern Porcupine Seabight west of Ireland. *Earth and Planetary Science Letters*, **89**, 387–397.

MARSDEN, G., YIELDING, G., ROBERTS, A.M. & KUSZNIR, N.J. 1990. Application of a flexural cantilever simple-shear/pure-shear model of continental lithosphere extension to the formation of the North Sea basin. *In*: BLUNDELL, D.J. & GIBBS, A.D. (eds) *Tectonic Evolution of North Sea Rifts*. Clarendon Press, Oxford, 240–261.

MASSON, D., DOBSON, M., AUZENDE, J.-M., COUSIN, M., COUTELLE, A., ROLET, J. & VAILLANT, P. 1989. Geology of Porcupine Bank and Goban Spur, northeastern Atlantic—preliminary results of CYAPORC submersible cruise. *Marine Geology*, **87**, 105–119.

MASSON, D.G., MONTADERT, L., SCRUTTON, R.A., *et al.* 1981. Regional geology of the Goban Spur continental margin. *In*: de Graciansky, P.C. & POAG, C.W. (eds) *Initial Reports of the Deep Sea Drilling Project, 80*. US Government Printing Office, Washington, DC, 1115–1139.

McDONNELL, A. & SHANNON, P.M. 2001. Comparative Tertiary stratigraphic evolution of the Porcupine and Rockall basins. *In*: SHANNON, P.M., HAUGHTON, P.D.W. & CORCORAN, D.V. (eds) *The Petroleum Exploration of Ireland's Offshore Basins*. Geological Society, London, Special Publications, **188**, 323–344.

McMAHON, N.A. & TURNER, J. 1998. The documentation of a latest Jurassic–earliest Cretaceous uplift throughout southern England and adjacent offshore areas. *In*: UNDERHILL, J.R. (ed.) *Development, Evolution and Petroleum Geology of the Wessex Basin*. Geological Society, London, Special Publications, **133**, 215–240.

MILLIMAN, J.D. & SYVITSKI, J.P.M. 1992. Geomorphic/tectonic control of sediment discharge to the ocean: the importance of small mountainous rivers. *Journal of Geology*, **100**, 525–544.

MOORE, J.G. & SHANNON, P.M. 1992. Palaeocene–Eocene deltaic sedimentation, Porcupine Basin, offshore Ireland—a sequence stratigraphic approach. *First Break*, **10**, 461–469.

MOORE, J.G. & SHANNON, P.M. 1995. The Cretaceous succession in the Porcupine Basin, offshore Ireland: facies distribution and hydrocarbon potential. *In*: CROKER, P.F. & SHANNON, P.M. (eds) *The Petroleum Geology of Ireland's Offshore Basins*. Geological Society, London, Special Publications, **93**, 345–370.

MURDOCH, L.M., MUSGROVE, F.W. & PERRY, J.S. 1995. Tertiary uplift and inversion history in the North Celtic Sea Basin and its influence on source rock maturity. *In*: CROKER, P.F. & SHANNON, P.M. (eds) *The Petroleum Geology of Ireland's Offshore Basins*. Geological Society, London, Special Publications, **93**, 297–319.

NADIN, P., KUSZNIR, N. & CHEADLE, M. 1997. Early Tertiary plume uplift of the North Sea and Faeroe–Shetland Basins. *Earth and Planetary Science Letters*, **148**, 109–127.

NAYLOR, D. & ANSTEY, N.A. 1987. A reflection seismic study of the Porcupine Basin, offshore west Ireland. *Irish Journal of Earth Sciences*, **8**, 187–210.

PARSONS, B. & SCLATER, J.G. 1977. An analysis of the variation of ocean floor bathymetry and heat flow with age. *Journal of Geophysical Research*, **82**, 5803–5827.

PETRIE, S.H., BROWN, J.R., GRANGER, P.J. & LOVELL, J.P.B. 1989. Mesozoic history of the Celtic Sea Basins. *In*: TANKARD, A.J. & BALKWILL, H.R. (eds) *Extensional Tectonics and Stratigraphy of the North Atlantic Margins*. American Association of Petroleum Geologists, Memoirs, **46**, 433–444.

RITCHIE, J.D. & HITCHEN, K. 1996. Early Paleogene offshore igneous activity to the northwest of the UK and its relationship to the North Atlantic Igneous Province. *In*: KNOX, R.W.O'B., CORFIELD, R.M. & DUNAY, R.E. (eds) *Correlation of the Early Paleogene in Northwest Europe*. Geological Society, London, Special Publications, **101**, 63–78.

ROBESON, D., BURNETT, R.D. & CLAYTON, G. 1988. The Upper Palaeozoic geology of the Porcupine, Erris and Donegal Basins, offshore Ireland. *Irish Journal of Earth Sciences*, **9**, 153–175.

SAUNDERS, A.D., FITTON, J.G., KERR, A.C., NORRY, M.J. & KENT, R.W. 1997. The North Atlantic Igneous Province. *In*: MAHONEY, J.J. & COFFIN, M.F. (eds) *Large Igneous Provinces: Continental, Oceanic, and Planetary Flood Volcanism*. Geophysical Monograph, American Geophysical Union, **100**, 45–93.

SCLATER, J.G. & CHRISTIE, P.A.F. 1980. Continental stretching: an explanation of the post-mid-Cretaceous subsidence of the Central North Sea basin. *Journal of Geophysical Research*, **85**, 3711–3739.

SHANNON, P.M., MOORE, J.G., JACOB, A.W.B. & MAKRIS, J. 1993. Cretaceous and Tertiary basin development west of Ireland. *In*: PARKER, J.R. (ed.) *Petroleum Geology of Northwest Europe: Proceedings of the 4th Conference*. Geological Society, London, 1057–1066.

SMALLWOOD, J.R., STAPLES, R.K., RICHARDSON, K.R. & WHITE, R.S. 1999. Crust generated above the Iceland mantle plume: from continental rift to oceanic spreading centre. *Journal of Geophysical Research*, **104**, 22885–22902.

SRIVASTAVA, S.P. & TAPSCOTT, C.R. 1986. Plate kinematics of the North Atlantic. *In*: VOGT, P.R. & TUCHOLKE, B.B. (eds) *The Geology of North America, The Western North Atlantic Region, Vol. M*. Geological Society of America, Boulder, CO, 379–404.

STOKER, M.S. 1997. Mid- to late Cenozoic sedimentation on the continental margin off NW Britain. *Journal of the Geological Society, London*, **154**, 509–515.

TATE, M.P. 1993. Structural framework and tectonostratigraphic evolution of the Porcupine Seabight Basin, offshore western Ireland. *Marine and Petroleum Geology*, **10**, 95–123.

TATE, M. & DOBSON, M.R. 1988. Syn- and post-rift igneous activity in the Porcupine Seabight Basin and adjacent continental margin W of Ireland. *In*: MORTON, A.C. & PARSON, L.M. (eds) *Early Tertiary Volcanism and the Opening of the NE*

Atlantic. Geological Society, London, Special Publications, **39**, 309–334.

TATE, M., WHITE, N. & CONROY, J.-J. 1993. Lithospheric extension and magmatism in the Porcupine Basin west of Ireland. *Journal of Geophysical Research*, **98** (B8), 13905–13923.

THOMPSON, R.N. 1974. Primary basalts and magma genesis, I: Skye, Northwest Scotland. *Contributions to Mineralogy and Petrology*, **45**, 317–341.

WHITE, N. 1994. An inverse method for determining lithospheric strain rate variation on geological timescales. *Earth and Planetary Science Letters*, **122**, 351–371.

WHITE, N. & LOVELL, B. 1997. Measuring the pulse of a plume with the sedimentary record. *Nature*, **387**, 888–891.

WHITE, R.S. 1997. Rift–plume interaction in the North Atlantic. *Philosophical Transactions of the Royal Society, London, Series A*, **355**, 319–339.

WHITE, R.S. & MCKENZIE, D.P. 1989. Magmatism at rift zones: the generation of volcanic continental margins and flood basalts. *Journal of Geophysical Research*, **94** (B2), 7685–7729.

WHITE, R.S., BOWN, J.W. & SMALLWOOD, J.R. 1995. The temperature of the Iceland plume and origin of outward-propagating V-shaped ridges. *Journal of the Geological Society, London*, **152**, 1039–1045.

WHITMARSH, R.B., LANGFORD, J.J., BUCKLEY, J.S., BAILEY, R.J. & BLUNDELL, D.J. 1974. The crustal structure beneath the Porcupine Ridge as determined by explosion seismology. *Earth and Planetary Science Letters*, **22**, 197–204.

Evidence of shallow gas above the Connemara oil accumulation, Block 26/28, Porcupine Basin

K. P. GAMES

Gardline Surveys Limited, Endeavour House, Admiralty Road, Great Yarmouth NR30 3NG, UK
(e-mail: ken.games@gardline.co.uk)

Abstract: The results of a site survey investigation, carried out in Block 26/28 of the Porcupine Basin, suggest the presence of shallow gas above the Connemara oil accumulation. The evidence for shallow gas at or close to the sea bed consists of a series of features interpreted as gas or fluid escape structures. The presence of gas at depth is suggested by two different features: an anomalously high-amplitude seismic reflector, interpreted as a gas-charged sand layer, and some isolated, seismically disturbed zones, which are identified as gas chimneys on both high-resolution 2D seismic and conventional 3D seismic data.

Block 26/28 lies at the northern end of the Porcupine Basin *c.* 165 km to the west of nearest landfall at Slyne Head (Fig. 1). A BP-led consortium discovered oil in this block, in 1979, with the drilling of well 26/28-1 (MacDonald *et al.* 1987). This well tested 5589 barrels per day of 32–38° API oil from sandstone reservoirs of Mid- to Late Jurassic age. A 3D survey, acquired by the BP consortium in 1982, revealed that the oil was trapped in a steeply dipping, complexly faulted, Late Jurassic structure sealed by mudstones of Early Cretaceous age. Drilling of four appraisal wells (26/28-2, 26/28-3, 26/28-4A and 26/28-5) confirmed the complex nature of the structure and the presence of modest reservoirs with potentially poor vertical communication between the sand bodies (MacDonald *et al.* 1987). The oil accumulation was subsequently named the Connemara Field (Earls 1995).

In 1996 Statoil Exploration (Ireland) Ltd acquired a new 380 km^2 3D survey over the Connemara accumulation with a view to further appraisal drilling and development of the field. This paper outlines the results of the interpretation of data from this 3D survey and the ensuing site investigation survey in Block 26/28. The primary objective of the study was to identify any safety hazards or constraints for drilling from a semi-submersible rig in this area. The investigation consisted of a detailed study of the sea-bed conditions using a swathe echo sounder and sidescan sonar. Near-sea-bed conditions were assessed using a hull-mounted pinger along with some mini airgun data. Deeper hazards, down to a depth of 2.5 km, were identified by means of high-resolution 2D seismic data in conjunction with the conventional 3D seismic dataset. The target site covered an area of 5 km × 7.5 km, but because of prolonged bad weather, only key lines were acquired.

Seismic evidence for a variety of shallow gas accumulation and migration features has been documented in a number of the basins of offshore Ireland. For example, gas fronts, 'bright' reflectors, gas plumes, breakthroughs, gas chimneys, pockmarks and seep mounds have been described in the western Irish Sea area (Croker 1995). In addition, there is evidence for the presence of an extensive development of Upper Tertiary to Recent carbonate knolls in the south–central part of the Porcupine Basin (Hovland *et al.* 1994; Henriet *et al.* 2001). Features identified in these papers include: acoustic voids, caused by the presence of methane in the shallow sediments; pockmarks, which are the result of gas venting through fine-grained sediment to the sea floor; ploughmarks, which are formed by iceberg abrasion of the sea floor; anomalously 'bright' reflectors caused by a gas-charged sand layer; gas chimneys, caused by focused gas venting through the sedimentary column (Hovland & Judd 1988).

The 2D site investigation data for this paper were acquired simultaneously by single-pass surveying. The lines of acquisition are shown in Fig. 1, along with the locations of the five exploration wells, drilled by the BP consortium, in Block 26/28. Bathymetry was recorded from a

From: SHANNON, P.M., HAUGHTON, P.D.W. & CORCORAN, D.V. (eds) 2001. *The Petroleum Exploration of Ireland's Offshore Basins*. Geological Society, London, Special Publications, **188**, 361–373.
0305-8719/01/$15.00

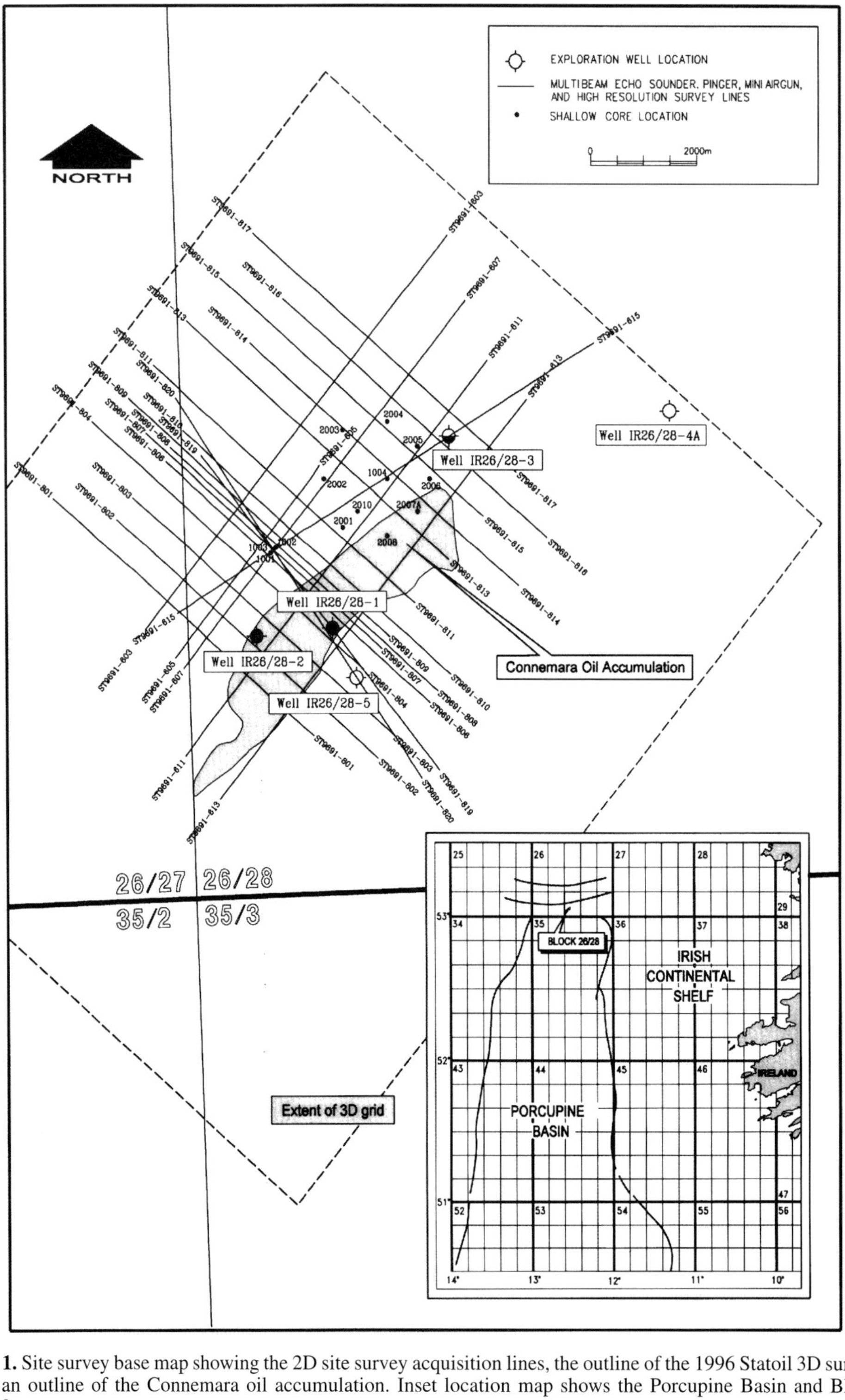

Fig. 1. Site survey base map showing the 2D site survey acquisition lines, the outline of the 1996 Statoil 3D survey and an outline of the Connemara oil accumulation. Inset location map shows the Porcupine Basin and Block 26/28.

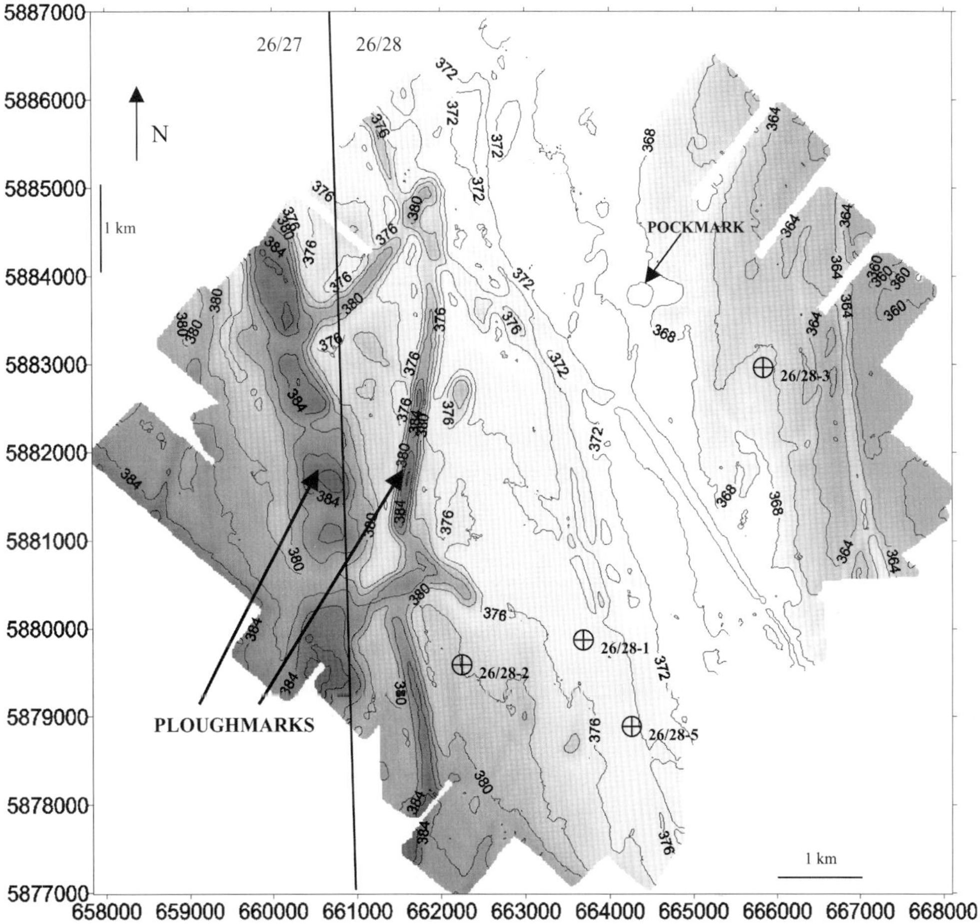

Fig. 2. Multi-beam bathymetric image of the sea bed covering the site survey area, illustrating a number of north–south-trending ploughmarks and some random pockmarks.

Simrad EM1000 multi-beam echo sounder, which has a 95 kHz transducer, and was run in the medium transmit mode. The data were processed into 10 m × 10 m cells using a Neptune post-processing suite. Sea-bed features were plotted from a GeoAcoustics 100 kHz sidescan sonar, using ranges of 200 m and 300 m per channel. Two sub-bottom profilers were used: a 16-element hull-mounted pinger array and a 10 cubic inch mini airgun. The pinger source has a central frequency of 3 kHz and the data were processed with a conventional swell filter, time variant gain (TVG) and band-pass filter. The mini airgun utilized a shot interval of 1 s and a recording sample interval of 0.5 ms. These data were processed using specially developed techniques to enhance the data quality, including event alignment, deconvolution and migration techniques. The high-resolution seismic data were acquired with a 160 cubic inch sleeve airgun array, which has a power level of 10 bar-metres, and a 96-channel streamer with a 12.5 m group length. These data were processed using conventional site survey parameters, which included a finite difference (FD) migration to collapse the many diffractions in the data. High cut filters varied from 240 Hz at the top of the section to 60 Hz at 2.5 s two-way time.

Sea-bed features

The sea bed over the area of the Connemara oil accumulation manifests a number of elongate features (Fig. 2). Multi-beam echo sounder data highlight what are interpreted to be two very distinct north–south-trending ploughmarks, the largest of which is c. 3 km long and up to 8 m deep. A number of smaller ploughmarks are also present. These ploughmarks are interpreted to represent the Quaternary abrasion of the continental shelf caused by floating ice grounding on the sea bed.

An abundance of pockmarks is also apparent on the sea bed in this area. Some of the larger pockmarks can be identified on the multi-beam bathymetry data, although they are at the limit of the resolution of this dataset (Fig. 2). The smaller pockmarks are observed only on the higher-resolution sidescan sonar data (Fig. 3). These pockmarks vary both in size and density of occurrence throughout the site. The density of distribution of the smaller pockmarks can be as great as 100 per km^2. These small pockmarks are often referred to as 'pits', or incipient or unit pockmarks (Harrington 1985). They are interpreted as being formed by the expulsion of fluid and/or shallow gas from the fine-grained seafloor sediment. The smaller pockmarks are typically 2–3 m in diameter and tens of centimetres deep. The larger pockmarks (Fig. 3), which average c. 20 m in diameter and up to 2 m in depth, often exhibit a high-reflectivity core or 'eye', indicative of carbonate concretions nucleating upon fluid or gas escape vents (Hovland & Judd 1988).

An anomalous zone of high reflectivity is also observed on the sidescan sonar data just outside the survey area. It consists of a linearly aligned series of seven minor point contacts, each about 80 cm high (Fig. 4). This zone can be interpreted as a series of carbonate concretions on the rims of small pockmarks. Alternatively, they may be interpreted as growths of the cold-water coral *Lophelia* sp., which have been observed to nucleate in linear bands in the south–central part of the Porcupine Basin (Hovland *et al.* 1994; Henriet *et al.* 2001). The latter is considered to be

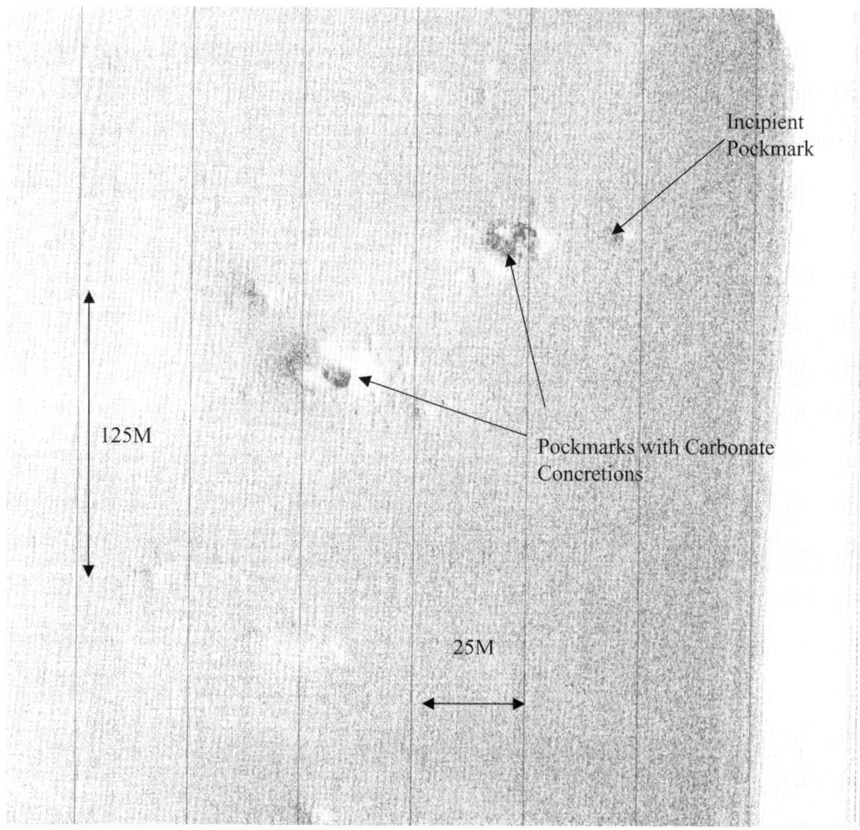

Fig. 3. Sidescan sonar example showing incipient pockmarks, which are typically a few metres in diameter and tens of centimetres deep. The larger pockmarks manifest a high-reflectivity core interpreted as carbonate concretion.

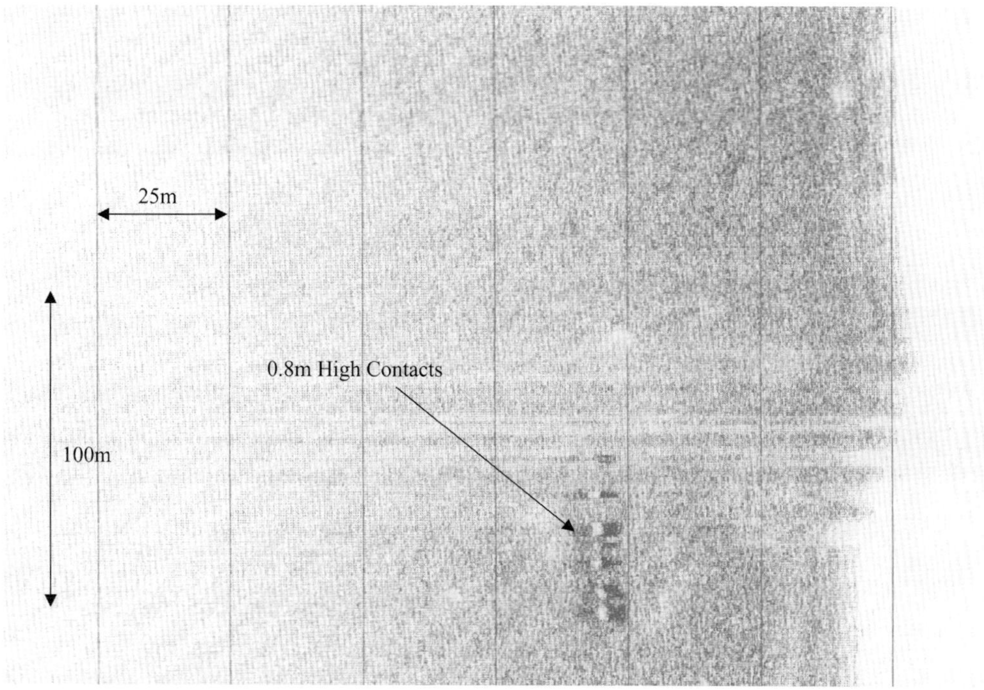

Fig. 4. Sidescan sonar example of linear zone of high reflectivity, which could be interpreted as carbonate concretions on the rims of small pockmarks. An alternative interpretation is that the seven contacts represent nucleations of the cold-water coral *Lophelia* sp.

the more favoured interpretation, as pockmarks very rarely form in such linear bands.

Near-sea-bed sediments

The pinger and mini airgun data have been interpreted with the aid of shallow sea-bed core data. The typical reflectivity sequence in the survey area is illustrated in Fig. 5. The upper 30 m of superficial sediment consists of a series of very soft clays interbedded with thin (10–40 cm) sand layers. Within this uppermost stratigraphic unit, numerous acoustic voids and diffraction hyperbolae are present (Fig. 5). These are interpreted as being caused by dewatering of the shallow sediments or by the presence of shallow gas within these sediments. The concentration of shallow gas is expected to be very low, as these features are not generally identified on the lower-frequency mini airgun data (Fig. 6). Locally there is some correlation between the acoustic voids and the ploughmarks described above (Fig. 7). In places, a definite loss of internal layering in the uppermost sediments is observed, which could be caused either by dewatering of the sediments or by diffuse gas. The observed lowering of the seismic frequency response in these areas is consistent with the absorption of the higher frequencies by the shallow gas (Fig. 7). However, no evidence of shallow gas was recorded in the shallow cores recovered from this area. Nevertheless, the presence of gas would be expected to reduce the shear strength of the near-surface sediments. Diffuse gasification is suggested by the measured variations in shear strength (from 10 to 30 kPa) that have been recorded, in similar sediment types, from different shallow cores.

Ploughmarks are also identified on the mini airgun data (Fig. 6). A major ploughmark is observed towards the SW of this section and there is a marked drop in seismic frequency below the indentation of the sea-bed reflector. A high-amplitude, low-frequency reflector is present at *c.* 40 m below sea bed. This reflector correlates with a sand horizon in the shallow core data. A relict ploughmark is interpreted on this horizon towards the NE of the line and a similar drop in frequency is observed beneath the ploughmark. In other basins, buried ploughmarks are recognized as potential drilling hazards because of their

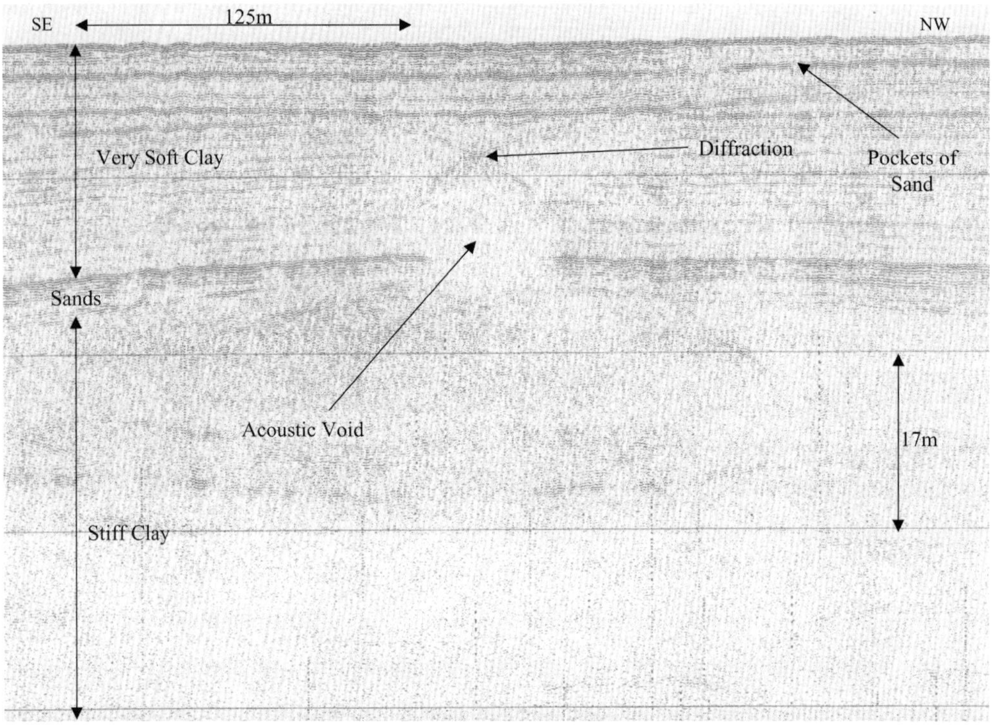

Fig. 5. Pinger profile illustrating the typical reflectivity sequence in the survey area. The near-sea-bed lithostratigraphy has been determined from nearby shallow core data. The presence of an acoustic void and associated diffractions are interpreted as indicating the presence of shallow gas.

tendency to accumulate and trap shallow gas (Gallagher *et al.* 1991; Gallagher & Heggland 1994).

Ploughmarks identified on both sea-bed and sub-sea-bed horizons are easily recognized on the 3D seismic data and manifest a similar trend across the site area. Figure 8 is a time–dip plot created on a Charisma workstation, which illustrates the similarity in trend of the ploughmarks identified on the sea-bed horizon (Fig. 8a) and the sub-sea-bed horizon (Fig. 8b). Impedance plots indicate that there is a decrease in impedance within the buried ploughmarks consistent with the presence of a gas-charged sand (Fig. 9).

Deeper sediments and associated gas indicators

Evidence for gas saturation in the deeper sediments was assessed using a combination of high-resolution 2D seismic data and conventional 3D seismic data, a technique developed by Statoil (Gallagher *et al.* 1991; Heggland *et al.* 1996). Logs for the five previously drilled exploration wells (26/28-1, 26/28-2, 26/28-3, 26/28-4A/ST, 26/28-5) were available for lithological identification. A number of major unconformities are identified on the high-resolution 2D seismic data (Fig. 10). In the area of the Connemara oil accumulation, the general stratigraphy consists of clays and sands of Quaternary age overlying a Tertiary succession of sandstones and mudstones with some lignites and limestone stringers to a depth of *c.* 1 km. These sediments in turn rest upon a Cretaceous and Jurassic succession that contains the source, reservoir and seal rocks for the Connemara oil accumulation.

Two indicators of the presence of shallow gas that could be hazardous to drilling operations have been identified on the high-resolution 2D records. The potential gas indicator identified in Fig. 10 occurs at *c.* 500 m below mean sea level. This amplitude anomaly covers an area of *c.* 6 km × 1 km, running NE–SW through the site survey

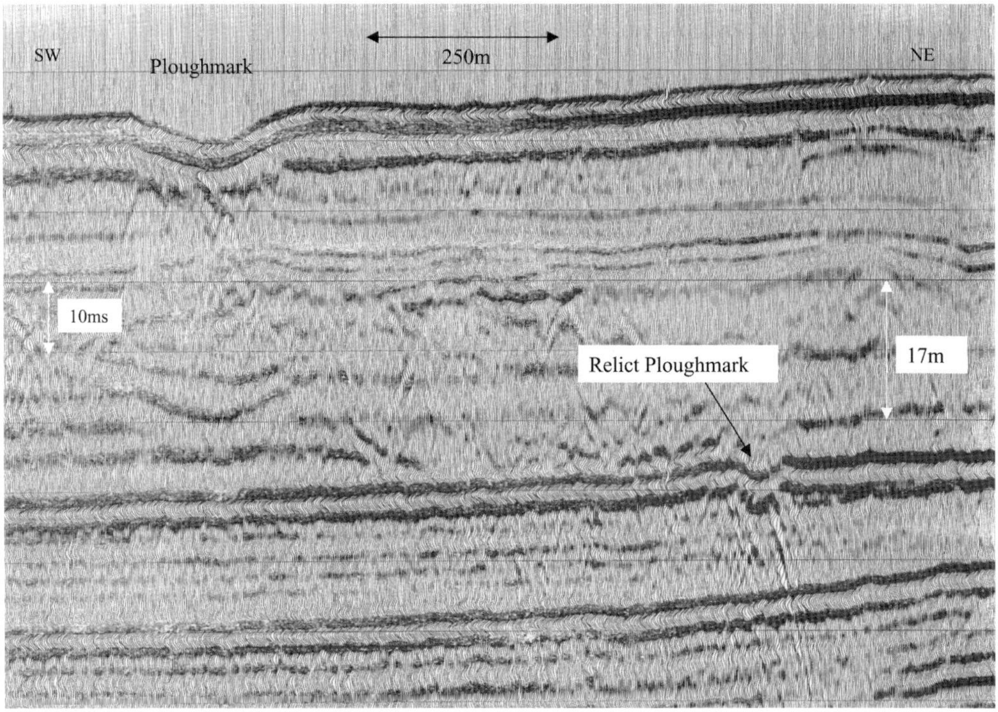

Fig. 6. Mini airgun record illustrating ploughmark with frequency reduction at its base. (Also note relict ploughmark at 40 m below sea bed, exhibiting similar frequency characteristics.)

area. Investigation of the 3D seismic data reveals that this anomaly extends significantly to the east of the site survey area. An attribute map of horizon amplitude indicates the extent of this anomalously high amplitude (Fig. 11). This horizon correlates with the top of a sand unit in the exploration wells, and the anomaly is interpreted as evidence of a gas accumulation within this sand, trapped by an overlying silty horizon. Other typical gasification characteristics, such as numerous diffractions and zones of frequency attenuation, are also associated with this anomaly and are best illustrated on amplitude and instantaneous frequency displays (Fig. 12). A decrease in frequency is observed below the gas-charged horizon, but is also observed extending upwards from it, suggesting leakage of gas vertically from the gas-charged layer. This is consistent with the fact that the overlying silty horizon may not act as an efficient seal. Only three of the exploration wells were drilled through this high-amplitude zone: well 26/28-1, where no data were logged for this interval, and wells 26/28-2 and 26/28-4A, where no gas shows were recorded. However, neither of these latter wells pass through the highest-amplitude area, so it is possible that the higher gas saturations have not been sampled by drilling.

The second indicator of deep-seated gas within the site area occurs in the form of isolated gas chimneys. These generally occur somewhat deeper in the succession, typically from 600 to 1200 ms. They appear as zones of acoustic masking often blanking out large areas of data. Figure 13 shows the appearance of one such chimney on a high-resolution 2D seismic line. The seismic characteristics of this chimney suggest that the rising gas migrates laterally at various levels as it passes through sand layers. This appears on the section as amplitude anomalies and minor diffractions on either side of the gas chimney.

These gas chimneys are also clearly identifiable on the 3D seismic data. Some typical features associated with vertical gas migration chimneys, which have been documented by Heggland (1997, 1998), are observed in Figs 13 and 14. These features include velocity pull-down, clearly shown on the high-resolution 2D example (Fig. 13), edge diffractions and brightening of abutting horizons (Fig. 14). Unpublished evidence from the 3D seismic time slice data indicates that these gas chimneys appear as

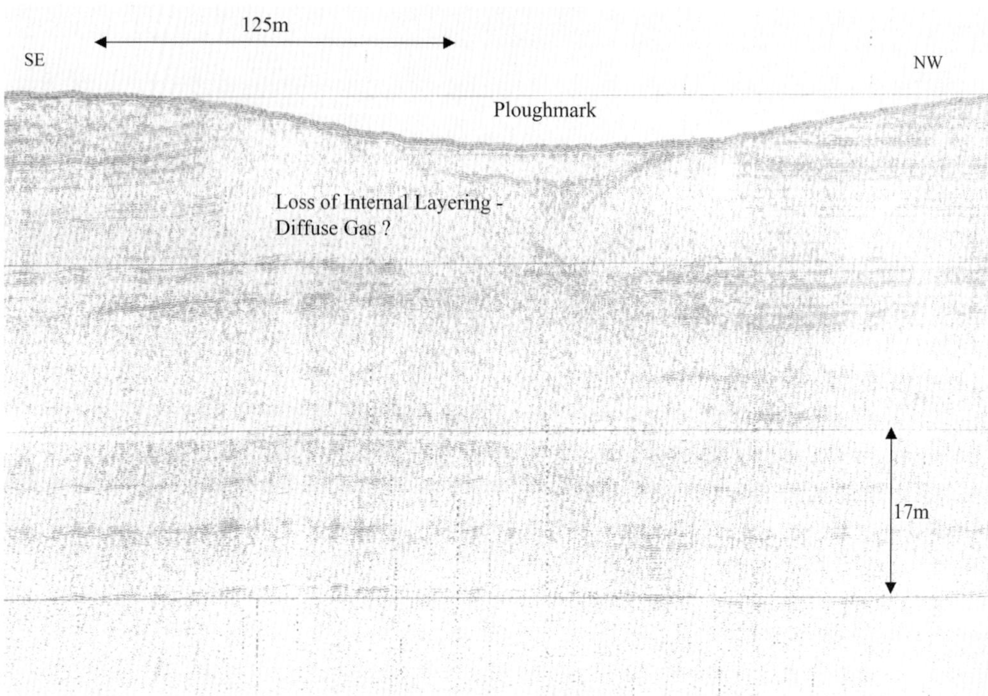

Fig. 7. Pinger profile demonstrating the typical seismic response to near-sea-bed sediments beneath ploughmark features in the site survey area. (Note the loss of internal layering, which is interpreted to represent dewatering or venting of shallow gas.)

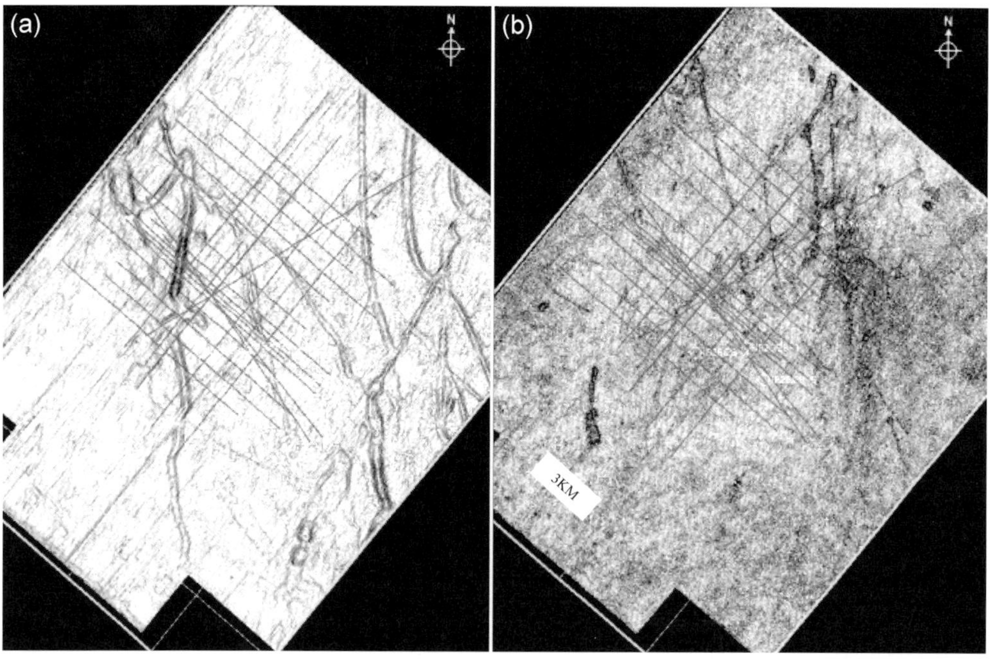

Fig. 8. A composite display, from the 3D seismic dataset, illustrating: (**a**) sea-bed time-structure and present-day ploughmarks; (**b**) time structure of sub-sea-bed horizon that contains the relict ploughmark identified in Fig. 9. It should be noted that the present-day and relict ploughmarks have a similar orientation. (Three-dimensional images based on original work performed by Statoil.)

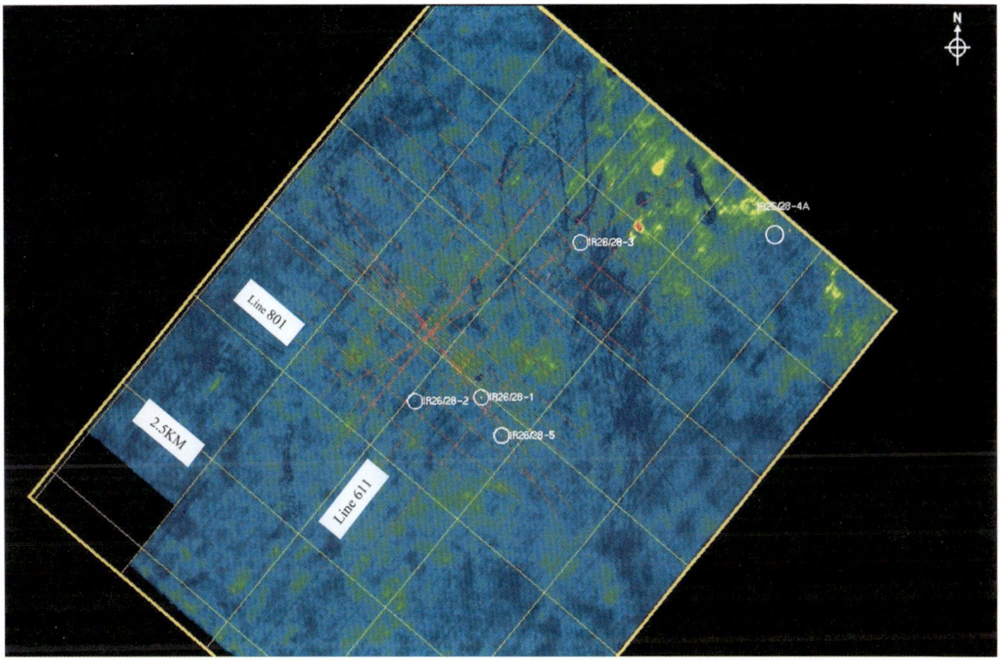

Fig. 9. Impedance plot of sub-sea-bed horizon containing a relict ploughmark, generated via 3D impedance attribute mapping with Charisma workstation. High-impedance area is represented by the darker image. Such features are often infilled with lower-impedance gas-charged sand. (Three-dimensional image based on original work performed by Statoil.)

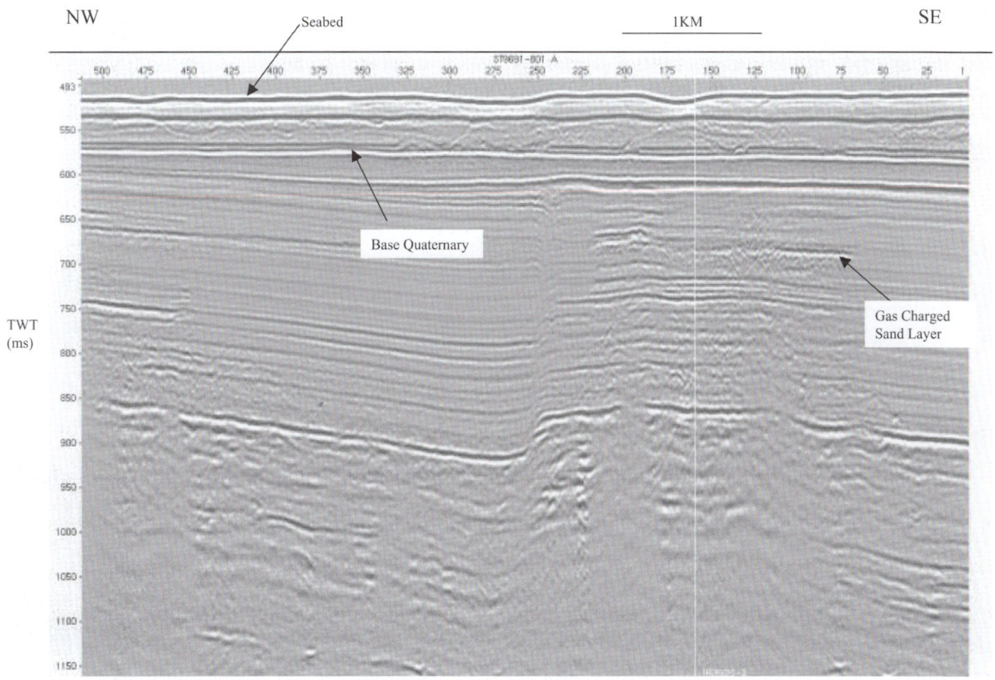

Fig. 10. High-resolution 2D seismic line, ST9691-801, through interpreted gas-charged sand layer at around 650 ms two-way travel time (TWT), shotpoint 100. (Note the high amplitudes and diffractions associated with this event.)

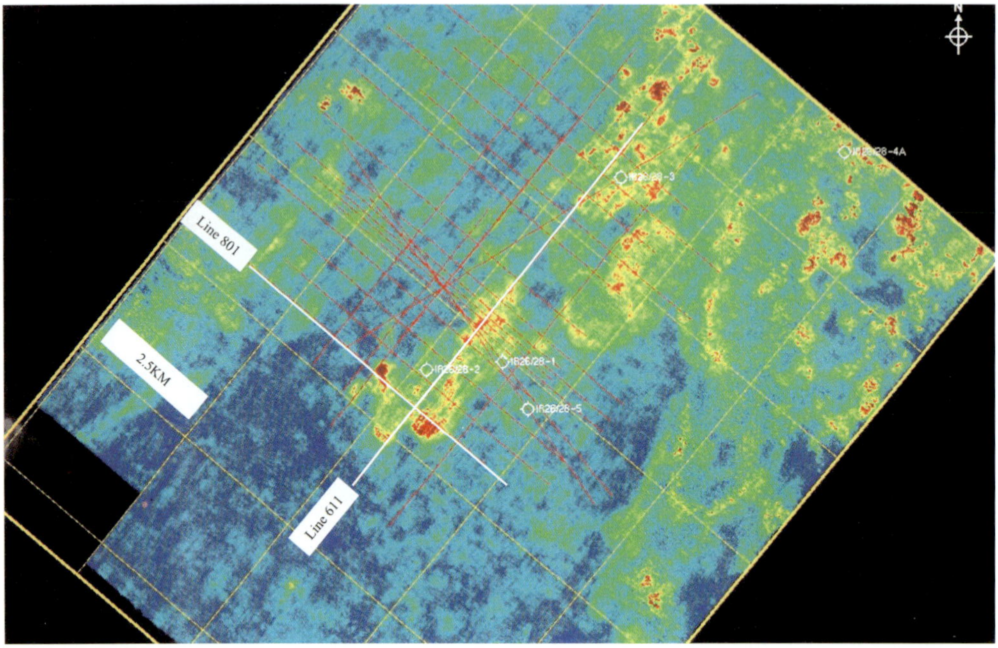

Fig. 11. Three-dimensional seismic amplitude map showing the extent of the anomalous gas-charged sand layer (3D image based on original work performed by Statoil).

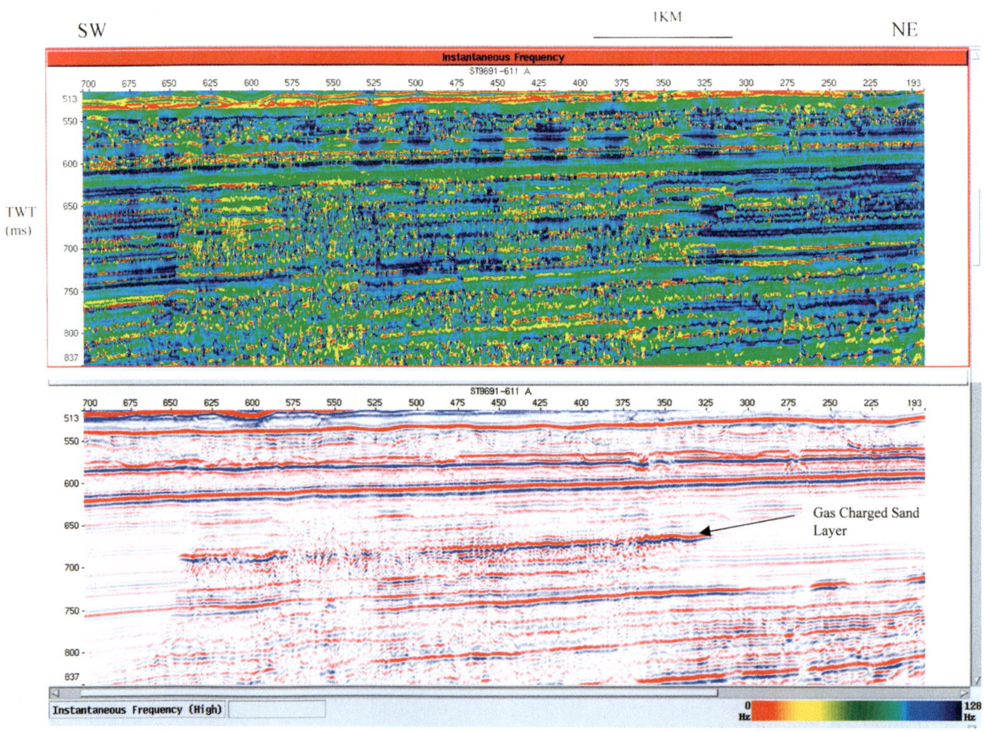

Fig. 12. High-resolution 2D seismic line, ST9691-611: (**a**) instantaneous frequency display; (**b**) amplitude display. It should be noted that the absorption of high frequencies is observed both below and above the interpreted gas sand, suggesting that vertical migration has occurred through an ineffective seal.

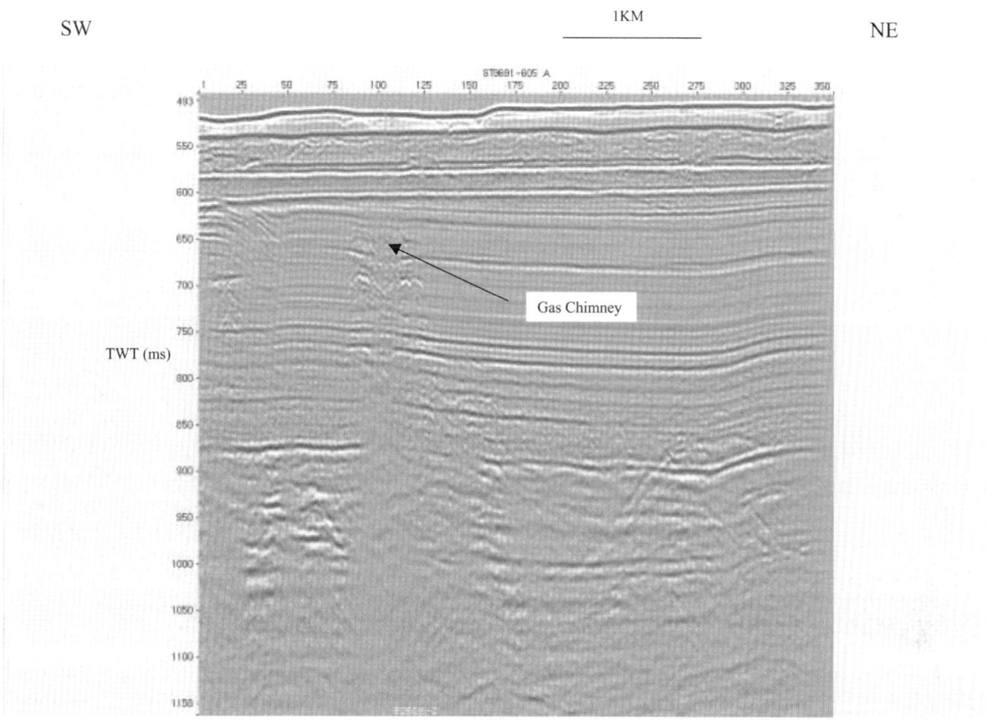

Fig. 13. Gas chimney as observed on high-resolution 2D seismic line. (Note the high amplitudes on the flanks of the chimney, which are indicative of lateral migration of gas into sand layers.)

circular anomalies in cross-section with a high-amplitude rim.

Discussion

The features described in this paper point to two potentially different sources for the shallow gas above the Connemara oil accumulation. In the near-sea-bed sediments, the evidence suggests a shallow source of diffuse gas, not directly linked to any deeper system. Sea-bed pockmarks, particularly those with carbonate concretions, are generally interpreted as being caused by venting gas. Evidence of diffuse gas in the near-sea-bed sediments is indicated by the presence of pockmarks, acoustic voids, diffraction hyperbolae and a low-frequency seismic response. The fact that this gas has not been detected by the shallow cores is not surprising, as the low quantities needed to produce these effects are not usually detectable, unless geochemical sampling is undertaken. A final pointer to the diffuse gasification is the variation in shear strength recorded in very similar sediment types at different locations in the site survey area. The 3D seismic data clearly identify the presence of relict ploughmarks, which are known to accumulate and trap shallow gas. Although the seismic characteristics of these features are similar to those encountered in other regions where gas shows have been confirmed, not enough evidence is available here to reach a definitive conclusion concerning the presence or absence of gas. In particular, none of the wells have penetrated these relict ploughmarks, so they remain unsampled by drilling in this area.

A second set of gasification indicators points to a deeper source of gas, which suggests that gas may be leaking from the Connemara trap or an adjacent source kitchen area. The horizon, which exhibits anomalously high amplitudes, together with masking of underlying reflectors, numerous diffractions and absorption of high frequencies, is most plausibly explained in terms of a gas-charged sand layer. Evidence from drilling is inconclusive, but the presence of a low saturation gas sand is a plausible geological model that could explain the seismic evidence. Finally, a number of potential gas chimneys have been identified in the area. Some of these chimneys are coincident with major faults, thus

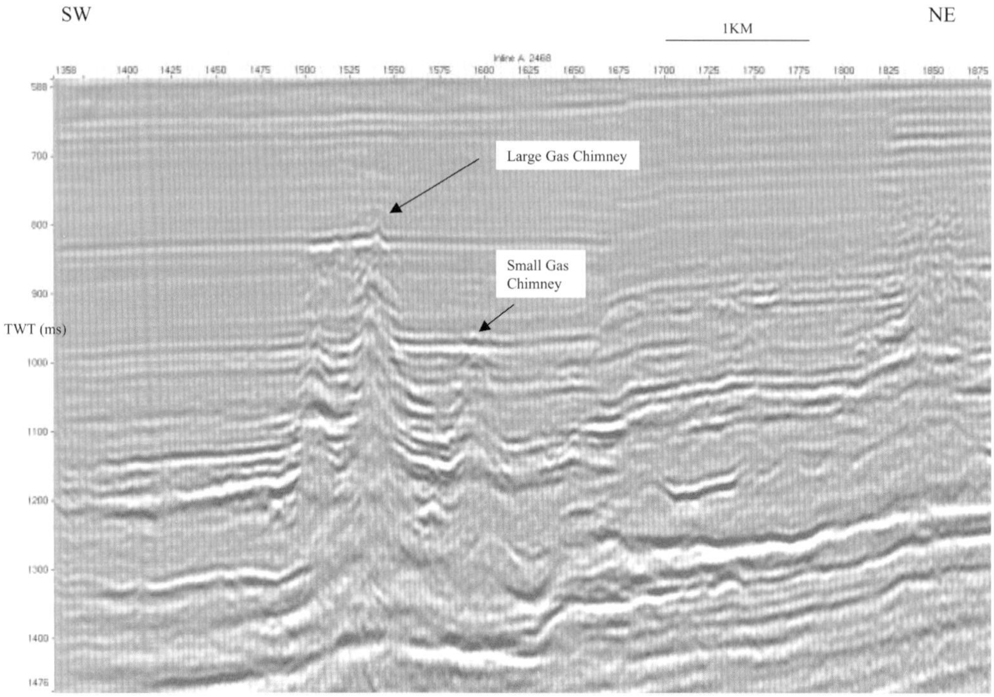

Fig. 14. Three-dimensional seismic inline 2468, illustrating a small gas chimney, which terminates just above 1000 ms, and a larger gas chimney, which rises to around 800 ms, where an apparent gas-charged horizon is present (based on original work performed by Statoil).

suggesting that the fault planes are providing the vertical migration pathways for the gas. These gas chimneys have been mapped by a combination of the high-resolution 2D seismic data, which possess frequencies of *c*. 100 Hz at the horizons of interest, and the 3D seismic data, which have a much lower resolution but give the spatial information that helps to elucidate the shape and distribution of these chimney features.

Conclusions

Results of the site survey investigation carried out in the area of the Connemara oil accumulation suggest the presence of shallow gas. The widespread occurrence of amplitude anomalies and interpreted gas accumulation and gas migration features over a relatively small area implies that shallow gas could pose serious problems for drilling in this part of the Porcupine Basin. The large gas chimneys interpreted in this area suggest that gas may be leaking from the Connemara trap or from a gas mature source kitchen adjacent to the trap.

I would like to thank Statoil Exploration (Ireland) Ltd for permission to use the data from this site survey. The comments of the two referees are gratefully acknowledged.

References

CROKER, P.F. 1995. Shallow gas accumulation and migration in the western Irish Sea. CROKER, P.F. & SHANNON, P.M. (eds) *The Petroleum Geology of Ireland's Offshore Basins*. Geological Society, London, Special Publications, **93**, 41–58.

EARLS, T.C. 1995. Potential for development of the Connemara field, Block 26/28. *In*: CROKER, P.F. & SHANNON, P.M. (ed.) *The Petroleum Geology of Ireland's Offshore Basins*. Geological Society, London, Special Publications, **93**, 343.

GALLAGHER, J.W. & HEGGLAND, R. 1994. Shallow gas evaluations based on conventional 3D seismic data. *EAEG 56th Meeting and Technical Exhibition, Vienna, 6–10 June, Extended Abstracts*.

GALLAGHER, J.W., HOVLAND, M., AMALIKSEN, K.G., BRAATEN, A.M., JACOBSEN, T., & GRANLI, J.R. 1991. *Shallow Gas on Haltenbanken: Summary of Study by the Shallow Gas Group; August 1986–December 1989*. Open report, Statoil.

HARRINGTON, P.K. 1985. Formation of pockmarks by pore-water escape. *Geo-Marine Letters*, **5**, 193–197.

HEGGLAND, R. 1997. Detection of gas migration from a deep source by the use of exploration 3D seismic data. *Marine Geology*, **137**, 41–47.

HEGGLAND, R. 1998. Gas seepage as an indicator of deeper prospective reservoirs: a study based on exploration 3D seismic data. *Marine and Petroleum Geology*, **15**, 1–9.

HEGGLAND, R., NYGAARD, E. & GALLAGHER, J.W. 1996. Techniques and experiences using exploration 3D seismic data to map drilling hazards. *Offshore Technology Conference Proceedings*, 119–127.

HENRIET, J.-P., DE MOL, B., VANNESTE, M., HUVENNE, V., VAN ROOIJ, D. & THE 'PORCUPINE–BELGICA' 97, 98 AND 99 SHIPBOARD PARTIES. 2001. Carbonate mounds and slope failures in the Porcupine Basin: a development model involving fluid venting. *In*: SHANNON, P.M., HAUGHTON, P.D.W. & CORCORAN, D.V. (eds) *The Petroleum Exploration of Ireland's Offshore Basins*. Geological Society, London, Special Publications, **188**, 375–383.

HOVLAND, M., JUDD, A.G. 1998. *Seabed Pockmarks and Seepages: Impact on Geology, Biology and the Marine Environment*. Graham & Trotman, London.

HOVLAND, M., CROKER, P.F. & MARTIN, M. 1994. Fault-associated seabed mounds (carbonate knolls?) off western Ireland and north-west Australia. *Marine and Petroleum Geology*, **11**, 232–246.

MACDONALD, H., ALLAN, P.M. & LOVELL, J.P.B. 1987. Geology of oil accumulation in Block 26/28, Porcupine Basin, offshore Ireland. *In*: BROOKS, J. & GLENNIE, K.W. (eds) *Petroleum Geology of North West Europe*. Graham & Trotman, London, 643–651.

Carbonate mounds and slope failures in the Porcupine Basin: a development model involving fluid venting

J. P. HENRIET, B. DE MOL, M. VANNESTE, V. HUVENNE, D. VAN ROOIJ & THE 'PORCUPINE–BELGICA' 97, 98 AND 99 SHIPBOARD PARTIES

Renard Centre of Marine Geology (RCMG), Gent University, Krijgslaan 281, S8, B-9000 Gent, Belgium (e-mail: jeanpierre.henriet@rug.ac.be)

Abstract: High-resolution reflection seismic investigations carried out in the Porcupine Basin, SW of Ireland, have shed light on the presence of several provinces of giant carbonate mounds. An intriguing setting is found on the northern slope of the basin. A cluster of surface mounds appears to be flanked by a large upslope, crescent-shaped province of buried mounds. Below the transitional zone, large imbricated slide scars suggest repeated failures. The buried mounds rise from an undisturbed basal horizon and seem to represent a single event, confined in time and space. Both high-resolution and industrial seismic data reveal a close vertical match of the mound cluster with a lower, buried sea-bed failure, where hydrate build-up may have played a role. The latter association may not be entirely fortuitous. It is suggested that gas venting may have triggered the formation of the mound clusters, and that the underlying sea-bed failure forms a previous but different expression of gas venting, on a common, episodic fluid migration pathway but under strongly contrasting bottom water temperature conditions.

Industrial seismic surveys and cruises of the research vessels *Belgica*, *Prof. Logachev* and *Pelagia* in 1997, 1998 and 1999 in the Porcupine and Rockall basins, west of Ireland, have revealed large provinces and clusters of impressive sea-bed mounds, up to 200 m high and 2 km in diameter (Akhmetzanov *et al.* 1998; Croker & O'Loughlin 1998; Henriet *et al.* 1998). A set of 31 large mounds was described by Hovland *et al.* (1994) on the northern slope of the Porcupine Basin. These 'Hovland' mounds are clustered in an area of *c.* 375 km^2 (15 km × 25 km) and lie in water depths between 600 and 900 m. Many of them are girdled by deep moats. A high-resolution seismic survey carried out by R.V. *Belgica* in May 1997 revealed that these large sea-bed mounds are only one of several mound types in the basin (Henriet *et al.* 1998). The Hovland mounds are flanked to the north by a large, crescent-shaped province (Fig. 1) of smaller, buried build-ups, which we have named the 'Magellan' province, after the commercial survey vessel from which they were first observed. Still another, fundamentally different, mound setting has been discovered on the eastern margin, the impressive range of 'Belgica' mounds, named after the R.V. *Belgica*.

The possible role of methane seeps in the genesis and growth of carbonate mounds in these settings and in hydrocarbon provinces in general is a debated issue (Hovland 1998; Ivanov *et al.* 1998; van Weering & Henriet 1998). The Hovland mounds seem to be associated with possible fault-controlled paths of migration of methane from deeper hydrocarbon reservoirs (Hovland *et al.* 1994). Pockmarks are observed south of them (R.V. *Belgica* 97 data). The Connemara oil field, some 90 km further upslope of the Magellan province (53°N), features large diapiric gas chimneys (Games 2001). High-resolution seismic data from site surveys and from the 1999 survey of R.V. *Belgica* in this oil field area have given ample evidence of shallow gas. Other evidence for fluid migration occurs on the eastern margin of Porcupine Basin: in the Belgica mounds area, where an intriguing mound structure (Henriet *et al.* 1999) may argue for fluid migration. Paradoxically, analyses of pore fluids and carbon isotopes in carbonates from shallow cores on the Porcupine mounds have hitherto failed to produce any conclusive evidence of the presence of hydrocarbons in sediment pores or on the possible role of methane in any carbonate precipitation in surficial layers or corals (De Mol *et al.* 1998; Ivanov *et al.* 1998).

The present paper adds a few more elements to the debate. It focuses on an intriguing relationship between the Hovland and Magellan mound

From: SHANNON, P.M., HAUGHTON, P.D.W. & CORCORAN, D.V. (eds) 2001. *The Petroleum Exploration of Ireland's Offshore Basins*. Geological Society, London, Special Publications, **188**, 375–383.
0305-8719/01/$15.00

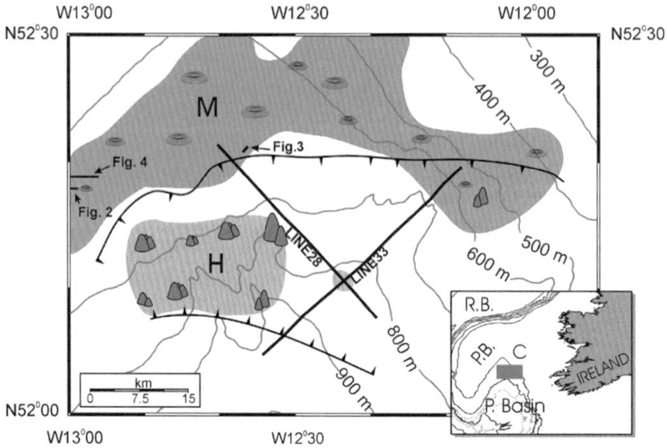

Fig. 1. Generalized map of the buried Magellan mound province (shaded area M, north) and the Hovland province of surface mounds (shaded area H, south) in the north of the Porcupine Basin, SW of Ireland. The headwall scars of the slope failures in the Hovland province are indicated. The sea-bed failure feature underlying the Magellan province is considered to largely match the areal extent of this province, and is not shown separately. Bathymetry is in metres. The location of the Connemara oil field (C) in the Porcupine Basin (P. Basin) is shown in the inset map: PB, Porcupine Bank; RB, Rockall Basin. The location of Figures 2–4, and of Profiles 28 and 33 (see Fig. 7), are shown.

settings, and on their spatial association with large buried slope failure features. A development model involving gas venting and the build-up and decay of gas hydrates may suggest a remarkable interplay between internal (geological) and external (climatic) controls on the coupling of mounds and slope failures on the northern slope of the Porcupine Basin through Quaternary times.

The Magellan mounds

Gravity cores, box cores and grab samples recovered by industrial surveys and by R.V. *Prof. Logachev* from the Hovland mounds and from a single outcropping Magellan mound contained carbonate-rich silt with extensive deep-water colonial coral debris (*Lophelia pertusa*, *Madrepora oculata*). These are identical to the fauna collected by Wyville Thomson in the same area during the historical cruise of the H.M.S. *Porcupine* in 1869 (Thomson 1873). Though none of these mounds have been drilled to date, their seismic character and the facies of the sediments sampled at their surface, coupled to a striking morphological analogy with some Palaeozoic carbonate build-ups (Monty 1995), suggest the presence of giant deep-sea carbonate mounds.

The Magellan province is crescent-shaped and closely matches the isobaths of the northern slope of the Porcupine Basin. The province was mapped in May 1997 and 1998 by R.V. *Belgica* and extends *c*. 90 km along the slope, covering an area of *c*. 1200 km^2. The width, in upslope direction, varies between 8 and 20 km.

The Magellan mound province appears as a kind of upslope 'halo' fringing the Hovland mound cluster, and separated from the latter by a moundless gap of 6–8 km. Swarms of sometimes very closely spaced mounds within this province typically rise from a single, undisturbed horizon, and are generally buried below a few tens of metres of drift sediments. This suggests a particular growth event, confined in space and time. Furthermore, there seems to be a distinct spatial pattern within this crescent, with mounds reaching their largest size at the western boundary, where a few mounds even extend to the present sea bed or rise high above the sea bed (Fig. 2, left). In a northerly direction, however, the mounds decrease progressively in size and finally fade out. Some detached patches of very small mounds displaying this upslope fade-out have been identified further north, midway between the Magellan province and the Connemara oil field. The downslope boundary is abrupt and it rather closely matches the trace of a large buried erosional or slump scar.

High-resolution reflection seismograms over the Magellan province show the frequent occurrence of lung- to butterfly-shaped twin structures. Examples are displayed in Figures 2 and 3. They might represent (axial to slant) cross-sections through ring-shaped build-ups. The ring hypothesis could be supported in a preliminary

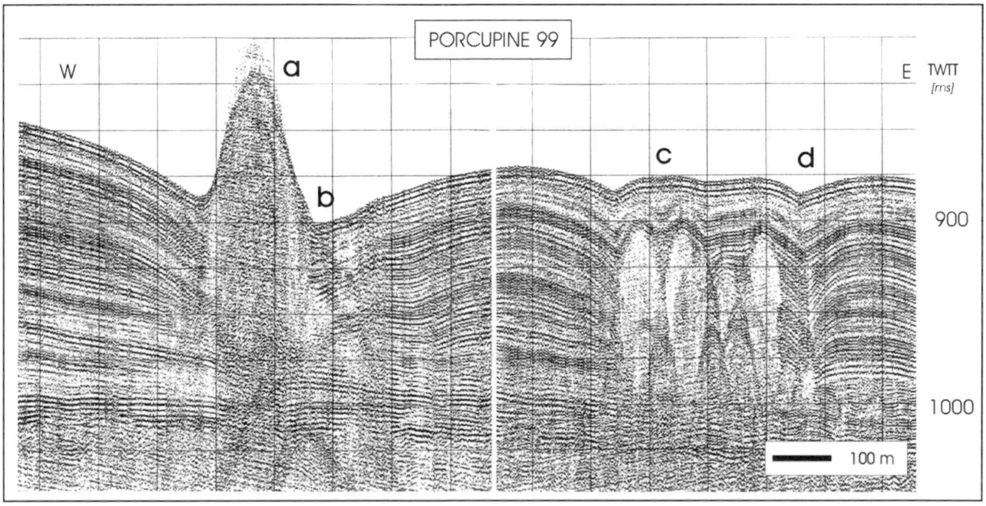

Fig. 2. High-resolution reflection seismic profile through a surfacing Magellan mound (a, left) and its moats (b) at the southwestern edge of the Magellan province, and of adjacent buried twin-shaped cross-sections, right (c). The cuspate depression, not directly flanking the mound structures (d), probably betrays the lateral proximity of an off-line mound. Location is shown in Figure 1.

way by two out of three carefully navigated perpendicular sections from the 1997 survey, but conclusive evidence requires further detailed investigations, which are also planned with deep-towed seismic devices to increase the lateral resolution.

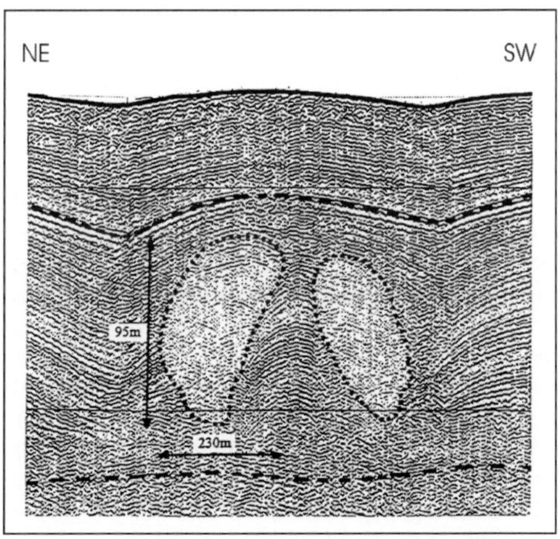

Fig. 3. High-resolution reflection seismic profile through a symmetrical twin structure, interpreted as a possible cross-section through a ring-shaped mound, in the northern part of the Magellan mound province. The top line is the sea bed; dashed lines are correlation lines. The moats fringing the mound are well expressed in the embedding drift sediment, and can be traced to the sea bed. Location is shown in Figure 1.

A count of the mounds along the high-resolution seismic lines, extrapolated to the whole area, suggests the presence of several hundreds of Magellan build-ups (Pillen 1998). Fully developed mounds have a height between 60 and 90 m, exceptionally up to 130 m. At least 40% of the medium-sized cross-sections in the northern part of the province suggest a possible ring shape. The outer diameter of twin structures generally ranges between 100 and 400 m, exceptionally up to 800 m. The central gap has a width between 25 and 75 m. Many butterfly- or lung-shaped cross-sections are remarkably symmetrical (Figs 2 and 3), other are asymmetric, and some display coalescing twin structures. Some sections also distinctly show twin elements merging towards the top, which might suggest, in the case of axial cross-sections, a possible terminal closure and capping of the central gap.

The setting of the Magellan mounds in the embedding drift sediments is most interesting. The flat and undisturbed basal set of reflectors caps a peculiar deeper horizon characterized by small-scale deformations and laterally varying amplitude anomalies (alternating white patches and bright spots), located some 15–20 m below the base of the mounds (Fig. 4), under a very low-angle unconformity. Above this flat sequence, the drift sediments show a wavy to hummocky reflector configuration, with cuspate depressions alternating with broad convex swells. This morphology extends to the sea bed.

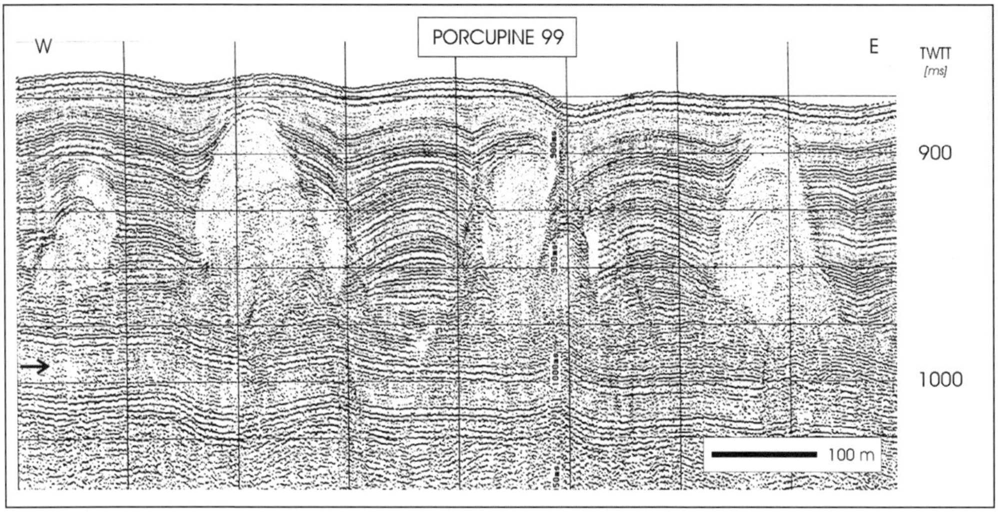

Fig. 4. High-resolution reflection seismic profile through buried mounds at the southwestern edge of the Magellan province, displaying the lower disturbed horizon (arrow), with small patches of laterally varying amplitudes. (Note the coincidence between some deformations in this horizon and overlying mounds or moats.) Location is shown in Figure 1.

On many sections, the V-shaped cusps do coincide with moats flanking the buried mounds (Fig. 2), suggesting a primary control from the mound build-up. At the surface, the grooves and ridges apparently display a remarkable parallel, north–south-trending pattern on the sea bed (Croker, pers. comm.). Not all mounds, however, are flanked by moats. In some profiles, the embedding sediments remain close to horizontal up to the flanks of the mounds. The top of virtually all mounds is capped by a convex sediment drape.

The hummocky to wavy reflector configuration of the drift sequence is generally confined to the Magellan province and stops abruptly at its boundaries. It is believed that many cuspate depressions, not directly flanking mound structures (Fig. 2, marked 'd'), nevertheless betray the lateral proximity of a mound.

Genetic model involving fluid venting

It is assumed that these mounds are largely built of carbonate sediments, with a carbonate concentration of 25–65% (measured on a proximal surface mound in the Hovland province; Mazurenko 1998). The carbonate at the surface of the Hovland mounds and in the shallow subsurface consists of a normal pelagic coccolith ooze. Foraminiferal assemblages are typical of the upper bathyal realm (Coles *et al.* 1996). This does not exclude a possibly wide compositional and textural range within the mounds, including recurrences of large coral debris, as observed in some cores.

Although large differences do exist, it is tempting to refer to fossil examples of large carbonate mud mounds, for instance the Upper Tournaisian 'Waulsortian reefs', which also developed in subphotic environments. Lees & Miller (1995) published an influential paper emphasizing the role of a bacterial biofilm in providing organic substrates for biogeochemical reactions leading to carbonate precipitation and early induration. Early induration can account for the often remarkable steep slopes of such primarily soft muds, preserving them from failing in strong bottom current regimes. Endemic or opportunist epifauna such as cold-water corals (hosting polychaetes), sponges, bryozoans, bivalves, gastropods and echinoids could be regarded as a guest fauna, not essential for the mound growth. Other examples of the possible mediating role of bacteria in carbonate build-up and/or early diagenesis have been reported (Peckman *et al.* 1998), in particular in Devonian mounds in Belgium (Bourque & Boulvain 1993), Algeria (Wendt *et al.* 1997), Italy (Cavagna *et al.* 1998) and Morocco (Belka 1998).

Any local proliferation of life on the sea bed, of whatever nature, requires a significant flux of nutrients. The way in which mounds display a spatial association with potential fluid migration sites in Porcupine Basin and the lack of any isotopic evidence in surface samples has been mentioned previously. It is suggested that the paradox may be solved by invoking short-lived 'triggering' controls of mound nucleation, which may have been of internal (seep) origin, followed by sustained 'growth' controls, which probably come from external fluxes. Seeps, whether on ridges or on active or passive margins, are by nature transient phenomena, both through the source dynamics controlling the advection and the sealing of conduits by mineral precipitation.

The dense and well-confined Magellan mound province of Porcupine Basin may bear witness to such a methane venting 'spike', of internal (seep) origin but in its effect probably modulated by external (climatic) factors. The interpreted ring shape of a large number of buried build-ups, if further corroborated, may argue for focused venting. The undisturbed nature of the basal horizons systematically observed immediately below the Magellan mounds (and above the disturbed horizon with variable reflection amplitudes), however, rules out large-scale mud volcanism or diapirism. Venting could have facilitated the formation of a fringe of authigenic carbonates, like those widely reported as primary deposits fringing seep sites (Hovland & Judd 1988; Reilly *et al.* 1996; Bohrmann *et al.* 1998). Ambient epifauna could have settled on such solid substrate, thus possibly providing a nucleus for mound growth.

A role for hydrates?

The spatial match between the mound swarm and the deeper, disturbed horizon characterized by a laterally varying acoustic reflectivity suggests a link. The vertical projection at the sea bed of the upper boundary of this disturbed horizon spatially matches the upslope boundary of the overlying Magellan mound province, which appears to fit the 500–600 m isobath. The control of water depth and hence pressure may, in turn, under suitable conditions of temperature and methane flux, involve a possible role of methane hydrates.

Gas hydrates are ice-like crystalline compounds that occur when water molecules form a cage structure around guest molecules under conditions of high pressure and low temperature (Sloan 1998). Natural gas hydrates are mostly found in marine or lacustrine sediments where water depths exceed 300–500 m (Kvenvolden 1998). Their generation generally requires the presence of a prolific methane source.

Recent industrial data covering the westernmost part of the Magellan province shed a new light on the nature of the buried deformed layer under the Magellan mounds, characterized by laterally varying amplitudes, and allow us to identify it with a large sea-bed failure, resembling a shattered windscreen with large angular

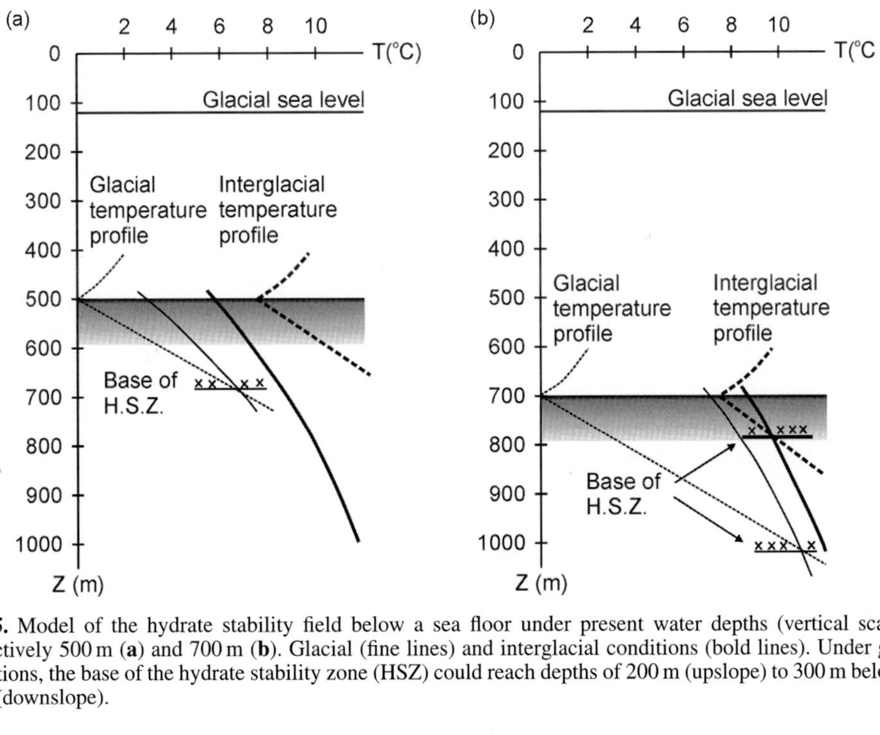

Fig. 5. Model of the hydrate stability field below a sea floor under present water depths (vertical scale) of respectively 500 m (**a**) and 700 m (**b**). Glacial (fine lines) and interglacial conditions (bold lines). Under glacial conditions, the base of the hydrate stability zone (HSZ) could reach depths of 200 m (upslope) to 300 m below sea floor (downslope).

blocks (Croker, pers. comm.). If the spatial match between the Magellan mound province and the underlying past sea-bed failure can be corroborated over the full areal extent of the Magellan mounds crescent, it means that the upper boundary of the slope failure also closely fits (palaeo)depth contour lines. Although further control is needed on palaeodepths and their evolution in this part of the Porcupine Basin, the sea-bed failure area could originally have been situated in a palaeodepth range between 500 and 700 m under interglacial sea-level conditions. This depth interval, when exposed to cold-water currents in glacial seas, moves into the stability field of gas hydrates, as verified by a simple model calculation below.

Evidence of glacial conditions in Porcupine Basin rely primarily on the record of Irish land ice, which has been found up to the Munsterian Cold Stage (300–130 ka BP; Mitchell & Ryan 1997). Earlier glacial conditions cannot be ruled out. The Connemara Field 3D seismic survey, upslope of the Magellan mound province, has highlighted examples of buried iceberg ploughmarks (Games 2001). A prominent horizon scoured by icebergs is also clearly visible on site survey data and on the high-resolution reflection data shot by R.V. *Belgica* in June 1999 on the Connemara Field. Under such glacial edge conditions, bottom water temperatures on the northern slope of Porcupine Basin probably did not exceed 0 °C.

Figure 5 shows the stability conditions of methane hydrates below the sea bed, under present water depths of 500 m (Fig. 5a) and 700 m (Fig. 5b). The methane hydrate stability in sea water (Dickens & Quinby-Hunt 1994) is depicted by the phase boundary (bold continuous line). The model assumes an arbitrarily set minimal post-glacial bottom water temperature value of 7.5 °C. The NOAA Levitus *World Ocean Atlas* refers to a present sea-floor temperature of 10.2 °C in the near vicinity of the Hovland mounds (Levitus *et al.* 1994). A thermal steady-state profile with a gradient of 0.03 °C m^{-1} is assumed (Croker & Shannon 1987), as well as current values for the thermal properties of sea-bed sediments. Under glacial conditions, a bottom water temperature of 0 °C is assumed, with a sea level lowered by 100–120 m. The temperature profile is drawn as a fine dashed line. Under such glacial conditions and below a slope situated between water depths of 580 and 380 m, the lower boundary of the hydrate stability zone lies between 300 m (downslope) and 200 m (upslope) below sea floor.

We next consider a postglacial situation, 10 ka after a sudden water temperature rise of +7.5 °C (bold dashed line). At the upper end of the slope, under interglacial water depths of 500 m, we clearly leave the hydrate stability window. Downslope, below a sea floor at 700 m, a residual stability zone of 100 m depth is still found, but it disappears if temperature is raised to 9 °C. This argues for the absence of gas hydrates in the northern part of the Porcupine Basin,

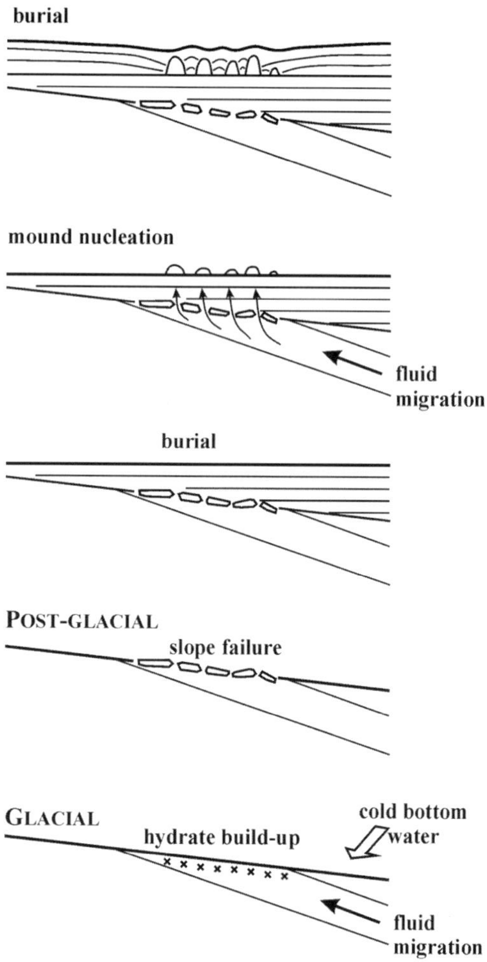

Fig. 6. Proposed development model for the genesis of the swarm of Magellan mounds on a site of episodic fluid migration, under strongly varying bottom water temperatures. Under glacial conditions, a horizon of gas hydrates could build up in regions of prolific (even transient) methane flux. Decay of the hydrated horizons could generate slope failure. In renewed methane flux conditions, in warmer waters, the disrupted horizon funnelled the migrating fluids to the sea bed, possibly contributing to venting and mound nucleation. The mounds and associated moats influence the morphology of the burying sediments.

upslope of the Hovland mound cluster, under present conditions.

Obviously, many factors have not been taken into account in this simplified model, e.g. burial depth, basin subsidence, the isostatic response of the margin in glacial–interglacial cycles, etc. The model, however, suggests the possible role of a critical depth zone on the continental slope of glaciated margins, in which hydrates may have developed. Provided a prolific methane source is present, such sea-bed depth zone is prone to failure under the influence of the waxing and waning of hydrates, paced by glacial forcing cycles. This could have been the fate of the past sea bed, found some 15–20 m below the Magellan mounds.

This failed sea bed was consequently buried, and just above it, under later and probably strongly contrasting oceanic conditions, a dense and strongly delineated swarm of mounds started growing. Is this spatial coincidence purely fortuitous, or can these superimposed sea-bed processes (the possibly hydrate-controlled sea-bed failure and the growth of a dense swarm of mounds) have a common origin? In other words, could the Magellan mound clusters and the underlying sea-bed failure form two different sea-floor expressions, at different times and under contrasting oceanic conditions, of a common fluid migration and gas venting pathway? This development model, displayed in Figure 6, is a topic for further research, in particular within the framework of European projects.

Major slope failures in the Hovland province

As shown in Figure 1, the downslope boundary of the Magellan province coincides with or slightly straddles (in the northeastern region) a complex, major buried slope failure. Profiles through the western slide scar (Fig. 7, top) show two imbricated slide events (a, b), each followed by a phase of fill, first by high-energy deposits and next by drift drape. The maximal scar height (at the western extremity) is *c.* 200 m. Further downslope, the slope failure suddenly abuts against another steep erosional slope, shaping a bowl rather than an open-ended slide track.

Such observations support past events of possibly seep- or hydrate-related slope failures in the Hovland area. The proposed match of the Hovland mounds with deeper faults has been interpreted in terms of a possible near-vertical pathway of fluids under recent conditions (Hovland *et al.* 1994). A glacial setting, with possible hydrate build-up, may have caused deflected flow. In such a model of deflected pathways, the possible role of the buried slide scars and of stratigraphic migration pathways, funnelling fluids further upslope towards the Magellan area, deserves further research. As the hydrate stability model shows, the depth of the crater below the Hovland area fits computed depths for the lower boundary of a hydrate stability zone under full glacial conditions. Sediment mobilization may have occurred either through blow-out or through scouring of

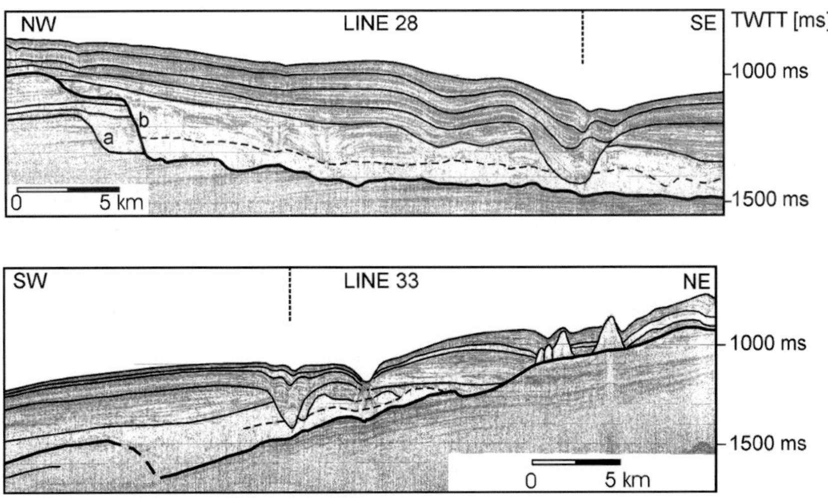

Fig. 7. Interpreted cross-sections through the slope failure area under the area of Hovland mounds (see Fig. 1 for line location). Note the imbricated slope failures (upper profile, a, b) at the northwestern extremity of the depression.

gas-charged sediments in an environment characterized by vigorous currents over the recent geological past. Sequences of high-energy deposits observed on the high-resolution seismic profiles argue for such strong currents.

Conclusions

Recent high-resolution reflection seismic investigations in Porcupine Basin, SW of Ireland, have unveiled a variety of carbonate mound provinces. The most intriguing province is an extensive region of hundreds of buried mounds: the Magellan mounds. This province fringes a large and complex buried slope failure feature, shaped as a crater-like depression and filled with high-energy deposits capped by a cluster of large sea-bed mounds: the Hovland mounds. The extent of the Magellan region itself largely coincides with a past sea-bed failure, located some 15–20 m below the base of the mounds.

Both mounds and slope failures in this region may be related to episodic fluid venting from deeper reservoirs, in a marine environment characterized by important variations in temperature and current regime. During Quaternary times, with repeated fluctuations from polar conditions in front of an Irish ice-sheet to temperate conditions, extreme variations in bottom water temperatures (up to 10 °C) may have translated into cycles of local growth of gas hydrates, fuelled by methane from deeper hydrocarbon reservoirs, and their subsequent decay. The Magellan mounds and the underlying sea-bed failure may form two different sea-floor expressions, at different times and under contrasting oceanic conditions, of a common fluid migration and gas venting pathway.

If this hypothesis can be substantiated by further evidence, it may shed light on the importance, in those hydrocarbon basins that have experienced glacial conditions, of a transient hydrate zone on the slope, between 500 and 700 m depth. Here, the waxing and waning of hydrate horizons, controlled by glacial cycles, may have played a significant role within a coupled system of fluid flow, slope destabilization and response of the biosphere.

The seismic investigations in Porcupine Basin have been carried out within the framework of the EU MAST III project ENAM II (European North Atlantic Margins), the MAST III concerted action CORSAIRES, the IOC/UNESCO programme Training Through Research and the University of Ghent BOF projects 'Deep water geophysics' and 'GOA Porcupine–Belgica'. Three of the authors (M.V., B.D.M., D.V.R.) are preparing a PhD supported by an IWT grant (Vlaams Instituut voor de bevordering van het Wetenschappelijk-Technologisch Onderzoek in de Industrie), and one (V.H.) with support of FWO (Fonds voor Wetenchappelijk Onderzoek—Vlaanderen). Support is gratefully acknowledged from the Petroleum Affairs Division (Dublin); Statoil Exploration (Ireland) Ltd and its partners Conoco (I.K.) Limited, Enterprise Energy Ireland Limited and Dana Petroleum plc; DWTC (Antarctic research programme, Federal Government, Brussels); and the Management Unit of the Mathematical Model of the North Sea (Brussels), for awarding R.V. *Belgica* ship-time access.

References

AKHMETZANOV, A.M., van Weering, Tj., KENYON, N.H. & IVANOV, M.K. 1998. Carbonate mounds and reefs at the Rockall Trough and Porcupine margins. *In*: DE MOL, B. (ed.) *Geosphere–Biosphere Coupling: Carbonate Mud Mounds and Cold Water Reefs*. UNESCO, Intergovernmental Oceanographic Commission Workshop Report, **143**, 45–46.

BELKA, Z. 1998. The origin of the Early Devonian Kess-Kess mud mounds of the Eastern Anti-Atlas (Morocco): evidence for submarine venting of methane-rich fluids. *In*: DE MOL, B. (ed.) *Geosphere–Biosphere Coupling: Carbonate Mud Mounds and Cold Water Reefs*. UNESCO, Intergovernmental Oceanographic Commission Workshop Report, **143**, 17–18.

BOHRMANN, G., GREINERT, J., SUESS, E. & TORRES, M. 1998. Authigenic carbonates from the Cascadia subduction zone and their relation to gas hydrate stability. *Geology*, **26**, 647–650.

BOURQUE, P.A. & BOULVAIN, F. 1993. A model for the origin and petrogenesis of the red stromatactis limestone of Palaeozoic carbonate mounds. *Journal of Sedimentary Petrology*, **63**, 607–619.

CAVAGNA, S., CLARI, P. & MARTIRE, L. 1998. The fossil cold seep carbonates of Monferrato (NW Italy): an example of localized carbonate precipitation induced by methane-degrading bacteria. *In*: DE MOL, B. (ed.) *Geosphere–Biosphere Coupling: Carbonate Mud Mounds and Cold Water Reefs*. UNESCO, Intergovernmental Oceanographic Commission Workshop Report, **143**, 14–15.

COLES, G., AINSWORTH, N.R., WHATLEY, R.C. & JONES, R.W. 1996. Foraminifera and Ostracoda from Quaternary carbonate mounds associated with gas seepage in the Porcupine Basin, offshore western Ireland. *Revista Española de Micropaleontologia*, **28**, 113–151.

CROKER, P.F. & O'LOUGHLIN, O. 1998. A catalogue of Irish offshore carbonate mud mounds. *In*: DE MOL, B. (ed.) *Geosphere–Biosphere Coupling: Carbonate Mud Mounds and Cold Water Reefs*. UNESCO, Intergovernmental Oceanographic Commission Workshop Report, **143**, 11.

CROKER, P.F. & SHANNON, P.M. 1987. The evolution and hydrocarbon prospectivity of the Porcupine Basin, offshore Ireland. *In*: BROOKS, J. & GLENNIE,

K.W. (eds) *Petroleum Geology of North West Europe*. Graham & Trotman, London, 633–642.

DE MOL, B., KEPPENS, E., SWENNEN, R. & HENRIET, J.P. 1998. Isotopic characterisation of ahermatypic coral on a 'Hovland' mound. *In*: DE MOL, B. (ed.) *Geosphere–Biosphere Coupling: Carbonate Mud Mounds and Cold Water Reefs*. UNESCO, Intergovernmental Oceanographic Commission Workshop Report, **143**, 31–32.

DICKENS, G.R. & QUINBY-HUNT, M.S. 1994. Methane hydrate stability in seawater. *Geophysical Research Letters*, **21**, 2115–2118.

GAMES, K.P. 2001. Evidence of shallow gas above the Connemara oil accumulation, Block 26/28, Porcupine Basin. *In*: SHANNON, P.M., HAUGHTON, P.D.W. & CORCORAN, D.V. (eds) *The Petroleum Exploration of Ireland's Offshore Basins*. Geological Society, London, Special Publications, **188**, 361–373.

HENRIET, J.P., DE MOL, B., PILLEN, S. *et al.* 1998. Gas hydrate crystals may help build reefs. *Nature*, **391**, 648–649.

HENRIET, J.P., DE MOL, B. & THE PORCUPINE–BELGICA '97 AND '98 SHIPBOARD PARTIES. 1999. Carbonate mounds, ring bioherms and past slope failures in the Porcupine Basin: prologue to a far-reaching story? *In*: CROKER, P.F. & O'LOUGHLIN, O. (eds) *The Petroleum Exploration of Ireland's Offshore Basins*. Petroleum Affairs Division, Department of the Marine and Natural Resources, Dublin, 90–92.

HOVLAND, M.T. 1998. Do carbonate reefs form due to fluid seepage? DE MOL, B. (ed.) *Geosphere–Biosphere Coupling: Carbonate Mud Mounds and Cold Water Reefs*. UNESCO, Intergovernmental Oceanographic Commission Workshop Report, **143**, 11–12.

HOVLAND, M., JUDD, A.G. 1988. *In*: *Seabed Pockmarks and Seepages*. Graham & Trotman, London.

HOVLAND, M., CROKER, P.F. & MARTIN, M. 1994. Fault-associated seabed mounds (carbonate knolls?) off western Ireland and north-west Australia. *Marine and Petroleum Geology*, **11**, 232–246.

IVANOV, M., KENYON, N.H., HENRIET, J.P., SWENNEN, R., LIMONOV, A. & TTR-7 SHIPBOARD PARTY. 1998. Carbonate mud mounds and cold water corals in the Porcupine Seabight and Rockall Bank: are they methane related? *In*: DE MOL, B. (ed.) *Geosphere–Biosphere Coupling: Carbonate Mud Mounds and Cold Water Reefs*. UNESCO, Intergovernmental Oceanographic Commission Workshop Report, **143**, 22–23.

KVENVOLDEN, K.A. 1998. A primer on the geological occurrence of gas hydrates. *In*: HENRIET, J.P. & MIENERT, J. (eds) *Gas Hydrates: Relevance to World Margin Stability and Climate Change*. Geological Society, London, Special Publications, **137**, 9–30.

LEES, A. & MILLER, J. 1995. Waulsortian banks. *In*: MONTY, C.L.V., BOSENCE, D.W.J., BRIDGES, P.H. & PRATT, B.R. (eds) *Carbonate Mud-Mounds. Their Origin and Evolution*. International Association of Sedimentologists, Special Publications, **23**, 191–271.

LEVITUS, S., BURGETT, R., BOYER, T.P., *et al.* 1994. *World Ocean Atlas 1994: NOAA Atlas NESDIS4. Vol. 4: Temperature*. US Department of Commerce, Washington, DC.

MAZURENKO, L.L. 1998. Fine sediment from different morphological features of the Porcupine Seabight basin floor. *In*: DE MOL, B. (ed.) *Geosphere–Biosphere Coupling: Carbonate Mud Mounds and Cold Water Reefs*. UNESCO, Intergovernmental Oceanographic Commission Workshop Report, **143**, 34–35.

MITCHELL, F. & RYAN, M. 1997. *Reading the Irish Landscape*. Town House, Dublin.

MONTY, C.L.V. 1995. The rise and nature of carbonate mud-mounds: an introductory actualistic approach. *In*: MONTY, C.L.V., BOSENCE, D.W.J., BRIDGES, P.H. & PRATT, B.R. (eds) *Carbonate Mud-Mounds. Their Origin and Evolution*. International Association of Sedimentologists, Special Publications, **23**, 11–48.

PECKMAN, J., REITNER, J. & NEUWEILER, F. 1998. Seepage related or not? Comparative analyses of Phanerozoic deep-water carbonates. *In*: DE MOL, B. (ed.) *Geosphere–Biosphere Coupling: Carbonate Mud Mounds and Cold Water Reefs*. UNESCO, Intergovernmental Oceanographic Commission Workshop Report, **143**, 18–19.

PILLEN, S. 1998. *Detailkartering en seismische analyse van de Magellan-mounds in het Porcupine Bezzen, ten zuidwesten van Ierland*. MSc thesis, Gent University.

REILLY, F.J. JR, MACDONALD, I.R. JR, BIEGERT, E.K. JR & BROOKS, J.M. JR. 1996. Geological controls on the distribution of chemosynthetic communities in the Gulf of Mexico. *In*: SCHUMACHER, D. & ABRAMS, M.A. (eds) *Hydrocarbon Migration and its Near-surface Expression*. American Association of Petroleum Geologists, Memoirs, **66**, 39–62.

SLOAN, E.D. JR 1998. Physical/chemical properties of gas hydrates and application to world margin stability and climatic change. *In*: HENRIET, J.P. & MIENERT, J. (eds) *Gas Hydrates: Relevance to World Margin Stability and Climate Change*. Geological Society, London, Special Publications, **137**, 9–30.

THOMSON, C.W. 1873. *In*: *The Depths of the Sea*. MacMillan, London.

VAN WEERING, T.C.E. & HENRIET, J.P. 1998. Fluid migration, gas hydrates and biogenic carbonate mud mounds: fundamental questions and outlook on research actions. *In*: DE MOL, B. (ed.) *Geosphere–Biosphere Coupling: Carbonate Mud Mounds and Cold Water Reefs*. UNESCO, Intergovernmental Oceanographic Commission Workshop Report, **143**, 13–14.

WENDT, J., BELKA, Z., KAUFMANN, B., KOSTREWA, R. & HAYER, J. 1998. The world's most spectacular carbonate mud mounds (Middle Devonian, Algerian Sahara). *Journal of Sedimentary Research*, **67**, 424–436.

Thermally driven porosity reduction: impact on basin subsidence

P. A. BJØRKUM, O. WALDERHAUG & P. H. NADEAU

Statoil a.s, 4035 Stavanger, Norway (e-mail: pab@statoil.com)

Abstract: At temperatures higher than *c.* 80 °C, thermally driven, isochemical diagenetic porosity loss in siliciclastic sediments leads to a thinning of the sediment column. This process, termed thermochemical compaction, results in surface subsidence and generation of sediment accommodation space. The diagenetic reactions driving thermochemical compaction will operate regardless of the initial mechanism of basin formation and in addition to any externally controlled processes causing continued subsidence. Thermochemical subsidence rates are an inverse function of geothermal gradients. The total rates of subsidence at the surface, including isostatic and mechanical compaction effects, may reach several tens of metres per million years, and may have been important in driving Tertiary subsidence in basins west of Ireland. Unlike the exponentially decaying subsidence caused by tectonic thinning and rifting of the crust, thermochemical basin subsidence is a self-regulated intrabasinal process, which proceeds at a relatively high constant rate over geological time. If not arrested by extrabasinal or tectonic events, the overall effect can ultimately result in sediment metamorphism and granitization.

It is generally agreed that the formation of sedimentary basins starts with crustal subsidence. Subsidence is generally related to stretching and thinning of the crust, resulting in a transient increased heat flux and hence increased temperature gradients. The period of thinning is followed by a thermo-tectonic subsidence of the crust (McKenzie 1978). According to this model, evolution of a sedimentary basin following a rifting event will not last for more than some 200 Ma, with subsidence rates decaying exponentially with time. Only some 10% of the basin infill will take place during the last 100 Ma according to this model. However, many basins, including the offshore basins along the Irish Atlantic margin, do not show a clear exponential decay in subsidence rates (Fig. 1), and sedimentation often persists at relatively high rates, in conflict with the McKenzie model (Einsele 1992). Moreover, many long-lived basins lack major extensional faulting during these periods, which is difficult to explain if the subsidence was caused by extensional rifting events. For instance, intracratonic sag basins are known from all continents and are a type of basin that may accumulate undeformed or little deformed sedimentary sequences of 10–15 km thickness over long periods of time, i.e. of the order of 200–1000 Ma (Einsele 1992). Although some of these basins may be underlain by rift structures, their post-rift subsidence history appears to be much longer than that which can be attributed to a rift-induced thermal anomaly (Quinlan 1987). In the North Sea, for example, the Tertiary post rift sequences of 2–5 km thickness, which correspond to an average sedimentation rate of 15–100 m Ma^{-1} (Nielsen *et al.* 1986), remain unexplained unless a new (third) rifting phase in Late Cretaceous to Early Tertiary time is assumed (Einsele 1992). However, empirical evidence for a rifting phase within this period of time has been difficult to obtain. Here, we propose that intrabasinal diagenetic processes exert a major and previously overlooked effect on basin subsidence, and that these processes cause steady-state basin subsidence after basin initiation by rifting or other processes.

Thermochemical porosity loss

In most existing compaction models porosity reduction is described as a physical/mechanical response to loading (see Hermanrud (1993) for a review), and hence reduction of porosity will take place only if accommodation space is already available. Other factors must create the accommodation space that makes deposition of sediment and mechanical compaction possible. According to recent models for cementation in sediments, however, porosity reduction also takes place within the sedimentary package independent of sediment loading (Oelkers *et al.* 1996; Walderhaug 1996; Bjørkum *et al.* 1998). At temperatures higher than *c.* 80 °C, porosity reduction in sandstones can occur by thermally

From: SHANNON, P.M., HAUGHTON, P.D.W. & CORCORAN, D.V. (eds) 2001. *The Petroleum Exploration of Ireland's Offshore Basins*. Geological Society, London, Special Publications, **188**, 385–392.
0305-8719/01/$15.00

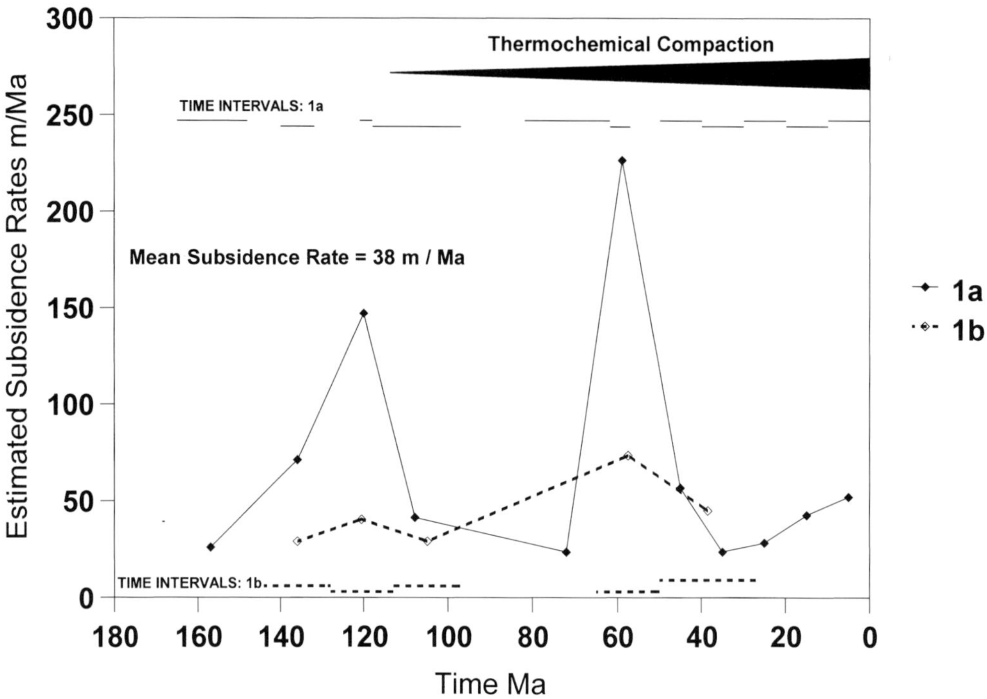

Fig. 1. Estimated subsidence rates from Porcupine Basin well 35/19-1 using two sets of biostratigraphic data (1a, shown by continuous curve with ◆; 1b, shown by dashed curve with ◇; bars indicate stratigraphic intervals used). The rates are corrected for palaeowater depth and both conventional mechanical as well as thermochemical compaction. Periods of inversion during earliest Cretaceous and Eocene time have been omitted for simplicity. Noteworthy features are the lack of an exponential decrease in subsidence with decreasing age, as anticipated from conventional tectonic subsidence models (see text), and the increasing contribution of thermochemical compaction-related subsidence, particularly during the last *c.* 40 Ma, with estimated present-day rates of *c.* 20 m Ma^{-1}. This value is estimated from the *c.* 2 km of sediments within the thermochemical compaction 'window' of 80–200 °C, using an average rate of 0.1% Ma^{-1} (see text).

driven, effective stress-insensitive, and kinetically controlled sequentially coupled dissolution/precipitation mineral reactions (Oelkers *et al.* 1992; Walderhaug 1994; Bjørkum 1996). The available evidence indicates that thermochemically driven diagenetic reactions are also responsible for porosity loss in shales (Bradley 1975; Bjørlykke 1998). Despite the lack of calibrated quantitative models of the type available for sandstones, the magnitude of the porosity loss caused by thermochemically driven dissolution of load-bearing reactants and precipitation of mineral cements in shales can be estimated from shale porosity v. depth curves (Bradley 1976; Magara 1980). The combined effect of the thermally driven diagenetic reactions in sandstones and shales is a thinning of the sediment column located between the 80 °C isotherm and the top of the inert zone where the sediments have effectively lost all porosity, typically around 180–220 °C. This process, hereafter referred to as 'thermochemical compaction', will cause subsidence at the surface.

Sandstones

At temperatures of around 80 °C (i.e. burial depths of typically 2–3 km) most quartzose sandstones have mechanically compacted to porosities of around 25–30% (Lander & Walderhaug (1999) and references therein). Further reduction in porosity is mainly due to quartz dissolution and precipitation of quartz overgrowths on available quartz surfaces (e.g. Bjørlykke *et al.* 1986; Ehrenberg 1990; Giles *et al.* 1992). The sequentially coupled process is characterized by the following three steps: (1) dissolution: quartz is dissolved along stylolites and micro-stylolites within the sandstones by a stress-insensitive dissolution process; (2) diffusion: dissolved silica is transported into

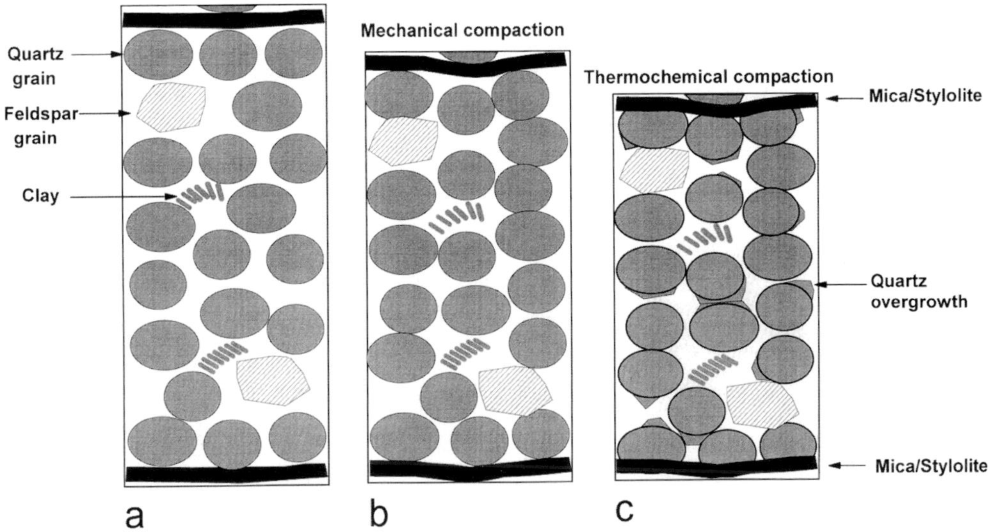

Fig. 2. Thinning of sedimentary rocks as a result of mechanical compaction, (**a**) to (**b**), and quartz cementational chemical compaction, (**b**) to (**c**). Sandstone chemical compaction takes place at temperatures higher than *c*. 80 °C and results in a thinning as a result of dissolution of minerals at the stylolite and reprecipitation as cement overgrowths in the interstylolite zone.

interstylolite zones by diffusion through the formation water; (3) precipitation: silica is precipitated as quartz overgrowths on available quartz surfaces, i.e. detrital quartz grains and overgrowths.

As pointed out many years ago by Heald (1955), this isochemical reduction in porosity causes a rock volume reduction of the sandstones, mainly a thinning (Fig. 2). The overall rate control on this sequentially coupled dissolution–transport–precipitation process is

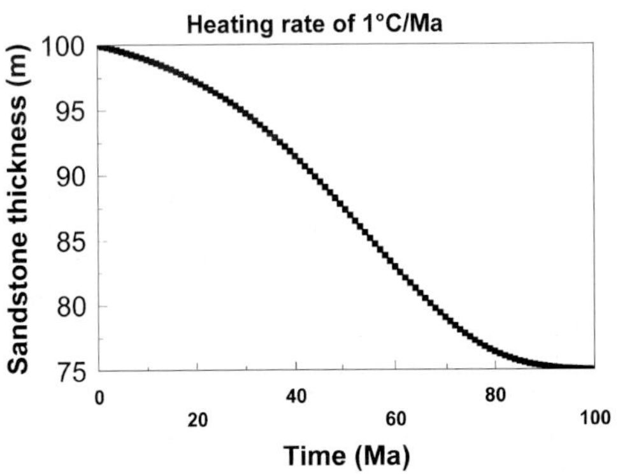

Fig. 3. Modelled vertical shortening of a sandstone of 100 m thickness as a result of thermochemical compaction as shown in Figure 2, assuming a constant heating rate of 1 °C Ma^{-1} (after Walderhaug *et al.* 2001).

the precipitation step. Because the rate of quartz precipitation per unit quartz surface area increases exponentially with temperature, the rate of porosity loss also increases with temperature. At some point the available quartz surface area becomes so small that the rate of porosity reduction slows down. This turning point, where the porosity loss rate reduces from its maximum, depends on several factors, and typically occurs in the range 120–140 °C (Fig. 3). At 180–200 °C, most of the porosity has been lost, implying that porosity reduction by quartz cementation predominantly takes place within a temperature window of around 120 °C (80–200 °C). The average porosity reduction rate for a sandstone that subsides at a constant rate within this temperature window is around 0.25% Ma^{-1} if the heating rates are around 1 °C Ma^{-1}, implying that given a porosity of 25% at the 80 °C isotherm, it takes *c*. 100 Ma to reach zero porosity. Average porosity loss rates allow us to also estimate the rate of thermochemical compaction and subsidence for sandstones undergoing quartz cementation.

Shales

Estimates of the thermochemically driven porosity loss rates for shales are also required to estimate the total magnitude of the thermochemical compaction of the sediment column. These estimates can be based on porosity v. depth observations. Shales are typically deposited with very high porosity, usually >50%, and most of the porosity is lost by mechanical compaction during the early stages of burial. After some 2–3 km of burial, shale porosities are usually 15–30% (Rieke & Chilingarian 1974), which implies that shales and sandstones often enter the thermochemical window with approximately similar porosities, and both lithologies generally have very little remaining porosity at temperatures of *c.* 200 °C. Consequently, the rate of thermochemical compaction for shales can also be estimated at *c.* 0.25% Ma^{-1}. Because the calculations of subsidence rates presented here are based on average porosity loss rates, the shape of the porosity v. depth curve for each stratum is not crucial for calculation of the thermochemical compaction and resulting basin subsidence, and the similar average porosity loss rates for sandstones and shales implies that the sand-to-shale ratios of the basins do not seriously affect our simplified calculations.

Although clay diagenetic reactions in shales starting at *c.* 60 °C may cause major permeability reduction below 80 °C (Bjørkum & Nadeau 1998), these reactions probably have little volumetric impact until temperatures higher than *c.* 80 °C are encountered. Thermochemical shale compaction in the temperature range 60–80 °C is therefore not considered here. Further research is required, however, to more accurately describe thermochemical compaction and the rate-limiting steps for specific diagenetic reactions occurring within shales.

Subsidence model

With an estimated average thermochemical porosity loss rate of 0.25% Ma^{-1}, i.e. a heating rate of 1 °C Ma^{-1}, each kilometre of sediment will shorten at a rate of 2.5 m Ma^{-1}, and the total thermochemical shortening of a sedimentary package will be controlled by the total sediment thickness within the 80–200 °C interval. The sedimentary column thickness between these isotherms is determined by the geothermal gradient. In a basin with a geothermal gradient of 20 °C km^{-1}, 6 km of sediments will be within the 80–200 °C temperature range, which will result in a shortening rate of 15 m Ma^{-1}. In a sedimentary basin with a geothermal gradient of 40 °C km^{-1}, there will be 3 km of sediments between the same isotherms, and the rate of thermochemical shortening will therefore be 7.5 m Ma^{-1}. If no sedimentation occurs, the rate of vertical accommodation space generated at the surface is equal to the rate of thermochemical shortening. If the accommodation space generated by thermochemical compaction is filled with sediments, then the sediments above the 80 °C isotherm will compact mechanically as a result of the increased load. This will cause thinning of the sedimentary package above the 80 °C isotherm. The average mechanical compaction rate can be estimated by assuming a porosity at the surface of 50%, and 25% at the 80 °C isotherm. On the basis of this assumption, approximately half of the initial sediment porosity is lost from deposition to the 80 °C isotherm. Because the depth interval where mechanical compaction operates is normally less than the depth interval where thermochemical compaction operates, the time taken to mechanically compact the sediments is shorter. Hence, the average rate of porosity loss by mechanical compaction is approximately two times greater than 0.25% Ma^{-1}. Although the zone of mechanical compaction tends to be thinner than the zone of thermochemical compaction, the faster compaction in this zone will result in a similar cumulative rate of vertical shortening for the sediment column within the thermochemical compaction zone.

Given sufficient sediment supply, the effects of loading and isostatic subsidence combine with the thermochemical shortening of the sediments located between the 80 °C and 200 °C isotherms, inducing surface subsidence and generation of additional accommodation space at a rate between three and four times greater than the rate of thermochemical shortening (Walderhaug *et al.* 2001). This implies that sedimentary basins with geothermal gradients of 20 °C km^{-1} and 40 °C km^{-1} will have thermochemically induced average subsidence rates approaching 60 m Ma^{-1} and 30 m Ma^{-1}, respectively (Fig. 4), and that these rates, in contrast to subsidence as a result of the McKenzie rifting model, may be essentially constant as long as the sediment supply and distribution systems are adequate to fill the generated accommodation space. Time-limited extrabasinal impulses or processes may drive the basin out of this dynamic equilibrium state, imposing transient effects that may increase, decrease or invert (producing uplift and erosion) the subsidence rate. The long-term response to the transient effects will be different for these three situations as follows.

When the overall subsidence rate is increased to values greater than that of intrabasinal equilibrium subsidence, the increased extrabasinally induced subsidence rate will reduce the sediment residence time within a given thermal zone, and can allow sediments to pass beyond the 200 °C isotherm with some porosity preserved. Hence, the effect of an extrabasinally induced,

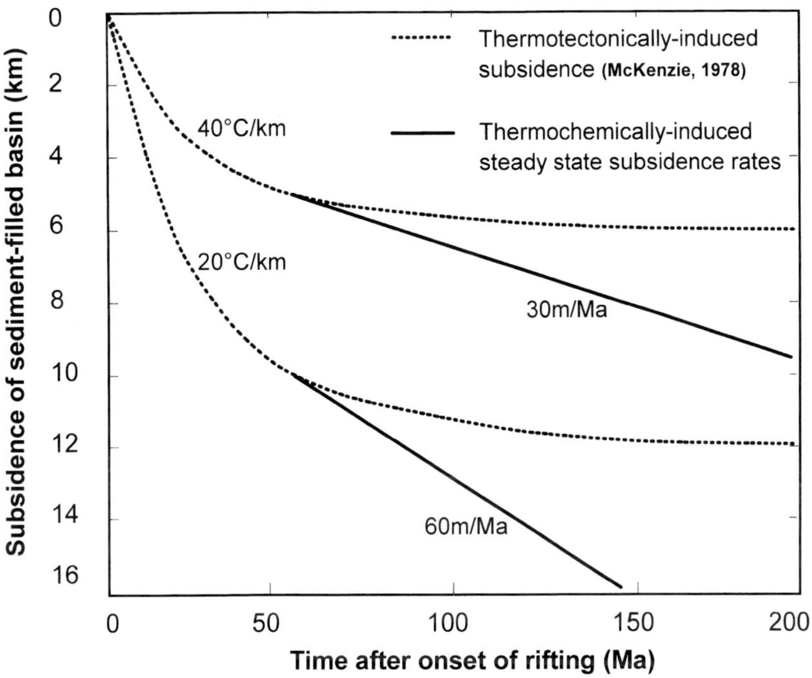

Fig. 4. Rates of subsidence for two different geothermal gradients with and without thermochemical subsidence.

increased subsidence rate is to increase the vertical thickness of sediments undergoing thermochemical compaction relative to the intrabasinally driven steady-state situation. This in turn increases the overall subsidence rates, and the increased rates will be maintained after the extrabasinal causes cease to operate if sediment supply is sufficient to keep the basin filled.

In cases where the subsidence rate is slowed down relative to the dynamic equilibrium rate, a slower than the previous dynamic equilibrium rate will be established when the extrabasinal causes cease. This results because porosity is continuously being reduced in sediments present at temperatures higher than 80 °C despite periods of reduced overall subsidence. In effect, the sediment residence time is increased, and hence, the zero-porosity isotherm, i.e. the isotherm beyond which the sediment porosity is effectively zero, will gradually shift to lower temperatures. Thus, the vertical thickness of sediments undergoing thermochemical compaction is reduced, resulting in a gradual lowering of the equilibrium subsidence rate. To resume the previous higher dynamic equilibrium subsidence rate, extrabasinal causes must be introduced. Reduced subsidence rates as a result of extrabasinal causes therefore influence the intrabasinal equilibrium subsidence rate in a way similar to that achieved by increasing the geothermal gradient.

In cases where the subsidence rate is inverted, i.e. when uplift and erosion occur, porosity reduction will still take place within those porous sediments located at temperatures higher than 80 °C. During erosion, the basin will be deprived of porous sediment that would otherwise contribute to thermochemically driven subsidence. If erosion results in a situation where there are no porous sediments present within the thermochemically active zone, i.e. when the prior-to-uplift (palaeo) zero-porosity isotherm has been uplifted shallower than the 80 °C isotherm, then the thermochemical drive for subsidence is terminated. The magnitude of 'terminal uplift and erosion' for this to occur corresponds to the thickness of the thermochemical window, i.e. 3–6 km, given typical geothermal gradients. In such circumstances, the erosional unconformity truncates the palaeo 120 °C isotherm, and the basin will resume subsidence only if some extrabasinal process reinitiates it. From a purely tectonic or extrabasinal perspective, the likelihood of having 'terminal erosion' shut off thermochemical compaction is greater in basins with high geothermal gradients, because less uplift and erosion is required to place the palaeo 200 °C

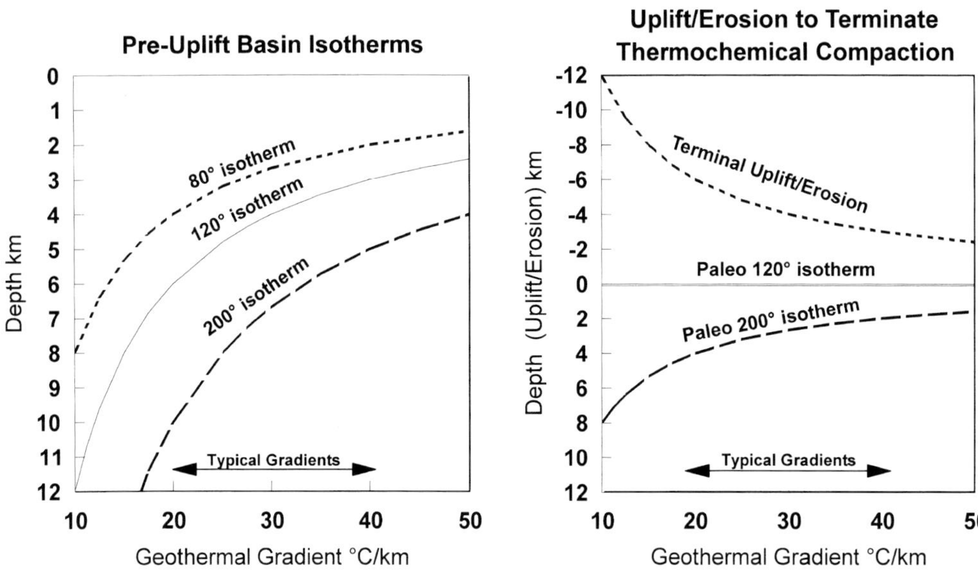

Fig. 5. Magnitude of uplift and erosion required to terminate thermochemical-related subsidence as a function of geothermal gradient. It should be noted that in all cases the erosional unconformity truncates the palaeo 120 °C isotherm.

(zero-porosity) isotherm above the 80 °C isotherm (Fig. 5).

Sedimentary basins as self-organized systems

According to the proposed model, a sedimentary basin may be considered a self-organized system. As long as sediment is supplied, for example by isostatic uplift along the flanks of the basin, the same sediment will cause an isostatic subsidence under the basin. Furthermore, the continuously operating thermochemically induced diagenetic subsidence mechanism may also contribute to maintain sediment supply by inducing uplift of the flanks of the basin via underlying mantle-level displacements. The overall process can be considered a self-organized geological convection cell. The energy that maintains this process is derived in part from the upward heat flux in the basin.

In the absence of extrabasinal factors affecting the convection cell, there is no theoretical time limit over which the process may continue. Sediments in the deeper part of the basin may readily pass into metamorphic pressure and temperature conditions, which is unlikely given the McKenzie stretching model. Hence, thermochemically induced subsidence can produce metamorphosed sedimentary rocks and ultimately granitic rocks derived from partial or total melting of the metasedimentary complex. It follows that metamorphic and granitic rocks may be the common culmination of sedimentary basin formation. Metamorphosed sedimentary rocks are often assumed to have formed at subduction zones, or during collision between continental plates. According to the proposed model, metamorphism and melting of sedimentary rocks are also the likely result of sedimentary basin formation given sufficient sediment supply and provided that the basin is not terminally inverted as a result of tectonic uplift. The rates for thermochemical compaction, basin subsidence and related uplift along the basin margins contribute to an overall geological equilibrium between weathering, erosion, sedimentation, diagenesis, metamorphism and granitization processes, which form the Earth's rock cycle.

Bjørkum & Nadeau (1998) have related thermochemical porosity reduction to fluid flow and hydrocarbon migration in sedimentary basins. In this paper we have concluded that basin subsidence and dynamics are strongly controlled by thermochemical porosity reduction in the sediment fill. Hence, both fluid migration and subsidence are controlled by intrabasinal thermally driven and stress-insensitive (Bjørkum 1996) diagenetic processes, which in turn make fluid migration and subsidence related in a way that could not be previously recognized. These relationships also allow evaluation of re-migration of reservoired hydrocarbons from structurally or stratigraphically deeper synrift levels, to create new shallower post-rift play models in response to continuing basin subsidence. The relationships can also be used to estimate the timing and rates of re-migration, which have important petroleum exploration applications along the Atlantic margin. As our understanding of diagenetic processes operating within sedimentary basins increases, other processes and phenomena that are currently considered as unrelated or poorly related may in fact turn out to be related via intrabasinal thermal controls on geological rates. Methods of sedimentary basin analysis are emerging based on self-organized processes, with internal thermal drives. These processes can contribute to or control rates of basin subsidence, sediment accommodation or distribution patterns, fluid migration and rock deformation.

Conclusions

Porosity loss in siliciclastic sediments at temperatures higher than *c.* 80 °C is dominated by thermally driven and stress-insensitive dissolution/precipitation diagenetic reactions, including quartz cementation. This thermochemical compaction results in a thinning of the sediment column, which in turn will lead to surface subsidence and generation of accommodation space. If the generated accommodation space is filled by sediment, subsidence is enhanced by isostatic effects and by mechanical compaction of the shallowest part of the sediment column. Hence, thermochemical subsidence does not depend upon extrabasinal processes, other than sufficient sediment supply, but takes place as a result of thermally driven diagenetic processes operating within the basin.

Rates of thermochemical compaction for the sediment column located beneath the 80 °C isotherm may reach values of the order of 10 m Ma^{-1}, implying a total surface subsidence rate of several tens of metres per million years when mechanical compaction and isostasy are taken into account. The subsidence caused by thermochemical compaction will be greatest in basins with low geothermal gradients because the thickness of sediments located within the zone of active thermochemical compaction (*c.* 80–200 °C) is inversely proportional to the geothermal gradient.

Unlike the McKenzie rifting model, thermochemical compaction can, if not interrupted by extrabasinal factors, continue to generate accommodation space for hundreds of millions of years at relatively constant rates provided there is a sufficient sediment supply. A sedimentary basin, therefore, can be considered as part of a large-scale geological convection cell, where the upward lithological flux is represented by the isostatically uplifted basin flank sediment source areas, and where the sedimentary basin subsidence represents the downward lithological flux. The rate of convection is driven in part by the thermochemical dissolution–precipitation porosity loss mechanisms operating within the porous part of the sedimentary basin. The energy required to perpetuate the resulting self-organized convection cell is taken in part from the heat flux and is consumed by the porosity-reducing mineral reactions operating within the sediments. The thermochemically induced subsidence may continue uninterrupted over geological time, and the sediments in the deeper part of the basin ultimately may undergo metamorphic reactions, which is unlikely to occur when subsidence is attributed solely to McKenzie-type stretching models.

The authors wish to thank Statoil a.s for support and permission to publish this work.

References

BJØRKUM, P.A. 1996. How important is pressure in causing dissolution of quartz in sandstones? *Journal of Sedimentary Research*, **66**, 147–154.

BJØRKUM, P.A. & NADEAU, P.H. 1998. Temperature controlled porosity/permeability reduction, fluid migration, and petroleum exploration in sedimentary basins. *Australian Petroleum Production and Exploration Association Journal*, **38**, 453–465.

BJØRKUM, P.A., OELKERS, E.H., NADEAU, P.H., WALDERHAUG, O. & MURPHY, W.M. 1998. Porosity prediction in quartzose sandstones as a function of time, temperature, depth, stylolite frequency, and hydrocarbon saturation. *AAPG Bulletin*, **82**, 637–648.

BJØRLYKKE, K. 1998. Clay mineral diagenesis in sedimentary basins—a key to the prediction of rock properties. Examples from the North Sea basin. *Clay Minerals*, **33**, 15–34.

BJØRLYKKE, K., AAGAARD, P., DYPVIK, H., HASTINGS, D.S. & HARPER, A.S. 1986. Diagenesis and reservoir properties of Jurassic sandstones from the Haltenbanken area, offshore mid-Norway. *In*: SPENCER, A.M., HOLTER, E., CAMPBELL, C.J., HANSLIEN, S.H., NELSON, P.H.H., NYSÆTHER, E. & ORMAASEN, E.G. (eds) *Habitat of Hydrocarbons on the Norwegian Continental Shelf*. Graham & Trotman, London, 275–286.

BRADLEY, J.S. 1975. Abnormal formation pressure. *AAPG Bulletin*, **59**, 957–973.

BRADLEY, J.S. 1976. Abnormal formation pressure: reply. *AAPG Bulletin*, **60**, 1127–1128.

EHRENBERG, S.N. 1990. Relationship between diagenesis and reservoir quality in sandstones of the Garn Formation, Haltenbanken, mid-Norwegian continental shelf. *AAPG Bulletin*, **74**, 1538–1558.

EINSELE, G. 1992. *Sedimentary Basins*. Springer, Berlin.

GILES, M.R., STEVENSON, S., MARTIN, S.V., CANNON, S.J.C., HAMILTON, P.J., MARSHALL, J.D. & SAMWAYS, G.M. 1992. The reservoir properties and diagenesis of the Brent Group: a regional perspective. *In*: MORTON, A.C., HASZELDINE, R.S., GILES, M.R. & BROWN, S. (eds) *Geology of the Brent Group*. Geological Society, London, Special Publications, **61**, 289–327.

HEALD, M.T. 1955. Stylolites in sandstones. *Journal of Geology*, **63**, 101–114.

HERMANRUD, C. 1993. Basin modelling techniques—an overview. *In*: DORÉ, A.G., AUGUSTON, J.H., HERMANRUD, C., STEWART, D.J. & SYLTA, Ø. (eds) *Basin Modelling: Advances and Applications*. Norwegian Petroleum Society (NPF), Special Publications, **3**, 1–34.

LANDER, R.H. & WALDERHAUG, O. 1999. Predicting porosity through simulating sandstone compaction and quartz cementation. *AAPG Bulletin*, **83**, 433–449.

MAGARA, K. 1980. Comparison of porosity–depth relationships of shale and sandstone. *Journal of Petroleum Geology*, **3**, 175–185.

MCKENZIE, D.P. 1978. Some remarks on the development of sedimentary basins. *Earth and Planetary Science Letters*, **40**, 25–32.

NIELSEN, O.B., SØRENSEN, S., THIEDE, J. & SKARBØ, O. 1986. Cenozoic differential subsidence of the North Sea. *AAPG Bulletin*, **70**, 276–298.

OELKERS, E.H., BJØRKUM, P.A. & MURPHY, W.M. 1992. The mechanism of porosity reduction, stylolite development and quartz cementation in North Sea sandstones. *In*: KHARAKA, Y.K. & MAEST, A.S. (eds) *Water–Rock Interaction*. Balkema, Rotterdam, 1183–1186.

OELKERS, E.H., BJØRKUM, P.A. & MURPHY, W.M. 1996. A petrographic and computational investigation of quartz cementation and porosity reduction in North Sea sandstones. *American Journal of Science*, **296**, 420–452.

QUINLAN, G. 1987. Models of subsidence mechanism in intercratonic basins and their applicability to North American examples. *In*: BEAUMONT, C. & TANKARD, A.J. (eds) *Sedimentary Basins and Basin-forming Mechanisms*. Canadian Society of Petroleum Geologists, Memoirs, **12**, 463–481.

RIEKE, H.H., CHILINGARIAN, G.V., *Compaction of Argillaceous Sediments. Developments in Sedimentology*, **16**, Elsevier, Amsterdam.

WALDERHAUG, O. 1994. Precipitation rates for quartz cement in sandstones determined by fluid-inclusion microthermometry and temperature-history modeling. *Journal of Sedimentary Research*, **A64**, 324–333.

WALDERHAUG, O. 1996. Kinetic modelling of quartz cementation and porosity loss in deeply buried sandstone reservoirs. *AAPG Bulletin*, **80**, 731–745.

WALDERHAUG, O., BJØRKUM, P.A., NADEAU, P.H. & LANGNES, O. 2001. Quantitative modelling of basin subsidence caused by temperature-driven silica dissolution and reprecipitation. *Petroleum Geoscience*, **7**, 107–113.

Interpretation of transverse gravity lineaments in the Rockall Basin

K. McGRANE, P. W. READMAN & B. M. O'REILLY
Dublin Institute for Advanced Studies, 5 Merrion Square, Dublin 2, Ireland
(e-mail: kmg@cp.dias.ie)

Abstract: A number of regional transverse gravity lineaments crosscutting the Rockall Basin are interpreted from satellite gravity data. Euler deconvolution carried out on gravity data along a wide-angle seismic profile indicates that a major NW–SE-trending lineament within the basin reflects pronounced variations in crustal structure and sedimentary thickness. These thickness variations are interpreted as the result of cross-basin faulting along a zone defined by this lineament. Transverse gravity lineaments to the north of this feature are similarly interpreted as major cross-basin fault zones.

Gravity gradient techniques have been used to identify and resolve major structural trends and fabrics in the Irish offshore region (Readman *et al.* 1995; O'Reilly *et al.* 1996). Application of these techniques accentuates subtle trends in the gravity field, and is used in this study to image regional transverse gravity trends within the Rockall Basin. The objective of this paper is to investigate the relationship between these trends and crustal structure by focusing on a major NW–SE-trending transverse gravity lineament that crosscuts the basin.

The study uses marine free-air anomaly data derived from satellite altimetry (Smith & Sandwell 1995), together with land gravity data collected by the Dublin Institute for Advanced Studies and the Geological Survey of Northern Ireland (Readman *et al.* 1997). The satellite gravity data have an accuracy of about 5 mGal (Sandwell & Smith 1997) and a typical anomaly resolution of 20–30 km, which is sufficient for interpreting regional structural relationships within sedimentary basins. A potential field technique known as Euler deconvolution is applied to a satellite gravity profile along a section of an axial wide-angle seismic line in the Rockall Basin. This profile is used to estimate source depth solutions for the observed NW–SE-trending lineament.

The free-air anomaly map

The marine satellite free-air anomaly map for offshore Ireland, integrated with the onshore Bouguer anomaly, is shown in Figure 1. The satellite gravity field over the Irish sector of the Rockall Basin is dominated by a NE–SW-trending fabric that swings SSW towards the Porcupine Abyssal Plain. A general NE–SW-trending regional fabric is also observed onshore Ireland and in the Irish Sea and Celtic Sea basins. This trend in the gravity field is attributed to the underlying Caledonoid basement structure, which may have influenced the geometry and evolution of the Rockall Basin (Readman *et al.* 1997).

A number of transverse lineaments crosscut the main NE–SW-trending fabric of the basin. These lineaments are large-scale regional features that trend approximately perpendicular to the basin axis (Fig. 1). The southern boundary of the Rockall Basin is defined by a major gravity lineament, the Charlie–Gibbs Fracture Zone (CGFZ). This lineament, which represents an eastwards extension of a Mid-Atlantic Ridge transform fault, has an approximate east–west trend and is characterized by steep gravity gradients. At *c.* 52°N, 17°W it appears to connect to an ESE–WNW-trending gravity lineament known as the Clare Lineament (CL). This feature has been interpreted as a Mesozoic precursor to the CGFZ (Bentley & Scrutton 1987; Megson 1987) and is believed to extend across the Porcupine High and Porcupine Basin to just east of 10°W (Tate 1992).

To the north of the CGFZ a prominent NW–SE-trending gravity lineament (labelled A in Fig. 1), transects the central part of the basin. This feature marks a major break in the dominant NE–SW-trending gravity fabric and represents one of the most striking gravity features in the basin. This lineament is characterized by steep gradients and large wavelengths (up to 80 km in places) and extends for a distance of *c.* 250 km across the basin. It appears to continue

From: SHANNON, P.M., HAUGHTON, P.D.W. & CORCORAN, D.V. (eds). 2001. *The Petroleum Exploration of Ireland's Offshore Basins*. Geological Society, London, Special Publications, **188**, 393–399. 0305-8719/01/$15.00

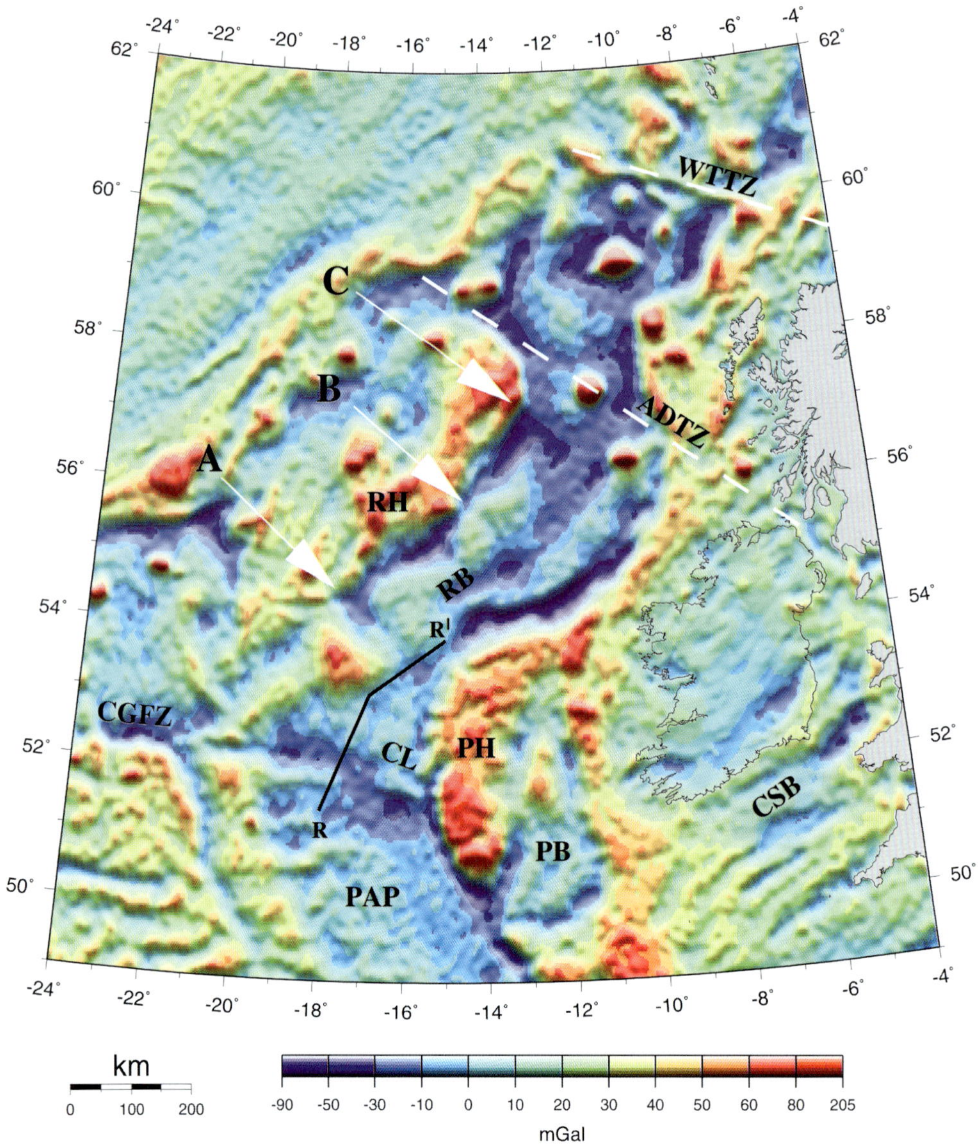

Fig. 1. Marine free-air gravity map for offshore Ireland, incorporating gridded onshore Bouguer data. Simulated illumination is from the north. The gridding interval is 2′ and the map projection UTM Zone 29. The section of the axial RAPIDS seismic profile used in this study, labelled RR′, is indicated. White arrows labelled A to C indicate transverse gravity lineaments. ADTZ, Anton Dohrn Transfer Zone; WTTZ, Wyville-Thomson Transfer Zone; CGFZ, Charlie–Gibbs Fracture Zone; CL, Clare Lineament; PAP, Porcupine Abyssal Plain; PB, Porcupine Basin; PH, Porcupine High; CSB, Celtic Sea Basin; RB, Rockall Basin; RH, Rockall High.

southeastwards across the Porcupine High, Porcupine Basin and into the Celtic Sea basins (see also Readman *et al.* 1995) becoming less well defined east of 10°W. A series of similarly trending transverse lineaments are observed to the north of this feature (labelled B and C in Fig. 1). Although these gravity lineaments are characterized by more subtle gradients they also mark distinct breaks in the dominant NE–SW-trending gravity fabric of the basin. Further north, the Anton Dohrn Transfer Zone (ADTZ) and the Wyville-Thomson Transfer Zone (WTTZ) define major offsets in the Rockall Basin (Doré *et al.* 1997; Waddams & Cordingley

1999; Corfield *et al.* 1999), but further discussion of these lineaments is beyond the scope of this paper.

Euler deconvolution

During the RAPIDS (Rockall and Porcupine Irish Deep Seismic) reflection experiments a wide-angle seismic profile was acquired along the axis of the Rockall Trough (Hauser *et al.* 1995). A section of this seismic profile, comprising line 24 and most of line 23, crosses the CGFZ, CL and lineament A (Fig. 1). A free-air gravity anomaly profile was extracted from the satellite data along this section. The effect of the topography of the eastern margin of the Rockall Trough was estimated by using simple Airy-type isostasy models at several positions along the profile. The magnitude of this topographic effect was then subtracted from the observed values to give the gravity variation along the profile

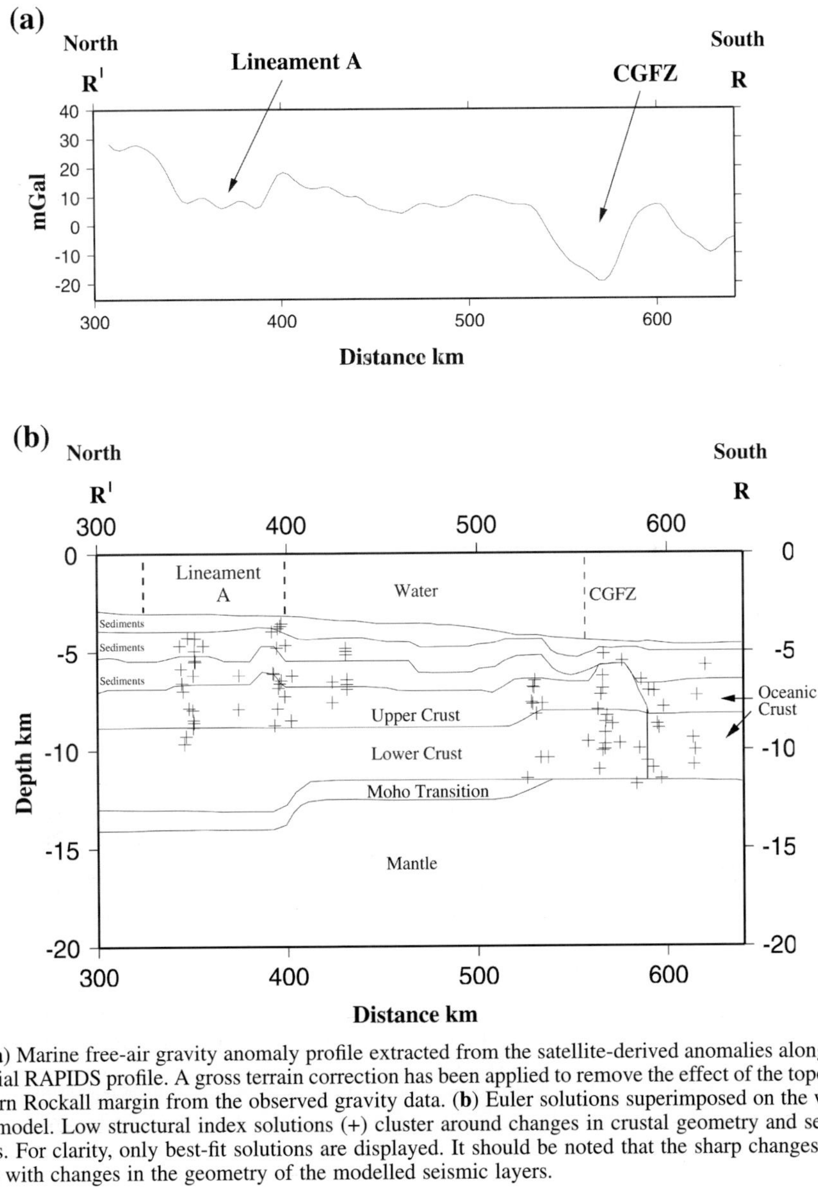

Fig. 2. (**a**) Marine free-air gravity anomaly profile extracted from the satellite-derived anomalies along a section of the axial RAPIDS profile. A gross terrain correction has been applied to remove the effect of the topography of the eastern Rockall margin from the observed gravity data. (**b**) Euler solutions superimposed on the wide-angle seismic model. Low structural index solutions (+) cluster around changes in crustal geometry and sedimentary thickness. For clarity, only best-fit solutions are displayed. It should be noted that the sharp changes in gravity correlate with changes in the geometry of the modelled seismic layers.

caused by a 2D structure (Fig. 2a). Euler deconvolution was then applied to the data and source depth estimates for the lineament anomalies were compared with the modelled structure developed from the wide-angle seismic data (Fig. 2b).

Euler deconvolution is a potential field technique that can be applied to gravity anomaly data to estimate source depths of causative bodies. Source depths along profile data can be calculated using Euler's 1D homogeneity relationship (Thompson 1982):

$$(x - x_0)\frac{\delta \Delta T}{\delta x} - z_0\frac{\delta \Delta T}{\delta z} = -n\Delta T(x)$$

The parameters x_0 and z_0 represent the source position of the body whose gravity field T is measured at (x, z). As the gravity field T is known, solving the above equation for various values of n will provide solutions for the depth of the source body. The quantity n is known as the structural index and is a measure of the rate of 'fall-off' of an anomaly pattern caused by the source body. This fall-off rate is a function of the geometry and depth of the causative body and consequently the structural index can be used to constrain the geological interpretation of the source body. Thompson (1982) and Reid *et al.* (1990) have investigated which structural indices provide the best-fit depth solutions for anomalies caused by typical geological structures. Source depth estimates for anomalies caused by faults, steeply dipping contacts and basin edges are best defined by low structural indices ranging from zero to one, whereas higher indices, ranging from two to three, provide best-fit depth estimates for features such as volcanic plugs, salt diapirs or kimberlite pipes. The most appropriate depth estimate and structural index for any given anomaly is chosen as that which displays the tightest degree of clustering of the solutions.

A range of structural indices between zero and three were tested along the extracted profile and depth solutions were generated for each index. Those that showed the highest degree of clustering were selected as the most reliable estimates of source depth. Interpretation of the solutions was constrained by the RAPIDS wide-angle seismic model. Euler solutions that displayed a scattered distribution and that bore no tangible relationship to modelled structure, derived from wide-angle seismic data, were considered as false solutions.

Results

The results of the 2D Euler deconvolution analysis on the profile are shown in Figure 2b, where the solutions are superimposed on the crustal model derived from the wide-angle seismic line. Only the best-fit solutions are displayed. Low structural indices provide the best-fit Euler solutions consistent with the geometries observed in the seismic model. The best-fit solutions (denoted by crosses) cluster in areas of pronounced changes in crustal and basement structure, generally at depths of between 5 and 9 km.

The anomaly pattern across the lineament that marks the CGFZ is characterized by a large-wavelength (65–70 km) gravity low, which correlates with a basement horst flanked by a region of thickened sediments (Fig. 2b). The best-fit Euler solutions for the lineament cluster within and around the horst block, defining a source depth of between 6 and 8 km, i.e. upper-crust level. Sets of solutions between 8 and 12 km could indicate even deeper sources in the lower crust, possibly related to the CGFZ and the boundary between the oceanic and continental crust. Hauser *et al.* (1995) have suggested that the gravity anomalies in this region may be detecting structure that has not been well resolved by the wide-angle seismic model.

The major NW–SE-trending lineament, labelled A in Fig. 1, is spatially associated with steps and offsets in the seismic crustal structure at *c.* 390 km, and with a significant change in Moho depth (Fig. 2b). The best-fit Euler solutions for this lineament are clustered adjacent to changes in crustal structure and sedimentary thickness. A set of solutions cluster tightly along the southern flank of a horst block structure in the upper crust, defining a source depth of *c.* 6–7 km. In addition, some shallow Euler solutions display modest clustering where sedimentary units exhibit minor changes in topography and isopach. These solutions, which correspond to higher frequency variations in the gravity field, provide depth estimates of between 4 and 5 km for these variations in isopach. These depth estimates suggest that the gravity signature of lineament A reflects contributions from the upper crust and the sedimentary cover. A change of 1–2 km in the depth to the Moho is not detected by the Euler method, possibly as a result of, in part, the damping effects of the Moho transition zone. This change in Moho depth contributes a broad regional variation to the gravity field of very large wavelength and has not been investigated by this Euler deconvolution analysis.

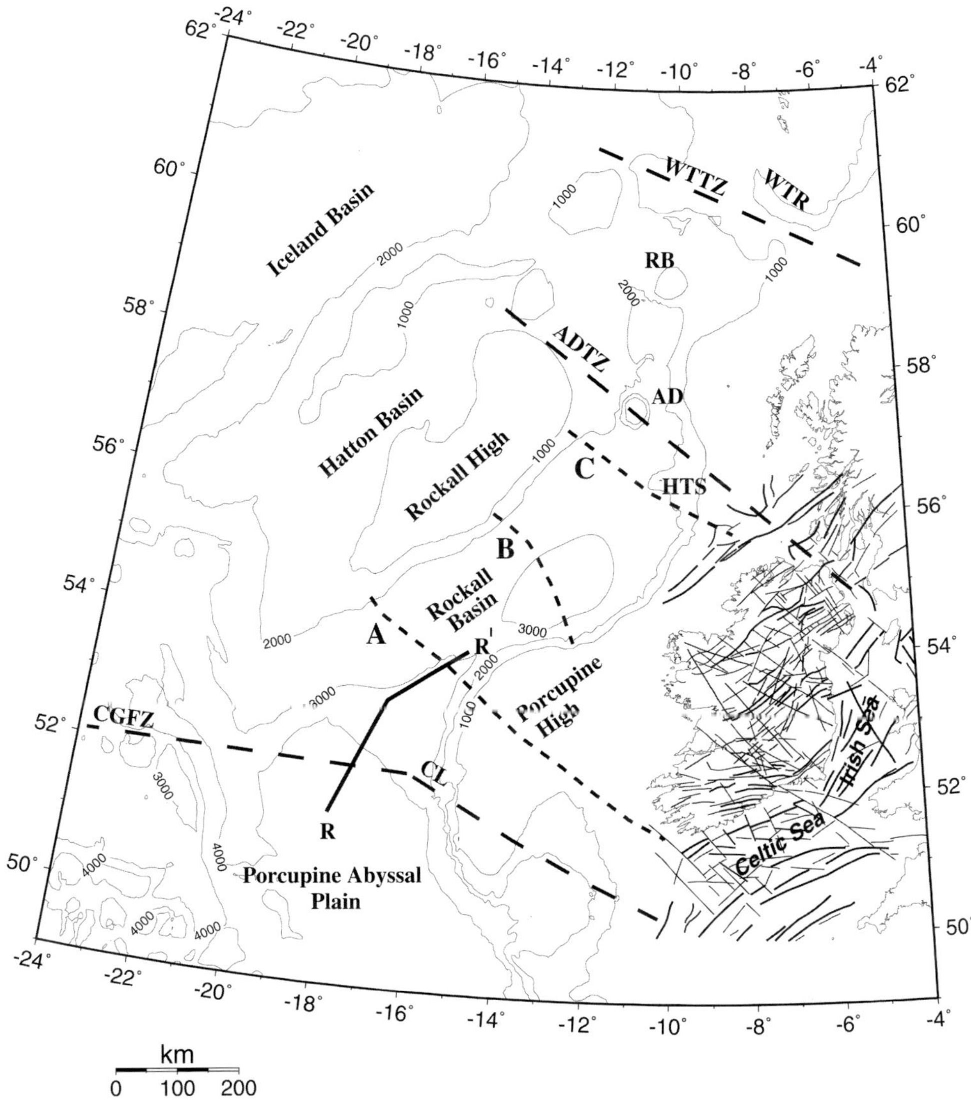

Fig. 3. Regional setting of interpreted cross-basin zones within the NE Atlantic region, including gravity lineaments onshore and near-offshore interpreted by Readman *et al.* (1995). Lineaments A, B and C are interpreted as major cross-basin fault zones (short-dashed lines). The bathymetry is contoured at 1000 m intervals and outlines prominent bathymetric features in the NE Atlantic region. Also shown, as a continuous bold line labelled RR′, is the section of the axial RAPIDS profile used in this study. Abbreviations as in Figure 1, and: AD, Anton Dohrn Seamount; HTS, Hebrides Terrace Seamount; WTR, Wyville-Thomson Ridge; RB, Rosemary Bank.

Regional interpretation

NW–SE-trending gravity lineaments have been described in a number of basins along the NE Atlantic margin (Musgrove & Mitchener 1996; Doré *et al.* 1997). They are prevalent onshore Ireland and in the Celtic Sea Basin (Fig. 3), where they coincide with a series of major offsets along the basin margins (Readman *et al.* 1995). In the Porcupine Basin they have been interpreted as strike-slip faults (Lefort & Max 1984; Masson & Miles 1986), whereas to the north in the Faroe–Shetland Basin they are attributed to closely spaced transfer zones, which strongly influenced sedimentation patterns in the basin (Rumph *et al.* 1993).

The transverse gravity lineaments described in this study suggest the presence of major cross-basin fault zones in the Rockall Basin. The Euler solutions and the wide-angle seismic data clearly illustrate that lineament A is coincident with changes in crustal topography and the morphology of the overlying sedimentary units. Variation in the isopachs of the sedimentary units is interpreted as the effect of faulting along a major cross-basin fault zone defined by this lineament. Transverse gravity lineaments (labelled B and C in Fig. 1), to the north of lineament A, are also interpreted as major cross-basin fault zones (Fig. 1).

The major variations in structure and isopach, coincident with lineament A, are generally confined to the upper-crustal and lower and middle sedimentary layers (Fig. 2b). Although the age of basement beneath the Rockall Basin is unknown, the basal sediments are interpreted as being of Late Carboniferous to Jurassic age, with the upper and middle sedimentary layers interpreted as being of Cretaceous to Paleocene and Eocene to Recent age, respectively (Shannon *et al.* 1995). This suggests that faulting may have occurred during Late Carboniferous to Jurassic times, with reactivation occurring during Cretaceous to Paleocene times to account for the observed change in isopach of the middle sedimentary layer. Minor changes in sedimentary thickness along the upper sedimentary layer may reflect differential compaction or some later movement along the fault zone marked by lineament A. Evidence from the northeastern margin of the Rockall Basin, including the inboard basins to the east, supports a recent age for transverse fault movement (Cunningham & Shannon, 1998; Walsh *et al.* 1999). The points of entry of lineament A onto the eastern and western margins of the Rockall Trough are coincident with pronounced variations in the free-air gravity field (Fig. 1). These variations in gravity are directly related to deep bathymetric embayments in the present-day margin of the Rockall Trough, and are considered as basin margin offsets caused by movements along this fault zone in Tertiary to Recent times. Although the dominant direction of the proposed faulting is interpreted to be cross-basin, no inferences regarding the style of faulting (i.e. dip-slip or strike-slip) can be made solely from the gravity data.

Conclusions

A gravity gradient technique, Euler deconvolution, coupled with wide-angle seismic data supports the presence of a major cross-basin fault zone within the Rockall Basin. This fault zone is defined by a major NW–SE-trending transverse gravity lineament that reflects pronounced variations in crustal structure and thickness of sedimentary units. Transverse gravity lineaments to the north of this feature are also interpreted as major cross-basin fault zones.

We thank S. M. Russell of the Southampton Oceanography Centre for the Euler deconvolution programs. P. Wessel and W. Smith are thanked for providing the Generic Mapping Tools software used to produce the figures in this paper. D. Inamdar and S. Smith are thanked for their helpful reviews.

References

BENTLEY, P.A.D. & SCRUTTON, R.A. 1987. Seismic investigations into the basement structure of the southern Rockall Trough. *In*: BROOKS, J. & GLENNIE, K.W. (eds) *Petroleum Geology of North West Europe*. Graham & Trotman, London, 667–675.

CORFIELD, S., MURPHY, N. & PARKER, S. 1999. The structural and stratigraphic framework of the Irish Rockall Trough. *In*: FLEET, A.J. & BOLDY, S.A.R. (eds) *Petroleum Geology of Northwest Europe: Proceedings of the 5th Conference*. Geological Society, London, 407–420.

CUNNINGHAM, G.A. & SHANNON, P.M. 1998. The Erris Ridge: a major geological feature in the NW Irish offshore basins. *Journal of the Geological Society, London*, **154**, 503–508.

DORÉ, A.G., LUNDIN, E.R., FICHLER, C. & OLESEN, O. 1997. Patterns of basement structure and reactivation along the NE Atlantic margin. *Journal of the Geological Society, London*, **154**, 85–92.

HAUSER, F., O'REILLY, B.M., JACOB, A.W.B., SHANNON, P.M., MAKRIS, J. & VOGT, U. 1995. The crustal structure of the Rockall Trough: differential stretching without underplating. *Journal of Geophysical Research*, **100**, 4097–4116.

LEFORT, J.P. & MAX, M.D. 1984. Development of the Porcupine Seabight: use of magnetic data to show the direct relationship between early oceanic and continental structures. *Journal of the Geological Society, London*, **141**, 663–674.

MASSON, D.G. & MILES, P.R. 1986. Structure and development of Porcupine Seabight sedimentary basin, offshore southwest Ireland. *AAPG Bulletin*, **70**, 536–548.

MEGSON, J.B. 1987. The evolution of the Rockall Trough and implications for the Faeroe–Shetland Trough. *In*: BROOKS, J. & GLENNIE, K.W. (eds) *Petroleum Geology of North West Europe*. Graham & Trotman, London, 653–665.

MUSGROVE, F.W. & MITCHENER, B. 1996. Analysis of the pre-Tertiary rifting history of the Rockall Trough. *Petroleum Geoscience*, **2**, 353–360.

O'REILLY, B.M., READMAN, P.W. & MURPHY, T. 1996. The gravity signature of Caledonian and Variscan tectonics in Ireland. *Physics and Chemistry of the Earth*, **21**, 299–304.

READMAN, P.W., O'REILLY, B.M., EDWARDS, J.W.F. & SANKEY, M.J. 1995. A gravity map of Ireland and surrounding waters. *In*: CROKER, P.F. & SHANNON, P.M. (eds) *The Petroleum Geology of Ireland's Offshore Basins*. Geological Society, London, Special Publications, **93**, 9–16.

READMAN, P.W., O'REILLY, B.M. & MURPHY, T. 1997. Gravity gradients and upper-crustal tectonic fabrics, Ireland. *Journal of the Geological Society, London*, **154**, 817–828.

REID, A.B., ALLSOP, J.M., GRANSER, H., MILLETT, A.J. & SOMERTON, I.W. 1990. Magnetic interpretation in three dimensions using Euler deconvolution. *Geophysics*, **55**, 80–91.

RUMPH, B., REAVES, C.M., ORANGE, V.G. & ROBINSON, D.L. 1993. Structuring and transfer zones in the Faeroe Basin in a regional tectonic context. *In*: PARKER, J.R. (ed.) *Petroleum Geology of Northwest Europe: Proceedings of the 4th Conference*. Geological Society, London, 999–1009.

SANDWELL, D.T. & SMITH, W.H.F. 1997. Marine gravity anomaly from Geosat and ERS-1 satellite altimetry. *Journal of Geophysical Research*, **102**, 10039–10054.

SHANNON, P.M., JACOBS, A.W.B., MAKRIS, J., O'REILLY, B., HAUSER, F. & VOGT, U. 1995. Basin development and petroleum prospectivity of the Rockall and Hatton region. *In*: CROKER, P.F. & SHANNON, P.M. (eds) *The Petroleum Geology of Ireland's Offshore Basins*. Geological Society, London, Special Publications, **93**, 435–457.

SMITH, W.H.F. & SANDWELL, D.T. 1995. Marine gravity field from declassified Geosat and ERS-1 altimetry. *EOS Transactions, American Geophysical Union*, **76**, 156.

TATE, M.P. 1992. The Clare Lineament: a relic transform fault west of Ireland. *In*: PARNELL, J. (ed.) *Basins on the Atlantic Seaboard: Petroleum Geology, Sedimentology and Basin Evolution*. Geological Society, London, Special Publications, **62**, 375–384.

THOMPSON, D.T. 1982. EULDPH: a new technique for making computer-assisted depth estimates from magnetic data. *Geophysics*, **47**, 31–37.

WADDAMS, P. & CORDINGLEY, T. 1999. The regional geology and exploration potential of the NE Rockall Basin. *In*: FLEET, A.J. & BOLDY, S.A.R. (eds) *Petroleum Geology of Northwest Europe: Proceedings of the 5th Conference*. Geological Society, London, 379–390.

WALSH, A., KNAG, G., MORRIS, M., QUINQUIS, H., TRICKER, P., BIRD, C. & BOWER, S. 1999. Petroleum geology of the Irish Rockall Trough—a frontier challenge. *In*: FLEET, A.J. & BOLDY, S.A.R. (eds) *Petroleum Geology of Northwest Europe: Proceedings of the 5th Conference*. Geological Society, London, 433–444.

The structural style and evolution of the Bróna Basin

A. THOMSON[1,2] & A. McWILLIAM[1,3]

[1]*Saga Petroleum Ireland, 150 Victoria Street, London SW1E 5LB, UK*

[2]*Present address: Talisman Energy (UK) Limited, 163 Holburn Street, Aberdeen AB10 6BZ, UK (e-mail: alastairt@talisman.co.uk)*

[3]*Present address: 33 Artillery Terrace, Guildford GU1 4NL, UK*

Abstract: The Bróna Basin lies on the eastern margin of the Rockall Basin and has a NNE–SSW trend. Although at present undrilled it is believed to contain Carboniferous, Triassic and Jurassic to Lower Cretaceous strata. It probably formed as a product of Late Jurassic rifting, which ultimately resulted in the creation of the Rockall Basin. The basin history is complex and its present form is the result of pre-Jurassic structuring, Late Jurassic rifting, later (latest Jurassic to earliest Cretaceous, and Early Tertiary) compressional pulses, and the effects of thermal subsidence in the adjacent Rockall Basin. The development of the Bróna Basin is of direct relevance to understanding the geological evolution of the other marginal basins lying on the flanks of the Rockall Basin, and to the Rockall region as a whole.

The Bróna Basin lies on the west side of the Porcupine High. It was first described by Tate (1993), who called it the Bean Basin. It was renamed by Naylor *et al.* (1999), who defined two sub-basins (the North Bróna Basin and the South Bróna Basin), separated by a basement high, the Clíona High (Fig. 1). The North Bróna Basin is situated on the eastern margin of the Rockall Basin in water depths ranging from 1000 to 1600 m. It is *c.* 50 km long by 20 km wide and is located between 52°20′ and 52°50′N and 15°00′ and 15°30′W, in Quadrant 83 of the Irish offshore region. Saga Petroleum Ireland operates Licence 10/97, which comprises blocks 83/13, 18, 19 and 20 in the southern half of the North Bróna Basin. The operatorship was awarded in 1997, in the Irish Rockall Frontier Licence Round. The partners in the group are Shell EP Ireland B.V., Statoil Exploration (Ireland) Ltd and Total Oil Marine plc. Following the licensing, 1600 km of 2D seismic data were acquired and processed for the group. This paper is primarily the result of the interpretation of these seismic data. No wells have been drilled in the basin. As a result, the interpretation has been kept as generalized as possible and the seismic stratigraphy has been assigned on the basis of structural and seismic stratigraphic comparison with adjacent basins, including the Slyne Basin (Dancer *et al.* 1999) and the Porcupine Basin (Croker & Shannon 1987; Tate & Dobson 1989). The regional geology in these basins, and the extent of the stratigraphic units, provides evidence supporting our proposed stratigraphic succession for the Bróna Basin. The objectives of the current paper are to describe the structural style of the Bróna Basin and to propose a model for the geological evolution of the basin.

Regional tectonic setting

The Bróna Basin is one of a series of marginal (perched) basins on the eastern margin of the Rockall Basin, which contain an assumed Carboniferous to Jurassic fill (Naylor *et al.* 1999). Other basins along the margin include the Pádraig and Macdara basins. These basins are thought to have similar Mesozoic successions but differ in the thickness of their preserved sedimentary succession. Regionally the Rockall Basin trends NE–SW but in the Bróna Basin area the orientation of the Rockall Basin is north–south, shaped largely by the location and orientation of the basement-cored Porcupine High (Fig. 1). The Bróna Basin trends NNE–SSW and we suggest that this is due to its development during the rifting phases that ultimately resulted in the formation of the Rockall Basin. As such, the rifting events that formed the Bróna Basin were likely to have been controlled by opening about a common pole of rotation, involving reactivation of Caledonian lineaments (Shannon 1991). The opening about a pole of rotation is illustrated by the clay model, constructed to replicate the development of a rifted margin during continental rift and drift

From: SHANNON, P.M., HAUGHTON, P.D.W. & CORCORAN, D.V. (eds). 2001. *The Petroleum Exploration of Ireland's Offshore Basins*. Geological Society, London, Special Publications, **188**, 401–410. 0305-8719/01/$15.00

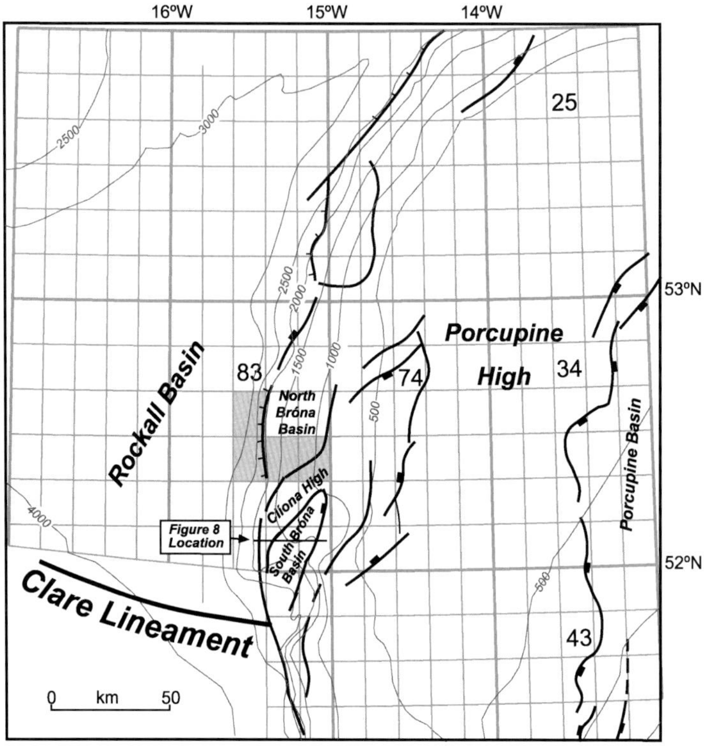

Fig. 1. Structural elements map of the eastern margin of the Rockall Basin, showing the location of the North and South Bróna basins. Bathymetry is in metres and quadrant numbers are shown in bold. The four blocks operated by Saga Petroleum Ireland Ltd are shaded. The location of Figure 8 is indicated.

phases (Fig. 2). The model demonstrates the tectonic setting of the Rockall Basin during Jurassic rifting when the continents separated about a pole of rotation (with the European plate on the right) and the resultant extensional faults developed as radii about this pole. This can explain the observation that, whereas the Rockall Basin trends north–south in the study area, the trends of the Jurassic marginal basins are oblique to the Rockall Basin margin and lie on an approximate NE–SW trend. This is a complementary explanation to that which identifies the older NE–SW trend as being solely a Caledonian overprint. Rifting ceased in earliest Cretaceous time and was followed by thermal subsidence of the Rockall Basin in Mid- to Late Cretaceous and Tertiary time. The present form of the Rockall marginal basins is therefore the result of the Jurassic rifting event, followed by a later phase of rapid subsidence in the main Rockall Basin that truncated the western portions of the marginal basins via fault movement and/or catastrophic slope failure. In addition, at least two compressional phases have been identified in the area.

Structural trends

The North Bróna Basin is a half-graben feature, bounded on its eastern margin by a NNE–SSW-trending fault against the Porcupine High (Fig. 3). Its western boundary is defined by the Rockall Basin margin, which cuts the basin obliquely on a north–south trend. The basin deepens and the strata increase in thickness to the south.

Four main structural trends are identified within the Bróna Basin. These are readily identified on the gravity residuals map (Fig. 4) and can also be mapped with confidence on the seismic data.

(1) NNE–SSW: this is the main fault trend within the basin and includes the steep basin-bounding fault, Fault 1 (Figs 3, 5 and 6), and intra-basinal faults including Fault 2.

(2) NE–SW: this trend defines the orientation of the Clíona High, the basement-cored ridge that separates the North and South Bróna basins (Fig. 1).

(3) North–south: this trend is represented by the orientation of the Rockall Basin margin and

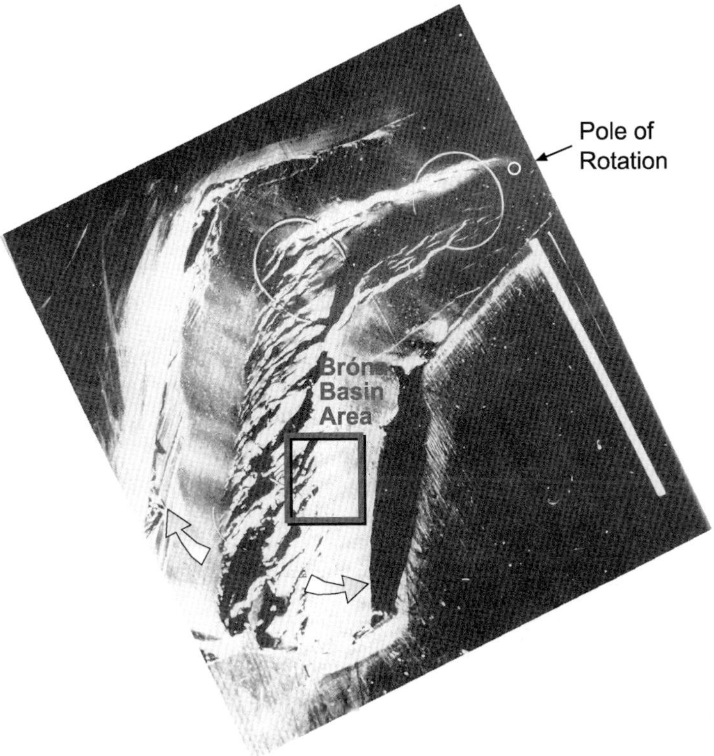

Fig. 2. Rotational clay model showing the development of structural orientations similar to those of the Bróna Basin and its general environs. The proposed location of the Bróna Basin (boxed area), the NE–SW trend of the synrift faults and the north–south trend of the region corresponding to the Rockall Basin should be noted. The eastern margin of the small boxed area is oriented north–south. From Petrecon Australia Pty Ltd.

by the associated low-angle slump faults that dip westwards into the Rockall Basin. It is also seen in the trend of the anticline intersected by line SG9805-406A (Figs 3 and 5).

(4) NW–SE: this trend is most apparent as offsets on the gravity residuals map (Fig. 4). Where faults of this orientation occur they exhibit lateral offset. Seismic mapping suggests that they can be interpreted as a series of transfer faults. This trend is also strongly developed in other parts of the Rockall Basin in Irish and UK waters (Rumph *et al.* 1993; McGrane *et al.* 2001), as revealed by regional gravity and magnetic data, and probably relates to deep-seated basement heterogeneity.

Stratigraphic evolution

The stratigraphy of the sedimentary fill in the Bróna Basin is speculative in view of the lack of drilling. The proposed stratigraphy is based largely on a comparison of the seismic character and regional stratigraphy with adjacent basins, including the Slyne Basin (Dancer *et al.* 1999) and the Porcupine Basin (Tate & Dobson 1989; Shannon 1991; Sinclair *et al.* 1994), where more robust well control exists.

The pre-rift succession in the Bróna Basin is interpreted to consist of Palaeozoic sedimentary strata (possibly Upper Carboniferous units) on account of its weakly reflective nature. Evidence for a regionally extensive Carboniferous unit in the pre-rift sequence is widespread. Wells drilled in adjacent basins that reach total depth in the pre-Triassic sediments almost invariably encountered Carboniferous strata ranging from late Namurian to Stephanian age (Croker & Shannon 1987; Robeson *et al.* 1988; Tate & Dobson 1989). For example, well 34/5-1, which was drilled on the NE of the Porcupine High, encountered Westphalian C–D sediments directly beneath a thin Tertiary section. Furthermore, reworked Carboniferous sediments are commonly found in the basal Middle Jurassic

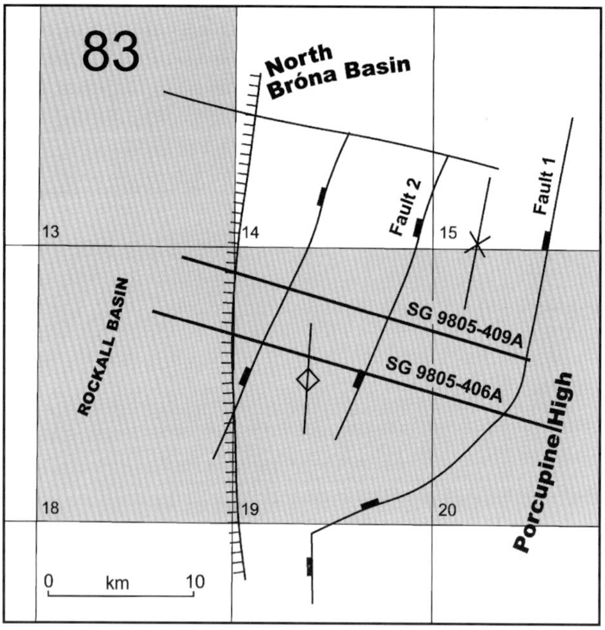

Fig. 3. North Bróna Basin structural elements. The locations of the seismic lines shown in Figures 4 and 5 are indicated. The four blocks (see Fig. 1) operated by Saga Petroleum Ireland Ltd are shaded.

sequence (Shannon 1991) and extensively reworked Carboniferous palynomorphs are reported in the Jurassic to Tertiary sediments within the Porcupine, Slyne and Erris basins.

The seismic unit overlying the pre-rift sequence is poorly reflective, suggesting relatively homogeneous strata. The unit thickens towards Fault 2, indicating synsedimentary tectonism and movement on this fault. We have interpreted this early rift phase unit as being of Triassic, rather than Carboniferous, age (Fig. 5). This is based on its thickness pattern, which indicates deposition in NNE-trending, fault-controlled sub-basins, and on the regional seismic and well evidence interpretation that an early NE- to NNE-oriented pre-Late Jurassic rift phase developed along reactivated Caledonian lineaments during Triassic times (Shannon 1991). There is the possibility that a basal, seismically banded section is of Permian age and, on the basis of its seismic facies reflection configuration, may consist of a heterogeneous Zechstein-equivalent unit. There is no seismic evidence for any pronounced halokinesis in the Bróna Basin, although the low-angle and listric nature of some of the major faults could be interpreted as indicating thin salt layers, which facilitated a detachment layer within which these faults sole out. An alternative interpretation, which we consider more likely because of the seismic facies comparison with the Slyne Basin and with published regional data, is that this early synrift basal unit consists of a Lower Triassic Sherwood Sandstone equivalent section. Further evidence for the existence of Triassic strata in the Bróna Basin comes from a submersible dredge sample taken in a bathymetric embayment on the western margin of the Porcupine High to the south of the South Bróna Basin (Auzende *et al.* 1989). The embayment is situated at the convergence of the southern end of the Cillian Basin (Naylor *et al.* 1999) with the steeply inclined fault scarp that marks the western edge of the Porcupine High. The dredge recovered barren red beds of a tentative Permo-Triassic age (Auzende *et al.* 1989). Although this is not direct evidence of Triassic sediments within the Bróna Basin itself, because of the lack of a seismic tie, it adds confidence to our interpretation regarding the presence of a regionally extensive early synrift Triassic sequence on the western side of the Porcupine High.

The seismic unit interpreted as being of Early to Mid-Jurassic age (Fig. 5) consists of parallel reflections. This seismic character is interpreted as a mixed energy (mudstones, sandstones and carbonates) succession. Mid-Jurassic time was a period of tectonic quiescence and this apparent heterogeneity may be similar to that encountered

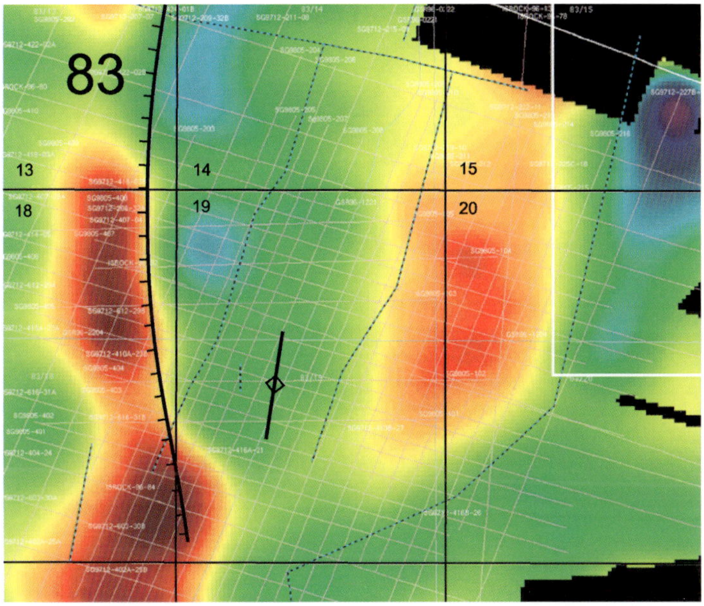

Fig. 4. Gravity residuals map; 20 km filtered. The Rockall Basin edge and main Bróna Basin faults are displayed. (See Fig. 3 for location.)

in the Middle Jurassic sediments in the Porcupine Basin. There the succession exhibits similar seismic characteristics and comprises marginal marine, fluvial and lacustrine facies of interbedded shales, sandstones and carbonates deposited in a late thermal subsidence and early onset warp phase of basin development (Sinclair *et al.* 1994). We have inferred similar lithologies for the Lower to Middle Jurassic section in the Bróna Basin. The fluvial systems responsible for the deposition of the Middle Jurassic sediments in the Porcupine Basin are thought to have flowed from the NNE (Shannon 1991), possibly along the Slyne and Erris basin axes. However, it is likely that the Mid-Jurassic system in the Bróna Basin is separate from that in the Porcupine Basin given the occurrence of the intervening Porcupine High.

The interpreted Upper Jurassic section comprises a seismic unit that increases in thickness towards the basin-bounding fault in the east (Figs 5 and 6). Again, its reflective seismic character and structural setting resemble those of the Upper Jurassic succession encountered in the Porcupine Basin. As such, this similarity with the Porcupine Basin has led us to tentatively interpret the Upper Jurassic section as comprising marine and marginal marine facies interbedded with turbiditic and alluvial sediments derived from the Porcupine High to the east. We therefore cautiously suggest that the Upper Jurassic unit in the Bróna Basin grades upwards from the continental deposits of Mid-Jurassic time through marginal to marine deposits related to the regional Late Jurassic marine transgression. It is suggested that the seismic heterogeneity seen in this unit is due to tectonically induced pulses of sediment flowing into the basin as a result of movements on the main basin-bounding fault. This is similar to the Late Jurassic tectonosedimentary development proposed for the Porcupine Basin by various workers (e.g. Croker & Shannon 1987; Sinclair *et al.* 1994).

The interpretation of the Base Cretaceous Unconformity (BCU) shown in Figs 5 and 6 is given added credence by the correlation of a BCU-lookalike event (regional post-rift unconformity) from the Rockall Basin into the south of the North Bróna Basin. It is therefore likely that Lower Cretaceous strata are preserved in the syncline above the BCU (Fig. 6).

We have interpreted the proposed Lower Cretaceous section in the Rockall Basin to consist of marine strata onlapping the Jurassic basin margins before the onset of Late Cretaceous sea-floor spreading in the North Atlantic. However, the interpreted Cretaceous seismic unit within the Bróna Basin contains a number of westerly prograding clinoforms, which resemble similar units of deltaic origin in the Porcupine Basin (Croker & Shannon 1987).

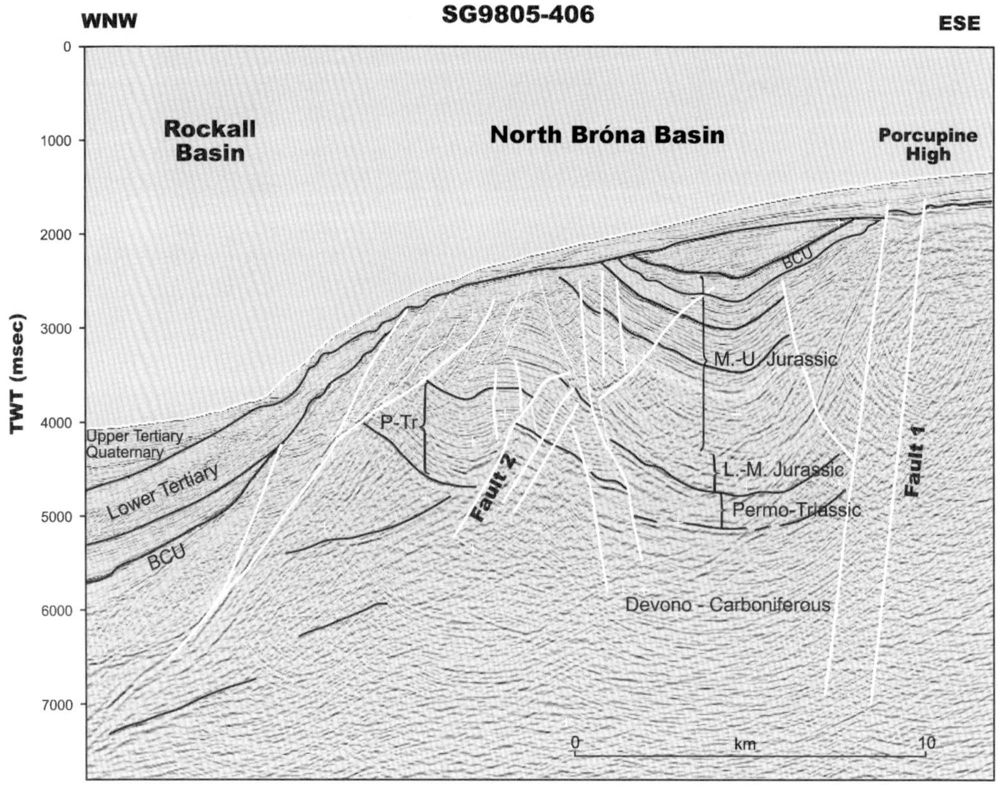

Fig. 5. WNW–ESE North Bróna Basin seismic dip line (SG9805-406A) illustrating the basin-bounding fault, the western basin margin and a compressional anticline. BCU, Base Cretaceous Unconformity. (See Fig. 3 for location.)

These clinoforms may indicate the presence of fan deltas, with progradation into the basin induced by local uplift along the margins of the Porcupine High in Aptian–Albian times (Shannon 1991).

A thin covering of possible Upper Cretaceous to Quaternary strata overlies the eroded Jurassic and Cretaceous sequences (Figs 5 and 6). From the west, approaching the edge of the Rockall Basin, the interpreted mid-Miocene unconformity is seen to onlap the interpreted Base Oligocene unconformity.

The southerly plunge of the basin is confirmed by the Bouguer gravity anomaly map (Fig. 7), which shows an increase in gravity towards the north (yellow to red on map), at the western margin of the North Bróna Basin. This gravity high may result from a northerly rising metamorphic basement. It is distinct from the Bouguer gravity high associated with the Rockall Basin. The tilting to the south of the Bróna Basin is interpreted to have continued from the Jurassic synrift period into Tertiary time. This early tilting is evidenced by the apparent development of thin-skinned tectonics over an uplifted basement décollement zone, adjacent to Fault 2 (Fig. 6) in the north of the study area. Later tilting resulted in progessively older strata being eroded northwards at the major regional post-'BCU' unconformity.

The South Bróna Basin (Fig. 8), like the North Bróna Basin, is a tilted half-graben structure that deepens to the south. It is bounded on its eastern margin by the Porcupine High and by a NNE–SSW-trending growth fault, and on its western margin by the Clíona High. The southern margins of both basins have been exploited by east–west-trending ?Plio-Pleistocene submarine channels, which lie above the hanging wall of the old southern basin margin faults.

Structural evolution

Minor extension in Triassic time is deduced from the thickening of the Permo-Triassic unit into the NNE–SSW-trending Fault 2 (Fig. 5). This unit

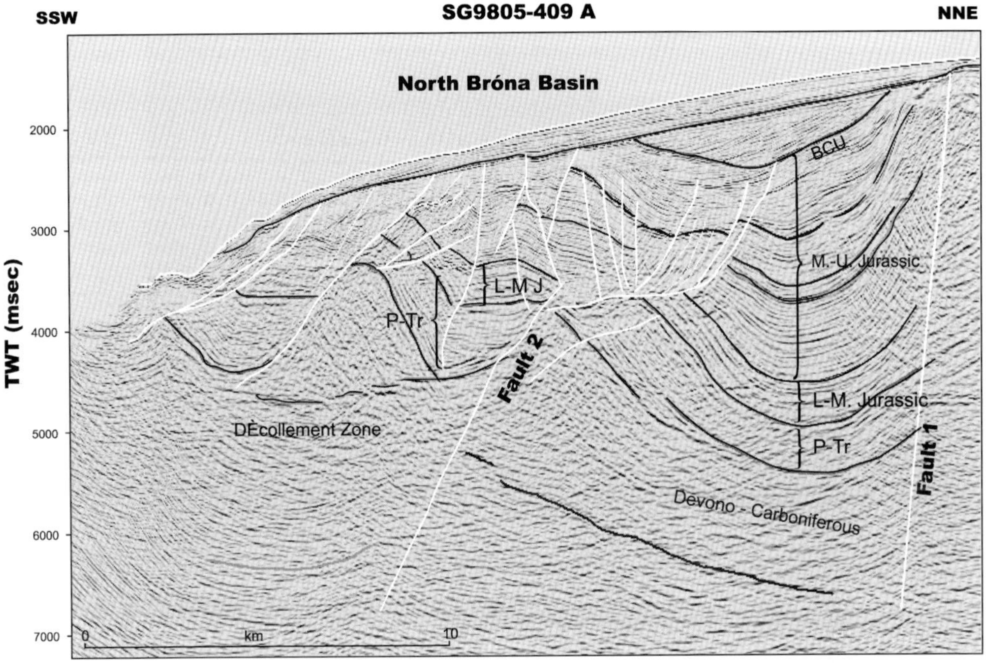

Fig. 6. SSW–NNE North Bróna Basin seismic dip line (SG9805-409A) illustrating basin form in the north and the relatively shallow basement. (See Fig. 3 for location.)

thins towards the eastern basin margin (Fig. 3), suggesting that the Triassic basin margin lay to the east of the preserved Jurassic margin. Fault 2 becomes shallower to the north, where it acts as a plane of décollement along the top of the shallowing basement (Fig. 6). Lower to Middle Jurassic strata show little change in thickness within the basin and were deposited during a period of relative tectonic quiescence. The main rifting phase is likely to have occurred in Late Jurassic to earliest Cretaceous time, and resulted in the development of the Bróna Basin half-graben. Most of the extension was taken up by the NNE–SSW graben margin fault (Fault 1). This rifting phase was followed by subsidence and continued deposition in the basin.

Two compressive periods can be deduced from the obvious bed shortening and folding seen in the section. These compressive phases were separated by a final, minor, late phase of extension, accommodated by movement on the low-angle Fault 2, which may have been the result of Rockall Basin subsidence. The older of these two compressive phases is (on the basis of the stratigraphy assigned herein) probably of latest Jurassic to earliest Cretaceous age, and the younger phase is interpreted to be of Early Tertiary age. The older compressive phase resulted in reversal of Fault 2 (Fig. 5) and the development of north–south-trending anticlines (Fig. 3). It should be noted that Cretaceous inversion periods have been documented in the Porcupine Basin and in the Celtic Sea basins, whereas a pronounced Early Tertiary phase on inversion is also identified in the Celtic Sea region (Shannon 1991).

The interpreted Lower Cretaceous sediments were deposited as a prograding, downlapping wedge onto the likely BCU surface, indicating that the Porcupine High was a positive feature during the early post-rift phase. The unit is now preserved within the synclinal axis (Fig. 6). The second compressive phase resulted in continued north–south folding and possible steepening of the basin margin fault (Fig. 6). The observations of compression are confirmed in the South Bróna Basin where, because of the presence of the Clíona High, a rigid basement high to the west (between the Rockall Basin and the South Bróna Basin), the results of this compression are much more pronounced. The folding events of both phases are the result of approximate east–west-directed compression, i.e. perpendicular to the Cretaceous Rockall Margin. It is not possible to date this second compressive phase within the Bróna Basin more precisely than to Early

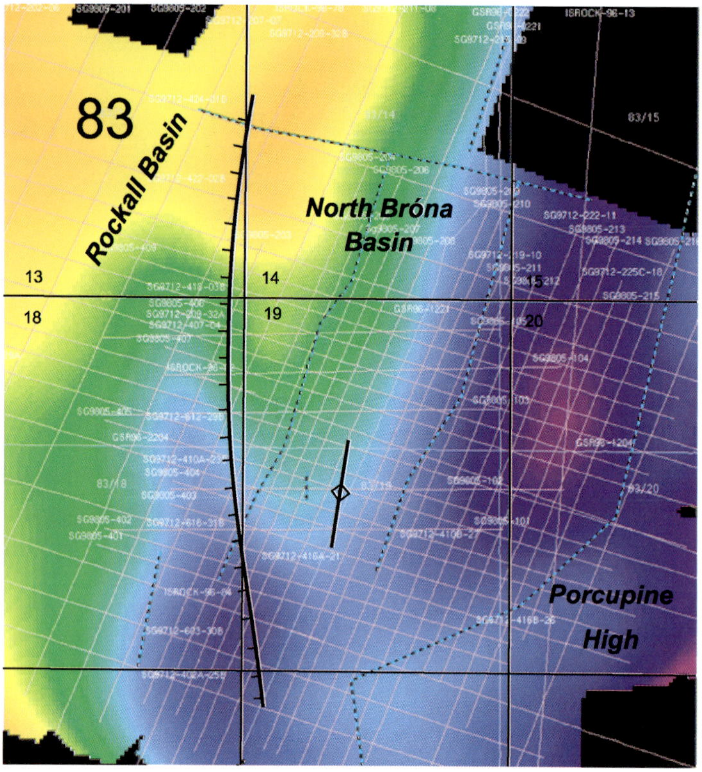

Fig. 7. Bouguer gravity anomaly map. The Rockall Basin edge and main Bróna Basin faults are displayed. Gravity highs are displayed as yellow to orange and gravity lows as blue. The gravity highs include the Rockall Basin, because of its thinned crust, and the northern part of the Bróna Basin, probably because of the occurrence of relatively shallow basement.

Cretaceous to Early Tertiary time. However, Paleocene to Eocene compressional phases have been documented within the axis of the Rockall Basin and in the Porcupine Basin (Shannon *et al.* 1993). It is therefore considered likely that this second phase of Bróna Basin folding was controlled by the same tectonic event. Tertiary compressive events have been noted in the Faroe–Rockall region by Boldreel & Andersen (1993), and in the North Sea basin systems, where typically the orientation of maximum horizontal stress is perpendicular to the basin axes. These compressional features are commonly observed along, and close to, plate boundaries, and are typically controlled by major zones of weakness in the crust (Davidson 1997). The north–south orientation of the folding axis suggests that the compression is unlikely to have been the result of African (Alpine) or Bay of Biscay movements, as this would have been unlikely to have resulted in north–south-oriented fold axes.

Conclusions

The main structural trends of the North Bróna Basin are NNE–SSW, with ancillary NE–SW, north–south and NW–SE trends apparent. The NNE–SSW trend is the result of Triassic and Jurassic rift events, which ultimately led to the development of the Rockall Basin. We suggest the presence of a Triassic rift basin succession, on the basis of basin alignment, seismic facies and the proven occurrence of Triassic sediments from drilling in adjacent basins and dredge samples in the adjacent area.

A Lower to Middle Jurassic sequence overlies this unit. By analogy with the Porcupine Basin to the east, this is suggested to comprise marginal marine, fluvial and lacustrine facies of interbedded shales, sandstones and carbonates.

The main rift phase within the North Bróna Basin occurred in Late Jurassic time and comprises a reflective seismic unit that increases in thickness towards the basin-bounding fault in the east. This seismic unit section resembles that

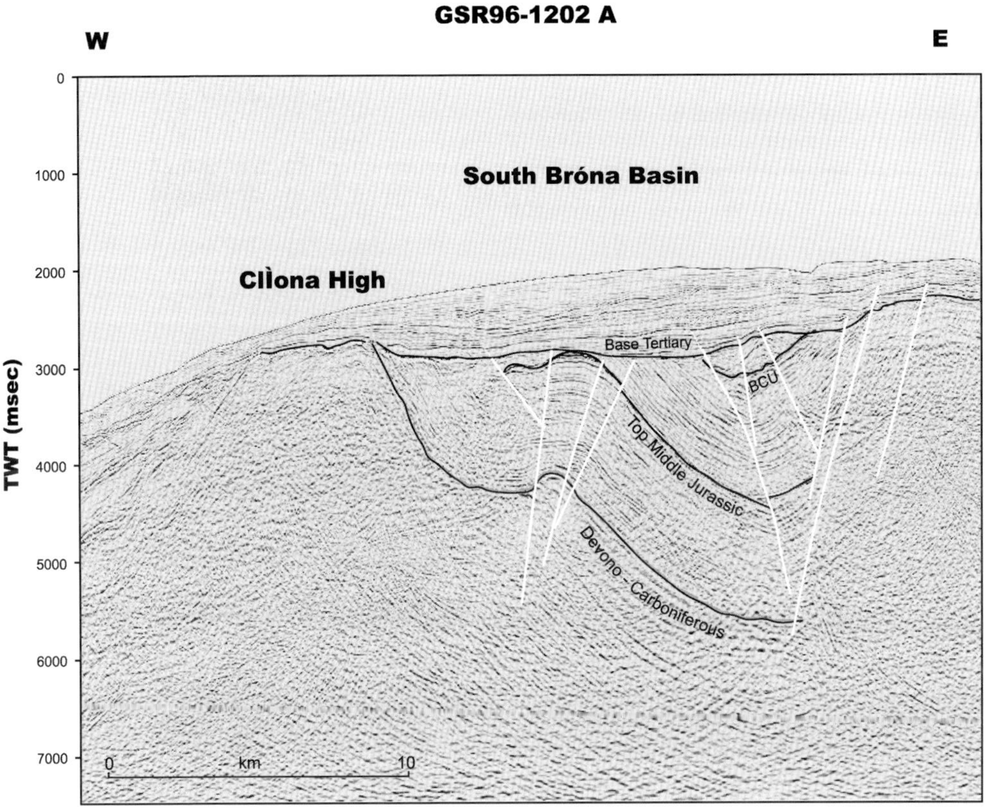

Fig. 8. East–west seismic dip line through the South Bróna Basin. It illustrates the general basin form, including the pronounced compressional anticline, within the basin. (See Fig. 1 for location.)

of the Upper Jurassic sequence seen in the Porcupine Basin and is interpreted as comprising marine and marginal marine facies interbedded with turbiditic and alluvial sediments derived from the Porcupine High to the east.

At least two compressive pulses, believed to have occurred in latest Jurassic to earliest Cretaceous time, and early Tertiary time, respectively, were responsible for giving the basin its characteristic appearance with anticlines developed along the western margin of the basin. This is most clearly seen in the South Bróna Basin, where, as a result of the presence of a rigid basement block between the Bróna Basin and the Rockall Basin, the results of the compression are very pronounced. The folding events of both phases are the result of approximate east–west-directed compression, i.e. perpendicular to the Cretaceous Rockall Margin.

The authors would like to thank the following for help and useful comments during the preparation of this paper: T. Rochester, Total Oil Marine plc; S. Smith, Shell EP Ireland B.V.; I. MacKenzie, Shell Exploration & Production UK Ltd; A. Walsh, Statoil Exploration (Ireland) Ltd; J. Davidson, Petrecon Australia Pty Ltd. We would also like to thank M. Golden for drafting the figures. Line GSR96-202 is reproduced by permission of Schlumberger Geco-Prakla. The other seismic examples were processed for the group by BiPS.

References

AUZENDE, J.-M., COUSIN, M., COUTELLE, A. & 5 OTHERS 1989. Stratigraphie des escarpements encadrant la baie de Porcupine: résultants préliminaries de la campagne *Cyaporc* (juillet–août). *Oceanologica Acta*, **12**(3), 117–131.

BOLDREEL, L.O. & ANDERSEN, M.S. 1993. Late Palaeocene to Miocene compression in the Faeroe–Rockall area. *In*: PARKER, J.R. (ed.) *Petroleum Geology of Northwest Europe: Proceedings of the 4th Conference*. Geological Society, London, 1025–1034.

CROKER, P.F. & SHANNON, P.M. 1987. The evolution and hydrocarbon prospectivity of the Porcupine Basin, offshore Ireland. *In*: BROOKS, J. & GLENNIE,

K.W. (eds) *Petroleum Geology of North West Europe*. Graham & Trotman, London, 633–642.

DANCER, P.N., ALGAR, S.T. & WILSON, I.R. 1999. Structural evolution of the Slyne Trough. *In*: FLEET, A.J. & BOLDY, S.A.R. (eds) *Petroleum Geology of Northwest Europe: Proceedings of the 5th Conference*. Geological Society, London, 445–453.

DAVIDSON, J.K. 1997. Synchronous compressional pulses in extensional basins. *Marine and Petroleum Geology*, **14**, 513–549.

MCGRANE, K., READMAN, P.W. & O'REILLY, B.M. 2001. Interpretation of transverse gravity lineaments in the Rockall Basin. *In*: SHANNON, P.M., HAUGHTON, P.D.W. & CORCORAN, D.V. (eds) *The Petroleum Exploration of Ireland's Offshore Basins*. Geological Society, London, Special Publications, **188**, 393–399.

NAYLOR, D., SHANNON, P. & MURPHY, N. 1999. *Irish Rockall Basin Region—a Standard Structural Nomenclature System*. Petroleum Affairs Division, Special Publication, **1/99**.

ROBESON, D., BURNETT, R.D. & CLAYTON, G. 1988. The Upper Palaeozoic geology of the Porcupine, Erris and Donegal basins, offshore Ireland. *Irish Journal of Earth Sciences*, **9**, 153–175.

RUMPH, B., REAVES, C.M., ORANGE, V.G. & ROBINSON, D.L. 1993. Structuring and transfer zones in the Faeroe Basin in a regional tectonic context. *In*: PARKER, J.R. (ed.) *Petroleum Geology of Northwest Europe: Proceedings of the 4th Conference*. Geological Society, London, 999–1009.

SHANNON, P.M. 1991. The development of Irish offshore sedimentary basins. *Journal of the Geological Society, London*, **148**, 181–189.

SHANNON, P.M., MOORE, J.G., JACOB, A.W.B. & MAKRIS, J. 1993. Cretaceous and Tertiary basin development west of Ireland. *In*: PARKER, J.R. (ed.) *Petroleum Geology of Northwest Europe: Proceedings of the 4th Conference*. Geological Society, London, 1057–1066.

SINCLAIR, I.K., SHANNON, P.M., WILLIAMS, B.P.J., HARKER, S.D. & MOORE, J.G. 1994. Tectonic control on sedimentary evolution of three North Atlantic borderland Mesozoic basins. *Basin Research*, **6**, 193–217.

TATE, M.P. 1993. Structural framework and tectonostratigraphic evolution of the Porcupine Seabight Basin, offshore western Ireland. *Marine and Petroleum Geology*, **10**, 95–123.

TATE, M.P. & DOBSON, M.R. 1989. Late Permian to early Mesozoic rifting and sedimentation offshore NW Ireland. *Marine and Petroleum Geology*, **6**, 49–59.

A Mid- to Late Cenozoic tectonostratigraphic framework for the Rockall Trough

M. S. STOKER[1], T. C. E. VAN WEERING[2] & T. SVAERDBORG[3]

[1]*British Geological Survey, Murchison House, West Mains Road, Edinburgh EH9 3LA, UK (e-mail: mss@bgs.ac.uk)*

[2]*Netherlands Institute for Sea Research, P.O. Box 59, 1790 AB Den Burg, Texel, Netherlands*

[3]*Department of Earth Sciences, University of Aarhus, DK-8000 Aarhus, Denmark*

Abstract: Regional subsidence in late Eocene time resulted in a marked change in the style of sedimentation in the Rockall Trough, coincident with the onset of bottom-current activity in the region. This subsidence is manifest by a shut-down of shelf-derived sediment supply, and the formation of a deep-water unconformity (reflector C30) caused by bottom-current erosion. The unconformity is particularly enhanced on the flanks of the basin where the downwarped and eroded surface of Eocene and older strata is onlapped by middle to upper Cenozoic sediments. The latter comprise three megasequences, RTc (of late Eocene to early Miocene age), RTb (of early Miocene to early Pliocene age) and RTa (of early Pliocene to Holocene age), which consist predominantly of deep-marine contourites, although a prograding clastic wedge has built out along the Hebrides–Malin margin since early Pliocene time. These megasequences reflect a gross three-stage depositional history; predominantly a response to intra-plate tectonism that modified sedimentation patterns and palaeoceanographic circulation. Megasequences RTc and RTb are separated by reflector C20, a reflective zone formed by lithification and diagenetic changes in lower Miocene strata, whereas RTb and RTa are separated by reflector C10, an early Pliocene angular unconformity. The development of C20 and C10 reflects major phases of Neogene basin evolution and can be linked, respectively, to the submergence of the Greenland–Scotland Ridge and the uplift and erosion of northern Britain and Ireland.

The Rockall Trough is a NE–SW-trending intracontinental basin located on the continental margin off NW Britain and Ireland (Fig. 1). It is bounded to the NE by the Wyville-Thomson Ridge, but deepens to the SW, where it opens out into the Porcupine Abyssal Plain. Water depths in the trough increase from 1000 m in the NE to 4000 m in the SW. The width of the trough (at the 1000 m isobath) is between 200 and 300 km. In the central and northern part of the trough, the continuity of the basin floor is interrupted by the seamounts of Rosemary Bank, Anton Dohrn and Hebrides Terrace. This configuration largely reflects late Mesozoic–early Cenozoic continental rifting across the margin, related to the processes that led to the opening of the NE Atlantic Ocean (e.g. Knott *et al.* 1993; Doré *et al.* 1999; Roberts *et al.* 1999).

The post-break-up history of the Rockall Trough has mainly been dominated by subsidence outpacing sedimentation, resulting in the present-day, starved, deep-water basin. However, sedimentation has not necessarily occurred amidst a background of tectonic quiescence. Information from British Geological Survey (BGS) core 57/12-18 from the top of the Anton Dohrn Seamount (Fig. 2), which proved a middle to upper Eocene nearshore conglomerate, suggests that the Rockall Trough may have subsided by at least 600–700 m since late Eocene time (Stoker 1997). Part of this subsidence may have occurred in the late Eocene interval linked to regional submergence of the continental margin. Regional subsidence is implied by the shut-down of shelf-derived source areas, such as the Rockall and George Bligh banks and the Hebrides–Malin shelf, which preserve Eocene prograding wedges (Fig. 3), and the change to a predominantly bottom-current-influenced deep-marine sedimentary environment in the Rockall Trough (Stoker 1997, 1998). According to Vanneste *et al.* (1995), post-Eocene subsidence in the NW Rockall Trough became non-uniform with increased differentiation between the flank and floor of the basin. This differential subsidence may have

From: Shannon, P.M., Haughton, P.D.W. & Corcoran, D.V. (eds). 2001. *The Petroleum Exploration of Ireland's Offshore Basins*. Geological Society, London, Special Publications, **188**, 411–438. 0305-8719/01/$15.00 © The Geological Society of London 2001.

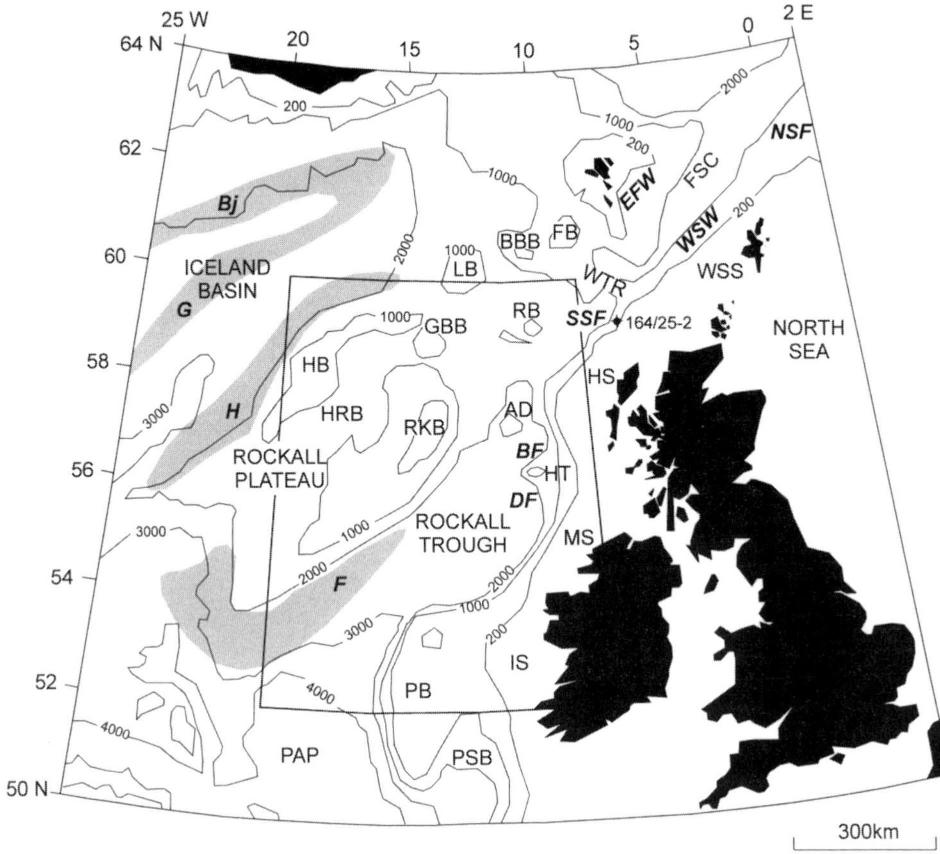

Fig. 1. Bathymetric setting of the continental margin off Britain and Ireland (contours in metres) showing location of main study area (boxed), well 164/25-2 referred to in text, and major sediment drifts (shaded). FSC, Faeroe–Shetland Channel; WSS, West Shetland Shelf; WTR, Wyville-Thomson Ridge; FB, Faroe Bank; BBB, Bill Bailey's Bank; LB, Lousy Bank; GBB, George Bligh Bank; RB, Rosemary Bank; AD, Anton Dohrn; HT, Hebrides Terrace; HS, Hebrides Shelf; RKB, Rockall Bank (*Rockall High*); HRB, Hatton–Rockall Basin (*Hatton Basin*); HB, Hatton Bank (*Hatton High*); MS, Malin Shelf; IS, Irish Shelf; PB, Porcupine Bank (*Porcupine High*); PSB, Porcupine Seabight (*Porcupine Basin*); PAP, Porcupine Abyssal Plain; ***NSF***, North Sea Fan; ***WSW***, West Shetland Wedge; ***EFW***, East Faroes Wedge; ***SSF***, Sula Sgeir Fan; ***BF***, Barra Fan; ***DF***, Donegal Fan; ***F***, Feni Ridge; ***H***, Hatton Drift; ***G***, Gardar Drift; ***Bj***, Bjorn Drift. It should be noted that the italicized names in parentheses represent the newly defined structural terminology for these features on the Irish Atlantic margin (Naylor *et al.* 1999), where the Rockall Trough is also renamed as the *Rockall Basin* (see text for further details).

been driven, at least in part, by a series of Oligo-Miocene compressional events that have been reported from around the northern margin of the Rockall Trough (e.g. Roberts 1989; Boldreel & Andersen 1993, 1995; Knott *et al.* 1993; Doré & Lundin 1996). Moreover, there is increasing evidence for uplift of the NW British hinterland during Neogene time (Stoker 1999), which, especially, modified the northeastern flank of the trough through significant shelf-margin progradation.

The ensuing changes in basin geometry and palaeobathymetry are likely to have modified palaeoceanographic circulation and deep-water sedimentation patterns. Consequently, the sedimentary response to these changes should be preserved in the mid- to late Cenozoic deep-water sediment fill of the Rockall Trough, expressed as regional unconformities and by changing styles of deposition. As the Rockall Trough was open to the NE Atlantic Ocean throughout this interval, it was particularly

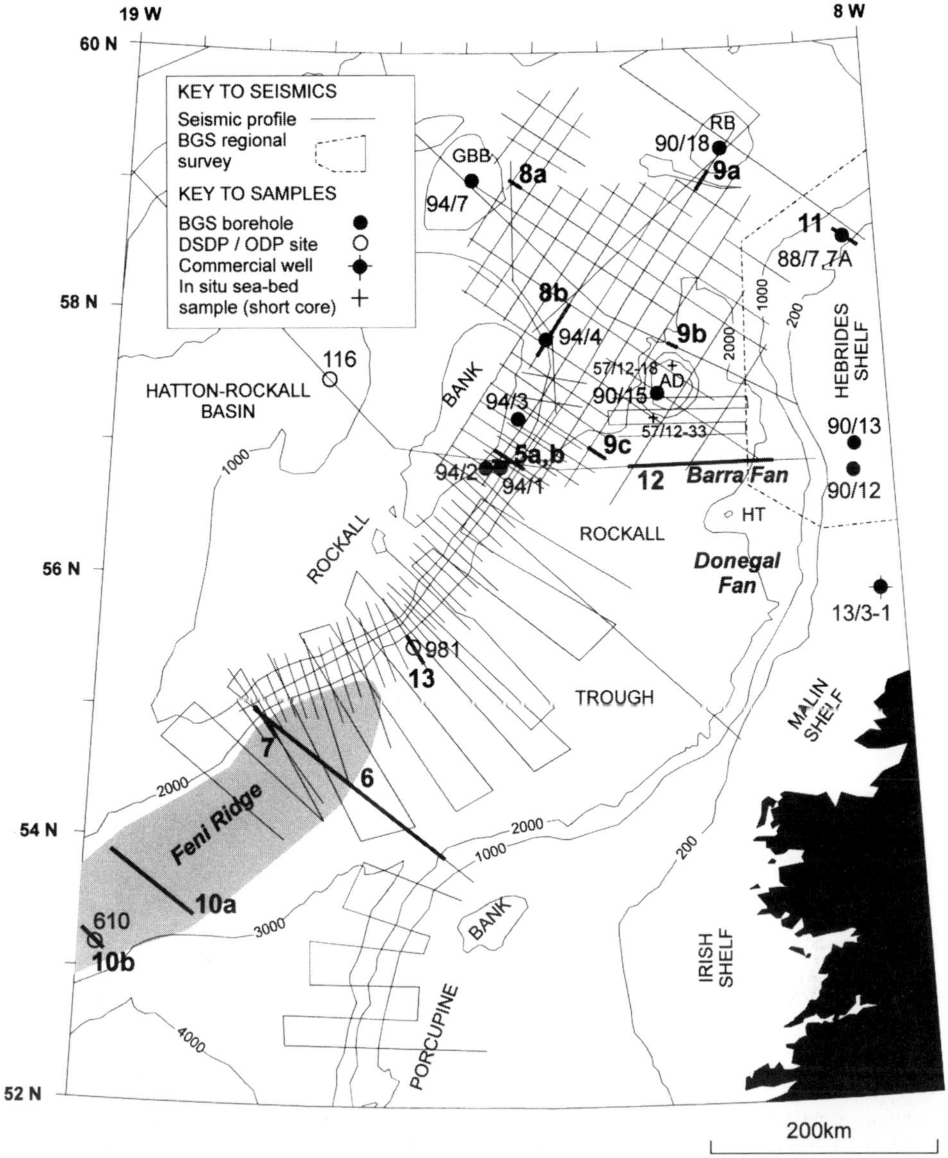

Fig. 2. Detailed map of Rockall Trough study area showing seismic-reflection database and positions of sample sites. Bold lines show locations of seismic sections illustrated in other figures (numbered). Shading indicates extent of Feni Ridge sediment drift.

sensitive to changes in bottom-current regime. Climate was an additional controlling factor, particularly during late Cenozoic time when glaciation greatly influenced sedimentation patterns (e.g. Stoker 1995). Recognizing the interplay and relative importance of these processes is crucial to developing a further understanding of the post-break-up history of the Rockall Trough.

The aim of this paper is to establish a mid- to late Cenozoic tectonostratigraphic framework for the Rockall Trough. Using seismic-stratigraphic techniques, emphasis is placed on the identification of regionally significant reflectors and megasequences that reflect major phases of basin evolution. The sedimentary record and history of palaeoenvironmental change preserved within

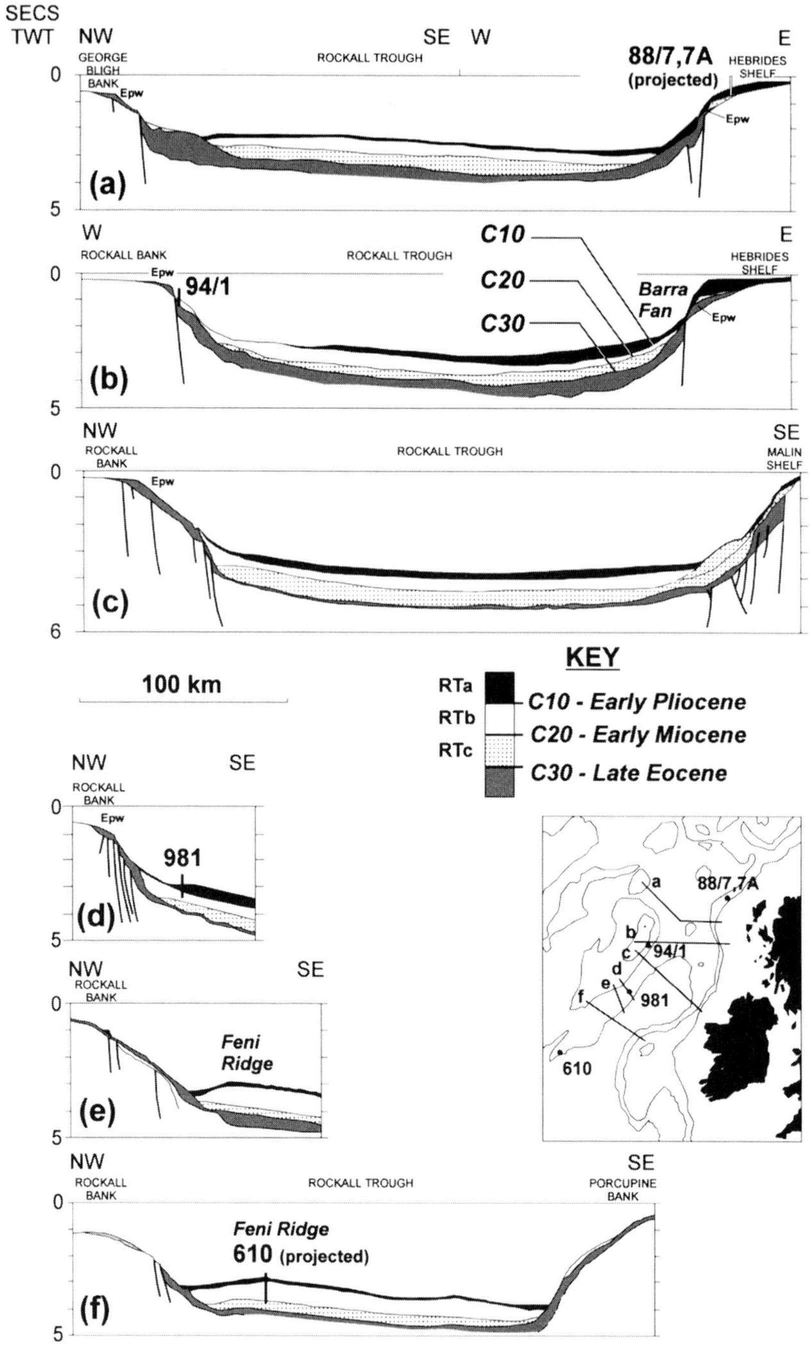

Fig. 3. Interpreted geoseismic sections across the Rockall Trough, focusing on the middle to upper Cenozoic stratigraphy and the relative setting of the key stratigraphic boreholes, BGS 88/7,7A and 94/1, DSDP site 610 and ODP site 981. Although 88/7,7A and 610 are projected onto sections (**a**) and (**f**), respectively, their calibration is constrained by regional seismic data. Sections (**a**) and (**b**) based on high-resolution and deep-seismic data acquired by the 'Rockall continental margin consortium' (see Acknowledgements); sections (**c**)–(**e**) based, respectively, on WRM profiles 96-103, 96-115 and 96-119; section (**f**) based on NIOZ profile 97-12. Sections located in the inset map. TWT, two-way time; Epw, well-preserved Eocene prograding wedge.

Table 1. *Correlation of regionally important mid- to late Cenozoic seismic reflectors*

Roberts (1975) & Roberts ***et al.*** (1981)	Miller & Tucholke (1983)	Masson & Kidd (1986)	Vanneste ***et al.*** (1995)	Bull & Masson (1996)	Jansen ***et al.*** (1996)	Stoker (1997)	Svaerdborg (1998)	This paper	Common age range
			BB5.0		R2		Yellow		Intra-Pliocene (?late)
			(BB5.0)		R3	***(A)***	Light green	C10	***Intra-Pliocene (early)***
	R1	Yellow		I					Latest late Miocene
			BB4.0				Dark green		Late mid-Miocene
		Purple							Mid-mid-Miocene
R4 (Rockall Trough)	R2	Green	BB3.0	II		A ***(B)***	***(Dark Green)***	C20	Late early Miocene
R4 (Rockall Plateau)	R4	Brown	R4	III		C	Blue	C30	Late Eocene

Bold represents more confident correlation between studies; bold italic and in parentheses represents revision to existing schemes (see text for details).

each megasequence enables us to directly assess the main controls on deep-water basin development west of Britain and Ireland during the mid- to late Cenozoic interval.

The study is based on the extensive geophysical and geological databases collected by the BGS and the Netherlands Institute for Sea Research (NIOZ), combined with commercial seismic data acquired by Fugro Geoteam (Fig. 2). The BGS data were acquired as part of its regional mapping programme of the UK continental margin, which is at present focused on the western frontier basins in the north Rockall region as part of a joint BGS–oil industry initiative (see Acknowledgements). The NIOZ data were mostly collected (1996–1998) as part of the European North Atlantic Margin (ENAM II) project, and provide extensive coverage of the central and southern Rockall Trough. Collectively, these datasets provide high-resolution (airgun, sparker and deep-tow boomer) and commercial deep-seismic reflection profiles. The widespread grid of seismic data has allowed, for the first time, the development of a unified seismic stratigraphy over most of the Rockall Trough. Geological calibration of the seismic stratigraphy was achieved with BGS boreholes and short rock cores, combined with other published material such as that of the Deep Sea Drilling Project (DSDP), the Ocean Drilling Program (ODP) and released commercial well data.

It should be noted that a new structural nomenclature system has recently been published for the Irish Atlantic margin (Naylor *et al.* 1999). Features such as the Rockall Bank, Rockall Trough, Hatton–Rockall Basin, Hatton Bank, Porcupine Bank and the Porcupine Seabight (Fig. 1), which are essentially bathymetric terms, have been renamed, respectively, as the Rockall High, Rockall Basin, Hatton Basin, Hatton High, Porcupine High and Porcupine Basin. However, these structural elements are more probably applicable to the syn- and early post-rift phases of margin development rather than the late post-rift expression of the margin, which is still being shaped today by bottom-current and sporadic mass-flow processes active since late Eocene time. For this reason, and the fact that the new nomenclature does not apply to UK waters, the traditional terminology is retained in the present study.

Seismic stratigraphy

The stratigraphic scheme proposed in this paper has developed through an iterative process of interpretation and reinterpretation ever since the

Table 2. *Correlation and revision of previous megasequence stratigraphies and key reflectors on basis of present study*

SOUTHERN ROCKALL TROUGH	NORTHERN ROCKALL TROUGH	ROCKALL TROUGH		
Shannon *et al.* (1993)	Stoker (1997)	This study		
		Reflector	*Megasequence*	*Age*
Sea Bed				
RT4	Middle Miocene to Holocene		RTa	
Yellow		C10	ANGULAR UNC.	Early Pliocene
RT3	A		RTb	
Green	Upper Eocene to Middle Miocene	C20	REFLECTIVE ZONE	Early Miocene
RT2	B		RTc	
Brown	C	C30	ANGULAR UNC.	Late Eocene
E O C E N E				

pioneering work by Roberts (1975) in the region. There have been a number of previous schemes that have considered different parts of the trough (Table 1). Although similarities between reflectors and acoustic facies of seismic units have been noted, regional correlation has largely remained ambiguous as a result of gaps in the seismic coverage, and the lack of cored geological information for the entire succession. Consequently, 'best guess' interpreted ages of reflectors based on long-range or regional correlation have infiltrated and stuck within the scientific literature. However, in the last 10 years or so the DSDP, BGS and ODP have collected long cores from the upper part of the infill within the Rockall Trough. These data have provided chronological control for several key reflectors that can now be established as regionally significant sequence boundaries. There remain, however, a number of reflectors (Table 1) for which regional correlation remains unattainable at present, partly through their restricted development and partly because of the quality and detailed resolution of the seismic data. For this reason, we have concentrated on the establishment of megasequences, which give options for future, more detailed, subdivision of the succession.

The seismic data have been interpreted according to the concepts, techniques and terminology of seismic-sequence stratigraphy and seismic facies analysis (e.g. Mitchum *et al.* 1977; Vail 1987; Emery & Myers 1996). Three key reflectors bounding three megasequences are identified throughout the study area; their characteristics are detailed below, and summarized in Table 2. The geometry and stratigraphical range of the middle to upper Cenozoic succession in the Rockall Trough is depicted in Figures 3 and 4. Stratigraphical information is based on a number of key sample sites, which are shown in Figures 2 and 3. For correlation purposes, the calcareous nannoplankton zonal scheme of Martini (1971) has been utilized: the prefix 'NP' applies to the Palaeogene zones, and 'NN' for the Neogene–Quaternary zones. The stratigraphic range of these biozones has been detailed by Harland *et al.* (1990).

Key reflectors

To simplify and draw together the existing plethora of reflector notation for the Rockall Trough, we propose a new, numbered, notation that allows for further subdivision if necessary (Table 1). The key Cenozoic reflectors described in this paper are numbered C10, C20 and C30. The numbering increases in descending stratigraphic order so that the same scheme may also be applied to the lower part of the Cenozoic succession if necessary.

In the Rockall Trough, the base of the middle to upper Cenozoic succession, reflector C30, is marked by a regional unconformity; three megasequences comprise the overlying strata bounded by reflectors C20 and C10. Each of these reflectors is described separately below, in ascending stratigraphic order.

Reflector C30. This boundary is expressed as a high-amplitude reflection at the margin of the basin, where it forms a prominent, angular unconformity (Fig. 3). A similar development of

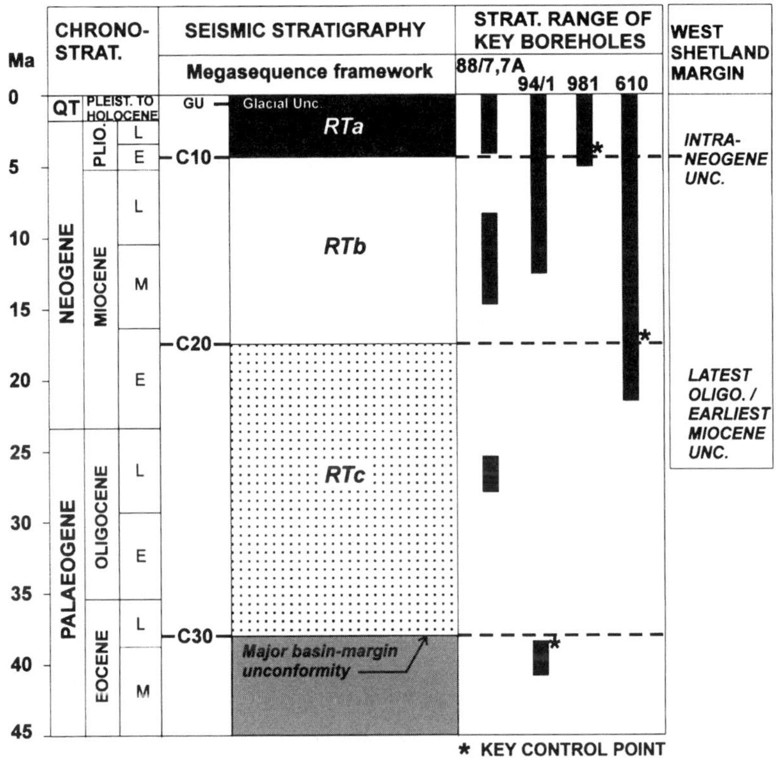

Fig. 4. Seismic stratigraphy and stratigraphical-range chart of key boreholes for the middle and upper Cenozoic succession in the Rockall Trough. Information from the West Shetland Margin is taken from Stoker (1999). The time scale is from Harland *et al.* (1990).

C30 is also observed adjacent to the seamounts of Rosemary Bank, Anton Dohrn and Hebrides Terrace (Stoker 1998). On both the western and eastern flanks of the Rockall Trough, C30 marks the top of a collapsed shelf-margin systems tract that has undergone much internal deformation, manifest as faulting, slumping and thickening (Figs 5–7). Consequently, the basin-margin expression of the reflector is commonly irregular, an attribute further enhanced through submarine erosion by deep-water bottom currents (Stoker 1997, 1998). On the Hebrides Slope, this surface is buried beneath a younger prograded shelf-margin succession. However, adjacent to the relatively starved slopes flanking the east Rockall Plateau and west Porcupine Bank the overlying slope apron is thin to absent, and C30 is locally exposed at the sea bed (Figs 3, 5–7 and 8a).

This reflector has also been imaged on seismic profiles from the NW part of the Rockall Trough, adjacent to Lousy, Bill Bailey and Faeroe banks (Vanneste *et al.* 1995; Boldreel *et al.* 1998). Traced basinwards, the reflector locally loses expression within the more parallel-bedded basin fill, although commonly it can be confidently traced across the basin. This boundary is broadly equivalent to reflector R4 as originally defined by Roberts (1975) from the Rockall Plateau. In the Rockall Trough, it correlates with the previously defined 'brown' reflector of Masson & Kidd (1986) in the southern Rockall Trough, reflector C of Stoker (1997) in the northern Rockall Trough, and the 'blue' reflector of Svaerdborg (1998) in the central Rockall Trough. It can also be correlated with reflector III of Bull & Masson (1996) from the southern margin of the Rockall Plateau.

On the western margin of the Rockall Trough, BGS borehole 94/1 (Fig. 5) penetrated below the C30 unconformity surface and recovered shallow-marine calcareous mudstones and sandstones of early late Eocene (NP16–18) age (Stoker 1997). Rocks of similar age (NP16–19) have also been proved by BGS boreholes 94/2 and 94/3 (Rockall Bank), and 94/7 (George Bligh Bank) where the unconformity surface occurs at or near the sea bed, and in well 164/25-2 on the NE flank of the basin

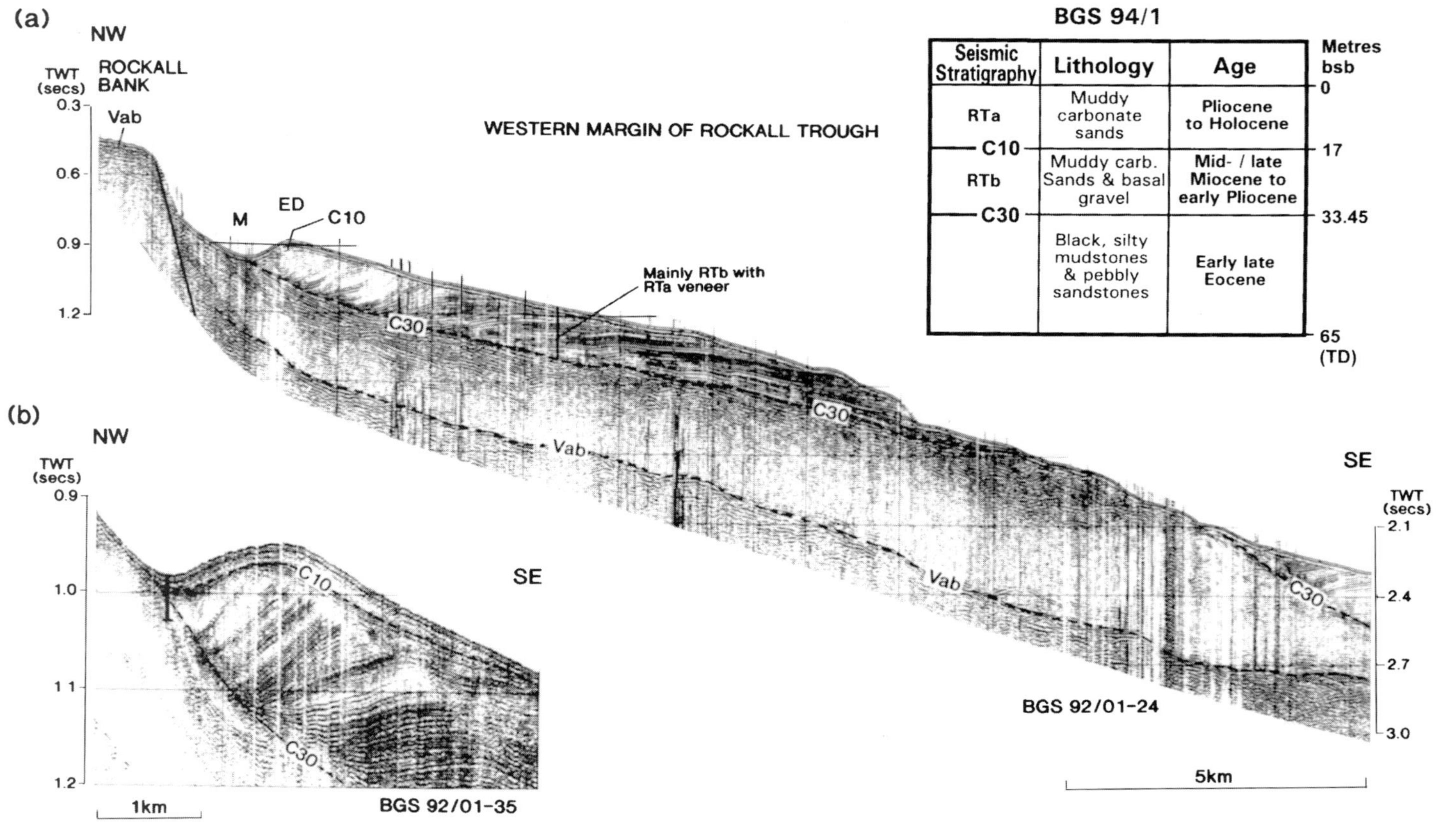

Fig. 5. (**a**) Airgun profile from the western margin of the Rockall Trough showing the seismic characteristics of late Cenozoic sediment-drift deposits (megasequences RTa and RTb), their modified geometry (by erosion), and progressive upslope migration across the late Eocene unconformity (reflector C30). (**b**) Sparker profile showing detail of lower, upslope-accreting unit (RTb), and upper, draped unit (RTa) where tested by BGS borehole 94/1. Inset shows details of borehole 94/1. Vab, volcanic acoustic basement (upper Paleocene–lower Eocene volcanic rocks); C30, late Eocene unconformity; C10, early Pliocene unconformity; ED, elongate mounded sediment drift; M, moat; bsb, below sea bed. Location of profiles is shown in Fig. 2. Estimates of scale on the seismic profiles can be calculated by assuming the velocity of sound in water is 1.45 km s^{-1}, and in sediments is from 1.55 to 1.8 km s^{-1}, based on increasing compaction with depth (Hamilton 1985). Modified after Stoker (1998).

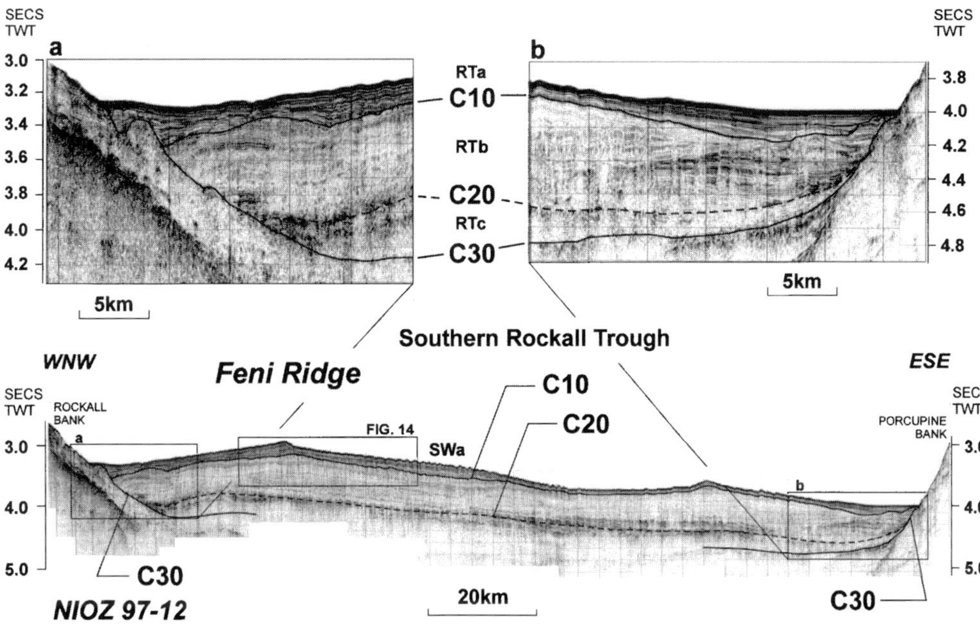

Fig. 6. Sleevegun profile across the southern Rockall Trough, between Rockall and Porcupine banks, showing the seismic characteristics of the middle and upper Cenozoic succession, including the Feni Ridge sediment drift, a smaller, subsidiary drift adjacent to Porcupine Bank, sediment waves at sea bed, and wide erosional moats adjacent to the banks. Insets (**a**) and (**b**) show details of the basin-margin and moated areas, particularly the erosional nature of C10, the onlapping character of C10 and C20, and the irregular (slumped) topography associated with C30. (**b**) also displays small-scale normal faults in RTb. Abbreviations as in Figure 5 except: SWA, sediment waves; C20, early Miocene reflector; RTa–RTc, middle–upper Cenozoic megasequences (see text for details). Profile is located in Figure 2. Location of Figure 14 is also indicated.

(Egerton 1998) (Figs 1 and 2). Regional stratigraphic evidence suggests that the oldest rocks overlying reflector C30 are of late Eocene–Oligocene age (Stoker 1997). BGS borehole 94/4 recovered upper Eocene–lower Oligocene mass-flow deposits from a lowstand fan on the NE slope of Rockall Bank (Fig. 8b). Similar deposits of equivalent age have been locally mapped adjacent to the Wyville-Thomson Ridge (Egerton 1998), and may also be present on the Irish Slope (Fig. 3c). On the upper Hebrides Slope and Shelf, the unconformity is overlain by carbonates (Jones *et al.* 1986) associated with the build-up of a large early Oligocene reef.

A late Eocene age for the development of C30 is consistent with the Hatton–Rockall Basin, where DSDP site 116 (Fig. 2) terminated in upper Eocene (NP19) limestones at about the level of reflector R4 (Laughton *et al.* 1972; Berggren & Schnitker 1983). It also supports the late Eocene age for the equivalent reflector III south of Rockall Plateau (Bull & Masson 1996). Characteristically, C30 is onlapped by the middle to upper Cenozoic section (Figs 5–9). This reflects the abrupt change in the style of deep-water sedimentation in late Eocene time noted above, and marks the onset of bottom-current activity in the Rockall Trough (Stoker 1998).

Reflector C20. In the Rockall Trough, this seismic horizon is characterized by a sub-horizontal, high-amplitude acoustic signature that varies from a sharp, single reflection to a reflective zone several tens of milliseconds thick (Figs 6–10). Around the margin of the Rockall Trough, reflector C20 generally onlaps reflector C30. However, in areas where downslope processes have contributed to the local development of a slope apron, this megasequence boundary is traced upslope along the top of the associated mass-flow deposits (Figs 3c and 8b). On the upper Hebrides Slope, the boundary is locally marked by an erosional hiatus (Stoker *et al.* 1994) (Fig. 11).

Reflector C20 correlates with the 'green' reflector of Masson & Kidd (1986) from the southern Rockall Trough, where DSDP site 610 proved a late early Miocene (NN4/5) age. Although C20 was not clearly imaged on the

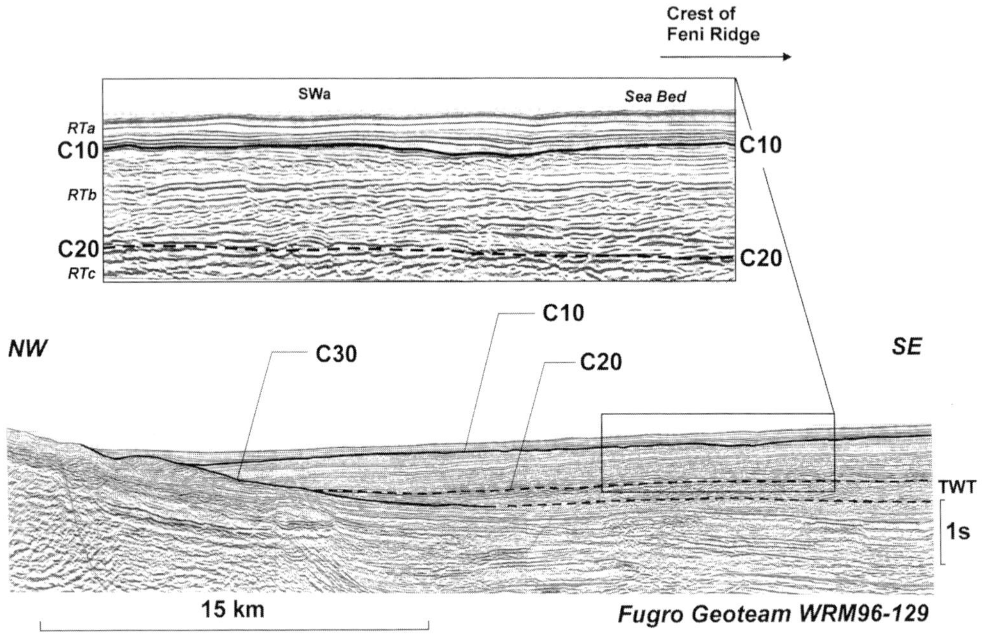

Fig. 7. Commercial seismic profile on western flank of Rockall Trough showing seismic characteristics of the middle and upper Cenozoic succession, particularly the onlapping configuration of the basinal strata, and the slumped nature of the Eocene basin-margin sediments (below C30). Inset shows erosional nature of C10, sediment waves in RTa, and occurrence of small-scale faults particularly in RTb. Abbreviations as in Figures 5 and 6. Profile is located in Figure 2.

high-resolution watergun record at the site 610, and may even be obscured by a sea-bed multiple, previously defined as 'acoustic unit c' by Ruddiman *et al.* (1986) (Fig. 10b), it does correlate with a highly reflective zone imaged on a nearby, lower-resolution, airgun profile (Fig. 10a). In the central Rockall Trough, the top of the reflective zone correlates with the 'dark green' reflector of Svaerdborg (1998). This reflector was originally assigned a late mid-Miocene age by Svaerdborg (1998) on the basis of correlation with the top of 'acoustic unit c' of Ruddiman *et al.* (1986). However, we have reinterpreted their unit to be an artefact, probably the sea-bed multiple, and propose a revised late early Miocene age for the dark green reflector consistent with all of the regional evidence presented herein. In the northern Rockall Trough, reflector C20 equates with reflector B of Stoker (1997). Although a latest Eocene age was previously inferred for reflector B, the revised interpretation is now confidently applied because of the extensive seismic coverage linking the two areas of the trough. Reflector A of Stoker (1997), originally correlated with the 'green' reflector (Masson & Kidd 1986), is now proved to be younger (see below). In the NW Rockall Trough, reflector C20 may correlate with reflector BB3.0 of Vanneste *et al.* (1995). In a regional context, reflector C20 is probably equivalent to NE Atlantic reflector R2 of Miller & Tucholke (1983), and reflector II of Bull & Masson (1996) from the southern margin of the Rockall Plateau.

According to Masson & Kidd (1986), the expression of reflector C20 results from a rapid increase in sonic velocity which, at DSDP site 610, occurs over a 50 m interval between 625 and 675 m sub-bottom (Fig. 10). Dolan (1986) attributed this increase in velocity to clay mineral cementation resulting in a zone of well-developed massive smectite (as distinct from the common crystalline detrital smectite that occurs throughout the sediment). The composition of the smectite, and the possible presence of minor zeolites, implies a derivation from the *in situ* alteration of volcanic glass. According to Dolan (1986), the parental volcanic glass may have been derived from a primary ash fall deposited in the Norwegian Basin. Its subsequent reworking and redeposition in the Rockall Trough may be related to an oceanographic event (Baldauf 1986). Miller & Tucholke (1983) reported a general intensification of bottom-current circulation in the NE Atlantic

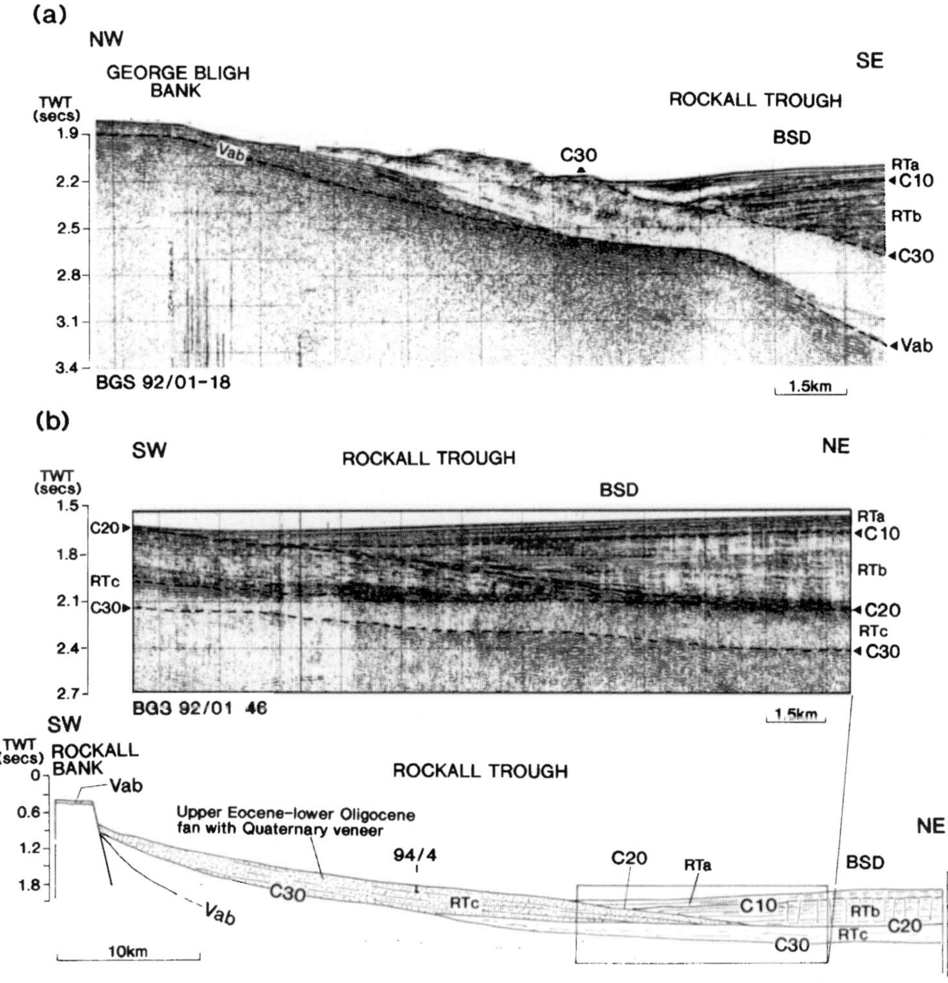

Fig. 8. Airgun profiles showing the seismic characteristics and morphology of the middle to upper Cenozoic sediment drift deposits in the NW Rockall Trough adjacent to (**a**) George Bligh Bank, and (**b**) NE Rockall Bank, and their progressive westward onlap and pinch-out onto the basin margin. In (**a**), the pre-sediment-drift strata exposed at the sea floor have been eroded and sculpted by bottom currents into an irregular sea-bed topography. In (**b**), a lowstand fan (stippled), calibrated by BGS borehole 94/4, is interbedded with the drift deposits; the stratigraphic relationships are clarified in the associated line drawing. Small-scale faults occur in the basinal strata in (**b**). Abbreviations as in Figures 5 and 6 except: BSD, broad-sheeted drift. Profiles are located in Figure 2. Modified after Stoker (1998).

region at about the early–mid-Miocene boundary, which created an unconformity associated with their R2 reflector. This is consistent with evidence for bottom-current erosion in the latest Oligocene–early Miocene interval in the Faeroe–Shetland Channel (Stoker 1999) and on the Hebrides Slope (Stoker *et al.* 1994). This vigorous pulse of deep-water circulation may have been responsible not only for erosion, but also for the transportation of the volcanic material southwards into the Rockall Trough. Although no unconformity is reported at site 610, there is a decrease in sediment accumulation rate at the level of the 'green' reflector (Masson & Kidd 1986). This probably reflects alternating periods of scouring and sporadic deposition that may be biostratigraphically unresolvable within the core at site 610 (Dolan 1986). The poor preservation of diatoms across this interval supports the likelihood of strong bottom currents and increased dissolution at this time (Baldauf 1986). Thus, the deep-water origin of reflector

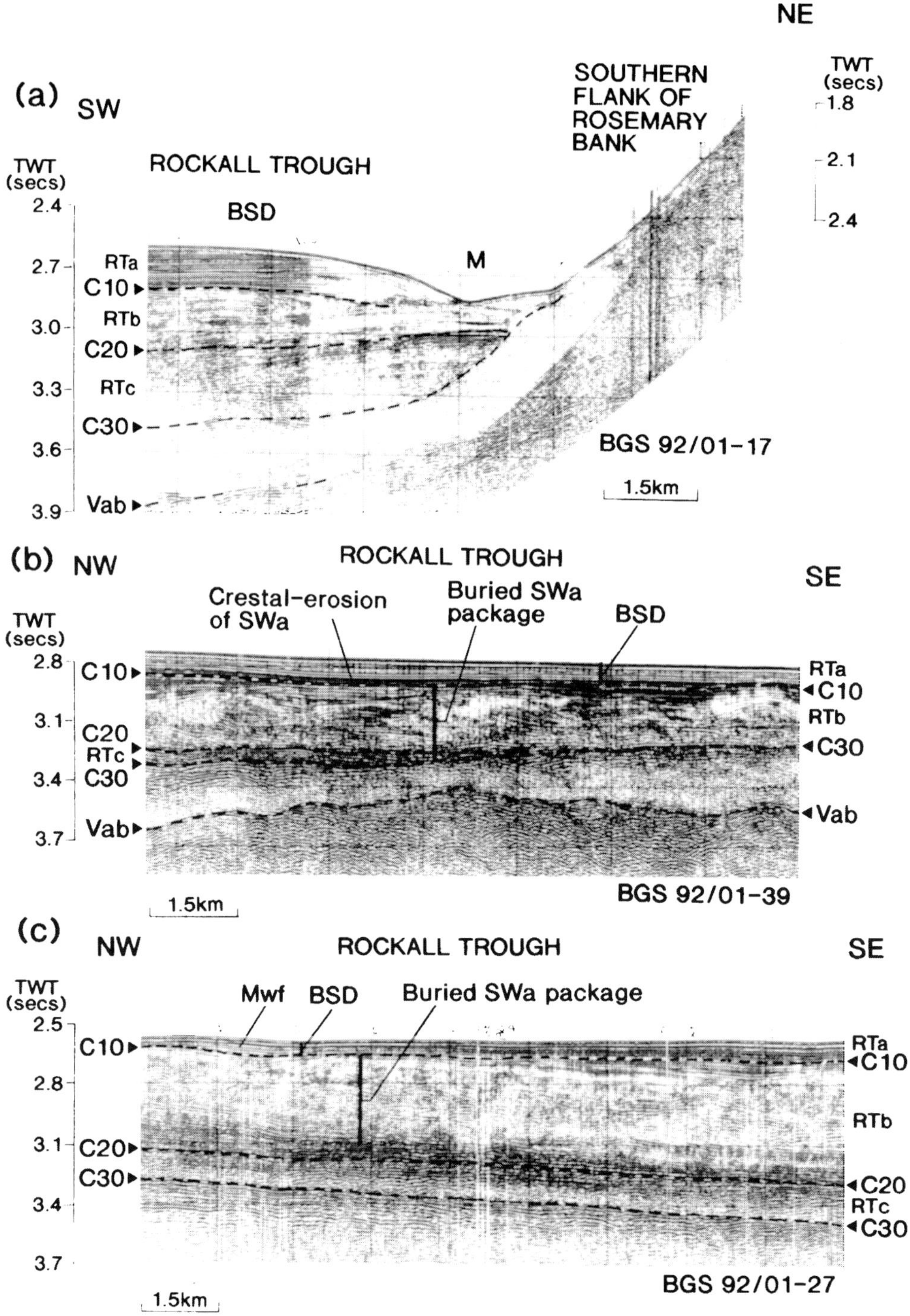

Fig. 9. Seismic profiles highlighting the variable seismic characteristics and morphology of the middle to upper Cenozoic sediment drift deposits in the north–central Rockall Trough. (**a**) Airgun profile showing broad sheeted drifts and a wide, deep, moat developed adjacent to Rosemary Bank Seamount. (**b**) and (**c**) Airgun profiles north and SW, respectively, of Anton Dohrn Seamount showing large-scale, buried, sediment waves preserved in megasequence RTb, buried beneath the predominantly flat-lying deposits of RTa. The crestal region of the buried waves has been locally eroded by erosion associated with the development of C10. Abbreviations as in Figures 5, 6 and 8 except: Mwf, migrating waveform. Profiles are located in Figure 2. Modified after Stoker (1998).

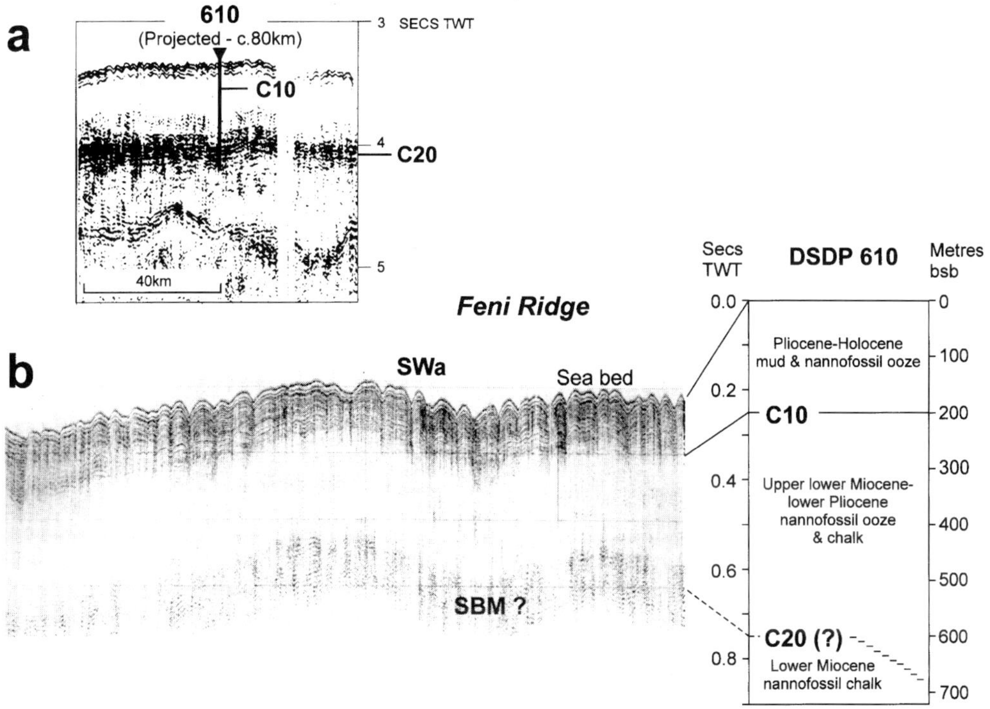

Fig. 10. (**a**) Airgun profile (Charcot 04/1969) from SW Rockall Trough in area of Feni Ridge showing projected location of DSDP site 610. (**b**) High-resolution watergun profile, collected by *Glomar Challenger*, from DSDP site 610 showing correlation with borehole, and abundant sediment sea-bed sediment waves. Profiles and log modified after Ruddiman *et al.* (1986). Abbreviations as in Figures 5 and 6 except: SBM, sea-bed multiple. Profiles located in Figure 2.

C20 probably reflects the integrated effects of bottom currents, resedimentation and diagenesis.

Reflector C10. This boundary is characteristically a high-amplitude, continuous reflection, locally offset by small-scale faults. It typically forms an erosional, angular unconformity truncating strata both on the flanks of the trough and in the basin (Figs 5–9, 11–13). On the basin margin, reflector C10 onlaps reflector C30, although locally it is eroded and truncated by the present-day sea-bed surface (Fig. 3). Adjacent to the Hebrides and NW Irish slopes, reflector C10 forms the base of the Upper Cenozoic prograding shelf-margin succession that includes the Barra–Donegal fan system (Figs 11 and 12). On most profiles, there is a change in acoustic facies across reflector C10 (described below), and there is localized onlap of the overlying strata, most notably at the location of ODP site 981 (Fig. 13).

In the central Rockall Trough, reflector C10 correlates with reflector R3 of Jansen *et al.* (1996) and the 'light green' reflector of Svaerdborg (1998). It can now also be confidently correlated with reflector A of Stoker (1997) in the northern Rockall Trough. On the basis of ODP site 981 (Fig. 13) this reflector is assigned an intra-early Pliocene age (Jansen *et al.* 1996), which revises the mid-Miocene age previously assigned by Stoker (1997) to the equivalent reflector to the north. This new dating is consistent with biostratigraphic information from BGS borehole 94/1, which penetrated reflector C10 on the western margin of the Rockall Trough (Fig. 5), and proved a mid-Miocene to early Pliocene (NN10/12–15) age range for sediments immediately underlying the unconformity. Supporting evidence for a major intra-Neogene regional unconformity is provided by BGS borehole 88/7,7A from the upper Hebrides Slope, which proved a phase of late Miocene–early Pliocene (NN10–12/13) erosion immediately before the development of the prograding shelf-margin succession (Fig. 11) (Stoker *et al.* 1994). This event has also been recognized on the West Shetland Margin, including the Faroe–Shetland Channel, where it has been informally termed the 'intra-Neogene unconformity' (Stoker 1999) (Fig. 4).

In the NW Rockall Trough, a similarly widespread, erosional unconformity was originally described by Smythe (1989) who inferred a late Miocene age for this boundary. This dating was subsequently adopted in the schemes of Boldreel & Andersen (1993); Kuijpers *et al.* (1998). In contrast, Vanneste *et al.* (1995) suggested an intra-late Pliocene age for this boundary (their reflector BB5.0). The new data suggest that some reinterpretation of these schemes may be necessary.

It is interesting to note that this unconformity was not reported at DSDP site 610, either in the log or on the seismic profile. Unfortunately, the recovery in the lower Pliocene section of the core is poor, and thus this event may have been missed in the sample gap. Inspection of the high-resolution seismic data (Fig. 10; see Masson & Kidd 1986, fig. 6), however, reveals a relatively sharp change in seismic character at about 0.25 s two-way time (TWT), sub-bottom, from an upper stratified section to a lower, reflection-free signature. By comparison with our data, we suggest that this change in seismic character coincides with reflector C10. Calibration of the core log with the seismic record indicates that this change occurred between earliest early Pliocene and latest early Pliocene time (i.e. between cores 610-9 and 610-10) and is thus an intra-early Pliocene event (Svaerdborg 1998).

The predominantly erosional character of reflector C10, on the slopes and basin floor, suggests that this boundary formed, at least in part, in response to an overall increase in the velocity and erosional power of bottom currents in the Rockall Trough.

Megasequences

Three megasequences (RTa, RTb and RTc) have been defined, comprising the middle to upper Cenozoic succession in the Rockall Trough

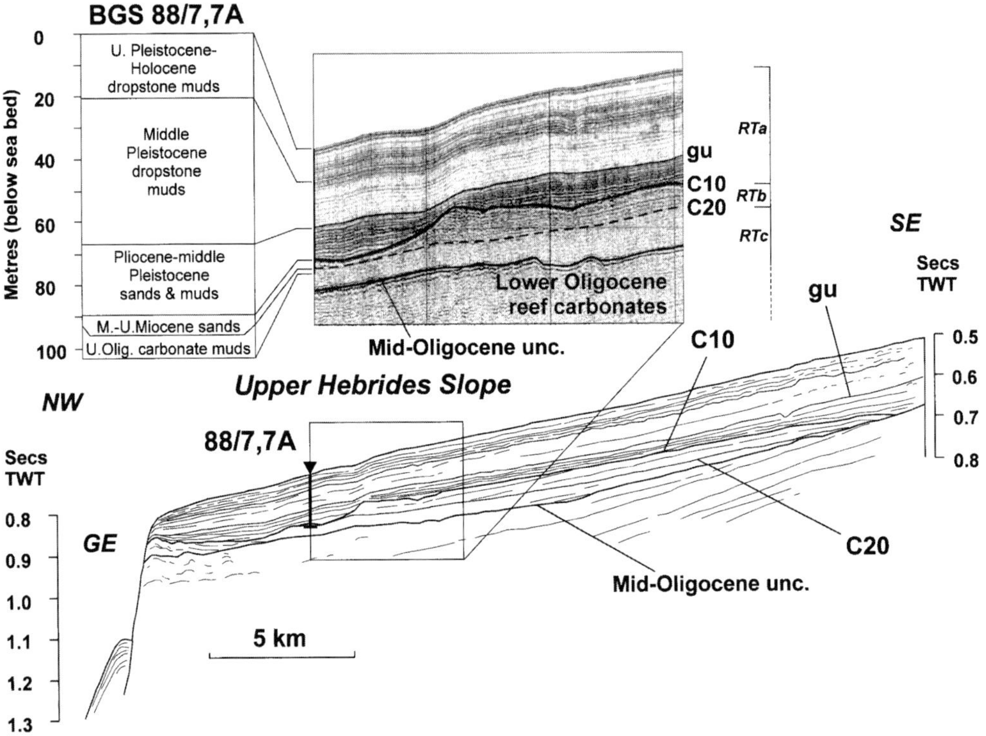

Fig. 11. BGS sparker profile 84/06-17 and interpreted line drawing across the Hebrides Slope showing the seismic characteristics of the middle to upper Cenozoic succession. The interpretation of mid-Oligocene and mid-Pleistocene (glacial) unconformities is based on the calibration with BGS borehole 88/7,7A. Abbreviations as in Figures 5 and 6 except: gu, glacial unconformity; GE, Geikie Escarpment. Profile is located in Figure 2. Modified after Stoker *et al.* (1994); Stoker (1995).

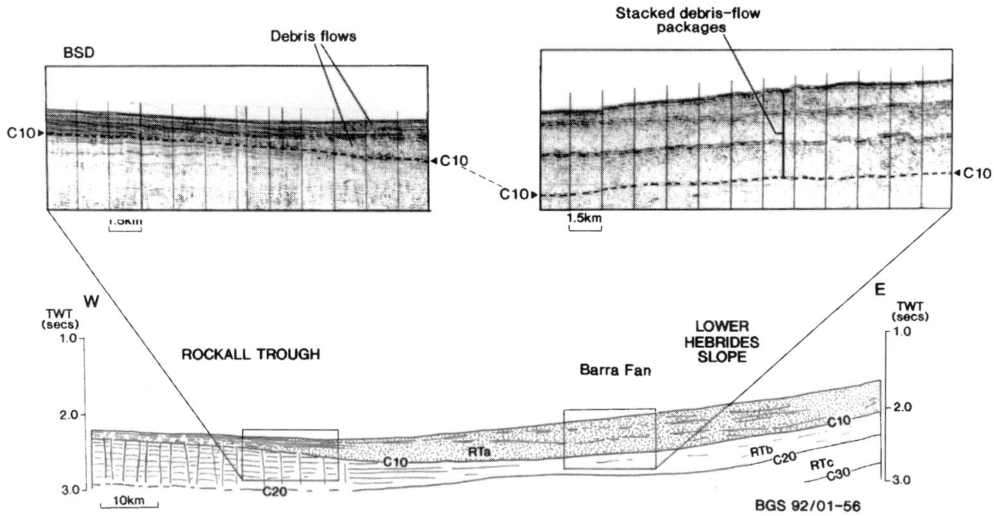

Fig. 12. Airgun profiles and interpreted line drawing across the lower part of the Barra Fan showing the contrasting seismic characteristics of, and the stratigraphical relationships between, the debris flows of the fan (stippled) and the basinal drift deposits, and the angular nature of the early Pliocene unconformity (C10). The basinal deposits in RTb are disrupted by numerous small-scale faults. Abbreviations as in Figures 5, 6 and 8. Profile is located in Figure 2. Modified after Stoker (1998).

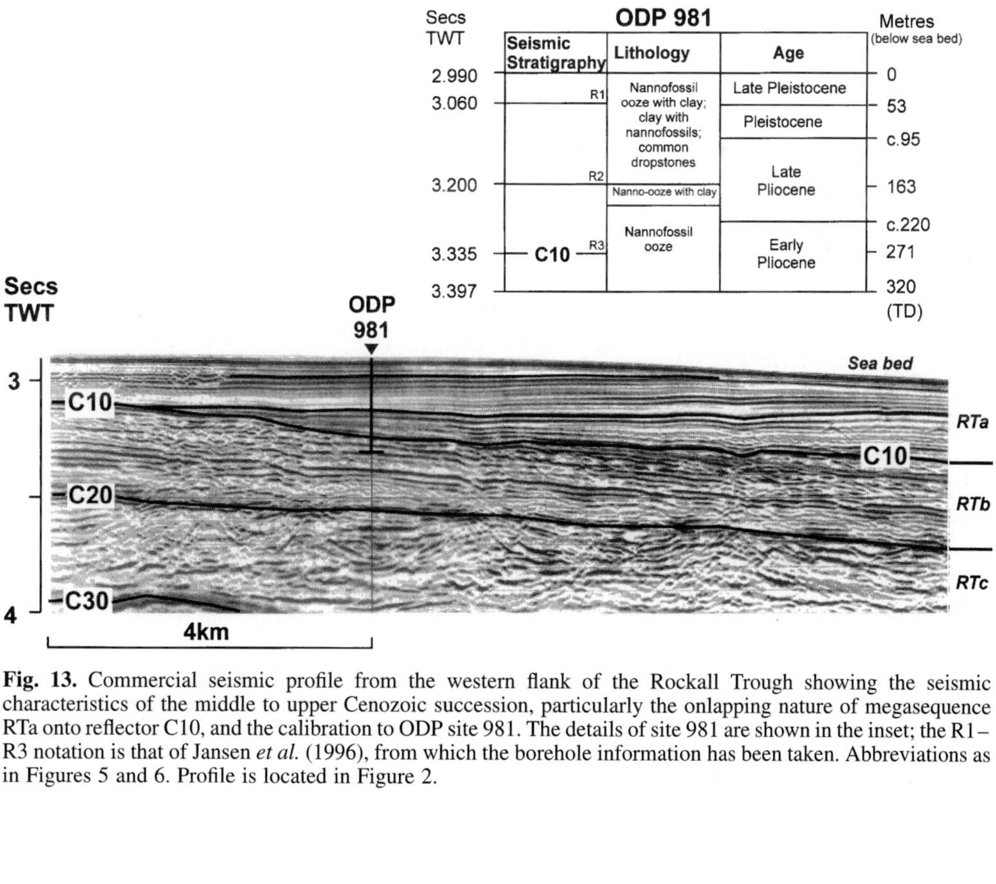

Fig. 13. Commercial seismic profile from the western flank of the Rockall Trough showing the seismic characteristics of the middle to upper Cenozoic succession, particularly the onlapping nature of megasequence RTa onto reflector C10, and the calibration to ODP site 981. The details of site 981 are shown in the inset; the R1–R3 notation is that of Jansen *et al.* (1996), from which the borehole information has been taken. Abbreviations as in Figures 5 and 6. Profile is located in Figure 2.

(Fig. 4; Table 2). The upper-case letters (RT) refer to the basin (Rockall Trough), and the lower-case letters (a–c) refer to the specific megasequence. On the evidence of the key reflectors, the stratigraphic range of the megasequences comprises upper Eocene to lower Miocene units (RTc), lower Miocene to lower Pliocene units (RTb), and lower Pliocene to Holocene units (RTa). This new scheme revises and replaces previous frameworks established by Shannon *et al.* (1993) and Stoker (1997) in, respectively, the southern and northern Rockall Trough (Table 2). On the Hebridean margin, the current subdivision of the Plio-Pleistocene succession has been described elsewhere (Stoker *et al.* 1993) and is beyond the scope of this paper. Each of the megasequences is summarized below, in ascending stratigraphic order.

Megasequence RTc (upper Eocene to lower Miocene units). Megasequence RTc is bounded by reflectors C30 and C20 and corresponds to the lower part of the previously designated 'Upper Eocene to Middle Miocene megasequence' (reflector C–B interval) of Stoker (1997) in the northern Rockall Trough (Table 2). This is a significant revision, as this seismic interval was previously assigned a late Eocene age. In the southern Rockall Trough, megasequence RTc is broadly equivalent to seismic sequence RT2 of Shannon *et al.* (1993).

This megasequence is largely restricted to the Rockall Trough, south of 59°N, but may locally extend onto the basin margin. It displays a predominant onlap fill that locally exceeds 0.4 s TWT in thickness (Fig. 3). On seismic profiles, the reflection configuration varies from reflection free to stratified (Figs 6–9). An onlapping-mounded form is locally preserved on the western margin of the basin and adjacent to the seamounts, e.g. Rosemary Bank (Fig. 9a), although a sheeted geometry is most common (Fig. 7). This depositional geometry has previously been attributed, at least in part, to bottom-current-influenced, deep-marine sedimentation resulting in the formation of sheeted contourites (Stoker 1997, 1998). The Feni Ridge (see below) may have been initiated during this interval (Masson & Kidd 1986), although its classic mounded morphology is predominantly the result of later Neogene deposition (Fig. 6).

Although the gross depositional environment of the Rockall Trough was dominated by deep-water sedimentation (see below), a number of stratigraphically equivalent deposits accumulated around the margin of the basin. Lowstand fans of late Eocene–Oligocene age occur locally on the NE slope of Rockall Bank (Fig. 8b), and on the southern flank of the Wyville-Thomson Ridge (Stoker 1997, 1998; Egerton 1998). Shallow-water carbonate-reef deposits of early Oligocene age, unconformably overlain by late Oligocene carbonate muds, are preserved at the Geikie Escarpment, on the edge of the Hebrides Shelf (Fig. 11) (Jones *et al.* 1986; Stoker *et al.* 1994; Stoker 1997). On the NW Irish margin, substantial slide deposits are preserved within this megasequence (Fig. 3c). All of these deposits reflect more variable, localized, and commonly more punctuated component sequences flanking the main basinal depocentre.

Megasequence RTb (lower Miocene to lower Pliocene units). Megasequence RTb is bounded by reflectors C20 and C10. It corresponds to the upper part of the previously defined 'Upper Eocene to Middle Miocene megasequence' (reflector B–A interval) of Stoker (1997) in the northern Rockall Trough (Table 2). The upper bounding reflector is as originally defined by Stoker, but its revised age necessitates the chronostratigraphic revision of the previous megasequence framework. In the southern Rockall Trough, megasequence RTb incorporates RT3 and the lower part of RT4 of Shannon *et al.* (1993). It should be noted that the purple and yellow reflectors of Masson & Kidd (1986), which occur within this megasequence, have not been recognized on a basinwide scale. Although with further detailed study such reflectors may form the basis for future subdivision of this unit, this is beyond the scope of the present work.

This megasequence is present throughout the Rockall Trough, extending onto the adjacent slopes, where it is mostly terminated by erosional pinchout. Outliers are locally preserved on the Hebrides Shelf. In the Rockall Trough, the basinal sediments commonly exceed 0.4 s TWT in thickness, with the thickest accumulation occurring in the western half of the basin. In general, these sediments display a ponded basin-floor fill external geometry but with locally significant upslope accretion by onlap onto the flanks of the basin. This sediment-body geometry reflects three main styles of sediment-drift accumulation: (1) broad, flat-lying to gently domed, sheeted drifts, which occupy a large part of the axial region of the basin floor (Figs 3a–c, 7–9); (2) elongate mounded drifts onlapping the margins of the basin (Fig. 5); (3) the giant, elongate, Feni Drift in the southern Rockall Trough (Fig. 6) (Stoker 1997, 1998). Adjacent to the Hebridean margin, the drift deposits are mostly buried beneath younger sediments

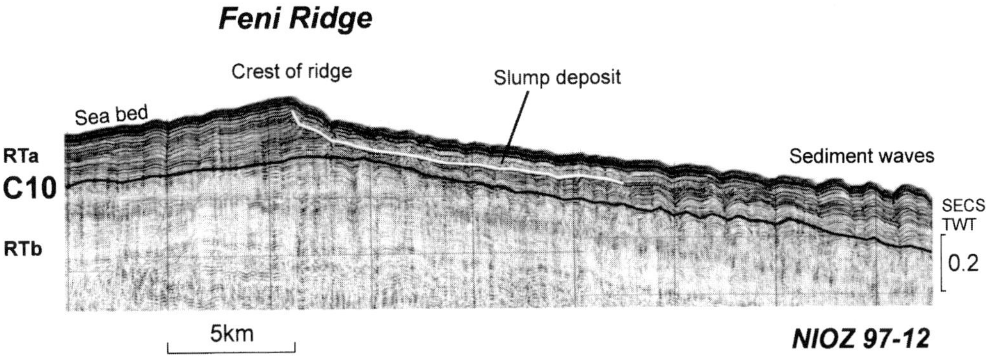

Fig. 14. Sleevegun profile across crest of Feni Ridge sediment drift showing localized failure of wavy, parallel-bedded drift-axis deposits and their redeposition as chaotic slump deposits which pinch-out downslope. Sediment waves have been destroyed in the area of the slump, but continued to develop farther downslope. The slump deposit is at present buried beneath a veneer of stratified sediments that display sediment-wave development. Abbreviations as in Figures 5 and 6. Profile is located in Figure 6.

(megasequence RTa), but along the western flank of the trough they are commonly exposed at the sea bed and have suffered extensive erosion since early Pliocene time (Fig. 3a–c). The basinal sediments also underwent extensive erosion during the formation of the early Pliocene upper bounding surface, reflector C10, which is commonly an erosional, angular unconformity (Figs 5–9, 12 and 13).

Seismic profiles have revealed a predominantly parallel-bedded internal reflection configuration over most of the area. The reflections are mostly continuous and vary in expression from flat-lying or gently domed to undulatory and waveform (Figs 7–9). The continuity of individual reflections is locally offset by small-scale normal faults (Figs 7, 8b and 12); however, some of the undulations represent large-scale sediment waves, up to 40 m high with a wavelength of 2–3 km (Fig. 9b and c) (Stoker 1998). The crest of these sediment waves has been locally eroded during the development of the upper bounding surface, reflector C10. The sheeted drifts onlap the margin of the basin, locally burying the older lowstand fans (Fig. 8b). In areas of upslope accretion, sediment-drift deposits have locally migrated from the basin floor onto the slope. These elongate, mounded drifts are characteristically separated from the adjacent slope by an erosional moat (Fig. 5). Internally, these drifts display both onlapping and downlapping reflector terminations onto the underlying strata. The relict nature of these sediments is evident from their erosional, irregular, upper surface with internal reflectors clearly truncated at, or near to, sea bed and mantled only by a veneer of younger deposits. In some areas, erosion has been so intense that parts of the section have been removed, exposing older strata (Fig. 5). This has resulted in parts of the upslope-accreting deposits now being detached from their basinal counterparts.

The Feni Ridge (Fig. 6) is a giant elongate drift, up to 80 km wide, ranging from 0.3 to 0.44 km in relief, and separated from Rockall Bank by a wide erosional moat. It can be traced for about 400 km parallel to the western margin of the southern Rockall Trough, but loses its classic morphological expression north of 55°N (Fig. 2). This may, in part, be an effect of erosion associated with the C10 reflector. A smaller drift occurs on the eastern margin of the trough, adjacent to the Porcupine Bank (Fig. 6), although the extent of this drift remains uncertain. These features have also been described by McDonnell & Shannon (2001).

Megasequence RTa (lower Pliocene to Holocene units). Megasequence RTa is bounded by reflector C10 and the present-day sea bed. It corresponds to the previously defined 'Middle Miocene to Holocene' megasequence of Stoker (1997) in the northern Rockall Trough (Table 2); the revised age of the lower bounding surface necessitates the revised chronostratigraphy. In the southern Rockall Trough, it corresponds to the upper part of RT4 of Shannon *et al.* (1993).

This megasequence is present across the study area, but displays a marked east–west asymmetry, being thickest along the Hebridean and NW Irish margins whereas it locally is below seismic resolution along the western flank of the Rockall Trough (Fig. 3). Nevertheless, short cores and

a

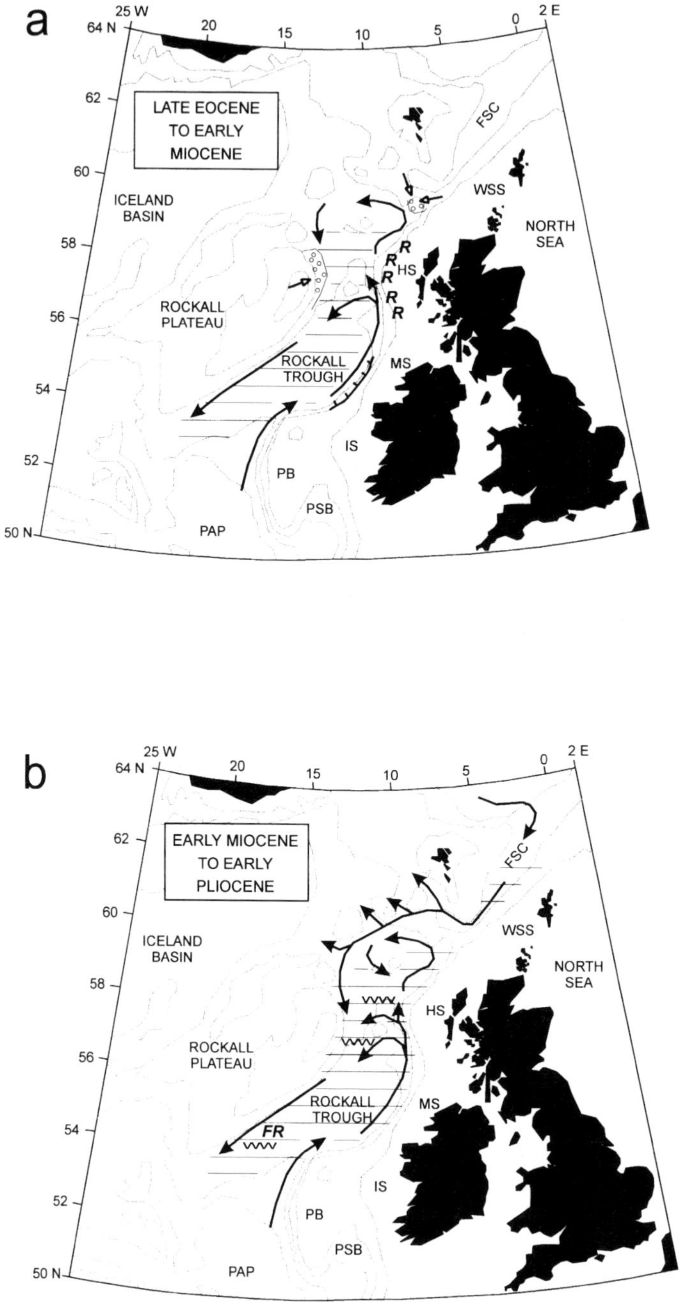

Fig. 15. Gross depositional environment maps summarizing the main depositional systems existing during the development of the mid- to late Cenozoic megasequences in the Rockall Trough. Abbreviations as in Figure 1. Present-day contours (as in Fig. 1) superimposed on maps to act as an approximate guide to the palaeo-morphology of the continental margin. Bottom-current circulation patterns inferred in (**a**) and modern in (**b**) and (**c**). (See text for details.)

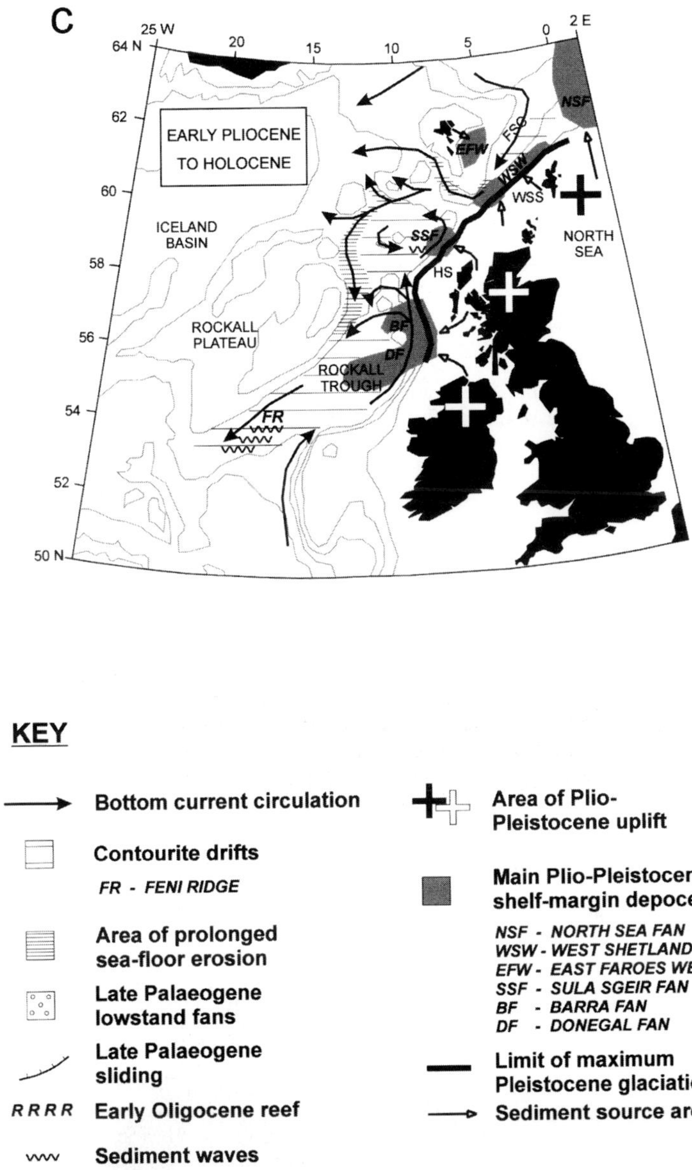

Fig. 15. Continued.

boreholes indicate that it is present on the top and slope of Rockall Bank, as well as the top of the axial seamounts, albeit as a veneer mostly less than 20 m thick. The thickest accumulations are associated with the Sula Sgeir and Barra–Donegal fans (Fig. 1), where the sediments locally exceed 0.8 s TWT in thickness. In the Rockall Trough, megasequence RTa is generally less than 0.2 s TWT in thickness north of Anton Dohrn Seamount, but commonly exceeds 0.3 s TWT in thickness to the south, largely as a result of the intercalation of debris-flow packages derived from the Barra–Donegal Fan and NW Irish margin. The long run-out distance of the debris flows has resulted in their being an integral part of the basinal section in the central Rockall Trough (Fig. 12). In the southern Rockall Trough, thicknesses are generally 0.2 s TWT or less (Figs 6 and 14). Along the western flank of the trough, north of 56°N, the seismically resolvable limit of the sediments is well defined on the middle to lower slope of Rockall Bank which, for the most part, was a zone of erosion throughout this interval (Fig. 3).

On seismic profiles, the Sula Sgeir and Barra–Donegal fans are dominated by packages characterized by a chaotic reflection configuration representing stacked debris flows. At the distal ends of the fans, the debris flows interdigitate with parallel-bedded sediment-drift deposits (Fig. 12). Major slide events identified within these fan systems (Baltzer *et al.* 1998; Holmes *et al.* 1998) may, in part, be responsible for debris flows extending well into the basin.

On the slope between the Sula Sgeir and Barra–Donegal fans (Fig. 1), the sediment-drift deposits display upslope accretion and elongate mounded drifts are locally developed on the Hebrides Slope. Farther north, Howe *et al.* (1994) and Stoker *et al.* (1998) have described single- and multi-crested elongate mounded drifts, broad-domed sheeted drifts and sediment waves forming a sediment-drift complex adjacent to the Wyville-Thomson Ridge. South of the Donegal Fan, the Irish margin is heavily canyoned and channelized, and there are fewer indications of shelf-margin progradation. Instead, the Irish and Porcupine margins are more probably zones of sediment starvation and bypass, with the canyons providing the main routes for the transfer of material from the shelf to the Rockall Trough.

North of Anton Dohrn Seamount, the basinal sediments predominantly consist of stratified, mostly parallel-bedded, broad-domed, sheeted drifts, which onlap the western flank of the Rockall Trough (Figs 3a and 8). Erosional moats are locally developed adjacent to the Rosemary Bank (Fig. 9a) and Anton Dohrn seamounts, with a relief in excess of 200 m below the general level of the sea bed (Stoker 1998). South of Anton Dohrn Seamount, drift deposits and reflection-free mass-flow sediments are commonly interbedded. Farther south, in the area of the Feni Ridge, parallel- to wavy-bedded drift deposits make up the bulk of the section, reflecting continued aggradation of the giant elongate sediment drift (Figs 6, 7 and 14). Migrating sediment waves are common on the flanks of this drift, and on the smaller drift adjacent to Porcupine Bank. Localized disruption of the drift deposits by mass-flow processes is common on the slopes of the Rockall Trough (e.g. Flood *et al.* 1979) and, more rarely, on the floor of the basin. An example of the latter is the failure of the crest of the Feni Ridge (Fig. 14).

Sedimentation and palaeogeography

Gross environment maps for the late Eocene to early Miocene, early Miocene to early Pliocene and early Pliocene to Holocene intervals, which correspond, respectively, to megasequences RTc, RTb and RTa, are presented in Figure 15, and the main points are summarized below. These maps supercede those presented by Stoker (1997).

Late Eocene–early Miocene sedimentation (Fig. 15a)

The generation of the early late Eocene unconformity, reflector C30, is interpreted to represent the onset of bottom-current activity in the Rockall Trough (Stoker 1997, 1998). The irregular nature of the unconformity reflects, at least in part, erosion of the sea bed by vigorous bottom currents concentrated along the margins of the basin. The predominantly sheeted character of the upper Eocene to lower Miocene sediments may imply a relatively high-energy, deep-marine environment (Stoker *et al.* 1998), in common with that proposed for the wider NE Atlantic for this time interval (Stow & Holbrook 1984). According to Stoker (1997, 1998, 1999), the bottom-current regime was southerly derived as there is no evidence for significant bottom-current activity in late Paleogene time in the Faroe–Shetland region. This implies that during this interval, the Wyville-Thomson Ridge was a barrier to deep-water flow.

In the basinal strata, DSDP site 610 proved lower Miocene, pelagic biogenic sediments (cyclic nannofossil chalk; Hill 1986; Stow *et al.* 1998) below reflector C20 (Fig. 10). This suggests that pelagic rain-out of biogenic material was an important source of deep-water sediment. The lowstand fans provided additional point sources of sediment input into the Rockall Trough. BGS borehole 94/4 recovered upper Eocene–lower Oligocene mass-flow calcarenites adjacent to Rockall Bank (Stoker 1997) (Fig. 8b); these are coarse-grained, porous, and massive to crudely bedded, with bioclastic material making up about 98% of the sediment. The nature of the reworked material implies a high-energy, shallow-marine, carbonate-dominated environment on Rockall Bank at this time. This is consistent with the early Oligocene reefal development on the Hebrides Shelf, which fringed the eastern margin of the Rockall Trough between about 57°N and 58°40′N (Jones *et al.* 1986; Stoker 1997, 1998). Although the reef was largely detached from the deep-water basin, there are indications of reworking of the seaward margin of the reef with mass-flow deposits interbedded with the basinal strata. Localized failure on the NW Irish margin (Fig. 3c) further provided a transfer mechanism for the transport of outer shelf and slope material into deep water by slides and slumps. Thus, at the margin of the Rockall Trough, megasequence RTc probably includes mixed mass-flow–contourite depositional systems.

Early Miocene–early Pliocene sedimentation (Fig. 15b)

This interval witnessed the most significant development and accumulation of sediment drifts and waves in the Rockall Trough. It was characterized by a vigorous, deep-water bottom-current circulation pattern and extensive lateral migration of sediment by upslope accretion onto the flanks of the trough. This modification to the oceanographic regime probably developed in response to the mixing of existing southerly derived water masses with the newly instigated, northerly derived, Norwegian Sea Deep Water (NSDW). The latter probably coincided with the onset of deep-water overflow across the Wyville-Thomson Ridge (Stoker 1998). The deep-water circulation pattern in the Rockall Trough has not changed substantially since this time (Stow & Holbrook 1984).

The nature of the deep-water sediments has been tested at several sites. DSDP site 610 sampled the Feni Ridge and proved Miocene nannofossil chalk overlain by lower Pliocene nannofossil ooze (Hill 1986) (Fig. 10). Lower Pliocene nannofossil ooze was also recovered from basinal, sheeted drifts in the upper part of the section at ODP site 981 (Jansen *et al.* 1996) (Fig. 13). Farther north, BGS borehole 94/1 proved middle Miocene–lower Pliocene carbonate sand and gravel from the moat of an elongate mounded drift on the slope of Rockall Bank (Fig. 5). In the NE Rockall Trough, a mud-dominated sheeted-drift succession was reported from well 164/25-2 (Fig. 1) (Egerton 1998). At shallower depths, BGS borehole 90/15 proved middle Miocene bioclastic sands and muds from the top of Anton Dohrn Seamount, and 90/18 recovered middle to upper Miocene nannofossil ooze from the top of Rosemary Bank (Fig. 2) (Stoker *et al.* 1993). On the upper Hebrides Slope and Shelf, middle to upper Miocene glauconitic sandstones were recovered in BGS boreholes 88/7,7A, 90/12,12A and 90/13; a similar facies was sampled on the Malin Shelf in well 13/3-1 (Stoker *et al.* 1993, 1994).

There are no indications of significant downslope input into the Rockall Trough during this interval, in terms of either submarine fans or mass failure. This begs the question as to the source of the basinal sediments. The flanks of the trough were eroded as the elongate mounded drifts migrated upslope, thus providing a local source of sediment. In contrast, the thick biogenic section preserved in the southern Rockall Trough implies high biological productivity and pelagic sedimentation.

Early Pliocene to Holocene sedimentation (Fig. 15c)

Bottom-current activity continued to dominate deep-water sedimentation in the Rockall Trough. However, the sedimentary regime was modified in two ways: (1) sediment drifts continued to accumulate in the basin and along the eastern flank of the trough, but the western flank of the basin underwent widespread erosion; (2) a significant shelf-margin wedge developed on the margin to the NW of Britain and Ireland, with sediment pathways focused on the Sula Sgeir and Barra–Donegal fans. The progradation of the shelf margin implies an abrupt increase in sediment supply to the continental margin during late Neogene time. This has been related initially to hinterland uplift and erosion, and subsequently enhanced through glaciation (Stoker 1997). This is consistent with the parallel development of the West Shetland Margin (Stoker 1999).

The interplay between downslope and along-slope processes indicates a dynamic, deep-water environment. The eastern flank of the Rockall Trough north of 55°N can be described as an overlapping or interdigitating fan–drift system; to the south, the slope is relatively starved and canyon and channel development is more predominant. The axial part (excepting the central trough) and western flank of the trough have been variably subjected to bottom-current erosion and deposition.

The prograding shelf-margin deposits are predominantly terrigenous in character. A sand-dominated lower Pliocene to lower middle Pleistocene section unconformably overlain by mud-dominated glacigenic deposits was proved in BGS borehole 88/7,7A (Stoker *et al.* 1994) (Fig. 11). In deeper water, BGS short cores from the distal edge of the Barra Fan recovered interbedded, thin-bedded turbidites and hemipelagites (Howe 1996; Knutz 1999). In contrast, pelagic, middle Pliocene chalk was recovered from BGS short core 57/12–33 in the central part of the Rockall Trough. The basinal section was also penetrated farther south, where DSDP site 610 and ODP sites 980 and 981 proved nannofossil ooze increasingly interbedded with calcareous terrigenous clay in the upper part of the section. The influx of clay and dropstones indicates that ice-rafted detritus reached these sites in late Pliocene time (Figs 10 and 13) (Ruddiman *et al.* 1986; Jansen *et al.* 1996). On the western flank of the trough, north of 56°N, the sediments are relatively condensed and generally reflect reworking of the underlying Miocene drifts and older strata, as in BGS boreholes 94/1 and 94/4 (Figs 5 and 8b). Bioclastic sands, gravelly sands, muds and gravel-lag deposits are common on the slopes of Rockall and George Bligh banks, together with localized occurrences of Pliocene bioclastic limestones. Similar thin sequences occur on the top of the seamounts (Stoker *et al.* 1993).

Tectonostratigraphic framework

Any attempt to understand the mid- to late Cenozoic stratigraphic development of the Rockall Trough and adjacent margins has to take into consideration its regional tectonic setting (Fig. 16). Megasequence development and regional unconformities (megasequence boundaries) tend to reflect major phases of basin evolution, commonly in response to regional tectonic events that modified sedimentation patterns and palaeoceanographic circulation. On the scale of a continental margin, plate-tectonic processes probably drive such changes. It is worth considering that the initiation of sea-floor spreading in the NE Atlantic placed the continental margin off NW Britain and Ireland in a buffer zone between the spreading ridge to the west and the Africa–Europe convergence zone to the SE (Knott *et al.* 1993), thus making it susceptible to any changes in the intra-plate stress field. Intra-plate tectonism may cause differential uplift and subsidence across a margin, and may lead to enhanced erosion of uplifted areas, increased basinal subsidence, and changes in the water circulation pattern (Cloetingh *et al.* 1990).

A number of tectonic events spanning the entire mid- to late Cenozoic interval are indicated in Fig. 16, although not all of these events necessarily have regional seismic-stratigraphic expression. However, the megasequence boundaries (key reflectors C10–C30) described in this study do represent a regional response to at least three tectonic events that affected the entire Rockall Trough. These events, which occurred in the late Eocene, early Miocene and early Pliocene intervals, are summarized below.

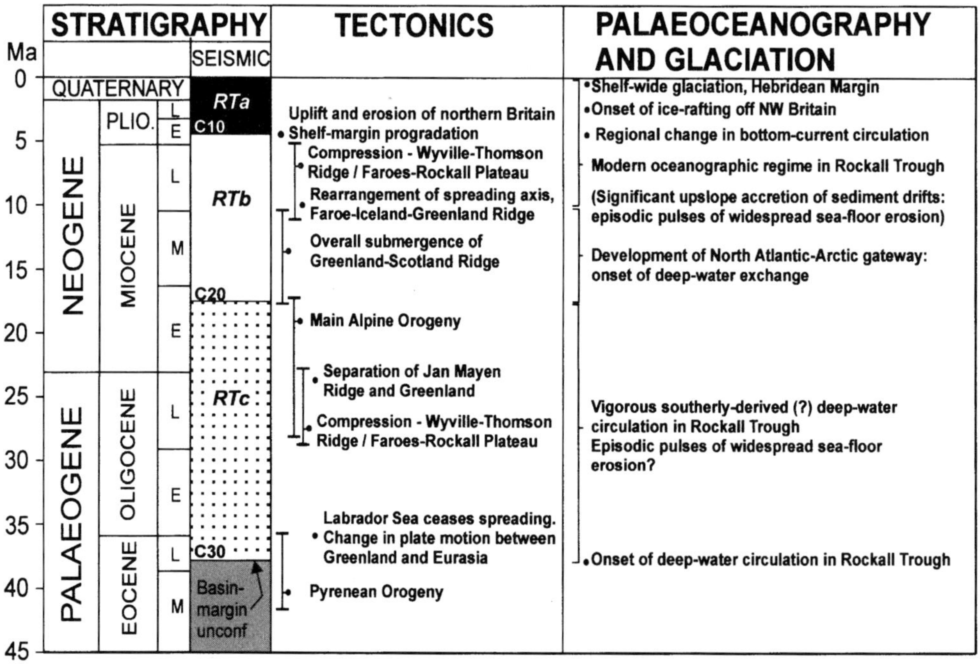

Fig. 16. Mid- to late Cenozoic tectonostratigraphical framework for the Rockall Trough. Tectonic and other events derived from a variety of sources (see text for references). Time scale from Harland *et al.* (1990).

Late Eocene event

Megasequence boundary C30 is a submarine erosion surface cut by deep-water bottom currents. The development of this unconformity marked a major change in the style of deep-water sedimentation, coincident with the initiation of bottom-current circulation in the Rockall Trough. The incursion of bottom waters into the Rockall Trough may have been a response to regional subsidence of the continental margin off NW Britain and Ireland. Bull & Masson (1996) also recorded a phase of subsidence at this time from the southern margin of the Rockall Plateau. It is perhaps no coincidence that the timing of this event was broadly coeval with the culmination of the Pyrenean orogeny (Knott *et al.* 1993), and with a major reorganization of spreading rate and direction in the North Atlantic, possibly caused by the cessation of sea-floor spreading in the Labrador Sea (Nunns 1983; Hinz *et al.* 1993). The combination of these factors may have affected the continental margin off NW Britain and Ireland such that the Rockall Trough subsided to a point that allowed bottom currents to enter the basin. According to Stoker (1997, 1998), the newly developed palaeocirculation pattern was anti-clockwise and predominantly southerly derived. This is consistent with the suggestion that deep-water exchange with northerly derived water masses was not fully established until Neogene time (Eldholm 1990; Jansen & Raymo 1996) (see below). This implies that the Greenland–Scotland Ridge, including the Wyville-Thomson Ridge, was an active barrier to deep-water flow at this time (Andersen & Boldreel 1995; Boldreel & Andersen 1995).

Early Miocene event

Megasequence boundary C20 resulted from an early Miocene deep-water oceanographic event that reworked and transported volcanic ash from the Norwegian Basin into the Rockall Trough before its alteration through processes of lithification and diagenesis (Baldauf 1986; Dolan 1986). It is proposed that this event is linked primarily to the submergence of the Wyville-Thomson Ridge, which, until this point, had been a substantial barrier to the exchange of northerly and southerly derived intermediate- and deep-water currents. On a larger scale, this is most probably related to the establishment of deep-water pathways linking the Arctic and North Atlantic oceans, via the Norwegian–Greenland Sea (Tucholke & Mountain 1986). It is the plate-tectonic evolution of the latter area, possibly also incorporating the distant effects of the early Miocene culmination of the Alpine orogeny (Knott *et al.* 1993), that strongly influenced the intra-plate stress field and the development of deep-ocean connections. Specifically, the latter involved the opening of the Fram Strait (the Northern Gateway), and the subsidence of the Greenland–Scotland Ridge (the Southern Gateway), including the Wyville-Thomson Ridge (Jansen & Raymo 1996; Thiede & Mhyre 1996). Although it has been suggested that deep-water circulation in the Norwegian–Greenland Sea may have been initiated during late Eocene–early Oligocene time (Berggren & Schnitker 1983; Ziegler 1988), a fully established pattern of deep-water exchange is probably a Neogene–Quaternary phenomenon, initiated during Miocene time as the Fram Strait developed a true deep connection, and the Greenland–Scotland Ridge became fully submerged (Eldholm 1990; Jansen & Raymo 1996). The Faroe–Shetland and Faroe Bank channel is the deepest passageway across the southern gateway, and thus represents a major transport route into the Rockall Trough for sediment derived from the Norwegian Basin. A major phase of deep-water erosion occurred along this passageway during latest Oligocene–earliest Miocene time (Stoker 1999).

Early Pliocene event

Megasequence boundary C10 is a regional, early Pliocene angular unconformity extending from the continental margin off NW Britain and Ireland into the Rockall Trough. At the shelf margin, it commonly forms the base of the prograding Plio-Pleistocene succession; in the trough, it is a submarine erosion surface cut by deep-water bottom currents. The formation of C10 links the instigation of shelf-margin progradation with widespread deep-water erosion.

In general, early Pliocene shelf-margin progradation off NW Britain was accommodated by regional seaward tilting of the Hebrides and West Shetland shelves (Stoker *et al.* 1993; Stoker 1999). The ensuing widespread development of depocentres, such as the Barra–Donegal fan system, the Sula Sgeir Fan, the West Shetland Wedge and the North Sea Fan (Fig. 15c), suggests that the tilting and sedimentation was linked to regional Neogene exhumation of Britain and Ireland. Onshore studies in Britain and Ireland (e.g. Japsen 1997; Galewsky *et al.* 1998) and the Irish Sea Basin (Green *et al.* 1999) support evidence for Neogene exhumation, although little has been done in terms of a regional synthesis of the onshore and offshore data. It is clear, however, that Britain and Ireland were part of a North Atlantic-scale exhumation (Japsen *et al.* 1998). Comparable prograding sedimentary wedges have been described from the eastern Faeroes region (Nielsen 1998) (Fig. 15c), the northern North Sea (Gregersen *et al.* 1997), the mid-Norwegian margin (Rokoengen *et al.* 1995; Henriksen & Vorren 1996; McNeill *et al.* 1998), and the East Greenland and Svalbard–Barents Sea margins (Solheim *et al.* 1998).

The synchronous development of depocentres stretching from Ireland to the Barents Sea is attributed to a general late Cenozoic uplift of the NW European margin (Doré *et al.* 1999), and suggests that the mechanics behind this phenomenon must be of a plate-tectonic nature. According to Doré (1992); Jensen & Doré (1998), the geometry of the vertical movements on a regional scale is a gentle folding (flexural uplift) with little or no faulting or fault reactivation, although localized brittle faulting cannot be discounted (Cunningham *et al.* 1999). Horizontal intra-plate stress or ridge push may have been the initial (and sustaining) cause of uplift, whereby a general uplift of cratonic areas occurs as a long-wavelength effect of plate-wide compression (e.g. Cloetingh *et al.* 1990). On the adjacent continental margins, the rapid and extensive progradation of the Plio-Pleistocene shelf-margin succession represents the main expression of uplift and erosion. Subsequently, an isostatic response to erosion and sediment loading in the adjacent basin may have continued to drive vertical movements. Flexural warping is also likely to have resulted in changes in the shape and palaeobathymetry of the adjacent basins, thus modifying the water circulation pattern (Cloetingh *et al.* 1990), which may account for the widespread deep-water erosion in the Rockall Trough.

Other events

Other events that have had important, albeit more restricted, consequences for stratigraphic development of the Rockall Trough include Cenozoic compression in the north Rockall–Wyville-Thomson Ridge–Faroes region (Boldreel & Andersen 1993, 1995) and shelf glaciation (Stoker 1995) (Fig. 16). Although the regional seismic-stratigraphic expression of these events cannot be demonstrated at present, they may form the basis for future subdivision.

The Cenozoic compressional events described by Boldreel & Andersen (1993, 1995) are attributed primarily to changes in the intra-plate stress field caused by North Atlantic plate reorganization. As the timing of these events is poorly constrained, their regional significance remains uncertain. The development of the deep-water passageway that is the Faroe–Shetland and Faroe Bank channel was influenced to some extent by Oligocene and mid- to late Miocene compression focused on a number of compressional ridges, including the Wyville-Thomson Ridge. Vertical movements of the latter may have caused fluctuations in the bottom-current regime, north and south of the ridge (Andersen & Boldreel 1995). Such fluctuations may be preserved as erosional unconformities in the deep-water sediment record in the northern Rockall Trough. If and when more precise constraints are placed on the timing and recognition of these events, it is conceivable that they may be more intimately linked to the events that created the C20 and C10 megasequence boundaries.

A consequence of the late Cenozoic uplift, in combination with climatic cooling, may have been the triggering of glaciation, with elevated plateaux becoming sites for snowfields, which ultimately developed into ice sheets (Eyles 1996). On the Hebridean margin, the base of the glacial succession is marked by a very distinct, irregular, shelf-wide, early mid-Pleistocene unconformity (Stoker 1995). Although this surface can be locally traced into the slope apron, it cannot be confidently correlated with the basinal succession for any great distance. Although this surface provides a potential marker for the future subdivision of megasequence RTa, at present it remains unassigned and informally referred to as the 'glacial unconformity' (Fig. 4).

Conclusions

(1) Three regionally significant reflectors, C30, C20 and C10, have been dated, respectively, to late Eocene, early Miocene and early Pliocene time. In the Rockall Trough, reflectors C30 and C10 are unconformities shaped, in part, by deep-water submarine erosion, whereas C20 is commonly a reflective zone formed through processes of lithification and diagenesis (Dolan 1986). These reflectors form the bounding surfaces of three megasequences (RTc, RTb and RTa in ascending stratigraphic order) that make up the middle and upper Cenozoic succession.

(2) The development of C30 marked a major change in the style of deep-water sedimentation in the Rockall Trough. This may have occurred in

response to regional subsidence of the continental margin during late Eocene time, together with the initiation of a vigorous bottom-current regime. The timing of this event was broadly coeval with the culmination of the Pyrenean orogeny and changes in North Atlantic plate motion. The origin of C20 is possibly related to a major oceanographic event, most probably linked to the submergence of the Greenland–Scotland Ridge (including the Wyville-Thomson Ridge) which, until this time, may have been a substantial barrier to deep-water exchange between the Arctic and North Atlantic oceans. The formation of C10 links the instigation of shelf-margin progradation with widespread deep-water erosion. The former is attributed to late Cenozoic uplift and erosion of Britain and Ireland, which is part of a North Atlantic-scale exhumation. It is proposed that exhumation may have been triggered by flexural warping initiated by intra-plate compression; complementary changes in the basinal geometry of the Rockall Trough may have modified the water circulation pattern, causing widespread erosion.

(3) The tectonostratigraphic framework provides a demonstrable link between intra-plate tectonics and changing basin geometry, palaeoceanography and sedimentation patterns. The formation of the key megasequence boundaries at times of significant plate rearrangement suggests that these events may be correlatable across the whole NW European margin.

We would like to thank K. Hitchen, D. Evans and the two referees, A. Densmore and M.S. Andersen, for their critical review of the paper; E.J. Gillespie for help in drafting the diagrams; P. Broad of Fugro Geoteam for allowing access to, and use of data from, the WRM96 survey; and the following oil companies who, together with the BGS, make up the 'Rockall' continental margin consortium, and without whose support this work could not have been undertaken: Agip, Amerada Hess, Arco, BG, BP-Amoco, Conoco, Elf, Enterprise, Esso, Mobil, Phillips, Shell, Statoil and Texaco. The NIOZ data were collected as part of the ENAM II programme funded by grant MAS 3-CT95. We wish to thank H. L. Andersen and P. Trinhammer of the Department of Geophysics, University of Aarhus, for their support. This paper is published with the permission of the Director of the British Geological Survey (NERC), and also forms NIOZ Contribution 3393.

References

ANDERSEN, M.S. & BOLDREEL, L.O. 1995. Effect of Eocene–Miocene compression structures on bottom-water currents in the Faeroe–Rockall area. *In*: SCRUTTON, R.A., STOKER, M.S., SHIMMIELD, G.B. & TUDHOPE, A.W. (eds) *The Tectonics, Sedimentation and Palaeoceanography of the North Atlantic Region*. Geological Society, London, Special Publications, **90**, 141–143.

BALDAUF, J.G. 1986. Biostratigraphic and palaeoceanographic interpretation of lower and middle Miocene sediments, Rockall Plateau region, North Atlantic Ocean. *In*: RUDDIMAN, W.F., KIDD, R.B., THOMAS, E. *et al.* (eds) *Initial Reports of the Deep Sea Drilling Project, 94*. US Government Printing Office, Washington, DC, 1033–1043.

BALTZER, A., HOLMES, R. & EVANS, D. 1998. Debris flows on the Sula Sgeir Fan, NW of Scotland. *In*: STOKER, M.S., EVANS, D. & CRAMP, A. (eds) *Geological Processes on Continental Margins: Sedimentation, Mass-Wasting and Stability*. Geological Society, London, Special Publications, **129**, 105–115.

BERGGREN, W.A. & SCHNITKER, D. 1983. Cenozoic marine environments in the North Atlantic and Norwegian–Greenland Sea. *In*: BOTT, M.H.P., SAXOV, S., TALWANI, M. & THIEDE, J. (eds) *Structure and Development of the Greenland–Scotland Ridge: New Methods and Concepts*. Plenum, New York, 495–548.

BOLDREEL, L.O. & ANDERSEN, M.S. 1993. Late Paleocene to Miocene compression in the Faeroe–Rockall area. *In*: PARKER, J.R. (ed.) *Petroleum Geology of Northwest Europe: Proceedings of the 4th Conference*. Geological Society, London, 1025–1034.

BOLDREEL, L.O. & ANDERSEN, M.S. 1995. The relationship between the distribution of Tertiary sediments, tectonic processes and deep-water circulation around the Faeroe Islands. *In*: SCRUTTON, R.A., SHIMMIELD, G.B., STOKER, M.S. & TUDHOPE, A.W. (eds) *The Tectonics, Sedimentation and Palaeoceanography of the North Atlantic Region*. Geological Society, London, Special Publications, **90**, 145–158.

BOLDREEL, L.O., ANDERSEN, M.S. & KUIJPERS, A. 1998. Neogene seismic facies and deep-water gateways in the Faeroe Bank area, NE Atlantic. *Marine Geology*, **152**, 129–140.

BULL, J.M. & MASSON, D.G. 1996. The southern margin of the Rockall Plateau: stratigraphy, Tertiary volcanism and plate tectonic evolution. *Journal of the Geological Society, London*, **153**, 601–612.

CLOETINGH, S., GRADSTEIN, F.M., KOOI, H., GRANT, A.C. & KAMINSKI, M. 1990. Plate reorganisation: a cause of rapid late Neogene subsidence and sedimentation around the North Atlantic. *Journal of the Geological Society, London*, **147**, 495–506.

CUNNINGHAM, M., PHILLIPS, A., DENSMORE, A., ALLEN, P. & GALLAGHER, K. 1999. Onshore Tertiary tectonism and sediment transport routes to the Porcupine and Donegal basins. *In*: CROKER, P.F. & O'LOUGHLIN, O. (eds) *The Petroleum Exploration of Ireland's Offshore Basins, Extended Abstracts*. Petroleum Affairs Division, Department of the Marine and Natural Resources, Dublin, 20–22.

DOLAN, J.F. 1986. The relationship between the 'R2' seismic reflector and a zone of abundant detrital and authigenic smectites, Deep Sea Drilling Project Hole 610, Rockall Plateau region, North Atlantic. *In*: RUDDIMAN, W.F., KIDD, R.B., THOMAS, E. *et al.* (eds) *Initial Reports of the Deep Sea Drilling Project, 94*. US Government Printing Office, Washington, DC, 1109–1115.

DORÉ, A.G. 1992. The base Tertiary surface of southern Norway and the northern North Sea. *Norsk Geologisk Tidsskrift*, **72**, 250–265.

DORÉ, A.G. & LUNDIN, E.R. 1996. Cenozoic compressional structures on the NE Atlantic margin: nature, origin and potential significance for hydrocarbon exploration. *Petroleum Geoscience*, **2**, 299–311.

DORÉ, A.G., LUNDIN, E.R., JENSEN, L.N., BIRKELAND, Ø., ELIASSEN, P.E. & FICHLER, C. 1999. Principal tectonic events in the evolution of the northwest European Atlantic margin. *In*: FLEET, A.J. & BOLDY, S.A.R. (eds) *Petroleum Geology of Northwest Europe: Proceedings of the 5th Conference*. Geological Society, London, 41–61.

EGERTON, P.D. 1998. Seismic characterisation of Palaeogene depositional sequences: northeastern Rockall Trough. *In*: STOKER, M.S., EVANS, E. & CRAMP, A. (eds) *Geological Processes on Continental Margins: Sedimentation, Mass-Wasting and Stability*. Geological Society, London, Special Publications, **129**, 217–228.

ELDHOLM, O. 1990. Paleogene North Atlantic magmatic–tectonic events: environmental implications. *Memorie della Società Geologica Italiana*, **44**, 13–28.

EMERY, D. & MYERS, K.J. 1996. *Sequence Stratigraphy*. Blackwell Science, London.

EYLES, N. 1996. Passive margin uplift around the North Atlantic region and its role in Northern Hemisphere late Cenozoic glaciation. *Geology*, **24**, 103–106.

FLOOD, R.D., HOLLISTER, C.D. & LONSDALE, P. 1979. Disruption of the Feni sediment drift by debris flows from Rockall Bank. *Marine Geology*, **32**, 311–334.

GALEWSKY, J., ALLEN, P.A. & DENSMORE, A.L. 1998. The Tertiary uplift of Ireland: the regional setting. *In*: BOLDREEL, L.O. & JAPSEN, P. (eds) *Neogene Uplift and Tectonics around the North Atlantic, International Workshop, Copenhagen*. Geological Survey of Denmark and Greenland, Copenhagen, 37–38.

GREEN, P.F., DUDDY, I.R., BRAY, R.J., DUNCAN, W.I. & CORCORAN, D. 1999. Thermal history of potential hydrocarbon source rocks in the central Irish Sea Basin. *In*: CROKER, P.F. & O'LOUGHLIN, O. (eds) *The Petroleum Exploration of Ireland's Offshore Basins, Extended Abstracts*. Petroleum Affairs Division, Department of the Marine and Natural Resources, Dublin, 40–42.

GREGERSEN, U., MICHELSEN, O. & SØRENSEN, J.C. 1997. Stratigraphy and facies distribution of the Utsira Formation and the Pliocene sequences in the northern North Sea. *Marine and Petroleum Geology*, **14**, 893–914.

HAMILTON, E.L. 1985. Sound velocity as a function of depth in marine sediments. *Journal of the Acoustical Society of America*, **78**, 1348–1355.

HARLAND, W.B., ARMSTRONG, R.L., COX, A.V., CRAIG, L.E., SMITH, A.G. & SMITH, D.G. 1990. *A Geologic Time Scale*. Cambridge University Press, Cambridge.

HENRIKSEN, S. & VORREN, T.O. 1996. Late Cenozoic sedimentation and uplift history on the mid-Norwegian continental shelf. *Global and Planetary Change*, **12**, 171–199.

HILL, P.R. 1986. Characteristics of sediments from Feni and Gardar drifts, sites 610 and 611, Deep Sea Drilling Project Leg 94. *In*: RUDDIMAN, W.F., KIDD, R.B., THOMAS, E. *et al.* (eds) *Initial Reports of the Deep Sea Drilling Project, 94*. US Government Printing Office, Washington, DC, 1075–1082.

HINZ, K., ELDHOLM, O., BLOCK, M. & SKOGSEID, K. 1993. Evolution of North Atlantic volcanic continental margins. *In*: PARKER, J.R. (ed.) *Petroleum Geology of Northwest Europe: Proceedings of the 4th Conference*. Geological Society, London, 901–913.

HOLMES, R., LONG, D. & dodd, L.R. 1998. Large-scale debrites and submarine landslides on the Barra Fan, west of Britain. *In*: STOKER, M.S., EVANS, D. & CRAMP, A. (eds) *Geological Processes on Continental Margins: Sedimentation, Mass-Wasting and Stability*. Geological Society, London, Special Publications, **129**, 67–79.

HOWE, J.A. 1996. Turbidite and contourite sediment waves in the northern Rockall Trough, North Atlantic Ocean. *Sedimentology*, **43**, 219–234.

HOWE, J.A., STOKER, M.S. & STOW, D.A.V. 1994. Late Cenozoic sediment drift complex, northeast Rockall Trough, North Atlantic. *Paleoceanography*, **9**, 989–999.

JANSEN, E. & RAYMO, M.E. 1996. Leg 162: new frontiers on past climates. *In*: JANSEN, E., RAYMO, M.E., BLUM, P. *et al.* (eds) *Proceedings of the Ocean Drilling Program, Initial Reports, 162*. Ocean Drilling Program, College Station, TX, 5–20.

JANSEN, E., RAYMO, M.E. & BLUM, P. 1996. Sites 980/981. *In*: JANSEN, E., RAYMO, M.E., BLUM, P. *et al.* (eds) *Proceedings of the Ocean Drilling Program, Initial Reports, 162*. Ocean Drilling Program, College Station, TX, 49–90.

JAPSEN, P. 1997. Regional Neogene exhumation of Britain and the western North Sea. *Journal of the Geological Society, London*, **154**, 239–247.

JAPSEN, P., BOLDREEL, L.O. & CHALMERS, J.A. 1998. Neogene uplift and tectonics around the North Atlantic: overview. *In*: BOLDREEL, L.O. & JAPSEN, P. (eds) *Neogene Uplift and Tectonics around the North Atlantic, International Workshop, Copenhagen*. Geological Survey of Denmark and Greenland, Copenhagen, 9–12.

JENSEN, L.N. & DORÉ, A.G. 1998. Cenozoic uplift in the North Atlantic area: magnitude, timing and mechanisms. *In*: BOLDREEL, L.O. & JAPSEN, P.

(eds) *Neogene Uplift and Tectonics around the North Atlantic, International Workshop, Copenhagen*. Geological Survey of Denmark and Greenland, Copenhagen, 75–76.

JONES, E.J.W., PERRY, R.G. & WILD, J.L. 1986. Geology of the Hebridean margin of the Rockall Trough. *Proceedings of the Royal Society of Edinburgh, 88B*, 27–51.

KNOTT, S.D., BURCHELL, M.T., JOLLEY, E.J. & FRASER, A.J. 1993. Mesozoic to Cenozoic plate reconstructions of the North Atlantic and hydrocarbon plays of the Atlantic margins. *In*: PARKER, J.R. (ed.) *Petroleum Geology of Northwest Europe: Proceedings of the 4th Conference*. Geological Society, London, 953–974.

KNUTZ, P.C. 1999. Contourite deposition, mass flows and ice-rafting events on the lower Barra Fan, Rockall Trough. *In*: *North-East Atlantic Slope Processes: Multi-disciplinary Approaches—Incorporating TTR-8 Post Cruise Conference, 4th ENAM Workshop and IGCP Workshop 432 (Contourites and Bottom Currents), Southampton Oceanography Centre, Abstract Book*, 48.

KUIJPERS, A., ANDERSEN, M.S., KENYON, N.H., KUNZENDORF, H. & VAN WEERING, T.C.E. 1998. Quaternary sedimentation and Norwegian Sea overflow pathways around Bill Bailey Bank, northeastern Atlantic. *Marine Geology*, **152**, 101–127.

LAUGHTON, A.S., BERGGREN, W.A. et al. (eds) *Initial Reports of the Deep Sea Drilling Project, 12*. US Government Printing Office, Washington, DC.

MARTINI, E. 1971. Standard Tertiary and Quaternary calcareous nannoplankton zonation. *In*: FARINACCI, A. (ed.) *Proceedings of the II Planktonic Conference, Rome, 1969*. Tecnoscienza, Rome, 739–785.

MASSON, D.G. & KIDD, R.B. 1986. Revised Tertiary seismic stratigraphy of the southern Rockall Trough. *In*: RUDDIMAN, W.F., KIDD, R.B. *et al.* (eds) *Initial Reports of the Deep Sea Drilling Project, 94*. US Government Printing Office, Washington, DC, 1117–1126.

MCDONNELL, A. & SHANNON, P.M. 2001. Comparative Tertiary stratigraphic evolution of the Porcupine and Rockall basins. *In*: SHANNON, P.M., HAUGHTON, P.D.W. & CORCORAN, D.V. (eds) *The Petroleum Exploration of Ireland's Offshore Basins*. Geological Society, London, Special Publications, **188**, 323–344.

MCNEILL, A.E., SALISBURY, R.S.K., ØSTMO, S.R., LIEN, R. & EVANS, D. 1998. A regional shallow stratigraphic framework off Mid Norway and observations of deep water 'special features'. *1998 Offshore Technology Conference, Houston, TX*, 1–13.

MILLER, K.G. & TUCHOLKE, B.E. 1983. Development of Cenozoic abyssal circulation south of the Greenland–Scotland Ridge. *In*: BOTT, M.H.P., SAXOV, S., TALWANI, M. & THIEDE, J. (eds) *Structure and Development of the Greenland–Scotland Ridge*. Plenum, New York, 549–589.

MITCHUM, R.M., VAIL, P.R. & SANGREE, J.B. 1977. Seismic stratigraphy and global changes of sea level, part 6: stratigraphic interpretation of seismic reflections patterns in depositional sequences. *In*: PAYTON, C.E. (eds) *Seismic Stratigraphy—Applications to Hydrocarbon Exploration*. American Association of Petroleum Geologists, Memoirs, **26**, 117–133.

NAYLOR, D., SHANNON, P. & MURPHY, N. 1999. *Irish Rockall Basin Region—a Standard Structural Nomenclature System*. Petroleum Affairs Division, Special Publication, **1/99**.

NIELSEN, T. 1998. Slope stability and sediment accumulations in the North Atlantic: evidence of Cenozoic tectonic episodes. *In*: BOLDREEL, L.E. & JAPSEN, P. (eds) *Neogene Uplift and Tectonics around the North Atlantic, International Workshop, Copenhagen*. Geological Survey of Denmark and Greenland, Copenhagen, 83–84.

NUNNS, A.G. 1983. Plate tectonic evolution of the Greenland–Scotland Ridge and surrounding areas. *In*: BOTT, M.H.P., SAXOV, S., TALWANI, M. & THIEDE, J. (eds) *Structure and Development of the Greenland–Scotland Ridge*. Plenum, New York, 11–30.

ROBERTS, D.G. 1975. Marine geology of the Rockall Plateau and Trough. *Philosophical Transactions of the Royal Society of London, Series A*, **278**, 447–509.

ROBERTS, D.G. 1989. Basin inversion in and around the British Isles. *In*: COOPER, M.A. & WILLIAMS, G.D. (eds) *Inversion Tectonics*. Geological Society, London, Special Publications, **44**, 131–150.

ROBERTS, D.G., MASSON, D.G. & MILES, P.R. 1981. Age and structure of the southern Rockall Trough – new evidence. *Earth and Planetary Science Letters*, **52**, 115–128.

ROBERTS, D.G., THOMSON, M., MICHENER, B., HOSSACK, J., CARMICHAEL, S. & BJØRNSETH, H.M. 1999. Palaeozoic to Tertiary rift and basin dynamics: mid-Norway to the Bay of Biscay—a new context for hydrocarbon prospectivity in the deep water frontier. *In*: FLEET, A.J. & BOLDY, S.A.R. (eds) *Petroleum Geology of Northwest Europe: Proceedings of the 5th Conference*. Geological Society, London, 7–40.

ROKOENGEN, K., RISE, L., BRYN, P., FRENGSTAD, B., GUSTAVSEN, B., NYGAARD, E. & SAETTEM, J. 1995. Upper Cenozoic stratigraphy on the Mid-Norwegian continental shelf. *Norsk Geologisk Tidsskrift*, **75**, 88–104.

RUDDIMAN, W.F., KIDD, R.B. & THOMAS, E. 1986. Site 610. *In*: RUDDIMAN, W.F., KIDD, R.B., THOMAS, E. *et al.* (eds) *Initial Reports of the Deep Sea Drilling Project, 94*. US Government Printing Office, Washington, DC, 351–470.

SHANNON, P.M., MOORE, J.G., JACOB, A.W.B. & MAKRIS, J. 1993. Cretaceous and Tertiary basin development west of Ireland. *In*: PARKER, J.R. (ed.) *Petroleum Geology of Northwest Europe: Proceedings of the 4th Conference*. Geological Society, London, 1057–1066.

SMYTHE, D.K. 1989. Rockall Trough—Cretaceous or Late Palaeozoic? *Scottish Journal of Geology*, **25**, 5–43.

SOLHEIM, A., FALEIDE, J.I., ANDERSEN, E.S. & 5 OTHERS 1998. Late Cenozoic seismic stratigraphy of the East Greenland and Svalbard–Barents Sea continental margins. *Quaternary Science Reviews*, **17**, 155–184.

STOKER, M.S. 1995. The influence of glacigenic sedimentation on slope-apron development on the continental margin off Northwest Britain. *In*: SCRUTTON, R.A., STOKER, M.S., SHIMMIELD, G.B. & TUDHOPE, A.W. (eds) *The Tectonics, Sedimentation and Palaeoceanography of the North Atlantic Region*. Geological Society, London, Special Publications, **90**, 159–177.

STOKER, M.S. 1997. Mid- to late Cenozoic sedimentation on the continental margin off NW Britain. *Journal of the Geological Society, London*, **154**, 509–515.

STOKER, M.S. 1998. Sediment-drift development on the continental margin off NW Britain. *In*: STOKER, M.S., EVANS, D. & CRAMP, A. (eds) *Geological Processes on Continental Margins: Sedimentation, Mass-Wasting and Stability*. Geological Society, London, Special Publications, **129**, 229–254.

STOKER, M.S. 1999. *Stratigraphic Nomenclature of the UK North West Margin. 3. Mid- to Late Cenozoic Stratigraphy*. British Geological Survey, Edinburgh.

STOKER, M.S., AKHURST, M.C., HOWE, J.A. & STOW, D.A.V. 1998. Sediment drifts and contourites on the continental margin off northwest Britain. *Sedimentary Geology*, **115**, 33–51.

STOKER, M.S., HITCHEN, K. & GRAHAM, C.C. 1993. *United Kingdom Offshore Regional Report: the Geology of the Hebrides and West Shetland Shelves, and Adjacent Deep-water Areas*. HMSO for the British Geological Survey, London.

STOKER, M.S., LESLIE, A.B., SCOTT, W.D. & 6 OTHERS 1994. A record of late Cenozoic stratigraphy, sedimentation and climate change from the Hebrides Slope, NE Atlantic Ocean. *Journal of the Geological Society, London*, **151**, 235–249.

STOW, D.A.V. & HOLBROOK, J.A. 1984. Hatton Drift contourites, Northeast Atlantic. *In*: ROBERTS, D.G. & SCHNITKER, D. (eds) *Initial Reports of the Deep Sea Drilling Program, 81*. US Government Printing Office, Washington, DC, 695–699.

STOW, D.A.V., FAUGRES, J.-C., VIANA, A. & GONTHIER, E. 1998. Fossil contourites: a critical review. *Sedimentary Geology*, **115**, 3–31.

SVAERDBORG, T. 1998. *A study of the Eocene to Recent deposits in the Rockall Trough, continental margin of the North East Atlantic Ocean: processing and interpretation of reflection seismic data*. MSc thesis, University of Aarhus.

THIEDE, J. & MHYRE, A.M. 1996. Introduction to the North Atlantic–Arctic gateways: plate tectonic–paleoceanographic history and significance. *In*: THIEDE, J., MHYRE, A.M., FIRTH, J.V., JOHNSON, G.L. & RUDDIMAN, W.F. (eds) *Proceedings of the Ocean Drilling Program, Scientific Results, 151*. Ocean Drilling Program, College Station, TX, 3–23.

TUCHOLKE, B.E. & MOUNTAIN, G.S. 1986. Tertiary paleoceanography of the western North Atlantic Ocean. *In*: VOGT, P.R. & TUCHOLKE, B.E. (eds) *The Geology of North America, Volume M, The Western North Atlantic Region*. Geological Society of America, Boulder, CO, 631–650.

VAIL, P.R. 1987. Seismic stratigraphy interpretation using sequence stratigraphy, part 1: Seismic stratigraphy interpretation procedure. *In*: BALLY, A.W. (ed.) *Atlas of Seismic Stratigraphy*. American Association of Petroleum Geologists, Studies in Geology, **27**, 1–10.

VANNESTE, K., HENRIET, J.-P., POSEWANG, J. & THEILEN, F. 1995. Seismic stratigraphy of the Bill Bailey and Lousy Bank area: implications for subsidence history. *In*: SCRUTTON, R.A., STOKER, M.S., SHIMMIELD, G.B. & TUDHOPE, A.W. (eds) *The Tectonics, Sedimentation and Palaeoceanography of the North Atlantic Region*. Geological Society, London, Special Publications, **90**, 125–139.

ZIEGLER, P.A. (ed.) *Evolution of the Arctic–North Atlantic and the Western Tethys*. American Association of Petroleum Geologist, Memoirs, **43**.

Slope instability and sediment redistribution in the Rockall Trough: constraints from GLORIA

V. UNNITHAN[1], P. M. SHANNON[1], K. McGRANE[2], P. W. READMAN[2], A. W. B. JACOB[2], R. KEARY[3] & N. H. KENYON[4]

[1]*Department of Geology, University College Dublin, Belfield, Dublin 4, Ireland (e-mail: vikram.unnithan@ucd.ie)*

[2]*Dublin Institute for Advanced Studies, 5 Merrion Square, Dublin 2, Ireland*

[3]*Geological Survey of Ireland, Haddington Road, Dublin 2, Ireland*

[4]*Southampton Oceanography Centre, Empress Docks, Southampton SO14 3ZH, UK*

Abstract: A recent GLORIA (Geological LOng Range Inclined Asdic) sidescan survey covered 200 000 km^2 of the sea bed in the Irish Rockall Trough. It revealed a range of sedimentary features on the trough floor and its steep (>6°) margins. The western margin is characterized by large-scale (of the order of hundreds of kilometres in length) downslope mass movement. Smaller-scale slides and slumps (tens of kilometres across) occur on the eastern margin, but they are subordinate to canyon, channel and fan systems. The western and central parts of the trough floor contain the Feni Sediment Ridge, a 600 km long contourite sediment build-up covered by large sediment waves trending sub-parallel to the dominant modern current pattern. Strong, northward-flowing bottom currents are thought to have eroded the base of the slope in the east and redeposited the sediments on the western margin and the trough floor. Mass wasting and terrigenous sediment input through canyons is regarded as the primary source of sediment in the region. The increase in the degree and frequency of canyon incision along the NE margin of the trough reflects increased terrigenous input from the Irish mainland and a possible glacial influence on the basin margin. The GLORIA images reflect a broad interplay of alongslope and downslope sediment transport processes in the Rockall Trough with sediments sourced from the NE margin and redistributed by currents along the western margin. Although alongslope and downslope processes are the major controlling factors, basin subsidence, Quaternary glaciations and glacio-eustatic sea-level fluctuations have also influenced the pattern of sedimentation in the Rockall Trough.

The Atlantic Irish Regional Survey (AIRS96), a GLORIA (Geological LOng Range Inclined Asdic) sidescan sonar survey, was carried out in August 1996. It covered an area of 200 000 km^2 of the continental margin and basin floor in the Irish sector of the Rockall Trough (Fig. 1). A major aim of the survey was to delineate slope instability features and sediment transport pathways along the margins of the bathymetric trough and to examine the relationship between sedimentation patterns and ocean bottom currents. Although various parts of the Rockall Trough have been subject to detailed studies (e.g. Roberts 1972; Flood *et al.* 1979; Lonsdale & Hollister 1979), this GLORIA survey provides the first integrated regional view of the sea-bed sedimentology. This paper illustrates some of the major sedimentary features identified during the survey, discusses factors influencing deep-sea sedimentary processes and attempts to place these processes in a regional geomorphological framework.

Within the last few years there has been a major increase in exploration interest in the frontier basins of the Atlantic margin (Shannon & Spencer 1999). This has led to the licensing of a significant number of blocks on the margin of the Irish Rockall Trough. A detailed understanding of the sea-floor morphology, and of the constraints on slope stability and ocean-bottom current patterns will be essential in the event of exploration success resulting in the installation of sea-bed production facilities. In these areas, a better understanding of the interplay between erosional and depositional processes in large basins such as the Rockall Trough will also benefit lithology prediction in basins with active bottom-current systems.

From: SHANNON, P.M., HAUGHTON, P.D.W. & CORCORAN, D.V. (eds) 2001. *The Petroleum Exploration of Ireland's Offshore Basins*. Geological Society, London, Special Publications, **188**, 439–454.
0305-8719/01/$15.00

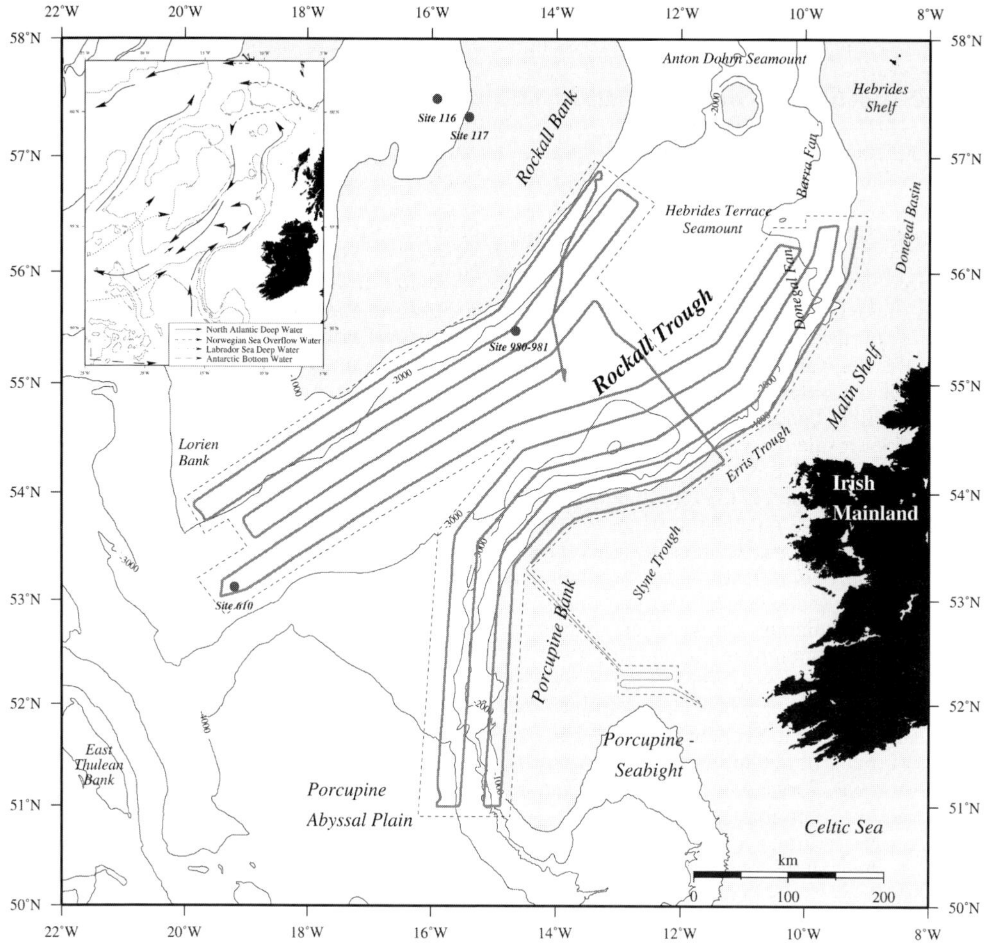

Fig. 1. (**a**) Location of the Rockall Trough, with dashed lines representing the limit of sonar coverage of the AIRS96 survey. The bold continuous lines are the ship's tracks. The location of DSDP and ODP drilling sites are indicated. (**b**) Generalized ocean-bottom circulation pattern in the NE Atlantic (after Lonsdale & Hollister (1979); Stoker (1998)) highlighting anticlockwise circulation in the northern and central Rockall Trough.

Regional setting and geological evolution

The Rockall Trough is an elongate bathymetric depression trending approximately NNE–SSW, 1000 km long and up to 250 km wide (Fig. 1). It is underlain by the Rockall Basin (Naylor *et al.* 1999), which forms part of a chain of late Palaeozoic–Cenozoic sedimentary basins extending from Norway to Ireland along the North Atlantic seaboard. The Rockall Trough extends from the Charlie–Gibbs Fracture Zone and Porcupine Abyssal Plain in the SW, to the Wyville-Thomson Ridge in the NE. Water depths along the axis of the trough decrease from 4500 m in the Porcupine Abyssal Plain to 1250 m near the Wyville-Thomson Ridge. The basin floor is punctuated by a number of shallow (200–500 m) igneous seamounts such as Hebrides Terrace and Anton Dohrn (Fig. 1). The margins of the Rockall Trough are characterized by relatively narrow (10–40 km) zones of relatively steep slopes with an average dip of 6°. The steepest slopes (locally >20°) occur along the SE margin and the lowest dips (4°) are along the western margin. The shelf break occurs in water depths ranging from 300 to 600 m and is easily identified as a sharp increase in gradient. The slope margins are arbitrarily subdivided into upper slope (shelf break to 1200 m), mid-slope (1000–2000 m) and lower slope zones (1800–3000 m). The continental rise occupies the deeper reaches

(>3000 m) of the Rockall Trough and is characterized by dips of less than 1°. Although both margins have a general NNE–SSW orientation, the eastern margin is subdivided into a north–south-striking segment (west of Porcupine Bank), an east–west-striking portion (north of Porcupine Bank), and a NNE–SSW-striking segment (west of Malin Shelf).

The structural and tectonic evolution of the Rockall Basin reflects a long history of Late Palaeozoic to early Cenozoic continental rifting. This is linked to major plate reorganization processes in the NE Atlantic (Stoker 1997). Pulsed extensional episodes with intervening periods of subsidence and inversion during Mesozoic and Cenozoic times (Shannon & Naylor 1998) resulted in extreme crustal thinning beneath the Rockall Basin. Eventual continental break-up was confined to the region between the Rockall Plateau and Greenland, where sea-floor spreading during Tertiary times resulted in the opening of the NE Atlantic. The onset of this spreading was preceded by thermally driven uplift and voluminous intrusive and extrusive igneous activity along the margins of the NE Atlantic. The last major tectonic phase in the NE Atlantic was one of regional Neogene uplift of the British Isles and Scandinavia (Doré *et al.* 1999). Estimates of net uplifts are in the range of 2–3 km for the Norwegian margin (Doré *et al.* 1999), and 1–2 km for the Irish offshore and northern UK regions (Stoker *et al.* 2001). However, the overall post-break-up history of the Rockall Basin was dominated by subsidence outpacing sedimentation (Stoker 1997).

Post-break-up subsidence commenced during early Neogene time (Shannon & Naylor 1998) in the Rockall Trough and was marked by the onset of deep-water circulation and drift sedimentation. Climatic changes and sea-level fluctuations influenced the onset of Antarctic glaciation during late Palaeogene–early Neogene and late Neogene–Quaternary Arctic glaciations (Tucholke & Mountain 1986). In NW Europe, the first evidence for major cooling is found at the base of the Pleistocene sequences. The western ice limit is characterized by repeated Late Quaternary ice advances across the Hebrides shelf and the NE Rockall Trough margin (Stoker 1995). The Armorican and southern Celtic Sea shelf margins were, however, relatively unaffected by these glacial advances. Extrapolation of the extent of Quaternary glaciations from the adjacent onshore areas, and from the UK offshore, suggests that the area investigated by the AIRS96 survey is likely to straddle the glaciated–non-glaciated transition to the west of Ireland.

The current oceanographic regime of counter-clockwise circulation in the Rockall Trough (see Fig. 1b) with a predominantly northern deep-water source is thought to have been established during Pliocene time (Stoker 1997). The modern circulation pattern in the Rockall Trough is still poorly known. Lonsdale & Hollister (1979) suggested three possible principal bottom current components: (1) Norwegian Sea Overflow Water (NSOW), which flows across the Wyville-Thomson Ridge southward along the western margin of the Rockall Trough; (2) the cold Labrador Current, which flows east along the Charlie–Gibbs Fracture Zone and then north along the eastern margin of the Hatton Basin and Rockall Trough; (3) Antarctic Bottom Water (AABW), which occupies the deepest parts of the Porcupine Abyssal Plain. North Atlantic Deep Water (NADW) includes these three major bottom waters in addition to intermediate waters such as the Slope Current (Huthnance 1986), which flows northwards along the eastern margin of the Rockall Trough. The modern deep-sea circulation pattern in the Rockall Trough is characterized by northward drift and westward deflection of the NADW as the trough shallows towards the north. This anti-cyclonic gyre formed by the NADW and the NSOW in the Rockall Trough is considered to be one of the most important factor controlling sedimentation on the basin floor (Lonsdale & Hollister 1979; Dickson & Kidd 1986; Dowling & McCave 1993).

Previous work

The Rockall Trough and underlying Rockall Basin are still relatively unexplored and poorly understood. Using seismic, sonar and bathymetry data, Stride *et al.* (1969) and Roberts (1975) provided the first detailed geomorphological analysis of the Atlantic continental margins of Europe. Roberts (1972, 1975), in particular, highlighted the main features in the Rockall Trough and recognized features such as the Donegal Fan in the northeastern part of the Trough. The contrasting physiography and stratigraphy of the margins of the Trough were thought by Roberts (1975) to reflect progressive or intermittent subsidence and differential deposition of oozes under the influence of bottom currents. Lonsdale & Hollister (1979) mapped and highlighted the importance of the strong anticlockwise bottom current pattern in the Trough using bottom photographs. Large-scale, episodic slope failure on Rockall Bank (the Rockall Bank Mass Flow), first identified on seismic lines by Roberts (1975), was further investigated and mapped by Flood *et al.* (1979) and Faugeres *et al.* (1981) using cores and echosounder data. Kenyon (1987) identified numerous slope failure features and canyons along the eastern margin of the Trough, and suggested slope gradient, contour currents and sediment influx as possible controls on down-slope sediment failure. Iceberg scour marks in water depths of 140–500 m on the northern margins (Belderson *et al.* 1973) provide the first clues to the extent of the glacial influence on these margins. Stoker (1997, 1998) used seismic profiles and shallow core data to document Quaternary glacial advances across the Hebrides shelf and the aggradation of glacial sediments at the shelf edge. In spite of such detailed work, the region still lacks an integrated framework for the understanding of sedimentary processes.

GLORIA and the AIRS96 survey

Since the late 1960s, the GLORIA sidescan sonar tool has been used to survey 10% of the world's deep oceans. It is a reconnaissance deep-sea mapping tool with the ability to cover in excess of 10 000 km^2 daily (depending on water depth and cruise speed). The GLORIA system is a shallow-towed, conventional sidescan sonar system operating at around 6.5 kHz. The sonar tool has been described by Laughton (1981); Somers (1996). Data-processing techniques have been detailed by Chavez (1986); Searle *et al.* (1990); Le Bas & Masson (1994).

GLORIA sidescan sonar images have a pixel resolution of 50 m and are displayed as monochrome images. The grey-scale variations represent energy received from the sea bed by the transducers. This 'backscatter' energy is expressed on GLORIA images as tonal variations, with strong backscatter producing lighter tones and weak backscatter producing darker hues. This variation is caused by a complex range of factors, such as sea-floor slope, topographic variability, grazing angle of insonification (geometry of the sensor–target system), physical characteristics of the target surface (e.g. surface roughness), and the intrinsic nature of the target (variations in sediment composition; Ulrick 1983). Scattering from volume inhomogeneities and sub-bottom interfaces associated with stacked sediment layers also affects GLORIA backscatter intensity. The implication of these lithological properties is that any reliable ground-truthing of GLORIA images by sampling is not feasible in complex areas. Individual samples (a few square metres) would not be representative of backscatter intensity of a single pixel (2500 m^2). Only in morphologically simple regions (flat with no lateral and vertical lithological variation), close to the ship's track (at far range even micro-topography affects backscatter intensity), would bottom sampling aid ground-truthing of the sonar images. This impracticality of ground-truthing GLORIA sonar images adds to the degree of ambiguity of sonar data interpretation.

The AIRS96 survey initially concentrated on the margins of the Rockall Trough with survey lines parallel to the margins. The swath width averaged 32 km with a nominal track separation of 25 km, to ensure overlap between adjacent swaths. The analysis was complicated by the very steep slopes (locally in excess of 20°) along the SE trough margin in the survey area. Tracklines were oriented parallel to the bathymetric contours (slope margin), which meant that signal response levels were higher from the upslope regions than the downslope side. A further complication, in terms of image resolution and interpretation, is the degree of acoustic penetration. The signal-to-noise ratio was fairly low during the survey, especially in the areas with thick unconsolidated sediments on the trough floor. Various studies (Gardner *et al.* 1991) have shown that the GLORIA signal penetrates the sea floor to varied depths, depending on the nature of the sub-bottom materials.

In addition to the GLORIA tool, a 3.5 kHz profiler was also deployed. This has a vertical resolution of 0.8 m and in areas of soft unconsolidated sediments penetrated to a maximum depth of 80 m. The echosounder data are useful as they compensate for the lack of GLORIA coverage directly under the ship's tracks, and provide accurate depth values needed for the sonar processing.

Before presenting the sonar images, it is prudent to point out that, as a large number of factors contribute to the backscatter energy, sidescan sonar interpretation is qualitative rather than quantitative. The technique adopted in this study is similar to that of Gardner *et al.* (1996). The sonar mosaic is described using descriptive backscatter terms such as high, low, mottled, speckled, etc. The second stage involves the interpretation of trackline data such as 3.5 kHz data and seismic profiles in terms of reflectivity based on the scheme devised by Damuth (1978) (i.e. strong surface, sub-bottom reflectors, etc.) and bathymetric features in terms of geomorphological characteristics (channel, canyon, levee, outcrops, etc.). The trackline analysis along successive lines was mapped and compared with the sonar mosaic. The final map, in combination with freely available satellite-derived free-air gravity data, forms the basis for the interpretation of surface morphology of the region.

Sedimentary features

The sonar mosaic of the entire AIRS96 survey is shown in Figure 2. The mosaic can be broadly categorized into three major acoustic facies. Relatively high backscatter punctuated by linear features extends from the upper slope to the edge of the rise along the eastern margin. This is associated with abrupt changes in topography such as the steep walls of canyons and/or changes in lithology. Patchy, high backscatter interspersed by regions of texturally homogeneous, low backscatter on the western margin is associated with changes in the micro-topography such as those observed in submarine mass movement. A low to very low, monotonous, homogeneous backscatter pattern (dark grey to black tones) is caused by absorption of the acoustic pulse by thick, unconsolidated hemipelagic muds and biogenic oozes. Homogeneous, fine-grained, well-sorted sands also cause low backscatter (Gardner *et al.* 1991).

Combining sidescan sonar characteristics, 3.5 kHz profiler data and other trackline information, three broad geomorphological domains can be distinguished: (1) canyons, channels and fan systems; (2) slumps, slides and debris flows; (3) contourite and drift-related deposits.

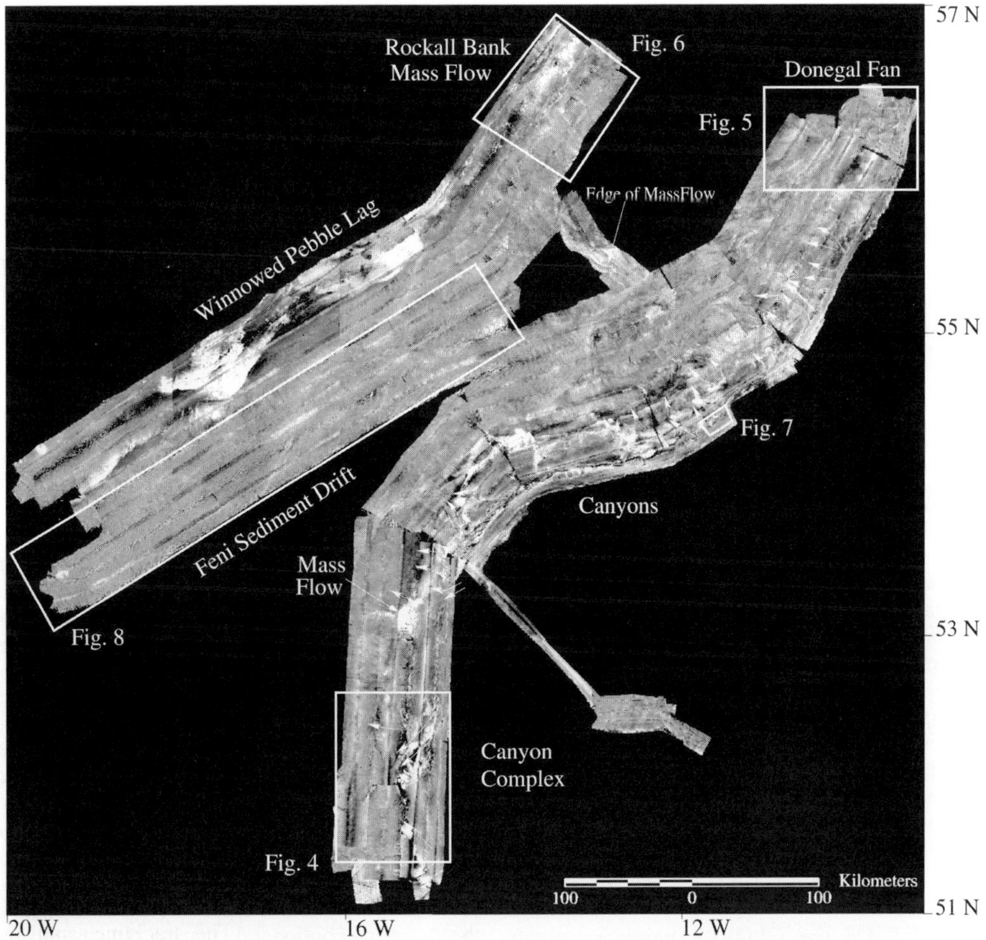

Fig. 2. GLORIA sonar mosaic of the AIRS96 survey (dotted region in Fig. 1a) highlighting the distribution of large-scale features and the location of subsequent figures described in the text. The white arrows indicate sites of major canyon incision.

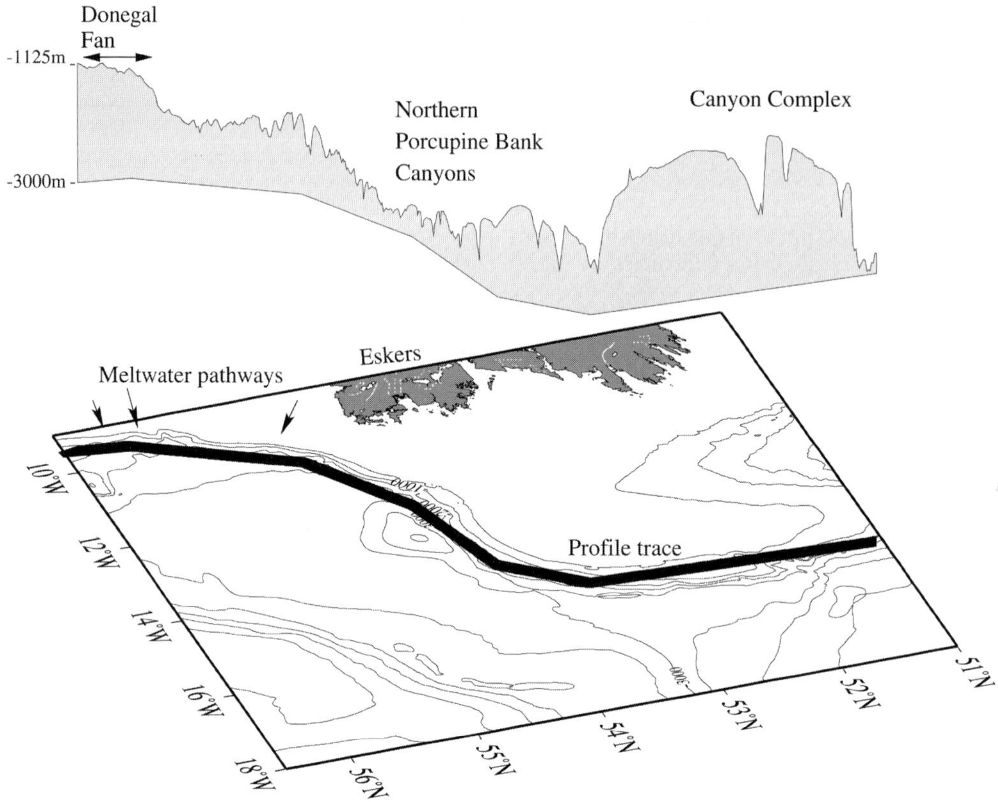

Fig. 3. A 3D perspective view of a bathymetric profile along the eastern margin of the Rockall Trough. The bathymetry data is based on 3.5 kHz profiler data, sampled at 1 km. Eskers and meltwater pathways are inferred from Warren & Ashley (1994).

Canyons, channels and fan systems

Deep marine canyons are defined in a similar fashion to their terrestrial counterparts as being a deep incision into the underlying bedrock or sediment. They serve as focused and long-lived sediment transport conduits. In contrast to channels, the steep canyon walls prevent sediment spillage and the formation of levees or overbank deposits. Channels are depositional features with characteristic channel-levee morphology. They are generally an order of magnitude smaller than canyons. Sediment fans are lobate, typically mounded, sediment accumulations at the base of slope generally formed at the distal termination of channel and canyon systems.

Canyons. On the GLORIA mosaics canyon systems are imaged across the upper and mid-slope of the eastern margin of the Rockall Trough, where they are the most pronounced feature on the slope. They are absent from the western margin of the Rockall Trough. On the eastern margin, they appear as linear to curved features with associated high and low backscatter. In profile, they have steeply incised walls and flat floors. A thin (5–15 m) layer of sediment covers the canyon floors. Canyon frequency increases from the SE margin to the NE, with the highest density along the northern margin of the Porcupine Bank (Fig. 3). Spatial variation in the density of these features is also accompanied by changes in canyon morphology. Canyons in the SE are predominately oblique to the basin margin. The most spectacular system occurs on the western flank of the Porcupine Bank (Fig. 4). This canyon complex is 20 km wide and extends for *c*. 70 km from the shelf edge to the basin floor. It is comparable in size with some of the large-scale canyon systems mapped by GLORIA along the eastern continental margin of the USA (Popenoe & Dillon 1996). The canyon originates in 500–700 m water depth on the upper slope and consists of two broad (20 km wide) canyon

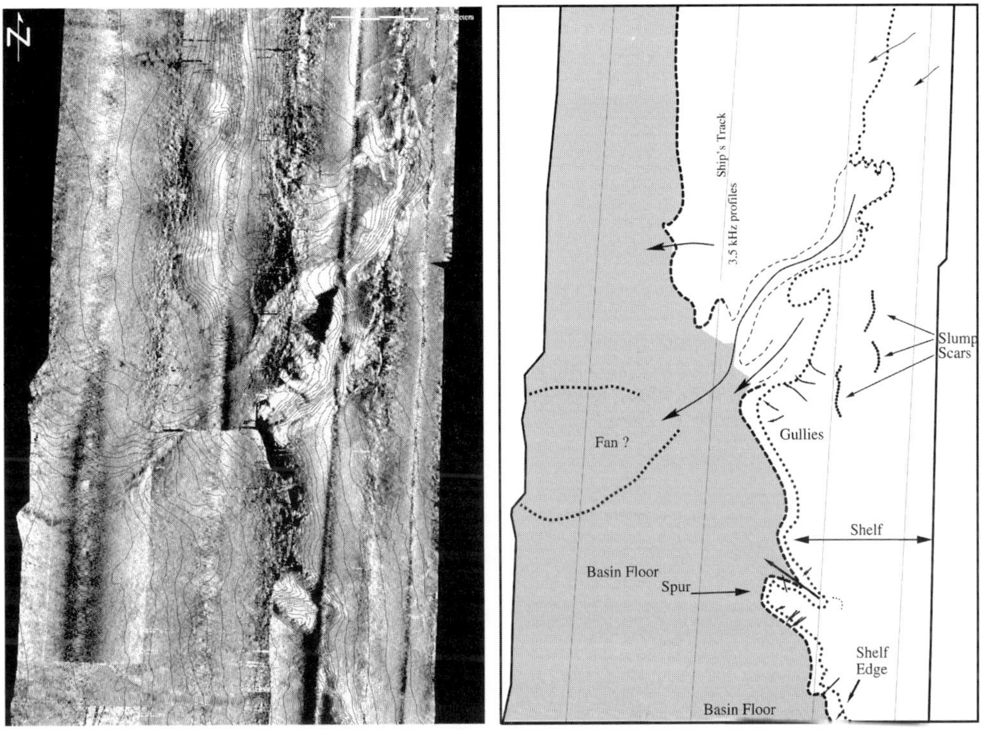

Fig. 4. Section of the GLORIA mosaic (left) from the SE margin of the Rockall Trough highlighting the major NE–SW trending canyon complex. Bathymetric contours (interval 100 m and depth range 500–4400 m) are superimposed. Locations of the channel and sediment fan associated with canyon complex are indicated with dashed lines (right). The arrows indicate canyon axes. The dotted lines demarcate the shelf region and the shaded area is the basin floor. The high dips (<20) to the south of the canyon complex should be noted. Figure location shown in Figure 2.

heads, which open individually into NE–SW-trending canyons. The canyon walls and the inter-canyon areas appears as high-backscattering targets on GLORIA, whereas the canyon floor is a low to medium backscattering region. In profile, the canyons are U-shaped, flat-floored and steep-walled. Gradients along the walls exceed 20°. Downslope they merge at a depth of 2000 m to form a single, low-sinuosity, poorly backscattering channel. Subtle variations within the predominantly low-backscattering region near the canyon mouth suggest the presence of a lobate fan measuring 45–60 km long and 25 km wide.

Along the eastern trough margin to the north and NE of Porcupine Bank, there are several regions where canyon incision and channel formation are focused (Fig. 3). In contrast to the canyons along the SE margin, most canyons are perpendicular to the basin margin and appear to originate on the mid- to lower slope (1500–2000 m). Both U-shaped and V-shaped canyons are observed in this region. These straight canyons typically measure 3 km in width and can be traced on the sonar images for short distances of about 10–15 km. Some of the canyons are inferred from local bathymetry and 3.5 kHz records rather than sidescan imagery. Such canyons do not have a high-backscatter component associated with the steep gradients along the canyon walls and hence could not be identified on the sonar records. The angle of insonification and the presence of sediments along the flanks of the canyons are probably the cause of the poor backscatter. The north–south-oriented canyons to the north of Porcupine Bank are partly imaged on older GLORIA records (Kenyon 1987).

Channels. In the Rockall Trough, channel systems are poorly imaged on the GLORIA mosaics. Where observed, channels are confined to the lower slopes and rise along the eastern margin of the Trough. On the sonar images they appear as sinuous, lineated backscatter trends. Recognition of channels at the base of slope is complicated by lack of relief and/or poor lithological contrast.

Channels are observed along the east–west and NNE–SSW margin segments of the eastern Rockall Trough. They are not as frequent as the canyon systems. They are generally small features (i.e. 5–10 m deep and 20–100 m wide) bordering the maximum resolution of both the 3.5 kHz profiler and sonar records. The channels to the north of the Porcupine Bank (55°N, 12°30′W) are observed on the sonar records as a series of curved tributaries coalescing downslope to form a single, discontinuous channel 10 km long and *c.* 100 m wide.

Sediment fans. GLORIA images from the NE margin of the Rockall Trough are characterized by low to medium, grainy, lineated backscatter. The most notable feature of this area is the lack of major canyon incision. The regional bathymetry to the south of Hebrides Terrace Seamount decreases radially towards the shelf, suggesting the presence of a low-relief, mounded submarine fan complex. On the basis of bathymetry and a few shallow seismic profiles, Roberts (1975) called this feature the Donegal Fan. Several arcuate and sinuous features observed at the base of slope region are interpreted as turbidity current pathways or mid-fan sinuous channels (Fig. 5). The lineated pattern reflects lithological and topographic contrasts between the channel and inter-channel areas of turbidite pathways

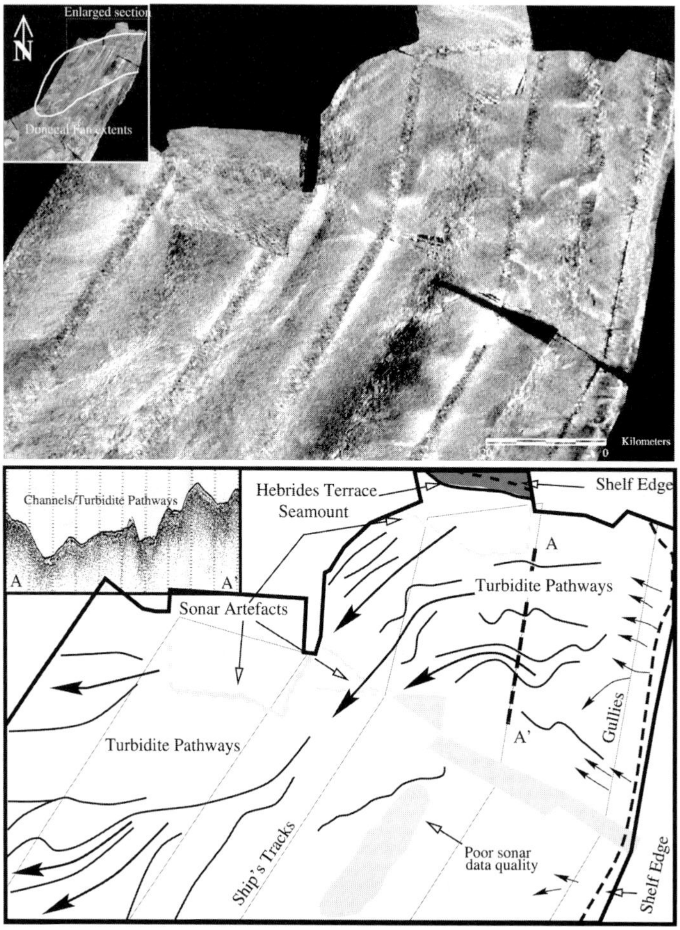

Fig. 5. Sonograms and interpretation of the Donegal Fan. The streaky lineations and backscatter pattern are typical of turbidity currents and reworking by vigorous bottom currents. The sediment transport direction is SW. Inset (left top), section of sonar mosaic, shows the regional extent of the Donegal Fan. Figure location shown in Figure 2.

(Masson *et al.* 1992, 1993). Along with the Barra Fan to the north of the Hebrides Terrace Seamount, the Donegal Fan is considered to be part of a large Neogene–Holocene trough-mouth fan complex (Armishaw *et al.* 1998) and forms the southernmost glacial fan along the NE Atlantic seaboard. On the profiler data, the inferred turbidity current deposits appear as acoustically transparent lenses (type II; Damuth 1978) and can be traced into the central region of the Rockall Trough. Streaky, NE–SW-trending backscatter patterns on the rise and east–west backscatter lineations on the slope appear to converge towards the apex of the Donegal Fan.

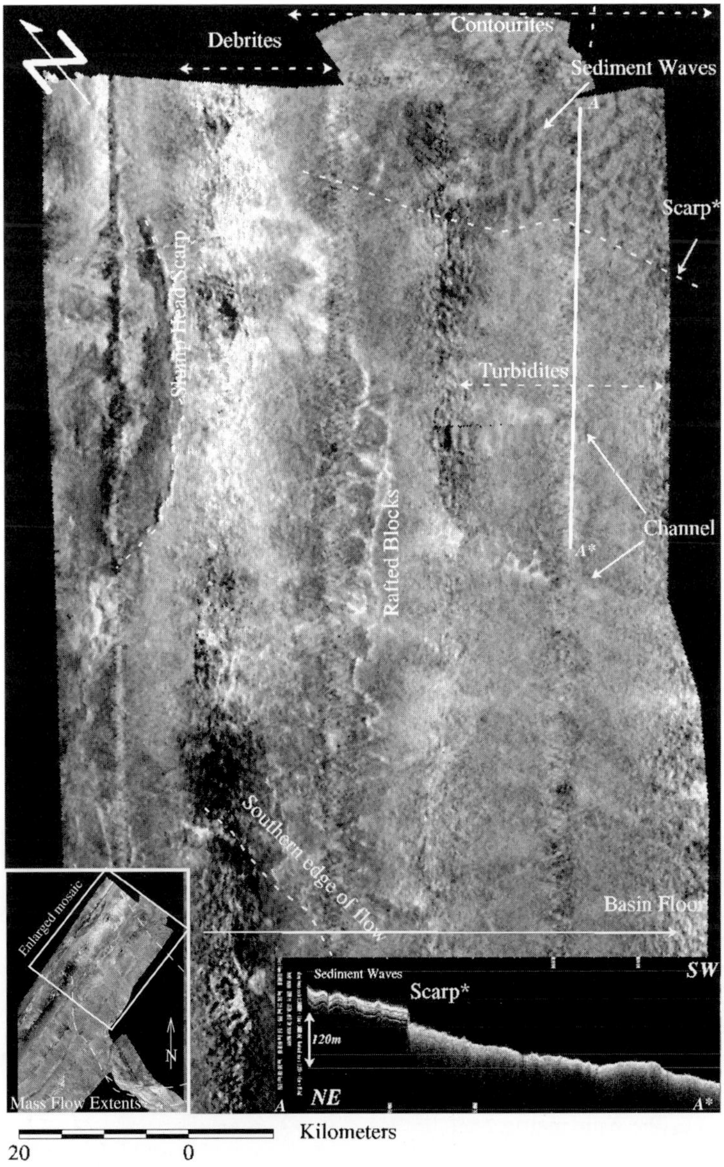

Fig. 6. Sonar mosaic of the Rockall Bank Mass Flow. The mottled bright backscatter is interpreted as debrites, whereas the low-backscatter, uniform texture on the trough floor is interpreted as turbidite flows possibly triggered by failures on the upper slope. Inset (bottom right) is an echosounder profile highlighting a 20 m high scarp bounding the mass flow and truncating a sediment wave field. The area affected by mass wasting is shown on the inset (bottom left) sonar mosaic. Figure location shown in Figure 2.

On the basis of acoustic facies mapping, the fan covers an area of *c.* 12 000 km^2, about twice the area covered by the Barra Fan.

Slumps, slides and debris flows

There is a large amount of literature on gravity-driven processes and the terminology varies greatly based on the approach used. In this paper, gravity-driven downslope processes will be described using the scheme of Stow *et al.* (1996). Mass flow, in this context, is a generic term referring to the downslope transfer of sediments in a coherent, incoherent or fluidized fashion by gravity-driven processes. Slumps and slides are gravity-driven cohesive sediment movement, whereas debris flows and turbidity currents refer to increasing downslope sediment disaggregation and fluid content, in which the initial stratification or structure is lost during transport and emplacement. Slumps and slides are distinguished on the basis of internal deformation or the lack of it. They represent a continuum of increasing downslope sediment movement and thus are difficult to differentiate. To minimize confusion, any reference to slumping or sliding in this paper strictly implies coherent downslope sediment movement. Debris flows are differentiated from turbidity currents on the basis of their inferred grain size and the presence or absence of cohesive sediment blocks.

The western and eastern margins of the Rockall Trough differ in terms of geometry, size and distribution of downslope mass wasting features. The western margin is characterized by a single large-scale mass wasting structure, whereeas there are numerous small slump, slide and debris flow deposits along the eastern margin of the Trough.

Western margin. GLORIA sidescan sonar images from the NW margin of Rockall Trough show clear evidence for large-scale downslope gravitational sediment movement. The AIRS96 survey imaged the mid- to lower slope of the Rockall Bank Mass Flow (Fig. 6). The mid-slope region is characterized by a low-backscattering arcuate ridge. Further downslope from the ridge, an area of mottled, high, saturated backscatter pattern with large (1–2 km diameter) blocks aligned parallel to the bathymetric contours is thought to represent the debris flow stage of this mass wasting feature. Fine, streaky, low to medium backscatter patterns on the basin floor reflect the turbidity flow deposits possibly related to the slumping and sliding events on the upper slopes of the margin. To the north, 3.5 kHz profiles across the debris and turbidity flow deposits show that large lateral scarps, in excess of 20 m high, truncate a sediment wave field associated with the northward continuation of the Feni Sediment Drift (Fig. 6). The scarps are less pronounced along the southern margin of the failed slope margin.

The deposits of turbidity currents downslope from the failures extend to at least 100 km from the base of slope. The total volume of sediment estimated to be redeposited is of the order of 3000 km^3 (Flood *et al.* 1979). The presence of two stacked, lobate backscatter structures on the basin floor (Fig. 2) suggests that there were probably two major flow episodes. Using bathymetric and seismic profiles, Roberts (1972) identified two distinct slip surfaces on the upper slopes and associated these with the Rockall Bank Mass Flow events. Whether these flows are related to the two sliding events as suggested by Roberts (1972) is a matter of speculation. Two channel-like features within the area of inferred turbidite deposition indicate that flow was probably confined through two funnel-shaped entrant zones. These zones are partially filled with sediments, indicating that smaller flows, pelagic sedimentation and strong bottom currents are modifying the turbidite deposits.

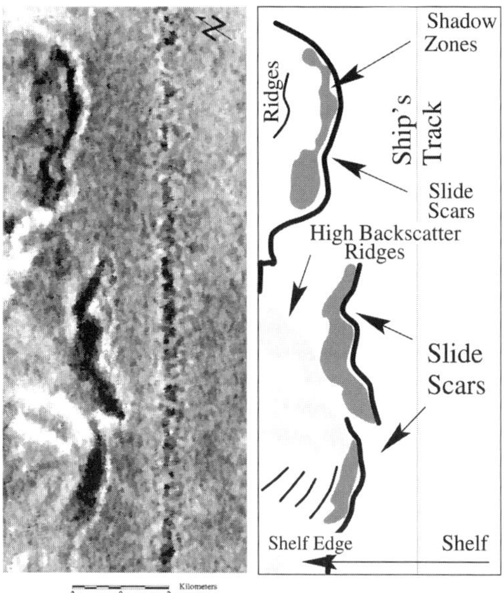

Fig. 7. Sonogram and interpretation of cuspate, extensional slide scars on Porcupine Bank *c.* 1–2 km long with downslope sediment transport of a few hundred metres. Figure location shown in Figure 2.

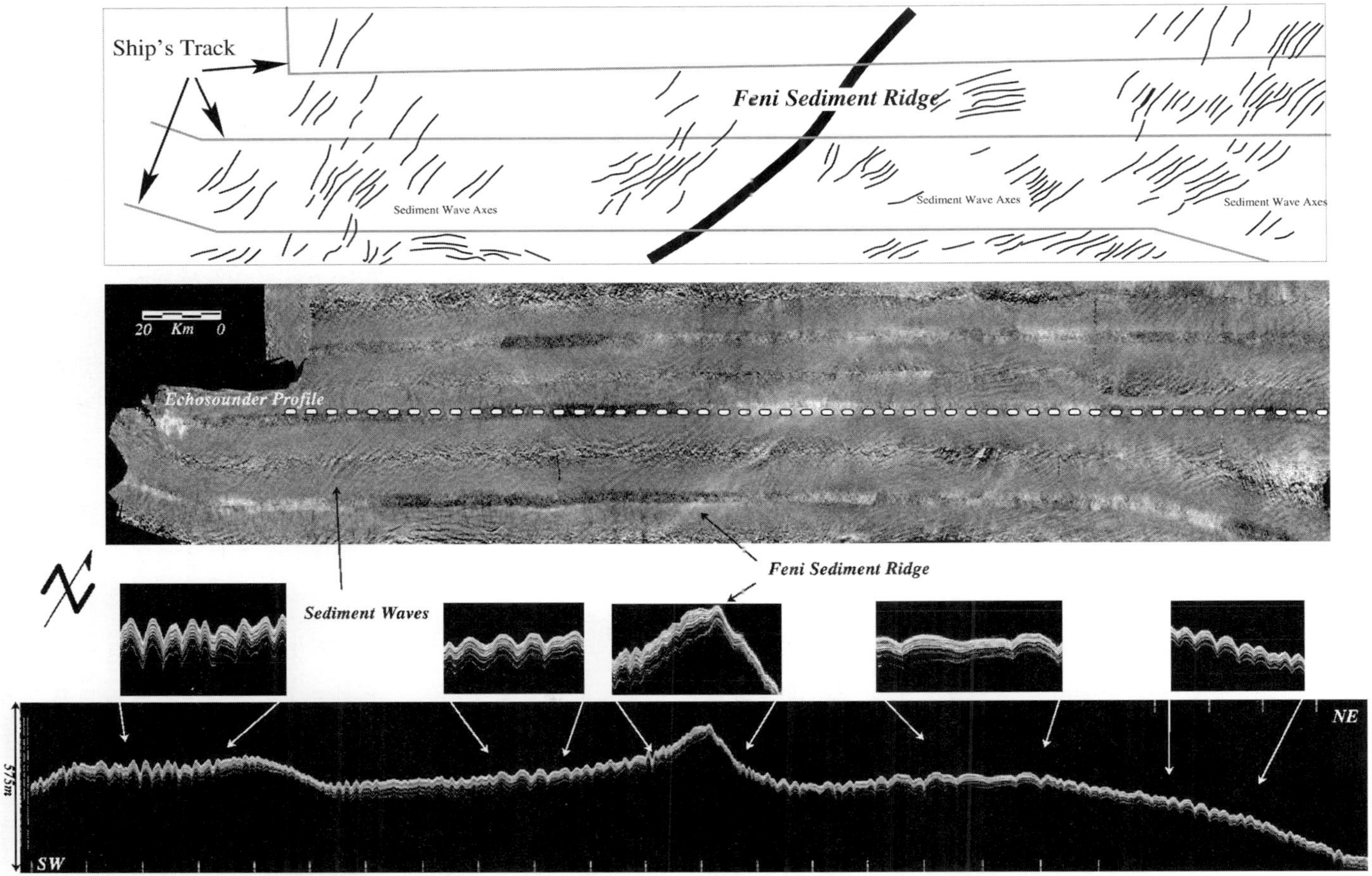

Fig. 8. GLORIA sidescan sonar mosaic highlighting subtle backscatter caused by the sediment waves of the Feni Sediment Drift. The crest of the sediment drift is seen as a diffuse band of stronger backscatter. A 3.5 kHz echosounder profile across the Feni Drift shows the internal geometry of the wave field and the change in character of the drift along the profile line from a mounded, sediment wave dominated drift in the SW to a relatively smooth, sedimented drift in the NE. Figure location shown in Figure 2.

Eastern margin. A wide spectrum of slope failure features occurs along the eastern margin of the Rockall Trough. Cuspate to linear, NE–SW-trending, approximately slope-parallel failure scars incise the upper slope of the eastern margin. The GLORIA images resolve these failure zones as narrow curvilinear belts of high reflectivity (Fig. 7). They are typically 2–8 km long and represent downslope displacements of *c.* 100–2000 m. These failures produce a progressive downstepping of the margin. They define a general area of collapse along a large section of the slope (about 100 km long), which is attributed to the steep and unstable nature of this part of the margin.

Along the SE margin, an area covering over 600 km^2 (53°N, 15.5°W) characterized by saturated levels of bright backscatter, is interpreted as a mass failure feature (Fig. 2). This feature occupies the mid- to lower slope region and is characterized by slumping and debris flows on the mid-slope and turbidite deposits on basin floor.

A large range of smaller instability features occur concentrated on the mid-slope region along the northern margin of the Porcupine Bank. A submarine slide, associated with a 'scallop'-shaped failure scar, shows a general SE–NW transport direction. Coherent mass movement features are defined by a series of bright backscattering ridges, and probably represent extensional tensional gashes. In addition along the mid-slope region, areas of high reflectivity (typically 50 km long and 5 km wide) with oval-shaped poorly backscattering spots (1 km diameter) probably represent consolidated strata exhumed by multiple failure events.

Contour current deposits

Although the exact pattern of deep-sea circulation in the Rockall Trough is unknown, the morphology and asymmetry of sedimentary deposits along the basin margins and floor highlights the importance of currents in redistributing sediments within the basin.

The Feni Sediment Drift. The most prominent sedimentary feature, covering the central and western parts of the Rockall Trough basin floor, is a giant elongate, mounded sediment drift, the Feni Sediment Drift. This is the oldest and one of the largest sediment drifts in the NE Atlantic (Wold 1994). It is a sinuous, NE–SW-oriented axial ridge with dips along the flanks in the range of 0.5–1° and is characterized by elevated topography. The surface of the drift is covered with large sediment waves, with dominant slope sub-parallel crests 2–20 km long and with 0.5–4 km wavelength. The amplitude of these waves varies between 20 and 60 m.

On sonar records these sediment waves appear as faint linear backscatter features towards the far range of the sonar swath (Fig. 8). AIRS96 GLORIA data show that wave trends are dominantly parallel to the regional slope. On the basis of a single GLORIA traverse across the Feni Drift, Roberts & Kidd (1979) recognized not only sub-parallel sediment wave trends but also a markedly transverse trend to the regional bathymetric contours. Without the advantage of adjacent GLORIA tracks, it is likely that Roberts & Kidd (1979) in some cases interpreted far range defraction patterns as sediment wave axes. A diffuse zone of high backscatter about 100 km in length represents the main axis of the drift. The overall low backscatter intensity and lack of features is mainly due to the absence of significant relief and the thick, fine-grained nature of the basin-floor sediments. Pelagic ooze, calcareous muds and terrigenous silts (Kidd & Hill 1986) of the Feni Drift have a low microscale roughness and acoustic impedance necessary for enhanced backscatter levels.

On echosounder profiles (Fig. 8), the sediment waves have a symmetric to asymmetric fold-like geometry with broad, single, gently rolling hyperbolae with a sharp bottom echo and continuous and conformable sub-bottom echoes (type IB; Damuth 1978). The symmetry of the sediment waves changes from symmetric away from the ridge axes to asymmetric closer to the axes. Flood (1994) suggested that this asymmetry could be caused by preferential deposition on the upstream or upslope wave flank. As wave migration ceased at *c.* 2.4 Ma (Kidd & Hill 1986), sediment loading, compaction and gravitational creep are likely causative mechanisms for the present-day wave asymmetry.

Winnowed pebble or gravel lag field. South of the debris flows on NW Rockall Trough, along the mid- to upper slope of the Rockall Bank, an area of very high backscatter (white to light grey hues) is observed on the sidescan sonar images (Fig. 2). On the 3.5 kHz echograms, continuous, sharp and prolonged bottom echoes with no sub-bottom reflections (type 1; Damuth 1978) and irregular relief indicate the presence of a hard, rough surface covered by an intermittent, thin veneer of unconsolidated sediments. Although nearly all detailed structure is obscured by the saturated high-backscatter patches, convoluted ridges on the upper to mid-slope are suggestive of sediment failure. Further downslope, linear high- and low-backscatter ridges parallel to the bathymetric contours are also indicative of sliding and slumping.

In the absence of geological samples and high-resolution seismic data, the origin of this bright backscatter region is speculative. Outcropping basement, consolidated sedimentary strata or a winnowed glacial pebble or gravel lag deposit are the most likely causes of the high acoustic response. Conventional seismic data (Roberts 1975) from this region do not support the interpretation of basement at or near the sea bed. The reflectors are coherent, laterally continuous and indicate the presence of thick sedimentary cover. A recent sidescan sonar and shallow bottom sampling survey (Kenyon *et al.* 1998) along the NW margin of this region shows evidence for strong currents and the active transport of coarse sandy material. In addition, Roberts (1975) also identified Flandrian beach conglomerates in water depths of 200 m along the margins of Rockall Bank. Hence, on the basis of the limited regional studies and acoustic characteristics, the bright backscattering region imaged on the GLORIA records is likely to be caused by coarse, current-winnowed sediments such as glacial gravel or pebble lags.

Discussion

The spatial distribution of sedimentary features along the margins of the Rockall Trough highlights the difference between the two margins. The western margin is characterized by a general absence of canyon incision and channelized sediment transport. Mass wasting episodes appear to be the main mechanism of downslope sediment transport on this margin. In contrast, the eastern margin has numerous canyon and channel systems. The frequency, styles and orientations of canyon erosion vary along the eastern margin. In the SE, the canyons are large, infrequent and oriented predominantly oblique to the basin margin, whereas the canyons along the northern Porcupine Bank and Malin Shelf are small, frequent and straight erosional forms trending perpendicular to slope. The oblique orientation of canyons in the SE is believed to reflect the influence of underlying basement faults (McGrane *et al.* 2001). Canyon systems documented by Kenyon *et al.* (1978) along the Armorican margin trend obliquely across the continental slope and appear to follow lines of tectonic weakness. In contrast, canyons along the northern part of the Porcupine Bank and to Malin Shelf trend perpendicular to slope, suggesting no underlying basement structural control on canyon formation. The higher frequency of canyon incision to the north of the Porcupine Bank is interpreted as a response to enhanced terrigenous supply from the Irish mainland. The western and SW Porcupine Bank margins are sediment starved. The Porcupine Basin shows a similar pattern with respect to the distribution of channel–canyon systems and mass flows features. The eastern margin, closest to the Irish mainland, is dominated by canyon and channel systems whereas mass wasting is confined to the sediment-starved western margin of Porcupine Basin (Roberts 1975).

If increased canyon frequency reflects enhanced sediment supply, then large sediment accumulations in fans or slope aprons would be expected along the eastern margin. From the AIRS96 survey, it is obvious that such large sediment accumulations do not exist along the eastern margin of the Rockall Trough. We believe that deep-sea thermohaline circulation is responsible for erosion (Dickson & Kidd 1986) and entrainment of sediments transported by the canyon and channel systems along the eastern margin and re-sedimentation along the western margin as the elongate Feni Sediment Drift (Roberts & Kidd 1979, van Weering & de Rijk 1991). Hence, continuous sediment supply and accommodation space need not result in sediment accumulation at the distal end of canyon or channel systems.

A wide range of slope failure features from very large (100 km wide) to small (2 km wide) is observed along the margins of the Rockall Trough. Large- and medium-scale catastrophic flows such as the Rockall Bank Mass Flow are likely to be initiated by a complex set of factors such as overloading (under- or over-consolidation) and oversteepening of the margins. The nature of sediments (cohesive, non-cohesive) and external factors such as seismic activity enhance the probability of slope instability. It is interesting to point out that both the Rockall and Porcupine banks are relatively starved of continuous sediment supply. Smaller-scale mass wasting along the upper slopes of the eastern margin appears to be spatially related to the canyon and channel systems. Areas of extensive canyon incision occur where a pronounced and significant amount of slope failure and mass movement has taken place. For instance, the NE–SW-trending canyon system on the SE margin has clearly undergone mass movements, evidenced by gullying across the lower canyon wall and slope failure within and adjacent to the canyon heads. However, O'Leary (1996) noted that the relationship between mass movements and canyon erosion is complicated, as the two processes might not be complementary. This is illustrated in the SE area further north from the canyon complex, where massive sediment failure involving a large section of the basin slope is devoid of canyon incision.

Constraints on the ages of mass movements and canyon erosion are problematic, because of the absence of core and detailed seismic data. Flood *et al.* (1979) dated the Rockall Bank Mass flows to 15–16 ka BP, which coincides with the end of the last glacial maximum. Pleistocene glaciation or deglaciation effects are likely to have had a significant impact on northern margins of the Rockall Trough. The combined result of glaciation or deglaciation (i.e. lowered sea level, increased meltwater transport, sediment creep and possible seismic activity induced by isostatic post-glacial rebound) are thought to have played a major role in triggering canyon erosion and mass movement. Major glacial meltwater pathways, mapped on the basis of eskers in Ireland by Warren & Ashley (1994), suggest increasing meltwater transport across the NE shelf margins during the last glacial maximum. In addition to supplying sediment to the Donegal Fan, ice streams from the Irish mainland were probably responsible for increased sediment and canyon incision along the north margin of the Porcupine Bank. The low frequency of canyon incision and slope failure along the SE margin suggests that this margin was largely unaffected by glaciation.

In addition to glaciation or deglaciation, factors such as seismic activity, increased wave action, storm surges and internal waves may have played an important subsidiary role in the development of the margin. Seismic activity, although sparsely documented in the literature from the Rockall Trough (Jacob *et al.* 1983), could provide the horizontal acceleration needed to initiate downslope mass movement. Studies along the Norwegian margin have highlighted the importance of earthquakes. All three stages of the giant Storegga Slide are attributed to overloading and earthquake activity (Bugge *et al.* 1988). Lowering of the storm wave base during glacio-eustatic sea-level fall would affect large parts of the margins of the Rockall Trough, bringing them within the zone of influence of storm surges and wave action.

In summary, the observed sedimentary features in the Rockall Trough can be explained by a simple morphogenetic model. Downslope processes such as gravity-driven mass wasting and alongslope processes involving sediment redistribution by ocean bottom currents are the two main mechanisms controlling sedimentation and erosional patterns. Sediment is supplied by episodic mass wasting along the upper slopes of both Trough margins and by sediments sourced from the NE shelf margin. The later flux is focused through canyons on the eastern margin and feeds turbidity currents which run across the Trough floor. Strong bottom currents rework and redeposit sediments derived from the eastern margin on the western margin. At longer time scales (>10 ka), downslope and alongslope processes are controlled by factors such as climate (glaciation or deglaciation), glacio-eustatic sea-level fluctuations and tectonics. Continued subsidence of the Rockall Trough and Neogene uplift of onshore Ireland and the UK provides the general tectonic framework. Quaternary glaciations and glacio-eustatic sea-level fluctuations modulate both downslope and alongslope processes. Hence, the static conceptual model of coupled downslope and alongslope processes must be placed in climatic, sea-level and tectonic framework to fully appreciate the complexity of this dynamic system.

Conclusions

The GLORIA mosaics from the AIRS96 survey in the Rockall Trough provide information and constraints on a wide range of erosional and depositional features. The main findings are as follows.

(1) The survey highlights the contrast in sedimentary features along the eastern and western margin of the Rockall Trough. A combination of canyons, channels, sediment fans and mass wasting features dominate the eastern margin. The western margin is characterized by large-scale mass wasting and lacks evidence for significant canyon development and channelized sediment transport.

(2) Contour currents are responsible for reworking and deposition of sediments on the basin floor. The lack of prominent canyon mouth lobes and channels along the eastern margin attests to the strong erosive northern current along this margin. Sediments are redeposited by SW-flowing contour-hugging currents along the western margin.

(3) The spatial variability of features observed along the margins and basin floor of the Rockall Trough is caused by the interaction of downslope and alongslope processes. The increase of canyon frequency along the NE margin of the Rockall Trough reflects increased sediment flux from the margins. Factors influencing these processes at longer time scales are likely to be basin subsidence, Quaternary glaciations and glacio-eustatic sea-level fluctuations.

(4) Although there are few constraints on the ages of these features, Tertiary uplift and Quaternary glaciations of the Irish margin are likely to have had a major impact on late Cenozoic sedimentary processes in the Rockall Trough.

This work was supported under contract IR.95.MR.021 of the Marine Research Measure (Operational Programme for Fisheries 1994–1999) administered by the Marine Institute, and part funded by their European Union's Regional Development Fund. The GLORIA system was provided under the EC's Human Capital and Mobility, Access to Large-scale Facilities Programme under their contract ERBCHGECT 930029 with Southampton Oceanography Centre.

References

ARMISHAW, J.E., HOLMES, R.W. & STOW, D.A.V. 1998. Morphology and sedimentation on the Hebrides Slope and Barra Fan, NW UK continental margin. *In*: STOKER, M.S., EVANS, D. & CRAMP, A. (eds) *Geological Processes on Continental Margins: Sedimentation, Mass-Wasting and Stability*. Geological Society, London, Special Publications, **129**, 81–104.

BELDERSON, R.H., KENYON, N.H. & WILSON, J.B. 1973. Iceberg plough marks in the northeast Atlantic. *Palaeogeography, Palaeoclimatology, Palaeoecology*, **13**, 215–224.

BUGGE, T., BELDERSON, R.H. & KENYON, N.H. 1988. The Storegga Slide. *Philosophical Transactions of the Royal Society of London*, **325**, 357–388.

CHAVEZ, P.S. 1986. Processing techniques for digital sonar images from GLORIA. *Photogrammetry, Engineering and Remote Sensing*, **52**, 1133–1145.

DAMUTH, J.E. 1978. Echo characteristics of the Norwegian–Greenland Sea: relationship to Quaternary sedimentation. *Marine Geology*, **28**, 1–36.

DICKSON, R.R. & KIDD, R.B. 1986. Deep circulation in the southern Rockall Trough, the oceanographic setting of Site 610. *In*: RUDDIMAN, W.F., KIDD, R.B., THOMAS, E. *et al.* (eds) *Initial Reports of the Deep Sea Drilling Project, 94*. US Government Printing Office, Washington, DC, 1061–1074.

DORÉ, A.G., LUNDIN, E.R., JENSEN, L.N., BIRKELAND, Ø., ELIASSEN, P.E. & FICHLER, C. 1999. Principal tectonic events in the evolution of the northwest European Atlantic margin. *In*: FLEET, A.J. & BOLDY, S.A.R. (eds) *Petroleum Geology of Northwest Europe: Proceedings of the 5th Conference*. Geological Society, London, 41–61.

DOWLING, L.M. & MCCAVE, I.N. 1993. Sedimentation on the Feni Drift and Late Glacial bottom water production in the northern Rockall Trough. *Sedimentary Geology*, **82**, 79–87.

FAUGERES, J.-C., GONTHIER, E., GROUSSET, F. & POUTIERS, J. 1981. The Feni Drift: the importance and meaning of slump deposits on the eastern slope of the Rockall Bank. *Marine Geology*, **40**, M49–M57.

FLOOD, R.D. 1994. Abyssal bedforms as indicators of changing bottom current flow: examples from the US East Coast continental rise. *Paleoceanography*, **9**, 1049–1060.

FLOOD, R.D., HOLLISTER, C.D. & LONSDALE, P. 1979. Disruption of the Feni sediment drift by debris flows from Rockall Bank. *Marine Geology*, **32**, 311–334.

GARDNER, J.V., BOHANNON, R.G., FIELD, M.E. & MASSON, D.G. 1996. The morphology, processes, and evolution of Monterey Fan: a revisit. *In*: GARDNER, J.V., FIELD, M.E. & TWICHELL, D.C. (eds) *Geology of the United States Seafloor: The View from GLORIA*. Cambridge University Press, Cambridge, 193–220.

GARDNER, J.V., FIELD, M.E., LEE, H., EDWARDS, B.E., MASSON, D.G., KENYON, N.H. & KIDD, R.B. 1991. Ground-truthing 6.5 kHz side scan sonographs: what are we really imaging? *Journal of Geophysical Research, B, Solid Earth and Planets*, **96**, 5955–5974.

HUTHNANCE, J.M. 1986. The Rockall slope current and shelf-edge processes. *Proceedings of the Royal Society of Edinburgh*, **88B**, 83–101.

JACOB, A.W.B., NEILSON, G. & WARD, V. 1983. A seismic event near the Hebrides Terrace Seamount. *Scottish Journal of Geology*, **19**, 287–296.

KENYON, N.H. 1987. Mass-wasting features on the continental slope of Northwest Europe. *Marine Geology*, **74**, 57–77.

KENYON, N.H., BELDERSON, R.H. & STRIDE, A.H. 1978. Channels, canyons and slump folds on the continental slope between South-West Ireland and Spain. *Oceanologica Acta*, **1**, 369–380.

KENYON, N.H., IVANOV, M.K., AKHMETZHANOV, A.M. 1998. *In*: *Cold Water Carbonate Mounds and Sediment Transport on the Northeast Atlantic Margin*. UNESCO Intergovernmental Commission Technical Series, **52**, 178.

KIDD, R.B. & HILL, P.R. 1986. Sedimentation on Feni and Gardar sediment drifts. *In*: RUDDIMAN, W.F., KIDD, R.B., THOMAS, E. *et al.* (eds) *Initial Reports of the Deep Sea Drilling Project, 94*. US Government Printing Office, Washington, DC, 1217–1244.

LAUGHTON, A.S. 1981. The first decade of GLORIA. *Journal of Geophysical Research*, **86**, 511–534.

LE BAS, T.P. & MASSON, D.G. 1994. Suppression of the multiple reflections in GLORIA side-scan sonar images. *Geophysical Research Letters*, **21**, 549–552.

LONSDALE, P. & HOLLISTER, C.D. 1979. A near-bottom traverse of Rockall Trough: hydrographic and geologic inferences. *Oceanologica Acta*, **2**, 91–105.

MASSON, D.G., KIDD, R.B., GARDNER, J.V., HUGGETT, Q. & WEAVER, P.P.E. 1992. Saharan continental rise and facies distribution and sediment slides. *In*: POAG, C.W. & DE GRACIANSKY, P.C. (eds) *Geologic Evolution of Atlantic Continental Rises*. Van Nostrand Reinhold, New York, 327–343.

MASSON, D.G., HUGGETT, Q.J. & BRUNSDEN, D. 1993. The surface texture of the Saharan Debris Flow deposits and some speculations on submarine debris flow processes. *Sedimentology*, **40**, 583–598.

McGrane, K., Readman, P.W. & O'Reilly, B.M. 2001. Interpretation of transverse gravity lineaments in the Rockall Basin. *In*: Shannon, P.M. & Haughton, P.D.W. & Corcoran, D.V. (eds) *The Petroleum Exploration of Ireland's Offshore Basins*. Geological Society, London, Special Publications, **188**, 393–399.

Naylor, D., Shannon, P. & Murphy, N. 1999. *Irish Rockall Basin Region—a Standard Structural Nomenclature System*. Petroleum Affairs Division, Special Publication, **1/99**.

O'Leary, D.W. 1996. The timing and spatial relations of submarine canyon erosion and mass movement on the New England continental slope and rise. *In*: Gardner, J.V., Field, M.E. & Twichell, D.C. (eds) *Geology of the United States Seafloor: the View from GLORIA*. Cambridge University Press, Cambridge, 47–58.

Popenoe, P. & Dillon, W.P. 1996. Characteristics of the continental slope and rise off North Carolina from GLORIA and seismic-reflection data: the interaction of downslope and contour processes. *In*: Gardner, J.V., Field, M.E. & Twichell, D.C. (eds) *Geology of the United States Seafloor: the View from GLORIA*. Cambridge University Press, Cambridge, 59–83.

Roberts, D.G. 1972. Slumping on the eastern margin of the Rockall Bank, North Atlantic Ocean. *Marine Geology*, **13**, 225–237.

Roberts, D.G. 1975. Marine geology of the Rockall Plateau and Trough. *Philosophical Transactions of the Royal Society of London, Series A*, **278**, 447–509.

Roberts, D.G. & Kidd, R.B. 1979. Abyssal sediment wave fields on Feni Ridge, Rockall Trough: long-range sonar studies. *Marine Geology*, **33**, 175–191.

Searle, R.C., Le bas, T.P., Mitchell, N.C., Somers, M.L., Parson, L.M. & Patriat, P.H. 1990. GLORIA image processing: the state of the art. *Marine Geophysical Researches*, **12**, 21–39.

Shannon, P.M. & Naylor, D. 1998. An assessment of Irish offshore basins and petroleum plays. *Journal of Petroleum Geology*, **21**, 125–152.

Shannon, P.M. & Spencer, A.M. 1999. Atlantic margin: offshore Norway to offshore Ireland. Introduction and review. *In*: Fleet, A.J. & Boldy, S.A.R. (eds) *Petroleum Geology of Northwest Europe. Proceedings of the 5th Conference*. Geological Society, London, 229–230.

Somers, M.L. 1996. The GLORIA system and data processing. *In*: Gardner, J.V., Field, M.E. & Twichell, D.C. (eds) *Geology of the United States Seafloor: the View from GLORIA*. Cambridge University Press, Cambridge, 29–42.

Stoker, M.S. 1995. The influence of glacigenic sedimentation on slope-apron development on the continental margin off Northwest Britain. *In*: Scrutton, R.A., Stoker, M.S., Shimmield, G.B. & Tudhope, A.W. (eds) *The Tectonics, Sedimentation and Palaeoceanography of the North Atlantic Region*. Geological Society, London, Special Publications, **90**, 159–177.

Stoker, M.S. 1997. Mid to late Cenozoic sedimentation on the continental margin off NW Britain. *Journal of the Geological Society, London*, **154**, 509–515.

Stoker, M.S. 1998. Sediment-drift development on the continental margin off NW Britain. *In*: Stoker, M.S., Evans, D. & Cramp, A. (eds) *Geological Processes on Continental Margins: Sedimentation, Mass-Wasting and Stability*. Geological Society, London, Special Publications, **129**, 229–254.

Stoker, M.S., van Weering, T.C.E. & Svaerdborg, T. 2001. A Mid- to Late Cenozoic tectonostratigraphic framework for the Rockall Trough. *In*: Shannon, P.M., Haughton, P.D.W. & Corcoran, D.V. (eds) *The Petroleum Exploration of Ireland's Offshore Basins*. Geological Society, London, Special Publications, **188**, 411–438.

Stow, D.A.V., Reading, H.G. & Collinson, J.D. 1996. Deep seas. *In*: Reading, H.G. (ed.) *Sedimentary Environments: Processes, Facies and Stratigraphy*. Blackwell Science, Oxford, 395–453.

Stride, A.H., Curray, J.R., Moore, D.G. & Belderson, R.H. 1969. Marine geology of the Atlantic continental margin of Europe. *Philosophical Transactions of the Royal Society of London, Series A*, **26** (1148), 31–75.

Tucholke, B.E. & Mountain, G.S. 1986. Tertiary paleoceanography of the western North Atlantic Ocean. *In*: Vogt, P.R. & Tucholke, B.E. (eds) *The Geology of North America: M, the Western North Atlantic Region*. Geological Society of America, Boulder, CO, 631–650.

Ulrick, R.J. 1983. *Principles of Underwater Sound*. 3rd edn. McGraw–Hill, New York.

van Weering, T.C.E. & de Rijk, S. 1991. Sedimentation and climate induced sediments on Feni Ridge, NE Atlantic Ocean. *Marine Geology*, **101**, 49–69.

Warren, W.P. & Ashley, G.M. 1994. Origins of the ice-contact stratified ridges (eskers) of Ireland. *Journal of Sedimentary Research*, **A64**, 433–449.

Wold, C.N. 1994. Cenozoic sediment accumulation on drifts in the northern North Atlantic. *Paleoceanography*, **9**, 917–941.

Slope failure features on the margins of the Rockall Trough

P. M. SHANNON[1], B. M. O'REILLY[2], P. W. READMAN[2], A. W. B. JACOB[2] & N. KENYON[3]

[1]*Department of Geology, University College Dublin, Belfield, Dublin 4, Ireland (e-mail: p.shannon@ucd.ie)*

[2]*Geophysics Section, Dublin Institute for Advanced Studies, 5 Merrion Square, Dublin 2, Ireland*

[3]*Southampton Oceanography Centre, Southampton SO14 3ZH, UK*

Abstract: A TOBI sidescan sonar survey in the Irish sector of the Rockall Trough reveals the presence of a range of slope failure features of various sizes and extent along both the eastern and western margins. A number of different types are identified. These include incipient cuspate slides, slab failures and evolved slides, and debris flows. It is suggested that the incipient cuspate slides, slab failures and evolved slides represent slope failure of muddy sediments whereas the failures that gave rise to debris flows lie on steeper slopes and may be of less muddy composition. Many of the slope failure features are relatively recent (probably <15 ka), although some evidence points towards either a prolonged period of movement or a number of phases of slope movement locally along the margins. A comprehensive understanding of the nature, distribution, age and controls on the formation of the slope failure features will be necessary in planning the likely location of sea-bed structures in the event of petroleum development in the region.

The Rockall Trough, a deep steep-sided bathymetric feature, lies on the continental shelf to the west of Ireland and the UK (Fig. 1). Water depths in the region increase from a few hundred metres on the margins to >4000 m in the southern abyssal parts of the Trough. The Rockall Trough is underlain by a sedimentary basin, the Rockall Basin (Naylor *et al.* 1999), which contains *c.* 6 km of presumed Upper Palaeozoic to Recent sediments and rests upon thinned continental crust (Shannon *et al.* 1994, 1995; Hauser *et al.* 1995; O'Reilly *et al.* 1996). During the past few years this deep-water area has become a frontier region of significant petroleum exploration interest, with the main prospects located towards the margins of the basin.

Both the eastern and western margins of the Rockall Trough display evidence of significant current activity (see Vermeulen 1997). The water column in the region is highly stratified and consists of waters with different salinities, oxygen contents and velocities (Lonsdale & Speiss 1977; Lonsdale & Hollister 1979). Along the eastern margin of the Rockall Trough northward-flowing currents have been recorded in various water depths. Kenyon (1986) documented longitudinal and transverse bedforms in water depths of up to 1000 m, which indicate the existence of a northward-flowing current with velocities sometimes in excess of 40 cm s^{-1}. This current flows along the upper continental slope west of Ireland, Scotland and Norway (see Howe & Humphery 1995). In a detailed study of current dynamics, Dickson & McCave (1986) documented northward-flowing currents with maximum velocities ranging from *c.* 30 cm s^{-1} to almost 50 cm s^{-1} at various depths from 500 m to *c.* 2500 m along the eastern margin of the Rockall Trough. These probably represent a combination of the shallower current described above, and a deeper water current that flows northwards along the eastern margin and deflects westwards where the Rockall Trough shallows at *c.* 56°N (see Howe & Humphery 1995). Currents were also recorded in even deeper waters along the eastern margin slope by Thorpe & White (1998), who described a deep intermediate nepheloid layer at a depth of 2550 m in the region. A current on the lower part of the slope, 10–30 m off the bottom, reached velocities of >15 cm s^{-1}. Along the western margin of the Rockall Trough currents flow southwards. Strong to moderate currents were reported in 2000–2400 m water depth, along the western margin (Lonsdale & Hollister 1979). Thermohaline currents were recorded at 2400–2900 m (Lonsdale & Speiss 1977), and

From: SHANNON, P.M., HAUGHTON, P.D.W. & CORCORAN, D.V. (eds) 2001. *The Petroleum Exploration of Ireland's Offshore Basins*. Geological Society, London, Special Publications, **188**, 455–464.
0305-8719/01/$15.00

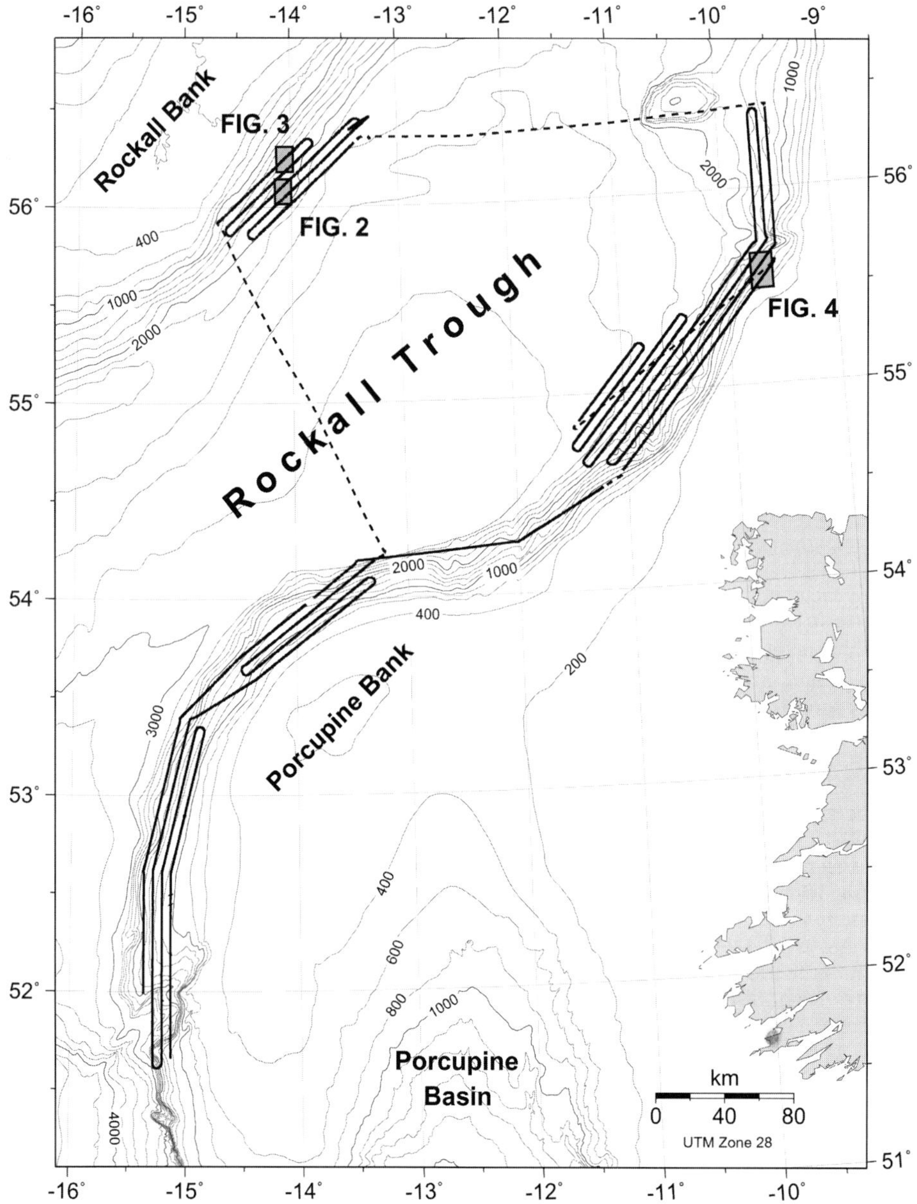

Fig. 1. Track chart for TOBI and 3.5 kHz survey (dashed lines are 3.5 kHz data only) with 200 m isobath contours. The location of Figs. 2–4 is shown in the boxes.

significant bedload transport and reworking at a depth of *c.* 1300 m was also noted. Velocities of up to 30 cm s^{-1} have been recorded along the western margin (see Vermeulen 1997). These currents probably represent a combination of the anticlockwise circulatory current recorded on the eastern margin together with the southerly directed Norwegian Sea Overflow current (see Howe & Humphery 1995). The latter is probably the most significant current in the Rockall Trough, determining both sediment distribution and the sea-floor bedforms present. In addition to the predominant southward-directed currents along the western margin, some WNW currents with velocities up to 13.5 cm s^{-1} were reported by Lonsdale & Hollister (1979). In summary, currents predominantly flow northwards along the eastern margin and southwards along the western margin of the Rockall Trough, thereby effecting a cyclonic pattern of sediment transport. This has the overall effect of transporting sediment from the delivery points on the eastern margin to the western side of the Rockall Trough.

The margins of the Rockall Trough are typically of the order of 10–40 km wide. They have relatively steep slopes (Unnithan *et al.* 2001). Dips on the eastern margin are on average 6° and are occasionally in excess of 20°, whereas those on the western margin average close to 4°. Existing seismic data and regional sidescan sonar information suggest the presence of erosional and depositional features including major slope failure and mass flow events (Unnithan *et al.* 2001). However, existing data only outline the general location and broad shape of such features, with detection of the detailed geometry and characteristics beyond the limits of resolution of commercial seismic and reconnaissance (GLORIA; Geological Long Range Inclined Asdic) sidescan sonar technology. The location, type and behaviour of such mass failure features is likely to have a considerable impact on the design, planning and location of sea-bed structures in the event of petroleum development and production in the region.

A number of large mass failure features have already been documented further north in the Rockall Trough and on the Norwegian continental margin (Flood *et al.* 1979; Kenyon 1987; Bugge *et al.* 1988; Dowdeswell & Kenyon 1997; van Weering *et al.* 1998). The age constraint on the failure features in the Rockall Trough is poor; mass flows on the eastern margin of the Rockall Bank have been biostratigraphically dated to 15–16 ka BP, coinciding with the last glacial maximum (Flood *et al.* 1979). The effects of Quaternary glaciation and post-glacial sedimentation (e.g. large sediment fans) are less pronounced further south, in the Irish Rockall Trough, than in the UK sector. Analysis of the features along the Irish margin of the Rockall Trough may therefore help to unravel the Late Quaternary and Recent history of the region in the context of a regional North Atlantic framework.

An extensive high-resolution sidescan sonar survey was carried out in the Irish sector of the Rockall Trough in June–July 1998. One of the major objectives of the TRIM (TOBI (Towed Ocean Bottom Instrument) Rockall Irish Margins) project, which was funded by the Irish Petroleum Infrastructure Programme (Rockall Studies Group), was to map and assess the type and extent of Recent slope failure features in the region. A total of 3100 line km of TOBI (30 kHz) sidescan sonar and 3700 km of 3.5 kHz profiler data were collected along the margins of the Rockall Trough (Fig. 1). The objective of this paper is to illustrate some of the preliminary results from the project. The present paper addresses only the slope failure features and is thus not intended to be a comprehensive demonstration of the wide range of features observed in the TOBI data. In particular, the paper does not describe the various turbidite systems, the deposits of which occur throughout the region. In addition, the spectacular large canyon systems, which occur along the eastern margin, are not described in detail but instead only some of the slope failure features associated with the smaller canyons are illustrated.

In view of the importance, both scientifically and industrially, of mass movements on continental slopes, there exists a vast literature describing such features and processes. However, the literature on slope failure terminology is often confusing, as studies have typically concentrated upon specific types of deposits such as sediment gravity flows (e.g. Middleton & Hampton 1973; Lowe 1979, 1982). Few publications have addressed the entire spectrum of changes that may occur during a gravity-induced failure between initiation and final deposition. An exception is the classification of offshore mass movements described and discussed by Mulder & Cochonat (1996). They divided the processes involved in offshore mass movement deposits into slides or slumps, plastic flows and turbidity currents. These three types are distinguished mainly by the style of motion, the architecture of the deposit and the shape of failure surface. This static classification system is related to a dynamic system that underlines the morphological transformation that may occur to a gravity event where a slide can evolve into a plastic flow and then into a turbidity current. The classification and terminology adopted in the present paper largely follows that of Mulder & Cochonat (1996). The features described comprise various types of slides or slumps and plastic flows. Slides and slumps are movements of coherent masses of sediments bounded on all sides by distinct failure planes. The displacement is limited and the internal structure of the removed material is largely undisturbed. The failure plane usually follows the stratification. The difference between a slide and a slump is based on the value of the Skempton ratio h/l, where h is the depth and l is the length of the failure. Slides are translational and have Skempton ratios of <0.15 whereas slumps are rotational and have Skempton ratios of >0.33

(Skempton & Hutchinson 1969). Plastic flows usually result from the movement of typically unconsolidated masses of sediments and may evolve from the disintegration of material that detached during slide movement. Debris flows are supported by their matrix strength and may be regarded as transitional between slides and turbidity currents (Hampton 1972). Shear is typically distributed throughout the sediment mass and movement occurs along innumerable shear planes within the body of the material. The mass therefore effectively flows rather than slides. Debris flows may be cohesive or cohesionless, and the flow characteristics of the different types have been discussed by Postma (1986).

TOBI

TOBI (Towed Ocean Bottom Instrument), designed, built and operated by the Southampton Oceanography Centre (SOC), is a deep-towed 30 kHz sidescan sonar instrument. It is towed at an altitude of 300–400 m above the sea bed at a speed of 1.5–3 knots depending on conditions and water depth. The swath width is about 6 km.

TOBI weighs about 2.5 tons and requires a ship with suitable aft deck space and an A-frame for deployment and retrieval. The R.V. *Pelagia*, operated by the Netherlands Instituut voor Onderzoek der Zee (NIOZ), was used for the TRIM survey. This is a modern research vessel 66.05 m in length. The TOBI instrument is of open frame construction into which the sidescan transducers are fixed and onto which other instruments can be attached. A large part of its weight (and volume) is made up of buoyancy material added to make it neutrally buoyant. TOBI is a two-body system employing a 600 kg depressor weight attached to the TOBI vehicle by means of a 200 m slightly buoyant umbilical connecting rope. This two-body arrangement gives the vehicle excellent stability and allows its use in fairly high sea states.

TOBI can operate in water depths of up to 6000 m. The depth of the instrument is controlled by varying the length of the tow cable and the speed of the ship. Attitude sensors on the vehicle record roll, pitch and yaw, and the heading of the vehicle is recorded by a gyrocompass.

A hull-mounted ORETECH 3.5 kHz echo-sounder was, in addition, used during this cruise. Depth information was logged from this sounder onto the ship's data logging system at 10 s intervals.

The intensity of the reflected backscatter on the TOBI mosaics is controlled by the nature of the sea bed and therefore it is generally possible to distinguish rock outcrop (high backscatter) from unconsolidated sediments (containing many grains producing lower backscatter). TOBI imagery may sometimes allow distinction between various unconsolidated sediment lithologies or between deposits formed by different processes (e.g. thinly bedded sands or muds from reworked and mixed sands or muds). In general, sandy sediments give rise to high acoustic backscatter, whereas muddy sediments typically result in a low backscatter on TOBI mosaics. However, muddy sediments with thin silt or sand layers can give rise to high acoustic backscatter, whereas sandy sediments with thin intercalated mud layers may produce low backscatter, possibly as a result of interference effects (Gardner *et al.* 1991). Consequently, it is inadvisable to always associate sandy or coarse sediments with high backscatter without extensive bottom sampling. Damuth (1980) recognized three basic echo types on sub-bottom profilers, which show a qualitative correlation with the relative abundance of coarse, bedded sediments (and thus with bed roughness) in the top few metres below the sea floor: areas with little or no coarse sediment display distinct echoes with continuous sub-bottom reflections; sediments with moderate amounts of sand or silt produce semi-prolonged echoes with transient or discontinuous sub-bottom events; sediments with large amounts of sand or silt and gravel have very prolonged echoes with no sub-bottom structure. In addition, the TOBI sonar penetrates into the sediments below the sea-bed surface and the backscatter may be related to sediments below the sea bed rather than only on the surface. Therefore, correlation of the TOBI sonar backscatter intensity with the character of the vertical incidence 3.5 kHz profiler data helps to constrain the overall sediment facies interpretation. Resolution of features on TOBI mosaics is normally down to *c.* 5–10 m.

The TRIM survey

The survey tracks are shown in Figure 1 with TOBI tracks indicated as continuous lines, and transit lines, where only 3.5 kHz profiler data were collected, shown as dashed lines. Preliminary interpretation of the TOBI mosaics, integrated with the 3.5 kHz echo-sounder profiles, reveals the presence of an extensive array of dramatic slope failure features on both margins of the Rockall Trough. A number of different types of features are seen and three are described in the present paper to illustrate the range of morphology and internal character. These are illustrated in the mosaics and the accompanying

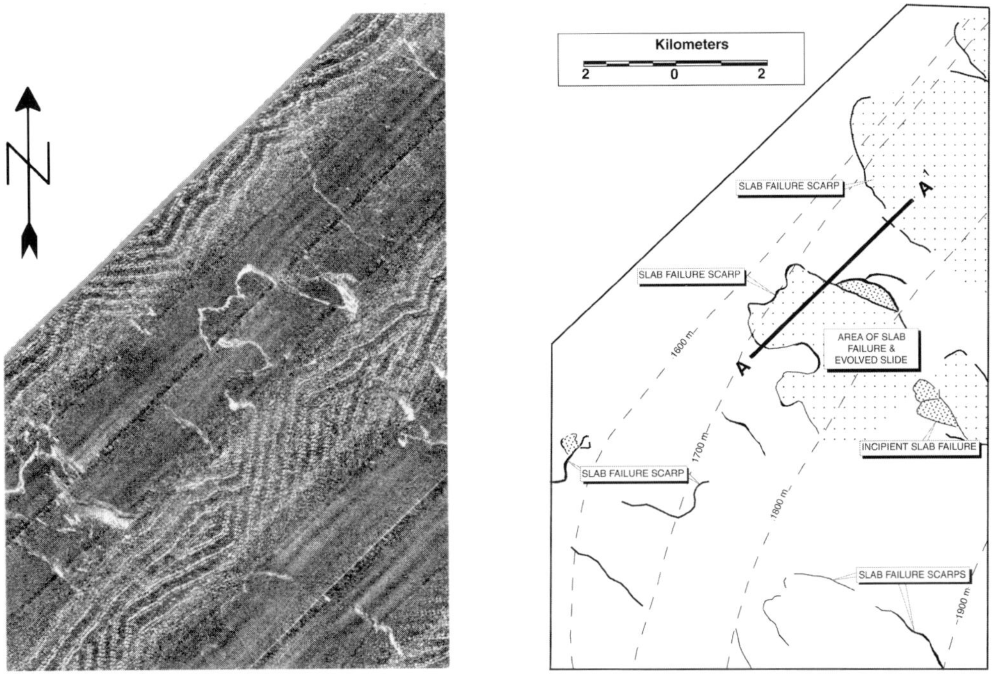

Fig. 2. Mosaic (left-hand side) and interpretation (right-hand side) of slab failures and evolved slides on the western margin of Rockall Trough. Water depths are in metres. The scarp features on the upslope sides of the slab failures but the absence of scarps on the downslope sides should be noted. This is interpreted as the result of disintegration and removal of the slab material. The scarps, together with the bathymetric lows in the area of slab removal, are clearly seen on the 3.5 kHz echo-sounder profile (Fig. 5). (See Fig. 1 for location.)

interpretations (Figs 2–4), together with an example of the 3.5 kHz profiler data. However, many slope failure features throughout the region show characteristics intermediate between these types, indicative of a continuum and a gradation between all types.

Incipient cuspate slides

These generally occur on the flanks and margins of the large slab failures described below and are well developed on the western margin of the Rockall Trough. They are less common than the other two types of slope failure discussed in this paper. They are typically 2–3 km wide and can be mapped for 5–10 km in a downslope direction. They have a low acoustic backscatter and the sub-bottom character suggests a mud-prone lithology. Within the headwall area of the slides a series of cuspate normal faults occur (Fig. 3) producing dramatic scarp features. Individual fault scarps can be mapped for up to a few hundred metres and rarely for up to a few kilometres (Fig. 3). The amount of downslope movement is typically small in the extensional heads, with the cohesive strata behaving in a semi-brittle manner facilitating the development of a linked series of broadly slope-parallel, curved scarps. Unlike classical slump structures (Moore & Shannon 1991), no evidence is seen of compressional toe areas associated with the incipient cuspate slides on the western margin of the Rockall Trough. This, together with the general absence of slide structures in the lower slope regions of the basin margins, may be due to detachment and remobilization of the central and lower parts of the slides to form turbidity currents that transported material several tens of kilometres into the basin centre.

Incipient cuspate slides are not always associated with large slab failures. Along the southeastern part of the margin, west of the Porcupine Bank, a region of major canyon incision occurs. Upslope of the canyons a complex region of sediment sliding is present. Canyon incision commences at a depth of *c.* 1500 m throughout this area. Crescent-shaped slide and slump structures are present in the region upslope of the canyons and sometimes an internal slump structure is resolved. The patterns

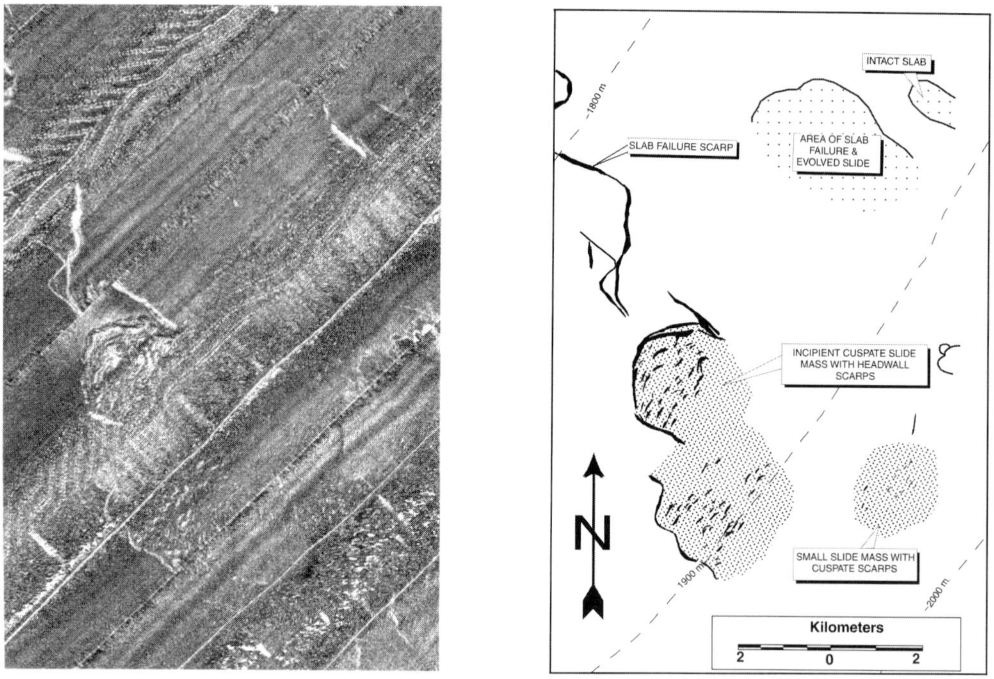

Fig. 3. Mosaic (left-hand side) and interpretation (right-hand side) of slope failure features on the western margin of Rockall Trough. A slide mass, with cuspate headwall scarps, and a series of slab failure scarps, are indicated. Water depths are in metres. (See Fig. 1 for location.)

on the TOBI mosaic become more complex and fragmented in the mid-slope region (typically very bright backscatter). Further downslope of this region the backscattering pattern becomes very strong and the surface smoother. This change in properties down the slope may represent a change from upper slope mass wastage to mid-slope net sediment accumulation.

Slab failures and evolved slides

These range in dimensions from several hundred metres to tens of kilometres across and are slightly to noticeably elongate in a downslope direction. They typically occur as clusters and are well developed on both margins of the Rockall Trough. In some examples the sediment slabs show a low degree of block disintegration and slab gliding. These slabs have moved slightly downslope, with scarps indicating the separation from the intact slope. In most instances, however, the slides have evolved more fully, with terraces or scarps preserved on the upslope and sides of the basin slope (Figs 2 and 5). In these instances the glide slabs have probably been removed by disintegration and extensive downslope movement. In general, the size of the fragmented slabs and residual slide scarps within a region of extensive mass wastage become smaller and more numerous in a downslope direction before disintegrating into probable debris flow or turbidite deposits.

The 3.5 kHz sub-bottom character and the generally low acoustic backscatter within the slab and scarp areas are consistent with mud-prone lithologies. The tabular geometries of the slab failure areas suggest that they are controlled by bedding plane detachments. Individual scarps are hundreds of metres to a few kilometres in length and displacements measured from 3.5 kHz profile data vary from several metres to tens of metres (Fig. 5). Slope failure escarpments are commonly oriented both parallel and perpendicular to bathymetric contours. In many instances the scarps are sharp, suggesting a relatively recent formation (probably within the last 10–15 ka) as there is little evidence of current reworking despite the strong currents reported by other workers (see above) in the region.

West of the Porcupine Bank a close spatial relationship is noted between slope failure escarpments associated with tabular slab failure and a region of probable carbonate mounds in the upper slope region. The mounds occur in elongate linear trends parallel and frequently

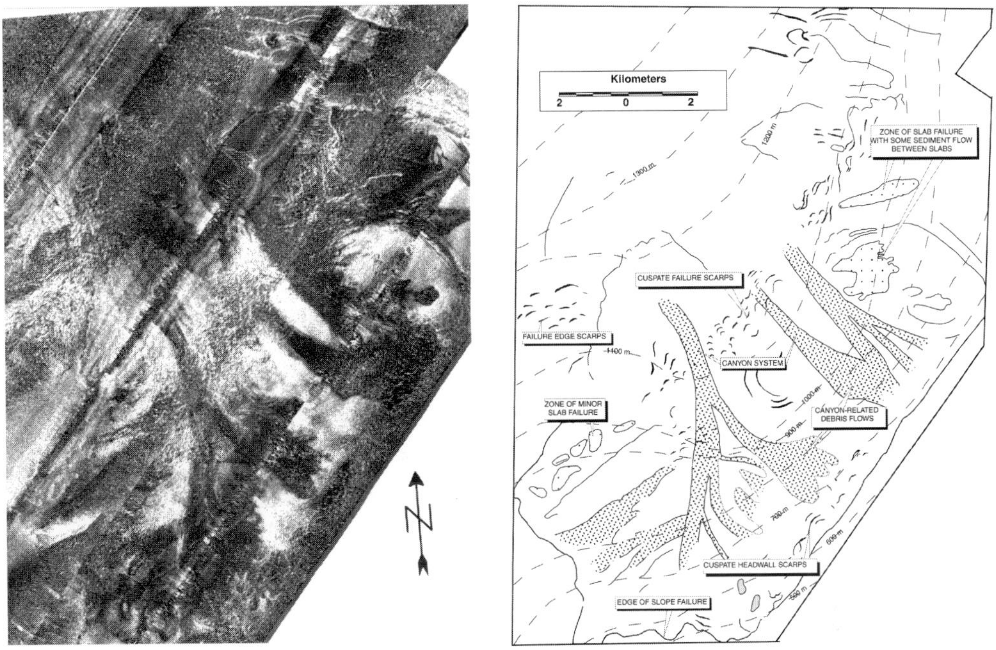

Fig. 4. Mosaic (left-hand side) and interpretation (right-hand side) of complex slope failure features along the eastern margin of the Rockall Trough. A series of cauliform gullies and canyons are interpreted as debris flow deposits that tap sediments fed onto the upper slope and result in a progressive updip migration of the area of slope failure. Zones of slab failure and evolved slides are also seen. Water depths are in metres. (See Fig. 1 for location.)

adjacent to fault scarps. It is tempting to speculate that the growth of the mounds may be associated with, or controlled by, the evolution of the slab failure features, in the light of a gas hydrate collapse model suggested for such mounds in the Porcupine Basin (Henriet *et al.* 1998). In this model the gas acts both as a slope failure trigger within the sediment and as a nutrient source for carbonate mound growth.

Debris flows

These are typically complex failure features, which are well developed in a number of areas

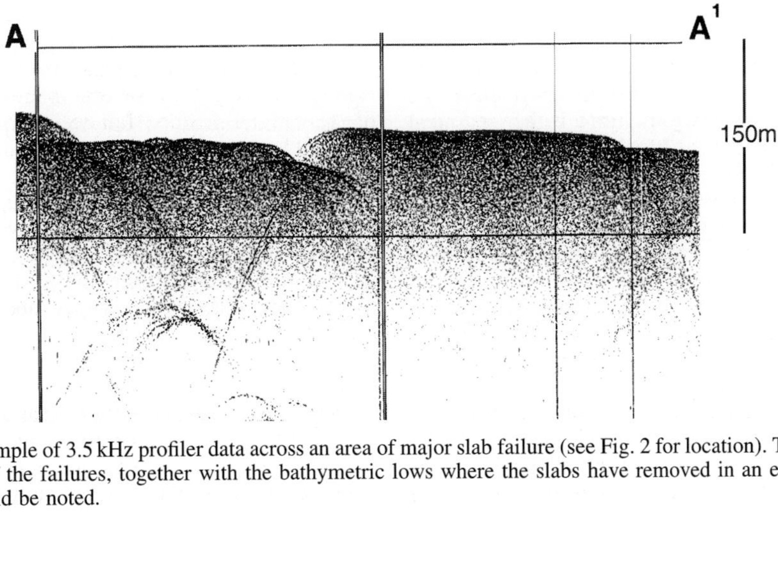

Fig. 5. Example of 3.5 kHz profiler data across an area of major slab failure (see Fig. 2 for location). The scarps at the edge of the failures, together with the bathymetric lows where the slabs have removed in an evolved slide mass, should be noted.

along the upper and mid-slope of the eastern margin of the Rockall Trough. They are best illustrated where they occur in association with some of the smaller canyon systems (Fig. 4).

The detailed morphology of the canyon-associated debris flow failures, from the upper slope at *c.* 400 m depth to the lower slope at *c.* 1200 m depth, typically consists of an extensive gully system incised into the slope (Fig. 4). This feeds downslope into the main broader canyon systems. The interconnection of the gullies on the upper slope defines a cauliform or finger-like pattern where the broad 'fingers' represent the effects of downslope sediment movement, which becomes more focused and channelized in a downslope direction. The incised tributary channels generally have a V-shaped profile. Along parts of the Porcupine Bank margin the channels coalesce into broader U-shaped canyons in the mid-slope region, which widen into aprons of high backscatter with a prolonged sub-bottom echo. Damuth (1980) demonstrated that such echo character is typical of high-energy coarse clastic sediments. As such, the present deposits are interpreted as probable sandy to gravelly flows. They also have similar sidescan sonar patterns to the gravel and sand swales described from underwater Holocene fan deltas by Prior & Bornhold (1990).

The transparent echo-sounder character and the broad, slightly lensoid geometry are very similar to the characteristics of debris flows documented by Damuth (1980) in a comprehensive review of deep-sea echograms. The 3.5 kHz profiler data across the present deposits show them to be generally confined within the channels and canyons. They lack any indications of pronounced convex-up geometries or of large entrained blocks standing proud of the deposit surface. They are therefore consistent with an interpretation of deposition as cohesionless debris flows (Postma 1986).

Within the debris flow failure systems it is suggested that canyons funnel the sediment catchment from a broad, upslope area of sediment failure. They are interpreted as upslope-migrating regions of sediment failure, which progressively tap sediments fed onto the upper slope. The backscatter pattern and sub-bottom echo suggest that the lobate fingers feeding the canyons are consistent with a less muddy composition than elsewhere in the other failure types.

Discussion

Although the geometry of the slope failures is well constrained by the TOBI sidescan sonar and echo-sounder data, there is still relatively little direct evidence of the lithologies involved in the different failure types. The backscatter pattern and the sub-bottom echo in the incipient slides and areas of slab failure are consistent with a muddy sediment, and this interpretation is also supported by data from a small number of unpublished oil industry shallow gravity cores. These failure features are therefore likely to reflect consolidated or semi-consolidated muds. The canyon-associated debris flow failures, on the other hand, appear to have developed on steeper slopes and may have developed in slightly less muddy sediments or where the degree of cohesiveness of the sediments was less. The incipient cuspate slides are likely to represent the initial stages of slope failure whereas the debris flow failure reflects the effects of more advanced mass wastage where the sediment has suffered more complete disintegration. The slab failures and evolved slides represent an intermediate phase, with occasional preserved slabs and areas of slide movement and slab disintegration.

The age of the slope failure features is also unknown. The sharp scarps on most of the failure features, lacking obvious evidence of reworking (despite the strong currents reported in the region), point towards a recent origin. Unfortunately, there is little precise information available on the likely ages of the sediments involved. The work of Flood *et al.* (1979) would point to the cessation of slumping on the Feni Drift at 10–15 ka. However, within regions of extensive debris failure some feeder chute or canyon systems appear to cross-cut older systems (Fig. 4), suggesting an older or an extended period of slope failure.

Comparison of the canyons with others along the Atlantic margin suggests that they have a significant history. Along parts of the southern Porcupine Bank margin there is little evidence of significant slope failure features, except for occasional terraces, which may represent the seaward boundary of a submarine slide zone. In the middle slope (1000–2500 m water depth) some changes in sub-bottom character (increase in backscatter downslope) occur and these may be due to remobilization and winnowing of sediments by bottom currents. Such reworking may explain the lack of evidence for slope failure structures, which are so common elsewhere along the eastern Rockall margin.

Conclusions

(1) Slope failure features are widespread along both the western and eastern margins of the

Rockall Trough. A variety of such features, of various sizes, have been identified on recent sidescan sonar data. They include incipient cuspate slides, slab failures and evolved slides, and debris flow failures.

(2) The composition of the sediment in the slope failure areas is largely unknown. It is suggested that the debris flow failures may be somewhat less muddy, or less consolidated, than the interpreted mud-prone nature of the incipient cuspate slides and the slab failures.

(3) It is suggested that many of the slope failure features are relatively recent, although some evidence is seen to suggest either a prolonged period of movement or a number of phases of slope movement, particularly in those features associated with the canyon systems.

(4) Further detailed mapping and analysis of the features, integrated with the results from shallow cores, will help to build up a more comprehensive understanding of the nature, distribution, age and controls on the formation of the slope failure features.

This project, including data and survey results acquired for the purpose, has been undertaken on behalf of the Rockall Studies Group (RSG) of the Irish Petroleum Infrastructure Programme Group 2, which was established by the Petroleum Affairs Division of the Department of the Marine and Natural Resources on 4 June 1997 in conjunction with the award of exploration licences under the Rockall Trough Frontier Licensing Round. The RSG comprises: Agip (UK) Ltd, Anadarko Ireland Company, ARCO Ireland Offshore Inc., BG Exploration & Production Ltd, BP Exploration Operating Company Ltd, British-Borneo International Ltd, Elf Petroleum Ireland B.V., Enterprise Oil plc, Mobil North Sea Ltd, Murphy Offshore Ireland Ltd, Phillips Petroleum Exploration Ireland Ltd, Saga Petroleum Ireland Ltd, Shell EP Ireland B.V., Statoil Exploration (Ireland) Ltd, Total Oil Marine plc, Union Texas Petroleum Ltd and the Petroleum Affairs Division of the Department of the Marine and Natural Resources.

References

BUGGE, T., BELDERSON, R.H. & KENYON, N.H. 1988. The Storegga Slide. *Philosophical Transactions of the Royal Society of London*, **325**, 357–388.

DAMUTH, J.E. 1980. Use of high-frequency (3.5–12 kHz) echograms in the study of near-bottom sedimentation processes in the deep-sea: a review. *Marine Geology*, **38**, 51–75.

DICKSON, R.R. & MCCAVE, I.N. 1986. Nepheloid layers on the continental slope west of Porcupine Bank. *Deep-Sea Research*, **33**, 791–818.

DOWDESWELL, J.A. & KENYON, N.H. 1997. Long-range side-scan sonar (GLORIA) imagery of the eastern continental margin of the glaciated polar North Atlantic. *In*: DAVIES, T.A., BELL, T., COOPER, A.K. *et al.* (eds) *Glaciated Continental Margins: an Atlas of Acoustic Images*. Chapman & Hall, London, 260–263.

FLOOD, R.D., HOLLISTER, C.D. & LONSDALE, P. 1979. Disruption of the Feni sediment drift by debris flows from Rockall Bank. *Marine Geology*, **32**, 311–334.

GARDNER, J.V., FIELD, M.E., LEE, H. *et al.* 1991. Ground-truthing 6.5 kHz side scan sonographs: what are we really imaging? *Journal of Geophysical Research*, **96**, 5955–5974.

HAMPTON, M.A. 1972. The role of subaqueous debris flows in generating turbidity currents. *Journal of Sedimentary Petrology*, **42**, 775–793.

HAUSER, F., O'REILLY, B.M., JACOB, A.W.B., SHANNON, P.M., MAKRIS, J. & VOGT, U. 1995. The crustal structure of the Rockall Trough: differential stretching without underplating. *Journal of Geophysical Research*, **100**, 4097–4116.

HENRIET, J.P., DE MOL, B., PILLEN, S. & 9 OTHERS 1998. Gas hydrate crystals may help build reefs. *Nature*, **391**, 648–649.

HOWE, J.A. & HUMPHERY, J.D. 1995. Photographic evidence for slope-current activity, Hebrides Slope, NE Atlantic Ocean. *Scottish Journal of Geology*, **30**, 107–115.

KENYON, N.H. 1986. Evidence from bedforms for a strong poleward current along the continental slope of Northwest Europe. *Marine Geology*, **72**, 187–198.

KENYON, N.H. 1987. Mass-wasting features on the continental slope of Northwest Europe. *Marine Geology*, **74**, 57–77.

LONSDALE, P. & HOLLISTER, C.D. 1979. A near bottom traverse of Rockall Trough: hydrographic and geologic inferences. *Oceanologica Acta*, **2**, 91–105.

LONSDALE, P. & SPIESS, F.N. 1977. Abyssal bedforms explored with a deeply towed instrument package. *Marine Geology*, **23**, 57–75.

LOWE, D.R. 1979. Sediment gravity flows: their classification and some problems of application to natural flows and deposits. *In*: DOYLE, L.J. & PILKEY, O.H. (eds) *Geology of Continental Slopes*. Society of Economic Paleontologists and Mineralogists, Special Publications, **27**, 75–82.

LOWE, D.R. 1982. Sediment gravity flows: II. Depositional models with special reference to the deposits of high-density turbidity currents. *Journal of Sedimentary Petrology*, **52**, 279–297.

MIDDLETON, G.V. & HAMPTON, M.A. 1973. Sediment gravity flows: mechanics of flow and deposition. *In*: MIDDLETON, G.V. & BOUMA, A.H. (Chairmen) *Turbidites and Deep Water Sedimentation*. Society of Economic Paleontologists and Mineralogists, Pacific Section, Short Course Lecture Notes, 1–38.

MOORE, J.G. & SHANNON, P.M. 1991. Late Tertiary slump structures in the Porcupine Basin, offshore Ireland. *Marine and Petroleum Geology*, **8**, 184–197.

MULDER, T. & COCHONAT, P. 1996. Classification of offshore mass movements. *Journal of Sedimentary Research*, **66**, 43–57.

NAYLOR, D., SHANNON, P. & MURPHY, N. 1999. *Irish Rockall Basin Region—a Standard Structural Nomenclature System*. Petroleum Affairs Division, Special Publication, **1/99**.

O'REILLY, B.M., HAUSER, F., JACOB, A.W.B. & SHANNON, P.M. 1996. The lithosphere below the Rockall Trough: wide-angle seismic evidence for extensive serpentinisation. *Tectonophysics*, **255**, 1–23.

POSTMA, G. 1986. Classification for sediment gravity-flow deposits based on flow conditions during sedimentation. *Geology*, **14**, 291–294.

PRIOR, D.B. & BORNHOLD, B.D. 1990. The underwater development of Holocene fan deltas. *In*: COLELLA, A. & PRIOR, D.B. (eds) *Coarse-Grained Deltas*. Special Publication of the International Association of Sedimentologists, **10**, 75–90.

SHANNON, P.M., JACOB, A.W.B., MAKRIS, J., O'REILLY, B., HAUSER, F. & VOGT, U. 1994. Basin evolution in the Rockall Region, North Atlantic. *First Break*, **12**, 515–522.

SHANNON, P.M., JACOB, A.W.B., MAKRIS, J., O'REILLY, B., HAUSER, F. & VOGT, U. 1995. Basin development and petroleum prospectivity of the Rockall and Hatton region. *In*: CROKER, P.F. & SHANNON, P.M. (eds) *The Petroleum Geology of Ireland's Offshore Basins*. Geological Society, London, Special Publications, **93**, 435–457.

SKEMPTON, A.W. & HUTCHINSON, J.N. 1969. Stability of natural slopes and embankment foundations. State-of-the-Art Report. *In*: *7th International Conference on Soil Mechanics and Foundation Engineering, Proceedings, Mexico City*, **2**, 291–335.

THORPE, S.A. & WHITE, M. 1998. A deep intermediate nepheloid layer. *Deep-Sea Research*, **35**, 1665–1671.

UNNITHAN, V., SHANNON, P.M., MCGRANE, K., READMAN, P.W., JACOB, A.W.B., KEARY, R. & KENYON, N.H. 2001. Slope instability and sediment redistribution in the Rockall Trough: constraints from GLORIA. *In*: SHANNON, P.M., HAUGHTON, P.D.W. & CORCORAN, D.V. (eds) *The Petroleum Exploration of Ireland's Offshore Basins*. Geological Society, London, Special Publications, **188**, 439–454.

VAN WEERING, T.C.E., NIELSEN, T., KENYON, N.H., AKENTIEVA, K. & KUIJPERS, A.H. 1998. Sediments and sedimentation at the NE Faeroe continental margin; contourites and large-scale sliding. *Marine Geology*, **152**, 159–176.

VERMEULEN, N.J. 1997. Hydrography, Surface Geology and Geomorphology of the Deep Water Sedimentary Basins to the West of Ireland. *Marine Institute, Marine Resources Series*, **241** pp.

Appendix: A list of common abbreviations

°	Oil gravity, degrees
AFTA	Apatite fission track analysis
AOM	Amorphous organic matter
API	American Petroleum Institute
BBL	Barrels
BCF	Billion cubic feet
BCM	Billion cubic metres
BCPD	Billion cubic feet per day
BHT	Bottom hole temperature
BOPD	Barrels of oil per day
CMP	Common mid-point
CPI	Carbon preference index
DMO	Dip move-out
DST	Drill stem test
FD	Finite difference
GC	Gas chromatography
GC-MS	Gas chromatography mass spectrometer
GIIP	Gas initially in place
GOR	Gas–oil ratio
GWC	Gas–water contact
HI	Hydrogen index
MDBRT	Measured depth below rotary table
MMBBL	Million barrels
MMBO	Million barrels of oil
MMSCFD	Million standard cubic feet of gas per day
MMSCMD	Million standard cubic metres of gas per day
NMO	Normal moveout
OIP	Oil in place
OWC	Oil–water contact
RFT	Repeat formation test
RKB	Rotary kelly bushing
r.m.s.	Root mean square
SEM	Scanning electron microscope
SP	Shot point
TCF	Trillion cubic feet
TD	Total depth
THR	Thermal history reconstruction
TOC	Total organic carbon
TVD	True vertical depth
TVD sub SB	True vertical depth below seabed
TVD sub GL	True vertical depth below ground level
TVDSS	True vertical depth subsea
TVG	Time variant gain
TWT	Two way travel time
VR	Vitrinite reflectance
ZFTA	Zircon fission track analysis

From: SHANNON, P.M., HAUGHTON, P.D.W. & CORCORAN, D.V. (eds). 2001. *The Petroleum Exploration of Ireland's Offshore Basins*. Geological Society, London, Special Publications, **188**, 465.

Index

Page numbers in italic refer to figures. Page numbers in bold refer to tables.